THE HIDDEN GAME
OF BASEBALL

THE HIDDEN GAME OF BASEBALL

A Revolutionary Approach to Baseball and Its Statistics

John Thorn and Pete Palmer
with David Reuther

Foreword by Keith Law

The University of Chicago Press
Chicago and London

John Thorn, a sports historian and author, has been the official baseball historian for Major League Baseball since 2011. Pete Palmer is a statistician, a baseball analyst, and a former consultant to Sports Information Center. Together Thorn and Palmer were the lead editors of *Total Baseball: The Official Encyclopedia of Major League Baseball.*

The University of Chicago Press, Chicago 60637
The University of Chicago Press, Ltd., London
© 1984, 1985 by John Thorn, Pete Palmer, and David Reuther
All rights reserved. Originally published by Doubleday & Company, Inc. in 1984.
Printed in the United States of America

24 23 22 21 20 19 18 17 16 15   1 2 3 4 5

ISBN-13: 978-0-226-24248-4 (paper)
ISBN-13: 978-0-226-27683-0 (e-book)
DOI: 10.7208/chicago/9780226276830.001.0001

Library of Congress Cataloging-in-Publication Data

Thorn, John, 1947– author.
    The hidden game of baseball : a revolutionary approach to baseball and its statistics / John Thorn and Pete Palmer with David Reuther ; foreword by Keith Law. — Third edition, enlarged.
        pages cm
    "Originally published by Doubleday & Company, Inc. in 1984."—Title page verso.
    Includes bibliographical references.
        ISBN 978-0-226-24248-4 (paperback : alkaline paper) — ISBN 0-226-24248-X (paperback : alkaline paper) 1. Baseball. 2. Baseball—Statistics. 3. Baseball—Miscellanea. I. Palmer, Pete, author. II. Reuther, David, author. III. Law, Keith, 1973– writer of preface. IV. Title.
    GV867.T49 2015
    796.357′021—dc23
                                                                            2014037002

♾ This paper meets the requirements of ANSI/NISO Z39.48–1992 (Permanence of Paper).

*For Sharon, Maureen, and Margie,
baseball widows no more.*

CONTENTS

FOREWORD

Keith Law

When I first joined the front office of the Toronto Blue Jays in January 2002, tasked with becoming the team's first full-time analytics employee, I decided to learn the fundamentals of the craft. I tracked down all of the available *Bill James Baseball Abstracts.* I bought Craig Wright and Tom House's *The Diamond Appraised.* And, for just $2, I found a used copy of the first edition of the book you now hold in your hands, *The Hidden Game of Baseball*, still the iconic book on thinking critically about the sport.

When those titles were all initially published in the 1980s, the market for prose about baseball analysis was thin. Baseball cards still promulgated myths like, "Good pitchers accumulate lots of wins" and "Good batters have lots of RBIs." And while "OBP" and "SLG" might have appeared as columns on the backs of the cards, good luck finding explanations of their meaning, let alone their relative importance. In three short decades, the hierarchy of baseball insight has been flipped on its head. Outsiders proved adept at developing new metrics and concepts in analyzing players, and eventually many of them moved into front offices to join the insiders. The voices in the media who once held a monopoly on telling you which players were good have found themselves drowned out by an egalitarian tsunami of new writers and experts, armed with granular data that didn't exist a decade earlier. We now take OBP and SLG for granted, and our tools and measurements have become increasingly refined: WAR, FIP, and UZR are now common terms of art. In 2013, when the Phillies hired a full-time analyst for their front office, it was newsworthy not because a baseball team employing a "quant" was novel but because they were the *last* team to do so.

If you've checked a player's wOBA or WAR total on Baseball-Reference or Fangraphs or any other site, you've benefited from a concept first laid out in these pages 30 years ago, that of linear weights—adding up the weighted values of individual events to come up with a single number that represents the total value of a player's contributions. It formed the basis for Pete Palmer's Total Player Rating, a direct ancestor of today's total metric of choice, Wins Above Replacement, and it drives continuing efforts inside and outside of front offices to find more accurate ways to measure defense or to separate the contributions of pitchers and fielders. A few years ago, when Mike Fast, then of Baseball Prospectus and now with the Houston Astros, first isolated and attempted to measure the effects of catcher framing—a catcher gaining or losing called strikes based on how well he receives pitches—he measured the values in runs, which are the basic units in these total-value metrics. This allowed us to add catcher-framing value to offensive production, something that wasn't even conceivable in decades past. That concept first went mainstream here.

Thorn and Palmer were among the first to think about the division of responsibility for each win between the run-scoring unit (the offense) and the run-prevention unit (the starting pitcher, any relief pitchers used, and the defense), and they even attempted to tackle the question of valuing defense, something Branch Rickey had previously labeled a fool's errand. Their efforts were limited by the poor quality of data available at the time, but by estimating how important defense was and trying to put a number on it, they began a line of inquiry that continues today with Ultimate Zone Rating and Defensive Runs Saved.

The seminal final chapter, "Rumblings in the Pantheon," also presaged the stark and often acrimonious debates surrounding the Hall of Fame and even annual player awards today, such as the decade-long argument over whether Jack Morris was a Hall of Famer. Thorn and Palmer would likely have scoffed at the notion had their book appeared in 1994, right after Morris retired, and their spiritual descendants carried the torches to argue that Morris (43.8 career WAR) was not worthy of induction while Bert Blyleven (96.5 career WAR) was.

Yet the brilliance of *The Hidden Game of Baseball* lies in its prose, not its formulae. Presenting information to an unwilling audience requires tact, diplomacy, and clarity, something Thorn and Palmer provided in spades. *Hidden Game* was, and remains, an eminently readable book, accessible to the true lay reader who's never taken a statistics class and would rather not think about confidence intervals or multivariate regressions. They begin by eviscerating sacred cows like RBIs, but gently, so that the cow barely knows what hit her, tearing down the reader's resistance to such heretical ideas as ditching batting average for on-base percentage or valuing a pitcher's performance without considering whether his team won the game.

All the while, the authors' love of and veneration for the national pastime shines through on every page. The next time an angry dinosaur, fearing imminent extinction from the impact of a Hit F/X comet, tells you that stat geeks "don't even like baseball," hand them a copy of *The Hidden Game of Baseball* and encourage them to read it. Love for the sport is what drove Palmer, Thorn, James, Rickey, and others to try to deepen their understanding of the game—that and the search for an edge on the field, something any general manager in 1984 would have had if only he'd picked up this book.

PREFACE

John Thorn and Pete Palmer

The statistical side of baseball has always gripped me. I believed that in numbers one might uncover truths not visible to the naked eye, in the way that flying at night a pilot will learn things from the instrument panel that his senses can't show him. In the summer of 1981, I was on assignment for the *Sporting News*. I went to my first convention of SABR (the Society for American Baseball Research), walked into a reception area, and met Pete Palmer. Pete, I quickly realized, was the best at what he did, which was to think hard about baseball and its numbers. Pete became my dear friend and more or less constant collaborator over the next 20 years.

But our first collaboration was not this book. With David Reuther, Pete and I developed an idea for a new sort of encyclopedia that would provide more revealing stats and tell better stories than the landmark books in the field at the time, which were known as ICI/Macmillan (1969) and Turkin/Thompson (first published in 1951). We called it *Complete Baseball*, I think, and we received a handsome bid for it, but the schedule demanded by the publisher was unworkable. So we walked away from what was at that time very big money and took much less to create *The Hidden Game of Baseball*, which came out in 1984. (The sort of encyclopedia we proposed did not come out until 1989, as *Total Baseball*).

We had no idea what impact *Hidden Game* might have, but our publisher certainly hoped we would enjoy some measure of the success Bill James was having with his first commercially published *Baseball Abstract*. Bill, of course, was one of the pioneers of what came to be known as sabermetrics. He had been releasing his *Abstract* annually, focusing on the season just past and the prospects for the next and including essays that articulated his inimitable take on baseball's statistics and how they might be improved. Like Bill, we had been interested in developing measures that tied runs scored and allowed to player performance—those numbers were demonstrably related to the outcome of a game or a season. Bill's best measure, modified over the years, was called Runs Created. Pete's was Linear Weights, which you can read all about in this book.

I say "Pete's" rather than "ours" because he was the statistician while I was the historian; he was the genius, I was the explainer. The conventional wisdom about *Hidden Game* has been that Pete did the numbers and I did the writing. That notion is more right than wrong, but Pete's words are presented and reflected throughout the book and, oddly, so is some of my statistical noodling. As with any successful collaboration, presumed areas of specialty don't stay sharply defined for long. Still, none of the innovative measures in *Hidden Game* may be called mine. I have never been a statistician, though I have been called one. All the same, Thorn and Palmer or Palmer and Thorn have endured as a pioneering sabermetric tandem because of *Hidden Game* and our subsequent work together.

The hidden game is the one played with statistics. It raises important questions about why we measure, what we think we are measuring, what we are truly measuring, and, most importantly, what the measurement means. Such questions informed our thinking throughout this book more than thirty years ago, and, even as Big Data and refined statistics sharpen our focus with each new season, sabermetricians today still cannot stray far from them. We were not the first to think unconventionally about baseball statistics, and we were careful to lay out their history from the 1840s on and to credit those who had innovated in our field long before us. In the original acknowledgments, we even invoke Bernard of Chartres.

Bill James has remarked that a meeting of sabermetricians at, say, a SABR convention in the early 1980s could have been—and more or less was—held in a hotel room. We were barely a tributary, miles from the mainstream. The chapter titles we chose then reflect the windmills we felt compelled to tilt at. It was much harder back then to convince baseball professionals and beat writers that what we were saying held any water. And yet now it's hard to find a baseball professional who does not see the value of analyzing all the data that are available to us.

As general managers and managers came to understand that outs and runs are the currency of the game, as they always have been, they began to value on-base percentage, which measures not just the hits that a batter gets but all the ways he gets on base—and the hidden value of not using up an out and permitting another man to bat with a runner(s) on base. Keeping track of pitch counts was not merely a way to preserve your own pitchers' arms—it was also a weapon: By having his batters work counts, a manager might force the hand of his opposing number and sooner get to the middle relievers, who are the soft underbelly of every pitching staff.

Today, the thinking in baseball has changed so much from thirty years ago that it is probable that we now overvalue walks where formerly they had been undervalued. Similarly, we scorn risky base running, when once it was the prime delight of players and fans. The charm of the grand old game is that it appears to be the same as it ever was, or at least the same as in President McKinley's day, but of course it has changed radically. In terms of strategy the game is now hardly about base running and fielding at all, though recent sabermetric work in these areas may alter the balance yet again.

As much as things have changed, we do think this book can still boast of its own achievements and lasting contributions. Tying individual statistics to team accomplishment—restating batting, pitching, and fielding records in runs scored or saved—still seems worthwhile. Restoring baseball statistical thinking to the 1860s core of the game—securing or conserving outs—was good. Pete came up with the first "Unified Field Theory" of baseball: the Total Player Rating, with all players' offensive and defensive contributions measured in runs above or below average, with league average performance defined as that which, when aggregated, would produce a .500 record for a team. This baseline troubled some of our colleagues, who contended that Hall of Fame players like Lloyd Waner or Tommy McCarthy could not possibly have been worse than league average over their long careers, as our calculations revealed. The current sabermetric standard is Wins Above Replacement, with some differing notions of what a replacement player (i.e., a somewhat below average one that any team might employ) might look like. Call us old fogies, but Pete and I still think a team of league-average players producing a league-average result (81–81 over the course of a modern season) sounds about right.

We have entertained offers over time to update and revise the original edition of this book, but we think it is better to leave it as it was, a stone along the road to a much greater understanding of how the game might best be played and who has played it best. (Pete has provided a list of the top 500 players of all time as of 2013, though, which appears as an appendix.) The updating, revising, and improving has been better left to the formerly tiny but now vast sabermetric community.

Still, how might we have approached *Hidden Game* differently—say, if we were to write it afresh today? We would say a good deal about F. C. Lane, a sabermetric pioneer and critic of the batting average whom we unfairly neglected. His had been the first attempt to estimate the run value of batting events, beginning with an article in the March 1916 issue of *Baseball Magazine*. When we wrote this book, play-by-play data were only beginning to be kept by the Elias Sports Bureau, and retrospective play-by-play had not yet been compiled by Retrosheet. We were compelled to develop our measures based on computer simulations and partial play-by-play. We would benefit from the work reflected at Baseball-Reference.com, Baseball Prospectus, FanGraphs, MLB.com, SABR.org, and so many other websites. We could not ignore the advances of the digital age: live data capture through time-stamped video. PITCHf/x provides pitch trajectory, velocity, and location data, and FIELDf/x tracks all moving objects on the field: fielders, runners, umpires, the ball. Our run values were the product of simulations; today those values may be tested against reams of play-by-play data, and they would be slightly different—not so different, however, as to alter any of our basic findings and tenets. More data bits may be available after a single game today than were available to us in 1984 for all baseball history, but is our understanding of the game radically altered? Or is the way we play it substantially different? Unbalanced defensive alignments—shifting infielders around to compensate for hitters' directional tendencies—are a novel reaction to data, for which in time there will be a counterreaction. Baseball is an entropic game.

Yet analytics are here to stay, and it is fair to say that the best constructed clubs—the ones that are in contention year after year—are not just the teams with the most money to lavish upon talent but the teams that spend wisely and exhibit patience with their young players. It has been ever thus. The backlash against sabermetrics, present to some degree as soon as Bill James began to be widely read, is different from the one we experienced in the 1980s.

Most fans believe the game's useful history begins with when they first started playing it or watching it. In my household, as my three sons grew up in the game, there was always talk at the dinner table about Ken Griffey, Jr. and Greg Maddux and Mike Schmidt—and

Babe Ruth and Cy Young and Ty Cobb, too. They were all part of the game. Indeed, they were all part of the family—more so than distant cousins and aunts and uncles. We talked about who was better than whom, what Cobb might do if he had to face Maddux, how many homers Ruth would hit today, what Griffey's OPS might have been against 1920s pitching staffs, that sort of thing.

Baseball fans of earlier generations had fewer statistics at their disposal, but a simpler game perhaps had less need of them. Ultimately, the statistical fragments that were once saved in scrapbooks, or the new measures devised by ingenious fans, become relics that remind us at every moment that our youth was a wonderful, if remote, time.

Cory Schwartz of Major League Baseball Advanced Media has said, "I'm old enough to remember when we had to wait two days to find West Coast box scores in the newspaper and wait until the Monday and Tuesday editions of *USA Today*." Pete and I are older than that, and we recall some of the individuals who were tilling this field before us. We are in a bold new Age of Enlightenment, but fans and writers are not unanimous in believing that we are in a new Age of Enjoyment.

Stats contain and crystallize stories but are not stories in themselves. They are something of a fetish, an encapsulation of a thing once alive. A stat serves to recall and revivify the past and sometimes to transform the future. As fans, Pete and I both follow baseball as closely as we ever did. But sabermetric writing lies more behind us than ahead and not only because we are nearer to life's ninth inning. Amid today's mix of straight-on game account and metric analysis of who is better than whom, we miss the fun that made us come to love the game in the first place.

For this we could blame Bill James, and ourselves too. Early on, what interested us more than fiddling with formulas or lobbying for Dick Allen to enter the Hall of Fame was the web of illusion that stats created for fans and players alike, evading more interesting theoretical or philosophical questions. Read *Hidden Game* in that spirit, the one that spurred us thirty years ago, and we think you will be rewarded. Others may say better than Pete and I what *Hidden Game* has meant, but for us it may be simply that it continues to be sought and cited, all these years later. With this reissue, no longer will fans need to scour antiquarian book sites to luck upon a copy.

ACKNOWLEDGMENTS

Pete and David have delegated the writing of this section to me, and, apart from my obvious debt to them, I would like to thank the following for their tangible and intangible contributions:

The Society for American Baseball Research, that singular group of some five thousand individuals who take their baseball seriously and in whose publications much of the best statistical analysis of the past decade has appeared. We are grateful for SABR's permission to reprint portions of articles by Dallas Adams, Dick Cramer, William Rubinstein, and George Wiley, and we thank the many SABR writers whose work has furthered the state of the art and heightened our understanding of this wondrously complex game. We wish to express our gratitude particularly to, in alphabetical order, Dallas Adams, Bob Carroll, Merritt Clifton, Dick Cramer, Bob Davids, Bill Deane, Alex Haas, John Holway, Bill James, Bob McConnell, David Shoebotham, the late John Tattersall, Frank Williams, and Craig Wright (and to anyone we have inadvertently neglected to mention—well, you know who you are). Other SABR-ites who provided valuable assistance along the way are Cliff Kachline, Tom Heitz, and Mark Rucker.

Steve Mann of the Baseball Analysis Company provided vital help in the modification of the tabular material to accommodate 1983 data.

We availed ourselves of the research facilities of the National Baseball Library in Cooperstown, New York; the New York Public Library; and the Interlibrary Loan System, courtesy of the always helpful people at the Public Library of Saugerties, New York.

For the wise and adroitly delivered counsel which moved this project beyond the realm of idea, where it threatened to linger and die, profound appreciation goes to George Greenberg. For bringing the apparent order of a neatly typed manuscript to the chaotic jumble of musings and jottings which was the first draft, thanks to Andrea Miller.

And to bring this book from manuscript to your eyes, we were compelled to make extraordinary demands of many individuals, who came through in extraordinary fashion: our editor, Paul Aron; designer Leslie Bauman; copy editor Scott Kurtz; production advisor Elaine Chubb; proofreader Kathy Guthmuller; Jim Gibson, Dennis Gibbons, and Phyllis Robertson of Waldman Graphics; and Steve Fisher, Jeff Koelbel, and Connie Sohodski of Fisher Composition. A special thanks as well to Lindy Hess, who brought the project to Doubleday in the first place.

My wife, Sharon, not only typed the final draft under difficult circumstances but also did her utmost to preserve an air of normalcy about our family life while this book dominated our every moment. To my sons, Jed and Isaac, I owe many months of bedtime stories.

And last, in recognition of those who have also tilled this soil, like Henry Chadwick, Ernie Lanigan, Branch Rickey, George Lindsey, and especially Bill James, who made the statistical analysis of baseball not only respectable but entertaining as well, we echo the words of Bernard of Chartres, who said nearly 900 years ago, "We are like dwarfs on the shoulders of giants, so that we can see more than they, and things at a greater distance, not by virtue of any sharpness of sight on our part, or any physical distinction, but because we are carried high and raised up by their giant size."

JOHN THORN
Saugerties, New York

1

THE MUSIC OF THE SPHERE AND ASH

On April 27, 1983, the Montreal Expos came to bat in the bottom of the eighth inning trailing the Houston Astros 4-2. First up to face pitcher Nolan Ryan was Tim Blackwell, a lifetime .228 hitter who had struck out in his first time at bat. At this routine juncture of this commonplace game, Ryan stared down at Blackwell, but his invisible—yet, for all that, more substantial—opponent was a man who had died the month before Ryan was born, a man about whom Ryan knew nothing, he confessed, except his statistical line. For at this moment of his seventeenth big-league year, Ryan had a career total of 3,507 strikeouts, only one short of the mark Walter Johnson set over twenty-one seasons, from 1907 to 1927. Long thought invulnerable, in 1983 Johnson's record was in imminent danger of falling not only to Ryan but also to Steve Carlton and Gaylord Perry.

Ryan fanned Blackwell and then froze the next batter, pinch-hitter Brad Mills, with a 1-and-2 curveball. The pinnacle was his. Johnson had been baseball's all-time strikeout leader since 1921, when he surpassed Cy Young. Ryan would hold that title only for a few weeks, then would have to eat Carlton's dust. During his brief tenure at the top, baseball savants scurried to assess the meaning of 3,509 for both the deposed King of K and the new.

What's in a number? The answer to "How many?" and sometimes a great deal more. In this case, 3,509 men had come to the plate against Ryan and failed to put the ball in play, one more man than Johnson had returned to the dugout, cursing. So what's the big deal? That Ryan was .0002849 faster, scarier, tougher—better—than Johnson? An absolute number like 3,509, or 715 (the home-run record once thought invulnerable, too), does not resound with meaning unless it is placed into some context which will give it life.

In the aftermath of Ryan's feat, writers pointed out that he only needed sixteen full seasons, plus fractions of two others, in which to record 3,509 strikeouts while Johnson needed twenty-one, or that Johnson pitched over 2,500 more innings than Ryan. Coming into the 1983 season, Ryan had fanned 9.44 men per nine innings, while Johnson was way down the list at 5.33. And Ryan allowed fewer hits per nine innings than Johnson, or, for that matter, anyone in the history of the game. So, it would seem 3,509 was not just one batter better than Johnson, but rather was mere confirmation for the masses of a superiority that was clear to the cognoscenti years before.

However, other writers introduced mitigating factors on Johnson's behalf, much as Ruth found supporters as the home-run king even after Aaron hit number 715. These champions of the old order cited Johnson's won-lost record of 417-279[1] and earned run average of 2.37 while scoffing at Ryan's mark, entering 1983, of 205-186 with an ERA of 3.11. This tack led to further argument in print, bringing in the quality of the teams each man pitched for and against, the resiliency of the ball, the attitudes of the batters in each era toward the strikeout, the advent of night ball, integration, expansion, the designated hitter, the overall talent pool, competition from other professional sports . . . and on down into the black hole of subjectivism.

Why were so many things dragged into that discussion? Because the underlying question about 3,509 was: Does this total make Ryan better than Johnson, or even a better *strikeout* pitcher than Johnson? At the least, does it make him a great pitcher? In our drive to identify excellence on the baseball field (or off it), we inevitably look to the numbers as a means of encapsulating and comprehending experience. This quantifying habit is at the heart of baseball's hidden game, the one ceaselessly played by Ryan and Johnson and Ruth and Aaron—and, thanks to baseball's voluminous records, nearly 13,000 other players—in a stadium bounded only by the imagination.

The hidden game is played with statistics (and, it could be said, by them), but it extends beyond the record books. One enters the game whenever one attempts to evaluate performance, which is possible only through comparison, implied or explicit. How good a hitter is

Eddie Murray? How would Rogers Hornsby do if he were active today? Why can't the Red Sox get themselves some decent starting pitchers? What value does Oakland receive from Rickey Henderson's stolen bases? When is an intentional base on balls advisable? The answers to these and countless other questions are of concern to those who play the hidden game, for which this book serves as a guide.

In the eternal Hot Stove League, statistics stand in for their creators, and the better the statistics, the more "real" (i.e., reasonable) the results. In recent years baseball's already copious traditional stats have been supplemented, though not supplanted, by a variety of new formulations—some of them official, like the save and the game winning RBI, most of them outlaws, like Runs Created or Total Average, though not without adherents. And with the explosion of new stats has come an outspoken antistatistical camp, with the two sides aligning themselves along battle lines that were drawn almost at the dawn of baseball.

The antis might argue that baseball is an elementally simple game: pitch, hit, run, throw, catch—what else is there that matters? Playing it or watching it is deeply satisfying without examination of any sort, let alone rigorous statistical analysis. So why do we need new stats? Don't we have enough ways to measure performance? Don't we have *too many?* Why subject every incident on the field to such maniacal ledger-book accounting?

How can baseball's beauty fail to wither under the glare of intense mathematical scrutiny? For those of an antistatistical bent, baseball, like a butterfly, is poetry in motion and a cold, dead thing when pinned to the page. If we subject the game to ever more intricate analysis, in hope that it will yield up its mysteries, are we not breaking the butterfly upon a wheel, in Pope's phrase?

For the statistician, too, baseball is indeed like a butterfly, whose grace can be glimpsed while it is in flight . . . but then it is gone, having scarcely registered upon the memory. One doesn't truly know any longer what it looked like, where it came from, how it vanished in an instant. The butterfly's coloring, its detail, cannot be absorbed while it is in flight; it must be examined to appreciate its complexity. One may love its simplicity in flight as one may love the simplicity of baseball while standing in the outfield or sitting in the grandstand. But the complex texture of the game, which for many is its real delight—the thing that pleases the mind as well as the eye—cannot be fully grasped while the game is in progress.

And that's what statistical analysis allows us to do. Statistics are not the instruments of vivisection, taking the life out of a thing in order to examine it; rather, statistics are themselves the vital part of baseball, the only tangible and imperishable remains of games played yesterday or a hundred years ago.

Baseball may be loved without statistics, but it cannot be understood without them. Statistics are what make baseball a sport rather than a spectacle, what make its past worthy of our interest as well as its present.

For those who view baseball statistics in this way—that as they increase our understanding of the game, they deepen our enjoyment—the numbers of the hidden game take on reality and, sometimes, beauty, in the way that the circumference of a circle may be described arithmetically or aesthetically. The Pythagoreans and Cabalists may have had it right in believing that numbers are at the core of creation.

Without sinking into a morass of Philosophy 101 disputation about whether statistics reside in the things we observe or whether we impose them, let's look at the "reality" of the thing itself, which for our purposes is the game of baseball. The form in which it comes to most of us is a telecast, which flattens the game into two dimensions, transforming baseball into ambulatory chess or Pac-Man; to restore contours to the game we have to imagine it even as we watch it. The televised game offers signposts of what baseball is like for those on the field or at the park; to recreate that feeling, the viewer relies upon his experience of playing the game or of seeing it in the open. This act of imagination, this restructuring of the video image, progresses from what is seen to what is unseen. Disorientingly, in this instance the game that is seen is the abstraction while the unseen game is concrete, or "real."

This movement from the seen to the unseen describes the impulse and the activity of the game's statisticians, too. For them, plumbing the meaning of numbers is not mere accounting; to bring the hidden game of baseball into the open is an act of imagination, an apprehension and approximation of truth, and perhaps even a pursuit of beauty and justice.

Baseball offers a model of perfection, a utopian, zero-sum system in which every action by the offense has a corresponding and inverse act by the defense, and everything balances in the end. The box score reads like the Book of Life held by St. Peter at the pearly gates; no action is left unaccounted. Although it has been written that baseball is a microcosm of American life, in no place in society at large can the harmony of the ball field be matched.

Many people who find "real life" too much for them, or at least a source of turmoil and anxiety, derive immense satisfaction from the order, regularity, justice, and essential stability of baseball, and this goes a long way toward explaining its continuing appeal to adults. We're all pursuing something that we can't quite identify; we all would like to think there is a simple answer to a multitude of complex problems. And a good many of us who may, by the world's standards, be entirely sane take great pains to investigate the clockwork mechanism, the mathematical construct that is baseball, because of its seeming offer of such an answer. The lure, the tease, for baseball statisticians is that the mathematical universe in which the game is played can be fully comprehended. And if this game can be fathomed, might not others? Those who analyze baseball by its numbers may, sometimes, hear the music of the spheres.

Even those who profess to abominate statistics—among whom are included several baseball managers, general managers, league officials, ballplayers—are statisticians despite themselves, for we are all, all of us humans, intuitive statisticians. We base our actions upon a quick assimilation of similar experiences, weigh the results, and decide what to do. An example: With men on first and second bases and one out, a ground ball is hit into the hole; the shortstop stabs the ball and makes the play at third base. This move was a product not of sudden inspiration (except for the shortstop who made the play for the first time in baseball history) but of calculated risk. While he is racing to the hole, the shortstop is figuring: Based upon the speed of the runners and how hard the ball is hit, he probably has no chance of a double play; he may have little chance of a play at second; and he almost certainly has no play at first. He throws to third because the distance from the hole to the bag is short, and his calculation of the various probabilities led him to conclude that this was his "percentage play."

Now not so much as a glimmer of any number entered the shortstop's head in this time, yet he *was* thinking statistically. In much the same way, a manager who pinch-hits with a righthanded batter when the opposing pilot brings in a lefty reliever is said to be playing the percentages. Surely he never calculated them, nor did he perform any empirical study of the question, but based upon his thirty or forty years of observation and upon folk wisdom handed down to him, he assumes that *statistically* the righthanded batter has a better chance of reaching base against the lefthanded pitcher. We are all statisticians whenever we generalize from a group of specific, similar experiences; those of us who work with the numbers, though, get more accurate results.

Most players and managers feel that they can do just fine without relying upon statisticians. Haven't they always done splendid, as Casey Stengel used to say? Well, no. Stable as it is by society's standards, the game does change. The rug was pulled out from under the managers' feet some time ago yet they don't know it: Dead ball era strategies continue to be employed sixty years beyond the point at which they outlived their usefulness. Front office decisions are made on the basis of player-performance measures which tell next to nothing of a man's value to his team. Teams tailor their personnel to their home-park peculiarities in such a way that they are left vulnerable at home and impotent on the road. The Hall of Fame becomes stuffed with players from between the wars simply because their stats are misunderstood.

This book was written not in a spirit of crusade, to right wrongs across the board, but in an attempt to see the old ball game in some new ways which both illuminate and entertain. Even if they (baseball management, players, Hall of Fame electors) don't want to know the score, you can. And baseball is, after all, a game, not real life, about which a tirade might be more apt.

Moreover, this book was written not only for those already converted to the statistical persuasion. If you have found the analysis of baseball through its statistics confusing or off-putting, give us a chance to show how powerful and elegant it can be at its best—which is what this book delivers. The hidden game will be out in the open, revealing the true stars, the honored impostors, the real percentages behind "percentage plays," what makes teams win, the statistical effect of every home park and more—and all this in a historical context that goes back to Alexander Cartwright, with revolutionary statistics applied from the beginning of major-league play in 1876 through the 1983 season.

Playing the hidden game of baseball—the interior stadium peopled with memories and images and numbers—you can position Nap Lajoie at second base for the 1983 Yankees and figure what difference he might have made in the team's won-lost record. Or if manipulating history is not irresistible, you can trade Andre Dawson to the Braves, just for fun, and see if Atlanta would have won the National League West. This book will give you the tools to evaluate such a move, or to see who would benefit more from a trade of, say, Jesse Orosco for Terry Kennedy.

Were players better in 1930? Or 1960? Or 1975? We'll take a stab at answering that statistically, too. Does clutch play exist or is it, as the curveball was once thought to be, an optical illusion? Why don't they steal home anymore?

And more . . . though we ask that you read the chapters in order, not because there are shocking revelations at the end (there are) nor because we are vain enough to wish each word appreciated (we are), but because the later chapters and the data in the tables which conclude the book build upon principles set forth in sequence, principally in the next five chapters. This book will not settle all the arguments that rage in the Hot Stove League, but it will elevate the discussion and provide some new understanding of baseball truisms that are no longer true, or never were. We may even move the powers that be to adopt some of our new statistics as official, or prompt newspapers to carry them on a daily basis; but baseball is an institution heartily resistant to change, and we are not optimistic. The RBI, for example, was introduced in 1879 yet did not become an officially recorded stat until 1920; the ERA and slugging percentage were both in use before 1876, yet not accepted for forty to fifty years; and the On Base Average, familiar through twenty-five to thirty years of mention in the press, is still not official. We will wait our turn.

We believe player performance can be measured in a better fashion than it is today, even by those in the vanguard of statistical analysis. We view performance—batting, pitching, fielding, baserunning—in terms of its runs contributed or saved and within a context formed by the average level of performance prevailing at the time and the effect of the home park upon run scoring.

These statistics are marvelous tools for cross-era comparisons, enabling us to determine if baseball's history is truly a seamless web or if its seams are real enough, only camouflaged by traditional statistics. These stats, like the batting average, were designed to identify individual accomplishments divorced from their effects on or by the team, and thus can promote values contrary to those of the team. With the aid of a computer simulation of the 130,000 major-league games played in this century, we have been able to break the won-lost record of a team into its individual components. Did a team finish 90-72? Then we can take the nine games above average (81-81) that the team won and attribute them: .7 wins to the third baseman, 2.4 wins to the top starting pitcher; 1.2 losses to the weak-hitting first baseman, and so on. These individuals' wins are predicated upon the run values of their every action on the ball field. How can a batter with a .250 average be better than another who bats .320? See Chapter 4. Why is Dave Winfield one of the top fifty batters in the history of the game? See Chapter 5. Who was the better pitcher, Bert Blyleven or Juan Marichal? See Chapter 6. Who is the best player of all time? The best player today? See Chapter 15 (last, please).

In identifying the very best performances of all time by our new, more accurate measures, we will inevitably confront the Cooperstown question. We will not agitate to rip Roger Bresnahan's plaque off the wall, nor to enshrine Dick Allen, although both actions are defensible. We will simply present our roster of the 140 top players and pitchers in history (corresponding to the number of major-league players selected for the Hall, excluding executives, umpires, etc.) as revealed by their run contributions, and leave the creation of a rival Hall of Fame to those so inclined.

We'll provide as well lists of the top 100 lifetime and single-season performances in each of our new measures, plus a season-by-season record of the top three in each category and a full statistical profile of each team, indicating why it finished where it did (or why it should have done much better or worse). We'll even supply the tools for you to work up your own variations of our stats, or to develop new ones altogether.

Promises, promises. Onward!

---

[1] The encyclopedias erroneously list Johnson at 416-279. The unearthed extra win is a product of research by Frank Williams, reported in "All the Encyclopedias Are Wrong," *The National Pastime,* 1982. All records mentioned in this book reflect the best available data, much of which has not yet gained "official" acceptance.

2

WHAT'S WRONG WITH
TRADITIONAL BASEBALL STATISTICS

Before we assess where baseball statistics are headed, we ought first to see where they've been.

In the beginning, baseball knew numbers and was not ashamed. The game's Eden dates ca. 1845, the year in which Alexander Cartwright and his Knickerbocker teammates codified the first set of rules and the year in which the *New York Herald* printed the primal box score. The infant game became quantified in part to ape the custom of its big brother, cricket; yet the larger explanation is that numbers served to legitimize men's concern with a boys' pastime. The pioneers of baseball reporting—William Cauldwell of the Sunday *Mercury,* William Porter of *Spirit of the Times,* the unknown ink-stained wretch at the *Herald,* and later Father Chadwick—may indeed have reflected that if they did not cloak the game in the "importance" of statistics, it might not seem worthwhile for adults to read about, let alone play.

Americans of that somewhat grim period were blind to the virtue of play (much to the befuddlement of Europeans) and could take their amusements only with a chaser of purposefulness. Baseball, though simple in its essence (a ball game with antecedents in the Egypt of the pharaohs), was intricate in its detail and thus peculiarly suited to quantification; statistics elevated baseball from other boys' field games of the 1840s and '50s to make it somehow "serious," like business or the stock market.

In the long romance of baseball and numbers, no figure was more important than that of Henry Chadwick. Born in England in 1824, he came to these shores at age thirteen steeped in the tradition of cricket. In his teens he played the English game and in his twenties he reported on it, for a variety of newspapers including the Long Island *Star* and the New York *Times*. In the early 1840s, before the Knickerbocker rules eliminated the practice of retiring a baserunner by throwing the ball *at* him rather than to the base, Chadwick occasionally played baseball too, but he was not favorably impressed, having received "some hard hits in the ribs." Not until 1856, by which time he had been a cricket reporter for a decade, were Chadwick's eyes opened to the possibilities in the American game, which had improved dramatically since his youth. Writing in 1868: "On returning from the early close of a cricket match on Fox Hill, I chanced to go through the Elysian Fields during the progress of a contest between the noted Eagle and Gotham clubs. The game was being sharply played on both sides, and I watched it with deeper interest than any previous ball match between clubs that I had seen. It was not long before I was struck with the idea that base ball was just the game for a national sport for Americans . . . as much so as cricket in England. At the time I refer to I had been reporting cricket for years, and, in my method of taking notes of contests, I had a plan peculiarly my own. It was not long, therefore, after I had become interested in base ball, before I began to invent a method of giving detailed reports of leading contests at base ball. . . ."

Thus Chadwick's cricket background was largely the impetus to his method of scoring a baseball game, the format of his early box scores, and the copious if primitive statistics that appeared in his year-end summaries in the New York *Clipper,* Beadle's *Dime Base-Ball Player,* and other publications.

Actually, cricket had begun to shape baseball statistics even before Chadwick's conversion. The first box score (see Table II, 1) reported on only two categories, outs and runs: Outs, or "hands out," counted both unsuccessful times at bat and outs run into on the basepaths; "runs" were runs scored, not those driven in. The reason for not recording hits in the early years, when coverage of baseball matches appeared alongside that of cricket matches, was that, unlike baseball, cricket had no such category as the successful hit which did not produce a run. To reach "base" in cricket is to run to the opposite wicket, which tallies a run; if you hit the ball and do *not* score a run, you have been put out.

Table II, 1. *The First Box Score: October 25, 1845*					
NEW YORK BALL CLUB			BROOKLYN CLUB		
	Hands out	Runs		Hands out	Runs
Davis	2	4	Hunt	1	3
Murphy	0	6	Hines	2	2
Vail	2	4	Gilmore	3	2
Kline	1	4	Hardy	2	2
Miller	2	5	Sharp	2	2
Case	2	4	Meyers	0	3
Tucker	2	4	Whaley	2	2
Winslow	1	6	Forman	1	3
	12	37		12*	19

*The "Hands out" total 13 rather than 12—a 139-year-old typo now beyond rectification.

Cricket box scores were virtual play-by-plays (see Table II, 2), a fact made possible by the lesser number of innings (two) and the more limited number of possible events. This play-by-play aspect was applied to a baseball box score as early as 1858 in the New York *Tribune* (see Table II, 3); interestingly, despite the abundance of detail, hits were still not accounted. Nor did they appear in Chadwick's own box scores, not until 1867 (see Table II, 4 for 1863 box of Chadwick's), and his year-end averages to that time also reflected a cricket mind-set. The batting champion as declared by Chadwick, whose computations were immediately and universally accepted as "official," was the man with the highest average of Runs Per Game. An inverse though imprecise measure of batting quality was Outs Per Game. After 1863, when a fair ball caught on one bounce was no longer an out, fielding leaders were those with the greatest total of fly catches, assists, and "foul bounds" (fouls caught on one bounce). Pitching effectiveness was based purely on control, with the leader recognized as the one whose delivery offered the most opportunities for outs at first base and the fewest passed balls.

In a sense, Chadwick's measuring of baseball as if it were cricket can be viewed as correct in that when you strip the game to its basic elements, those that determine victory or defeat, outs and runs *are* all that count in the end. No individual statistic is meaningful to the team unless it relates directly to the scoring of runs. Chadwick's blind spot in his early years of baseball reporting lay in not recognizing the linear character of the game, the sequential nature whereby a string of base

Table II, 2. *A Cricket Box Score of 1856*

BROOKLYN CLUB vs. THE LONG ISLAND CLUB
LONG ISLAND

1st Innings		2nd Innings	
Walker, b Stevens	14	b Mack	7
Curry, b Stevens	2	b Stevens	8
Hartshorn, c Ely, b Stevens	0	not out	6
Henry, c Walden, b Stevens	0	b Mack	7
James, b Mack	8	c Ely, b Stevens	3
Brooks, c and b Stevens	1	b Mack	0
D. Clear, b Mack	4	b Mack	3
J. Eastmead, b Stevens	0	c Mack, b Stevens	0
Pick, b Mack	2	c Spencer, b Stevens	4
Clear, Jun., c Mack, b Stevens	4	b Stevens	16
Hollely, not out	0	c Bainbridge, b Stevens	0
Leg bye 1, wide 1	2	Byes 7, lb 2, w 2	11
Total	37	Total	65

BROOKLYN

1st Innings		2nd Innings	
Mack, b Brooks	5	b Peck	3
Spencer, b Brooks	10	c and b Brooks	26
Bainbridge, b Brooks	0	run out	3
Douglass, b Hartshorn	0	b Brooks	0
Stevens, b Brooks	11	b Peck	9
Grant, c Pick, b Hartshorn	4	c Curry, b Peck	3
Jingle, b Brooks	0	not out	3
Alexander, c Hartshorn, b Brooks	2	not out	5
Ely, not out	0	b Brooks	3
Sweetland, run out	4	b Brooks	2
Byes 6, lb 1	7	Leg byes	3
Total	43	Total	60

hits or men reaching base on error (there were no walks then) was necessary in most cases to produce a run. In cricket each successful hit must produce at least one run, while in baseball, more of a team game on offense, a successful hit may produce none.

Early player stats were of the most primitive kind, the counting kind. They'd tell you *how many* runs, or outs, or fly catches; later, how many hits or total bases. Counting is the most basic of all statistical processes; the next step up is averaging, and Chadwick was the first to put this into practice.

BROOKLYN

| | Outs | Runs | FIELDING | | | | HOW PUT OUT | | | | | |
			Fly	Bound	Base	Total	Fly	Bound	1b	2b	3b	Foul
Leggett, c	5	1	0	7	0	7	1	1	1	0	0	2
Holder, 2b	4	2	0	0	1	1	1	1	0	0	0	2
Pidgeon, ss	4	1	2	2	1	5	1	0	1	1	0	1
Grum, cf	2	4	0	0	0	0	1	0	0	0	0	1
P. O'Brien, lf	3	2	0	3	0	3	1	0	0	1	0	1
Price, 1b	1	3	0	0	4	4	0	0	0	1	0	0
M. O'Brien, p	2	3	2	1	1	4	0	1	0	0	0	1
Masten, 3b	4	1	2	1	0	3	1	2	1	0	0	0
A. Burr, rf	2	1	0	0	0	0	0	1	1	0	0	0
	27	18	6	14	7	27	6	6	4	3	0	8

NEW YORK

| | Outs | Runs | FIELDING | | | | HOW PUT OUT | | | | | |
			Fly	Bound	Base	Total	Fly	Bound	1b	2b	3b	Foul
Pinckney, 2b	2	3	0	2	5	7	0	0	1	0	0	1
Benson, cf	3	3	1	0	0	1	1	1	0	0	0	1
Bixby, 3b	3	1	0	0	0	0	0	1	1	0	0	1
De Bost, c	3	2	1	7	0	8	1	2	0	0	0	0
Gelston, ss	4	2	0	0	0	0	1	1	0	0	0	2
Wadsworth, 1b	3	3	2	1	2	5	0	0	0	0	0	3
Hoyt, lf	2	4	0	0	0	0	0	0	1	1	0	0
Van Cott, p	2	4	1	2	0	3	0	0	2	0	0	0
Wright, rf	5	0	2	1	0	3	2	0	1	0	0	2
	27	22	7	13	7	27	5	5	6	1	0	10

As professionalism infiltrated the game, teams began to bid for star-caliber players. Stars were known not by their stats but by their style: Every boy would emulate the flair of a George Wright at shortstop, the whip motion of a Jim Creighton pitching, the nonchalance of a John Chapman making over-the-shoulder one-handed catches in the out-field (this in the days before the glove!). But Chadwick recognized the need for more individual accountability, the need to form objective credentials for those perceived as stars (or, in the parlance of the

BATTING

EXCELSIOR	H.L.*	RUNS	UNION	H.L.*	RUNS
Flanly, 2d b	0	5	Nicholson, 1st b	5	0
Smith, p	5	0	E. Durell, lf	3	2
Masten, 3d b	2	1	Abrams, 2d b	4	2
Whiting, 1st b	3	1	Hannegan, p	3	3
McKenzie, lf	4	0	Birdsall, c	3	3
H. Brainard, ss	2	2	Hyatt, 3d b	2	3
Cline, c	4	0	Gaynor, ss	4	2
Fairbanks, rf	3	0	Collins, cf	1	2
Leggett, cf	4	0	F. Durell, rf	2	3
Total		9	Total		20

* Hands lost, same as Hands out in Table II, 1.

RUNS MADE IN EACH INNING

	1st	2d	3d	4th	5th	6th	7th	8th	9th
Excelsior	2	0	2	1	1	1	1	0	1-9
Union	4	1	2	1	3	1	3	2	3-20

Umpire—Mr. Pearce, of the Atlantic club.
Scorers—Messrs. Holt and Travers.
Home runs—Hyatt, 1.
Struck out—Smith, 1.
Catches missed—Leggett, 1; Masten, 1; Smith, 1; Fairbanks, 1; Birdsall, 2; E. Durell, 1.
Put out at first base—Excelsiors, twice; Unions, 9 times.
Put out at home base—E. Durell, by Masten.
Fly catches made—Masten, 2; Fairbanks, 1; H. Brainard, 1; McKenzie, 1; Flanly, 2; Cline, 1; Abrams, 3; Birdsall, 4; E. Durell, 1; Hannegan, 1; Hyatt, 2; Nicholson, 2.
Put out on foul balls—Excelsiors, 16 times; Unions, 8 times.
Time of the game—three hours and thirty minutes.

period, "aces"). The creation of popular heroes is a product of the post-Civil War period, with a few notable exceptions (Creighton, Joe Start, Dickey Pearce, J.B. Leggett). So in 1865, in the *Clipper,* Chadwick began to record a form of batting average taken from the cricket pages—Runs Per Game. Two years later, in his newly founded baseball weekly, *The Ball Players' Chronicle,* Chadwick began to record not only average runs and outs per game, but also home runs, total bases, total bases per game—and hits per game. The averages were expressed not with decimal places but in the standard cricket format of "average and over": Thus a batter with 23 hits in six games would have an average expressed not as 3.83 but as "3-5"—an average of 3 with an overage, or remainder, of 5. Another innovation was to remove from the individual accounting all bases gained through errors. Runs scored by a team, beginning in 1867, were divided between those scored after a man reached base on a clean hit and those arising from a runner's having reached base on an error.

In 1868, despite Chadwick's derision, the *Clipper* continued to award the prize for the batting championship to the player with the greatest average of Runs Per Game. Actually, the old yardstick had been less preposterous a measure of batsmanship than one might imagine today, because team defenses were so much poorer and the pitcher, with severe restrictions on his method of delivery, was so much less important. If you reached first base, whether by a hit or by an error, your chances of scoring were excellent; indeed, teams of the mid-1860s registered more runs than hits! By the 1876 season, the first of National League play, the caliber of both pitching and defense had improved to the extent that the ratio of runs to hits was about 6.5 to 10; today the ratio stands at roughly 4 to 10. To illustrate the futility of applying this ancient measure of offensive ability to the present day, here are the top ten major-league players of 1983, as measured by Runs Per Game (100 games minimum).

Table II, 5. *Runs Per Game Leaders, 1983*

	R/G	Avg.
1. Tim Raines, Montreal	133/156	.853
2. Dale Murphy, Atlanta	131/162	.809
3. Steve Garvey, San Diego	76/100	.760
4. Cal Ripken, Baltimore	121/162	.747
5. Eddie Murray, Baltimore	115/156	.737
6. George Brett, Kansas City	90/123	.732
7. Rickey Henderson, Oakland	105/145	.724
8. Bob Horner, Atlanta	75/104	.721
9. Gary Redus, Cincinnati	90/125	.720
10. Willie Randolph, NY (A)	73/104	.702

There are some fine players here, to be sure, but where are Wade Boggs, Dickie Thon, Mike Schmidt, Robin Yount? They may not have scored as many runs as the fellows on this list, but they had superior 1983 seasons to such as Redus, Randolph, Garvey, et al.

The Outs Per Game figure was tainted as a measure of batting skill because it might reflect as easily a strikeout or a double unsuccessfully stretched to a triple. Or, in a ridiculous but true example, a man might get on base with a single, then be forced out at second base on a ground ball. The runner who was forced out is debited with the out; not only does the man who hit the grounder fail to register a notch in the out column—if he comes around to score, he'll get a counter in the run column.

In the late 1860s Chadwick was recording total bases and home runs, but he placed little stock in either, as conscious attempts at slugging violated his cricket-bred image of "form." Just as cricket aficionados watch the game for the many opportunities for fine fielding it affords, so was baseball from its inception perceived as a fielders' sport. The original Cartwright rules of 1845, in fact, specified that a ball hit out of the field—in fair territory or foul—was a foul ball! "Long hits are showy," Chadwick wrote in the *Clipper* in 1868, "but they do not pay in the long run. Sharp grounders insuring the first-base certain, and sometimes the second-base easily, are worth all the hits made for home-runs which players strive for."

If it was so easy to score once having reached first base, why exert oneself to the utmost simply to hear the oohs and ahs for a towering fly ball? Batters in the 1860s held their hands apart some two to four inches on the handle, so that when one did swing "for the fences," it meant a mighty sweep of the upper torso rather than a less demanding if more powerful snap of the wrists. (It was with one of these mighty upper-body contortions that Jim Creighton, in 1862, belted a home run and ruptured his bladder, leading to his death a few days later. The long hit was not only "showy" but, as practiced then, unnatural.)

Chadwick's bias against the long ball was in large measure responsible for the game that evolved and for the absence of a hitter like Babe Ruth until 1919. When lively balls were introduced—as they were periodically from the very infancy of baseball—and long drives were being belted willy-nilly, and scores were mounting, Chadwick would ridicule such games in the press. What he valued most in the early days was the low scoring game marked by brilliant fielding. In the early annual guides, he listed all the "notable" games between significant teams—i.e., those in which the winner scored under *ten* runs!

Chadwick prevailed, and Hits Per Game became the criterion for the *Clipper* batting championship and remained so until 1876, when the problem with using games as the denominator in the average at last became clear. If you were playing for a successful team, and thus were surrounded by good batters, or if your team played several weak rivals who committed many errors, the number of at bats for each individual in that lineup would increase. The more at bats one is granted in a game, the more hits one is likely to have. So if Player A had 10 at bats in a game, which was not unusual in the '60s, he might have 4 base hits. In a more cleanly played game, Player B might bat only 6 times, and get 3 base hits. Yet Player A, with his 4-for-10, would achieve an average of 4.00; the average of Player B, who went 3-for-6, would be only 3.00. By modern standards, of course, Player A would be batting .400 while Player B would be batting .500.

In short, the batting average used in the 1860s is the same as that used today except in its denominator, with at bats replacing games. Moreover, Chadwick posited a primitive version of the slugging percentage in the 1860s, with total bases divided by number of games; change the denominator from games to at bats and you have today's slugging percentage—which, incidentally, was not accepted by the National League as an official statistic until 1923 and the American until 1946 (the game was born conservative). Chadwick's "total bases average" represents the game's first attempt at a *weighted* average—an average in which the elements collected together in the numerator or the denominator are recognized numerically as being unequal. In this instance, a single is the unweighted unit, the double weighted by a factor of two, the triple by three, the home run by four. Statistically, this is a distinct leap forward from, first, counting, and next, averaging. The weighted average is in fact the cornerstone of today's statistical innovations.

The 1870s gave rise to some new batting stats and to the first attempt to quantify thoroughly the other principal facets of the game, pitching and fielding. Although the *Clipper* recorded base hits and total bases as early as 1868, a significant wrinkle was added in 1870 when at bats were listed as well. This is a critical introduction because it permitted the improvement of the batting average, first introduced in its current form in the Boston press on August 10, 1874, and first computed officially—that is, for the National League—in 1876. Since then the BA has not changed.[1]

The objections to the batting average are well known, but to date have not dislodged the BA from its place as the most popular measure of hitting ability. First of all, the batting average makes no distinction between the single, the double, the triple, and the home run, treating all as the same unit—a base hit—just as its prototype, Runs Per Game, treated the run as its unvarying, indivisible unit. This objection was met in the 1860s with Total Bases Per Game. Second, it gives no indication of the effect of that base hit; in other words, it gives no indication of the value of the hit to the team. This was probably the objection that Chadwick had to tabulating base hits originally, because it is not likely that the idea just popped into his head in 1867, upon which he decided to act immediately; he must have thought of a hit-constructed batting average earlier and rejected it.

A third objection to the batting average is that it does not take into account times first is reached via base on balls, hit by pitch, or error. This, too, was addressed at a surprisingly early date. In 1879 the National League adopted as an official statistic a forerunner of the On Base Average; it was called "Reached First Base." Paul Hines was the leader that year with 193, which included times reached by error as well as base on balls and base hits.[2] But the figure was dropped after that year.

What happened to Runs Per Game? Chadwick—"that demon eliminator," in the words of statistician Ernie Lanigan—dropped it like the plague from his Official Guide, *along with* the figures for runs and hits; he was not one to convert halfheartedly. But RPG continued to appear in Al Spalding's publication, the Constitution and Playing Rules (commonly referred to as "the league book"), along with the number of runs and hits; runs were not restored to the Chadwick-edited guides (Beadle's, DeWitt's, and later Spalding's) until 1882, hits till 1880.

The year 1876 was significant not only for the founding of the National League and the official debut of the batting average in its current form, it was also the Centennial of the United States, which was marked by a giant exposition in Philadelphia celebrating the mechanical marvels of the day. American ingenuity reigned, and technology was seen as the new handmaiden of democracy. Baseball, that mirror of American life, reflected the fervor for things scientific with an explosion of statistics far more complex than those seen before, particularly in the previously neglected areas of pitching and fielding. The increasingly minute statistical examination of the game met a responsive audience, one primed to view complexity as an indication of quality.

When the rule against the wrist-snap was removed in 1872, permitting curve pitching, and as the number of errors declined through the early 1870s—thanks to the heightened level of competition provided by baseball's first professional league, the National Association—the number of runs scored dropped off markedly.

With the pitcher unshackled—transformed from a mere delivery boy of medium paced, straight balls to a formidable adversary—the need to identify excellence, to plot the stars, arose just as it had for batters in the 1860s. Likewise, as fielding errors became more the exception than the rule, they became at last worth counting and contrasting with chances accepted cleanly, in other words, the fielding percentage. Fielding skill was still the most highly sought after attribute of a ballplayer, but the balance of fielding, batting, and pitching was in flux; by the 1880s pitching and batting would begin their long rise to domination of the game, Chadwick's tastes notwithstanding.

The crossroads of 1876 highlights how the game had changed to that point, and how it has changed since.

In that year, the number of offensive stats tabulated at season's end (in either the Chadwick publications or the Spalding-Reach) was six: games, at bats, runs, hits, runs per game, and batting average. Of these, only runs and runs per game were common in the 1860s, while that decade's tabulation of total bases vanished. The number of offensive stats a hundred years later? Twenty. (Today the number is twenty-one, with the addition of the game winning RBI.)

The number of pitching categories in 1876 was eleven, and there were some surprises, such as earned run average, hits allowed, hits per game, and opponents' batting average. Strikeouts were not recorded, for Chadwick saw them strictly as a sign of poor batting rather than good pitching (his view had such an impact that pitchers' K's were not kept officially until 1887). The number of pitching stats today? Twenty-four.

The number of fielding categories in 1876 was six. One hundred years later it was still six (with the exception of the catcher, who gets a seventh: passed balls), dramatizing how the game—at least the hidden game of statistics—had passed fielding by. The fielding stats of 1876 were combined to form an *average,* the "percentage of chances accepted," or fielding percentage. A "missing link" variant, devised by Al Wright in 1875, was to form averages by dividing the putouts by the number of games to yield a "putout average"; dividing the assists similarly to arrive at an "assist average"; and to divide putouts plus assists by games to get "fielding average." These averages took no account of errors. (Does Wright's "fielding average" look familiar? You may have recognized it as Bill James's Range Factor! Everything old is new again.)

This is all testimony to the changing nature of the game—not just to the evolving approaches of statisticians, but to fundamental changes in the game. These will be detailed in Chapter 7, "The Good Old Days Are Now."

The public's appetite for new statistics was not sated by the outburst of 1876. New measures were introduced in dizzying profusion in the remaining years of the century. Some of these did not catch on and were soon dropped, some for all time, others only to reappear with renewed vigor in the twentieth century.

The statistic which never resurfaced after its solitary appearance in 1880 was "Total Bases Run," a wonderfully silly figure which signified virtually nothing about either an individual's ability in isolation or his value to his team. It was sort of an RBI in reverse, or from the baserunner's perspective. Get on with a single, proceed to score in whatever manner, and you've touched four bases. Abner Dalrymple of Chicago was baseball history's only recorded leader in the category, with 501. Now there's a *major league* trivia question.

Another stat which was stillborn in the 1870s was times reached base on error (it was computed again in 1917-19 by the NL, then dropped for all time). Its twentieth-century companion piece, equally short-lived after its introduction in the 1910s, was runs allowed by fielders. Lanigan records this lovely bit of doggerel written to "honor" Chicago shortstop Red Corriden, whose errors in 1914 let in 20 runs:

> *Red Corriden was figuring the cost of livelihood.*
> *"'Tis plain," he said, "I do not get the money I should.*
> *According to my figurin', I'd be a millionaire*
> *If I could sell the boots I make for 30 cents a pair."*

Previously mentioned was another stat which blossomed in only one year (1879), Reached First Base. This resurfaced, however, in the early 1950s in an improved form called On Base Average, which may be the most widely familiar of all unofficial statistics. In the same manner, the "total bases per game" tabulation of the 1860s vanished only to be named an official stat decades later in its modified version of slugging percentage. And yet another 1860s stat, earned run average, dropped from sight in the 1880s only to return triumphant to the NL in 1912 and the AL in 1913, when Ban Johnson not only proclaimed it official but also dictated that the AL compile *no* official won-lost records (this state of affairs lasted for seven years, 1913-19.)

Another stat which was "sent back to the minors" before settling in for good in 1920 was the RBI. Introduced by a Buffalo newspaper in 1879, the stat was picked up the following year by the Chicago *Tribune* which, in the words of Preston D. Orem, "proudly presented the 'Runs Batted In' record of the Chicago players for the season, showing Anson and Kelly in the lead. Readers were unimpressed. Objections were that the men who led off, Dalrymple and Gore, did not have the same opportunities to knock in runs. The paper actually wound up almost apologizing for the computation." Even then astute fans knew the principal weakness of the statistic to be its extreme dependence on situation—in a particular at bat, whether or not men are on base; over a season or career, one's position in the batting order and the overall batting strength of one's team. It is a curious bit of logical relativism to observe that the fans of the nineteenth century rejected ribbies because of their poor relation to run-producing ability while twentieth-century fans embrace the stat for its presumed indication of that same quality.

Other statistics introduced before the turn of the century were stolen bases (though not caught stealing), sacrifice bunts, doubles, triples, homers, strikeouts for batters and for pitchers, bases on balls, hit by pitch (HBP), and, erratically, grounded into double play (GIDP). Caught stealing figures are available on a very sketchy basis in some of the later years of the century, as some newspapers carried the data in the box scores of home-team games. From 1907 on, Lanigan recorded CS in the box scores of the New York *Press,* but the leagues did not keep the figure officially until 1920. The AL has CS from that year to the present, excepting 1927, which members of the Society for American Baseball Research are now engaged in reconstructing from newspaper box scores. The NL kept CS from 1920 to 1925, then not again until 1951. League officials and managing editors evidently believed the stat to be of little value—newspapers still do not print it in their daily tabulation of league leaders or in the weekly summaries—but as will be demonstrated in a later chapter, the failed steal attempt has twice the impact on a team of the successful one.

The sacrifice bunt became a prime offensive weapon of the 1880s and began appearing as a statistical entry in box scores by 1889. The totals piled up in the years when a single run was precious—that is, from '89 to '93, then again from 1901 to 1920—were stupendous by modern standards (sacrifices counted as at bats until the early 1890s). Hardy Richardson had 68 sacrifice hits in 1891 (in 74 games!), Ray Chapman 67 in 1917; today it is unusual to see a player with as many as 20.

Batter bases on balls (and strikeouts) were recorded for the last year of the American Association, 1891, by Boston's Clarence Dow, and for some years of the mid-'90s in the National League, but didn't become an official statistic until 1910 in the NL, 1913 in the AL. Caught stealing, hit by pitch and grounded into double plays were not kept steadily in the nineteenth century, making it impossible for modern statisticians to apply the most sophisticated versions of their measures to early players.

The new century has added little in the way of *new* official statistics—ERA and RBIs and SLG are better regarded as revivals despite their respective adoption dates of 1912, 1920, and 1923. These are significant measures, to be sure, but they represent official baseball's classically conservative response to innovation: Wait forty or fifty years, then "make it new." Running counter to that trend have been baseball's two most interesting new stats of the century, the save and the game winning RBI. Both followed in fairly close relationship to a perception that something was occurring on the field yet, because it was not being measured, it had no verifiable reality. (Another such stat which did not survive, alas, was stolen bases off pitchers, which the American League recorded only in 1920-24.)

The same could have been said back in 1908, in a classic case of a statistic rushing in to fill a void, as Phillies' manager Billy Murray observed that his outfielder Sherry Magee had the happy facility of providing a long fly ball whenever presented with a situation of man on third, fewer than two outs. Taking up the cudgels on his player's behalf, Murray protested to the National League office that it was unfair to charge Magee with an unsuccessful time at bat when he was in fact succeeding, doing precisely what the situation demanded. Murray won his point, but baseball flipflopped a couple of times on this stat, in some years reverting to calling it a time at bat, in other years not even crediting an RBI.

The most delightfully loony stat of the century (though the GWRBI is giving it a run for the money) was unofficial: the "All-American Pitcher" award, given to Giants' reliever Otis Crandall after the 1910 season, then sinking into deserved oblivion. It went like this: Add together a pitcher's won-lost percentage, fielding percentage, and batting average, and voilà, you get an All-American. Crandall's combined figures of .810, .984, and .342, respectively, gave him 2,136 points and, according to those in the know, the best mark of all time, surpassing Al Spalding's 2,096 points of 1875. Who's the all-time All-American since 1910? You tell us. But seriously, folks, the idea wasn't a bad one—measuring the overall ability of pitchers—it was just that the inadequacies of the individual statistics were magnified by lumping them in this way.

There have been other new statistical tabulations in this century, but of a generally innocuous sort: counting intentional bases on balls, balks, wild pitches, shutouts, and sacrifice bunts and sacrifice flies against pitchers. Other new stats of a far superior quality appeared in the 1940s and '50s but have not yet gained the official stamp of approval. But unofficial new statistics, while they are what this book is about, are not the subject of this chapter.

Now that the genealogy of the more significant official measures has been described, it's time to evaluate the important ones you saw in the newspapers over breakfast, and a few which are tabulated officially only at year's end, or are found in the weekly *Sporting News*.

The first offensive statistic to consider will be that venerable, uncannily durable fraud, the batting average. What's wrong with it? What's right with it? We've recited the objections for the record, but we know as well as anyone else that this monument just won't topple; the best that can be hoped is that in time fans and officials will recognize it as a bit of nostalgia, a throwback to the period of its invention when power counted for naught, bases on balls were scarce, and no one wanted to place a statistical accomplishment in historical context because there wasn't much history as yet.

Time has given the batting average a powerful hold on the American baseball public; everyone knows that a man who hits .300 is a good hitter while one who hits .250 is not. Everyone knows that, no matter that it is not true. You want to trade Bill Madlock for Mike Schmidt? Bill Buckner for Darrell Evans? BA treats all hits in egalitarian fashion. A two-out bunt single in the ninth with no one on base and your team trailing by six runs counts the same as Bobby Thomson's "shot heard 'round the world." And what about a walk? Say you foul off four 3-2 pitches, then watch a close one go by to take your base. Where's your credit for a neat bit of offensive work? Not in the BA. And a .250 batting average may have represented a distinct accomplishment in certain years, like 1968 when the American League average was .230. That .250 hitter stood in the same relation to an average hitter of his season as a .277 hitter did in the National League in 1983—or a *.329* hitter in the NL of 1930! If .329 and .250 mean the same thing, roughly, what good is the measure?

So in attempting to assess batting excellence with the solitary yardstick of the batting average, we tend to diminish the accomplishments of (a) the extra-base hitter, whose blows have greater run-scoring potential, both for himself and for whatever men may be on base; (b) the batter whose talent it is to extract walks from pitchers who do not wish to put him on base, or whose power is such that pitchers will take their chances working the corners of the plate rather than risk an extra-base hit; (c) the batter whose misfortune it is to be playing in a period dominated by pitching, either because of the game's evolutionary cycles or because of rules-tinkering to stem a previous domination by the batters; and (d) the man whose hits are few but are well-timed, or clutch—they score runs. In brief, the BA is an unweighted average; it fails to account for at least one significant offensive category (not to mention hit by pitch, sacrifices, steals, and grounded into double play); it does not permit cross-era comparison; and it does not indicate value to the team.

And yet, the batting champion each year is declared to be the one with the highest batting average, and this will not soon change. And the Hall of Fame is filled with .300 hitters who couldn't carry the pine tar of many who will stay forever on the outside looking in. Knowledgeable fans have long realized that the ability to reach base and to produce runs are not adequately measured by batting average, and they have looked to other measures, for example, the other two components of the "triple crown," home runs and RBIs. Still more sophisticated fans have looked to the slugging percentage or On Base Average (and recently to the new statistics proposed by such men as Tom Boswell and Bill James; these will be discussed in the following chapter).

How well do these other stats compensate for the weaknesses of the BA when viewed in conjunction with it or in isolation? The slugging percentage does acknowledge the role of the man whose talent is for the long ball and who may, with management's blessing, be sacrificing bat control and thus batting average in order to let 'er rip. (The slugging percentage is the number of total bases divided by at bats rather than hits divided by at bats, which is the BA.) But the slugging percentage has its problems, too.

It declares that a double is worth two singles, that a triple is worth one and a half doubles, and that a home run is worth four singles. All of these proportions are intuitively pleasing, for they relate to the number of bases touched on each hit, but in terms of the hits' value in generating runs, the proportions are wrong. A home run in four at bats is not worth as much as four singles, for instance, in part because the run potential of the four singles is greater, in part because the man who hit the four singles did not also make three outs; yet the man who goes one for four at the plate, that one being a homer, has the same slugging percentage of 1.000 as a man who singles four times in four at bats.

Moreover, it is possible to attain a high slugging percentage without being a slugger. In other words, if you have a high batting average, you must have a decent slugging percentage; it's difficult to hit .350 and have a slugging percentage of only .400. Even a bunt single boosts not only your batting average but also your slugging percentage. (The attempt to counteract this problem is a statistic called Isolated Power, which will be discussed in Chapter 3.)

Other things the slugging percentage does not do are: indicate how many runs were produced by the hits; give any credit for other offensive categories, such as walks, hit by pitch, or steals; permit the comparison of sluggers from different eras (if Jimmie Foxx had a slugging percentage of .749 in 1932 and Mickey Mantle had one of .705 in 1957, was Foxx 7 percent superior? The answer is no, and we'll tell you why in Chapter 5).

Well, how about On Base Average? It has been around for quite a while and is still not an official statistic of the major leagues. But it does appear on a daily basis in some newspapers' leaders section, weekly in *The Sporting News,* and annually in the American League's averages book (since 1979, when Pete Palmer put it there). The OBA has the advantage of giving credit for walks and hit by pitch, but is an unweighted average and thus makes no distinction between those two events and, say, a grand-slam homer. A fellow like Eddie Yost, who drew nearly a walk a game in some years in which he hit under .250, gets his credit with this stat, as does a Gene Tenace, one of those guys whose statistical line looks like zip without his walks. Similarly, players like Mickey Rivers or Mookie Wilson, leadoff hitters with a lot of speed, no power, and no patience are exposed by the OBA as distinctly marginal major leaguers, even in years when their batting averages look respectable or excellent. In short, OBA does tell you more about a man's ability to get on than BA does, and thus is a better indicator of run generation, but it's not enough by itself to separate "good" hitters from "average" or "poor" ones.

RBIs? Don't they indicate run production and clutch ability? Yes and no. They tell how many runs a batter pushed across the plate, all right, but they don't tell how many fewer he might have driven in had he batted eighth rather than fourth, or how many more he might have driven in on a team that put more men on base. They don't even tell how many more runs a batter might have driven in if he had delivered a higher proportion of his hits with men on base. (The American League kept RBI Opportunities—men on base presented to each batter—as an official stat for the first three weeks of 1918, then saw how much work was involved and ditched it.)

If you've got George Foster batting cleanup for the Mets, and the men batting ahead of him are Mookie Wilson, Bob Bailor, and Dave Kingman, there is no way he's going to drive in 140 or 150 runs the way he did when he had Pete Rose, Davey Concepcion, and Joe Morgan on base all the time.

How to credit clutch hitters on teams that are last in the league or division in runs scored? You could find the percentage of a team's runs driven in by the individual, as Bob Carroll did in the 1982 issue of *The National Pastime.* He found that only nine times in history had a man driven in 20 percent or more of his team's runs (the all-time best was 22.75 percent by Nate Colbert of the 1972 Padres). That's an interesting stat, but it has the same problem as RBIs. It's situation-dependent. To get a high percentage of team runs batted in, you must play for a lousy club. The RBI champions of the major leagues on an all-time basis, Hack Wilson with 191 and Lou Gehrig with 184, couldn't make the 20 percent level because they were playing for good ballclubs; yet Bill Buckner did, in 1981.

The RBI does tell you something about run-producing ability, but not enough: It's a situation-dependent statistic, inextricably tied to factors which vary wildly for individuals on the same team or on others. And the RBI makes no distinction between being hit by a pitch to drive in the twelfth run of a game that concludes 14-3 and, again for comparison, the Thomson blast.

And so we come to the newly formulated game winning RBI—a noble attempt at describing the value of a hit to the team, its "clutchness"—but a measure which was misconceived in its presumption that a game could be won with a hit in the first inning. A man who drives in a run in the first inning is simply doing his job, not performing an extraordinary feat; if the pitcher makes that run hold up by throwing a shutout, bully for him, but why credit the hitter? Were he to drive in the lone run of the game in the seventh inning or later, that would be different. Nonetheless, the current formulation of the stat would give the man who drove in that first-inning run a GWRBI even if his team eventually won 22-0, since it gave the team a lead that was never relinquished.

Worst, the GWRBI is situation-dependent to an even greater degree than the RBI. You can't play for a lousy team and lead the league in GWRBIs because there aren't enough GWs to go around. And it's as hard to accumulate GWRBIs from the eighth place in the batting order as it is to accumulate RBIs. Last, if you put your team ahead with an RBI in the bottom of the eighth, why should you lose your GWRBI simply because the pitcher allows the lead to be lost? Wasn't your hit "clutch"? Say the pitcher allows the score to be tied, then a teammate might pick up the GWRBI that should have been safely tucked away for you. Nicely motivated, the GWRBI, but we liked Bases Touched better.

The third jewel in the triple crown, home runs, is an important event on the baseball field because it can deliver several runs at a time: It is the offense's most productive weapon. But statistically, it's as dull as dishwater because it is a counter stat—it simply tells you *how many*, not when or how. For comparison purposes we are forced to assume that more home runs is better than fewer, but to know that one man hit 37 homers and another 9 would not tell us which of the two was the better hitter (produced more runs for his team); in fact, these are the 1982 figures for Dave Kingman and Keith Hernandez. It would not tell us whose hits produced more extra bases, who made the fewest outs proportionate to his times at bat, and so on. Other counter stats are doubles, triples, steals, strikeouts, bases on balls, and so on. There's nothing wrong with them if all we ask them to do is tell us how many. It's when we attempt to make more of them that we overextend their usefulness. A man with a lot of stolen bases is not necessarily the best baserunner; he might have been caught as often as he stole and thus have cost his team many runs on balance. A man with the most triples is not necessarily a slugger or a speed merchant; he probably plays half his games in a park conducive to triples, like the Astrodome.

Hits? High numbers are largely a function of at bats (presuming ability), so this stat tells little not indicated by the batting average, presuming regular play.

Runs? They're situation-dependent, just as they were 140 years ago when they were first recorded. Runs don't tell you anything useful any more, though. In the early 1860s, when hits were not recorded, runs served very nearly in their stead. Today you still have to get to first in order to score a run, but if you're playing for a poor offensive team, you get left on base a lot. Scoring a large number of runs is a function not only of the team you play for, but also of the position you occupy in the batting order. You're not going to find anybody leading the league in runs (or any other offensive category) batting seventh or eighth.

It's an odd fact that from being the most interesting stat in the early days of baseball, runs has become the least interesting stat of today; it's odd in that runs remain the essence of baseball, remain the key to victory. What has happened over the years is that the correlation between runs and times reached base has been almost constantly widening. In 1875 the number of hits allowed per nine innings was, incredibly, not much different from what it is today. Tommy Bond of Hartford allowed only 7.95 hits per nine innings (facing underhand pitching was easy?). Bases on balls were in force at this time, but eight balls were required to get one, which accounts for their scarcity in the 1870s. Today, with walks greatly increased and hits only somewhat reduced, the number of *runs* per nine innings has dropped dramatically, although not the number of earned runs. Indeed, as the ratio of hits to runs has diminished through the years, the ratio of earned runs to total runs has increased. In 1876, for example, the National League scored 3,066 runs, of which only 1,201—39.2 percent—were earned. By the early 1890s this figure reached *70* percent, an extraordinary advance. It took until 1920 to reach 80 percent, and by the late 1940s it leveled off in the 87-89 percent range, where it remains.

On to the pitching statistics, the ones you commonly see. First is wins, with its correlated average of won-lost percentage. Wins are a team statistic, obviously, as are losses, but we credit a win entirely to one pitcher in each game. Why not to the shortstop? Or the left fielder? Or some combination of the three? In a 13-11 game, several players may have had more to do with the win than any pitcher. No matter. We're not going to change this custom, though Ban Johnson gave it a good try.

To win many games a pitcher generally must play for a *team* which wins many games (we discount relievers from this discussion because they rarely win 15 or more) or must benefit from extraordinary support in his starts or must allow so few runs that even his team's meager offense will be enough, as Tom Seaver and Steve Carlton did in the early 1970s. Verdict on both wins and the won-lost percentage: situation-dependent. Look at Red Ruffing's W-L record with the miserable Red Sox of the 1930s, then his mark with the Yankees. Or Mike Cuellar with Houston, then with Baltimore. Conversely, look at Ron Davis with the Yanks and then with the Twins. There is an endless list of good pitchers traded up in the standings by a tailender to "emerge" as stars.

The recognition of the weakness of this statistic came early. Originally it was not computed by such men as Chadwick because most teams leaned heavily, if not exclusively, on one starter, and relievers as we know them today did not exist. As the season schedules lengthened and pitchers began to throw breaking balls—by 1884, overhand—the need for a pitching staff became evident, and separating out the team's record on the basis of who was in the box seemed a good idea. It was not and is not a good statistic, however, for the simple reason that one may pitch poorly and win, or pitch well and lose.

The natural corrective to this deficiency of the won-lost percentage is the earned run average—which, strangely, preceded it, gave way to it in the 1880s, and then returned in 1913. Originally, the ERA was computed as earned runs per game because pitchers almost invariably went nine innings. In this century it has been calculated as ER times 9 divided by innings pitched.

The purpose of the earned run average is noble: to give a pitcher credit for doing what *he* can to prevent runs from scoring, aside from his own fielding lapses and those of the men around him. It succeeds to a remarkable extent in isolating the performance of the pitcher from his situation, but objections to the statistic remain. Say a pitcher retires the first two men in an inning, then has the shortstop kick a ground ball to allow the batter to reach first base. Six runs follow before the third out is secured. How many of these runs are earned? None. (Exception: If a reliever comes on in mid-inning, any men he puts on base who come in to score would be classified as earned for the reliever, though unearned for the team statistic. This peculiarity accounts for the occasional case in which a team's unearned runs will exceed the individual totals of its staff.) Is this reasonable? Yes. Is it a fair depiction of the pitcher's performance in that inning? No.

The prime difficulty with the ERA in the early days, say 1913, when one of every four runs scored was unearned, was that a pitcher got a lot of credit in his ERA for playing with a bad defensive club. The errors would serve to cover up in the ERA a good many runs which probably should not have scored. Those runs would hurt the team, but not the pitcher's ERA. This situation is aggravated further by use of the newly computed ERAs for pitchers prior to 1913, the first year of its official status. Example: Bobby Mathews, sole pitcher for the New York Mutuals of 1876, allowed 7.19 runs per game, yet his ERA was only 2.86, almost a perfect illustration of the league's 40 percent proportion of earned runs.

In modern baseball, post-1946, with 88 of every 100 runs being earned, the problem has shifted. The pitcher with a bad defense behind him is going to be hurt less by errors than by the balls that wind up recorded as base hits which a superior defensive team might have stopped. Bottom line: You pitch for a bad club, you get hurt. There is no way to isolate pitching skill completely unless it is through play-by-play observation and meticulous, consistent bookkeeping.

In a column in *The Sporting News* on October 9, 1976, Leonard Koppett, in an overall condemnation of earned run average as a misleading statistic, suggested that total runs allowed per game would be a better measure. It is a proposition worth considering, now that the proportion of unearned runs has been level for some forty years; one can reasonably assume that further improvements in fielding would be of an infinitesimal nature. However, when you look at the spread in fielding percentage between the worst team and the best, and then examine the number of additional unearned runs scored, pitchers on low-fielding-percentage teams probably still have a good case for continuing to have their effectiveness computed through the ERA. In 1982, for example, in the American League, only 39 of the runs scored against Baltimore were the result of errors; yet Oakland, with the most error-prone defense in the league, allowed 84 unearned runs.

What gave rise to the ERA, and what we appreciate about it, is that like batting average it is an attempt at an isolating stat, a measure of individual performance not dependent upon one's own team. While the ERA is a far more accurate reflection of a pitcher's value than the BA is of a hitter's, it fails to a greater degree than BA in offering an isolated measure. For a truly unalloyed individual pitching measure we must look to the glamor statistic of strikeouts, the pitcher's mate to the home run (though home runs are hugely dependent upon home park, strikeouts to only a slight degree).

Is a strikeout artist a good pitcher? Maybe yes, maybe no, as indicated in the discussion of the Carlton-Ryan-Johnson triad; an analogue would be to ask whether a home-run slugger was a good hitter. The two stats run together: Periods of high home-run activity (as a percentage of all hits) invariably are accompanied by high strikeout totals. Strikeout totals, however, may soar even in the absence of overzealous swingers, say, as the result of a rules change such as the legalization of overhand pitching in 1884, the introduction of the foul strike (NL, 1901; AL, 1903), or the expansion of the strike zone in 1963.

Just as home-run totals are a function of the era in which one plays, so are strikeouts. The great nineteenth-century totals—Matches Kilroy's 513, Toad Ramsey's 499, One Arm Daily's 483—were achieved under different rules and fashions. No one in the century fanned batters at the rate of one per inning; indeed, among regular pitchers (154 innings or more), only Herb Score did until 1960. In the next five years the barrier was passed by Sandy Koufax, Jim Maloney, Bob Veale, Sam McDowell, and Sonny Siebert. Walter Johnson, Rube Waddell, and Bob Feller didn't run up numbers like that. Were they slower, or easier to hit, than Sonny Siebert?

Even in today's game, which lends itself to the accumulation of, by historic standards, high strikeout totals for a good many pitchers and batters, the strikeout is, as it always has been, just another way to make an out. Yes, it is a sure way to register an out without the risk of advancing baserunners and so is highly useful in a situation like man on third with fewer than two outs; otherwise, it is a vastly overrated stat because it has nothing to do with victory or defeat—it is mere spectacle. A high total indicates raw talent and overpowering stuff, but the imperative of the pitcher is simply to retire the batter, not to crush him. What's not listed in your daily averages are strikeouts by batters—fans are not as interested in that because it's a negative measure—yet the strikeout may be a more significant stat for batters than it is for pitchers.

On second thought, maybe it's just the same. So few errors are being made these days—2 in 100 chances, on average—maybe there's not a great premium on putting the ball into play anymore. Sure, you might move a runner up with a grounder hit behind him or with a long fly, but on the other hand, with a strikeout you do avoid hitting into a double play. At least that's what Darryl Strawberry said in his rookie season when asked why he was unperturbed about striking out every third time he came to the plate!

Bases on balls will drive a manager crazy and put lead in fielders' feet, but it is possible to survive, even to excel, without first-rate control, provided your stuff is good enough to hold down the number of hits. Occasionally you will see a stat called opponents' Batting Average, or opponents' On Base Average, or opponents' Slugging Percentage, all of which seem at first blush more revealing than they are. In fact these calculations are all academic, in that it doesn't matter how many men a pitcher puts on base. Theoretically he can put three men on every inning, leave twenty-seven on base, and pitch a shutout. A man who gives up one hit over nine innings can lose 1-0; it's even possible to allow no hits and lose. Who is the better pitcher? The man with the shutout and twenty-seven baserunners allowed, or the man who allows one hit? No matter how sophisticated your measurements for pitchers, the only really significant one is runs.

The nature of baseball at all points is one man against nine. It's the pitcher against a series of batters. With that situation prevailing, we have tended to examine batting with intricate, ingenious stats, while viewing pitching through generally much weaker, though perhaps more copious, measurements. What if the game were to be turned around so that we had a "pitching order"—nine pitchers facing one batter? Think of that one for a minute. The nature of the statistics would change, too, so that your batting stats would be vastly simplified. You wouldn't care about all the individual components of the batter's performance, all combining in some obscure fashion to reveal run production. You'd care only about *runs*. Yet what each of the nine pitchers did would bear intense scrutiny, and over the course of a year each pitcher's opponents' BA, opponents' OBA, opponents' SLG, and so forth, would be recorded and turned this way and that to come up with a sense of how many runs saved each pitcher achieved.

A stat with an interesting history is completed games. This is your basic counter stat, but it's taken to mean more than most of those measurements by baseball people and knowledgeable baseball fans. When everyone was completing 90-100 percent of his starts, the stat was without meaning and thus was not kept. As relief pitchers crept into the game after 1905, the percentage of complete games declined rapidly, as illustrated in Table II, 6. By the 1920s it became a point of honor to complete three quarters of one's starts; today the man who completes half is quite likely to lead his league. So with these shifting standards, what do CGs mean? Well, it's useful to know that of a pitcher's 37 starts, he completed 18. That he *accepted* no assistance in 18 of his 37 games is indisputable; that he required none is a judgment for others such as fans or press to make. There is managerial discretion involved: It is seldom a pitcher's decision whether to go nine innings or not, and there are different managerial styles and philosophies. There are the pilots who will say give me a good six or seven, fire as hard as you can as long as you can, and I'll bring in The Goose to wrap it up. There are others who encourage their starting pitchers to go nine, feeling that it builds team morale, staff morale, and individual confidence. Verdict: situation-dependent, to a fatal degree. CGs tell you as much about the manager and his evaluation of his bullpen as they tell you about the arm or heart of the pitcher.

Table II, 6. *Percentage of Games Completed and Saved, 1876-1982*

	CG Pct.	Save Pct.
1876-1904	90.5	1.3
1905-1923	63.3	5.4
1924-1946	45.9	8.7
1947-1958	35.1	12.7
1959-1973	25.8	18.7
1974-1982	22.9	19.2

Can we say that a pitcher with 18 complete games out of 37 starts is better than one with 12 complete games in 35 starts? Not without a lot of supporting help we can't, not without a store of knowledge about the individuals, the teams, and especially the eras involved. The more uses to which we attempt to put the stat, the weaker it becomes, the more attenuated its force. If we declare the hurler with 18 CG's "better," how are we to compare him with another pitcher from, say, fifty years earlier who completed 27 out of 30 starts? Or another pitcher of eighty years ago who completed all the games he started? (Jack Taylor completed every one of the 187 games he started *over five years*.) Or what about Will White, who in 1880 started 76 games and completed 75 of them? But the rules were different, you say, or the ball was less resilient, or they pitched from a different distance, with a different motion, or this, or that. The point is, there are limits to what a traditional baseball statistic can tell you about a player's performance in any given year, let alone compare his efforts to those of a player from a different era.

Perhaps the most interesting new statistic of this century is the one associated with the most significant strategic element since the advent of the gopher ball—saves. Now shown in the papers on a daily basis, saves were not officially recorded at all until 1960; it was at the instigation of Jerry Holtzman of the Chicago *Sun-Times,* with the cooperation of *The Sporting News,* that this statistic was finally accepted. The need arose because relievers operated at a disadvantage when it came to picking up wins, and at an advantage in ERA. The bullpenners were a new breed, and as their role increased, the need arose to identify excellence, as it had long ago for batters, starting pitchers, and fielders.

The save is, clearly, another stat that hinges on game situation and managerial discretion. If you are a Ron Davis on a team that has a Goose Gossage, the best you can hope for is to have a great won-lost record, as Davis did in 1979 and '80. To pile up a lot of saves, you have to be saved for save situations, as Martin reserves Gossage; Howser, Quisenberry; or Herzog, Sutter. These relief stars are not brought in with their teams trailing; the game must be tied or preferably the lead in hand. The prime statistical drawback is that there is no negative to counteract the positive, no stat for saves blown (except, all too often, a victory for the "fireman").

In April 1982, *Sports Illustrated* produced a battery of well-conceived, thought-provoking new measurements for relief pitchers which at last attempted, among other things, to give middle and long relievers their due. Alas, the *SI* method was too rigorous for the average fan, and the scheme dropped from sight. It was a worthy attempt, but perhaps the perfect example of breaking a butterfly on the wheel. The "Rolaids Formula," which at least takes games lost and games won into account, is a mild improvement over simply counting saves or adding saves and wins. It awards two points for a save or a win and deducts one point for a loss. The reasoning, we suppose, is that a reliever is a high-wire walker without a net—one slip may have fatal consequences. His chances of drawing a loss are far greater than his chances of picking up a win, which requires the intervention of forces not his own.

Briefly, fielding stats. The central weakness of fielding percentage is well known: You can't make an error on a ball you don't touch. To counter the weakness in fielding percentage and to credit the plays made as well as the plays not made, total chances per game is a useful statistic—and when errors are deducted from chances, you have Range Factor.

Another difficulty with the fielding percentage is that to understand what figure represents average performance (and thus be able to identify inferior and superior fielders), one must adjust for position: A shortstop who fields .980 has done extremely well, but a first baseman, catcher, or outfielder with that figure would have been far below average. Thus the fan must bring to the fielding percentage a great deal of background knowledge—the average percentage for each position. This is a demand not created by the batting average (all men stepping to the plate occupy the same position—batter) . . . and yet the sophisticated fan knows that a batting line of .267, 10 HR, 80 RBI will mean different things when applied to a shortstop or to a left fielder. In other words, just as any evaluation of fielding performance carries an inherent positional bias, so does batting performance.

High double-play totals are believed to indicate excellence among middle infielders, but the more double plays a club turns, as a rule, the lousier the pitching. Which teams had the most double plays in major league history? In the 154-game season, the Philadelphia A's of 1949 and the Dodgers of 1958; in the 162-game season, Toronto and Boston of 1980 and Pittsburgh of 1966. Of these, only the last mentioned had a team ERA better than the league average. If the pitchers are putting a lot of men on base, the team can get a lot of double plays even without a great fielding shortstop and second baseman.

The idea of crediting stellar fielding plays individually has been proposed occasionally ever since 1868, when Chadwick wrote: "The best player in a nine is he who makes the most good plays in a match, not the one who commits the fewest errors, and it is in the record of his good plays that we are to look for the most correct data for an estimate of his skill in the position he occupies."

Chadwick never got anywhere with this rating system because of the subjectivity that would contaminate any such stat. At a meeting of the Society for American Baseball Research a few years back, an individual advocated the creation of a "game saving play" stat to credit fielders for clutch performance in the manner that the game winning RBI is intended to credit hitters. As with the GWRBI, the intentions are noble but the implementation awful. In fact, the fielding equivalent would be far worse, as hometown boosterism might infect the official scorers, who would be called upon to make judgments about the difficulty of particular plays.

Although we pilloried the GWRBI earlier in this chapter, it does bespeak a new concern for integrating individual performance and team success, an intention 180 degrees removed from the practice of the past hundred years, in which the aim was to isolate the play of the individual from his surroundings. This new concern is, ironically, the same as the old concern, the same vision of baseball statistics as Henry

Chadwick's. He and his predecessors originally measured simply outs and runs, not even hits; the emphasis then was the team game, not the individual accomplishment.

This orientation today forms the basis not only of the GWRBI but of the entire new statistics movement. Whether the new measure be Steve Mann's Run Productivity Average, Bill James's Runs Created, or Pete Palmer's Linear Weights, all attempt to express a man's performance in terms of his contribution to victory.

The batting average has retained its primacy for so long because it affords the illusion that what is being measured in isolation is pure ability, ability apart from situation. But because such traditional stats are not normalized to league average, nor adjusted for home park, nor weighted properly, all you have is the *illusion* of purely individual accountability, while in fact the stats are extremely situation-dependent.

The new statistics come around full circle to the game as it was originally understood. And what's remarkable about this is that in order to be led back to the primordial simplicity of the 1840s, '50s, and '60s, we are availing ourselves of information produced from computer simulations, and we are employing some complicated, if not complex, modes of computation. In other words, what we have with the new statistics is simplicity arising from complexity, while what we have had for the last hundred years or so has been complexity as a product of simplicity. We had the aura of simplicity, but in fact we were using statistics—such as the BA, the RBI, the W-L Pct.—so fraught with bias, so antithetical to the nature of the game, so demanding of special knowledge about historical context—that they were in reality highly complicated. If you compare Ty Cobb's .382 batting average of 1910 with Ted Williams's .388 batting average of 1957, the difference appears to be six thousandths (six "percentage points"). To find out the true difference without benefit of a new statistical approach would involve you in a series of fairly convoluted assumptions if not serpentine computations.

The computer has made possible the rapid analysis of mountains of raw data based upon observed cases or mathematically accurate, probabilistic computer simulations. What is a single worth, or a walk, or a homer, in terms of its run-producing capacity? How valuable is a stolen base? These questions were once thought to be unanswerable, but they are mysteries no longer. Now, for example, the slugging percentage can be reweighted, baserunning can be viewed in conjunction with other offensive weapons, and more. The computer has reversed the thought process represented by traditional baseball statistics, and the new statisticians, rather than wrapping the game in layer ·upon layer of newfound complexity, are peeling the existing layers away.

Let's go on and look at the grand old game through brand-new glasses.

------

[1] In 1876, bases on balls were counted as outs; in 1887, they were counted as hits. *The Baseball Encyclopedia* corrected all the 1887 averages to remove walks from consideration, but failed to change all 1876 BAs.

[2] Being hit by a pitch did not send the batter to first base until 1884 in the American Association, 1887 in the National League.

3

THE NEW STATISTICS

"There's no such thing as a new statistic," quoth Earl Weaver. The former (and, in all likelihood, future) manager *par excellence* is worth listening to on most any baseball matter, but not this one. Wasn't the batting average once new? Or the ERA, the RBI, the save? Each came into being to account for something real that was happening on the baseball field but was not being measured. It is odd for a man proclaimed as a baseball visionary to presume that all that can ever be done—in any area of the game—has already been done.

Weaver's remark illustrates perfectly how baseball's inbred resistance to change affects even its best minds. One can try to close the door on innovation, but it will occur all the same, whenever conditions are ripe. And for innovation in baseball statistics, conditions are ripe right now.

In the 1980s the computer has become less mystifying and thus less frightening to the American public, to the extent that the family with a microcomputer in its home today is no more peculiar than a family possessing a television set in 1950. Although the computer applications to baseball were clear to a handful of individuals as early as the '50s, it was not until the present decade that a major-league organization sought to augment its analysis of game situations and personnel with the aid of a computer. As this is written, five clubs—Oakland, Chicago (AL), New York (AL), Texas, and Atlanta—have committed to the twentieth century; by the time you read this, there may well be more. The 1980s give every sign of being as receptive to technological advance as the 1880s, the last great period of statistical experimentation and innovation in baseball.

The 1980s also brought unprecedented media attention to the efforts of two men in particular to redefine the measures of individual performance. Tom Boswell, not a statistician by trade or inclination but rather a sportswriter for the Washington *Post*, has become widely known for his statistic Total Average, a measure of offensive proficiency which takes into account not only batting but also baserunning skills. Bill James has become celebrated not for any one new statistic but for a bundle of them: Range Factor, Runs Created, Value Approximation Method, and more. James has popularized a different approach to the whole question of what baseball statistics are for—that they are not brass knuckles to beat a barroom adversary with, but tools for achieving a better understanding of the game and heightening one's pleasure in it.

To this method he has given the name "sabermetrics," a neologism combining the acronym of the Society for American Baseball Research (SABR) with the suffix indicating measurement. It's not a euphonious coinage, but it may be too late to turn back its incursion into baseball argot; already the business card of the statistical analyst of the Texas Rangers reads, "Craig Wright, Sabermetrician."

So what is sabermetrics? Simply, in James's words, "Sabermetrics is the mathematical and statistical analysis of baseball records." It differs from conventional sportswriting, he wrote in the *1981 Baseball Abstract*, the last of his five self-published editions: "Sportswriting draws on the available evidence, and forces conclusions by selecting and arranging that evidence so that it points in the direction desired. [Note: This is what Disraeli had in mind when he said, 'There are three kinds of lies: lies, damned lies, and statistics.'] Sabermetrics introduces *new* evidence, previously unknown data derived from original source material." Also, "sabermetrics puts into place formulas, schematic designs, or theories of relationship which would compare not only this player to that one, but to any other player who might be introduced into the discussion."

James gave a name to this method of baseball analysis, but he didn't invent it, as he indicated in the 1981 edition: "I could have said [in response to the question, 'What is sabermetrics?'] . . . like Louis XIV, '*Sabermetrics? C'est moi.*' But people have been doing sabermetric-type things for at least sixty years, and I am only thirty-one." So where did sabermetrics come from? James may have had in mind the analytical work of Ernie Lanigan, who once confided to Fred Lieb, "I really don't care much about baseball, or looking at ball games. . . All my interest in baseball is in its statistics" (Lanigan might have served as a model for J. Henry Waugh, the proprietor of Robert Coover's *The Universal Baseball Association*). But the sabermetric method as defined above could be said to have been employed sixty years earlier still, when Chadwick formulated the prototypical slugging average or when he filched from cricket the statistic Runs Per Game.

The pioneers of the New Statistics, men like Chadwick, Dow, and Lanigan, all recognized at least some of the weaknesses in the measures then prevailing and were spurred by a vision of how the game's yardsticks might be improved. Each new statistic or adjustment flowed from an earlier one, until the long skein of invention came to a close, or at least a break, after the adoption of the slugging percentage in 1923. The next significant event on the New Statistical trail was the private publication by Ted Oliver, in 1944, of a booklet entitled *Kings of the Mound*. Little known then or now, its new statistic, the Weighted Rating System for pitchers, was motivated by the inadequacies of both the won-lost percentage and the ERA when it came to evaluating pitchers laboring for poor teams. The Oliver formula, ingenious if flawed, was: pitcher's won-lost percentage minus the team's won-lost percentage—after removing the pitcher's decisions from the team's record—then multiplying the difference by the pitcher's number of decisions. (With a slight modification, we discuss this statistic as Wins Above Team in Chapter 10 and record its all-time and seasonal leaders in the tables at the end of the book.) Here is an example of the Oliver method as applied to Bobby Castillo, who in 1982 pitched very well in going 13-11 for a very bad Minnesota club (60-102; without him, 47-91).

$$\left(\frac{13}{24} - \frac{47}{138} \right) \times 24$$

or

$$.542 - .341 \times (24)$$

or

$$.201 \times 24 = 4.824$$

The figure of 4.824 would have been represented by Ted Oliver as "4,824 points"; he did not seem to recognize that had he retained the decimal point, his rating would have been expressed in *wins*. Thus the number of wins Castillo accounted for in his 24 decisions that an average Minnesota pitcher would *not* have gained was 4.8. This performance, by the way, ranked Castillo second (to Charlie Hough of Texas) among AL pitchers that year *as judged by this stat alone;* it is by no means the sole tool or the best one to analyze pitching performance. The weaknesses of the Oliver method include a mathematical bias in favor of good pitchers on poor teams and against good pitchers on great staffs, and a basic flaw is that it is denominated in actual victories, which are not purely the accomplishment of a hurler.

Another "alternate" statistic came up in 1951, a pitcher's stat as well and even less heralded than Oliver's. A fellow named Alfred P. Berry came up with the invention of Average Bases Allowed or ABA. This too was designed to supplement won-lost pct. and earned run average. The ABA was very simple: Total bases allowed divided by innings pitched. The ABA, according to Berry, made the pitcher rating truly individual, in the same way as the batter ratings (BA or SLG) freed the batter's accomplishment from those of his teammates. The earned run average, Berry reasoned, deceives because the poorer the team's defense, the more earned runs, as well as unearned runs, will be charged to the pitcher. His analysis of the ERA was correct, but the ABA was not the answer for, as discussed earlier, the task of the pitcher is not to deny basehits or baserunners, but to deny runs.

Also accruing to Berry's credit was the recognition that old-time pitchers, working at a time when errors were plentiful, had a good many of their own pitching mistakes covered up by the presence of an error in the inning to wipe the ERA slate clean. Grover Cleveland Alexander, for example, posted a remarkable ERA of 1.22 in 1915, allowing only 51 earned runs in 376 innings. However, he allowed an additional 35 of the unearned variety. Sandy Koufax, in 1963, posted an ERA of 1.88, allowing 65 earned runs in 311 innings, presumably 50 percent "worse" than Alexander's mark, but Koufax allowed only *3* unearned runs! If we were to adopt Leonard Koppett's idea of maintaining pitcher records on a runs-allowed basis, Alexander would have a mark of 2.06, and Koufax 1.97. Not quite fair, we know—surely more than a few of Alex's unearned runs scored without his complicity—but provocative nonetheless.

Although the impulse to improve our understanding and appreciation of baseball through the laying on of numbers had been present from the game's beginnings, it was not until August 2, 1954, in of all places *Life* magazine, that the New Statistics movement was truly born. On that date there appeared an article by the game's designated guru Branch Rickey, supported considerably by statistician Allan Roth, which was optimistically titled "Goodby to Some Old Baseball Ideas." With the aid of some new mathematical tools, it sought to puncture long-held misconceptions about how the game was divided among its elements (batting, baserunning, pitching, fielding), who was best at playing it, and what caused one team to win and another to lose. This is a pretty fair statement of what the New Statistics is about.

Although the old ideas remained in place despite his efforts, Rickey had shaken them to their foundations. He attacked the batting average and proposed in its place the On Base Average; advocated the use of Isolated Power (extra bases beyond singles, divided by at bats) as a better measure than slugging percentage; introduced a "clutch" measure of run-scoring efficiency for teams, and a similar concept for pitchers (earned runs divided by baserunners allowed); reaffirmed the basic validity of the ERA and saw the strikeout for the insubstantial stat it was; and more. But the most important thing Rickey did for baseball statistics was to pull it back along the wrong path it had taken at the crossroads long ago: to strip the game and its stats to their essentials and start again, this time remembering that individual stats came into being as an attempt to apportion the players' contributions to achieving victory, for that is what the game is about.

"Baseball people generally are allergic to new ideas," Rickey wrote. "We are slow to change. For fifty-one years I have judged baseball by personal observation, by considered opinion and by accepted statistical methods. But recently I have come upon a device for measuring baseball which has compelled me to put different values on some of my oldest and most cherished theories. It reveals some new and startling truths about the nature of the game. It is a means of gauging with a high degree of accuracy important factors which contribute to winning and losing baseball games. . . . The formula, for so I designate it, is what mathematicians call a simple, additive equation:

$$\left(\frac{H + BB + HP}{AB + BB + HP} + \frac{3\,(TB\text{-}H)}{4\,AB} - \frac{R}{H + BB + HP} \right) - \left(\frac{H}{AB} + \frac{BB + HB}{AB + BB + HB} + \frac{ER}{H + BB + HB} - \frac{SO}{8(AB + H + HB)} - F \right) = G$$

"The part of the equation in the first parenthesis stands for a baseball's team offense. The part in the second parenthesis represents defense. The difference between the two—G, for game or games—represents a team's efficiency."

What we have here is the first attempt to represent the totality of the game through its statistical component parts. Another way of stating the formula above is to say that if the first part—the offense, or runs scored—exceeds the second part—the defense, or runs allowed—then G, the team efficiency or won-lost percentage, should exceed .500. This is a startlingly simple (or rather, seemingly simple) realization, that just as the team which scores more runs in a game gets the win, so a team which over the course of a season scores more runs than it allows should win more games than it loses—and by an extent correlated to its run differential!

How did Rickey and Roth come up with the formula? "Only after reverting to bare ABC's was any progress noted. We knew, of course, that all baseball was divided into two parts—offense and defense. We concluded further that weakness or strength in either of these departments could be measured in terms of runs." Once mathematicians at M.I.T. confirmed for them that the correlation of team standings with run differential was 96.2 percent accurate over the past twenty years, the task became to identify the component parts of runs.

In the formula on page 41, the first segment of the offense (H + BB + HP) ÷ (AB + BB + HP), is the On Base Average. The second segment is Isolated Power, multiplied by .75. The third segment, applicable to teams but not to individuals, is percentage of baserunners scoring, or run-scoring efficiency ("clutch"); RBIs were not, Rickey stated, a suitable measure of individuals' clutch ability.

In the defensive half of the formula, the first segment is simply opponents' batting average. The second is opponents reaching base through pitcher's wildness. (Rickey divided the opponents' On Base Average into these constituent parts in an attempt to isolate "stuff" from control.) The third segment indicates a pitcher's "clutch" ability, and the fourth, his strikeout ability, multiplied by only .125 because it was not very important. The fifth segment of the defense, F for fielding, was deemed unmeasurable. "There is nothing on earth anyone can do with fielding," Rickey declared, but he did indicate that fielding was far less significant than pitching as a proportion of total defense: He ventured that while good fielding might account for the critical run in four or five games a year, it was worth only about half as much as pitching.

Rickey and Roth's fundamental contribution to the advancement of baseball statistics comes from their conceptual revisionism, their willingness to strip the game down to its basic unit, the run, and reconstruct its statistics accordingly. The Rickey formula (though perhaps Roth deserves even more credit) has been superseded in terms of accuracy. The method of correlating runs with wins has been improved in recent years, and the formula for analyzing runs in terms of their individual components has, too. But the existence of the space shuttle does not tarnish the accomplishment of the Wright brothers (Orville and Wilbur, not Harry and George).

In recognizing that traditional baseball statistics did not give an adequate sense of an individual's worth or of a team's prospects of victory, Rickey anticipated the future. Twenty-eight years later, a writer for *Discover* magazine, surely unaware that he was echoing baseball's Mahatma, described the impetus to the New Statistics: "Sabermetricians have tackled this problem [the inadequacy of traditional offensive measures] by devising a new statistic, one that directly measures a player's ability both to score and to drive in runs. The number has been calculated by various analysts under various designations: batting rating, run productivity average, runs created, and batter's run average, to name a few. It usually comes down to this simple fact: The total number of runs a team scores in a season is proportional to some combination of its hits, walks, steals, and other factors that result in batters getting on base or advancing other runners. Although the number of runs scored by a particular hit depends on how many men were on base, the differences tend to cancel themselves out over a season."

This understanding did not evaporate in the years between Rickey's article and the dawn of sabermetrics by that name. In 1959 the scholarly *Operations Research Journal* published an article by George R. Lindsey titled "Statistical Data Useful for the Operation of a Baseball Team." As far as baseball people were concerned, Lindsey might as well have been writing in Icelandic. Lindsey and his father had recorded play-by-play data of several hundred baseball games in order to evaluate such long-standing perplexities as whether in facing a right-handed pitcher, a lefthanded hitter did possess an advantage over his righthanded counterpart, and if so to precisely what extent (he did, by about 15 percent); whether a team in the field should set its infielders for an attempted double play with the bases loaded early in the game and no outs (it should); whether a man's batting average can serve as a predictor of future performance in a given at bat or game or season (at bat and game, no, season, yes); and more.

Lindsey followed this article with one that is even more central to the issues raised by Rickey and revived by the New Statisticians. In 1963, again in *Operations Research,* he published "An Investigation of

Strategies in Baseball." He wrote in his abstract, or summary, of the article: "The advisability of a particular strategy must be judged not only in terms of the situation on the bases and the number of men out, but also with regard to the inning and score. Two sets of data taken from a large number of major league games are used to give (1) the dependence of the probability of winning the game on the score and the inning, and (2) the distribution of runs scored between the arrival of a new batter at the plate in each of twenty-four situations and the end of the half-inning. . . .[Note: the twenty-four situations are all the combinations of baserunners, from none to three, and outs, none, one, and two.] By combining the two sets of data, the situations are determined in which an intentional base on balls, a double play allowing a run to score, a sacrifice, and an attempted steal are advisable strategies, if average players are concerned. *An index of batting effectiveness based on the contribution to run production in average situations is developed.*" [Emphasis ours.]

Where Rickey had added the On Base Average and Isolated Power to arrive at a batter rating—and it was a good one, far more accurate in its correlation to run production than was the batting average—Lindsey employed an additive formula based on the run values of each event: .41 runs for a single, .82 for a double, 1.06 for a triple, 1.42 for a home run. (These values are not quite right, but they're close; more on this in the next chapter.) To illustrate how Lindsey's method was applied, let's look at the 1983 records of three substantial National League players, Dale Murphy, Mike Schmidt, and Andre Dawson. Note that Lindsey's method is to express all *hits* in terms of runs, but not the *outs;* these he brings into the picture through the traditional averaging process, dividing the run total by at bats. Yet an out has a run value, too, though it is a negative one.

Table III, 1. *The Lindsey Additive Formula*

	1B	2B	3B	HR	Runs/AB	Run Avg.
Andre Dawson	111(.41) +	36(.82) +	10(1.06) +	32(1.42) =	$\frac{131.07}{633 \text{ AB}}$	= .207
Dale Murphy	114(.41) +	24(.82) +	4(1.06) +	36(1.42) =	$\frac{121.78}{589 \text{ AB}}$	= .207
Mike Schmidt	76(.41) +	16(.82) +	4(1.06) +	40(1.42) =	$\frac{105.32}{534 \text{ AB}}$	= .197

How did Lindsey arrive at these values? It is a bit complicated for the general reader, but those with the appetite for probability theory we refer to the bibliographical citations at the back of the book. In brief, Lindsey devised a table, based on observation of 6,399 half innings (all or part of 373 games in 1959-60); he recorded how many times a batter came to the plate in any one of the twenty-four basic situations. Moreover, he deduced what the run-scoring probability became after the batter had hit a single, double, whatever, by computing the difference between the run-scoring value of the situation that *confronted* the batter—for example, man on first and nobody out—and that of the situation which prevailed *after* the batter's successful contribution. That difference represents the run-scoring value of that contribution.

With these new values, proper weighting became possible, in, say, the slugging percentage. A home run was demonstrably not worth as much as four singles, nor a triple as much as a single and a double, and so on. What Lindsey did not account for was such offensive elements as the base on balls or hit by pitch; this had been done the year before in a formula proposed at a conference at Stanford University by Donato A. D'Esopo and Benjamin Lefkowitz. This formula, which they called the Scoring Index, is too complicated to go into, but in any event it was only marginally an improvement on Rickey's, which similarly had accounted for walks and hit by pitch as well as total bases. The Scoring Index overcredited these events, to the extent that in ranking the top hitters of the National League in 1959, Joe Cunningham, whose slugging percentage was .478 to Henry Aaron's .636, rated higher than Aaron, just as he did in On Base Average.

The term Scoring Index reappeared in 1964, but was defined differently by Earnshaw Cook in *Percentage Baseball,* a book which created considerable media stir for its controversial suggestions to revise baseball strategy in line with probability theory. Among these suggestions was to start the game with a relief pitcher and pinch-hit for him his first time up; to realign the batting lineup in descending order of ability; to restrict severely the use of the intentional base on balls and sacrifice bunt, etc. Indeed, Cook's Scoring Index did not appear in a form intelligible to the layman until the appearance of his next book, *Percentage Baseball and the Computer* (1971), in which the "DX," as he abbreviated it, was represented by:

$$\frac{\text{Hits} + \text{Walks} + \text{HBP}}{\text{At Bats} + \text{Walks} + \text{HBP}} \times \frac{\text{Total Bases} + \text{Steals} - \text{Caught Stealing}}{\text{At Bats} + \text{Walks} + \text{HBP}}$$

The first component is simply On Base Average; the latter is a bizarre amalgam of power and speed in which, in effect, baserunning exploits are averaged by plate appearances in the same manner as total bases are. The rationale, evidently, is that net stolen bases (steals minus times caught stealing) adds extra bases in the way that doubles do to singles. This is not quite so, but in any event, the formula works pretty well in spite of its logical shortcomings. At the time of its introduction, the DX was the most accurate measure of total offensive production yet seen and the first to combine ability to get on base in all manners; to move baserunners around efficiently through extra-base hits; and to gain extra bases through daring running.

The original Cook book was highly abstruse in its detail and, despite the hubbub which met its publication in 1964, it is regarded today as perhaps a setback to the cause of improving baseball's statistics. If the job was going to be *that* much trouble, why bother?

If *Percentage Baseball,* despite its brilliance, was not an open sesame to the unlocking of baseball's secrets, a genie came forth in 1969 with the appearance of *The Baseball Encyclopedia,* compiled for Macmillan by Information Concepts, Inc. (ICI). A battalion of researchers commanded by David Neft foraged through baseball history to provide for those who had no ERAs, RBIs, slugging percentages, saves, and all manner of wonderful things. There had never been anything like this mammoth ledger book of the major leagues. But what place does it have in the present discussion of the New Statistics movement? *The Baseball Encyclopedia* is important because the researchers not only found new data to correct old inaccuracies but also applied the new yardsticks to men who had gone to their graves never having heard of an RBI or a save. ICI did not create new stats, it created new stars. Sam Thompson, Addie Joss, Roger Connor, Amos Rusie—their phenomenal level of play was hidden simply because statisticians back then were not recording the particular numbers which would show them off to best advantage. If sabermetrics consists of finding things in the existing data that were not seen before, or collecting the data which makes possible the application of new statistics to old performances, *The Baseball Encyclopedia* is a true monument of the New Statistics movement.

What is more, *The Baseball Encyclopedia* forms a monument in computer technology. Not only was all the material which finally appeared in the tome entered into a data bank—along with much that was withheld for reasons of space—but the book was one of the first typeset entirely by computer, now a common practice. David Neft, a partner in ICI and the head of the research group, is a professional statistician who went on to develop a baseball table game for *Sports Illustrated* which embodied the myriad probabilities of the game.

With the aid of this game, in 1973 R. Allan Freeze of the University of British Columbia ran a "Monte Carlo" style computer simulation of 200,000 complete baseball games. His object was to determine the validity of Earnshaw Cook's hypothesis about how best to structure the batting order—Cook proposed placing the best batter in the lead-off spot, the second best in the second spot, and so on (more of this in Chapter 8). The computer simulation was based on the model of the roulette wheel at Monte Carlo in its ability to generate random numbers which would produce various outcomes, dependent only upon probability and not upon external conditions, such as clutch ability or loss of concentration. Any baseball game, of course, is mightily affected by "external conditions," as is any particular at bat or play, but over the course of 200,000 games these variables tend to cancel each other so that valid results can be obtained. Freeze's study, incidentally, was a victory for baseball's traditional batting order: Cook's provided more at bats over a season for the best hitters, but fewer runs for the team.

This was not the first time a computer simulation model had been applied to the elaborate probabilistic construct that is the game of baseball. In 1959, R.E. Trueman had constructed a model based on 5,000 games (a smaller number than Freeze's, reflecting the smaller capacity of the computers of that time), and in 1960 W.G. Briggs and others espoused the value of a Monte Carlo simulation to determine the true values of the articles of faith that comprise "The Book," such as the desirability of a sacrifice bunt with a man on first and none out in the bottom of the ninth, trailing by a run.

In 1965 the General Electric Company, in order to promote its GE-235 computer, digested the complete batting records of the American League and spat out the top clutch hitters, based upon percentages assigned for each of many possible pressure situations (for example, 5 percent to "first inning, nobody on or out, team ahead by two runs"; 120 percent to "ninth inning, one run down, one out, runners on first and third"). The results of this study—that as pressure increased, for example, Harmon Killebrew became awesome and Tony Oliva helpless, in 1965 at least—impressed Gabe Paul, the general manager of Cleveland, but prompted him to say, "Computers are coming. They are ready for us, but we are not quite ready for them." Nearly twenty years later, all but five of the big-league clubs would say the same thing. Yet the computer has proved its usefulness in statistical analysis, and its employment in baseball circles can only increase.

In 1969 and 1970, the Mills brothers (the nonsinging variety, in this case Eldon and Harlan), who were partners in a self-started enterprise called Computer Research in Sports, tracked two entire major-league seasons on a play-by-play basis. Then they applied to that record the probabilities of winning which derived from each possible outcome of a plate appearance, as determined by a computer simulation incorporating nearly 8,000 possibilities. What, for example, was the visiting team's chance of winning the game before the first pitch was thrown? Fifty percent, if we are pitting two theoretical teams of equal or unknown ability on a neutral site. If that first man fails to get on base, the chances of the visiting team winning are reduced to 49.8 percent; should he hit a double, the visiting team's chance of victory is raised to 55.9 percent, as determined by the probabilistic simulation. Every possible situation—combining half inning, score, men on base, and men out—was tested by the simulator to arrive at "Win Points." The Millses' purpose was to determine the clutch value of, say, hitting a homer with two men on and two men out in the bottom of the ninth, with the team trailing by two runs, the situation that Bobby Thomson faced in the climactic National League game of 1951. (It gained for him 1,472 Win Points; had it come with no one on in the eighth inning of a game in which his team led 4-0, the homer would have been worth only 12 Win Points. More on this in Chapter 9.)

What the Mills brothers were attempting to do was to evaluate not only the *what* of a performance, which traditional statistics indicate, but the *when*, or clutch factor, which no statistic to that time could provide. If this project, detailed in a small book issued in 1970 called *Player Win Averages*, sounds similar to the GE-235 program discussed above, it may be because mathematician Harlan Mills at one time served on the technical staff at General Electric.

August 10, 1971, marked another milestone, the founding of the Society for American Baseball Research, the group in whose annual publications most of the New Statisticians have cut their analytical teeth. At first a band of only sixteen men brought together in Cooperstown by Bob Davids, a veteran researcher of the game's arcana, SABR today numbers some 4,500 fans and students of the game. Its statistical analysis research committee, headed for the last decade by Pete Palmer, has served as a sounding board for the inventive approaches of such men as Dallas Adams, Dick Cramer, Bill James, and more. Their accomplishments, as demonstrated most notably in the *Baseball Research Journal, The Baseball Abstract,* and *The Baseball Analyst,* define the state of the art, although they by no means represent the whole of the statistical vanguard. The work of Tom Boswell, Barry Codell, Merritt Clifton, Steve Mann, and others is also significant and will come under discussion.

The most prominent of the New Statisticians, the man whose annual *Baseball Abstract* had done so much to make the statistical analysis of baseball accessible—even *popular,* a prospect unimagined a decade ago—is Bill James. The tall, bearded sabermetrician reminds one of the young Chadwick, whose stature he seems destined to attain. James published the *Abstract* from his home in Lawrence, Kansas, for five years to a minute if appreciative audience (1977 publication budget: $112.73). In 1982 Ballantine Books assumed publication of the *Abstract* and the audience became sizeable indeed by commercial standards and positively enormous by the standards of earlier sabermetric-type books such as Cook's and the Mills brothers'.

We will not devote to James's work the space which it clearly merits. Chadwick, Lanigan, and Rickey are not around to expound their views, but Bill is. The scope and ingenuity of his approach to the game are unique and best experienced direct. If you are not already a reader of the *Abstract,* we suggest you become one immediately—provided you haven't come across this paragraph while browsing in the book-store; in that event, let Bill's book sit on the shelf another day and bring this one to the register.

Perhaps the most widely known of the many James indices is Runs Created, his formula for defining the number of runs a batter accounts for through his various offensive contributions. In its basic expression, it is:

$$\frac{(\text{Hits } + \text{ Walks}) (\text{Total Bases})}{\text{At Bats } + \text{ Walks}}$$

The essence of this formulation is that the ability to get on base and the ability to push baserunners around fairly describes offensive ability, and this is so, except that in Runs Created the effect of total bases is diminished by walks, a point of seeming unfairness; the accuracy of the formula would be improved by separating the two components, dividing (Hits + Walks) by (At Bats + Walks) and Total Bases by only At Bats. Also, no factor exists to represent base stealing ability, which is surely an offensive contribution (as are the hit by pitch, sacrifice hit, sacrifice fly and, in a negative way, the grounded into double play). So James refined the formula to:

$$\frac{(\text{Hits } + \text{ Walks } - \text{ Caught Stealing}) (\text{Total Bases } + .7 \text{ Stolen Bases})}{\text{At Bats } + \text{ Walks } + \text{ Caught Stealing}}$$

In its rudimentary form, Runs Created is a rough predictor of offensive contribution by individuals and becomes more accurate as the sampling base widens to teams or leagues. Accuracy, by the way, may be measured by correlating a team's or league's predicted Runs Created with its actual runs scored. (A single season's record is not really enough, however, to rightly assess the accuracy of a statistic; for this we need a larger sample and a basic statistical tool, the standard deviation, which we will apply at the conclusion of this chapter.) The extended form of Runs Created is superior, and its accuracy surpasses that of any measure discussed thus far; however, it cannot be applied to performances prior to 1951, when caught stealing data was recorded erratically.

Why did James multiply stolen bases by .7 in the formula above—and by .65 in a 1983 adjusted formula which incorporated HBP, SH, SF, and GIDP for the first time—rather than weight them equally with singles or walks? In the 1982 *Abstract,* he wrote, "A stolen base advances only the runner; each base of a hit advances the batter as a runner and anyone else who happens to be aboard." He is correct, of course, but is a walk as good as a hit? No. A walk can never advance a baserunner two bases, as a single can—most particularly the two bases from second to home—and a walk cannot drive in a runner from third, as a single can, unless the bases are loaded. If stolen bases can be fractionally weighted, why not walks? Moreover, Lindsey and Cook both confirmed that a home run was not worth as much as four singles, so why not adjust the erroneous weights that go into the slugging percentage? Simplicity is a virtue, but once having sinned by making the formula more accurate, why not go all the way? Bottom line: Runs Created in its extended version could be more theoretically appealing and thus more accurate, and its simple version is less accurate than other run-denominated measures of batting prowess; all the same, it is a fine statistic and we offer tables at the end of this chapter detailing lifetime and seasonal leaders in this category.

James also employs the Rickey measure called Isolated Power, which he went to the trouble of devising independently as "Power Percentage" before he'd read the Rickey article. It's a useful corrective to one of the inherent flaws of the slugging percentage, namely, that it is boosted by singles. Isolated Power is total bases minus hits, divided by at bats, or slugging percentage minus batting average. A low-average, high-power hitter like Dave Kingman will appear high on these all-time lists (see the tables at the rear of the book), but in the middle of the pack of the slugging percentage leaders.

Unlike most other practitioners of the New Statistics, James does not limit his investigations to the offense. In the opinion of many, it is on the question of fielding, that stepchild of statistical progress since 1876, that he is at his best. He has converted thousands to his view that errors don't count for much anymore, and thus the number of chances cleanly handled per game is the best measure of fielding prowess. He has put forth, too, a Defensive Efficiency Record for teams, and a Defensive Spectrum construct to explain the games' perpetual abundance of designated-hitter types and dearth of good middle infielders. Pitching seems to interest him less, but he and his helpers keep track of pitcher run support in order to explain at least in part the won-lost failure of a Ross Baumgarten or the won-lost success of a Bob Mc-Clure.

As we have seen, the first of the efforts to pare offensive statistics to their essence, runs, and then reconstruct them for individuals so as to reflect their run-producing ability, were Rickey and Roth's. Next came Lindsey, followed by Cook. Between Cook and James arose a formula called the Batter's Run Average devised in 1972 by Dick Cramer, a research scientist from the Philadelphia area who nearly a decade later prepared and sold to the Oakland A's the first computerized statistical analysis service, "Edge 1.000." The BRA, a perhaps infelicitous choice for an acronym, was simply the On Base Average times the slugging percentage, or

$$\left(\frac{\text{Hits} + \text{BB} + \text{HBP}}{\text{AB} + \text{BB} + \text{HBP}} \right) \times \left(\frac{\text{Total Bases}}{\text{At Bats}} \right)$$

It turned out that this correlated beautifully with actual team scoring data, though it was a less effective measure for individuals.

In 1977 Cramer refined the BRA to become the Batter Win Average, a measure of a player's runs contributed as subtracted from the league average. The formula was rather daunting, however. It may be enough to say that the BWA stands in the same relationship to the BRA as James's complex version of Runs Created does to its simple prototype.

Cramer's accomplishments include not only Edge 1.000, which will be discussed in Chapter 8, but also a brilliant analysis of what had been the two great statistical chimeras of the Hot Stove League: whether clutch hitters existed, and if so how they could be identified; and how average batting skill could be compared across time. How could we compare a .320 hitter of 1893 with a .320 hitter of 1983 besides comparing each mark to the league batting average? The former point will be discussed at length in Chapter 9, the latter in Chapter 6.

Cramer's 1980 article on average batting skill in the *Baseball Re-*

search Journal was followed in the 1982 *BRJ* by an article from Dallas Adams, a globetrotting engineer of great statistical acumen, on average pitching and fielding skill. Adams has also contributed stimulating papers to *The Baseball Analyst* on the distribution of runs scored and the relationship of run differentials to won-lost percentage.

The discussion of average batting, pitching, and fielding skill to which Adams and Cramer devoted themselves advanced a discussion which had begun in 1976 with the first *BRJ* article on cross-era comparison, in which David Shoebotham proposed a new statistic called the Relative Batting Average. Shoebotham recognized that the .320 batting average in 1893, when the National League batted .280, did not represent the same level of accomplishment as that average did in 1968 when, for a number of reasons, the National League batted a measly .243. His solution? To normalize the players' averages to their respective league averages through the simple formula:

$$\text{Relative BA} = \frac{\text{Individual Hits/At Bats}}{(\text{League Hits} - \text{Individual Hits})/(\text{League At Bats} - \text{Individual At Bats})}$$

In this fashion he demonstrated, for example, that Pete Rose, who led the NL with a .335 BA in 1968, had a Relative BA of 1.38; while Ed Delahanty, who led the NL with a BA of .380 in 1893, had a Relative BA of only 1.36. Another way of stating this conclusion is that Rose's .335 was 38 percent above the average batting performance in the NL of 1968, while Delahanty exceeded his league's norm by 36 percent. The inferences that might be drawn from this approach are many: that batting skill has not declined since the days of Ruth, Gehrig, Foxx, et al., but that pitching skill might have increased; that no batting average of the years around 1930 ought to be taken without a carload of salt; that some of the most notable batting performances of all time, as measured by the batting average, have occurred right under our noses, unbeknownst to us.

Independently from Shoebotham, at about the same time, Merritt Clifton of southern Quebec was performing the same sort of relativist calculations, eventually self-publishing his findings in 1979. (Clifton was not then a member of SABR.) His booklet *Relative Baseball* extended the relativity theory to slugging percentage and home runs, and he employed an alternate method of normalizing as well, not only to league average but also to league leaders. In *The Baseball Analyst,* Ward Larkin compared BAs across eras by means of their standard deviations and showed that Rod Carew and Ty Cobb had nearly identical batting marks.

Clifton's method was applied to fielding percentages in the 1983 issue of *The National Pastime,* another SABR publication, by Bill Deane. Also, David Maywar viewed pitcher strikeout records through a relativistic prism in the 1980 *BRJ.* The work of Shoebotham, Clifton, Maywar, and Deane—as well as the more involved theories of Cramer and Adams on the subject—will be addressed fully in Chapter 6.

Normalizing a statistic to its league average is a valuable analytical tool if employed logically. A Relative Batting Average, for example, tells a good deal more, and tells it more straightforwardly, than Relative Homers or Relative Strikeouts. The relativist approach works better with ratios such as the BA, OBA, or SLG—or for that matter, Runs Created or Isolated Power—than it does for simple counter stats.

Another worthwhile adjustment to various averages is for home-park effects, a park factor, if you will. The pioneering work in this area was done by Robert Kingsley, particularly in regard to why homers flew out of Atlanta's park despite its "normal" dimensions, but Pete Palmer was the first to measure the effects of home parks on run totals and then to devise a park adjustment for the records of batters and pitchers. These will be reviewed in Chapter 5.

In 1977 Steve Mann, like Cramer from the Philadelphia area, devised the Run Productivity Average, which assigned run production values to each of the offensive events in the manner of Lindsey, but with radically different run values because their basis was runs *produced,* defined as runs scored plus runs driven in. Where Lindsey's observations had led to a run value of 1.42 for a home run, for example, Mann's observations of some 325 complete games (or their equivalent in innings) led him to credit a home run as 2.63 runs; the two figures really differ by only 0.21, inasmuch as Mann counts the home run twice, once for its run driven in, another time for its run scored. The flaw in Mann's approach is in its concept, the evident belief that runs are produced only by those who score them or drive them home, when in fact the "man in the middle," whose offensive contribution is to advance baserunners through a hit, walk, or HBP, slips between the statistical cracks. This problem was repaired substantially by adding a correcting factor for On Base Average. The formula does represent an important step forward in the New Statistical movement's returning attention to the basic activity of the game, the creation/prevention of runs. (More on Mann's RPA in the next chapter.)

In early 1978 Barry Codell of Chicago wrote a paper describing his new statistic, the Base-Out Percentage, which he distributed to major-league executives, fellow statisticians, and figures in the various sports

Table III, 2. *Best Lifetime Batters, Runs Created*

NO	PLAYER	YRS	SPAN	G	RC
1	Babe Ruth	22	1914-1935	2503	2767.7
2	Hank Aaron	23	1954-1976	3298	2601.8
3	Stan Musial	22	1941-1963	3026	2582.8
4	Ty Cobb	24	1905-1928	3034	2546.2
5	Ted Williams	19	1939-1960	2292	2366.7
6	Willie Mays	22	1951-1973	2992	2357.0
7	Lou Gehrig	17	1923-1939	2164	2276.3
8	Tris Speaker	22	1907-1928	2789	2200.1
9	Jimmie Foxx	20	1925-1945	2317	2149.8
10	Carl Yastrzemski	23	1961-1983	3308	2129.2
11	Frank Robinson	21	1956-1976	2808	2117.1
12	Mel Ott	22	1926-1947	2730	2101.9
13	Rogers Hornsby	23	1915-1937	2259	2079.9
14	Pete Rose	21	1963-1983	3250	2065.3
15	Mickey Mantle	18	1951-1968	2401	1923.3
16	Honus Wagner	21	1897-1917	2789	1913.0
17	Al Kaline	22	1953-1974	2834	1848.8
18	Paul Waner	20	1926-1945	2549	1824.0
19	Eddie Collins	25	1906-1930	2826	1811.2
20	Al Simmons	20	1924-1944	2215	1810.2
21	Charlie Gehringer	19	1924-1942	2323	1727.7
22	Nap Lajoie	21	1896-1916	2474	1706.7
23	Harry Heilmann	17	1914-1932	2146	1686.9
24	Goose Goslin	18	1921-1938	2287	1686.0
25	Billy Williams	18	1959-1976	2488	1683.1

media. An aesthetically appealing formulation because it incorporated all offensive events under one "roof," it looked like this:

$$\frac{\text{Total Bases} + \text{Walks} + \text{HBP} + \text{Steals} + \text{Sacrifices} + \text{Sac. Flies}}{\text{At Bats} - \text{Hits} + \text{Caught Stealing} + \text{GIDP} + \text{Sacrifices} + \text{Sac. Flies}}$$

The elements of the numerator represented bases gained, while the events in the denominator represented outs produced (sacrifices and sacrifice flies appeared in both because they achieved both, gaining a base for the team while producing an out). Codell's paper was published in the *Baseball Research Journal* in 1979. He termed the BOP baseball's "most complete and informative offensive statistic."

At about the same time, Tom Boswell, a sportswriter for the Washington *Post* and self-confessed "stat freak," devised a formula he called Total Average:

Table III, 3. *Best Lifetime Batters, Total Average*

NO	PLAYER	YRS	SPAN	G	TA
1	Babe Ruth	22	1914-1935	2503	1.428
2	Ted Williams	19	1939-1960	2292	1.374
3	Lou Gehrig	17	1923-1939	2164	1.252
4	Jimmie Foxx	20	1925-1945	2317	1.170
5	Hank Greenberg	13	1930-1947	1394	1.125
6	Rogers Hornsby	23	1915-1937	2259	1.106
7	Mickey Mantle	18	1951-1968	2401	1.100
8	Stan Musial	22	1941-1963	3026	1.060
9	Joe DiMaggio	13	1936-1951	1736	1.038
10	Ralph Kiner	10	1946-1955	1472	1.035
11	Mel Ott	22	1926-1947	2730	1.035
12	Johnny Mize	15	1936-1953	1884	1.022
13	Charlie Keller	13	1939-1952	1170	1.019
14	Mike Schmidt	12	1972-1983	1638	1.001
15	Willie Mays	22	1951-1973	2992	.997
16	Hack Wilson	12	1923-1934	1348	.996
17	Ty Cobb	24	1905-1928	3034	.992
18	Frank Robinson	21	1956-1976	2808	.990
19	Tris Speaker	22	1907-1928	2789	.984
20	Joe Jackson	13	1908-1920	1331	.983
21	Earl Averill	13	1929-1941	1669	.968
22	Harry Heilmann	17	1914-1932	2146	.965
23	Hank Aaron	23	1954-1976	3298	.965
24	Duke Snider	18	1947-1964	2143	.963
25	Ken Williams	14	1915-1929	1397	.959

$$\frac{\text{Total Bases} + \text{Walks} + \text{HBP} + \text{Steals}}{\text{At Bats} + \text{Walks} + \text{HBP} + \text{Steals} + \text{Caught Stealing}}$$

By the time TA appeared in *Inside Sports* in January 1981, to considerable attention, it had been improved both aesthetically and mathematically (in its first incarnation, a home run in four at bats counted the same as four singles in four at bats with no penalty for contributing three outs). Boswell altered the concept of the denominator from "opportunities" to "failed opportunities" or outs:

$$\frac{\text{Total Bases} + \text{Steals} + \text{Walks} + \text{HBP}}{\text{At Bats} - \text{Hits} + \text{Caught Stealing} + \text{GIDP}}$$

In this regard Total Average is identical with the Base-Out Percentage except that it eliminates sacrifice hits and flies. In Codell's formula these had only a marginal impact anyway, slightly diminishing the

Table III, 4. *Best Batting Seasons Post-1900, Runs Created*

NO	YEAR	PLAYER	CLUB	LG	G	RC
1	1921	Babe Ruth	NY	A	152	233.8
2	1923	Babe Ruth	NY	A	152	217.3
3	1927	Lou Gehrig	NY	A	155	211.9
4	1922	Rogers Hornsby	STL	N	154	206.4
5	1920	Babe Ruth	NY	A	142	205.8
6	1932	Jimmie Foxx	PHI	A	154	205.6
7	1927	Babe Ruth	NY	A	151	203.0
8	1924	Babe Ruth	NY	A	153	200.4
9	1930	Lou Gehrig	NY	A	154	198.2
10	1930	Chuck Klein	PHI	N	156	194.1
11	1948	Stan Musial	STL	N	155	193.1
12	1936	Lou Gehrig	NY	A	155	192.5
13	1930	Hack Wilson	CHI	N	155	192.2
14	1934	Lou Gehrig	NY	A	154	190.3
15	1924	Rogers Hornsby	STL	N	143	189.2
16	1930	Babe Herman	BKN	N	153	189.1
17	1926	Babe Ruth	NY	A	152	188.2
18	1929	Rogers Hornsby	CHI	N	156	187.9
19	1930	Babe Ruth	NY	A	145	186.9
20	1925	Rogers Hornsby	STL	N	138	186.3
21	1931	Babe Ruth	NY	A	145	185.0
22	1941	Ted Williams	BOS	A	143	184.7
23	1929	Lefty O'Doul	PHI	N	154	184.7
24	1938	Jimmie Foxx	BOS	A	149	183.9
25	1931	Lou Gehrig	NY	A	155	182.7

BOP of the great players who achieved a higher number of bases than outs, and slightly augmenting the BOP of those whose ledger holds more outs than bases.

The TA is a user-friendly stat, and it has other virtues, too. Like Rickey's, Cook's, and James's offensive measures, TA incorporates the whole world of offense in its confines, and it holds no built-in advantage for sluggers or singles hitters or runners or walkers. But it is not as accurate as Runs Created or Cook's Scoring Index—or Batter's Run Average (On Base times Slugging), which is both more accurate and simpler. Tables III, 2-5 illustrate the differences between RC and TA.

Boswell admitted to other drawbacks of the Total Average. It is not useful for comparing players across eras (though in this characteristic it is no worse than the batting average, Runs Created, or any other absolute measure), and it may be inaccurate in its weighting of the various bases in the numerator. "A fellow scrivener who thought that,

Table III, 5. *Best Batting Seasons Post-1900, Total Average*

NO	YEAR	PLAYER	CLUB	LG	G	TA
1	1920	Babe Ruth	NY	A	142	1.885
2	1923	Babe Ruth	NY	A	152	1.808
3	1921	Babe Ruth	NY	A	152	1.801
4	1941	Ted Williams	BOS	A	143	1.782
5	1957	Ted Williams	BOS	A	132	1.677
6	1926	Babe Ruth	NY	A	152	1.646
7	1924	Babe Ruth	NY	A	153	1.632
8	1927	Babe Ruth	NY	A	151	1.595
9	1930	Babe Ruth	NY	A	145	1.554
10	1925	Rogers Hornsby	STL	N	138	1.548
11	1957	Mickey Mantle	NY	A	144	1.532
12	1927	Lou Gehrig	NY	A	155	1.527
13	1954	Ted Williams	BOS	A	117	1.510
14	1924	Rogers Hornsby	STL	N	143	1.502
15	1931	Babe Ruth	NY	A	145	1.501
16	1932	Jimmie Foxx	PHI	A	154	1.489
17	1946	Ted Williams	BOS	A	150	1.482
18	1942	Ted Williams	BOS	A	150	1.449
19	1930	Lou Gehrig	NY	A	154	1.449
20	1936	Lou Gehrig	NY	A	155	1.444
21	1932	Babe Ruth	NY	A	133	1.442
22	1947	Ted Williams	BOS	A	156	1.438
23	1949	Ted Williams	BOS	A	155	1.430
24	1928	Babe Ruth	NY	A	154	1.427
25	1961	Norm Cash	DET	A	159	1.424

for instance, a walk or a steal ought only count four-fifths as much as a base hit might find a sympathetic ear," he wrote in 1981.

Yet one year later Boswell wrote: "We should start with Total Average and modify our appraisal by looking at pertinent stats. After all, no system of offensive measurement could ever really approach the ideal too closely. Why? Because all bases are not created truly equal. Is a walk 100 percent as good as a hit? Or only 85 percent, or 75 percent as good? And is a home run worth exactly four times as much as a steal, or only three and a half times as much? This question is a cosmic stumper in the baseball universe."

Now, you can't blame Boswell for not having a lifetime subscription to *Operations Research* or for not slogging through Cook's *Percentage Baseball*, but this question was answered in the 1960s. If Boswell had known these values, would he have weighted his Total Average accordingly? We think not. Many people, he once wrote, including James and Palmer and others in SABR, are "dedicated to concocting a perfect offensive stat. However, the more ambitious the stat, the more complex and arbitrary it almost always becomes. What it gains in sophistication and the intuitive wisdom of its creator, it loses in simplicity and objectivity. How can you love a stat, or use it in arguments, if you can't really explain it?"

How many baseball fans can compute an ERA? One in ten would be a guess on the high side; perhaps that many can compute slugging percentage. That doesn't mean they can't cite the numbers for legions of players and understand perfectly well what the numbers *represent,* that the pitcher with the lowest ERA or the batter with the highest BA is the best around. If any of the New Statistics is to take hold—including Total Average—it will be because its concept is broadly appealing, not its computation.

So, even if RC, TA, DX, BA, and all the other initials leave something to be desired, as any baseball statistic must, how good are they in relation to each other? Throughout this chapter we have talked vaguely of statistical accuracy, but now that all important measures of offensive production have been reviewed, it is time to assess their accuracy precisely. How? By correlating each measure with actual batting performance for all teams in all years from 1946 through 1982. If a team had, say, a batting average that was 10 percent above the league average in a given season, and it also scored 10 percent more runs than the league average, the correlation would be perfect. However, that correlation would not be likely (to say the least) to remain perfect for all teams over thirty-seven seasons. Once we have obtained the annual differences between the runs scored "predicted" by the statistic, like batting average, and the actual runs, we can calculate the standard deviation of the two and thus identify the accuracy of the statistic. The smaller the standard deviation, the more accurate the statistic.[1]

Here's the report card.

Table III, 6. *Accuracy of Various Offensive Statistics*

STATISTIC	STANDARD DEVIATION (in runs)
Batting Average	54.8
On Base Average	53.0
Isolated Power	50.8
On Base Average *plus* (adding a corrector for SB, CS, and Outs On Base, inc. GIDP, pickoffs, out stretching)	48.1

Rickey	41.0
(without clutch measure, R/ [H + BB + HBP])	
Slugging Percentage	39.9
D'Esopo and Lefkowitz	39.6
Run Productivity Average	36.0
(without OBA corrector)	
Rickey	34.8
(with R/ [H + BB + HBP)	
Total Average	31.1
Total Average	30.3
(with a corrector for Outs On Base)	
Scoring Index (Cook's DX)	26.4
Runs Created	25.8
Runs Created	24.6
(with corrector for Outs On Base and HB)	
Scoring Index (Cook's DX with Outs On Base corrector)	24.6
Batter's Run Average	24.4
(OBA × SLG)	
Run Productivity Average	22.5
(with OBA corrector)	
On Base Average *plus* ×	20.4
SLG (with corrector for SB, CS, and Outs On Base)	
Linear Weights	19.8
(with SB, CS, and Outs On Base corrector)	

What does this evaluation tell us? That batting average bears little relationship to run production. That of traditional statistics, slugging percentage is considerably superior and much better than On Base Average, too, which leads one to the conclusion that for a team, extra-base power is more important than the ability to get on base . . . not exactly a surprise. That Total Average and Runs Created are not as accurate, even in their most inclusive versions, as Batter's Run Average, which is simply a multiplication of On Base times Slugging *not even bothering with stolen bases, caught stealing, or grounded into double play*. (Maybe stealing is not much of an offensive weapon?) That the most accurate of all offensive statistics is . . . Linear Weights? What's that?

Read on.

[1] For those of you who are interested, here is the full account of the correlation method and of the calculation of the standard deviation. Each offensive measure—let's take batting average as an example—was calculated for each league season and for each team in the period 1946-82. The team BA was then multiplied by the number of innings batted by the team and again times the number of runs scored per inning at the league average.

An example of the approach in action: Let's say that in 1983 the Mudville Excelsiors batted in 1450 innings, and that the league averaged half a run per inning (roughly 4.5 runs per game for the average team). If the Excelsiors' batting average was 10 percent higher than the league average, and if BA had a straight-line correlation with offensive production, then they figured to score 10 percent more runs than the average team. In fact, based on its BA, the Excelsiors figured to score 798 runs in 1983: 1450 Innings × 0.5 Runs Per Inning × 1.10 = 798. If the Excelsiors in fact scored 758 runs, or 838, BA was 40 runs off in its prediction.

But one season's results for one team provide too small a sample, so we used this approach across 37 years, for each team in each league—746 cases—and then calculated the *standard deviation* of the predicted and actual runs. This is done by taking the square root of the sum of the squares of the 746 differences. For example, assume the sample consisted of one team over five years. Let's say the differences between predicted runs and actual runs were 20, 6, 3, 11, and 5. Squaring each difference, we get 400, 36, 9, 121, and 25, which add to 591. Dividing 591 by 5 (the number of samples), we get 118.2. The square root of 118.2 is 10.87—the standard deviation of the BA for this sampling.

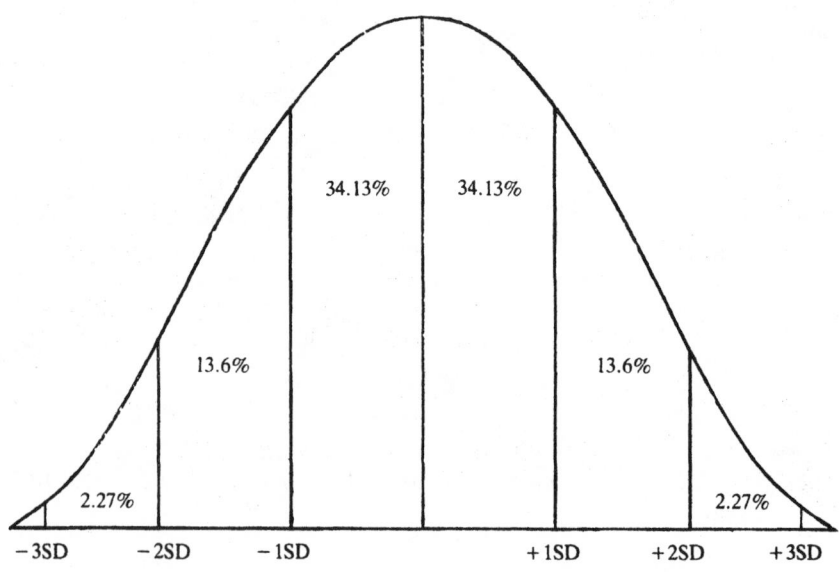

What the standard deviation tells us is that two thirds (actually 68.26 percent) of all the differences will fall within the range of one standard deviation, 95 percent of all differences will fall within two standard deviations, and 99.7 percent will be within three standard deviations. These values are characteristic of any normal distribution, as indicated by the classic bell curve below, which has its mean at the center and its variations, plus and minus, in absolute symmetry. Baseball's pattern of run scoring fits under this curve as snug as a bug in a rug.

The smaller the standard deviation, the more accurate the statistic. An offensive statistic with an SD of 50 will have one of three predictions off by 50 runs or more; a stat with an SD of 25 will miss its predictions by more than 50 runs only one time in twenty.

Actually, the standard deviations listed in the "report card" for offensive measures have one more complicating factor—linear curve fitting. In the formula $y = mx + b$, y is the result (in this case, runs per inning), and x is the prediction (team batting average/league average); m and b are chosen so as to minimize the error between x and y. For batting average, y, or the

$$\text{prediction} = 1.15 \left(\frac{\text{Team BA} \times \text{Runs/Inning, League}}{\text{League BA}} \right) - 102.$$

Without the slope corrector, the SD of the batting average would be even worse—58.9 rather than 54.8.

THE LINEAR WEIGHTS SYSTEM

In 1982, Milwaukee's Robin Yount had the year of his life, batting .331 with 29 homers, 114 RBIs and 129 runs scored; he led the American League in hits, doubles, total bases, and slugging percentage, while finishing just one point behind the league leader in batting average. He was voted the Most Valuable Player in the American League, being named first on all but one of the twenty-eight ballots cast by the baseball writers.

Over in the other league, Mike Schmidt of the Phillies was having an off year, batting only .280 with 35 homers and 87 RBIs; the previous year, when he was awarded the MVP, in only 102 games played he had totaled 31 homers and 91 RBIs. He did lead the league once again in 1982 in slugging percentage, and he did win the Gold Glove at third base for the seventh straight year, yet in the MVP balloting none of the ballots listed him higher than fourth; ten ballots were cast without listing him at all.

For Yount, 1982 was a crowning achievement; for Schmidt, a disappointment: That is the verdict reached by the baseball writers and conventional baseball statistics. Yet in terms of actual performance, as determined by the number of runs contributed, Schmidt's "off year" was scarcely different from Yount's. With the bat, Yount accounted for 55.7 runs beyond what an average batter might have contributed;

Schmidt, 51.0. Through base stealing, Yount added 2.4; Schmidt none. With the glove, Yount saved 12.2 runs; Schmidt, 14.5. Total runs contributed: Yount 70.3, Schmidt 65.5. Total wins contributed beyond average by each: Yount 7.0, Schmidt 6.5.

How do we know these things are so? By applying Linear Weights. The Linear Weights system provides not only the best batting statistic but also the most accurate measures of proficiency in fielding, base stealing, and pitching. Its back-to-basics foundation is the same underlying the Rickey formula and most of the new statistics since: that wins and losses are what the game is about; that wins and losses are proportional in some way to runs scored and runs allowed; and that runs in turn are proportional to the events which go into their making.

With Linear Weights, these events are expressed not in the familiar yet deceptive ratios—base hits to at bats, wins to decisions, etc.—but *in runs themselves,* the runs contributed (batting, stealing) or saved (pitching, fielding). Normalizing factors (to league average) built into the formulas for all but base stealing, where league average is not a shaping force, allow us to compute the number of runs that Mike Schmidt's bat provided last year in excess of those an average hitter might have produced in an equivalent number of plate appearances. And, by adjusting for home-park influences, the Linear Weights comparison may be extended to how many runs Schmidt accounted for beyond what an average player might have produced in the same number of at bats *had he too played half his games in Veterans' Stadium.*

Furthermore, having determined the number of runs required to transform a loss to a win in the final standings (generally around 10, historically in the range 9-11; more on this later) we can convert a player's Linear Weights record, expressed in runs, to the number of wins above average he alone contributed—and what are individual statistics for if not to achieve some understanding of this? Last, by reviewing the win contributions of all a team's batters, pitchers, fielders, and base stealers, we may establish a solid assessment of that team's strengths and weaknesses for the upcoming season whether, for example, it figures to be a pennant contender without any personnel changes, or whether it will have to import some new bodies just to stand in place.

HOW RUNS ARE MADE

George Lindsey, in his previously mentioned 1963 article, was the first to assign run values to the various offensive events which lead to runs: Runs = (.41) 1B + (.82) 2B + (1.06) 3B + (1.42) HR. He based these values on recorded play-by-play data and basic probability theory. Unlike Earnshaw Cook, who in the following year assigned run values on the basis of the sum of the individual scoring probabilities—that is, the *direct* run potential of the hit or walk plus those of the baserunners set in motion—Lindsey recognized that a substantial part of the run value of any non-out is that it brings another man to the plate. This additional batter has a one-in-three chance of reaching base and thus bringing another man to the plate with the same chance, as do the batters to follow. The *indirect* run potential of these batters cannot be ignored.

Steve Mann's Run Productivity Average assigned these values based on observation of some 12,000 plate appearances: RPA = (.51) 1B + (.82) 2B + (1.38) 3B + (2.63) HR + (.25) BB + (.15) SB − (.25)CS, all divided by plate appearances, then plus .016. His values, as mentioned before, were denominated in terms of the number of runs *and RBIs* each event produced. Bill James, at about the same time, came up with a similar formula, since shunned, with values based on runs plus RBIs *minus home runs.* The drawbacks to the approaches of Mann and James are the drawbacks of the RBI, which gives the entire credit for producing a run to the man who plates it, and of the run scored, which gives credit only to the man who touches home, no matter how he came to do so. For example, with no outs, a man reaches first on an error; the next batter hits a double, placing runners on second and third; the following batter taps a roller to short and is thrown out at first, with the run scoring from third. The man who produced the out is given the credit for producing a run, while the man who started the sequence by reaching first on an error is likewise credited with a run. The man who hit the double, which was surely the key event in the sequence which produced the run, and the only one reflecting batting skill, receives no credit whatsoever. In this regard, any formula based on "Runs Produced" (whether R + RBI or R + RBI − HR) is philosophically inferior to the formula Lindsey proposed, despite his failure to account for walks, steals, and other events.

Pete devised a corrector for Mann's RPA, adding (On Base Average − .330) to the result to reflect the run contribution of batters who advance baserunners without producing RBIs or runs; this corrector represents the extent to which the batter's OBA exceeds the normal OBA of about .330. While the accuracy of the RPA as measured by its standard deviation was thus improved dramatically, the resulting formula is an ungainly, jury-rigged thing, corrected first by adding .016, then by adding the on-base factor.

The run values in the Linear Weights formula for identifying batters' real contribution are derived from Pete's 1978 computer simulation of all major-league games played since 1901. All the data available concerning the frequencies of the various events was collected; following a test run, these were tabulated. Unmeasured quantities, such as the probability of a man going from first to third on a single vs. that of his advancing only one base, were assigned values based on play-by-play analysis of over 100 World Series contests. The goal was to get all the measured quantities very nearly equal to the league statistics; then the simulation would provide run values of each event in terms of *net runs produced above average*. Expressing the values in these terms would give a meaningful base line to individual performances, because if you are told that a player contributed 87 runs you don't know what that signifies unless you know the average level of run contribution in that year: 87 may sound like a lot, but if the norm was 80, then you know the player contributed only 7 runs beyond average.

The values obtained from the simulation are remarkably similar from one era to the next, confounding expectations that the home run would prove more valuable today than in the dead-ball era, or that the steal was once a primary offensive weapon. These values are expressed in beyond-average runs.

Table IV, 1. Run Values of Various Events, by Periods

EVENT	PERIOD			
	1901-20	1921-40	1941-60	1961-77
home run	1.36	1.40	1.42	1.42
triple	1.02	1.05	1.03	1.00
double	.82	.83	.80	.77
single	.46	.50	.47	.45
walk/HBP	.32	.35	.35	.33
stolen base	.20	.22	.19	.19
caught stealing	−.33	−.39	−.36	−.32
out*	−.24	−.30	−.27	−.25

*An out is considered to be a hitless at bat and its value is set so that the sum of all events times their frequency is zero, thus establishing zero as the base line, or norm, for performance.

In the years since this simulation was conducted, statistician Dave Smith ("Maury Wills and the Value of the Stolen Base," *Baseball Research Journal,* 1980) convinced Pete to adjust the values of the stolen base and caught stealing because of their situation-dependent, elective nature: Attempts are apt to occur more frequently in close games, where they would be worth more than if they were distributed randomly the way an event like a single or a home run would be. Pete revised the value for the steal upward to .30 runs, while for the caught stealing it becomes − .60 runs.

THE FORMULA

Just as these run values change marginally with changing conditions of play, they differ slightly up and down the batting order (a homer is not worth as much to the leadoff hitter as it is to the fifth-place batter; a walk is worth more for the man batting second than for the man batting eighth); however, these differences have been averaged out in the figures above. For evaluating runs contributed by any batter at any time, there is no better method than this Linear Weights formula derived from the computer simulation which is the basis of Table IV, 1.

$$\text{Runs} = (.46)1B + (.80)2B + (1.02)3B + (1.40)HR + (.33)(BB + HB) + (.30)SB − (.60)CS − (.25)(AB − H) − .50(OOB).$$

The events not included in the formula that you might have thought to see are sacrifices, sacrifice hits, grounded into double plays, and reached on error. The last is not known for most years and in the official statistics is indistinguishable from outs on base (OOB). The sacrifice has essentially canceling values, trading an out for an advanced base which, often as not, leaves the team in a situation with poorer run potential than it had before the sacrifice (more on this in Chapter 8). The sacrifice fly has dubious run value because it is entirely dependent upon a situation not under the batter's control: While a single or a walk always has a potential run value, a long fly does not unless a man happens to be poised at third base (whether it is achieved by accident or design is open to question, as well, but that is beside the question—getting hit by a pitch is not a product of intent, either). Last, the grounded into double play is to a far greater extent a function of one's place in the batting order than it is of poor speed or failure in the clutch, and thus it does not find a home in a formula applicable to all batters. It is no accident that Henry Aaron, who ran well for most of his long career and wasn't too bad in the clutch, hit into more DP's than anyone else, nor that Roberto Clemente, Al Kaline, and Frank Robinson, who fit the same description, are also among the ten "worst" in this department. If Boston's Glenn Hoffman doesn't hit into many twin killings, it's not because of adept bat handling or blazing speed but because he bats ninth.

The Linear Weights formula can be condensed by eliminating the components for steals, caught stealing, and outs on base. In fact, this abbreviated version is the one employed in all the leaders tables in the back of the book because data for caught stealing is not available for so many of baseball's 108 years. To include it for some years and not others would make comparisons across time unfair, while outs on base—calculated as Hits + Walks + Hit Batsmen − Left on Base − Runs − Caught Stealing—is meaningless for individuals. A further condensation we have availed ourselves of for our historical data is to set the value of the single at .47 runs and each extra base at .31, making a double .78, a triple 1.09, and a homer 1.40. This way one need not ascertain the precise number of doubles, triples, and home runs for each of the nearly 13,000 men who have played major-league ball. All that one need know is the number of hits and the total bases. Subtract the hits from the total bases and multiply the resulting extra bases by .31 and the hits by .47. This may introduce small variations from the rigorous formula, generally a fraction of a run, but the calculation is much snappier for those without a complete computer database at the ready. Of course, if you want to fiddle with some of these calculations yourself, perhaps for this year's players, you may use either version of the formula with good results.

The Linear Weights formula for batters may be long, even in its condensed form, but it calls for only addition, subtraction, and multiplication and thus is as simple as the slugging percentage, whose incorrect weights (1, 2, 3, and 4) it revises and expands upon. Each event has a value and a frequency, just as in slugging percentage, yet as in no batting statistic you have ever seen, outs are treated as offensive events with a run value of their own (albeit a negative one), a truth so obvious it somehow escaped notice. Just as the run potential for a team in a given half inning is boosted by a man reaching base, it is diminished by a man being retired; not only has he failed to change the situation on the bases but he has deprived his team of the services of a man further down the order who might have come up in this half inning, either with men on base and/or with scores already in.

What Linear Weights does is to take every offensive event and treat it in terms of its impact upon the team—an average team, so that a man does not benefit in his individual record for having the good fortune to bat cleanup with the Brewers or suffer for batting cleanup with the Mets. The relationship of individual performance to team play is stated poorly or not at all in conventional baseball statistics. In Linear Weights it is crystal clear: The linear progression, the sum, of the various offensive events, when weighted by their accurately predicted run values, will total the runs contributed by that batter or that team beyond the league average. Let's take as an example Wade Boggs, American League batting champion in 1983.

AB	H	2B	3B	HR	BB	HBP	SB	CS
582	210	44	7	5	92	1	3	3

Once we have determined his number of singles (154) by subtracting extra-base hits from hits, we can calculate his Linear Weights:

Runs = 154 (.46) + 44 (.80) + 7 (1.02) + 5 (1.40) + 92 (.33) +
       1 (.33) + 3 (.30) − 3 (.60) − 372 (.25) =
       70.8 + 35.2 + 7.14 + 7.0 + 30.4 + 0.3 + 0.9 − 1.8 −
       93.0 = 56.9

By these calculations, we see that Wade Boggs contributed 56.9 runs *beyond* what an average player might have done in his stead. This figure differs from that which appears for Boggs in the seasonal leaders tables at the back of the book for five reasons: (1) to make comparisons more fair, especially for the years since the AL introduced the designated hitter, we have eliminated from consideration all batting performance by pitchers, thus increasing the measured level of production, which in turn makes an out more of a negative event, say, from − .25 to − .27 (− .27 was the value of the out in the American League of 1983); (2) the impact of the out fluctuates with conditions of play, rising slightly in years marked by heavy hitting, falling in years dominated by pitching; (3) stolen bases and caught stealing are not used so as to make comparisons across time more meaningful (the data are not uniformly available); (4) all extra bases are credited at + .31 runs per base; and (5) the computer calculations include more significant digits while the above calculation employs rounded figures.

Recognizing that the more dedicated readers will wish to keep track of batting performance by compiling Linear Weights themselves over the course of a season, and that they may be frustrated by the difficulty of separating out pitcher batting or of calculating the (At Bats − Hits) factor for the league, we advise that using a fixed value of − .25 for outs will tend to work quite well if you wish to include pitcher performance, and a fixed value of − .27 will serve if you wish to delete it. Actually, any fixed value will suffice in midseason; it's only when all the numbers are in and you care to compare this year's results with last year's (or with those of the 1927 Yankees) that more precision is desirable. At that point the value of the out may be calculated by the ambitious among you, but ideally, your newspaper or the sporting press will provide accurate Linear Weights figures. Who, after all, calculates ERA for himself?

For those to whom calculation is anathema, or at the least no pleasure, Batter Runs, or Linear Weights, has a "shadow stat" which tracks its accuracy to a remarkable degree and is a breeze to calculate: OPS, or On Base Average Plus Slugging Percentage.[1] While it is not expressed in runs and thus lacks the philosophical appeal of Linear Weights, the standard deviation of its most complete version is 20.4 runs compared to the 19.8 of Linear Weights. In other words, the correlation between Linear Weights and OPS over the course of an average team season is 99.7 percent.

OPS consists of two measures each of which is somewhat better than batting average, but not in a league with the newer statistics. On Base Average, as previously noted, brings the walk and hit batsman in from the statistical cold but treats all bases alike. Slugging percentage weights the hits according to the bases gained (intuitively sound if statistically false), but it does not take into account any base gained without a hit. These two one-legged men, when joined together, make for a very sturdy tandem. The weaknesses of the one are almost exactly compensated by the strengths of the other.

However, as an average or ratio, OPS measures the *rate* of batting success (efficiency), while Linear Weights measures the *amount* of success. For example, a batter who goes 2-for-5 with a walk in one game, those 2 hits being doubles, will have an OBA of .500 and a SLG of .800; his OPS will be 1.3. Another batter, who in 162 games gets 200 hits and 100 walks in 500 at bats, with 400 total bases, will have an identical OBA, SLG, and OPS. Which player has contributed more to his team? Clearly, longevity, or *amount* of production, is no less important than *rate* of production.

To cite a specific instance in which OPS and LWTS differ, take George Brett's remarkable 1980 season in which he batted .390, had 298 total bases, 75 bases through walks or HBP, and 118 RBIs—all in only 117 games played. In the table of all-time single-season leaders in OPS, the Kansas City third baseman ranks 34th when his OPS of 1.124 is normalized to the league average and adjusted for home-park effects. Yet in the table of park-adjusted Linear Weights, Brett's season ranks only 81st because he missed 45 games, in which his team derived no benefit from his high rate of performance. (Had Brett played 162 games and continued to perform at the same level, his Linear Weights would have been not 64.8 but 89.7, the 16th best mark in history.

What was the best mark? It's no surprise that Babe Ruth heads both the single-season and lifetime lists, but you may find interesting this list of the ten best Linear Weights batting performances since expansion in 1961 (these figures are not adjusted for home park; the table will be repeated in the next chapter with park factors incorporated).

Table IV, 2. *Batters' Linear Weights, Best Since 1961*

1.	1961 Norm Cash, DET	86.1
2.	1967 Carl Yastrzemski, BOS	76.4
3.	1961 Mickey Mantle, NY	76.3
4.	1969 Willie McCovey, SF	76.1
5.	1966 Frank Robinson, BAL	73.6
6.	1970 Carl Yastrzemski, BOS	71.7
7.	1977 Rod Carew, MIN	67.0
8.	1962 Frank Robinson, CIN	66.6
9.	1972 Dick Allen, CHI (A)	66.1
10.	1969 Harmon Killebrew, MIN	65.6

The OPS chart below looks somewhat different for reasons already explained. These numbers have been normalized to league average to make for a consistent comparison with Linear Weights, which has a built-in normalizing factor in its variably weighted out. While normalizing techniques and applications will be explored in full in Chapter 6, here's a brief example of how it works for OPS. In 1983 National League MVP Dale Murphy had an OBA of .396 and a slugging percentage of .540. The league OBA was .324 and the league SLG .376. The normalized OPS is calculated by adding the normalized OBA and the normalized SLG, then subtracting 1. For Murphy,

$$\frac{.396}{.324} + \frac{.540}{.376} - 1 = 1.22 + 1.44 - 1 = 1.66$$

For ease of expression, we will state Murphy's NOPS (Normalized OPS) simply as 166.[2]

Table IV, 3. *Normalized OPS, Best Since 1961*

1.	1969 Willie McCovey, SF	208.6
2.	1980 George Brett, KC	204.1
3.	1981 Mike Schmidt, PHI	204.0
4.	1961 Norm Cash, DET	204.0
5.	1972 Dick Allen, CHI (A)	202.6
6.	1971 Henry Aaron, ATL	201.9
7.	1961 Mickey Mantle, NY	199.4
8.	1966 Frank Robinson, BAL	197.3
9.	1962 Mickey Mantle, NY (A)	192.1
10.	1967 Frank Robinson, BAL	187.8

When OPS is adjusted in the next chapter to eliminate home-park bias, the composition of the list will shift once again.

RUNS AND WINS

Because OPS is not expressed in runs, it is less versatile than Linear Weights. For just as runs are proportional to the events which form them, so are they proportional to wins and losses. This statement, a truism today, was a novelty in 1954 when Rickey and Roth first stated the correlation between run differentials and team standings. But they did not take the next step, to recognize that not only a team's standing but even its won-lost record could be predicted from the run totals.

"The initial published attempt on this subject," Pete wrote in the 1982 issue of the SABR annual *The National Pastime*, "was Earnshaw Cook's *Percentage Baseball*, in 1964. Examining major-league results from 1950 through 1960 he found winning percentage equal to .484 times runs scored divided by runs allowed. . . . Arnold Soolman, in an unpublished paper which received some media attention, looked at results from 1901 through 1970 and came up with winning percentage equal to .102 times runs scored per game minus .103 times runs allowed per game plus .505. . . . Bill James, in the *Baseball Abstract*, developed winning percentage equal to runs scored raised to the power x, divided by the sum of runs scored and runs allowed each raised to the power x. Originally, x was equal to two but then better results were obtained when a value of 1.83 was used. . . .

"My work showed that as a rough rule of thumb, each additional ten runs scored (or ten less runs allowed) produced one extra win, essentially the same as the Soolman study. However, breaking the teams into groups showed that high-scoring teams needed more runs to produce a win. This runs-per-win factor I determined to be ten times the square root of the average number of runs scored per inning by both teams. Thus in normal play, when 4.5 runs per game are scored by each club, the factor comes out equal to ten on the button. (When 4.5 runs are scored by each club, each team scores .5 runs per inning—totaling one run, the square root of which is one, times ten.)"[3]

Note that when Pete refers to the need for approximately ten additional runs scored (or ten fewer allowed) to provide a team with an additional win, he does *not* mean that it takes ten runs to win any given game. Obviously, in a specific case, a one-run margin is all that is required; but statistics are designed for the long haul, not the short.

What does this have to do with Linear Weights? Remembering that LWTS is expressed not simply in runs but in beyond-average runs, the conversion from a batter's Linear Weights to his *wins* is a snap; taking the aforementioned exploits of Wade Boggs in 1983, we see that he contributed 56.9 runs,[4] or 5.7 wins, since in the American League in '83 it took 10.02 runs to produce an additional win. If every other player on the Red Sox had performed at the league average, the Boston record should have been 87-75; if each of the eight other batters had performed as well as Boggs (discounting reserves, pitchers, fielders, and stealers, whom we shall presume for this discussion to have been average), the Red Sox would have finished 9 × 5 wins over average or 132-30.

Alas for Red Sox fans, this was not so. Just as Linear Weights will show an above-average hitter to have contributed beyond-average runs and thus wins to his team, it will also show below-average hitters to have negative run marks, which result whenever the runs lost through outs exceed the runs gained through times reached base. Take the rest of the Boston infield . . . please. In 1983 first baseman Dave Stapleton cost his team 13 runs with his bat, second baseman Jerry Remy 17, and shortstop Glenn Hoffman 13. (Their gloves were no help—Remy alone was −22 runs in the field).

BASE STEALING RUNS

The quick-witted will remember that stolen bases are not used in these computations, but had we included them, the Boston infield's record would have looked better by only 1.2 runs. The extended Linear Weights formula for batters contains a factor for base stealers, expressed in runs. Although it is not used in the Batter Runs listings in the tables at the rear, it *is* presented as a separate Linear Weights formula in all years since 1951, when caught-stealing records began to be kept on an uninterrupted basis in both leagues. How do you judge the effectiveness of a base stealer? Conventional baseball statistics will lead you to the conclusion that whoever has the most steals is the best thief; that is the sole criterion for *The Sporting News* annual "Golden Shoe Award" in each league. How often the man with the most steals may have been thrown out is of no concern.

An article in the 1981 *Baseball Research Journal* by Bob Davids offered something more sophisticated yet utterly simple: a stolen base percentage, which is simply stolen bases divided by attempts. The best stolen base average of all time, insofar as we know and based on a minimum of 20 attempts, is Max Carey's in 1922 when he stole 51 bases in 53 attempts. The most times caught stealing in the course of a season was Ty Cobb's 38 in 1915, until 1982 when Rickey Henderson was nabbed 42 times. But the best method yet devised, and one that is pleasingly simple, is to apply the Linear Weights method to get "stealer's runs." One multiplies the steals by their run value of .30 and the failed attempts by −.60, and adds the two products. The implication for such men as Cobb, Henderson, Raines, et al., is clear: It takes a fabulous stealing performance to produce as much as one extra win for the team.

In 1915 Ty Cobb, when he established the modern stolen-base record of 96, can be seen to have contributed to his team 28.8 runs, while his 38 foiled larcenies cost 22.8. Thus Cobb, for all his whirling dervish activity, accounted for only 6 non-par runs—not even a single win. Whoa! You mean that not a single one of Cobb's steals produced a victory? That's not what's being said: The fact is that while the gain from the stolen base is entirely visible—an extra base which may be followed by a hit that would otherwise not have produced a run—the cost of the caught stealing is entirely invisible, or conjectural, except with the aid of statistics. How many big innings did Cobb run his team out of? How many batters reached base in ensuing innings who might, in an earlier inning, have had their contributions count for runs? What Linear Weights indicates is that, *on balance,* not on a specific-case basis, the stolen base is at best a dubious method of increasing a team's run production.

Now let's take a look at what Henderson did. Henderson's 130 stolen bases in 1982 produced 39 runs for his team. His 42 failed attempts took away 25.2 possible runs. Net effect: approximately 14 runs, or one and a half wins, a performance nearly *three times* as good as Cobb's. In 1983, stealing 22 fewer bases, he was even better, accounting for 21.0 runs. However, the all-time best stealing record is that of Maury Wills in 1962, when he stole 104 bases and was caught only 13 times. Wills's 104 stolen bases produced 31.2 runs; his 13 steals cost only 7.8. So, his baserunning contribution was 23.4, or a little over two wins.[5]

DEFENSIVE RUNS

Let's get off the case of the Red Sox infield, and go back one year earlier to that of second baseman Doug Flynn. Surely the brain trust at Montreal (and earlier at Texas, the New York Mets, and Cincinnati) did not grant him employment for philanthropic reasons, and surely they know their business. These organizations must have reckoned that Flynn saved the team more runs with his glove than he cost at bat. Second base is a "skill" position, along with shortstop, catcher, and third base. For these positions the tradition has been to tolerate poor offense if it is accompanied by fine defense. So what, one might argue, if Flynn's bat cost his 1982 teams 4 wins they might have gained with an average hitter in his place? Mike Ivie batted at precisely the league average in '82, yet you couldn't have put him in Flynn's spot without costing the team a great many more than 4 wins.

This is beyond argument, but what about replacing Flynn with an average *fielding* second baseman? How many runs did Flynn save the Rangers and Expos beyond what an average second baseman might have saved? The Linear Weights formula for second basemen, third basemen, and shortstops begins by calculating a league average for the position, in this fashion:

$$\text{AVG.} \ \frac{\text{pos.}}{\text{lg.}} = \left(\frac{.20 \ (PO + 2A - E + DP) \ \text{league at position}}{PO \ \text{League total} - K \ \text{league total}} \right)$$

where A = assists, PO = putouts, E = errors, DP = double plays, and K = strikeouts. Then a rating is calculated for the team in question at that position:

$$\text{Team Runs} \ (\text{per pos.}) = .20 \ (PO + 2A - E + DP) \ \text{team at pos.} - \text{Avg. pos. lg.} \times \left(\frac{PO}{\text{team}} - \frac{K}{\text{team}} \right)$$

Assists are doubly weighted because more fielding skill is generally required to get one than to record a putout. To evaluate a particular player, prorate by putouts. In other words, if the team's second base rating was +30 runs and one man had 324 of the team's 360 putouts at second base, he would get credit for ($^{324}/_{360}$) or 90 percent of the team's +30 rating, or 27 runs. Calculating Doug Flynn's Defensive Linear Weights in this manner, we find that he contributed −3 runs.

For catchers, the formula was modified only by removing strikeouts from their putouts. For first base, because putouts and double plays require so little fielding skill in all but the odd case, they were eliminated, leaving only .20(2A − E) in the numerator. For outfielders, the formula becomes .20(PO + 4A − E + 2DP). The weighting for assists was boosted here because a good outfielder can prevent runs through the threat of assists that are never made; for outfielders, the assist is essentially an elective play.

To solve the problem of how to assign outfield putouts and assists to left, center, and right fields, we take the three outfielders on each team with the most putouts and select the one of these with the most putouts as the center fielder, then pool the other two, and get league averages for each group. Center fielders are compared to the average for center fielders, left and right fielders to the average for those positions. (Because late-inning substitution is common among outfielders, and the number of innings each man played is difficult to estimate, there could be a source of error here.) For pitchers' fielding, of course, no such problem presents itself: Innings are known.

Since expansion, the top ten single-season defensive performances have been:

Table IV, 4. *Defensive Linear Weights, Best Since 1961*

1.	1963	Bill Mazeroski, PIT	46.5
2.	1980	Ozzie Smith, SD	42.8
3.	1966	Bill Mazeroski, PIT	40.8
4.	1962	Bill Mazeroski, PIT	40.7
5.	1983	Ryne Sandberg, CHI (N)	39.9
6.	1971	Graig Nettles, CLE	39.5
7.	1982	Buddy Bell, TEX	36.9
8.	1964	Bobby Knoop, LA (A)	36.5
9.	1977	Ivan De Jesus, CHI (M)	36.1
10.	1977	Manny Trillo, CHI (N)	35.9

There is a great deal more to say on the subject of fielding skill and how it may be measured, and this will be attempted in Chapter 12.

PITCHING LINEAR WEIGHTS

Determining the run contributions of pitchers is much easier than determining those of fielders or batters, though not quite so simple as that of base stealers. Actual runs allowed are known, as are innings pitched. Let's assume that a pitcher is responsible only for earned runs. Then why, we hear some of you asking, is the ERA not measure enough of his ability? Because it tells only the pitcher's *rate* of efficiency, not his actual benefit to the team. In a league with an ERA of 3.50, a starter who throws 300 innings with an ERA of 2.50 must be worth twice as much to his team as a starter with the same ERA who appears in only 150 innings. Through Linear Weights, we seek to determine the number of beyond-average runs a pitcher saved—the number he prevented from scoring that an average pitcher would have allowed.

The formula for Earned Run Average is:

$$ERA = \frac{\text{Earned Runs} \times 9}{\text{Innings Pitched}}$$

The number of average, or par, runs for a pitcher, which is represented by a Linear Weight of zero, is equal to:

$$\frac{\text{League ERA} \times \text{IP}}{9}$$

If the league ERA is 3.64[6] (as the National League's was in 1983) and a pitcher's ERA is also 3.64 he will by definition have held batters in check at the league average no matter how many innings he pitched. If, however, his ERA was 2.69 and he hurled 274 innings (as Mario Soto did for the Reds in '83), he will have saved a certain number of runs that an average pitcher might have allowed in his place; to find that number we employ the Linear Weights formula:

$$\text{Pitcher's Runs} = \text{Innings Pitched} \times \left(\frac{\text{League ERA}}{9} \right) - \text{ER}$$

This represents the difference between the number of earned runs allowed at the league average for the innings pitched and the actual earned runs allowed. For the case of Soto, we get

$$\text{Runs} = 274 \times \frac{3.64}{9} - 82 = 28.8$$

Soto was 28.8 runs better than the average National League pitcher in 1983, and had he been transported to an average NL team—that mythical entity which scores as many runs as it allows while winning 81 and losing 81—he would have made that team's mark 84-78. An alternative way to calculate pitchers' Linear Weights, useful with oldtimers for whom you may have the ERA but not the number of earned runs allowed, is to use the pitcher's ERA, subtracted from the league's ERA, multiplying by the innings pitched, then dividing by nine. In Soto's case, this approach would look like:

$$(3.64 - 2.69) \times \frac{274}{9} = 28.9$$

The difference of a tenth of a point is accounted for because we are using the ERA of 2.69, which has been rounded off, rather than the absolute figure of the pitcher's earned runs allowed, 82.

The two parts of performance—efficiency and durability, or how well and how long—are incorporated into all Linear Weights measures. If you are performing at a better than average clip, the more regularly you do so, the more your team will benefit and thus the higher your Linear Weights measure. If you are stealing bases 9 times out of 10, your team will benefit more from 60 attempts than from 40; if you are batting at an above average clip, it's better to play in 160 games than 110; if you're allowing one earned run per game less than the average pitcher, your LWTS will increase with innings pitched.

A problem emerges in this regard when trying to compare the

LWTS of a pitcher from 1978 like Ron Guidry, with that of Hoss Radbourn in 1884. In the "efficiency" component of the formula, which may be understood as the league ERA minus the individual's ERA, the two compare this way:

$$\text{Guidry} = 3.76 - 1.74 = 2.02 \qquad \text{Radbourn} = 2.98 - 1.38 = 1.60$$

Guidry's differential is "unfairly" boosted by the higher league ERA of 1978; in fact, if we had compared the two by their normalized ERAs, which is logically sounder, the results would have been:

$$\text{Guidry} = \frac{3.76}{1.74} = 2.16 \qquad \text{Radbourn} = \frac{2.98}{1.38} = 2.16$$

Yet because rules and playing conditions allowed Radbourn to extend his efficiency over 679 innings, while Guidry hurled "only" 274, their LWTS look like this:

$$\text{Guidry} = 62.0 \qquad \text{Radbourn} = 120.0$$

For this reason we have separated the listings for single-season and lifetime leaders at 1901.

The ten best Pitchers' Linear Weights performances since 1961, not adjusted for home park, have been:

Table IV, 5. *Pitchers' Linear Weights, Best Since 1961*

1.	1966 Sandy Koufax, LA	67.4
2.	1968 Bob Gibson, STL	63.2
3.	1978 Ron Guidry, NY (A)	62.0
4.	1964 Dean Chance, LA (A)	61.0
5.	1975 Jim Palmer, BAL	61.0
6.	1971 Wilbur Wood, CHI (A)	57.7
7.	1971 Vida Blue, OAK	57.2
8.	1972 Steve Carlton, PHI	56.9
9.	1965 Sandy Koufax, LA	56.1
10.	1971 Tom Seaver, NY (N)	54.3

There is a great deal more to say on the subject of pitching and statistics: see Chapters 10 and 11.

LINEAR WEIGHTS IN PRACTICE

Having formulas for pitching, fielding, baserunning, and batting, we can assess the run-scoring contribution of every individual who has ever played the game, and thus the number of wins that he has contributed in a given season or over his career. The number of runs required to produce an additional win has varied over the years between 9 and 11 runs, with a very few league seasons outside those parameters. (The formula for obtaining the runs-per-win factor was cited earlier, but in the LWTS tables at the rear, which are rank-ordered by wins, these factors have been incorporated in any seasonal or lifetime calculation.)

Limited by conventional baseball statistics one might, in 1982, have uttered something like, "Dale Murphy hit .281 with 36 homers and 109 RBIs—the guy must have been worth 10 extra wins to Atlanta all by himself!" Or: "The White Sox are only one pitcher away from winning the division." Or: "The Mets are only three players away from being a contender." Or, in mid-1983: "Trading Keith Hernandez for Neal Allen was the worst move the Cardinals ever made—no way that a pitcher can be worth as much to them as a first-rate everyday player."

With Linear Weights, these statements, or rather the concerns they reflect, can be approached with some data and with some degree of objectivity. First: Dale Murphy had a fine year in 1982 and a better one in '83, but to have contributed 10 wins by himself he would have had to account for some 95 runs, a mark that has been attained by only 14 men in major-league history. In fact, Murphy contributed 3.2 wins in '82.

As to the White Sox, they finished 87-75 in 1982, while their Linear Weights projected them to finish at 88-74. The Angels, who won the AL West at 93-69, actually projected to finish with an even better record, 95-71. The other team which finished ahead of Chicago, Kansas City, went 90-72, but exceeded their projection of 88-74, the same as that of the White Sox. So, the Sox management might have asked, how to close ground on the Angels, if we can presume the Royals are in the same boat as the Sox? Could one pitcher—like Floyd Bannister, whom they picked up in the free-agent bazaar—make the difference? To do so, he would have to contribute 70 runs by the Linear Weights formula, a feat only four pitchers in this century have been able to accomplish, none since Lefty Grove in 1931 (even Guidry in 1978 saved only 62 runs). In 1982, pitching for Seattle—and remember, the LWTS formula is divorced from considerations of batter support—Bannister contributed 3.3 wins, second only to Toronto's Dave Stieb. So presuming that he pitched as well for the Sox as he did for the Mariners, or even slightly better, he would not be enough to "win" Chicago the flag on paper; Chicago would need help from other quarters, and perhaps the Angels' age would begin to tell. And of course both came to pass, as Carlton Fisk, Greg Luzinski, Rich Dotson, and Salome Barojas all came through, and Reggie Jackson and Tommy John didn't.

Regarding the statement, "The Mets are only three players away from being a contender," which was uttered in the spring of 1983: They closed the 1982 season with only two of their eight regulars having made positive run contributions, i.e., being better than average; their pitching staff had only three in the plus column. Replacing Hubie Brooks (−1.6 wins) with Mike Schmidt, Ron Gardenhire (−1.7) with Robin Yount, and Mike Scott (−2.7) with Steve Rogers would have produced a net gain of 22.6 wins, transforming the '82 Mets' record from 65-97 to 88-64, good enough for third place. If that's what was meant, yes, the Mets were three players away, but find us the G.M. who can swing those deals.

The Allen for Hernandez swap was complicated by such considerations as Hernandez's contractual status, Allen's psychological disarray in the first half of the 1983 season, and the question of how Whitey Herzog intended to use Allen (in the pen or in the rotation). Also pressing was the need to find a spot for David Green to play. But as to whether a pitcher can be worth as much as an everyday player, despite the conventional wisdom, the answer is yes indeed. A run is a run is a run, and runs contributed by pitchers, as measured by LWTS, have the same win value as runs contributed by batters. In 1982 Keith Hernandez had a typical year in which he accounted for 2.3 wins. That mark was topped on his own team by no other batter, but Joaquin Andujar contributed 4.2 wins, so it *was* possible for a pitcher to make the deal good from the St. Louis point of view. Neal Allen, however, pitching exclusively in relief for the Mets in 1982, had contributed only 0.4 wins above average, and that was as high a mark as he had ever posted. The Cards also got Rick Ownbey, whose track record was not of consequence. If Hernandez could be counted on to produce in 1983 as he had in '82, then Allen would have had to emerge as a "new man"—perhaps as a starter, where some in New York felt he belonged from the outset. Unless the Mets fail to sign Hernandez for '84, or Allen becomes a world-beater as a starter (he can't be of as much use in a bullpen headed by Bruce Sutter), this looks on paper like a one-sided deal favoring the Mets. Before the trade, Hernandez had contributed 0.3 wins, offensively and defensively, to the Cards, while Allen had been −0.7 wins for the Mets; afterward, Hernandez supplied 2.0 wins to the Mets and Allen −0.2 to the Cards.

At the end of each of the seasonal listings of league leaders at the end of the book is a section of team statistics, among which are team Linear Weights for batting and pitching (Fielding Wins are given for 1946 to the present, Base Stealing Wins not at all because they have so little impact). This data is useful for analyzing past results and predicting future outcomes. Let's look at the Braves and Dodgers of 1982 as an example:

Table IV, 6. *Atlanta & Los Angeles, 1982*

	W	L	R	OR	BAT RUNS (adj)*	BAT WINS	PIT RUNS (adj)*	PIT WINS	DIFF
ATL	89	73	739	702	−28.6	−3.0	19.1	2.0	9.0
LA	88	74	691	612	85.6	9.0	13.9	1.5	−3.5

*The run figures have been adjusted for home-park influences.

What these numbers tell us, basically, is that Atlanta did it with mirrors, and LA should have won in a cakewalk. First, look at the run differentials (R-OR) for the Braves, 37; for the Dodgers, 79. Simply using these figures to predict W-L record as outlined earlier, Atlanta should have finished at 85-77 while LA should have finished 89-73. Looking at their projected wins resulting from the run contributions of all their batters, Atlanta's offense was worse than the league average, while LA's was outstanding, the best in the league. The Atlanta pitching staff was its real source of strength, contributing 2 wins, which was the top mark in the division (though not approaching the 8 pitching wins of the Cards), while the Dodger staff contributed 1.5 wins. The Linear Weights totals for Atlanta thus projected to a W-L mark of 80-82 (3 games below average for the batting, 2 above for the pitching), while the Dodgers projected to a mark of 90-72 or 91-71; the Braves exceeded expectations by a whopping 9 games while the Dodgers fell below theirs by 3.5 games. These are the win contributions of the Atlanta Braves of 1982:

Table IV, 7. *Linear Weights in Wins, Atlanta 1982*

Benedict, c	−1.8	Niekro	0.9
Chambliss, 1b	0.5	Mahler	−0.6
Hubbard, 2b	−1.4	Walk	−1.8
Ramirez, ss	−1.2	Bedrosian	2.5
Horner, 3b	1.8	Garber	2.2
Butler, of	−2.0	Camp	0.6
Murphy, of	3.2	Perez	0.8
Washington, of	−0.1	Dayley	−0.5
Pocoroba, c	−0.4	Cowley	−0.3
Others	−0.4	Others	−1.7

This is your ordinary, average, .500 or thereabouts club. It was not going to repeat in 1983 (flukes seldom repeat) without some substantial improvements in offense or in its starting rotation (the bullpen was excellent). What happened in 1983 is precisely that. Benedict and Butler moved their offense onto the plus side of the ledger, Ramirez and Hubbard improved, and most notably Walk and Mahler gave way to Perez and McMurtry. And yet it wasn't enough. The Braves won 88 games in 1983, on merit this time, but the Dodgers' pitching was incredible, by far the best in the league; the Braves still needed another starting pitcher.

Are you wondering how the Braves' offense could have been "sub-par" in 1982 with 739 runs scored while the Dodgers were the cream of the league with only 691? Park Factor. It's enormously important in its effect on performance, and on perceptions, and it is what we will discuss next.

---

[1] Run scoring for teams is proportional to On Base Average *times* slugging percentage, while for individuals added to a lineup of average players, runs produced by that player are proportional to On Base Average *plus* slugging percentage.

[2] In the tables for the 1983 season, Murphy's mark will appear as 157 because there we have removed pitchers' batting performance from the league averages. The league OBA and SLG thus are substantially higher than the .324 and .376 cited in the example, making Murphy's NOPS accordingly lower.

[3] James handled this situation by adjusting his exponent to reflect run-scoring patterns; the power of 1.83 was correct for the case in which each team scored 4.5 runs per game. Bill's method and Pete's work about as well; Soolman's is slightly less accurate, Cook's considerably worse. About a year after Pete's article appeared, Bill Kross, a Purdue professor, devised an elegant little formula that was not only simpler than the others, but also very nearly as accurate, erring only when run differentials were extreme (± 200 runs). If a team is outscored by its opponents, Kross predicts its winning percentage by dividing runs scored by two times runs allowed; if a team outscores its opponents, the formula becomes

$$1 \; - \; \frac{\text{runs allowed}}{2 \,(\text{runs scored})} \, .$$

[4] This figure differs from that in the tables at the rear of the book; there it is 48.4, after adjustment for home park.

[5] From play-by-play data Dave Smith calculated that Wills produced 24 runs and 3 wins.

[6] This figure is calculated by summing up all the pitchers' IP and ER and then computing, rather than incorporating the "Team ER" category dictated since 1971 by Rule 10.18, which creates the bizarre result of a team allowing fewer earned runs than the teams' pitchers do.

5

THERE'S NO PLACE LIKE HOME

Familiarity breeds success. Every team is expected to win more games at home than it does on the road, to the extent that if it only breaks even on the road it is deemed to have a shot at the pennant. In 1983, for example, only three National League clubs had plus .500 records on the road (and none won more than 43 games); in 1981 Montreal finished 8 games *under* .500 away from home yet topped the National League East.

Hitters' park or pitchers' park, the home team should take advantage of its peculiarities better than the visiting team. The Houston Astros may score fewer runs at home than they do on the road, but their *differential* between runs scored and runs allowed will be greater than their run differential on the road. The Boston Red Sox may allow more runs at home than they do on the road, yet the result should be the same: Their run differential, and thus their won-lost record, should be better at Fenway than in the hinterlands. If it's not—and it was not in 1983—shake up that front office.

Why would a team, strong or weak, perform better in their own park than on the road? The players benefit from home stands of reasonable duration—say, eight to thirteen games—when they live in their own residences, sleep at more nearly regular times, play before appreciative fans, and benefit from the physical park conditions which to some degree may have made their organizations acquire them in the first place. It is difficult for fans to grasp the difficulty of playing on a travel day or of adjusting to jet lag and hotel "comforts."

In 1983, when, as noted above, only three NL clubs had winning records on the road, only one (the Reds) was a loser at home. Almost anybody, it seems—which is to say, the Mets and Cubs of '83—can play .500 ball at home. This is somewhat deceptive, for while the team that goes 81-81 on the season is by definition an average team, to be an average performer at home requires a team to win 54 percent of its games. Substantiation for this assertion rests in the table below, which gives the home won-lost percentages of the American and National (and Federal) Leagues for every decade since 1900. Totaling all the games played at home by all the teams, we come up with a record of 62,205 wins and 52,426 losses, a winning percentage of .543. The inverse, .457, is the average road record.

Table V, 1. *Home-Park Won-Lost Records*

	National League			American League		
	W	L	Pct.	W	L	Pct.
1900-10 (1901 AL)	3489	2995	.538	3345	2530	.569
1911-20	3189	2755	.537	3201	2754	.537
1921-30	3360	2770	.548	3344	2787	.545
1931-40	3353	2760	.549	3349	2753	.549
1941-50	3319	2823	.540	3383	2754	.551
1951-60 ('61 NL)	3681	3098	.543	3291	2863	.535
1961-68 ('62 NL)	3075	2591	.543	3462	3003	.535
1969-76	4088	3638	.529	4142	3568	.537
1977-82	3023	2473	.550	3451	2951	.539
	30,577 −	25,903 =	.541	30,968 −	25,963 =	.544

Federal League

1914-15	660 − 560 = .541
TOTAL	62,205 − 52,426 = .543

If the average home winning percentage is .543, then an average team (defined as 81-81) should be expected to go 45-36 at home and 36-45 on the road. What this means is that breaking even on the road (impossible in ordinary practice, but say 41-40 or 40-41) represents a performance that is distinctly *above* average. Only six teams in this century have won pennants with below-average road records. The worst on a percentage basis was the Expo team of 1981; the others were the 1902 Athletics, the 1944 Browns, the 1974 Pirates, and the Phillies and Royals of 1978. Not a single one won a World Series, and in the four cases in which the teams were divisional champions, not one made it to the World Series. There is no statistical reason for this; we offer simply a cautionary tale.

Just as in the previous chapter we indicated how runs scored and runs allowed might predict won-lost records, now we move backward from won-lost records—the actual home-park norm of 45-36, which is about 10 percent better than the theoretical norm of 41-40—to examine runs scored and runs allowed. It develops that individuals bat and pitch at a rate 10 percent higher at home, on average. That is, On Base Average and slugging percentage each tend to be 5 percent higher (when combined to create OPS, they are 10 percent higher); batting average will be 5 percent higher too. Linear Weights, because it is denominated in runs, will be 10 percent higher at home, while earned run average, for the same reason, will be 10 percent lower. These statements are true on average, but in some cases home-park variations may run considerably higher or lower: The ERAs of Red Sox pitchers may soar 20 or 30 percent, and the OPS of Astro batters may plummet by as much.

Keeping in mind that the home record of the average hitter, as reflected in his OPS, should be 1.10 times his OPS on the road, let's look at the lifetime ratio of Normalized OPS at home to NOPS on the road for some leading American League batters.

Table V, 2. *Lifetime Ratios of Normalized OPS at Home to NOPS on the Road*

Ty Cobb, 1.03	Mickey Mantle, 1.07
Joe DiMaggio, 0.88	Babe Ruth, 1.04
Jimmie Foxx, 1.24	Tris Speaker, 1.22
Lou Gehrig, 0.94	Ted Williams, 1.09

Speaker and Foxx derived more than average benefit from their home parks to a staggering degree while the others, notably DiMaggio and Gehrig, had their batting performance suffer for playing where they did (Williams took nearly average—1.10—benefit of Fenway). Thus the batting statistics of all eight men—like those of every man who ever played the game—reflect not only how they played, but *where* they played, with the latter proposition having enormous effect on those batters blessed to have played half their games in Shibe Park, Fenway, or Wrigley Field and on those cursed to have been denizens of Yankee Stadium, San Diego Stadium, or the Astrodome. For pitchers, naturally, the stigmata are reversed.

For hard luck in home parks, it is tough to top the record of Dave Winfield, who has had the misfortune to call both San Diego and Yankee Stadiums home. Through 1982, his lifetime OPS, normalized to league average but not adjusted for park effects, was 102nd best on the all-time list of those playing in 1000 games. Had he played his home games instead in Fenway Park, his NOPS would have projected to the 25th best of all time. (The statistical method by which such projections are made is explained below.)

If we desire to remove the silver spoon or the millstone that a home park can be, and measure individual ability alone, we must create a statistical balancer which diminishes the individual batting marks created in parks like Fenway and augments those created in San Diego. Pete has developed an adjustment which enables us, for the first time, to measure a player's accomplishments apart from the influence of his home park.

Parks differ in so many ways that it may be hard to imagine how their differences can be quantified. The most obvious way in which they differ is in their dimensions, from home plate to the outfield walls, and from the base lines to the stands. The older arenas—Fenway Park, Wrigley Field, Tiger Stadium—tend to favor hitters in both regards, with reachable fences and little room to pursue a foul pop. The exception among the older parks is Chicago's Comiskey which, in keeping with the theories of Charles Comiskey back in 1910 and the team's perceived strength, was built as a pitchers' park. Yet two parks can have nearly equal dimensions, like Pittsburgh's Three Rivers Stadium and Atlanta's Fulton County Stadium, yet have highly dissimilar impacts upon hitters because of climate (balls travel farther in hot weather), elevation (travel farther above sea level), and playing surface (travel faster and truer on artificial turf). Yet another factor is how well batters think they see the ball; Shea Stadium is notorious as a cause of complaints.

And perhaps more important than any of the objective park characteristics, suggested Robert Kingsley in a 1980 study of why so many homers were hit in Atlanta, is the attitude of the players, the way that the park changes their view of how the game must be played in order to win. Every team that comes into Atlanta in August knows that the ball is going to fly and, whether it is a team designed for power or not, it plays ball there as if it were the 1927 Yankees. In their own home park the Astros may peck and scratch for runs, but in Atlanta they will put the steal and hit-and-run in mothballs. Conversely, a team which comes into the Astrodome and plays for the big inning will generally get what it deserves—a loss. The successful team is one that can play its game at home—the game for which the team was constructed—yet is flexible enough to adapt when on the road. How to quantify attitude?

Rather than try to assign a numerical value to each of the six or more variables that might go into establishing an estimator of home-park impact, Pete looked to the single measure in which all these variables are reflected—runs. After all, why would we assign one value to dimensions, another to climate, and so on, except to identify their impact on scoring? If a stadium is a "hitters' park," it stands to reason that more runs would be scored there than in a park perceived as neutral, just as a "pitchers' park" could be expected to depress scoring.

To measure park impact, Pete looks not at the runs scored by the home team, which may have been put together specifically to take advantage of a park's peculiar features, but rather those scored by the visiting teams. By totaling the runs allowed at home for all teams in a league year and dividing that figure by the runs allowed by all teams in their road games, we take the first step in determining the Park Factor, which may be applied to a team's batters and pitchers (it might also be applied to base stealers, inasmuch as Craig Wright's studies have shown that it is 12 percent easier to steal on artificial surfaces, and fielders, who also benefit from the carpet, as shown by Paul Schwarzenbart's study in *The Baseball Analyst;* however, this task awaits another day).

The succeeding steps, alas, become increasingly complicated, and for this reason the full explanation for the computation of the Park Factor is left to the footnote, where hardy readers might consider taking a peek right now.[1] For most of us, though, it will be enough to understand that the Park Factor consists mainly of the team's home-road ratio of runs allowed, computed as it was above for the league, compared to the league's home-road ratio. The batter adjustment factor, or Batter Park Factor (BPF), consists of (1) the Park Factor and (2) an adjustment for the fact that the batter does not have to face his own pitchers. The pitcher adjustment factor, or Pitcher Park Factor (PPF), likewise consists of the Park Factor and an adjustment for the fact that the pitcher does not have to face his own team's batters.

The BPF and PPF are expressed in relation to the average home-park factor, which is defined mathematically as 1.00. A park which featured 5 percent more scoring than the average park would have a BPF of 1.05, while that same park's PPF might be 1.04 or 1.06, for instance, because it is adjusted differently (correcting for the absence of home team batters rather than that of home team pitchers). In practice, a BPF might be used in this way: To express the individual batting performance of, say, Joe Morgan in 1976, take his Normalized On Base Plus Slugging (NOPS) of 1.91 and simply divide that by his Batter Park Factor that year, which was 1.08: the result is 1.77, which is the NOPS that Morgan would have totaled had he played in an average home park, not deriving the 8 percent additional benefit of Riverfront Stadium. (Batter and Pitcher Park Factors for each year since 1901 are listed in the team stats section of the year-by-year record at the back of the book.) Analogously, the normalized earned run average of Cincinnati pitcher Pat Zachry in that year (league ERA of 3.51 over Zachry's ERA of 2.74, or 1.28) is bettered by *multiplying* the NERA by the Pitcher Park Factor of 1.06 (result: an NERA of 1.36, or an ERA of 2.58).

To apply Batter Park Factor to any other average—On Base, slugging, Isolated Power, batting average—use the square root of the BPF. This is done so that run scoring for teams, which is best mirrored by On Base Average *times* slugging percentage, can be represented clearly:

$$\frac{OBA}{\sqrt{BPF}} \times \frac{SLG}{\sqrt{BPF}} = \frac{O \times S}{BPF}$$

The application of the Batter Park Factor to Linear Weights, which is not an average, is more complicated—the explanation will be found in the footnotes—but Park Adjusted figures are offered in the tables for the top three in LWTS for batters and starting pitchers for all years since 1900 and for relievers since 1946.[2]

The previous chapter presented the top ten Linear Weights performances since 1961 in batting and pitching without adjustment for home park. Here are those same lists with Park Factors incorporated (the superscript numbers indicate ranking in unadjusted LWTS).

Table V, 3. *Batters' Linear Weights*

	LWTS	BPF	LWTS (Adjusted for park)
1. 1961 Norm Cash, DET[1]	86.1	1.003	85.8
2. 1961 Mickey Mantle, NY[3]	76.3	.908	83.4
3. 1969 Willie McCovey, SF[4]	76.1	1.004	75.8
4. 1962 Frank Robinson, CIN[8]	66.6	.921	73.2
5. 1966 Frank Robinson, BAL[5]	73.6	1.023	72.0

	LWTS	BPF	LWTS (Adjusted for park)
6. 1977 Rod Carew, MIN	67.0	.980	68.7
7. 1972 Dick Allen, CHI (A)[9]	66.1	.978	67.3
8. 1970 Carl Yastrzemski, BOS[6]	71.7	1.067	66.6
9. 1970 Willie McCovey, SF	62.0	.940	66.5
10. 1969 Harmon Killebrew, MIN[10]	65.6	.996	65.9

Note that the second place finisher in the earlier ranking drops off the list entirely (Carl Yastrzemski, 1967) thanks to a BPF of 1.174 in that year, which meant that playing half his games in Fenway boosted the totals of all visiting players by some 17 percent (who said it was a righthanded hitters' park?). However, that extreme BPF does not mean that Yaz in particular benefited by 17 percent; in fact, an analysis of his record reveals that he hit only 10 percent better than average at home. To perform this kind of home-road breakdown for every player in every season is beyond human capacity, so a Batter Park Factor remains the best method for adjusting the records of all hitters. An aside: Much of the reason for the dominance of such early 1960s types as Cash, Mantle, and Robinson lies in the higher run-scoring pattern of the period. If we divide the Park Adjusted LWTS by the number of runs required to produce an additional win, Dick Allen rises to fifth and Frank Robinson in 1962 falls to sixth; also, Reggie Jackson's 1969 season finds its way onto the list, along with Yaz's 1967.

Table V, 4. *Pitchers' Linear Weights with Park Factors*

	LWTS	PPF	LWTS (Adjusted)
1. 1973 Bert Blyleven, MIN	47.0	1.131	65.1
2. 1971 Vida Blue, OAK[7]	57.2	1.059	64.3
3. 1978 Ron Guidry, NY[3]	62.0	1.016	63.9
4. 1971 Wilbur Wood, CHI(A)[6]	57.7	1.040	62.8
5. 1965 Juan Marichal, SF	46.0	1.120	59.9
6. 1966 Sandy Koufax, LA[1]	67.4	.940	59.7
7. 1968 Bob Gibson, STL[2]	63.2	.948	57.9
8. 1972 Steve Carlton, PHI[8]	56.9	.990	55.6
9. 1969 Bob Gibson, STL	49.5	1.041	54.7
10. 1972 Gaylord Perry, CLE	44.0	1.079	53.3

This list underwent drastic revision once home-park influences were discounted. Bert Blyleven, whose unadjusted LWTS of 47.0 did not make the top ten, zooms to first on the list when the hardships of trying to hold down the score in the old Minnesota stadium are considered. On the other side of the coin, Dean Chance's LWTS of 61, which was the fourth best of the expansion era, dropped off the list entirely because of the Chavez Ravine Pitcher Park Factor of .887, by far the lowest for any pitcher in the list of top 100 season performances.

This is not to say Chance had anything but a marvelous year: 20 wins, a 1.65 ERA, and 11 shutouts are hard to argue with; yet he was aided considerably by park conditions which were not available to, say, Blyleven in 1973. In 81 home games in 1964, the Angels allowed 226 runs; in 81 games on the road, they allowed 325—44 percent more, where a 10 to 11 percent increase would have been normal. If you are to compare Chance and Blyleven fairly, you must deny one the benefit of his home park and remove from the other the onus of his. This is what Park Factor does.

For decades, the all-time scoring squelcher was Chicago's South Side Park, which saw service at the dawn of the American League. From 1901 through 1909, its last full year of service to the White Sox, this cavernous stadium produced home run totals like the 2 in 1904, 3 in 1906, and 4 in 1909; in two years the Sox failed to hit *any* homers at home, thus earning the nickname "Hitless Wonders." In 1906, Chicago pitchers held opponents to 180 runs at South Side Park, an average of 2.28 runs per game, earned *and* unearned, in a decade when 4 of every 10 runs were unearned. This mark held until 1981, when the Astrodome intimidated opposing hitters to such a point that in the 51 home dates of that strike-shortened season, Astro hurlers were touched for only 106 runs—2.08 per game. The Pitcher Park Factor of .817 for the Astrodome was the lowest ever. Those who suspected that men like Joe Niekro, Don Sutton, Vern Ruhle, et al., were perhaps not worldbeaters after all were right: Look at the ERAs the Astro starters registered that year, and what these ERAs might have been in an average park like Shea that year (BPF: 1.00) or a moderately difficult pitchers' park like San Francisco (BPF: 1.06).

Table V, 5. *Houston Pitchers, 1982*

	ERA	BPF:1.00	BPF:1.06
Nolan Ryan	1.69	2.07	2.19
Joe Niekro	2.82	3.43	3.64
Vern Ruhle	2.91	3.56	3.77
Bob Knepper	2.18	2.66	2.82
Don Sutton	2.60	3.17	3.36
HOUSTON (all)	2.66	3.24	3.44
SAN FRANCISCO (all)	3.28	3.09	3.28

Some observations prompted by this table: San Francisco with its team ERA of 3.28 had a better pitching staff than Houston with its 2.66; and

Houston batters, regarded as a Punch-and-Judy crew by all observers, must have been a lot more effective than heretofore suspected. In fact, when Houston batters' totals (eighth in runs scored, eighth in LWTS) are adjusted for park, the Astros emerge on ability as the *best* hitting team in the National League of 1981! Even without the application of Park Factor, one might have come to a similar conclusion by examining the runs scored totals for all NL clubs on the road in 1981. Houston's total was exceeded only by those of the Dodgers and Reds.

Proceeding from a similar hunch, we may look at the batting record of the "Hitless Wonders" of 1906, who won the pennant (and the World Series, in four straight over a Cubs team which went 116-36 during the season). Baseball lore has it that a magnificent pitching staff (Ed Walsh, Doc White, Nick Altrock, and others) overcame a puny batting attack (BA of .230, 6 homers, slugging percentage of .286). In fact, the Sox scored more runs on the road than all but one AL team, and their Batting Linear Weights, when adjusted for park, was third in the league—the same rank achieved by their pitching. (How they won the pennant remains a mystery, for both Cleveland and New York had vastly superior teams on paper.)

There have been nine other notable "pitchers' parks" since 1900, those that held scoring down at a rate 15 percent or more below normal.

Table V, 6. *Worst Hitters' Parks Since 1900*

1981	Houston	.817
1906	Chicago (A)	.820
1981	Texas	.821
1918	Boston (A)	.822
1958	Milwaukee	.825
1926	Boston (N)	.832
1976	Houston	.838
1950	Boston (N)	.843
1953	Cleveland	.844
1975	Oakland	.844

The great hitters' parks—those providing 15 percent greater run scoring than normal—have been more numerous, but these are the top ten, in order.

Table V, 7. *Best Hitters' Parks Since 1900*

1955	Boston	1.22
1970	Chicago (N)	1.19
1972	Detroit	1.19

1957 Brooklyn	1.18
1968 Cincinnati	1.17
1967 Boston	1.17
1977 Boston	1.17
1981 Toronto	1.17
1917 Cleveland	1.17
1911 New York (A)	1.16

Looking at the most recent of these, Exhibition Stadium in Toronto, one wonders how the Blue Jays' young staff managed to avoid a mass nervous breakdown in 1981, let alone post an ERA of 3.82 which, when adjusted for park, proved second best in the AL.

Illuminating as the application of Park Factor can be to team results, it is positively mind-bending when applied to individuals. Here is a sampling of the revelations which emerge from a casual perusal of the tables at the rear.

- In 1981, Houston's Art Howe was the third best hitter in the National League.
- The superstar numbers posted by Jim Rice would be hardly as impressive if the Fenway Park Factor were taken into account (through 1982, slugging percentage home/away for Rice, .584/.476; LWTS home/away, 211.2/85.2).
- Of the 815 men who have played in 1000 or more games since 1900, no batter has suffered more for his "choice" of home park than Houston's José Cruz (Winfield's home-park advantage is nearly as poor).
- Of the top 100 lifetime marks in park-adjusted Batters' LWTS, only three have been achieved despite Park Factors 5 percent below average, those of Lou Gehrig, Gene Tenace, and Dave Winfield. And Cesar Cedeno, whose career is universally regarded as one of failed expectations, occupies the 117th spot on the list, his accomplishments adjusted for a low Park Factor of .95.
- When measuring batting by OPS adjusted for park, the list of top 100 seasons includes such unexpected delights as: Harry Lumley, the Dodger first sacker whose 1906 season (PF: .91) ranks 76th; or Frank Howard, whose 1968 season (PF: .89) ranks 81st; or Bobby Murcer in 1971 (PF: .94), who takes the 88th spot. These men were stars of the first magnitude, but did not receive their due until now. Howard's lifetime NOPS, in fact, is the 35th best of all time. In the 1983 Hall of Fame election, he received no votes (Ray Sadecki got two).

- Among the top fifty seasons posted by pitchers, as measured by LWTS, the 4th best of all time was the 1944 campaign of Dizzy Trout, who had to contend with a Park Factor of 1.15. Other perhaps unexpected occupants on the list: Bert Blyleven in 1973 (11th best), Steve Rogers in 1982 (42nd best), and Frank Sullivan in 1955 (43rd).
- Of the top fifty lifetime LWTS by pitchers, the greatest park handicap had to be overcome by Phil Niekro (10th best); others who spent their careers in home parks 5 percent more conducive to hitting were Fergy Jenkins, Dizzy Trout, Hal Newhouser, and Virgil Trucks.
- Nolan Ryan's lifetime ERA, an impressive 3.11, has been hugely helped by the fact that he has pitched all his home games in pitchers' parks, first Shea, then Anaheim, now the Astrodome. His lifetime PF of .942 is lower than that of all the top 100 pitchers except Warren Spahn, Lefty Gomez, and Don Sutton. Ryan's ERA at home through 1982, in fact, was 2.41, while on the road it was 3.75—not even a league average performance.

Of the top thirty hitters of all time, as measured by their NOPS, it is strange that the two who played in the worst hitters' parks were men whose rankings did not need the boost of their Park Factors: Babe Ruth and Lou Gehrig. Their totals are so awesome that no matter what measure of offense you use, no matter what adjustments you make, Ruth is going to rank first and Gehrig third, with Ted Williams second. However, the gap between Williams and Gehrig narrows considerably once each record is adjusted for park. What would Gehrig have done in almost any other park in the American League at that time? In 1930 he drove in 117 runs *on the road,* with 27 homers and a .405 BA. Lifetime, his road BA was .351; at home—where his batting average should have been, with a normal home-park advantage, .372—it was "only" .329.

Oddly, just as Babe Ruth's star was dimming in the late '30s, Gehrig seems to have concentrated more on pulling the ball for homers, with the result that in 1934, '36, and '37, he hit 30, 27, and 24 homers at home, a level previously unreached. Indeed, in 1934 he had one of the great *home* records of all time, with a BA of .414, 98 RBIs, and a NOPS of 2.48. In recent years, one of the best home batting marks has been that of Fred Lynn in 1979, when he batted .386 with 28 homers, 83 RBIs, and a NOPS of 2.50; since moving to Anaheim in 1981, he has not accomplished in a full season what he did in that half season.

A few other notable home hitting records:

- In 1912 Joe Jackson hit .483 at home, which was Cleveland's League Park.
- In 1920 Babe Ruth slugged .985 and had a NOPS of 3.10—more than three times the league average!—at the Polo Grounds (Yankee Stadium was not built until 1923).
- Also in 1920, George Sisler hit .473 at Sportsman's Park in St. Louis. Two years later, when he hit .453 there, teammate Ken Williams chipped in with 32 homers and 103 RBIs at home, and the Browns made their first serious run at a pennant.
- In 1936 Cleveland's Hal Trosky hit 30 homers and drove in 99 runs at home.
- In 1938 Hank Greenberg hit an all-time-high 39 fourbaggers at Briggs Stadium and Jimmie Foxx hit 35 homers at Fenway to go with his all-time high 104 RBIs.
- In 1941 Ted Williams had an On Base Average of .541 in Boston.

Pitchers, too, have compiled unbelievable home records. In the early part of the century:

- In 1908 Ed Walsh went 23-5 in 241 innings pitched at Chicago's South Side Park with an ERA of 1.04. That's a full season's work and then some.
- Joe Wood was 18-2 at home for the 1912 Red Sox (of course, he was impartial, going 16-3 away).
- Chief Bender won 11 and lost only 1 for the 1914 A's.

In 1916, pitching in tiny Baker Bowl, Pete Alexander hurled 9 shutouts. But the most astounding record may be that of Lefty Grove, who showed his partiality to home cooking throughout his career, in two cities. Here is his record with the A's between 1929 and 1933:

Table V, 8. *Lefty Grove, Home/Away, 1929-1933*

	HOME	AWAY	TOTAL
1929	9-2	11-4	20-6
1930	17-2	11-3	28-5
1931	17-1	14-3	31-4
1932	16-4	9-6	25-10
1933	16-2	8-6	24-8

Obviously, the man's road record of 53-22 over the five years wasn't too shabby, but it pales before the otherworldly home mark of 75-11, a winning percentage of .872. How to explain it? Shibe Park did not favor pitchers except in 1933; if anything, it is conventionally regarded as having been a very friendly place for righthanded hitters like Jimmie Foxx or Al Simmons. Even in Grove's later years, when he lost his fastball and was traded to the Red Sox—to pitch in Fenway, another park congenial to righthanded batters—he maintained his mastery at home, going 18-0 there over a three-year period.

In recent years, the best home pitching record has been that of Ron Guidry in 1978, when he was 13-1; however, he was scarcely less effective on the road at 12-2. In between Grove and Guidry, a supposedly washed-up Billy Pierce pitched the Giants to the 1962 flag by starting 12 games at home and winning all of them, the most wins without a loss either at home or on the road. And Mel Parnell, a lefthander, was 16-3 at Fenway in 1949.

The best and worst records by teams in this century, home and away, are perhaps best presented in tabular form (see Table V, 9). With so many factors going into a team's winning percentage, its runs scored and allowed, its home runs, etc., it may be useful to think of the *road* record as the best index of ability pure and portable. The New York Giants may have hit 131 homers at home in 1947, but we ought to be more impressed by the 124 hit away from home by the Milwaukee Braves ten years later—or by the Milwaukee Brewers in 1982, with 127 on the road. The Phillies of 1930 may have scored 7.05 runs a game at home, but we rub our eyes in disbelief at the road record of the 1939 Yankees: 7.8 runs scored per game, only 3.9 allowed (and that was the year in which Gehrig was replaced in May by a .235-hitting Babe Dahlgren).

The Houston Astros have always done well at home, even before they were a .500 team overall and before there was an Astrodome. Visiting teams hated to play in the heat before 1965 and have hated to play in the air conditioning since. In 1969 the eventual champion Mets lost all their games in Texas; that year produced an Astro record of 52-29 at home, best in the league, and a symmetrical 29-52 on the road.

Occasionally an organization runs out of kilter and gathers, somehow, an overabundance of players ill-equipped to take advantage of its home park. Imagine Richie Zisk and Pat Putnam in Houston, or Terry Puhl and José Cruz in Boston; or reflect back to when San Diego paid big money to acquire first Gene Tenace, then Oscar Gamble. The customary way out of such a fix is to swing a deal, or to go out in the free-agent mart and buy what you need to redress the team's balance. Occasionally a franchise has opted to keep the personnel and change the park, as the Cleveland Indians did in 1970 when they moved the fences in to such an extent that the number of homers hit in Municipal

Table V, 9. *Team Home/Away Records, Top Five Since 1900**

Best Winning Percentage (Home)

1932	Yankees	.805
1961	Yankees	.802
1931	Athletics	.800
1949	Red Sox	.792
1946	Red Sox	.792

Best Winning Percentage (Away)

1939	Yankees	.730
1933	Senators	.697
1928	Cardinals	.688
1971	Athletics	.688
1923	Yankees	.684

Most Runs Scored Per Game (Home)

1950	Red Sox	8.12
1932	Athletics	7.43
1931	Yankees	7.08
1930	Athletics	7.05
1930	Cardinals	7.03

Most Runs Scored Per Game (Away)

1939	Yankees	7.80
1930	Yankees	7.57
1936	Yankees	7.35
1931	Yankees	6.69
1932	Yankees	6.58

Fewest Runs Allowed Per Game (Home)*

1981	Astros	2.08
1964	White Sox	2.63
1968	Dodgers	2.65
1966	White Sox	2.68
1958	Braves	2.69

Fewest Runs Allowed Per Game (Away)*

1968	Yankees	2.83
1972	Orioles	2.84
1968	Tigers	2.87
1972	Tigers	2.90
1954	White Sox	2.95

Worst Winning Percentage (Home)

1939	Browns	.234
1911	Braves	.260
1923	Phillies	.267
1915	Athletics	.267
1962	Mets	.275

Worst Winning Percentage (Away)

1935	Braves	.167
1916	Athletics	.169
1945	Athletics	.171
1909	Senators	.195
1904	Senators	.197

Fewest Runs Scored Per Game (Home)*

1942	Phillies	2.46
1968	Dodgers	2.62
1972	Padres	2.71
1968	Mets	2.73
1972	Angels	2.76

Fewest Runs Scored Per Game (Away)*

1972	Indians	2.65
1981	Cubs	2.73
1942	Phillies	2.75
1963	Mets	2.78
1963	Colt .45s	2.82

Most Runs Allowed Per Game (Home)

1930	Phillies	8.36
1923	Phillies	7.96
1929	Phillies	7.63
1939	Browns	7.19
1936	Browns	7.17

Most Runs Allowed Per Game (Away)

1930	Phillies	7.03
1932	White Sox	6.76
1936	Athletics	6.68
1901	Brewers	6.59
1950	Athletics	6.58

Most Homers Hit Per Game (Home)

1947	Giants	1.72
1956	Reds	1.66
1970	Indians	1.64
1954	Giants	1.58
1977	Red Sox	1.55

Most Homers Hit Per Game (Away)

1957	Braves	1.61
1982	Brewers	1.57
1961	Yankees	1.56
1980	Brewers	1.41
1963	Twins	1.41

Fewest Homers Hit Per Game (Home)*			*Fewest Homers Hit Per Game (Away)**		
1924	Senators	.013	1920	Athletics	.114
1945	Senators	.013	1920	Pirates	.130
1927	Reds	.038	1920	Dodgers	.143
1928	Reds	.038	1944	White Sox	.156
1921	Red Sox	.039	1928	Browns	.156
1924	Reds	.039			
1920	Red Sox	.039			

Best Batting Teams (Batter Run Rating)†			*Worst Batting Teams (Batter Run Rating)*		
1913	Athletics	1.43	1981	Blue Jays	.664
1947	Yankees	1.40	1942	Phillies	.672
1933	Yankees	1.40	1910	Braves	.682
1931	Yankees	1.38	1903	Senators	.719
1930	Yankees	1.37	1932	Red Sox	.720

Best Pitching Teams (Pitcher Run Rating)**			*Worst Pitching Teams (Pitcher Run Rating)*		
1906	Cubs	.644	1915	Athletics	1.42
1909	Cubs	.686	1911	Braves	1.38
1907	Cubs	.694	1953	Tigers	1.36
1926	Athletics	.700	1904	Senators	1.35
1905	Cubs	.722	1968	Senators	1.33

---

* In some cases—particularly, fewest runs scored and allowed per game and fewest homers hit—the tables present post-1920 data to avoid total dominance in these categories by dead-ball era teams.

† The Batter Run Rating is the team's runs scored per inning, normalized to league average and adjusted for home park.

** The Pitcher Run Rating is the team's runs allowed per inning, normalized to league and adjusted for park.

Stadium jumped from 116 to 236 (the Indians themselves jumped only one place, from last to next-to-last). An equally bizarre leap of the imagination was George Steinbrenner's decision in the winter of 1981 that speed was the wave of the future; overnight, he transformed the Yankees' traditional posture from power and pitching, which had been good enough since their park was built in 1923, to a team of jackrabbits and slap hitters. In came Dave Collins and Ken Griffey, out went Reggie Jackson and Bob Watson, among others. These moves made a fifth-place finisher of a pennant-winning club, and the Yankees' speed era came to a speedy conclusion, as the next winter's shopping expedition brought Don Baylor and Steve Kemp.

A home park with extreme characteristics—heavily favoring pitchers or batters, lefty or righty—can be a problem. In Fenway visiting teams almost never start a lefthander because the Red Sox have historically stacked their lineup with righthanded hitters who can pull 350-foot fly balls over The Wall. (Hall of Famer Whitey Ford had an ERA of 6.16 in Fenway Park—things got so bad that the Yanks eventually decided to skip his turn in the rotation if they happened to be in Boston.)

This dearth of lefthanded opposition negates much of the presumed advantage that the Sox front office has labored to construct. Likewise, the Sox have rarely had a first-rate lefthanded pitcher on their own staff for the same reason: dread of The Wall. However, this has left Boston hurlers very vulnerable to lefthanded-hitting lineups in their road games. In this regard, it will be interesting to see Boston's road record in 1984, considering the club's 1983 acquisition of yet another righthanded slugger in Tony Armas when they knew Yaz was on the way out, taking with him the only lefthanded power they had. Entering 1984 without acquiring at least one more lefthanded hitter will be perilous indeed. Overreliance upon a strength can become, in the end, a weakness.

Home park characteristics certainly are on the minds of management as they contemplate trades. They may even have been on the minds of Messrs. Ruppert and Huston back in 1920 when they brought Babe Ruth to New York by giving to cash-strapped Red Sox owner Harry Frazee $100,000 and a $350,000 mortgage on Fenway Park. Ruth had been a sensation in 1919, hitting 29 homers to set a new baseball record. What has not been examined until now, but may have been known to the Yankees, was that of Ruth's 29 homers, only 9 were hit in his 63 games in Fenway Park, while 4 were hit in the Polo Grounds in the 11 games he played against the Yankees. A simple projection from these figures would indicate a plausible home-run mark for Ruth in 1920, playing 77 games in the Polo Grounds, of 28 homers in N.Y. plus 20 more on the road, with Fenway replacing the Polo Grounds as a road park. In fact, Ruth hit 54, which has been universally attributed to the introduction of the lively ball that year. Had he played with the same ball used in 1919, however, he figured to hit about that many anyway.

The classic "What if?" proposition regarding home parks also involved New York and Boston, the fancied trade of Ted Williams, a lefthanded pull hitter in a park that was thought to benefit only righthanded hitters (in fact it benefits all hitters) for Joe DiMaggio, a righthanded power hitter playing in a stadium that was cavernous in left and cozy in right. The thinking behind the proposed deal was that Williams, playing in Yankee Stadium, would have a shot at Ruth's home run marks and would hit for an even higher average, while

DiMaggio would benefit in like fashion from Fenway. This hypothetical exchange of titans was very nearly consummated in 1949, long after the point at which it might have been a trade of equals, for DiMaggio's career would end in 1951 while Williams's would continue through 1960.

What if the deal had been completed a decade earlier? Park Factor, useful tool though it may be, should not be employed as if it were a magical button on your Betamax; it does not permit a "true" replay of seasons long past with the characters transported to different locales or new characters introduced. Park Factor offers a *suggestive* truth, one that is essentially and logically plausible, but not "verifiable" statistically (statistics never *prove*, anyway—they are estimations of truth). So, we really shouldn't be doing this, but what the hell. Here are the lifetime batting average and home run totals of Joe DiMaggio and Ted Williams as they might have looked if each had played his entire career in the other's uniform.

Table V, 10. *DiMaggio with Boston, Williams with New York*

Joe DiMaggio NY   BA: .325; HR: 361
Joe DiMaggio, BOS BA: .340; HR: 417

Ted Williams, BOS BA: .344 HR: 521
Ted Williams, NY   BA: .328* HR: 497*

---

*Williams's figures in Yankee pinstripes would be higher still—a .340 BA and 513 homers—if we adjust for the fact that he would have been batting against Red Sox pitching rather than against Yankee pitching. Yankee pitchers were 7 percent better than the league average during 1939-60, Williams's span, while the Red Sox hurlers performed at the league average for DiMaggio's span, the period 1936-51.

The basis of these calculations is not simply Park Factor, but the precise batting data for each player at Boston, at New York, and elsewhere. Williams hit much better at Fenway over his career than he did in Yankee Stadium (BA, .361 to .309; SLG, .652 to .543) but his homers at Boston came at about the same rate as they did in New York. DiMaggio hit no better at Fenway than he did in the average of all the other road parks, but he *did* hit much better there than at Yankee Stadium (BA: .334 to .315; SLG, .605 to .546).

Bottom line? The trade was better off not being made as far as the Red Sox were concerned. DiMaggio would have built up more impressive career totals had he come up with the Red Sox in 1936, but by 1949 it was too late.

1 PARK ADJUSTMENT: (Step 1) Add up runs allowed at home for all teams in the league (ROH—Runs, Opposition, Home) and runs allowed away for all teams in the league (ROA—Runs, Opposition, Away). Form a league average home/road ratio of HRL = ROH/ROA, where HRL is the home/road ratio of the league. (Step 2) Find games, losses, and runs allowed for each team at home and on the road. Take runs per game allowed at home over runs per game allowed on road, all over HRL. (Step 3) Make corrections for innings pitched at home and on the road. This is a bit complicated. First find the league average road winning percentage (wins on road over games on road). For each team compare its road winning percentage to the league average. If it is higher, this means the innings pitched on the road are higher because the other team is batting more often in the last of the ninth. This rating is divided by the Innings Pitched Corrector (IPC):

$$IPC = 1 + (\text{Road Win Percentage, League} - \text{Road Win Percentage, Team}) \times .113$$

(Step 4) Make corrections for the fact that the other road parks' total difference from the league average is offset by the park rating of the club which is being rated. Multiply rating by this Other Parks Corrector (OPC):

$$OPC = \frac{\text{No. of teams} - \text{Run Factor, team}}{\text{No. of teams} - 1}$$

Example: In 1982, the runs allowed at home by all teams in the National League was 3993; on the road it was 3954. Thus the home/road ratio (HRL) is 1.010. Atlanta allowed 387 runs at home in 81 games, 315 runs allowed on the road in 81 games. The initial factor is (387/81) / (315/81) / 1.010 = 1.216. The league road winning percentage was .487 (473 wins in 972 games). The Braves' road record was 47-34, or .580. Thus the IPC = 1 + (.487 − .580) × .113 = .989. The team rating is now 1.216/.989 = 1.230. The OPC = (12 − 1.230) / (12 − 1) = .979. The final rating is 1.230 × .979, or 1.204.

We warned you it wouldn't be easy.

The batter adjustment factor is composed of two parts, one the park factor and the other the fact that a batter does not have to face his own pitchers. The initial correction takes care of only the second factor. For the first start with the following:

SF = Scoring Factor, previously determined (for Atlanta, 1.204)

SF1 = Scoring Factor of the other clubs (NT = number of teams);

$$1 - \frac{SF - 1}{NT - 1}$$

Next is an iterative process in which the initial team pitching rating is assumed to be one, and the following factors are employed:

RHT, RAT = Runs per game scored at home (H), away (A) by team
OHT, OAT = Runs per game allowed at home, away, by team
RAL = Runs per game by both teams

Now, with the Team Pitching Rating (TPR) = 1, we proceed to calculate Team Bat Rating (TBR):

$$TBR = \left(\frac{RAT}{SF1} + \frac{RHT}{SF}\right)\left(1 + \frac{TPR - 1}{NT - 1}\right)\Big/RAL$$

$$TPR = \left(\frac{OAT}{SF1} + \frac{OHT}{SF}\right)\left(1 + \frac{TBR - 1}{NT - 1}\right)\Big/RAL$$

The last two steps are repeated three more times. The final batting corrector (BF) is:

$$BF = \frac{(SF + SF1)}{\left(2 \times \left[1 + \frac{TBR - 1}{NT - 1}\right]\right)}$$

Similarly, the final pitching corrector (PF) is:

$$PF = \frac{(SF + SF1)}{\left(2 \times \left[1 + \frac{TBR - 1}{NT - 1}\right]\right)}$$

Now an example, using the 1982 Atlanta Braves once more.

$$RHT = \frac{388}{81} = 4.79 \qquad\qquad RAT = \frac{351}{81} = 4.33$$

$$OHT = \frac{387}{81} = 4.78 \qquad\qquad OAT = \frac{315}{81} = 3.89$$

$$RAL = \frac{7947}{972} = 8.18 \qquad\qquad NT = 12$$

$$SF = 1.204 \qquad\qquad SF1 = 1 - \left(\frac{1.204 - 1}{11}\right) = .981$$

$$TBR = \left(\frac{4.33}{.981} + \frac{4.79}{1.20}\right)\left(1 + \frac{1 - 1}{11}\right)\Big/8.18 = 1.027$$

$$\text{TPR} = \left(\frac{3.89}{.981} + \frac{4.78}{1.20}\right)\left(1 + \frac{1.027 - 1}{11}\right)\Big/8.18 = .974$$

Repeating these steps gives a TBR of 1.02 and a TPR of .97. The batting corrector is:

$$\text{BF} = \frac{(1.204 + .981)}{\left(2 \times \left[1 + \frac{.97 - 1}{11}\right]\right)} = 1.09$$

This is not a great deal removed from taking the original ratio,

$$\frac{1.216 + 1}{2}, \text{ which is } 1.11.$$

The pitching corrector may be calculated in analogous fashion.
[2] To apply the Batter Park Factor to Linear Weights, one must use this formula:

$$\text{LWT}_{\text{corrected}} = \text{LWT}_{\text{uncorrected}} - \frac{\text{Runs (league)}}{(\text{AB} + \text{BB} + \text{HBP})_{\text{league}}} \times (\text{BF} - 1) \times (\text{AB} + \text{BB} + \text{HBP})_{\text{player or team}}$$

For example, if a player produces 20 runs above average in 700 plate appearances with a Batter Park Factor of 1.10, and the league average of runs produced per plate appearance was .11, this means that his uncorrected LWT was 20 over the zero point of 700 × .11 (77 runs). In other words, 77 runs is the average run contribution expected of this batter had he played in an average home park. But because his Batter Park Factor was 1.10, which means his home park was 10 percent kinder to hitters, you would really expect an average run production of 1.1 × 77, or 85 runs. Thus the player whose uncorrected LWT was 97 with a BF of 1.1 was only + 12 runs rather than + 20, and 12 is his Park Adjusted Linear Weights Runs.

$$12 = 20 - .11 \times (1.10 - 1) \times 700$$

[3] Some other great road batting marks: Harry Heilmann had a BA of .456 in his away games of 1925; Gehrig slugged .805 away in 1927; and Ted Williams had an OBA of .528 on the road in 1957, with a NOPS of 2.68. The last men to post batting averages over .400 on the road were Lou Boudreau and Stan Musial, both in 1948.

6

THE THEORY OF RELATIVITY
AND OTHER ABSOLUTE TRUTHS

"How'm I doin'?", Mae West used to ask, a question recently revived by New York City Mayor Ed Koch. Although they are probably referring to different kinds of performance, a response to either of them will be formed by preliminary answers to these questions: How well did you once do it? How well are others doing it? How well have others done it in the past? The answers supply a context for evaluating whether an achievement is inferior, superior, or acceptable (average). In baseball it is the same: If Batter A presented himself to you for approval with these statistics—.330 BA, 16 HR, 107 RBI—what would your reaction be? You'd like to have him on your team, right? And what to make of Batter B, who presents these numbers—.257 BA, 14 HR, 53 RBI? Not bad for a middle infielder with a good glove, you say, but otherwise undistinguished? In fact, the "impressive" figures of Batter A represent the *average* performance of a National League outfielder in 1930, while the "blah" figures of Batter B are those of the average American League outfielder of 1968: The former has more than twice the RBIs of the latter, along with a Batting Average 73 points higher, yet the two performed at identical levels, and an argument could be made that Batter B was superior.

In a similar comparison involving those two years of extremes, Bill Terry led the National League in 1930 with a BA of .401, a mark surpassed by Ted Williams in 1941 but not equaled since; Carl Yastrzemski led the American League of 1968 with a performance which oldtimers held to be a disgrace, a lowly BA of .301, the worst ever to win a batting championship. Terry's mark was achieved at a time when most pitchers had only two pitches, a fastball and a curve, and not enough confidence in the latter to throw it when behind in the count at 2-0 or 3-1. The parks were smaller; there was no night ball; the game was segregated racially; and you played 22 games with each team, none farther west of the Mississippi than St. Louis. Moreover, 1930 was the year in which National League officials, attempting to match the popularity of the slugging American League, juiced the ball to such an extent that the entire *league* batted .312.[1] In other words, the average nonpitcher in the NL of 1930 batted higher than the AL leader in 1968! When Yaz hit .301, pitchers dominated the game and the average American League player hit .238. How to compare Terry and Yaz, who played under such different conditions 38 years apart?

You could view Terry's .401 in relation to his league's BA of .312, concluding that Memphis Bill was a better hitter (by BA alone, which despite its previously cited deficiencies remains the most comfortable stat by which to introduce this technique) by 28.5 percent. You could compare Yaz's .301 to *his* league's BA of .238 and conclude that he was a better than average hitter by 26.5 percent. A mere 2 percentage points separate the men—had they both played in the National League of 1983, when the league average was .255, the Terry of 1930 might have hit .328, the Yaz of 1968, .323. A further refinement of this method would be to delete Terry's at bats and hits from his league's, and those of Yastrzemski from his league's, so that the batters are not in effect compared with themselves. This, however, necessitates the use of at bats and hits rather than simply the averages and does not significantly alter the results.

The method illustrated above—normalizing to league average, or measuring in relative fashion—was touched upon in earlier chapters out of necessity, since the Linear Weights formulas (all but base-stealing) contained built-in normalizing features; then, to illustrate the batting LWTS "shadow stat" of On Base Plus Slugging (OPS) and stack it up fairly against its big brother, we had to normalize *it* as well (to become NOPS).

Why do we need relative measures? Basically, for the same reason we need statistics altogether, to compare, to interpret, and to compre-hend, but in a more reasonable and accurate manner when the dis-parity of the data sources makes the use of absolute, unadjusted numbers illogical. If the analysis involves data produced under widely varying conditions, such as a sample including baseball performances 20, 50, or 100 years apart, any comparison will be meaningless without dragging in a series of rather complex historical understandings to modify the analysis—and in a highly subjective, unreliable manner. To compare Terry's .401 with Yastrzemski's .301 with no recognition of the *context* in which these marks were achieved, that is, to infer that Terry was 100 points *better* than Yaz, is equivalent to comparing Babe Ruth's salary of $80,000 in 1930 with Pete Rose's $806,250 of fifty years later and concluding that Rose was $726,250 richer. To understand those dollar figures we must place them within a context which includes such factors as I.R.S. regulations and inflation: We might think to re-express the two salaries in terms of their purchasing power, multiplying each by the Consumer Price Index of its time as expressed in 1967 dollars; doing this would be to compute a "relative salary" for Ruth and Rose, just as we computed a Relative Batting Average for Terry and Yaz. (And just as we discovered there was little difference between the BAs of the latter couple, we would discover there is little difference between the salaries of the former pair.)

Few are the fans who could cite the context of Ross Barnes's .429 batting average of 1876,[2] let alone evaluate its ingredients (these include considerations of equipment, schedule, travel, physiology, racial exclusion, daytime games, rules variations, attitudes, and customs). A statistic removed from its historical context can be as deceptive as a quotation pulled out of context. How, then, to compare Barnes's .429 with, say, Bill Madlock's league-leading figure of .339 a century later? Should we discount Barnes's average 10 percent because in his day batters could demand a pitch above the waist or below? Or should we augment it 17 percent because a pitcher could throw eight "balls" before allowing a walk?

We were confronted with a similar problem in the previous chapter when we wondered how to quantify the various differences between home parks; our solution then was to look at the single measure which reflected all the variables—runs—and from that measure we proceeded to devise a formula for Park Factor. Similarly, the many variables which supply the context for Barnes in 1876 supplied an identical context for every other batter *in that year*—and the context in which Bill Madlock hit .339 prevailed for every other National League batter in 1976 (except for home park, of course). Accordingly, if we form a ratio of Barnes's .429 to his league's average (.265)[3] and another of Madlock's to his league's average (.263) we obtain figures (1.62 for Barnes, 1.28 for Madlock) which may reasonably be compared with each other: Barnes was 62 percent better than his league in BA, while

Madlock was 28 percent better than *his;* these become the comparables, not the .429 and .339. The method will not become a time machine—putting Barnes on a modern club and Madlock on an old-time one—any more than the Park Factor was a place machine, switching DiMaggio to Beantown and Williams to the Bronx. However, the relativist approach offers suggestive truths and does measure precisely the extent to which Barnes's and Madlock's BAs dominated those of their contemporaries.

Until the 1970s, when David Shoebotham ("Relative Batting Averages," *Baseball Research Journal,* 1976) and Merritt Clifton ("Relative Baseball," *Samisdat,* 1979) introduced the relativist approach, all baseball stats were absolute. And for cross-era comparison, that favorite Hot Stove League activity, absolute stats were absolutely useless, generating plenty of heat and precious little light. What the theory of relativity, baseball-style, does beautifully is to eliminate the need for bringing historical baggage to statistical analysis. The normalized or relative versions of *any* statistic—BA, OPS, ERA, SLG, you name it; even homers or strikeouts, though there are problems with these—will be greater than 1.00 for all above-average performers (1.41, for example, means 41 percent better than average in the given category) while relative statistics less than 1.00 will indicate a below average level of play (0.88 means 12 percent below the norm).

It is as simple as can be. So Early Wynn had a 3.20 ERA in 1950? What does that *mean?* Well, the league ERA was 4.58, so Wynn did very well indeed. His normalized ERA thus was 1.43, a mark better than that earned by Tom Seaver in 1968, when he had an absolute ERA a full run lower at 2.20.

Has your appetite been whetted? Here are tables of the century's top 20 single season and lifetime relative performances in batting average, On Base Average, and slugging percentage, with a few observations interspersed. In the tables at the rear, you will also find Relative (Normalized) OPS and Isolated Power. Nineteenth century performers are listed separately, in Relative OBA, SLG, OPS, and ERA. (Indeed, even Linear Weights, with its built-in normalizing factor, can be made more "accurate," i.e., meaningful, by dividing the runs contributed through batting, fielding, base stealing, or pitching by the number of runs required for an additional win in the given year. For example, two batters in different years might have contributed the same number of runs but because one played in a year which featured more run scoring, more runs would be required to produce an extra win and his wins total would be less. Case in point: Ty Cobb in 1911 and Lou Gehrig in 1932 each contributed 75.2 batting runs beyond average, adjusted for park. Yet Cobb, who attained that total in a year in which 10.18 runs were required for an additional win, contributed 7.39 wins, while Gehrig, in a year in which 10.81 runs were required for an additional win, accounted for "only" 6.96 wins. Note that all the Relative Batting Averages in Tables VI, 1 and VI, 2 are adjusted for park; league BAs are provided for those who may wish to calculate their own RBAs without Park Factor. Also, decimal points have been shifted right for convenience of expression, so that a par figure is 100 rather than 1.00 and a normalized figure of 1.382 is expressed as 138.2.

Table VI, 1. *Relative Batting Average, Best Seasons Since 1901*

	TEAM	AVG.	LG.	REL. ADJ.
1. 1912 Ty Cobb	DET	.410	.272	155.2
2. 1910 Nap Lajoie	CLE	.384[4]	.250	151.9
3. 1916 Tris Speaker	BOS(A)	.386	.257	150.5
4. 1901 Nap Lajoie	PHI (A)	.426[4]	.284	149.5
5. 1924 Rogers Hornsby	STL (N)	.424	.290	148.9
6. 1910 Ty Cobb	DET	.383[4]	.250	148.6
7. 1913 Ty Cobb	DET	.390	.265	148.4
8. 1904 Nap Lajoie	CLE	.376	.251	147.9
9. 1918 Ty Cobb	DET	.382	.260	147.9
10. 1977 Rod Carew	MIN	.388	.266	147.2
11. 1919 Ty Cobb	DET	.384	.276	146.7
12. 1941 Ted Williams	BOS(A)	.406	.276	146.6
13. 1911 Ty Cobb	DET	.420	.281	146.5
14. 1917 Ty Cobb	DET	.383	.255	146.4
15. 1905 Cy Seymour	CIN	.377	.265	146.1
16. 1912 Joe Jackson	CLE	.395	.272	145.9
17. 1980 George Brett	KC	.390	.269	145.8
18. 1911 Joe Jackson	CLE	.408	.281	145.4
19. 1974 Rod Carew	MIN	.364	.258	144.2
20. 1909 Ty Cobb	DET	.377	.253	143.1

Table VI, 2. *Relative Batting Average, Lifetime Since 1901*

	AVG.	LG.	REL. ADJ.
1. Ty Cobb 1905-28	.366[5]	.272	133.6
2. Joe Jackson 1908-20	.356	.268	132.2
3. Nap Lajoie 1896-1916	.339	.265	128.1
4. Rod Carew 1967-82	.331	.258	127.9
5. Rogers Hornsby 1915-37	.358	.284	127.1
6. Ted Williams 1939-60	.344	.269	125.9
7. Willie Keeler 1892-1910	.343	.272	124.3

8. Tris Speaker 1907-28	.344	.275	123.8
9. Honus Wagner 1897-1917	.329	.265	122.9
10. Eddie Collins 1906-30	.333	.274	122.1
11. Stan Musial 1941-63	.331	.267	122.1
12. Roberto Clemente 1955-72	.317	.263	121.0
13. George Brett 1973-83	.316	.262	120.9
14. Babe Ruth 1914-35	.342	.288	120.7
15. Lou Gehrig 1923-39	.340	.290	120.1
16. Elmer Flick 1898-1910	.315	.265	120.0
17. Harry Heilmann 1914-32	.342	.286	119.3
18. Joe DiMaggio 1936-51	.325	.278	118.5
19. George Sisler 1915-30	.340	.287	118.5
20. Dale Mitchell 1946-56	.312	.269	118.4

These Relative Batting Average lists seem to confirm the oldtimers' notion that hitting, or at least hitting for average, ain't what it used to be: The dead-ball era heroes dominate the upper echelons here nearly to the same degree as they do with absolute BAs. Only Rod Carew and George Brett of modern (post-1945) players make the top twenty seasons chart, while Roberto Clemente and Dale Mitchell (!) join them among the lifetime leaders. Why is this so? For one, the best hitting talents in baseball in recent years have been applied to run production through power,[6] which necessitates a fuller cut, which in turn necessitates an earlier commitment to swing at a pitch and thus a greater chance of miscalculation. If a hitter with the ability of a Dave Winfield, or a Willie McCovey, or a Mike Schmidt would content himself with slapping and slashing at the ball in the old-style way, aiming simply to reach base, he might well have a higher BA. Besides, attempts to hit for power in the early days produced results like those of Home Run Baker or Tim Jordan: 10-15 homers at best and a lesser measure of prestige than that accorded to Cobb or Lajoie or Wagner. A Joe Jackson took a full swing, but the ball didn't go that far that often. Playing today, he might well compile numbers more like Schmidt's than like those of Carew.

A second and no doubt more significant reason for the lowered Relative Batting Averages of the postwar era is that the *overall* level of play today—especially among the third, fourth, and fifth starting pitchers and the relievers—is superior to that of 1910, and thus the star stands out to a lesser degree. The gap between the average and the peak performance has been steadily narrowing throughout baseball history, with only occasional blips such as the World War II years or those immediately following a league expansion.

Merritt Clifton attempted to bridge this generation gap—the ever narrowing divergence of peak and average levels of play—by normalizing the absolute stat not only to the league average but also to the league-leading figure. Take Joe Jackson's lifetime BA of .356 and normalize it to his league average in that period, .268; you get a Relative Batting Average of 1.329.[7] Next, take the average league-leading BA during Jackson's career, .383, and divide that by the league average of .268, arriving at 1.429. Then you divide Jackson's average normalized to the league by the leaders' average normalized to league. Assuming that the league-leading performance was perfection, inasmuch as it was the best anyone could do, you now have Jackson's "percentage of perfection": 93. (Had he led the league in BA every year he played, he would have had a mark of 100 percent.)

Clifton's ingenious technique—which, incidentally, places Ted Williams and Rod Carew (1, 2) rather than Cobb, Jackson, or Lajoie at the top of the heap—accords equal statistical value to the level of play of one individual, the leader, as to that of the hundreds of players in a league. The result can be, in the case of an aberrationally high performance like that of Cobb and Lajoie in 1910, when they each batted more than 53 percent better than the league, to depress, perhaps unreasonably, the ratings of players who may have batted at the same level in relation to the league as they did in years past.

A still more sophisticated approach to Relative Batting Average was proposed in 1982 by Ward Larkin in *The Baseball Analyst*. This method employs the standard deviation of batting average from the norm rather than a simple ratio of individual to league norm or to league leader.

Relative Slugging or Relative On Base Average is each a better measure of hitting ability in absolute or normalized form than batting average, and when added together they form an index nearly as good as LWTS. Also, these two categories are not dominated by the men who played in 1900-20 but rather show a nice chronological mix through the century.[8] In the lifetime tables opposite you will not be surprised by the names which occupy the first five to ten spots, but we daresay you will be surprised by the players in spots eleven-twenty and more so by the full list at the back of the book. Here are Park Adjusted Relative OBA and SLG, top twenty seasonal and lifetime.

Table VI, 3. *Relative OBA, Best Lifetime Since 1901*

		REL. ADJ.
1.	Ted Williams 1939-60	136.5
2.	Babe Ruth 1914-35	135.1
3.	Mickey Mantle 1951-68	135.1
4.	Rogers Hornsby 1915-37	127.8
5.	Lou Gehrig 1923-39	127.0
6.	Roy Thomas 1899-1911	127.0
7.	Ty Cobb 1905-28	126.8
8.	Eddie Collins 1905-30	124.6
9.	Joe Jackson 1908-20	124.5
10.	Tris Speaker 1907-28	123.0
11.	Topsy Hartsel 1898-1911	122.3
12.	Elmer Flick 1898-1910	122.0
13.	Stan Musial 1941-63	121.8
14.	Joe Morgan 1963-83	121.5
15.	Gene Tenace 1969-83	121.4
16.	Mike Hargrove 1974-83	121.1
17.	Mel Ott 1926-47	121.1
18.	Rod Carew 1967-83	120.8
19.	Ken Singleton 1970-83	120.5
20.	Eddie Stanky 1943-53	120.1

Table VI, 4. *Relative OBA, Best Seasons Since 1901*

		REL. ADJ.
1.	1941 Ted Williams, BOS (A)	155.8
2.	1957 Mickey Mantle, NY (A)	154.6
3.	1954 Ted Williams, BOS (A)	150.4
4.	1957 Ted Williams, BOS (A)	150.3
5.	1924 Rogers Hornsby, STL (N)	149.7
6.	1962 Mickey Mantle, NY (A)	148.5
7.	1920 Babe Ruth, NY (A)	147.6
8.	1942 Ted Williams, BOS (A)	145.8
9.	1926 Babe Ruth, NY (A)	145.5
10.	1923 Babe Ruth, NY (A)	144.6
11.	1947 Ted Williams, BOS (A)	144.5
12.	1948 Ted Williams, BOS (A)	143.0
13.	1961 Norm Cash, DET	142.6
14.	1930 Babe Ruth, NY (A)	142.6
15.	1932 Babe Ruth, NY (A)	142.5
16.	1916 Tris Speaker, CLE	142.0
17.	1931 Babe Ruth, NY (A)	142.0
18.	1977 Ken Singleton, BAL	141.8
19.	1946 Ted Williams, BOS (A)	141.7
20.	1915 Eddie Collins, CHI (A)	141.4

We can hear you saying: Since Rod Carew is in the lineup to get on base rather than to drive men in, are we to conclude that a team would be better off with Mike Hargrove or Gene Tenace? Joe Morgan was *that* good? Who are Roy Thomas and Topsy Hartsel? Of the top twelve seasons, Ted Williams had *six?* (Yes, and on a straight, unadjusted basis, Williams's lifetime OBA was .483—meaning he reached base nearly half the time—and his seasonal best of .551 in 1941 was baseball's best ever; unbelievably, he led the league in OBA every year he qualified except for his rookie season, and in three of his four injury-shortened seasons his OBA was higher than the league leader's!) And only one National Leaguer is among the top 20 seasons— one fewer than in the BA list, two fewer than in SLG below.

Table VI, 5. *Relative Slugging, Best Lifetime Since 1901*

	REL. ADJ.
1. Babe Ruth 1914-35	173.4
2. Ted Williams 1939-60	157.0
3. Lou Gehrig 1923-39	154.6
4. Rogers Hornsby 1915-37	148.1
5. Joe Jackson 1908-20	145.7
6. Jimmie Foxx 1925-45	145.3
7. Joe DiMaggio 1936-51	144.1
8. Mickey Mantle 1951-68	144.0
9. Hank Greenberg 1930-47	143.9
10. Johnny Mize 1936-53	142.3
11. Henry Aaron 1954-76	140.5
12. Willie Mays 1951-73	140.0
13. Ty Cobb 1905-28	139.9
14. Dick Allen 1963-77	139.6
15. Mike Schmidt 1972-83	137.8
16. Stan Musial 1941-63	137.5
17. Willie Stargell 1962-82	137.4
18. Frank Robinson 1956-76	136.0
19. Nap Lajoie 1896-1916	135.1
20. Gavvy Cravath 1908-20	134.7

Table VI, 6. *Relative Slugging, Best Seasons Since 1901*

	REL. ADJ.
1. 1920 Babe Ruth, NY (A)	209.7
2. 1921 Babe Ruth, NY (A)	198.5
3. 1927 Babe Ruth, NY (A)	188.5
4. 1927 Lou Gehrig, NY (A)	186.8
5. 1926 Babe Ruth, NY (A)	185.0
6. 1933 Jimmie Foxx, PHI (A)	183.1
7. 1923 Babe Ruth, NY (A)	182.2

8. 1924	Babe Ruth, NY (A)	181.1
9. 1919	Babe Ruth, BOS (A)	180.7
10. 1928	Babe Ruth, NY (A)	180.4
11. 1918	Babe Ruth, BOS (A)	180.3
12. 1941	Ted Williams, BOS (A)	180.1
13. 1925	Rogers Hornsby STL (N)	178.5
14. 1934	Lou Gehrig, NY (A)	178.2
15. 1957	Ted Williams, BOS (A)	177.3
16. 1948	Stan Musial, STL (N)	176.0
17. 1924	Rogers Hornsby, STL (N)	175.6
18. 1961	Mickey Mantle, NY (A)	175.5
19. 1930	Babe Ruth, NY (A)	175.2
20. 1956	Mickey Mantle, NY (A)	174.5

Did anyone out there still need convincing that Ruth was one of a kind? Nine of the eleven best seasons; a *lifetime* mark that would have been the 23rd best *season* ever; a 1920 in which his SLG was more than *twice* the league average. The lifetime list is more interesting: Dick Allen, 14th best slugger of the century, received 14 votes in the Hall of Fame balloting for 1983; and Mike Schmidt is, except for Nap Lajoie, the lone representative in the top twenty of a skill position—shortstop, second base, catcher, third base (Allen broke in at third but played more games at first base).

To date, only one study has been published which applies the relativity method to pitching performance, James P. Maywar's "Who Are the Most Impressive Strikeout Pitchers?" in the 1981 *Baseball Research Journal.* Maywar's study was somewhat flawed by the use of a strikeout differential—individual's K's per nine innings *minus* the league average—rather than a ratio of the two. Taking the five best strikeout-differential seasons of ten pitchers generally acknowledged as the top strikeout pitchers (minus Waddell and Johnson, whose differentials, surprisingly, were not that impressive), Maywar calculated a cumulative strikeout differential. In Nolan Ryan's five best years, he fanned 10.3 batters per nine innings while the league average was 5.1: thus a mark of 5.2 for Ryan, crowning him "the best strikeout pitcher of all time." Yet had Maywar *divided* Ryan's record by the league's to obtain his Relative Strikeouts, he would have pegged Ryan's top five seasons at 2.02, fourth on the list behind Dazzy Vance's 2.50 and the 2.14 registered by Bob Feller and Lefty Grove.

Whichever way such a study is conducted, the intrinsic problem is that pitchers find it easier to notch strikeout victims today because of several factors unrelated to their ability. Because big-swinging hitters (and homers) were less common in 1910 than they were in 1970, pitchers back then did not strive for K's the way they later did; in 1910, the hurler's objective was to make the batter hit the ball to a fielder, not to vanquish a batting order all by himself. The employment of relative "counter" stats of any sort—be they strikeouts, homers, triples, steals, whatever—is fraught with danger. With homers, for example, the man who led the National League in 1876—George Hall, with 5—may emerge as superior to Babe Ruth in 1921 as a relative home run hitter, which flies in the face of reason, for Hall was 5'7" and weighed 142 pounds and probably legged out all five of his homers; his high Relative Homer mark is merely the result of an extremely low league average in a category with a theoretically unlimited ceiling, unlike BA, OBA, or SLG. It is not reasonable, either, to presume that Gavvy Cravath was a superior home run hitter to Ralph Kiner or Henry Aaron on a lifetime basis, yet this is precisely what a Relative Homers measure can imply. This problem will not arise to the same degree with ratios, for their ranges throughout baseball history—both within a season and over time—have been much narrower. Hall's home run total of 5 exceeded his league's norm by 826 percent, while Jim Rice's 39 in 1977 exceeded his league's norm by "only" 144 percent—an enormous spread. Ty Cobb's league-leading BA of .410 in 1912, on the other hand, exceeded the norm by only 51 percent (unadjusted for park) while Rod Carew's .388 in 1977 exceeded the norm by a comparable 46 percent.

On to pitching. We cannot employ a Relative Won-Lost record, for the league average is every year the same: .500. (A logical corollary is that one cannot fruitfully use relative measures of any sort for a single season's analysis, as all like figures will be compared to the same league average. The numbers may be changed into normalized form, but the players' rankings will be unchanged: The top ten in BA in 1983, for example, will retain their ranks in RBA.) We have already normalized the Pitching Linear Weights formula after a fashion by building in a factor for comparison to the league ERA; the resulting run total can be further normalized by dividing it by the number of runs required in a given year to gain an additional win (usually 9 to 11, as previously explained). A normalizing technique we have not discussed is one modeled on a procedure developed by Vic Meyer: normalizing innings pitched. Vic's idea was to eliminate the advantage enjoyed by oldtime pitchers in many categories by taking the IP of those who pitched prior to 1881 (when the pitching distance was increased from 45 to 50 feet) and multiplying by one-half, and taking the IP of those who pitched after 1881 but before 1893 (when the distance became today's 60'6") and multiplying by two-thirds; these factors corresponded roughly to the decline in the number of innings pitched by the clubs' top starters in that period (approximately 550 in 1881, 320 in 1893, 280 since 1940). The ambitious among you may wish to normalize the individual's IP to the IP of the average of the clubs' top starters. This would bring the LWTS of Hoss Radbourn in line with the LWTS of Ron Guidry rather than leave Radbourn's record on another planet, where it now resides along with the nineteenth century marks of Guy Hecker and Amos Rusie.

Here are the top twenty Relative or Normalized ERAs, lifetime and seasonal. As with Relative Batting Average, the figures are park adjusted but the league averages are supplied for those who may wish to calculate unadjusted Relative ERAs.

Table VI, 7. *Relative ERA, Best Lifetime Since 1901*

	ERA	LG.	REL. ADJ.
1. Lefty Grove 1925-41	3.06	4.42	146.0
2. Walter Johnson 1907-27	2.17	3.24	145.5
3. Hoyt Wilhelm 1952-72	2.52	3.76	145.4
4. Ed Walsh 1904-17	1.82	2.76	143.6
5. Mordecai Brown 1903-16	2.06	2.89	143.2
6. Addie Joss 1902-10	1.89	2.72	140.8
7. Cy Young 1890-1911	2.63	3.54	136.3
8. Pete Alexander 1911-30	2.56	3.40	136.1
9. Christy Mathewson 1900-16	2.13	2.92	136.1
10. Rube Waddell 1897-1910	2.16	2.88	133.2
11. Harry Brecheen 1940-53	2.92	3.83	133.2
12. Whitey Ford 1950-67	2.74	3.84	132.3
13. Hal Newhouser 1939-55	3.06	3.83	132.1
14. Sandy Koufax 1955-66	2.76	3.70	131.6
15. Tom Seaver 1967-83	2.73	3.59	131.4
16. Dizzy Dean 1930-47	3.04	3.88	131.2
17. Carl Hubbell 1928-43	2.98	3.96	130.5
18. Bob Gibson 1959-75	2.91	3.59	129.2
19. Ron Guidry 1975-83	2.99	4.00	129.2
20. Stan Coveleski 1912-28	2.89	3.64	127.8

Table VI, 8. *Relative ERA, Best Season Since 1901*

	ERA	LG.	REL. ADJ.
1. 1913 Walter Johnson, WAS	1.14	2.93	285.4
2. 1914 Dutch (Hub) Leonard, BOS (A)	0.96	2.74	283.5
3. 1906 Mordecai Brown, CHI (N)	1.04	2.63	271.2
4. 1968 Bob Gibson, STL	1.12	2.99	252.7
5. 1909 Christy Mathewson, NY (N)	1.15	2.60	240.8
6. 1912 Walter Johnson, WAS	1.39	3.35	238.9
7. 1915 Pete Alexander, PHI (N)	1.22	2.75	237.1
8. 1918 Walter Johnson, WAS	1.27	2.77	235.5

Table VI, 8. *Relative ERA, Best Season Since 1901* continued

	ERA	LG.	REL. ADJ.
9. 1907 Jack Pfiester, CHI (N)	1.15	2.46	229.8
10. 1905 Christy Mathewson, NY (N)	1.27	2.99	228.7
11. 1907 Carl Lundgren, CHI (N)	1.17	2.46	225.8
12. 1978 Ron Guidry, NY (A)	1.74	3.78	220.6
13. 1919 Walter Johnson, WAS	1.49	3.22	219.1
14. 1902 Jack Taylor, CHI (N)	1.33	2.78	216.4
15. 1931 Lefty Grove, PHI (A)	2.06	4.38	213.7
16. 1908 Addie Joss, CLE	1.16	2.39	210.4
17. 1958 Whitey Ford, NY	2.01	3.77	204.4
18. 1901 Cy Young, BOS (A)	1.63	3.66	202.9
19. 1905 Ed Reulbach, CHI (N)	1.42	2.99	202.4
20. 1971 Vida Blue, OAK	1.82	3.47	201.8

Harry Brecheen eleventh best of all time in Relative ERA? Lefty Grove beating out Walter Johnson? (This was due to Park Factor; on an unadjusted basis, the all-time leader is Ed Walsh, followed by Johnson, Hoyt Wilhelm, and Grove.) We suppose that Wilhelm's ranking will not come as a surprise to baseball aficionados, though it may come as a surprise to some of the electors of Cooperstown. And look at all the Cubbies on the seasonal list—Brown, Pfiester, Lundgren, Taylor, and Reulbach; where are they now that Chicago *really* needs them? And with all those dead-ball flingers taking up space, where are the men whose names we recognize? Here's a list of the top ten lifetime and seasonal Relative ERAs achieved mostly after 1961 by those who have pitched in 1500+ innings, excepting those already represented in the top twenty (numbers indicate ranking on full post-1900 table at the back):

Table VI, 9. *Relative ERA, Best Lifetime Since 1961*

21. Bert Blyleven	127.0
22. Jim Palmer	127.0
32. Steve Rogers	121.9
38. Andy Messersmith	121.6
40. Juan Marichal	121.0
43. Steve Carlton	120.9
46. Don Drysdale	120.6
47. Rollie Fingers	120.0
49. Phil Niekro	119.6
56. Dean Chance	118.7

Table VI, 10. *Relative ERA, Best Seasons Since 1961*

26. 1968 Luis Tiant, CLE	196.0
27. 1966 Sandy Koufax, LA (N)	196.0
30. 1964 Dean Chance, LA (A)	194.9

33.	1964	Sandy Koufax, LA (N)	191.6
35.	1971	Tom Seaver, NY (N)	189.5
39.	1971	Wilbur Wood, CHI (A)	188.8
40.	1962	Hank Aguirre, DET	188.3
49.	1965	Juan Marichal, SF	185.2
51.	1967	Phil Niekro, ATL	184.9
78.	1964	Whitey Ford, NY (A)	175.1

The only published study of Relative Fielding Performance is that by Bill Deane, "The Best Fielders of the Century," in *The National Pastime* of 1983. Deane elected to base his study on Fielding Average (FA) rather than Total Chances/Game, but he did not normalize to the league average: "To compare against the average performances of an era," Deane wrote, "tends to favor stars of lower levels of competition. In 1910, for example, Terry Turner's .973 FA was over 4 percent higher than the .935 average of American League regular shortstops; to exceed 1982's league average by 4 percent, an AL shortstop would have needed an impossible 1.007 FA." This dilemma does not manifest itself with other measures because only FA measures so small a relative quantity, namely, errors. A batter who hits .300 is a star, while a man who fields .900 would not crack a high school lineup. ERA is theoretically unlimited, and the league averages for OBA, SLG, etc., are not proximate to 1.000 in the way that FA's is.

What Deane did was to employ Clifton's concept of measuring against the league leaders rather than the league average, obtaining a percentage of perfection for the lifetime leaders in all the positions (grouping all outfielders and not treating pitchers). All seasons of 50 or more games at a position were included in the study. Here is Bill's table of average league-leading FAs by decade:

Table VI, 11. *Average League-Leading Fielding Average by Decade and Position, 1900-79*

Decade	1B	2B	SS	3B	OF	C
1900s	.990	.966	.946	.946	.982	.982
1910s	.993	.972	.958	.960	.985	.984
1920s	.995	.976	.964	.969	.987	.988
1930s	.995	.979	.966	.968	.991	.992
1940s	.995	.982	.973	.968	.994	.992
1950s	.994	.987	.976	.969	.994	.995
1960s	.996	.987	.977	.971	.996	.996
1970s	.996	.989	.985	.975	.996	.995

It is interesting to note that of the distance between the league-leading figures of the first decade and perfection (1.000), outfielders closed 78 percent of the gap, shortstops and catchers 72 percent, second basemen 68 percent, first basemen 60 percent, and third basemen only 54 percent. Perhaps thanks to the installation of so many artificial

TABLE VI, 12. *Relative Fielding Average, Best by Position Since 1900*

POS	PLAYER	YEARS	G	FA	RFA
1B	Dan McGann	1900-08	1094	.9894	.9993
	Wes Parker	1965-72	1077	.9959	.9992
	Steve Garvey	1973-82	1468	.9957	.9991
	Joe Adcock	1953-66	1460	.9941	.9991
	Vic Power	1955-65	1283	.9943	.9990
2B	Frankie Frisch	1921-36	1687	.9741	.9995
	Hughie Critz	1924-35	1453	.9738	.9976
	Eddie Collins	1909-27	2600	.9706	.9970
	Red Schoendienst	1946-60	1776	.9834	.9970
	Bobby Doerr	1938-51	1805	.9804	.9968
SS	Lou Boudreau	1939-51	1538	.9725	1.0009
	Everett Scott	1914-24	1552	.9655	.9987
	Eddie Miller	1939-50	1362	.9726	.9975
	Leo Durocher	1929-40	1462	.9613	.9964
	Larry Bowa	1970-82	1870	.9802	.9960
3B	Brooks Robinson	1958-76	2788	.9715	.9968
	Jim Davenport	1958-69	1004	.9650	.9967
	Heinie Groh	1915-24	1256	.9677	.9965
	George Kell	1944-57	1691	.9686	.9956
	Pinky Whitney	1928-38	1392	.9614	.9950
OF	Pete Rose	1967-74	1220	.9921	.9964
	Amos Strunk	1911-22	1282	.9813	.9959
	Jimmy Piersall	1953-66	1467	.9898	.9951
	Tommy Holmes	1942-50	1222	.9889	.9948
	Gene Woodling	1949-61	1427	.9892	.9944
C	Bill Dickey	1929-43	1663	.9881	.9970
	Bill Freehan	1963-76	1578	.9933	.9969
	Jim Sundberg	1974-82	1255	.9919	.9969
	Johnny Edwards	1961-73	1360	.9916	.9966
	Sherm Lollar	1949-62	1505	.9918	.9960

YEARS = Refer to years in which player appeared in 50 or more games, in one league, at particular position; subsequent statistics (G, FA, RFA) are for those corresponding years only.

G = Games played; minimum 100 games played in rated seasons.

Averages carried out as many decimal points as necessary to break ties. Most statistics based upon those in *The Baseball Encyclopedia;* remaining stats provided by Ev Cope and Pete Palmer, mostly from Spalding annual guides.

surfaces, FAs at short and third rose dramatically in the 1970s. At the left is Bill's table of the top five at each position, through 1982. (The anomalous value of 1.0009 for Lou Boudreau is accounted for, Deane writes, by his having "led AL shortstops in FA in each of the eight seasons in which he played the required 100 games. Additionally, Lou had five seasons in which he played between 50 and 99 games at short, in three of which his FA *exceeded* that of the league leader.")

It may be time to recapitulate the strengths of relative baseball stats, by now perhaps obscured by the various caveats we have been duty bound to issue. Relativism in baseball echoes not only Einstein but also Shakespeare, whose words in *Hamlet* might be modified to read "There is nothing either good or bad, but context makes it so." No longer must we accept arbitrary assessments of performance or regard with awe such old-time figures as Hugh Duffy's BA of .438 in 1894 (not the accomplishment that Rod Carew's .388 was in 1977) or George Sisler's .407 in 1920 (not as good as Roberto Clemente's .357 in 1967). Conversely, a "mediocre" performance of recent years, such as Bobby Murcer's .292 of 1972, for instance, stacks up as the equal of Eddie Collins' .360 in 1923, while Charlie Grimm's seemingly solid .298 in 1929 compares unfavorably to Mike Cubbage's .260 in 1976. Relative measures permit comparisons across time where absolute figures do not, for it is reasonable to compare two figures however many years apart by their relation to those of their peers and/or their relation to the league leading performances.

Relativism redefines our understanding not only of particular accomplishments but also of baseball history itself. We see that the men who batted .400 with numbing regularity in the 1890s and 1920s were not supermen (would you swap Wade Boggs for Tuck Turner? George Brett for Harry Heilmann?), no more than the sub-2.00 ERA pitchers of the late 1960s (Gary Peters, Bob Bolin, Dave McNally, et al.). Absolute figures lie. Are hitters today worse because none has hit .400 since 1941? Or are they superior because a Dave Kingman can average nearly 30 homers a year while Cap Anson only averaged 4? Are infielders better today because they make fewer errors than their counterparts of 50, 75, or 100 years ago? Do modern outfielders have limp-noodle arms because their assist totals pale before those registered in the 1900s? Is baseball improving or declining, and has its rise or fall been steady? One can spit absolute stats on the hot stove all winter long and get no closer to the answer, but with relative statistics, the issues are clarified. The relative approach is not a panacea for all that ails absolute stats, but it is a substantial advance. And there is more progress to come, for no area of baseball statistics has been the focus of such stimulating work in recent years as that of cross-era comparison.

In the May 1983 issue of *The Coffin Corner,* the newsletter of the Professional Football Researchers Association, Bob Carroll offered a witty and perceptive dissection of the relative approach to football statistics. It was based upon a comparison of two great running backs, Tuffy Leemans of the New York Giants of the late 1930s and early '40s and George Rogers, currently with the New Orleans Saints. "I've always liked the story," Carroll wrote, "of the little old lady who scornfully toured a Picasso exhibit and then sniffed, 'If Rembrandt were alive today, *he* wouldn't paint this way!' To which a bystander replied, 'Ah, but if Rembrandt were alive today, he wouldn't be Rembrandt.'

"The bystander knew the truth that genius is unique to its own time and place. He knew better than to compare two artists from different times and places, because different circumstances produce different results, even with genius. Nowhere is this common-sense rule disobeyed more often than in the world of sports. . . . I should know better than to get into it, but then, at my age, I should know better than to do a lot of things. So, here goes."

Carroll then compared Tuffy Leemans' 830 yards, with which he led the NFL in 1936, to George Rogers's 1674 yards, with which Rogers led the NFL in 1981. He manipulated these numbers in sundry typical ways, normalizing to the length of the season (12 games for Leemans, 16 for Rogers); the number of carries; the number of yards per carry; the value of those yards as opposed to the value of the yards obtained by passing; the number of times Leemans carried the ball compared to the total offensive plays of the Giants versus the number of times Rogers carried the ball compared to the Saints' offensive plays. He also normalized Leemans' average number of yards per carry to that of all runners in 1936 and created the equivalent ratio for Rogers. For purposes of comparison, he dismissed the obvious disadvantage to Leemans of having to play offense *and* defense. What he came up with in the end was that Leemans' 830 yards were "more" than Rogers's 1674 yards, upon which he observed:

"For some strange reason my nose has grown so long in the last few minutes that I now can type with the tip. All right, the truth. As most of you know, anything can be proved with statistics so long as only certain statistics are used. (For my next number I'll prove that Jack Lambert is a better passer than Ken Anderson.) What this little exercise proves, really, is that there are too many variables to compare a great star from one era with an equal star of a different era. . . .

"Try this: Rogers ran against bigger defenders. But his blockers were also bigger. And so is Rogers. *But,* if Leemans was twenty-three today, he'd be bigger too. He'd also be stronger and faster. But that also applies to defenders. And to blockers. AAAAAAAARGH!!!

"We can't 'if' our way to an answer, but we're left with the imponderable: the difference between a 1936 yard and a 1981 yard is a million miles.

"We'll close by rephrasing what we started with: If Tuffy Leemans were running today, he wouldn't be Tuffy Leemans.

"But he just might be George Rogers."

Bob's bottom line about relative stats— AAAAAAAARGH!!!— may be the way you feel right now, too, but baseball is hugely different from football, which has changed radically over its briefer history. Also, each play in a football game requires the synchronization of eleven efforts on offense and eleven on defense, with responsibility for success or failure always, and most often falsely, applied to the man or men who get their hands on the ball. Baseball is a team game, too, but one in which individual effort stands out more clearly and individual credit or blame is doled out more fairly. What's more, the game of baseball as it was played in 1936, or 1896, is more like the game of today than it is unlike it; the same cannot be said of pro football. The best football club of 1936 wouldn't stand a chance in a game against the worst today, while it could be argued (and is) that the 1936 New York Yankees would beat the pants off any baseball team today.

There are things that relative baseball stats won't do, questions they won't answer. What would Ty Cobb bat if he were playing today? Lefty O'Doul was asked this question by a fan at an offseason baseball banquet in 1960. "Maybe .340," O'Doul answered. "Then why do you say Cobb was so great," the fan remarked, "if he could only hit .340 with the lively ball today?" "Well," O'Doul said, "you have to take into consideration that the man is now 74 years old." Relative Batting Average cannot tell with certainty what Cobb would hit today, for as Carroll wrote of Tuffy Leemans, if Cobb were playing today he wouldn't be the same Cobb; he would be bigger, stronger, and faster, and he might choose to steal less and go for the long ball more. What RBA can do is to state that Cobb's lifetime batting average was more than a third above the league average, from which one can infer that Cobb, had he played in 1961-83, the span enjoyed by Carl Yastrzemski, might have batted .346 (league average = .257, Cobb's unadjusted RBA = 1.347; the two multiplied = .346).

The trouble with this inference, reasonable though it is on its face, lies in a truth Einstein would appreciate: Everything is relative, including relativity. The National League batting average of .266 in 1902 does not mean the same thing as the American League BA of .266 in 1977, any more than Willie Keeler's .336 in 1902 means the same thing as Lyman Bostock's .336 in 1977: It does violence to common sense to suppose that, while athletes in every other sport today are measurably and vastly superior to those 50 or 75 years ago, in baseball alone the quality of play has been stagnant or in decline. Keeler's and Bostock's Relative Batting Averages are identical, which signifies that each player exceeded his league's performance to the same degree. But the question that is begged is "How do we measure *average* skill: What do the .266s of 1902 and 1977 mean?" That question, and other Hot Stove League stumpers that have not been approached statistically until now, will be addressed in the next chapter.

---

[1] Note that in this book we remove pitchers' batting performance from the league averages; as conventionally expressed, the NL of 1930 batted .303 and the AL of 1968 .230.

[2] *The Baseball Encyclopedia* shows Barnes at .404, which was the average credited to him in 1876 (actually, it was .403 because of improper rounding off); but the Special Baseball Records Committee of Organized Baseball ruled in 1968 that "Bases on balls shall always be treated as neither a time at bat nor a hit for the batter. (In 1887 bases on balls were scored as hits and in 1876 bases on balls were scored as outs.)" Barnes's average, when his 20 walks are no longer counted as outs, becomes .429.

[3] All league averages prior to 1900 have *not* been adjusted upward by the elimination of pitchers' batting. In the nineteenth century, pitchers were much better hitters than they are today (Guy Hecker led the AA in batting with a .342 mark in 1886, amassing 345 at bats while going 27-23 as a pitcher!). And hurlers often played regular positions on those days when they were not in the box.

[4] Lajoie's BA in 1901 has long been in dispute. All sources today list him at .422, but box-score research by Information Concepts, Inc., in 1968 revealed his correct BA to have been .426. The major reference works adhere to .422 for reasons of tradition rather than truth. The same holds for Lajoie's BA in 1910 and Cobb's as well: The discrepancies in these averages were discovered by Pete in 1980 and publicized by *The Sporting News* in 1981, yet Bowie Kuhn ruled that the incorrect figures would remain official.

[5] The figure for Cobb is one point less than what Organized Baseball recognizes as official; our figure reflects the recent research which revealed that clerical errors by the American League office falsified (unintentionally) the record, and not only in the 1910 season.

[6] Who are the modern players among the tops in lifetime RBA? They include such good but not great batters as Tony Oliva (21), Matty Alou (22), Manny Mota (23), Richie Ashburn (38), Thurman Munson (44), Mickey Rivers (46), Manny Sanguillen (50), and Ralph Garr (51).

[7] If you're doing the math yourself, you'll come up with 1.328, but we employ additional decimal places in both individual and league BA before rounding off.

[8] Nineteenth century performances are excluded from these lists because (a) we do not yet have park adjustments for years before 1900; (b) unique styles and rules of play created extreme conditions for both hitters and pitchers not seen in this century; and (c) the overall level of play was so poor as to give the stars of the period a statistical edge which would lead to their domination of many categories. A separate listing of pre-1900 Relative Slugging and Relative OBA will be found in the tables.

7

THE GOOD OLD DAYS ARE NOW

Much of the charm of baseball is its archaic quality, its *tableau vivant* of a simpler time gone by. The game of 1893 is recognizably and comfortingly the same as that of 1983, which cannot be said of much else in a country busy reinventing itself every twenty years or so. For many Americans, baseball has become the fixed point in a turning world; things may fall apart, but the center holds. If the America of our youth survives anywhere as anything more than an artifact—if there is a link between the generations, a reminder of the way we used to be—it is in baseball, that peculiar exercise in nostalgia in no matter what setting, domed stadium or Little League field.

Baseball seems to exist under a bell jar, oblivious and impervious to the stresses of the world outside. With America's institutions under assault from all directions, baseball remains a world unto itself with its small, slow changes arising only from its own mandate.

On its own terms, the game has changed a great deal, but you wouldn't know it to look at the statistical log, baseball's Doomsday Book in which good works and bad are recorded. In the Olympics of 1896, the winning time in the 1500-meter run was 4:33.2; in 1980 it was 3:38.4, a clear statement that in this event, the top runner of today is capable of performances 20 percent better. Baseball in 1896, however, saw four men hit over .390, a level of performance seemingly unattainable today. In truth, such a level can be attained whenever management deems it desirable. The rule in baseball as in the garment trade is change or die, for without periodic tinkering with the rules, the ball, the playing conditions, etc., it is certain that one force in baseball—the offense or the defense—will rule the game to such an extent that spectators will lose interest. Or at least that's the way the owners have always felt and acted.

Baseball is not a purely athletic event like a track meet, where the attraction is men and women performing to their maximums, but rather an entertainment, an exhibition tailored to the presumed preferences of the paying public, like horse racing. Just as horses are assigned different weights to produce a close race, so are pitchers or batters, in their turns, hampered or abetted so as to keep the game competitive and uniform with that of seasons past.

Although the contestants of today are very different in their abilities, physiologies, attitudes, and training, the game *looks* the same as that of 1893 or 1953, the scores are about the same, and the individual performances are about the same. The seamless web of baseball is an illusion, the seams smoothed over by statistics. If Jesse Burkett hit .410 to lead the National League in 1896, why does no one today bat .500, let alone .400? Or if Burkett was a superman, look at the league average of .290: Why would today's league averages be 10 percent lower rather than 10 percent higher? Was the average player better some ninety years ago? Have the numbers changed with the game, or has the game changed to keep the numbers more or less constant? These are some of the questions to be addressed in this chapter.

Where are the stars of yesteryear? Where today is a man like Hoss Radbourn, who pitched 27 consecutive Providence games and won 26 of them? Where is a Rogers Hornsby, who *averaged* over .400 for a five-year period? A Babe Ruth, who in 1920 hit more homers than fourteen of the fifteen other big-league *teams?* A Jack Taylor, who over five years completed 187 consecutive starts? Why were so many all-time pitching records set between 1900 and 1919 and so many batting records over the next two decades? Giants assuredly walked the earth in those days, as demonstrated in the previous chapter's tables of relative averages; but men of the same stature, or greater, are among us today, their abilities camouflaged by the comparative expertise of those around them.

We believe baseball today—by which we mean the 1970s and '80s, not any specific season—is better than it's ever been, and that baseball players are, too. We believe that the Golden Age is a period identified flexibly with one's youth: Octogenarians glow rapturous over the days of Cobb and Wagner, septuagenarians over the days of Ruth and Gehrig, our generation over Brooklyn's Boys of Summer or Stengel's Yanks or the Sox of Williams, Pesky, and Doerr. "Oh, they don't make 'em like that anymore!" all the generations cry in unison. In the year 2000, we will surely be joined by a gaggle of thirty-to-forty-year-olds lamenting the departed Murphy and Stieb and Dawson. Baseball is a backward-looking game.

As we grow older, we bear less and less outward resemblance to the child left behind at adulthood's door, yet that child lives on within us, remaining an essential part of our identity, if not *the* essential part. Life's cares make it more and more difficult to touch base with the child within, which needs attention if it is to sustain us. Watching a baseball game, or even thinking about baseball, steady, comfortable, and unchanging baseball, brings us into a unifying relationship with the child, the part of us that loves baseball, even if it is the adult that understands it. Because the game is so evocative, on the deepest level, of our childhood, it is not surprising that the sharply formed impressions of the game during that period are the ones that stay with us for all time. To suggest that baseball is better today than it was when we were young is heretical, akin to telling a child there is no Santa Claus. "Where are players like we had in the good old days?"

"All around you" is a response faintly depressing to write, as it surely is to read.

Isn't it odd that baseball owners and officials would like us to believe that the game has never been better than it is today, in order to put more fannies in the seats, yet their Rules Committee leaps into action whenever the game seems to be on the verge of change, whenever the traditional balance between offense and defense appears in danger of tilting lastingly one way or the other. Baseball's owners are in the entertainment business, where the customer is always right, and the successful impresario anticipates what the public wants. If pitching seems to be dominating the game, they make a seemingly minor adjustment like lowering the mound five inches and—poof—there is an explosion of hitting. If the batters are making major-league baseball look like Sunday picnic beerball, they make another minor adjustment, like decreasing the tension with which the wool in the ball is wound, and the former balance is restored. (Historically, by the way, the trend has always been for pitchers to take the upper hand in those periods not marked by manipulations from on high.)

Want more stolen bases? Tighten up on the balk rule. Too many infield hits and not enough outfield assists? How about 91-foot basepaths? Baseball is a delicate mechanism, and seemingly minuscule adjustments in its rules can have mammoth results.

Just how Organized Baseball determined what "balance" in baseball should be, i.e., what the fan "wants," we do not know; attendance figures do not provide the answer. But since major-league baseball began in 1876, whenever league batting averages dipped below .250,[1] the owners would be on the alert for possible intervention. If the BA dipped below .240, you could be sure that a significant rule change would follow.

The specific rules changes which have maintained baseball's "balance"—and which, in 1920, shattered and then redefined it—will be detailed later. But another force affecting the game has been fashion, that mass delusion which acts on baseball in the same way it does on hemlines and hair styles. An edict or aphorism delivered by a respected source like Henry Chadwick or John McGraw or Earl Weaver may shape the way the game is played for years. A World Series champion with a style of play perceived as unusual or "revolutionary" will spawn imitators throughout baseball, from the majors to the sandlots. When the Dodgers were winning championships with pitching, speed, and defense in the early 1960s, they set a trend for others to follow—often to no profit. A particular team's style of play cannot be appropriated by just anybody who admires it: It must be suited to a team's home park, for example, as George Steinbrenner learned to his chagrin in 1982. The St. Louis Cardinals won the '82 Series and the baseball publications were full of talk about "Whiteyball"—speed, defense, and pitching, especially from the bullpen. We are not convinced, however, that the Cardinals' style is the wave of the future—and not only because they backslid in 1983—for in their World Championship year they were 46-35 at home (only two games above average) and 46-35 on the road (nine games above average). Had the Brewers won Game Seven—and they certainly had a superior team before Fingers was hurt—would the pundits have proclaimed the era of "Harveyball"? Probably; and where is poor Harvey Kuenn now?

With individual, team, and league performance all captive to the rules tinkerers and the dictates of fashion gurus, as well as broad trends in society which have impact on the talent pool available to the major league clubs, it is quite a task to separate out and quantify the variables that combine to keep the statistics within so narrow a range. One who claims that baseball is better than ever, as we do, should have no expectation of being believed without evidence. Relative measures are a first step, permitting us to place individual performance in the context of league performance, but with the game in a constant state of minute agitation—a continuing series of corrections designed to maintain an implicit order or balance—how can we evaluate the context itself?

Has baseball been improving as it has changed, or has change been enforced to mask a deterioration of play, as was charged when the hitters were made beneficiaries of a bail-out plan in 1969? Was baseball at its peak in the 1920s, as the old fogies claim? Or in the early 1950s, when a massive minor-league network funneled talent into only sixteen big-league clubs, as the new fogies claim? Better, of course, can be an entirely subjective question. Do you like pitching and defense? Or batting orgies with lots of home runs? Or clever play— hitting behind the runner, trick defenses for bunt situations, outfielders who can hit the cutoff man? It's possible, maybe even likely, that young men coming into the big leagues in 1910 knew how to play the game better, had a better grounding in the fundamentals, than their counterparts today. But a fellow blessed with athletic talent back then who wanted to make his living in sports could look only to professional baseball, which he played and practiced single-mindedly. Professional football scarcely existed outside a three-state area in 1910, pro basketball was off in the distance. Tennis, golf, soccer, hockey were not ways to make a buck in the U.S. at that time. So, baseball was siphoning off almost all of the athletic talent. However, in 1910 the population of the country was 92 million; today it's 225 million. With a population 2.5 times as great, we should be producing at least 2.5 times the number of superior athletes and in all probability many times more, because of improved conditioning, training, and education. John McGraw recognized as much in 1923, when he wrote: "In thirty years I have seen much baseball. My greatest asset has been a good memory. There is no question in my mind but that present-day baseball is better. Also there are more good ballplayers today than there ever were before, simply because there are more people playing ball."

Was there more competition for baseball jobs in the old days? Sure there was. In 1949 there were 59 separate minor leagues with 448 teams. At that time the majors consisted of 16 teams, each carrying 25 men—thus the baseball "400," even more difficult to crack than the high society version. Today there are 26 teams: 250 more spots open than in 1949, and 42 fewer minor leagues with 284 fewer teams. Where 35 years ago about 22 professionals waited in the wings for each big-league place, today there are only 5. It would appear from this that the talent pool for major league baseball has shrunk dramatically. However, the colleges have lately been supplying many of the blue chippers coming into the major leagues—a Tom Seaver, a Reggie Jackson, a Dave Winfield, a Dave Stieb. And in 1949 the black minor leaguer was a comparative rarity.

George Bamberger pitched in the Pacific Coast League for 13 years. His total major league experience, though, came to only 10 games and 14 innings. Shortly before he resigned as manager of the Mets, he said in the New York *Times,* "Once I pitched in the Pacific Coast League against Elmer Singleton, and we knew that if either of us got the decision, we'd reach a combined total of 500 credits in the league— games won and lost. That's experience. But after the War everything changed. We had 16 teams in the big leagues then. Now we have 26. That's 100 more pitchers in the big leagues, 30 more catchers, 60 more infielders, 50 more outfielders, counting reserves. The result was that the talent got spread thin. A lot of the kids today wouldn't have made my old club in the Pacific Coast League. Many of them wouldn't have made Double A."

What Bamberger is saying is not that today's kids don't have the athletic ability, but that they come to the major leagues without sufficient experience. They can hit, they can run, they can field, they can throw—but they're not baseball-smart. Many of them hit the big time after only three years or less in the minors; if they are not placed on the parent club's 40-man roster after three minor-league seasons, any club in a higher league can claim them in the draft. But to say that the novices of today wouldn't have made it in Double A in 1949, or they wouldn't have made the Pacific Coast League—we can't agree. There may be some men on the rosters of the weaker teams in the major leagues who wouldn't have been in the major leagues of 1955, but Bamberger's view is inevitably colored by his age, by his life experience, and by the human tendency to form views when young that only harden with age.

Dick Williams, the manager of the San Diego Padres, who was in the Dodger chain for years before finally making the majors, was quoted in the same piece, saying, "You used to be able to keep a guy in the minors for seven years, but since the reserve clause was changed, you're force feeding a lot of guys. . . . You have guys coming along who can't really play in the big leagues, but you have to protect them. It spreads the talent thin. I know there were .210 hitters when Babe Ruth played, but there are more of them now." Williams, like Bamberger, may have forgotten more about baseball than we will ever know, but the .210 hitter of Babe Ruth's day—if there was one in that paradisical time for hitters—might not hit .150 today. The talent pool in Ruth's day, as from 1885 through 1946, included no blacks and only Latins of light hue.

Feeling outnumbered against both Bamberger and Williams, we're going to look for a bit of support from George "Specs" Toporcer, who was a reserve infielder with the Cardinals in the 1920s, later managed in the International League, and was in the front office of the Red Sox and the White Sox. He wrote a letter to Red Smith in the fall of 1981, which was published in the *Times*. Toporcer is that rare and refreshing individual who can look back upon his own days as a player and evaluate them in the light of what followed without the rosy tinge of nostalgia.

"It is impossible," Toporcer wrote, "to detail my convictions and conclusions of the relative merits of players of different eras in this letter. I will state, however, that anyone who does not think the game and the number of worthwhile players have improved from decade to decade—with the exception of the war and immediate postwar years—does not know whereof he speaks. Yes, outstanding stars of yesteryear would star today, but most people who try to rate them are misled by statistical evidence, which makes the old ones appear far greater than they actually were.

"They do not realize that the reason why Cy Young won 511 games with Mathewson, Alexander and Johnson winning well over 30 games in three different seasons, Ed Walsh winning 40 in one season, Ty Cobb batting for a lifetime average of .367, the New York Giants stealing a total of 347 bases in one season despite having only one really fast man on that 1911 team, and the fact that when the ball was enlivened in 1920, it led to having many teams batting for averages in excess of .300, and far more .400 individual batting marks than at any other period in the history of the game—seven in five years, if memory serves—that those amazing statistics were achieved only because the rank and file players were far inferior to those of today, thus enabling the stars to stick out like a sore thumb."

That's an amazing statement for a man now 84 years old to make. He continued: "In the last magazine article I wrote about three years ago, I ranked those I considered the 12 greatest of all time batters. I did not rank Cobb among those 12. You may think I did not do so because of my dislike of the man dating back to the 1920s, but that is not so. I think I judged the matter fairly and gave a lot more reasons than are contained in this letter for omitting his name from that list.

"I ranked them in this order: Ruth, Williams, Gehrig, Foxx, Di-Maggio, Hornsby, Musial, Mays, Aaron, Greenberg, Mantle, Frank Robinson."

Incredible. Of the top twelve in LWTS, of which Toporcer knew nothing, nine are on his list . . . and six played mostly after WWII. Each man on the list hit 300 homers. Not only is Toporcer free of the grip of nostalgia, he is also an intuitive statistician of the first water:

"Of course, no two experts, even the most knowledgeable, would rank players in exactly the same way, because they played in different eras against different opposition, in different ball parks and cities with winning or losing teams, and all these facts bear on the choices. As an example, I ranked Mays ahead of Aaron because Hank benefited greatly by playing most of his career in Atlanta. I think had he played in the Polo Grounds and Candlestick Park and Mays at Milwaukee and Atlanta, Mays, and not Aaron, would have been the one to break Ruth's home run mark, and the RBI's and runs scored by other players would have been altered."

Park Factor, anyone? But more interesting than reshuffling the deck of the top twelve in order to evaluate baseball long ago vs. baseball today is the question of average skill. In the 1980 *Baseball Research Journal,* Dick Cramer employed an ingenious method of evaluating average batting skill across time. He compared the batting performance of the same player in different seasons, avoiding the pitfall of environmental factors such as ball resilience, rule changes, racial mix, playing conditions, etc., by subtracting league averages before making any comparison. "Of course," Cramer wrote, "direct comparison cannot be made for seasons more than 20 years apart; few played much in both periods, say, 1950 and 1970. But these seasons can be compared indirectly, by comparing 1950 to 1955 to 1960, etc., and adding the results."

Cramer examined the question in terms of "Batter Win Average," a measure of runs contributed beyond the average divided by plate appearances. Fred Lynn's BWA of .120 in 1979, for example, signified that (a) every ten of his plate appearances generated 1.2 runs that an average player would not have produced, and (b) Lynn's run generation beyond the average was the same as that of a player who produced at the league average level in all measures—doubles, triples, homers, walks, HBP—except singles, of which he hit enough to have a batting average .120 higher than the league. Like Cramer's earlier Batter Run Average, discussed in Chapter 3, BWA is predicated on the strong correlation between On Base Average × slugging percentage and actual runs scored by a team or league. We quote Cramer's important study at length:

"The first stage in this study was a labor of love, using an HP67 calculator to obtain BWA's for every non-pitcher season batting record having at least 20 BFP (batter facing pitcher) in major league history. The second stage was merely labor, typing all those BFP's and BWA's into a computer and checking the entries for accuracy by comparing player BFP sums with those in the Macmillan Encyclopedia.

The final stage, performing all possible season-to-season comparisons player by player, took 90 minutes on a PDP10 computer. A season/season comparison involves the sum of the difference in BWA's for every player appearing in the two seasons, weighted by his smaller number of BFP's. Other weighting schemes tried seemed to add nothing to the results but complexity. . . .

"The results of my study are easiest to visualize from the graphical presentation on the next [spread]. (Because few readers will be familiar with the BWA units, I have not tabulated the individual numbers, but later convert them to relative BA's and slugging percentages.) Theories on the whys and wherefores of changes in average batting skill I leave to others with greater personal and historical knowledge of the game. But the major trends are clear:

(1) The average level of batting skill has improved steadily and substantially since 1876. The .120-point difference implies that a batter with 1979-average skills would in 1876 have had the value of an otherwise 1876-average batter who hit enough extra singles for a .385 batting average.

(2) The American and National Leagues were closely matched in average batting strength for the first four decades (although not in number of superstars, the AL usually having many more). About 1938 the National League began to pull ahead of the American, reaching its peak superiority in the early 60's. A resurgence during the 70's makes the American League somewhat the tougher today, mainly because of the DH rule.

(3) The recent and also the earliest expansions had only slight and short-lived effects on batting competitiveness. However, the blip around 1900 shows the substantial effect on competition that changing the number of teams from 12 to 8 to 16 can have!

(4) World War II greatly affected competitiveness in 1944 and 1945.

"Many baseball fans, myself included, like to imagine how a Ruth or a Wagner would do today. To help in these fantasies, I have compiled a table of batting average and slugging percentage corrections, based again on forcing differences in league batting skill overall into changes in the frequency of singles only. However, league batting averages and slugging percentages have been added back in, to reflect differences in playing conditions as well as in the competition. To convert a player's record in year A to an equivalent performance in season B, one should first add to his year A batting and slugging averages the corrections tabulated for season A and then subtract the corrections shown for season B. The frequency of such other events as walks or stolen bases then can, optionally, be corrected for any difference in league frequencies between seasons A and B.

"One interesting illustration might start with Honus Wagner's great 1908 season (BWA = +.145). What might Wagner have done in the 1979 American League, given a livelier ball but tougher competition? The Table yields a batting average correction of $-.059 - (+.003) = -.062$ and a slugging correction of $-.020 - (-.029) = +.009$, which applied to Wagner's 1908 stats gives a 1979 BA of .292 and SPct of .551. (In 600 ABs, he would have, say 30 HRs, 10 3BHs, 35 2BHs). Wagner's stolen base crown and tenth place tie in walks translate directly to similar positions in the 1979 stats. That's impressive batting production for any shortstop, and a '1979 Honus Wagner' would doubtless be an All-Star Game starter!

"These results are fairly typical. Any 20th century superstar would be a star today. Indeed a young Babe Ruth or Ted Williams would outbat any of today's stars. But of course, any of today's stars—Parker, Schmidt, Rice, Carew—would before 1955 have been a legendary superstar. Perhaps they almost deserve their heroic salaries!

"Facts are often hard on legends, and many may prefer to believe veterans belittling the technical competence of today's baseball as compared, say, to pre-World War II. Indeed, 'little things' may have been executed better by the average 1939 player. However, so great is the improvement in batting that if all other aspects of play were held constant, a lineup of average 1939 hitters would finish 20 to 30 games behind a lineup of average 1979 hitters, by scoring 200 to 300 fewer runs. This should hardly surprise an objective observer. Today's players are certainly taller and heavier, are drawn from a larger population, especially more countries and races, are more carefully taught at all levels of play. If a host of new track and field Olympic records established every four years are any indication, they can run faster and farther. Why shouldn't they hit a lot better?"

The implications of this study were troubling: That Ross Barnes's .429 BA of 1876 would have been only .306 in the National League of 100 years later; that Joe Torre's .363 of 1971 would have been .501 in 1894; that Nap Lajoie's .422 of 1901, had he played in the American League of 1977, would have been only .303 against Rod Carew's .390; that Rogers Hornsby's .424 of 1924 would have been only .339 in 1979, trailing Keith Hernandez; and that the *average* batter of 1894, like Louisville's Fred Pfeffer, a Hall of Fame caliber second baseman who batted .308, would have batted .163 in the American League of 1976!

Table VII, 1. *Average Batting Skill*

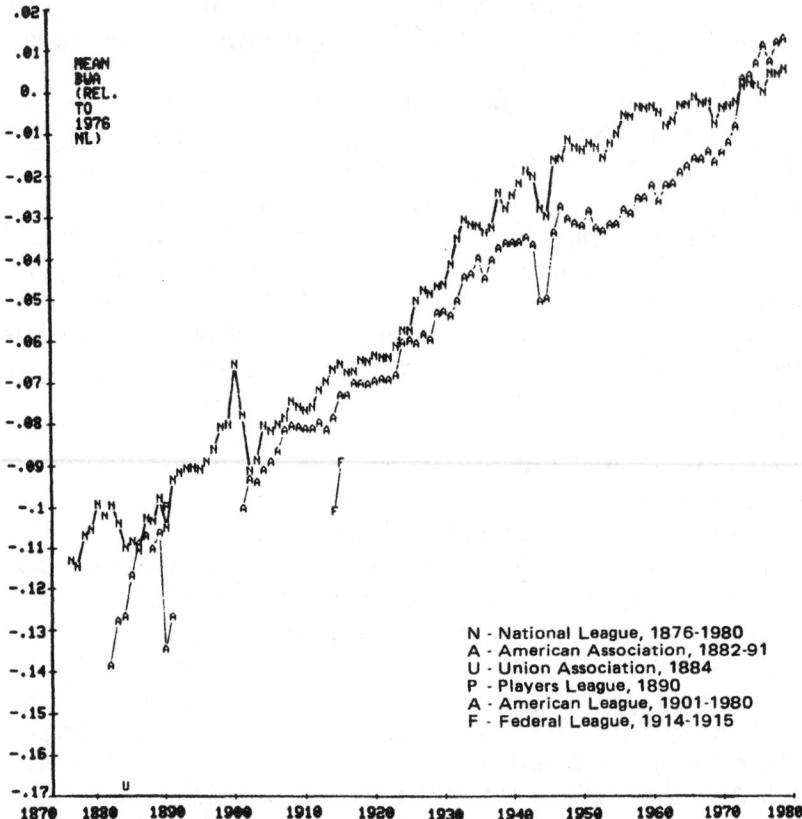

N - National League, 1876-1980
A - American Association, 1882-91
U - Union Association, 1884
P - Players League, 1890
A - American League, 1901-1980
F - Federal League, 1914-1915

Corrections Applied to a Player's Batting Average and Slugging Percentage in a Particular Season to Equate Them with 1976 National League Play.

	American BA	American SPct	National BA	National SPct		American BA	American SPct	National BA	National SPct
1979	+.003	−.029	−.001	−.017	1965	+.001	−.021	+.003	−.016
1978	+.011	−.007	+.005	−.004	1964	−.006	−.035	−.003	−.017
1977	+.002	−.031	−.004	−.033	1963	−.009	−.036	+.003	−.010
1976	+.016	+.017	(.000)	(.000)	1962	−.017	−.050	−.014	−.040
1975	+.009	−.006	.000	−.006	1961	−.022	−.055	−.012	−.049
1974	+.007	−.001	+.002	−.004	1960	−.017	−.044	−.004	−.031
1973	+.005	−.011	+.002	−.014	1959	−.018	−.043	−.009	−.043
1972	+.013	+.015	+.005	−.007	1958	−.019	−.042	−.011	−.048
1971	+.001	−.010	.000	−.008	1957	−.024	−.045	−.011	−.045
1970	−.004	−.027	−.007	−.035	1956	−.028	−.056	−.007	−.046
1969	−.002	−.019	−.003	−.016	1955	−.029	−.046	−.015	−.057
1968	+.016	+.013	+.010	+.018	1954	−.029	−.039	−.023	−.059
1967	+.008	−.001	+.003	−.005	1953	−.035	−.050	−.027	−.066
1966	+.004	−.019	−.002	−.024	1952	−.026	−.032	−.012	−.027

BA and SPct, continued

	American		National				American		National	
	BA	SPct	BA	SPct			BA	SPct	BA	SPct
1951	−.030	−.043	−.018	−.024		1913	−.078	−.052	−.077	−.063
1950	−.043	−.068	−.020	−.054		1912	−.085	−.062	−.089	−.080
1949	−.034	−.034	−.021	−.042		1911	−.094	−.073	−.082	−.072
1948	−.036	−.046	−.018	−.034		1910	−.064	−.029	−.078	−.054
1947	−.023	−.025	−.026	−.045		1909	−.065	−.024	−.065	−.029
1946	−.030	−.032	−.018	−.011		1908	−.060	−.019	−.059	−.020
1945	−.045	−.030	−.040	−.033		1907	−.069	−.026	−.067	−.027
1944	−.051	−.038	−.034	−.030		1906	−.075	−.038	−.070	−.030
1943	−.026	−.012	−.024	−.007		1905	−.070	−.037	−.082	−.053
1942	−.032	−.026	−.013	−.001		1904	−.075	−.046	−.075	−.042
1941	−.042	−.059	−.025	−.022		1903	−.078	−.072	−.103	−.076
1940	−.047	−.077	−.034	−.040		1902	−.108	−.096	−.096	−.050
1939	−.055	−.077	−.046	−.054		1901	−.117	−.105	−.077	−.065
1938	−.059	−.087	−.037	−.040		1900			−.090	−.071
1937	−.062	−.090	−.050	−.054		1899			−.108	−.086
1936	−.074	−.100	−.057	−.059		1898			−.097	−.067
1935	−.060	−.076	−.055	−.063		1897			−.123	−.111
1934	−.063	−.077	−.056	−.065		1896			−.124	−.115
1933	−.058	−.069	−.042	−.032		1895			−.132	−.129
1932	−.068	−.089	−.057	−.071		1894			−.145	−.165
1931	−.072	−.084	−.064	−.068		1893			−.103	−.109
1930	−.081	−.108	−.095	−.134		1892	Amer. Assoc.		−.082	−.042
1929	−.064	−.081	−.087	−.113		1891	−.127	−.110	−.091	−.075
1928	−.081	−.091	−.076	−.086		1890	−.132	−.105	−.104	−.086
1927	−.084	−.092	−.076	−.074		1889	−.112	−.098	−.107	−.096
1926	−.082	−.087	−.076	−.076		1888	−.092	−.063	−.088	−.068
1925	−.092	−.101	−.095	−.111		1887	−.124	−.112	−.117	−.123
1924	−.091	−.091	−.086	−.089		1886	−.096	−.070	−.107	−.092
1923	−.090	−.090	−.093	−.096		1885	−.107	−.083	−.095	−.070
1922	−.106	−.101	−.102	−.108		1884	−.111	−.090	−.102	−.089
1921	−.101	−.111	−.099	−.101		1883	−.124	−.097	−.111	−.103
1920	−.093	−.091	−.079	−.070		1882	−.127	−.089	−.096	−.082
1919	−.079	−.064	−.069	−.042		1881			−.107	−.079
1918	−.064	−.027	−.064	−.032		1880			−.089	−.058
1917	−.058	−.024	−.059	−.032		1879			−.106	−.074
1916	−.061	−.031	−.060	−.035		1878			−.111	−.065
1915	−.061	−.033	−.059	−.036		1877			−.131	−.092
1914	−.067	−.036	−.063	−.040		1876			−.123	−.073

Federal League

1915	−.090	−.069
1914	−.110	−.096

Union 1884   Players 1890

−.146 −.111 −.108 −.106

In the 1981 *Baseball Research Journal,* William D. Rubinstein drew additional conclusions from Cramer's study, conclusions "so astonishing that many SABR members will not believe them," he wrote.

"In the three tables below," Rubinstein continued, "I have detailed the following information: the 'real' lifetime batting averages (in terms of 1976 NL play, the standard used by Cramer) of nearly 50 of the top batting stars of past and present; secondly, the top 'real' single season batting averages, by era, since 1876, again in terms of 1976 NL play; and, thirdly, the top 'real' single season slugging averages between 1920 and 1979, in the same terms.

"Compiling the lifetime BA information is somewhat complicated, since each single season BA has to be weighed into the total by proportion of lifetime ABs. Nevertheless, its findings are remarkable in the extreme. Given that average batting skill rose by 120 points between 1876 and 1979, one would naturally expect today's lifetime BAs to rise, but these calculations reveal that if Ty Cobb's career had taken place under the conditions of the NL in 1976, his lifetime BA would have been only *.289.* Rogers Hornsby and Joe DiMaggio achieve identical *.280* marks. Bill Terry, Lou Gehrig, and Tris Speaker are average-to-mediocre hitters at .271, .269, and .265 respectively. The Babe's .262 is a disappointment, though he hits better than Al Simmons (.260) or Harry Heilmann (.257). If he had been playing today, Honus Wagner would certainly wear a 'good field–no hit' (lifetime .251) label! And as for the likes of Pete Browning (.248), Dan Brouthers (.238), or Billy Hamilton (.236)—it looks like Salt Lake City for them!

"On the other hand, today's players, even those who are not true superstars, appear to be veritable supermen in comparison. Tony Perez outbats Hornsby and DiMaggio by three points; Al Oliver (through 1979) outhits Ty Cobb by thirteen. Rod Carew (.341 through 1979), George Brett, Bill Madlock, and others among today's top hitters easily outhit those of yesterday by many points. Only Stan Musial (.315), Roberto Clemente (.313), and Ted Williams (.310) among recent stars of the past do really well.

TABLE VII, 2. *Some "Real" Lifetime Batting Averages Through 1979*

.340   Carew .341
       Brett .325
       Madlock .321
.320   J. Rice .319; Parker .317
       Musial .315; Clemente .313
       Rose .312

.310	T. Williams .310
	Garvey .305
	Oliver .302
.300	Aaron .301
	Mays .295
	J. Robinson .293
	F. Robinson .292; Brock .292; B. Williams .291
.290	Cobb .289; P. Waner .289; Yastrzemski .289
	G. Foster .288; Kaline .287
	Stargell .284; Perez .283
	Mize .282; Staub .282; J. Jackson .281; Mantle .281
.280	Hornsby .280; J. DiMaggio .280
	Snider .279
	Terry .271
.270	Gehrig .269; Bench .269
	Speaker .265; Foxx .264
	Ruth .262; Greenberg .262
.260	Simmons .260
	Heilmann .257; Lindstrom .255
	Wagner .251
.250	Keeler .250
	Browning .248; Youngs .248
.240	Brouthers .238
	Hamilton .236

"The problem with these statistics is that no one will believe them. I doubt if a single member of SABR can be persuaded by any amount of statistics that Tony Perez is a better hitter than Rogers Hornsby or Joe DiMaggio, or that Al Oliver is capable of outhitting Ty Cobb by 13 points or Honus Wagner by over 50 points. Certainly the Baseball Writers, who in 1969 voted Joe DiMaggio the greatest living player over anyone active more recently, do not accept this verdict. The statistics raise many questions which will be discussed below. Before doing this, the other tables presented here should be considered.

"The top 'real' single season BAs further shows that upward rise in batting averages since the 19th century. The top 'real' batting average of the 1876-1919 period, Tris Speaker's .327 in 1916, was far below the 15th highest BA of the period 1963-80. Ty Cobb still dominates the list of early hitters, but with averages which would be considered only good–excellent, rather than unbelievable, by today's standards. Only *one* 19th century batting average makes the list at all, while the incredible averages of the era, like Hugh Duffy's .438 in 1894, can be deflated into much lower figures—in Duffy's case, to an average of only .293 in 1976 terms. It will also be seen that 'true' .300+ averages were of the most extreme rarity, about one every other season in the early period.

"During the 'golden age' of baseball, 1920-42, top averages rose, but they are still lower than those of today. Ted Williams' 'real' .364 in 1941 outscores everyone else by a wide margin, with Hornsby's 'real' .338 in second place. It is interesting to note that no single hitter dominates the list, while many of the highest averages and best hitters of the period do not appear at all—like the top marks of Harry Heilmann, Bill Terry, and George Sisler. After the War, averages rose again, with Williams and Stan Musial in leading places. Joe Cunningham's .345 in 1959—which translates into a 'real' average of .336—turns out to be one of the highest in history, higher than any 'real' average ever achieved by Ty Cobb! In the contemporary period, averages rise again by a large amount, with Rod Carew the predominant figure—although George Brett may eventually equal his impact. Carew (three times) and Brett (once) hit for averages higher than anything ever seen in baseball history.

Table VII, 3. *Top "Real" Single Season Batting Averages*

1876-1919				1920-45		
1.	Speaker	.327	(1916)	T. Williams	.364	(1941)
2.	Cobb	.326	(1911)	Hornsby	.338	(1924)
3.	Cobb	.325	(1912)	Musial	.333	(1943)
4.	Cobb	.317	(1917)	Vaughan	.330	(1935)
5.	Cobb	.321	(1910)	Klein	.326	(1933)
6.	Lajoie	.320	(1910)	J. DiMaggio	.326	(1939)
7.	Cobb	.318	(1918)	Medwick	.324	(1937)
8.	Jackson	.314	(1911)	Williams	.324	(1942)
9.	Cobb	.312	(1909)	D. Walker	.323	(1944)
10.	Cobb	.312	(1913)	Garms	.321	(1940)
11.	Cobb	.312	(1916)	Simmons	.318	(1931)
12.	Jackson	.310	(1912)	Reiser	.318	(1941)
13.	Keeler	.309	(1897)	Lombardi	.317	(1942)
14.	Cobb	.308	(1915)	Travis	.317	(1941)
15.	Lajoie	.308	(1901)	P. Waner	.316	(1936)

1946-62				1963-80		
1.	T. Williams	.364	(1957)	Carew	.390	(1977)
2.	Musial	.358	(1948)	Brett	.385	(1980)
3.	Musial	.347	(1946)	Carew	.371	(1974)
4.	Aaron	.346	(1959)	Carew	.368	(1975)
5.	Mantle	.341	(1957)	Torre	.363	(1971)
6.	Musial	.340	(1957)	Clemente	.360	(1967)
7.	Ashburn	.339	(1958)	Carty	.359	(1970)
8.	Cash	.339	(1961)	Cooper	.357	(1980)
9.	Clemente	.339	(1961)	Carew	.355	(1973)
10.	H. Walker	.337	(1947)	Garr	.355	(1974)
11.	Musial	.337	(1951)	Madlock	.354	(1975)

12. Mays	.336	(1958)	Brett	.349	(1976)
13. Cunningham	.336	(1959)	Clemente	.348	(1969)
14. Kuenn	.335	(1959)	McRae	.348	(1976)
15. T. Williams	.333	(1948)	Carew	.347	(1976)

"Turning to 'real' top single season SAs achieved during the lively ball era, it will be seen that these 'deflate' to a much more limited extent than BAs. Babe Ruth is still the dominant figure, with top SAs superior to anything seen since. Top 'real' SAs have declined rather than risen since the War, as of course they have if one takes the record book figures at face value. This indicates that if Babe Ruth were active now, he would probably hit about as many homers as he actually did. In 1976 terms, his 'real' 1927 figures—a .272 BA, with a .698 SA— indicate that he probably would still have hit 55-60 home runs, but with many fewer singles and doubles.

Table VII, 4. *Top "Real" Single Season Slugging Percentages (1920-79)*

1920-45			1946-79		
1. Ruth	.756	(1920)	T. Williams	.686	(1957)
2. Ruth	.745	(1921)	Musial	.669	(1948)
3. Ruth	.698	(1927)	Aaron	.661	(1971)
4. Gehrig	.681	(1927)	Mantle	.649	(1956)
5. T. Williams	.676	(1941)	McCovey	.640	(1969)
6. Ruth	.670	(1923)	T. Williams	.635	(1946)
7. Ruth	.664	(1922)	Stargell	.632	(1973)
8. Ruth	.661	(1926)	Yastrzemski	.621	(1967)
9. Foxx	.660	(1931)	Mantle	.620	(1957)
10. Hornsby	.655	(1925)	Stargell	.620	(1971)

"Mr. Cramer's approach is very interesting, but I think there are very few SABR members who can accept its conclusions on face value. In the first place—as another SABR member, Dallas Adams, has pointed out to me—the table of 'corrections' Cramer has compiled (pp. [132-3]) apply to an entire league in a particular season, not to a particular player. And while one might well have to subtract 101 points from the BA of a mediocre AL hitter in 1921 to arrive at his 1976 BA, is this really also true of the Ty Cobbs and George Sislers? Evidence that it is not so seems implied in the well-documented fact that the gap between the average league BA and the league-leading BA has been consistently narrowing throughout the century: in 1911, for instance, Cobb's batting average exceeded the AL league BA by 137 points, but most recent batting champs have exceeded their league BA by only 60-70 points. It thus seems that the very best hitters of the past were very much better than the average hitters of their day. Because of this, I doubt that the 'corrections' provided by Cramer can be applied to *all* players of the past: they almost certainly cannot be applied to the very best players of former ages.

"How, then, would Ty Cobb do if he were young and active today? Based upon everything we know about his ability, incredible drive, speed, and willingness to learn, my guess is that if he were retiring next year, he would be carrying a .320-.325 lifetime BA with him. He would not hit for a lifetime .367—because of better fielding and pitching, modern ballparks, and jetlag, no one could under today's conditions— but he would still be among the very top active hitters, probably below Carew but above Rose and everyone else among today's senior stars, and far above the paltry .289 lifetime average suggested by the statistics."

Cramer responded, in the same issue of the *BRJ*, reminding Rubinstein that the corrections to BA and SLG detailed in the table were based upon the run-generating value of the player to his team, and could be rearranged in other ways to express the same run value. The BA correction contained a presumption that a player was drawing walks and hitting for extra bases *at the league average,* with any margin of above-average offense being reflected in his BA only as additional singles.

"Regarding the Batting Skill commentary by Dr. Rubinstein," Cramer replied, "I agree completely. Cobb probably would have a batting average higher than .289 if he were retiring today. However, I also think he would have a slugging percentage worse than the .465 that the tables by themselves suggest. To illustrate this point with an example, Cobb's 1911 heroics, when transformed as in the 1908 Wagner example, yield 193 hits (.328 BA), and 47 doubles, 23 triples, and 12 homers (.548 SPct); I agree that his actual BA would be higher, but I also suspect that his extra-base hits would be fewer. What the tables are supposed to mean is that Cobb's 'team batting value' to a 1979 team would be that of a .328 hitter with 47 doubles, 23 triples, 12 homers, and leading the league in stolen bases. Definitely a 1979 all-star, but possibly not what a young Cobb's stats 'should be' in 1979.

"To get more singles and fewer extra-base hits while retaining the same 'team batting value,' the following rough approximation to the relative values of hits may be useful: Converting an out to a single improves team run-scoring by as much as

> converting three doubles to three singles, or
> converting two triples to two singles, or
> converting a home run to a single.

"So, to fix up Cobb's '1979 record,' one could prune his extra-base hits down to, say, 38 doubles, 6 triples, and 6 homers. This would add respectively, 3 + 9 + 6 singles to his total of 193 hits, making a different (but value equivalent) BA of .356 on 211 hits.

"To model Cobb's 1911 teammate Crawford, one might, in contrast, change singles and triples to homers. The table would give a .284 BA and a .453 SPct, say 36 doubles, 14 triples, and 11 homers. If we assume Crawford would have hit .260 in 1979 with 5 triples, his homer total would then improve to 28. Exactly the same team value!"

Dallas Adams followed, in the 1982 *BRJ*, with an article entitled "Average Pitching and Fielding Skills Through Major League History." Based upon the presumption that Cramer's thesis is valid, it proceeds to show that pitching has improved at a rate commensurate with that of batting, and that fielding skill has been in nearly constant ascent since 1876.

Are all the oldtime accomplishments thus consigned to the trashbin? The previous chapter on relativity exposed Bill Terry's gaudy .401 of 1930 as scarcely different from Carl Yastrzemski's tacky .301 of 1968, and now Cramer's corrections for Average Batting Skill show that the .401 would only have been .290 in 1968, anyway. Is nothing sacred? Wouldn't it be nice to have someone hit .400 again, no matter what it means? Or win 35 games, or complete that many starts?

If we really want a .400 batting average, it can be arranged: Just let league BAs continue to drift upward, maybe juice the ball a little bit, and when a league's BA hits .293, voilà, you have a 50-50 shot at someone hitting .400. This prediction emerges from a neat little study by the aforementioned Adams of league-leading BAs as a function of league BA; from that he derived the probability of a .400 average arising from various league averages. The article, "The Probability of the League Leader Batting .400," appeared in the *BRJ* of 1981; it is brief, so it follows in full:

"In 1980 George Brett, while ultimately falling short, came close to hitting for a .400 average. The question naturally arises as to the probability of anyone hitting .400. The commonly held view nowadays is that night ball, transcontinental travel fatigue, the widespread use of top quality relief pitchers, big ballparks, large size fielders' gloves and other factors all act to a hitter's detriment and make a .400 average a near impossibility.

"But, surely, the above items will affect all batters, not only the potential .400 hitters; and, therefore, the net effect of all these factors will be reflected in the composite league batting average. If the league average is low, the chance of there being a .400 hitter is also low; a high league average means a higher chance of a .400 hitter.

"Consider the experimental data: Figure 1 shows, for each major league season from 1901 through 1980, the average for each league's batting champion plotted against the league batting average. Of particular interest is the dashed line which marks the rather well-defined upper boundary of the data points. This line represents the ultimate level of batting performance in 80 years of major league baseball. Note that this boundary crosses the .400 level of individual performance at a league average of .255, which can be considered the effective minimum league level from which a .400 hitter can, historically, emerge.

Figure VII, 1. *Batting Champion's Average as a Function of League Batting Average (1901-80)*

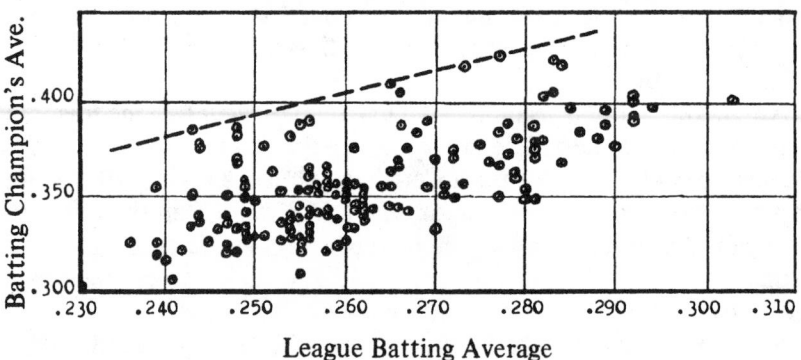

"For the present era, the minimum league level required is probably higher than .255. There is, for example, evidence that the gap in talent between the league's average and best players has been steadily lessening over time. Indeed, the points defining that upper boundary were all achieved in the deadball era.

"For any given league batting average, the experimental probability of an individual .400 hitter could, if there are sufficient data, be obtained directly off Figure 1 by counting. For example, at a league average of .265 there was one season with a .400 hitter and three seasons without. Unfortunately, the simple approach is inadequate because of sparseness of data: eleven .400 hitters spread over a range of .230 to .303 in league batting average. It is necessary, therefore, to group the data.

"For this study a moving average covering .009 points in league batting average was employed. This means that the experimental data for each specific league batting average was augmented by all the data within ±.004 points. Thus for a .265 league average, by way of example, the 29 data points in the range .261 through .269 are used, rather than only the four data points at exactly .265. Those ranges above .288 contained ten or fewer data points and were considered insufficiently populated to be included in the calculations. Despite the smoothing effect of the moving average technique, there remains some jumping about of the resultant experimentally determined probabilities but the general trend is apparent, as shown by the individual points on Figure 2.

Figure VII, 2. *Probability of a .400 Hitter as a Function of League Batting Average*

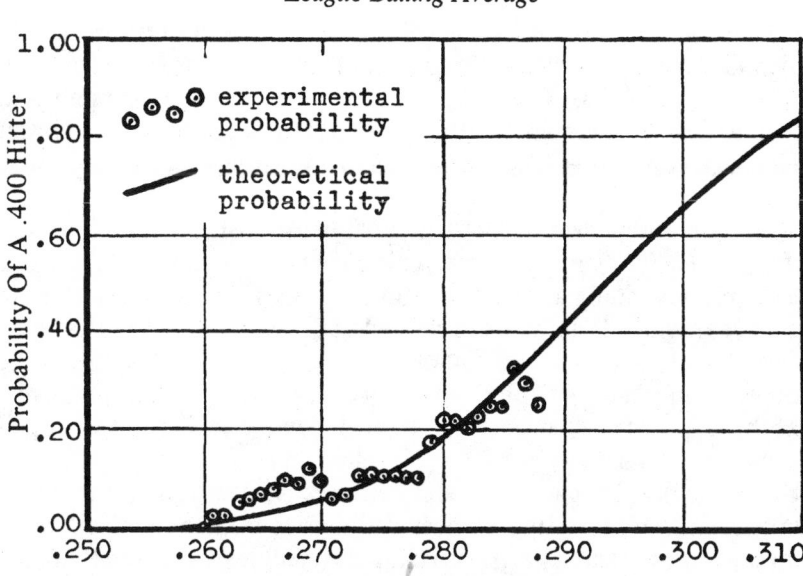

"From a more theoretical point of view: consider, for example, a league batting average of .265; .400 is 51% higher than .265. Thus the question: what is the probability of a player compiling a personal batting average which is at least 51% higher than his league's .265 average? At this juncture it is necessary to introduce the 'Relative Batting Average' concept of Shoebotham (1976 *Baseball Research Journal*, pages 37-42).

"In its simplest form, a relative batting average is a player's average divided by his league's average. If one calculates the relative batting average for all major league batting champions from 1901 through 1980, the results approximate a normal distribution (the familiar 'bell-shaped curve') with a mean (average) value of 1.361 and a standard deviation (a measure of the dispersion of data about the mean) of

0.075. Now, the useful thing about a normal distribution of known mean and standard deviation is that the probability of occurrence for any arbitrary value, above or below the mean, can be calculated. For a league average of .265, we want to calculate the probability of a player making an average 51% higher, a relative batting average of 1.51; the computations give a 2.4% probability for this.

"Similar computations have been made for a wide range of league batting averages, and the resulting theoretical probabilities are shown by the solid line on Figure 2. The theoretical and experimental results are in good agreement.

"Thus, we have two approaches for examining the historical odds which George Brett was challenging. The 1980 American League batting average was .269. From Figure 2, the experimental data points at and near .269 indicate that there's about a 10% chance for a .400 hitter under such conditions. The theoretical probability is even less optimistic; it shows a 4.7% chance.

"The long odds against Brett in 1980 help illustrate why there has not been a .400 average in the major leagues since 1941. The odds lengthen appreciably as league batting averages shrink below .269. In the 39 years since 1941, the American League batting average has only twice bettered .269 and the National League has never done it. If the theoretical probabilities of Figure 2 are used, the calculations reveal that there is a 51% chance of there NOT being a .400 hitter in any of the past 39 years."

As to winning 35 games or completing 35 starts, these accomplishments seem not to be duplicatable, no matter how the owners might monkey with the game short of ruling baseball a six-inning affair: No one has won 35 or more games since Walter Johnson in 1913, and only Bob Feller in 1946 has completed 35 or more since 1917, when Pete Alexander and Babe Ruth (!) did it.

There was a time, of course, when winning 35 was commonplace and completing less than 90 percent of one's starts unthinkable. Indeed, the early years of major league play provide records that, to one not familiar with the rules and conditions of play, are unfathomable: Will White completing all 75 of his starts in 1879 while pitching 680 innings; Jim Devlin of Louisville, Bobby Mathews of New York, and George Bradley of St. Louis each accounting for all his team's victories in 1876; and of course Hoss Radbourn winning 60 games in 1884. Were these men of iron, compared to the namby-pambies of today? Of course not. It was easier on the arm to pitch underhand, as they all were bound to do until 1883 and many, including Radbourn, continued to do at least some of the time into the 1890s. Another factor which promoted pitchers' staying power until 1920 was the staying power of the ball, which might last a whole game despite being scuffed or pounded out of shape.

Not having to fear the long ball and not having to worry much about control—walks were very infrequent, because it took nine balls to get one in 1876, diminishing by ones to four in 1887—pitchers could get by with one pitch or two. With few men reaching via the free pass, a pitcher could allow more base hits than a pitcher today can. The total number of baserunners allowed per nine innings has fluctuated within a very narrow range from 1876 to today, as has the number of earned runs allowed. (The total runs scored, however, are radically different, as discussed in Chapter 2.)

In the early 1870s, the years of the National Association, a pitcher could have an earned run average under 2.00 while allowing seven or eight runs a game. By 1876, the percentage of runs scored that was earned was about 40; by 1880, 50; and by the mid-1890s, 70. That figure increased steadily over the next forty years, hitting a plateau of 88 percent in 1946, where it has remained ever since. Meanwhile, over that same span of years, 1876-1946, the number of innings pitched by the top pitcher on each team was in steady decline. In the 1880s it was in the range of 500 innings pitched (up from the 400 of the 1870s strictly as a function of the expanded schedule); by 1893, when the pitching distance was increased to the present 60'6", the figure dipped to 400; and by 1905, below 300. It reached a level of 250 to 270 innings pitched in the 1930s and has stayed there ever since, with an upward blip in the first few years after the introduction of the DH.

Interestingly, the percentage of earned runs and the number of innings pitched are inversely tied to one another: The product of the two has been a virtual constant (225) since 1880. (See Figure VII, 3.) This suggests that as the fielding improved behind the pitcher, he bore more of a responsibility in the total defensive posture, that a higher proportion of opponents' runs was chalked up to his deficiencies. It seems likely that oldtimers pitched as many innings as they did because they relied on their fielders to a greater extent than is the case today and because the absence of the long-ball threat enabled them to pace themselves, or "coast," when their lead was not threatened.

A watershed year for baseball and baseball stats is 1893, when the pitching distance was moved back from 50 feet to 60'6", producing a lot of 55-foot curve balls and a quantum leap in offense (not only in '93 but also in the following two seasons). What few fans and writers realize is that the change in pitching distance was not 10'6", as indicated, but more like 5 feet. You see, prior to 1893 pitchers worked out of a box, the front line of which was, from 1881 to 1892, 50 feet from home plate; thus it was the pitcher's *front* foot that landed some 50 feet from home. When the distance was made 60'6" for the 1893 season, it was marked from the front of the rubber slab with which the pitcher's *back* foot had to make contact; taking into consideration a stride of about 5 feet, we see that the pitching distance was actually set back by the same amount as in 1881, when the shift was from 45 feet to 50.

There was a little-known foreshadowing of the 1893 move three years earlier. In 1890, the National League batting average was .254, and the American Association batting average was .253; yet the Players League, the product of the Brotherhood Rebellion, which attracted the best players from both the NL and the AA, had a league batting average of .274, and scored a run more per game. Why was that? Digging through some contemporary newspaper clippings at the Hall of Fame, we learned that the Players League teams, in an effort to rally fan support to their cause, decided that more hitting would help. They adopted a livelier ball, and they changed the pitching distance from 50 feet to 51½, only 4 feet short of today's distance.

After that burst of hitting in 1893-96, pitching resumed its historic tendency to dominate batting. Batting averages and runs scored totals declined through the late '90s all through the first decade of the twentieth century to the extent that in 1908 each league had a BA under .240 and an ERA under 2.40. A record number of shutouts was pitched in each league, and there were some remarkable performances, such as Walsh's 40 wins, Mathewson's 37, and a *staff* ERA of 2.02 by Cleveland. The frightened owners introduced the cork-centered ball in late 1909, and run scoring and home runs picked up dramatically in 1910-12: The NL, for example, hit 151 homers in 1909, then 314 in 1911—a 50 percent increase—surely attributable to the new ball.

Pitching seemed poised to take over again by 1920, when the transfer of Babe Ruth to New York and the introduction of a still more lively ball in the American League (not unrelated actions, one suspects) changed the game irrevocably—and, from Ring Lardner's point of view, irreparably. American League homers went from 240 to 369; the National League juiced its ball the following year, producing a rise in homers from 261 to 460.

Pitchers became so shell-shocked over the course of the next decade that, for the first time in baseball history, in several years of the 1930s the number of walks in the American League exceeded the number of strikeouts. In 1936, AL pitchers were roughed up for an average of 5.04 earned runs per game, the only time a league ERA has been over

Figure VII, 3. *Innings Pitched vs. Percentage of Earned Runs*

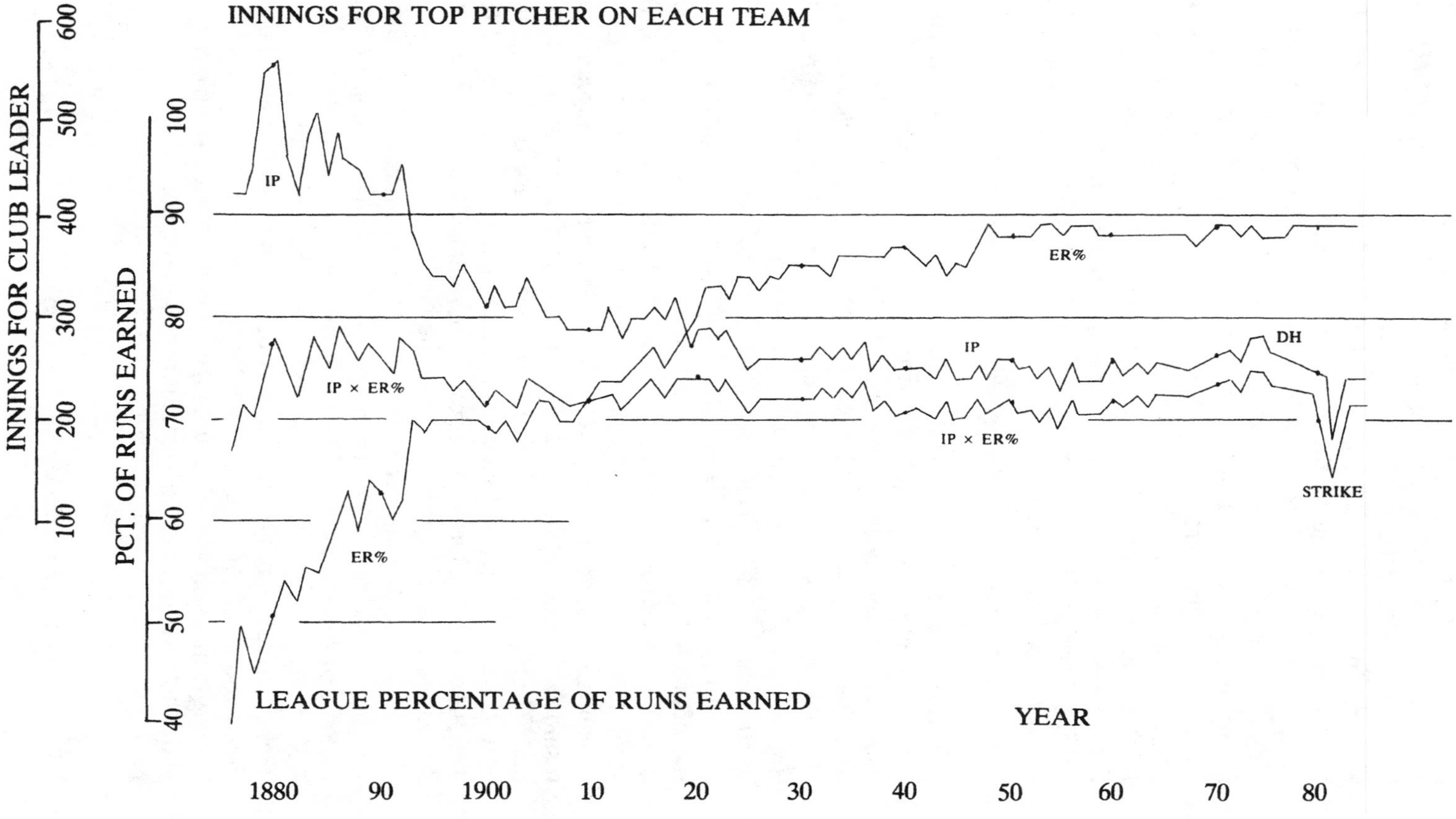

INNINGS FOR TOP PITCHER ON EACH TEAM

LEAGUE PERCENTAGE OF RUNS EARNED

YEAR

PCT. OF RUNS EARNED

INNINGS FOR CLUB LEADER

5.00 except for the 1894 season. In the period 1921-41, the ERA of the American League exceeded 4.00 in every year except one: 1923, and in that year it was 3.99.

During the War years, the quality of hitting declined. But so did that of pitching and fielding, so the numbers themselves look "normal." In the postwar revival of skills, as the returning veterans resumed their major league jobs, there was a notable upturn of hitting, particularly for power. And with it came an upturn of the 1930s trend of walking batters rather than grooving pitches. The level of bases on balls per league (prorated by number of teams) became the highest since the 1890s, which was the greatest previous period of adjustment. It was the culmination of the age of the home run. The National League in 1955, consisting of eight teams, hit 1263 home runs, 158 homers per team, the highest ever.

You would think that the next marked change would have come in the age of expansion, ushered into the AL in 1961, into the NL one year later. Yet, surprisingly, the dilution of talent had very little effect on league averages, though the dispersion of individual marks widened: The 1963 Mets, for example, had a batting average of .219, which was the lowest NL mark since 1908. In the mid-1960s pitching, after four decades of second-class status, reclaimed the game from the sluggers. By 1968 both league ERAs were under 3.00 for the first time in 50 years. The 1968 Yankees batted .214. Reflecting the axiom that you must have a strong franchise in New York and frightened by the rise of pro football as a rival to baseball as "the national pastime," the owners lowered the mound from 15″ to 10″ and tightened up the strike zone: Formerly the top of the shoulder to the bottom of the knee, it became the armpit to the top of the knee. When these changes seemed insufficient to revive hitting in the AL, which drooped badly in 1971 and '72 after a spurt in '69-70, management went for the Designated Hitter, an idea that had kicked around baseball since at least 1920. This innovation distorted American League stats to such an extent that AL run scoring is 10 percent above the National's.

In 1977 there was an explosion of home run hitting in the National League. Home runs went from 93 per team in 1976 to 136 in 1977, an increase of 46 percent; in 1977 as well, the ERA of the Atlanta staff hit 4.85, a postwar high. Then in the following year, 1978, homers slipped back down to 106 per team. The only thing that we can point to here is the introduction of a newly manufactured baseball. The balls were coming from Haiti and were no longer horsehide but instead cowhide, and with new production techniques the resilience of the balls may have been unintentionally altered.

We are in a hitters' period right now, and have been since 1973. In 1979 the American League batted .270 for the first time since 1950, and AL pitchers were blasted for the highest ERA since 1953. This was attributable to the hitter-friendly home parks of the two expansion teams, Toronto and Seattle, and to the new stadium in Minnesota, another hitters' haven. With a .270 league average behind us, can .400 be far away?

What are the tradeoffs? The basic seesaw action is between pitching and batting, with the balancing fulcrum ideal, which is to say mythical. It seems baseball officials are pretty happy with things the way they are now; that is, with the American League batting around .270 and the ERA crown being won each year by a pitcher whose mark is maybe 2.40 to 2.70 rather than 1.80. What makes them happier still is the high attendance figures of recent years. Nearly half the teams in 1983 drew 2 million and the "product" is deemed good enough by NBC for it to fork over a billion dollars for telecast rights.

Enhancing baseball's current attractiveness is its annihilation of the old power-speed dichotomy. It used to be that in a pitcher-dominated era, when runs came tough and all but the aficionados found the game boring, the steal would be an integral part of offensive strategy; and that when hitters dominated, and traditionalists found the pinball-game scoring tedious, the stealers would go into hibernation. Right now we are in a freak period featuring a combination of power and basepath daring never seen before. All the original sixteen major league franchises (with the exception of Oakland) established their historical highs in stolen bases between 1904 and 1917, with most of those highs clustered around 1910-13. This coincided with the dead-ball era when runs were scarce. When Ruth & Co. rewrote "The Book" in the 1920s, stolen bases vanished. Not right away—Max Carey nabbed 51 bases in 1922, but that was at the dawn of the new era and can be seen as a vestige of the previous one. Things got to the point in the 1940s and '50s where a man who had stolen fewer than 20 bases could lead his league.

Luis Aparicio returned the steal to the game, and he did so right in the middle of a hitting period. But Aparicio's particular situation was unusual: He played for a team that went against the trend deliberately, being *designed* for pitching and defense. The Chicago White Sox of 1959 were a poor hitting club with a BA below the league norm. They were matched in that characteristic by the Los Angeles Dodgers of that year. And how did the Dodgers join the White Sox in the World Series, the only one in which both teams had BAs worse than their leagues'? Through a September rush led by a rookie named Maury

Wills, who had been called up in midseason from Spokane.

Wills and Aparicio sparked weak-hitting teams through the 1960s, but Lou Brock did the same for a team that *could* hit, as in the last decade did Ron LeFlore, Joe Morgan, Tim Raines, Omar Moreno, and others. The combination of power and speed, or the attempt to steal many bases while playing for the big inning, is not a crowning achievement of baseball strategy, as will be shown in the next chapter. But it is new, and it is exciting, and we don't hear as much talk about that boring old has-been, baseball, as we once did.

Whether the game as it is played today is better than ever is mostly a matter of taste. What this chapter has been about is whether the *players* are better than ever, or worse, or the same. Using conventional statistics and simple comparisons we cannot tell, for the rules tinkerers have flattened out the differences that otherwise would have shown in the averages. But these gremlins in the baseball engine have done nothing to inhibit fielding, and it has enjoyed a steady ascent since 1876, as measured by the ratio of earned runs to total runs. (Total chances per game is weak for cross-era comparison.) Anyone who has been watching the game for 30 or 40 years and is of an unbiased cast of mind will tell you that the best fielders of all time, at almost any position you can think of, entered the game after World War II. Old-timers will tell you stories about Hal Chase or George Sisler, but were they better than Gil Hodges or Wes Parker or Keith Hernandez? Did Rabbit Maranville range farther and wider than Ozzie Smith? Did Tris Speaker cover more ground than Richie Ashburn?

The impulse to nostalgia is irresistible. Readers of this book who are now convinced that players today are the superiors of those who glisten in the mists will nonetheless find themselves, as they advance in years, tugged in the same way that everyone else has been. George Wright, who was the shortstop of the undefeated Cincinnati Red Stockings of 1869 and the greatest player of his day, left the ballpark in the mid-1890s and was heard to observe, "These modern players all wear gloves now. It wasn't like that in my day."

Ted Breitenstein, a very good pitcher of the 1890s who labored for some very poor clubs, told a reporter for *The Sporting News* in 1929: "I can name a dozen pitchers you couldn't touch today: Amos Rusie, Roaring Bill Kennedy, Frank Dwyer, Bill Rhines, and [Jouett] Meekin were some of them. Give them the same rest these boys get today, and they'd trim anyone."

In 1952 Ty Cobb wrote an article for *Life* magazine, referred to in Toporcer's letter, in which he declared that the only ballplayers of the modern era who could be compared with those of his day were Stan Musial and Phil Rizzuto. Lee Allen commented on the *Life* article:

"Cobb deplored the accent on slugging and the decline of bunting and the base stealing art, the absence of strategy, sign stealing, and inside baseball. He particularly singled out Ted Williams and Joe Di-Maggio as players who had not realized their potential ability. And he commented acidly on the brittleness of modern players. He was also disturbed about the lowly state of batting averages. The reaction was as violent as Cobb's original article. Players replied emotionally, many of them missing the point and citing gate receipts and attendance figures as proof of the superiority of the modern game. Thus was revived the old argument that can never be settled. Could Joe Louis beat Jack Dempsey? What about Ben Hogan and Bobby Jones? Was the fighting at Tarawa harder than at Verdun? There is not a barroom or living room in the land that has not provided a setting for this eternal battle of the generations."

On a personal level, I [Thorn] played high school basketball in the early 1960s. I was 5'10" and at that time weighed 145 pounds; if today I were in my prime, such as it was, I couldn't make a high school squad in the smallest hamlet in the most rural county of the most sparsely populated state. And others who played basketball or football decades ago would likewise acknowledge the superiority of players in those sports today. Yet, forty-year-old fans who once played some baseball themselves can look at the game on television or go out to the stadium and fancy themselves replacing a current major leaguer and doing a creditable job. That illusion is part of the lure of the game: It *looks* easy from the stands or from the other side of the television screen. That it was played better in the halcyon days of old, before The Calamity—free agency, expansion, WWII, the lively ball, the Flood (not Curt)—is the lore of the game, but it simply isn't so.

In *The Hot Stove League* (1955), one of the best baseball books ever and the source of the title for this chapter, Lee Allen wrote:

"Everything is bound to change, and the time is not far distant when some player will look back at the road he has traveled and say, 'Well, this is not the same. The game isn't as good as it used to be. This is not the way it was when I broke in in 1954. Those were the good old days."

The sentiment can be updated endlessly. If you'd like to substitute 1984 . . .

---

[1] For purposes of this discussion, league BA is expressed conventionally, including pitcher batting records.

8

THE BOOK... AND THE COMPUTER

Baseball, like poker, is a game in which the situations vary within a defined range and so may be modeled mathematically; within such a model, the probability of a particular tactic's being successful may be calculated. This is attested to by the time-honored tendency of managers to "play the percentages," not only because over the long haul percentage baseball is winning baseball but also because pilots, by calling upon higher authority in the form of "The Book," deflect much of the second-guessing that works against career longevity. But how can whole armies of managers claim to play the percentages if they don't know what they are? Maybe a man of genius like Casey Stengel can determine the right moves by the seat of his pants, but the ordinary guy will be looking primarily to cover his.

As Pete said to Joe Klein, who was preparing an article for *Sport* magazine, "You know, all these managers talk about playing by The Book, but they've never even *read* The Book. They don't know what's in it. They all use the same old strategies, many of which are ridiculous. Every mathematical analysis I've seen shows that the intentional walk is almost always a bad play, stolen bases are only marginally useful, and the sacrifice bunt is a relatively useless vestige of the dead-ball era when they didn't pinch-hit for pitchers."

Of course, when baseball people talk about The Book, they don't mean anything that's bound between hard covers (or, these days, soft). They're referring to the folk wisdom that has built up through trial and error, largely in the seventy-five years or so after Alexander Cartwright's Knickerbockers cavorted in Hoboken. Most of the significant elements of strategy go back before the turn of the century—the steal, the sacrifice, the hit and run, the defense giving up a run early to head off a possible big inning. The classic "percentage play" of pinch-hitting a lefthanded batter against a righthanded pitcher (or vice versa), or of platooning regulars depending upon the handedness of the pitcher—this goes back to John McGraw in the early 1900s.

Then, as now, percentage play consisted of nothing more than achieving the greatest possible gain in run scoring or run prevention while assuming the least possible risk. As the penalty for failure increases, so must the reward; otherwise the percentages are said to be working against you. Although the game has changed over the course of a century, most notably in the years following 1920, The Book has not, except in newly dictated situations such as those posed by artificial turf (whether to bunt, how deep to play the infield to cut off a runner at home, etc.).

The same maneuvers that Ned Hanlon, Connie Mack, and John McGraw used with so much success in the era of the dead ball have remained articles of faith for managers throughout the explosive hitting period between the wars and continue to be revered today. Like the Church, baseball is a conservative institution that does not reevaluate and revise its tenets lightly. In an everchanging society, it is a fundamentally unchanging force, which is a source of strength as well as weakness. Baseball, the game of the masses, is as slow to change its procedures as the Union Club or the Century Club. The old defense of outmoded practices, "We've always done it that way," bespeaks a conservative's preference to deal with the known rather than the unknown, until he is convinced of the former's utter, irredeemable inadequacy. The old code survives even when the circumstances which brought it into being vanished long ago.

Take the case of the sacrifice bunt. When this idea first came into the minds of baseball men in the mid-1880s, a time when league batting averages were in the .240s and slugging percentages in the .320s, it may have been a good idea. By 1908, when BAs had shrunk to the .230s in both leagues and SLGs to an all-time low of .304 in the American and .306 in the National, the sacrifice seemed even smarter. With every run dear and shutouts commonplace, playing for one run rather than for the big inning would appear to make sense. But what happened to this idea, born of a particular time and particular conditions, was that it became entrenched and grew, spreading itself into other times and other conditions which would not have been fertile for its invention. Managers in the 1930s or 1950s, hitting-dominated decades, sometimes instructed their fourth, fifth, or sixth batters to lay one down for the good of the team. A Gene Mauch in 1982, blessed with an Angelic batting order which included Reggie Jackson, Fred Lynn, Rod Carew, Don Baylor, and Bobby Grich (and cursed with a starting rotation which included nobody) would sac-bunt in the first inning. The last time that move really made sense was in 1968, and even then only if the team was the Cardinals and the man on the mound was Bob Gibson (who, with a 1.12 ERA and 13 shutouts, was a fair bet to make one run stand up).

Prove it, you say? OK. This is where the computer comes in to "rewrite" The Book, which was never written in the first place but transmitted across the generations like the lines of Homer (antiquity's augur of baseball?). The computer enables us to analyze masses of data, establish run values for situations and events, and evaluate the options available to a manager or player. This can be done with more tactics than we can possibly cover in one chapter, so we will examine only a handful of traditional plays (including the sacrifice bunt) which will serve to illustrate the technique, a technique that you, dear reader, may apply to other situations not detailed here.

As we earlier expressed the statistics of individuals to reflect their runs contributed or saved, so we can examine the elements of strategy to reflect their potential runs gained in the event of success, or lost in the event of failure. We will need to know: (a) the potential run-scoring situation that exists before a contemplated tactic is employed; (b) the run potential that would result if the move succeeded; and (c) the run potential remaining if it failed. Armed with this information, a manager (or fan) can weigh the possible gain against the possible loss. For the first time, he can determine objectively whether the tactic is indeed a percentage play or should be blue-penciled out of The Book.

Pete used his computer simulation of all baseball games played between 1900 and 1977 to calculate the expected run value of each possible strategy. He made two calculations: first, the run potential for the given situation regardless of score, and second, the probability of winning the game. These two calculations are different because a strategy may have far more consequence in the seventh, eighth, or ninth innings than it does early on; the sacrifice bunt or the intentional base on balls may not be distributed randomly over the course of a game.

Pete calculated the number of potential runs for each of the twenty-four base-out situations, for each of four periods of play: 1901–20,

1921–40, 1941–60, and 1961–77. The results for the last period (which would not change appreciably if updated to 1983) are shown in Table VIII, 1.

Table VIII, 1. *Potential Runs for Twenty-Four Base-Out Situations*

Runners	Number of Outs		
	0	1	2
None	.454	.249	.095
1st	.783	.478	.209
2nd	1.068	.699	.348
3rd	1.277	.897	.382
1st, 2nd	1.380	.888	.457
1st, 3rd	1.639	1.088	.494
2nd, 3rd	1.946	1.371	.661
Full	2.254	1.546	.798

At the beginning of a half inning, with nobody out and no runners on base, the run-scoring potential was .454 for the period 1961–77. In rough terms, over the last decade or so teams have tended to score about 4.5 runs, which breaks down to about half a run per inning. Why, then, is the figure in the table .454 and not .500? First, because a victorious home team does not bat in nine innings, but eight (except when the victory is gained in the ninth); second, because during most of the 1960s pitching dominated, so that the average team scored somewhat less than 4.5 runs. Had we provided a table for the period 1921–40, all the run values would have been higher.

If there is a man on third and one out, the team should score, on average, .897 runs. What does that mean? That 89.7 percent of the time, the man on third should score? No, not exactly: It means that the run-scoring potential is .897 as a function of there being a man on third *and* at least two additional batters in the half inning, barring a double play, pickoff, or failed steal attempt. Totaling the run potential of the man on third plus that of the two additional batters, who may get on base themselves, provides the .897. In the case of the first batter, let's say that no one was on base—then the run potential for the team would be .249 (see the table for the intersection of one out and no one on base). Thus we see that in the situation this batter confronts, .249 of the team's run value is attributable to the batter's possibility of reaching base, bringing up not only the next batter but perhaps several more, depending upon the outcomes. This means that of the run value inherent in the situation "man on third, one out" (namely, .897), .249 resides with the batter(s) and .648 with the baserunner. In other words, a runner on third with one out will score, on average, 64.8 percent of the time.

To quote from Pete's unpublished essay on baseball: "Once the distribution of potential runs for each of the game situations was found, the simulator calculated the probability of winning the game as a function of the half inning, score, outs, and baserunners. This was done by playing the game backwards. Given the last of the ninth with the score tied, the probability of the home team winning the game was equal to the probability of scoring at least one run from the given situation plus half the probability of not scoring any runs. [The latter factor is accounted for by the fact that the home team may score no runs in the bottom of the ninth yet still win the game in extra innings, when its chance of victory is the same as that of the visitor.] Similar calculations were made for being behind in the last of the ninth, and the result provided probabilities of winning at the start of the final half inning for each score possibility. These are the same as the win probabilities at the end of the top of the ninth. This procedure was continued until the top of the first inning was reached. Both home and visiting teams were given the same run distribution, so the probability of winning at the end of any inning when the score was tied was .500. [Table VIII, 2 shows win probabilities for the home team in the bottom of the seventh inning when trailing by one run.]"

Table VIII, 2. *Win Probabilities, Bottom of Seventh Inning, Down by a Run*

	Number of Outs			
	0	1	2	3
Runners				
None	.343	.298	.262	.239
1st	.413	.348	.289	"
2nd	.482	.403	.324	"
3rd	.537	.457	.334	"
1st, 2nd	.529	.432	.343	"
1st, 3rd	.594	.483	.353	"
2nd, 3rd	.654	.546	.393	"
Full	.683	.557	.411	"

Unlike the previous table, this one has a column for the three-out situation; Table VIII, 1 gave run values, of which with three outs there obviously is none, but Table VIII, 2 shows win probabilities in the bottom of the seventh, and the home team, down by a run, still has a 23.9 percent chance to win the game even if it is retired without a score. With these two tables, we can begin to pose some strategic alternatives and evaluate them. The leadoff batter in the seventh draws a walk; should you sacrifice? Say you do, and the batter pops the ball up and is retired. Now, do you give the runner a steal sign? Or should you hold him to his base because your best power hitter is up? Say you do, and the home-run threat strikes out; however, the next man singles, placing runners at first and third, with two out. Do you now send the runner on first, gambling that he will steal the base and set up a situation in which a single will drive in the lead run as well?

All these questions in turn beg a larger question: What is the break-even point? Where do risk and reward intersect, and what is the "percentage play"? To find the break-even point, we must identify the point at which the run value that exists *before* that strategy is employed equals the run value *after* the strategy has been employed. This may be expressed as an equation, for those mathematicians in the crowd:

$$P_b \times V_s + (1 - P_b) \times V_f = V_p$$

It's not as confusing as it looks. P_b is the probability of attaining the break-even point with a given strategy. V_s is the value of a success, while V_f is the value of a failure. V_p stands for the present value—i.e., before the strategy has been set in motion. Rearranging terms so as to set the break-even point off to one side, since this is what we are trying to find, we get:

$$P_b = (V_p - V_f) / (V_s - V_f)$$

And now let's put it to use for the multipart scenario above.

To begin with, our boys were batting in the bottom of the seventh, down by a run. With no one on and no one out, the win probability was .343, as indicated in the table.

The first batter walks, boosting the team's win probability to .413. Should we attempt a sacrifice? If we do and fail, we will have a man on first and one out—in rare cases, worse: no one on and two out—giving a win probability of .348. If we succeed, we have a man on second and one out, a win probability of .403—whoa! That's worse than what we started with! Whether the attempt to move the runner to second fails or succeeds, *it fails*. No need to calculate the break-even point here . . . there is none. Is the sacrifice bunt always a no-win proposition? No. More on this in a bit.

Moving on through our seventh-inning model, let's presume a failed attempt leaves us with a man on first and one out—win probability .348. Do we steal? A successful steal would increase our chances of winning to .403, while a failed attempt would drop us back to .262—

the risk outweighs the reward. But maybe the success is easy to attain: Using the break-even point formula, we get

$$P_b = (.348 - .262) / (.403 - .262) = .610$$

If you, the manager, have on first base a man who has been successful in 75 percent of his attempts this year and last, then send him. But a man who is by major league standards a below-average base stealer (that is, he steals about one base for every two attempts rather than the average of two for every three) is, because of the inning and the score, an overly risky proposition—unless a manager takes into account some other aspect of the specific situation (a pitcher with a slow release to the plate, for example) which balances the scales. Using Table VIII, 1, which listed run values for *random* base/out/inning situations, the break-even point would be higher: $P_b = (.478 - .095) / (.699 - .095) = .634$.

Still with us? Let's say we didn't send the runner, and the next batter fanned. Two outs, man on first—but then the next batter singles, sending the baserunner to third. Does the man on first steal second to try to bring in two men on a single? (Remember you're one run down.) Before we run through the numbers, what is your *impression* of the play? You might be inclined to like it because if the catcher throws through, the man on third (if he's fast) can break for the plate. It's a pressure play, which typically produces more mistakes than it does textbook executions. Now, the break-even point, presuming that the runner on third holds his base, is as follows:

$$P_b = (.353 - .239) / (.393 - .239) = .740$$

Wow! That play has to succeed 74 percent of the time just to break even. Not much of a play, not in the seventh inning, anyway, trailing by a run. In the bottom of the ninth, trailing by *two* runs, the break-even point would be lowered to 58 percent.

This little exercise gets easier the more you do it. Let's move on.

THE SACRIFICE

The sacrifice bunt, the discussion of which we so rudely interrupted, is a bad play, as several modern-day managers—but not enough of them—have concluded. Earl Weaver used it very sparingly with the Orioles, Dick Howser likewise with the Royals. Twenty years ago Paul Richards said, "The defense plays hitters better now. It has gotten to the point where they play you so well on the bunt that you can hardly afford to sacrifice any more. I just can't justify giving away a third of an inning trying to sacrifice."

Richards's observation was probably true forty years earlier as well. The potential run value is *always* lower after a successful sacrifice. With the introduction of the lively ball, the sacrifice bunt should have vanished, except perhaps for situations in which the pitcher is allowed to come to the plate in the late innings with a man on first. The sac bunt by any other man in the order should have become as infrequent a mode of strategy as the squeeze play. The squeeze has always been employed sparingly because of the high risk associated with it; as a result, on the whole it has been used wisely. In all likelihood, the sacrifice has been used promiscuously because the risk attached to it has not been as obvious (the statement holds true for the steal attempt, too). Managers have thought: Well, if it doesn't succeed, we'll lose the man on first through a force play, or the batter will be in the hole two strikes and then, four times in five, will be retired. But you can't give up an out. On average a runner on first with nobody out creates for his team a run-scoring potential of .783. A runner on second with one out—the situation that obtains after a successful sacrifice—is worth only .699 runs. The "successful" bunt reduces the potential offense for your team in that half inning by some 10 percent.

It may be argued that while a sacrifice decreases the possibilities for a two-run or three-run inning, it increases the chance of scoring one run, via the man already on base. But this is not so except in a very few cases. In the last of the ninth, with the score tied, no outs, and a runner on first, the run potential is .698; with only one run needed to win, a successful sacrifice here would increase the run potential to .715. A failed bunt, however, would decrease the run potential to .627—and the break-even point is a whopping 80.7 percent, making this a highly dubious stratagem. (Pete's analysis of a limited number of World Series and playoff games showed that of twenty sacrifice attempts, only ten succeeded.) The break-even point as calculated for a nine-man model, by the way, is lowered to 67 percent in the above situation if the pitcher is up.

Another of the few instances in which a successful bunt improves a team's run potential is a sacrifice with men on first and second, no one out, and the home team behind by one run in the seventh inning. Checking the table of win probabilities for the bottom of the seventh, we see that the successful bunt raises the win probability slightly, from .529 to .546; however, a failure, leaving men at first and second with one out, lowers the chance of winning substantially, to .432. Using the break-even point formula, we discover that such a play must succeed 85.1 percent of the time—an improbable rate of accomplishment for this difficult play.

THE STEAL

The stolen base, as indicated in the chapter on the Linear Weights System, is an overrated play, with even the best base stealers contributing few extra runs or wins to their teams. The reason for this is that the break-even point is so high, roughly two steals in three attempts. The precise figure can be obtained from Table VIII, 1 and the break-even point equation. A runner on first with no outs is worth .478 runs: A steal of second increases this to .699; a failure leaves no one on base and two out, worth .095.

$$P_b = (.478 - .095) / (.699 - .095) = .634$$

This figure will vary depending upon the situation but is valid for a discussion of the value of the stolen base irrespective of situation. What would the break-even point be in the last of the seventh, with the home team down by one? Why don't you figure out this one?[1]

As you would expect, the break-even point declines as time grows short for the team trailing by one or tied. In the last of the ninth, if the score is tied, two men are out, and you've got a runner on first faster than Cliff Johnson, then send him. Now, the astute fan probably doesn't need numerical support to justify that decision, but in this situation the break-even point is under 50 percent because the game will go on whether the attempt succeeds or fails.

What about stealing other bases? Television announcers will tell you that if you want to steal third, you'd better be sure you're going to make it. What's implicit in that remark is that your team will suffer far more for your being thrown out than it will benefit from your gaining third, because it stands a pretty good chance of scoring already, simply by there being a man at second base. With one out or none out, sure, it would be nice to get to third and perhaps score on an out. But stealing third requires a success rate of 80–90 percent to make it worthwhile.

Have you noticed that nobody steals home anymore? A few active players have done it, notably Rod Carew, but of his 17 lifetime steals of home, 7 came in 1969. If you see a steal of home at all today, it is likely to be off a delayed break from third after the runner on first has taken off for second. Managers and players avoid the play because they presume that, as with the steal of third, the break-even point is too high to make it worthwhile. They're wrong. Stealing home with two out is a good play, a far better percentage play than stealing third: Because of the enormous potential gain as compared to the risk, you only need a 35 percent probability of success in order to break even— (from Table VIII, 1: $P_b = (.382 - .000) / (1.095 - .000) = .349$. The break-even point dips below 30 percent if it's the last of the ninth, two out, and the score is tied. The two-out steal of home is the unknown great percentage play.

This 30–35 percent break-even range for stealing home with two outs similarly applies when a runner is on third and a fly ball produces the second out. If the third-base coach feels there's a one-in-three chance of the runner arriving at the plate successfully, he should send him. The same holds true for a man who's on second with two outs when the batter drives a single to the outfield. If the coach believes there is a one-in-three chance for the runner being safe at home, he should go for it.

THE INTENTIONAL BASE ON BALLS

Let's look at the free pass, a move widely condemned in recent years. Just as the DH accounts for the American League's lesser dependence on the sacrifice than the National's, it also accounts for the AL's lesser reliance on the IBB (37 per AL team in 1982, compared to 67 in the NL). And just as Pete's computer simulation showed that the sacrifice bunt never lifts the team's expected run value, so does it show that the intentional base on balls never reduces the expected number of runs scored. However, there are cases in which an IBB will lower the batting team's chance of winning the game. For example, with the score tied in the bottom of the ninth, one out, and a man on third, giving the batter a free pass reduces the batting team's win probability from .825 to .806. Similar but smaller gains were shown in the top of the ninth and the bottom of the eighth.

Because the pitcher is permitted to bat in the National League, an intentional base on balls is frequently issued when there are two out, one or two men on, a base open, and the eighth-place batter is at the plate. This is the classic use of the IBB—not to set up a force play but to work to a batter of lesser ability. This move reduces slightly the probability of a run scoring in that half inning—but the reduction is more than offset by the enhanced probability of the team scoring in its next turn at bat. This is because the next inning, instead of beginning with the pitcher batting and, eight times in ten, being retired, opens with the number-one hitter, who is likely to be retired not even seven times in ten. Over the course of many games—a season or seasons—extensive use of this strategy is the equivalent of playing with fire. You may escape unscathed this time or the next, but the total number of extra runs that will result from the IBB is, though invisible, nonetheless sizeable.

Treating this subject specifically, with the number-eight hitter coming to bat, a runner on second, and two outs, the team's run potential is .27. With the pitcher up, runners on first and second, and two outs—the situation that would obtain if you issued a free pass—the run potential is .26 runs. If the pitcher is retired, the defense may have saved a run or two, but the offense starts the next inning in a more beneficial situation. With its leadoff batter coming to the plate, the potential as that half inning opens is .55 runs, whereas if the pitcher had led off the inning, the run potential would have been .43.

Ready for a little quiz? With a man on second and no one out, a ground ball is hit to the shortstop. Should the runner attempt to go to third? The Book would lead you to believe this is a bad play. But now you've got the tools to determine the answer yourself.[2]

THE BATTING ORDER

Another chapter of The Book which has come under investigation in the age of the computer is the batting order. (A seminal article on the subject is R. Allan Freeze, "An Analysis of Baseball Batting Order by Monte Carlo Simulation," *Operations Research,* 1974.) The desirable traits for a leadoff batter have traditionally been: an ability to get on base; enough speed, should he reach base, to avoid the double play or to go to third base on a single to center or right; and not so much power that he wouldn't be more useful in the middle of the order. This last characteristic has occasionally been ignored: Such men as Willie Mays, Bobby Bonds, and Brian Downing have sometimes batted leadoff. A number-two man generally is: a good contact hitter, since a strikeout cannot advance a runner to scoring position for the big boys in the middle; willing and able to take a pitch to benefit a base-stealing threat; and able to manipulate the bat, to hit behind the runner or hit-and-run. Power is not essential here, according to The Book, yet in 1982 Robin Yount hit 29 homers while batting second.

Your number-three hitter is generally the best overall hitter on your team—the one with the highest OPS. The number four, or cleanup, batter is thought to be a clutch hitter with power, the team's top RBI man; his slugging percentage is lower than that of his immediate predecessor, generally as a result of his lower batting average. The fifth spot may be occupied by a power hitter with a lower BA and OBA than those of the two men ahead of him. Number six may be the same kind of hitter as number five, only not quite as proficient; or the fifth and sixth spots, if they are occupied by a lefty and a righty, may be reversed depending on the handedness of the opposing starting pitcher.

Your seventh and eighth positions, in the National League, are usually filled by men who are starters by virtue of their defensive ability—frequently the shortstop and catcher—and who will be your two weakest hitters. In the American League, where the DH is the "ninth" man who never bats ninth, the fashion has become for the seventh or eighth batter to be the team's weakest hitter, with the ninth batter sharing the characteristics of the leadoff batter. The seventh batter in each league generally shares the characteristics of the second (.260 BA, 5–10 HR, 50–60 RBI), but produces figures some 10–20 percent worse than the second batter. On an especially good-hitting club, the seventh spot might be filled by a batter who hits a lot of long balls but has a very low BA and doesn't draw many walks.

Table VIII, 3 depicts the traditional characteristics of the men who occupy the various spots in the batting order. Pete gathered data for all regulars in both major leagues in the period 1969–71, before the advent of the DH. (In those years, each league had twelve teams, so you might expect the three years to yield 72 regular player-seasons at each position; however, this is not the case. Variations from this figure are attributable to platooning, with more players being shuffled in and out of the lineup in the sixth through eighth slots than in the first five, as you also might expect.) The ninth spot in the order is broken down into two categories: 9P (pitchers) and 9PH (pinch-hitters). Pitchers comprise three quarters of ninth-spot batters, and pinch-hitters are one quarter.

Is the traditionally structured batting order the best one? Earnshaw Cook, in his *Percentage Baseball* of 1964, said no. He realized correctly that over the course of a season the leadoff batter had more at bats than the other eight spots; the second batter had more than the seven men below him; and so on. So, he reasoned, why not give the team's best hitters the maximum number of at bats so that they might achieve more hits and thus produce more runs? He proposed, in short, a batting order organized by batting strength which looked like this: 345612789, with the numbers corresponding to the traditional batting order prescribed by The Book. Alas, Cook's lineup was an example of a good idea gone wrong. With the aid of the computer simulation, Pete evaluated the run potential of Cook's lineup as well as that of the

Table VIII, 3. *Performance by Position in Batting Order*

	Batting-Order Position									
	1	2	3	4	5	6	7	8	9P	9PH
Min. plate app.	300	300	300	300	300	285	270	255		
Players	77	77	75	77	78	70	70	62		
Games	141	130	142	138	139	130	123	126		
At bats	539	468	516	490	477	427	400	393		
Runs	76	63	81	69	62	51	41	37		
Hits	146	125	148	137	128	110	100	92		
Doubles	21	19	24	22	20	18	15	13		
Triples	4	4	4	3	3	3	2	2		
Home runs	8	8	22	22	17	12	8	4		
Runs batted in	46	43	80	83	70	53	41	35		
Walks	51	43	67	60	51	45	37	36		
Hit by pitch	3	3	3	4	4	3	2	2		
Strikeouts	61	55	77	82	71	67	58	56		
Batting average	.270	.267	.287	.280	.268	.256	.250	.235	.155	.219
Slugging average	.374	.373	.476	.470	.430	.395	.359	.313	.194	.314
On Base Average	.337	.333	.373	.363	.344	.331	.318	.302	.206	.285

conventional order and some other possibilities. A key finding was that, as one might have expected, the run value of particular events (single, homer, walk, etc.) vary somewhat from position to position within the batting order: a home run hit by the leadoff batter in 1969–71 was worth not 1.40 runs, which is the average value applied to homers hit by all players in all periods, but 1.28, because the leadoff batter's homers tended to come with fewer men on base. The value of a homer hit by those occupying the fourth through eighth spots in 1969–71 was, by comparison, 1.46 runs.

So, while the third through sixth men in Cook's order *did* gain about 36 at bats per season, the increased run scoring they provided was more than offset by the decreased value of their extra-base hits. The traditional batting order in the period 1969–71 generated 4.141 runs per game for each team; Cook's order produced 4.130. Cook's other idea for revising the batting order was never to allow the pitcher to bat, but instead to pinch-hit for him and bring in a reliever, who would pitch until *his* turn at bat; this idea seems to have been adopted, after a fashion, by the American League, which increased its run production by 10 percent. Now imagine if *one* NL team tried it . . .

Pete noted a slight improvement in run scoring with a batting order of 134562789, to 4.154. Yet this gain is not terribly significant—less than two additional runs over the course of a season. In fact running through the 1969–71 model with a *backwards* order—987654321—was scarcely catastrophic, producing 4.026 runs. A reverse-strength order of 987216543—the worst possible—produced 4.003. Thus the difference between the best possible order and the worst is less than 25 runs, or 2 wins, per season. Conclusion? Since no manager will employ the worst batting order unless he has a death wish, the differences between all other possible configurations are not worth thinking about. All the time managers put into masterminding a winning lineup is so much thumb twiddling, and they are hereby granted an additional hour's sleep a night.

THE COUNT

The next element of strategy to be discussed is the impact of the count on the performance of both batter and pitcher. Managers have always known the importance for the pitcher of staying ahead in the count and the edge for the batter should the pitcher fall behind 2–0 or 3–1. It has been written that such counts make a .400 hitter of a .300 hitter. It has been said as well that if a pitcher could make his first toss to each batter a strike, he'd be awfully tough to beat. Until now all this has been guesswork; but through a pitch-by-pitch examination of over 100 World Series and League Championship games from 1974 to 1982, Pete has found the mathematical value of each ball-and-strike combination and has identified the implications for players, managers, and fans.

In Table VIII, 4 the vertical columns represent the strikes and the horizontal columns the balls. The values shown at each intersection of balls and strikes represent the average results of all batters who passed through each count. In other words, if a batter took a strike and a ball and then singled, he would have passed through a count of no strikes and no balls, a count of no balls and one strike, and a count of one ball and one strike. So he will be registered under each of these categories. (The run value here, by the way, is obtained through the Linear Weights method, expressed in runs beyond the average, as it was for earlier tables; On Base Plus Slugging is not quite as accurate in situational studies.) Not taken into consideration in this table are intentional walks, sacrifice hits, and sacrifice flies—all of which have proven to be statistically neutral over many seasons of simulated play. You can see that if the first pitch thrown is a ball, the batter produced runs at a rate 35 percent above average, nearly double that of batters who started 0-1. What is not evident from this table but is also true is that the eighth-place batter with a 1-0 count outperforms the cleanup batter at 0-0. Moreover, of all bases on balls allowed, 78 percent

Table VIII, 4. *Performance by Ball-Strike Count*

Strikes		Balls 0	1	2	3
	Samples	2,545	3,091	1,070	336
	BAT	.259	.267	.260	.250
0	SLG	.388	.415	.416	.321
	OBA	.317	.371	.477	.750
	RUNS	.002	.035	.085	.205
	Samples	3,196	2,689	1,450	583
	BAT	.240	.243	.265	.285
1	SLG	.347	.365	.418	.457
	OBA	.273	.306	.389	.600
	RUNS	− .031	− .010	− .044	.150
	Samples	1,208	1,733	1,322	750
	BAT	.198	.195	.208	.199
2	SLG	.279	.283	.310	.309
	OBA	.221	.239	.309	.479
	RUNS	− .076	− .065	− .022	.065

occurred when the first pitch was a ball. The message for pitchers, accordingly, is the traditional one: Get the first pitch over.

Table VIII, 4 also shows that batting average and On Base Average climb sizeably as a function of the count. Interesting, however, is the relative constancy of the slugging average. It seems that the extra-base hit is less a function of the count than we had imagined.

Clearly, the 3-1 and 2-0 counts are the great hitting situations, as expected. Oddly, the 3-0 situation, which carries a .750 likelihood of reaching base, bears a lower batting average and slugging percentage than the 3-1 count, and its SLG is only barely superior to that of the 2-0 count. The Book has dictated taking the 3-0 pitch in an attempt to draw the walk, but in the last fifteen years or so managers have allowed their power hitters to swing away, hoping to take advantage of a cripple pitch. The study of these postseason contests suggests that the batter in the 3-0 situation may be overanxious. In the 336 observed cases of this count, the batter swung at the next pitch 66 times, missing 34, hitting 16 fouls, making 11 outs, and gaining 5 singles. Much better results were obtained at 3-1 by the 141 batters who took the 3-0 pitch for a strike. The men who did *not* go on to reach base by a walk (129 of the 141) had a BA of .385 and an SLG of .654.[3]

The count study may also be used to assess the run value of a strategy like the pitchout (not a very good play) or to fine-tune the value of a sacrifice attempt (each missed or fouled bunt lowers the run potential by .04, beyond the run loss of the bunt that is put into play).

LEFT VS. RIGHT

In 1959 George Lindsey published an article in *Operations Research,* entitled "Statistical Data Useful for the Operation of a Baseball Team." Gathering play-by-play data from games of the National, American, and (Triple-A) International leagues, Lindsey set out to determine whether, and to what degree, lefthanded pitchers had an edge against lefthanded batters; how much these batters feasted on righthanded pitchers; and so forth. The seasons covered were 1951 and 1952, and the results were pooled. Lindsey recorded 5,197 at bats in which a righthanded batter faced a righthanded pitcher, which produced 1,201 hits, and a batting average of .231. For lefthanded pitchers against lefthanded batters there were 1,164 at bats, 270 base hits, and a .232 BA. When the pitcher and batter were of opposite orientations, the lefthanded batters batted .264 in 4,002 at bats and the righthanded batters .263 in 2,245 at bats. The total number of at bats, same-side and opposite, are quite close—and opposite-side situations produced batting averages 15 percent higher.

Pete did a study of 277 regular performers in the American League from 1974 through 1977 and used only players with at least 100 at bats each season against both lefthanded and righthanded hurlers. His results are in line with Lindsey's but may be regarded as more accurate because of the larger sampling. As illustrated in Table VIII, 5, a righthanded batter's BA was 7.5 percent higher against a lefthanded pitcher, while a lefthanded batter's BA was 11.5 percent higher against a righthanded pitcher.

Table VIII, 5. *Performance by Left-Right Combinations*

| | Pitcher | | | | | | | | | | |
| | Right | | | | Left | | | | Total | | |
Batter	PA*	BAT	SLG	OBA	PA	BAT	SLG	OBA	PA	BAT	SLG	OBA
Right	59	.255	.375	318	32	.274	.409	.342	91	.261	.387	.326
Left	29	.291	.450	.361	14	.261	.375	.322	43	.281	.425	.349
Switch	6	.264	.368	.354	3	.266	.370	.335	9	.265	.368	.347
Total	94	.266	.397	.334	49	.270	.397	.336	143	.268	.397	.334

*PA = Plate Appearances, in thousands.

Pete observed the substantial edge enjoyed by lefthanded batters, not only against opposite-handed pitchers but also overall. This he attributed to the facts that (a) lefthanded batters face favorable—i.e., righthanded—pitching situations two thirds of the time, while righthanders are favorably confronted only one third of the time; and (b) 70 percent of lefthanded batters come from the "hitting positions"—outfield and first base.

THE COMPUTER

Lindsey concluded his discussion of the left-right situation by writing: "It is probable that the sensitivity to same- or opposite-handedness varies from batter to batter. It is also probable that an individual may have widely different success in different [playing] fields, possibly on account of the location of the fences, by day and by night, according to whether a hit is vitally needed when he is at bat, and certainly against different pitchers and in different leagues. In order to remove all of these factors, it would be necessary to amass a large set of special averages for different situations (e.g., N in Detroit against pitcher A at night with none on base)."

It is the need for precisely this kind of data which is behind the contemporary applications of the computer by major league baseball. (The Chicago Cubs fooled around with a mainframe monster in the mid-1960s, but these were the same Cubs whose other innovations included the Manager-of-the-Month Club.) Thanks to the computer—now being used by Oakland, the Chicago White Sox, Atlanta, the New York Yankees, and Texas—it is at last worth the effort to record such data. Input is still a big job, but the manipulation of the data afterward for purposes of analysis is now a matter of proper programming, requiring far fewer man-hours than manual organization and calculation.

The computer enables play-by-play—indeed, pitch-by-pitch—recording and analysis, which would, among other things, permit us to improve the Linear Weights System. For now, we've had to estimate how many times a single drives a man on first over to third base, a factor that marginally affects the potential run value of a single; knowing the precise incidence in a large sampling may make it somewhat higher than its current .46 runs. By storing play-by-play data for future analysis, the computer permits managers to keep track of how individual batters fare against individual pitchers and vice versa, or how players perform under a myriad of varying conditions—day vs. night, artificial turf vs. grass, curveball vs. fastball, etc.

The Oakland program, which was provided by Edge 1.000, Dick Cramer and Matt Levine's outfit, has a provision for charting where balls are hit. No more "F9" for "fly ball to right"—now one can keyboard symbols for "bloop single over first base off righthanded pitcher, traveling 135 feet." This is data that a computer can easily store and later retrieve.

At this point, the use of the computer by managers to make spot decisions during ballgames is not imminent. Lindsey wrote in 1963: "The concept of the electronic computer in the dugout is a distasteful one, but, if progress demands it, this [the advisability of various strategies] is the type of calculation for which it could be programmed." Maybe so, but it is more likely that the computer will be used before and after games—for information storage and retrieval, for situational stats, for strategy analysis, for evaluating prospective roster changes—than in the dugout during games. A manager who has spent his entire adult life in baseball shouldn't need to consult the computer if he's trailing by one in the late innings with a man on first, one out, and a contact batter at the plate—that is, provided he has learned the *real* percentages beforehand.

With proper input, a computer can tell you how frequently steal attempts succeed against a particular catcher or, more interestingly, pitcher. Or it may reveal how a batter has fared against a particular pitcher: Boog Powell 1-for-61 against Mickey Lolich, Lenn Sakata 0-for-ever against the White Sox, etc. But the better managers have been keeping track of this stuff for years. Would Casey Stengel have skipped Whitey Ford's turn in Fenway Park any sooner if he had had a computer? That was intended as a rhetorical question, but maybe . . .

If the general manager of the Braves wanted to judge what would have happened to his team in 1983 had he picked up Rick Honeycutt or Sixto Lezcano rather than letting them go to other clubs, a computer model could tell him. A computer can run through an entire season with one new variable in a few minutes, so a front office can construct a scenario for a prospective trade that will predict the difference the new player(s) will make in the team's won-lost record. And that will tell the G.M., if he wants to dance, how much to pay the piper.

The computer can help a team adjust to the features of a road ballpark. It can help a manager decide whether to unleash his jackrabbits when playing in Toronto or Seattle, or restrain them in New York or Detroit. (The chances of stealing a base on artificial turf are about 12 percent better than on grass, but Toronto and Seattle favor hitters as well—a manager might run himself out of a big inning there as easily as in Detroit.)

We have seen how computer modeling of thousands of baseball games provides a true picture of the mathematical aspect of the game. Those managers who persist in the old seat-of-the-pants wisdom known as The Book are not playing the percentages, they are playing with dynamite. Unbeknownst to themselves, they are not percentage players but hunch players—no matter that their hunches are backed by tradition. In the long run, hunch players fail—that's why the folks in Las Vegas and Atlantic City entice them. They may beat the odds on a given night or stay "hot" for five or ten sessions; but if they're going to flout the laws of probability 162 times in half a year . . . well, it's not tough to get hot, but it is tough to stay hot.

Chuck Tanner rode the seat of his pants to a world championship in 1979 and retains much of his post–World Series aura. "Don't ever let stats get in your way of judgment," he proclaimed. "Figures don't always tell the truth. I'd rather use good judgment and common sense than cold stats." Who's saying stats are antithetical to good judgment or common sense? If we had a nickel for every man who aligned himself behind common sense while stating something absurd . . . "Sandy Koufax might beat a team twenty straight times," Tanner continued, "but there's no way to guarantee he'll beat you the twenty-first if you don't know how he woke up that morning." No, but even without that morning line, you'd sure know how to bet.

We believe we have shown that by analyzing events of all kinds in terms of their run-scoring potential and win probability, aided by the computer, the percentages need not be a matter for debate. The Book is no longer a figure of speech or a figment of the imagination, but a real book. It may even be the one in your hands.

---

[1] It's $(.348 - .262) / (.403 - .262) = .610$, or 61 percent.

[2] So you were lazy, eh? All right. The shortstop will not concede third base to the runner in this situation, so the question is how much does the runner help his team if he is safe, and how much does he hurt it if he is out? If he makes third base, you have men on first and third (1.639); if he is thrown out, you have a man on first and one out (.478); if he doesn't go, the batter will be retired and you will be left with a man on second and one out (.699). Thus if the runner on second makes third, the gain vs. holding second is .940 (1.639 − .699); if he fails, the loss is .221 (.699 − .478). All this makes the attempt look worthwhile—but what if it must succeed 80–90 percent of the time to make the risk equal to the reward? To answer this, we need to know the break-even point: $P_b = (.699 - .478) / (1.639 - .478) = .190$. The attempt to reach third looks like a terrific move. The break-even point of 19 percent is eminently reasonable to achieve so great a gain in run potential. Indeed, in 1982 the Oakland A's, according to Dick Cramer, were confronted 39 times with the situation of a man on second and fewer than two outs and a ground ball hit to shortstop or third; 16 times the A's tried to advance, and 16 times they succeeded!

[3] This includes those who swung and missed at 3-0 and those who took a ball at 2-1.

RISING TO THE OCCASION

An objection may be raised to the new statistics that while they are more accurate and logical than traditional stats, they still don't reveal what fans really want to know: the value of an accomplishment in a given time and place, its "clutchness." Just as not all bases are created equal—a walk is not as good as a hit, nor is a home run worth as much as four singles—so is it true that not all similar situations are of equal importance. A double with men on base, for example, is more valuable than a double with the bases empty. It is also worth more with two outs and runners on than with no one on and no one out—and perhaps more still, depending upon the score, the inning, the team's position in the standings, the point in the season, etc.

Most fans believe that *when* a player contributes something to his team is more important than, or at least equally important as, *what* he contributes. They defend their favorite ballplayer, in spite of ample statistical evidence that the man is a dud, with claims that he doesn't really bear down until the pressure is on—at which point he becomes a world beater. Thus one can argue that the player's puny OPS of .550 is not the true measure of the man: that he can only be appreciated by those who see him day in, day out, and who know his basehits to be particularly meaningful—he gets most of them in the late innings, or against first-division clubs, or he has a knack for starting rallies (those who follow him in the batting order presumably drawing inspiration from the rarity of his hits). This is rank subjectivism, of course, which statisticians have tried to combat almost since the beginning of major league play.

The first attempt to measure hitters' clutch ability came in 1879 with the introduction of the run batted in (RBI). Yet, as mentioned in Chapter 2, a central deficiency of this statistic was identified only one year later, when some Chicago White Stockings fans complained that RBI opportunities were not distributed randomly throughout the lineup, for which reason any ranking by this statistic would discriminate against their favorites, Abner Dalrymple and George Gore, who batted first and second. The other fundamental weakness of the RBI is even more damaging—that men on different teams do not have the same RBI opportunities; you can't drive in runners who aren't on base. If you're playing for a team that doesn't put many men on base, or you're left unprotected in the batting order—in other words, you are a good hitter followed by a comparative pushover—few fat pitches will be coming your way with men on base.

It is not unusual for a man who has been traded to see his RBIs increase or decrease by 20–30 percent while his other batting stats remain roughly the same. Has the batter whose RBIs decline left his clutch ability behind with his former club? The answer, certainly, is no. The test of how many runs he is contributing to his new team as opposed to what he did for his old team can be found through Linear Weights, or OPS. The RBI total is not a measure of intestinal fortitude any more than is the runs scored total: It is a measure of fortuity, not of clutch ability. An RBI total will tell you (a) how many runs a batter drove in, but not (b) how many baserunners he stranded, or what his "RBI Ratio"—a / (a + b)—was. In fact, an RBI Ratio has been recorded for the Boston Red Sox in recent years by Dick Bresciani, who advises that anything over .275 is pretty good, while ratios of .320–.330 are excellent. (He also keeps some other "clutch figures" for the Red Sox which we'll discuss later in the chapter.) Other clubs surely keep track of the RBI Ratio as well, but they regard it as proprietary material and will not release it to the press—which is not a great loss: An RBI Ratio may take into account the quality of the batters ahead of the player being evaluated, but it reveals nothing about the quality of the man batting *behind* him.

The RBI Ratio is kin to Branch Rickey's measure of team efficiency discussed in Chapter 2: runs scored divided by total baserunners. Rickey regarded this measure as being useful only for evaluating teams, for a player's runs were little reflection of his ability. And he did not transform the formula to an RBI Ratio because "RBIs were not only misleading but dishonest." And even if the Mahatma had felt different, no club was then interested in analyzing play-by-play data.

Once computers came on the scene, play-by-play analysis became not quite so onerous. In 1965 the General Electric Company, perhaps as a publicity stunt, had its GE-235 computer absorb all play-by-play data for the American League. GE programmers, using a point-by-point batting-pressure curve, determined weight values for all observed combinations of inning, out, baserunner, and score. These "pressure factors" ranged from 0 to 155 percent, with real pressure, as the fan understands it, coming on at about 75 percent.

Here are some of GE's clutch situations:

Table IX, 1. *Clutch Situations as Identified by General Electric, 1965*

Pressure Factor	Inning	Score	Outs	Bases Occupied
5%	1	−2	0	0
5%	9	+4	0	0
25%	9	+2	1	1
25%	9	−6	2	1, 2
50%	9	+1	0	3
50%	9	−4	1	3
75%	9	+1	1	2, 3
75%	9	−2	0	1
100%	9	0	0	3
100%	9	−2	1	3
120%	9	0	1	2, 3
120%	9	−1	0	1, 3

So far, so good . . . but all the GE wizards did was to record each player's batting average in each situation! A man coming to the plate with two men on base in the ninth, two out, and his team down by three runs, would receive the same credit for a game-tying homer or for a bunt single; a walk was of no measurable use.

Had this method recorded OPS instead, or Runs Created, or Cook's Scoring Index, it would have been better—but still not as good as the study conducted four years later by the Mills brothers, Eldon G. and Harlan D. Their method was put to use with the play-by-play record for all major league games of 1969, and their results were published the following year as a little book entitled *Player Win Averages*. The theory behind the PWA is the most sensible yet developed for measuring clutch hitting, though it is not without some remediable flaws.[1]

The Millses tried to identify winning or clutch players—those whose actions contributed most to team success. They ran a computer simulation of thousands of baseball games to identify the probable outcome of the game as dictated by every one of the almost 8,000 possible situations, and assigned a point value to each situation. (Win Points or Loss Points—each situation presents both, and they are inverses, so that a batter whose homer provides 500 Win Points is balanced by a pitcher who receives 500 Loss Points). The components of the situation were: number of outs, bases occupied, inning, visitor or home half of the inning, and so on. All of these components were taken into consideration, and then point values were assigned based on the league-average level of hitting for the particular season used in the computation, so that an average player at season's end would have as many Win Points as Loss Points. On a league basis, Win Points and Loss Points must be equal, in the same way that wins and losses must balance.

Next the play-by-play record was quantified. Each game presents 75–80 plays producing a baserunner, a base advanced, or an out. With 1,946 games played in 1969, the Millses had to score 155,000 plays for the offense and another 155,000 for the defense. Then the computer assigned point values to each of the 310,000 events. Each event merited the player a certain number of Win or Loss Points, according to how much his play affected the outcome for his team—that is, how much the win probability increased or decreased as a result of his action. If he hit a home run in the bottom of the ninth, and his team was leading by 8 runs, he would only get 5 Win Points. But Bobby Thomson's home run off Ralph Branca in the final game of the 1951 National League playoff—which, with one out in the bottom of the ninth and men on second and third, transformed a 4–2 defeat into a 5–4 victory—obtained for him 1,472 Win Points. No extra points were obtained by the fact the pennant swung with the outcome of the game. (The only blow that might gain more points would be a grand-slam homer in the bottom of the ninth, with two outs and the home team trailing by three runs: Bo Diaz of the Phils hit one of those in 1983.) And Ralph Branca was saddled with 1,472 Loss Points.

The Player Win Average (PWA) is simply an individual's Win Points divided by his total of Win and Loss Points, a computation analogous to the won-lost percentage. A player with 14,000 Win Points and 12,000 Loss Points will thus have a PWA of .538, which means that he was 38 percentage points better in the clutch than the average player. This figure could be attained by a batter or a pitcher. In 1969, the only year for which the Millses published their data (although the 1970 data is extant in an unpublished article), the league leaders in batting PWA were Willie McCovey in the NL, at .677, and

Mike Epstein in the AL, at .641, while the poorest clutch performers among the regulars were the NL's Hal Lanier (.348) and the AL's Zoilo Versalles (.330). Among pitchers, the upper range reached about the same level—25 to 35 percent above average—but the lower range extended only about half as far from the norm, some 10–15 percent. The reason for this may be that a poor clutch hitter like Lanier may still earn his keep through good fielding, but a poor clutch pitcher (one who gives up many runs, or allows them at inopportune moments) will have no redeeming quality, and so will not work enough innings to qualify as a regular.

The most recent measure of clutch ability is the game winning RBI, defined as the run which gives a team a lead it never relinquishes: For example, if the score is tied at 1–1 in the second inning, you are hit by a pitch with the bases loaded, and your team winds up holding the lead and winning 11–7, you are awarded the GWRBI. Is this a measure of clutch performance?

If a situation is clutch, a batter should sense its importance. The outcome of the game should hang in the balance, which it cannot very well do in the second inning. The GWRBI defines the clutch situation retroactively, which is unsatisfactory, an instance of hindsight being 20-20. Also, the GWRBI takes a situation-dependent stat—the RBI—and makes it contingent upon still further variables which are not randomly distributed. A batter for a team which wins few games will find it tougher to gain a GWRBI, just as a player with few RBIs, because of his spot in the order, will have commensurately few GWRBIs. For proof that the GWRBI is measuring clutch ability, advocates cite the caliber of player found at the top of the charts each year; the list of leaders, however, is nothing more than a list of good hitters under any circumstances, who happen to bat in the middle of the order for a plus-.500 team.

The GWRBI implies that a victory is equally attainable at any point in the game. The "when" doesn't count, only the "what." Yet, if clutch hitting exists at all, being able to hit in *timely* fashion is its very essence.

In adopting the game winning RBI as an official statistic, the owners have suggested that this is the best they can do. They sensed the public's appetite for a measurement of clutch performance and, feeling the need to do something, they came up with this. Better they should have done nothing, for if the GWRBI is a measure of clutch hitting, then it might be concluded that there is no such thing as clutch hitting.

Can clutch fielding performance be measured? In 1982, at a regional meeting of the Society for American Baseball Research, in Reading, Pa., a member advocated that the official scorer keep track of Game Saving Fielding Plays (GSFP). The proposed method was similar to the GWRBI in that one didn't have to contribute such a play in the seventh, eighth, or ninth inning: The play simply had to be an excellent one that prevented men on base from scoring. The Linear Weights System gives credit to infielders for runs saved, but these runs are statistical approximations. The GSFP, presumably, would be a superior measure because it would be based upon observed data. The difficulty, however, is that by putting this decision in the hands of an official scorer, who may have more than a smattering of hometown bias, rather ordinary fielding plays may be labeled outstanding (as a bloop single can look like a line drive in the next day's box score). Moreover, certain bumblers in the outfield can make an adventure of the most routine fly ball, and a hotdog infielder chasing a grounder may halt one step short of where he might have, and then reach dramatically across his body to stab at the ball.

How would you measure clutch pitching performance? One might keep track of situations in which the pitcher leaves men in scoring position. Or, taking a cue from Branch Rickey's formula, one might divide the pitcher's runs allowed by his baserunners allowed. The problem (actually, only one of several problems) with this approach is that it rewards a pitcher for putting himself in hot water. Really, all that matters for pitchers is runs, earned or not—unless you believe that a man can pitch "just well enough to win"—that is, allow 4 or 5 runs only when his team scores 6 or 7; then you might regard a pitcher's won-lost percentage over that of his team, minus his own efforts, as a clutch measure.

In a 1977 article in the *Baseball Research Journal,* entitled "Do Clutch Hitters Exist?", Dick Cramer used the Mills brothers' Player Win Average as the measure of *when* things happen and his own Batter Win Average (detailed in Chapter 7) as the measure of *what.* He confirmed a high degree of correlation between the two stats, originally noted by Pete: that the batter who has a high BWA will tend to have a high PWA as well. There were some exceptions, of course, but the correlations were sufficiently good to *predict* a PWA from the BWA, in this manner: PWA = BWA (1.37) + .484.

What this means is that most of the differences among batters' Player Win Averages, or clutch batting, are attributable to the quantity of times reached base (the On Base Average) and bases gained (slugging percentage). Willie McCovey had the top PWA in baseball in 1969 and 1970—*and,* not coincidentally, the highest BWA (and

OPS, too). In those cases which presented divergences between PWA and BWA, one might suppose that an unpredictably high PWA indicates an ability to hit in the clutch, and an unpredictably low PWA bespeaks a pattern of choking in the clutch. The greatest deviations from predicted PWA were for Carlos May of the Chicago White Sox in 1969, on the clutch side (.067 higher than his BWA predicted for him), and Tito Fuentes in 1970 on the choke side (.068 below BWA prediction).

The differences *can* be attributed to clutch ability or the lack of it . . . or to luck or the lack of it; determining which is a hoary dilemma. Statistically speaking, deviations of .067 and .068 from predicted values are not greater than one might expect to occur from mere luck. Furthermore, the distribution of *all* deviations from predicted clutch ability is normal; this is more consistent with an explanation of luck, or chance, than it is with the notion that some select players become titans when the chips are down.

The last test Cramer applied to this question of "chance or skill" was to say that if a player has clutch ability, it would be likely to manifest itself over time. "If clutch hitters really exist," he wrote, "one would certainly expect that a batter who was a clutch hitter in 1969 would tend also to be a clutch hitter in 1970. But if no such tendency exists, then 'clutch hitting' must surely be a matter of luck. After all, the only means of ever identifying a clutch hitter would be by his *consistency,* if not from situation to situation at least from season to season." Correlating the 1969 residuals or deviations in PWA with those of 1970, Cramer found a random pattern, with some of the "best" clutch hitters of 1969 becoming the "choke artists" of 1970, and vice versa. Pete found that the previous season's NOPS had a better correlation with the next year's PWA than the previous season's PWA did.

Does clutch hitting exist? The question is reminiscent of the dispute over the curveball that has simmered from the 1870s to the present day: Is it real, or is it illusion? With the curveball, the current answer is "both"—the ball does break, but not "late" as supposed. With clutch hitting, the answer appears to be "both" as well—that a batter can be "hot" in key situations for a period, brief or extended, but not over a span of many seasons. Those batters who are the most productive over the long haul are likely to be the best in the clutch as well, because the laws of chance have more opportunity to exert themselves over a greater number of hits.

"So fades a legend," wrote Cramer, "but after all, what was really meant when someone was called a 'clutch hitter'? Was he really a batter who didn't fold under pressure—*or* was he a lazy batter who bothered to try his hardest only when the game was on the line?''

Clutch, it seems, is in the eyes of the beholder, and questions relating to it may be better left to philosophers than to statisticians.

---

[1] The major flaw in the Mills brothers' system is that the Player Win Average weights a few events very heavily, many others quite lightly, so that it effectively has a smaller sample and is therefore less accurate. A combination of overall and situational data would be better.

Also, the way the Mills brothers handled errors was to count the error as an out for the batter and the pitcher—which moves the average PWA for batters over .500 and the average PWA for pitchers under .500.

10

44 PERCENT OF BASEBALL

Pitching is 70 percent of baseball, said Connie Mack. Pitching is 50 percent of baseball, said Branch Rickey. Pitching is 35 percent of baseball, said George Weiss. More recent guesstimates—for that is what they are, supported by impressions rather than evidence—have ranged from a high of 80 percent, by Herman Franks when he managed the Cubs, to a low of 15 percent, by Joe McDonald when he was G.M. of the Mets.[1] (If you reflect on the Cubs' chronic lack of pitching, and the Mets' chronic lack of hitting, and their equally chronic occupancy of the depths of the NL East, these extreme views take on a certain poignance.)

Everybody has an answer, but nobody, it seems, understands the question. How else to explain the extraordinarily divergent responses of those best in a position to know? To the question, "What part of baseball is pitching?", take an answer of, say, 70 percent. Does this mean that if your club's pitching is "better" than the other club's on a given day, or than the other clubs' in a given season, your chances of winning the game or pennant are 70 percent? Does it mean that however good your club's pitching is, the combined efforts of the batters, fielders, and baserunners (let's forget about intangibles like attitude) have only a 30 percent impact on the outcome? Does it mean that the essence of the game is the battle between pitcher and batter, and the pitcher will have his way 70 percent of the time? Or does it mean that pitching effectiveness has a 70 percent correlation with winning percentage? These do not exhaust the possible interpretations.

As far as we are concerned, the best answer to this confusing if not downright silly question is supplied in the title of this chapter. There is no getting around the fact that baseball is divided equally between offense and defense: Each run scored by Team A is a run allowed by Team B. With the game split down the middle at 50-50, the question becomes how much of the offense to attribute to batting and how much to baserunning, and how much of the defense is accounted for by pitching and how much by fielding. In 1910, when roughly 70 percent of all runs were earned, George Weiss's opinion that pitching was 35 percent of baseball would have been correct, by our lights, inasmuch as earned runs can be considered the responsibility of the pitchers and unearned runs the responsibility of the fielders (70 percent of the defense's 50 percent of the game equals 35 percent). As errors became less frequent—which is another way of saying, as the average level of fielding skill improved—pitchers took on a greater share of responsibility for all runs scored. In recent years, with only 12 percent of all runs being unearned, fielding thus accounts for 6 percent of the game and pitching 44 percent. If unearned runs continue to decline as a percentage of total runs, pitching will one day become as large a part of the game as batting; it is not as large now. However, it is undeniably important, and there are several unconventional, appealing ways to gauge its effectiveness.

The Linear Weights measure of runs saved by a pitcher beyond what an average pitcher would have allowed is the best pitching stat, and its accuracy may be heightened by adjusting for home-park effects and by converting the runs saved to wins gained. Rather than give a précis of the earlier discussion of Pitching LWTS, we refer you back to Chapter 4; a discussion of traditional pitching stats (notably won-lost percentage and ERA) will be found in Chapter 2.

Two measures of pitching effectiveness which may have originated with Branch Rickey—doesn't it seem that he invented *everything?*— are the relationship of hits to innings pitched, with a ratio of 1 : 1 or less representing quality; and the relationship of strikeouts to walks, for which a ratio of 2 : 1 or more is deemed outstanding. The former yardstick is not meaningful when applied to pitching before 1887, the first year in which only four wide pitches were required to obtain a base on balls. Prior to that time, walks were so infrequent that pitchers could yield 10 or 11 hits per game and still be in the top drawer of their trade. In modern baseball, however, in which average control has meant about 3 walks per 9 innings, it has been imperative to maintain a 1 : 1 ratio of hits to innings pitched. In fact, when an aging star pitcher first produces a season in which his hits exceed his innings pitched, his team will be looking to move him. If sympathy or wishful thinking intervene and the pitcher is retained, the results generally look like those of Juan Marichal, who lasted in the major leagues several years beyond the point when he lost his effectiveness. His hits exceeded his IP for the first time in 1970, and though his fortunes revived in 1971, the next four years produced a sad decline.

The hits-to-innings-pitched ratio is another way of looking at what has been called the Opponents' Batting Average. John Holway used to write articles for *Baseball Digest* ranking pitchers by Pitching Average, as he called it. When Luis Tiant held opponents to an all-time low batting average of .168, Holway speculated that "Looie" might have been the greatest pitcher ever, for one season. What to say, then, about the man whose record Tiant broke—Tommy Byrne, who for the 1948 Yanks held opponents to a .172 mark? Or such other men as Nolan Ryan, who in 1971 held batters to .171; or Tom Hall of Minnesota, who posted a Pitching Average of .173 in 1970? (Relievers Goose Gossage and Kevin Saucier did even better in 1981, with marks of, respectively, .144 and .160.)

All right, you say, Byrne was a wild man and so was Ryan. (Byrne's ERA in 1949, when opponents batted only .183 against him, was 3.72!) Would an Opponents' On Base Average stat be a better measure? We're off on a tangent here, whether we measure pitching effectiveness by Opponents' Batting Average or their On Base Average: The objective of the game is to win, and that objective is impeded only by *runs* allowed; the imperative for the pitcher is not to throw a no-hitter but to throw a shutout. Recording opponents' BA or OBA may be an interesting exercise, but it is no more meaningful than measuring a pitcher's ratio of strikes to balls—the latter is accounted for nicely by strikeouts and walks, as the former is by earned runs allowed.[2]

What about the ratio of strikeouts to bases on balls? Rickey posited a 2 : 1 ratio rather than the 1 : 1 of hits to innings pitched because the beneficial impact on the team of the strikeout is not nearly as great as the detriment of the base on balls. In fact the two events, while related, are not directly comparable, in that the strikeout has run value only when an out in the field may advance or score a baserunner, while a walk almost always increases the opponent's run potential.[3]

Strikeouts are surely overrated as an indicator of pitching ability. No one believed that Nolan Ryan, when he became, briefly, the all-time strikeout king, became the best pitcher in baseball history. The strikeout is romantic, glamorous—a pitcher's opportunity to do something all by himself, unaided by his fielders. As a glamor stat it resembles the home run, a similarly individual feat, similarly overrated. The analogy extends to the paranoid delusion of grandeur which sometimes seizes both the strikeout artist and the long-ball champ, that each has to "do it all by himself"—save runs, that is, or create them. A further extension may be the at best muddy correlation between home run or strikeout championships and overall batting or pitching prowess, as measured by Linear Weights.

A few years ago, Leonard Koppett proposed a triple crown for pitchers much like the triple crown for batters. The three jewels in the pitching crown would be won-lost percentage, earned run average, and strikeouts. A pitcher who wins such a crown has indeed had himself an outstanding year, but the three jewels are not equally precious: the ERA title is worth more than the won-lost title, which in turn is worth more than the strikeout leadership. It's possible to lead the league in strikeouts in an otherwise poor year, as Nolan Ryan did a couple of times with the Angels. Here are the men who would have won Koppett's triple crown:

Table X, 1. *Pitching Triple Crown Winners*

	W-L, Pct.	ERA	K
1877 Tommy Bond, BOS	40-17, .702	2.11	170
1884 Hoss Radbourn, PRO	60-12, .833	1.38	441
1888 Tim Keefe, NY (N)	35-12, .745	1.74	333
1889 John Clarkson, BOS (N)	49-19, .721	2.73	284
1913 Walter Johnson, WAS	36-7, .837	1.14	243
1915 Pete Alexander, PHI (N)	31-10, .756	1.22	241
1924 Walter Johnson, WAS	23-7, .767	2.72	158
1929 Lefty Grove, PHI (A)	20-6, .769	2.81	170
1930 Lefty Grove, PHI (A)	28-5, .848	2.54	209
1931 Lefty Grove, PHI (A)	31-4, .886	2.06	175
1934 Lefty Gomez, NY (A)	26-5, .839	2.63	158
1945 Hal Newhouser, DET	25-9, .735	1.81	212
1948 Harry Brecheen, STL (N)	20-7, .741	2.24	149
1965 Sandy Koufax, LA	26-8, .765	2.04	382

In 1965, when Sandy Koufax was closing out the above list, Tommy Holmes, sportswriter for the New York *Herald Tribune,* established a system based upon the beliefs of Branch Rickey, then entering his eighty-third year. Among these beliefs were: (1) both W-L Pct. and ERA are heavily influenced by the quality of the team one pitches for; (2) the lower the ratio of hits to innings pitched, the better the pitching; and (3) the higher the ratio of strikeouts to bases on balls, the better. Thus, "The Rickey Ratings": hit-inning differential plus strike-out-walk differential. The names on Holmes's list of all-time best single seasons are no pikers: Koufax, Johnson, Walsh, Mathewson, Alexander, Feller, et al. But this is, like Opponents' Batting Average and the Pitching Triple Crown, what Bill James has aptly called a "freak-show stat"—a meaningless correlation of this element with that, offering no particular insight or truth but only another list.

If Opponents' BA or OBA are not much use in measuring pitcher effectiveness, why not (you may wonder) Opponents' On Base Plus Slugging or Linear Weights? The salient feature of these assessments of batter proficiency is their excellent correlation with runs scored: LWTS, through proper weighting of each offensive element, re-expresses a batter's diverse accomplishments as his runs contributed above the norm. But pitchers' accomplishments don't need such translation—they are already denominated in runs.

Pitching LWTS, or Pitcher Runs, is arrived at by a fairly simple formula: Runs = Innings Pitched × (League ERA / 9) − Earned Runs Allowed. An alternate version is: IP / 9 × (Individual ERA − League ERA). From the alternate, it is clear that one aspect of Pitching LWTS is efficiency, as represented by the amount by which one betters the league ERA. Another way of stating this efficiency is to create a Normalized (or Relative) Earned Run Average, dividing the league figure by that of the individual. (For the Relative Batting Average, we divided the individual figure by the league's, but that was because a higher BA was the goal; with ERA, one aims at the lowest figure possible.) If the league ERA is 4.30 and yours is 2.15, your Normalized ERA (NERA) is 2.00.

There have been only thirty instances in the history of baseball in which a pitcher has attained an ERA of less than half that of his league. Here they are, without park adjustments (a complete table of the top hundred and lifetime NERAs will be found at the rear of the book, with park adjustments for twentieth-century pitchers):

Table X, 2. *Season ERA Half or Less That of League (NERA over 2.00)*

1.	1914 Dutch Leonard, BOS (A)	2.85
2.	1880 Tim Keefe, TRO	2.76
3.	1968 Bob Gibson, STL	2.67
4.	1913 Walter Johnson, WAS	2.57
5.	1906 Three-Finger Brown, CHI (N)	2.53
6.	1912 Walter Johnson, WAS	2.41
7.	1905 Christy Mathewson, NY (N)	2.36
8.	1915 Pete Alexander, PHI (N)	2.26
9.	1909 Christy Mathewson, NY (N)	2.26

10.	1882 Denny Driscoll, PIT	2.25
11.	1901 Cy Young, BOS (A)	2.25
12.	1964 Dean Chance, LA (A)	2.20
13.	1918 Walter Johnson, WAS	2.18
14.	1978 Ron Guidry, NY (A)	2.17
15.	1884 Hoss Radbourn, PRO	2.16
16.	1919 Walter Johnson, WAS	2.16
17.	1907 Jack Pfiester, CHI (N)	2.14
18.	1931 Lefty Grove, PHI (A)	2.13
19.	1905 Ed Reulbach, CHI (N)	2.11
20.	1907 Carl Lundgren, CHI (N)	2.11
21.	1882 Guy Hecker, LOU	2.09
22.	1902 Jack Taylor, CHI (N)	2.09
23.	1966 Sandy Koufax, LA	2.09
24.	1923 Dolf Luque, CIN	2.07
25.	1908 Addie Joss, CLE	2.06
26.	1953 Warren Spahn, MIL	2.04
27.	1964 Sandy Koufax, LA (N)	2.03
28.	1955 Billy Pierce, CHI (A)	2.01
29.	1933 Carl Hubbell, NY (N)	2.01
30.	1943 Spud Chandler, NY (A)	2.01

Another method of evaluating a pitcher is to compare his won-lost record with that of his team. This achieved some currency through its inclusion in *The Sports Encyclopedia: Baseball,* which David Neft, Dick Cohen, and Jordan Deutsch first published in 1974. They employed a formula which Branch Rickey had used: Individual W-L Pct. − Team W-L Pct., weighted by the number of decisions. An example of this formula, applied to the career of Sandy Koufax:

Table X, 3. *Koufax W-L vs. Dodger W-L*

Year	Team Pct. ×	Koufax Decisions =	Product
1955	.641	4	2.564
1956	.604	6	3.624
1957	.545	9	4.905
1958	.461	22	10.142
1959	.564	14	7.896
1960	.532	21	11.172
1961	.578	31	17.918
1962	.618	21	12.978
1963	.611	30	18.330
1964	.494	24	11.856
1965	.599	34	20.366
1966	.586	36	21.096
TOTAL		252	142.847

Take the product of 142.847 and divide it by Koufax's decisions, 252, to obtain a Dodger winning percentage for the years 1955–66 of .567. (This percentage, because it is weighted by Koufax's decisions, is different from the Dodgers' actual record, which was .570 based on 1,078 wins and 814 losses). What the weighting procedure does is to give you the winning percentage that an average Dodger pitcher—actually, an agglomeration of all the Dodger pitchers, *including Koufax*—might have been expected to attain in those games in which Koufax drew the decision. Over Koufax's career, by this method, an average Dodger pitcher with 252 decisions would have won 143 games; Koufax won 165. To obtain a percentage differential, subtract the Dodgers' weighted W-L Pct. of .567 from Koufax's W-L Pct. of .655 (165 wins, 87 losses) to get .088, which happens to be the seventh-best mark in baseball history.

This is, as you can see, not a quick computation. Moreover, it produces a tainted figure, because the Dodger won-lost percentage includes Koufax's substantial contribution. For example, in 1964 the Dodgers finished two games under .500 at 80-82, while Koufax was fourteen games over, at 19-5. Why not compare his W-L Pct. of .759 against the Dodgers' W-L Pct. without him, which would have been .442 rather than the .494 which in effect compares Koufax with himself?

This plainly more logical approach was employed in a booklet called *Kings of the Mound: A Pitcher's Rating Manual,* which was published privately by Ted Oliver in Los Angeles in 1944. Over the course of the previous four years, without the aid of a computer (obviously) or a baseball encyclopedia (the first to cover all players was the Turkin-Thompson tome of 1951), he compared the W-L Pct. of every pitcher from 1894 forward to that of his team, in this manner: He took the pitcher's won-lost percentage and subtracted from it the won-lost percentage of his team *in games not decided by that pitcher;* then he multiplied the resulting figure by the pitcher's number of decisions; the result is a number of points, either positive or negative, expressing the pitcher's ability to win compared to that of his team.

Oliver called this method the Weighted Rating System. It came into being to fill a perceived void: how to measure the effectiveness of pitchers shackled to a bad ballclub, like Red Ruffing, Claude Passeau, or Hugh ("Losing Pitcher") Mulcahy. Oliver wanted a stat like the batting average, in which anyone could excel regardless of the situation. Both the W-L Pct. and the ERA, he argued, were biased toward pitchers whose teams could hit or field proficiently. The Weighted Rating System revealed that a pitcher with a W-L Pct. of .600 for a club which played .600 ball was not a good pitcher, not .100 above average, but rather was no different from a .435 pitcher on a .435 club. Here is Oliver's example of the Weighted Rating System at work:

Table X, 4. *Sid Hudson, Washington, 1940*

Senators' W–L: 64-90
Hudson's W–L: 17-16, .515
Senators' W–L sans Hudson: 47-74, .388
Hudson's margin: .127
Points: .127 × 33 decisions = 4,191

The Oliver method was a significant step forward for baseball statistics: It was the first ratio to be weighted by a longevity factor, and it was the first to be normalized, in this instance to a team's record. However, *Kings of the Mound* received little publicity upon its wartime publication, and fewer sales; an updated edition appeared in 1947 (including coverage of the high-level minor leagues), but by the early 1950s both the Weighted Rating System and its inventor had faded from view.

Oliver did manage to thrust into the limelight such men as Bobo Newsom, Noodles Hahn, Jess Tannehill, Nap Rucker, and Urban Shocker, all of whom might have compiled Hall of Fame–level records with better clubs. And it remains a valuable tool for identifying the accomplishments of the pitcher apart from his team—a Floyd Bannister with Seattle, a Dave Stieb with pre-1983 Toronto, a Mario Soto with Cincinnati. For this reason we include a ranking of the top hundred lifetime and single-season performances in Oliver's statistic, and list the top three pitchers in each league season from 1876 on—*with one significant and curious difference.* When Oliver obtained his winning percentage margin (see Table X, 4 for Sid Hudson), it was expressed in thousandths (.127); yet when he multiplied that margin by the number of decisions, he dropped the decimal point to arrive at a statistic expressed in whole numbers (4,191). *Had he retained the decimal,* he would have identified the number of *wins* the pitcher achieved beyond what an average pitcher on his team might have gained (in Hudson's case, 4.191 wins). In the tables at the rear of this book, you will see the Oliver method expressed in this manner, as "Wins Above Team."

As with so many baseball statistics, the strength of the Oliver method also provides its prime weakness: It is easier for a good pitcher laboring for a poor club to compile an impressive figure, for the lower the team's W–L Pct., the easier it is to exceed by a sizeable margin. For example, in 1972 Steve Carlton was 27-10 for a Phillie club which was 32-87 without him; his Wins Above Team rate was 17.1, the highest mark of this century. Lefty Grove in 1931 was 27 games over .500, compared to Carlton's 17, yet he recorded only 8.3 Wins Above Team because he played for a pennant winner; to have matched Carlton's record, Grove would have had to go undefeated in 49 decisions— necessitating the extension of the A's schedule to 166 games. Carlton had a great year in 1972, but it wasn't *that* much better than Grove's— if it was better at all.

A further weakness of the Oliver method is that it normalizes to a narrow base (the team) rather than to a wide one (the league), and just as a summation of all the won-lost records in a league must result in .500, the sum of all individual records for a given team must result in the won-lost record of that team. On a league basis, the pitcher who goes 20-10 is balanced by other pitchers who finish ten games under .500; on a team basis, every success is likewise matched by a failure. Two pitchers will only rarely cancel each other, but take the case of Grove in 1931. The A's were a great team, playing .650 ball even without him; in calculating his Wins Above Team, we obtain a differential by subtracting .650 from his W-L Pct. of .886. No other pitcher on that staff had so low a figure subtracted from his own W-L Pct. because Grove's extraordinary performance boosted the team percentage to even greater heights. Rube Walberg, for instance, finished at 20-12 with an ERA 0.64 below the league's, not too shabby by conventional standards, yet the Oliver method subtracted a team W-L Pct. of .725 from his W-L Pct. of .625, producing a negative rating that was the third worst in the American League. And yet you know that Walberg was not the third worst pitcher in the league, or the pennant-winning A's would not have allowed him to pitch 291 innings—*more* than Grove or any AL pitcher! The upshot is that a good pitcher on a great staff—like Walberg with the A's, or Mike Garcia with the Indians when their rotation also included Early Wynn, Bob Lemon, and Bob Feller—will suffer by this stat, though not as much as, say, Tom Seaver suffered in his W-L Pct. for having worn a Mets uniform in 1983, and in so many years earlier. On balance, the virtues of the Oliver method, or Wins Above Team, outweigh its flaws.

A statistic which is somewhat similar—in that its intent is to provide recognition for the unrecognized, the top-flight pitcher laboring for a bottom-flight club—is Percentage of Team Wins. Its flaws are similar, too—the better the club, the harder it is for the pitcher to shine in this category, and extreme individual totals can be achieved only against the backdrop of extreme team ineptitude. Moreover, it is inherently weak because it is a derivative of the won-lost record, with all the bias that involves (see Chapter 2). The highest single-season mark in Percentage of Team Wins in this century is once again Steve Carlton's 1972 season: His 27 wins represented 45.8 percent of the Phils' total. And like Wins Above Team, this measure is nearly useless for the years before 1882 or so, when one pitcher might work 70 percent or more of his team's games. (Carlton's 45.8 percent would not make the list of top *100* single seasons from 1876 on!)

Sidelight: What would Carlton's epic 1972 totals have been had he played for an average club that year? Or for the NL West champs, the Reds? This may be predicted from his innings pitched, runs allowed, and runs scored (if the last figure is not known, one may multiply the team's runs scored per inning by the pitcher's innings). Pitching for a nearly average club (in 1972, say the Cardinals, who finished 75-81), Carlton figured to have won 25 games and lost 13. Pitching for the Reds, Carlton might have won 28 and lost 10.

It may be surprising that Carlton's record with vastly superior clubs would not itself have been vastly superior. Evidently the Phillies played much better ball behind Carlton, both in the field and at the bat, than they did behind Ken Reynolds (2-15), Billy Champion (4-14), or Dick Selma (2-9), for Carlton should, on the basis cited above, have gone 23-15 for the Phils.

It raises an interesting question about the relationship between a pitcher and his teammates. When a pitcher is going so well that his teammates, regardless of their abilities, feel they have a good chance to win any time he steps on the mound, they may perform with more confidence and manifest ability. This is a Norman Vincent Peale kind of argument, easy to scoff at, but it makes sense. Carlton's 1972 season was the best of this century when measured against the Phils' record without him, but in LWTS it stacks up as only the thirtieth best since 1900. On a park-adjusted basis, it was not even as good as Gaylord Perry's record in that same year.

The Park Factor or park adjustment for pitchers' performance is as daunting to discuss technically as it was for batters. For the precise method of calculation, see the footnotes to Chapter 5, but for purposes of this discussion, it should suffice to say that the Pitchers' Park Adjustment is calculated in the same way as the Batters' Park Adjustment, except that it adjusts for the fact that the pitcher does not have to face his own team's batters. Park Factors for pitchers thus may vary slightly from the Park Factors for that team's batters: In 1983, for example, Seattle's Batter Park Factor (BPF) was 102, while its PPF was even higher—104; this was because Mariner pitchers derived less benefit from not having to face Mariner batters than Mariner batters derived from not having to face Mariner pitchers.

Absent from the chapter to this point has been the relief pitcher, who presents a variety of statistical problems. The nature of the job is such that his won-lost record is not meaningful (even less so today than ten or fifteen years ago, with the ace in many bullpens being called upon only when his team has a lead in the eighth inning). A reliever may pick up a win with as little as a third of an inning's work, if he is lucky, while a starter must go five innings; a reliever may also pick up a loss more easily, for if he allows a run there may be little or no opportunity for his teammates to get it back, as they can for a starter. Earned run average is meaningful for the reliever, but it must be .15 to .25 lower to equate with that of a starter of comparable ability: A reliever frequently begins his work with a man or two already out, and thus can put men on base and strand them without having to register three outs.[4]

Ratios of hits to innings, strikeouts to innings, strikeouts to walks—all of these have their interest, but none is sufficient by itself to measure relief-pitcher effectiveness. Relievers may also have an edge in these ratios because they generally face each batter only once in a game, thus leading to fewer hits and more strikeouts per inning. Before discussing the modern alternatives of saves or Relief Points, let's review briefly the rise of the relief pitcher from the role of a mere hanger-on to, some would say, the most indispensable part of a winning team.

Relief pitching before 1891 was limited, with rare exceptions, to the starting pitcher exchanging places with one of the fielders, who was known as the "change pitcher." Substitutions from the bench were not permitted except in case of injury until 1889, when a tenth man became entitled to designation as a substitute for all positions; free substitution came in two years later, but no relief specialists emerged until Claude Elliott, Cecil Ferguson, and Otis Crandall in the first decade of this century.

The next decade's best relievers were starters doing double duty—notably Ed Walsh, Chief Bender, and Three Finger Brown. The 1920s, up to the end of World War II, brought the first firemen to be employed in the modern way, although they tended to work more innings and fewer games than today. These were men such as Firpo Marberry, Johnny Murphy, Ace Adams, and several other worthies. But because relievers were not yet a breed entirely apart from starters, as they are today, the tables in this book record seasonal relief marks only from 1946 on.[5]

When you think of a relief pitcher in the modern-day sense—that is, a man who can appear in 50 or more ballgames a year, all or nearly all in relief, and win/save 30 or more—you begin with Joe Page of the 1947–49 Yankees and Jim Konstanty of the 1950 Phils, though Marberry had one such season in 1926. None of the three, however, ever heard of a "save" in his playing days—this term wasn't introduced until 1960, the year after Larry Sherry's heroic World Series in which he finished all four Dodger victories, garnering two for himself and saving the others; 1959 was also the year fireman Roy Face went 18-1, not losing until September 11. Before Jerry Holtzman of the Chicago *Sun-Times* devised the save, baseball people were looking at really only one figure to measure a reliever's work, and that was the number of games in which he appeared; any other appreciation of his efforts was expressed impressionistically. A reliever did not work enough innings to qualify for an ERA title (Hoyt Wilhelm in 1952 being the exception), nor could he expect to win 20 games. The introduction of a specialized statistic for the fireman was acknowledgment of his specialized employment, and conferred upon it a status it had never enjoyed, not even after the exploits of Konstanty, Page, Wilhelm, and Face. Only when the save came into being did the majority of relievers take pride in their work and stop regarding their time in the bullpen as an extended audition for a starting role.

When *The Sporting News,* spurred by Holtzman, began recording saves in its weekly record of the 1960 season, the save was defined in a way different from today. Then, upon entering the game, a reliever had to confront the tying or winning run on base or at the plate, and of course finish the game with the lead. This definition later became eased, so that simply finishing a game would get the reliever a save; a memorably absurd result of the new ruling was that the Mets' Ron Taylor gained a save in 1969 by pitching the final inning of a 20–6 win over Atlanta. This outraged sportswriters and fans alike, so in 1973 the definition was changed yet again: A reliever had to work three innings *or* come in with the tying or winning run on base or at bat. This definition was relaxed yet again in 1975 so that the tying run could be *on deck,* thus giving the bullpenner license to allow a baserunner. It's a good thing for statisticians that Dan Quisenberry surpassed John Hiller's 1973 record of 38 saves by a decisive margin of 7.

There was a blip in the relievers' trend of rising importance when the American League introduced the designated hitter in 1973. The predicted outcome, based on the first few years' experience of the DH, was: increased offensive production, no more need to pinch-hit for the pitcher, and thus a greater number of complete games and fewer saves. All those things did happen in 1973–76, although not quite to the degree expected—and soon the American League's use of relief pitchers became as extensive as it had been in the early 1970s. In 1982, despite the DH, American League starters completed only 19.6 percent of their games, an all-time league low (though still substantially higher than the National League, where CGs have dropped below 15 percent the last few years).

The save, despite its varying definitions and the absence of a countervailing stat for blown saves, is firmly entrenched in the statistical pantheon, and basically on merit. However, it is nonsensical to look at the 18 wins of a Roy Face in 1959 and say that as a measure of relief effectiveness they count for nothing. Certainly many relief wins are accidents, "vultured" from a starter who departs with a lead, but a Bob Stanley or a Jesse Orosco is frequently called into a game that is tied. For them, the Rolaids Company's formula for its Relief Man Award is a godsend, in that it gives two points for a save *or* a win. And for fans and writers who wanted a negative stat for relievers, the Rolaids formula deducted one point for a loss. Why a win is twice as good as a loss is bad is a mystery—a study should have been done to establish that a reliever has more of a chance to be tagged with a defeat than he does to pick up a victory. This position is not without logic, but a statistical study might have shown that, say, a loss should be weighted at 1.5 points.

Relief Points *is* an improvement over saves, and in the tables at the rear, we provide the top fifty single-season and lifetime marks, as well as the top three in each league for each year after 1946 (saves are figured by the definition which prevailed at the time they were attained, post-1960; all years before are figured by the 1973 definition). In lifetime records, Rollie Fingers is the Relief Points leader with 718 through 1983; Hoyt Wilhelm is in second place with 600. The active relievers to watch are Rich Gossage with 505, Bruce Sutter with 483, and especially Dan Quisenberry, who has totaled 309 in little more than four seasons. "Quiz" is also the single-season leader, with 97 in 1983.

Some folks still have continued to feel the need for a "blown save" category, and others have longed for a measure of middle-relief effectiveness, that statistical no-man's-land. In April 1981 *Sports Illustrated* came up with an incredibly complicated series of tabulations to address these final injustices, and they were dazzling. However, the *SI* method dazzled in the same way that the Mills brothers' Player Win Average did—it was ingenious and well conceived, but involved too much work. Not only does it require play-by-play analysis, but it also reminds one (queasily) of the National Football League's quarterback-rating system. Quarterbacks are rated in four categories, variously weighted, to arrive at a number of "rating points." Not one fan in a thousand could tell you how the rating points are derived, and the same holds for the *SI* relievers' formulae.

The final relief statistic to be discussed is the one we think is the best—Linear Weights, or reliever's runs saved. How do we determine who is a relief pitcher? Our definition is a pitcher whose average number of innings per game is under three. The immediate objection may be: What about the guy who gets blasted out in the first or second inning time after time, and then is relegated to the bullpen? Chances are that his early exits from his starts will have produced so many runs against his record that he will not emerge among the league leaders no matter how successful he is afterward in relief. Our definition results in at least one aberration: Bob Stanley of the Red Sox in 1982 won 14 games and saved 14 more while posting the second-best ERA in the AL. He started not once in his 48 appearances, yet he worked 168⅓ innings—3.5 innings per outing, the most ever by a nonstarter—and thus is not awarded a Relief LWTS record. (Bob Stanley fans: His line was 18 runs—23 when adjusted for park.)

The advantage of Relief Linear Weights over Relief Points is much the same as the reasoning behind the *Sports Illustrated* formula: LWTS gives credit for good pitching no matter when in the game it occurs. It does not attempt to isolate clutch situations because relief pitchers, by and large, have no control over their use in clutch situations—that is in the hands of the manager. Another way of saying this is that saves and wins are situation-dependent, while LWTS is not: All that LWTS measures is how many runs a reliever prevents that an average pitcher, in the same number of innings, would not have prevented. You can say that the man who pitches the eighth and ninth is in a tougher position and his performance should in some way be credited at a higher rate than the man who pitches the so-called easy innings.[6] That's up to you. You might also consider using LWTS in conjunction with Relief Points.

Relievers' LWTS will be much lower than those notched by first-rate starters because the weighting factor is innings pitched. Victories don't count for anything; saves don't count for anything; all that does count is runs allowed and innings pitched. Yet occasionally a relief performance is so outstanding that it would make that season's list of top three marks by starters. In 1979 Jim Kern of Texas, despite working only 143 innings, prevented more runs from scoring than any other pitcher—starter or reliever—in either league. That, incidentally, was the best relief year anyone has ever had (Kern went 13-5 in 71 games, saving 29 with an ERA of 1.35). What would happen if we took Kern's innings and ERA and gave him a top starter's number of innings—say, 240? To be fairer, we could adjust his ERA upward by .20, to remove the reliever's advantage. Of course, this whole bit of dream-casting is unfair because Kern's performance may have been uniquely the product of his role—many games, few innings. All right, it *is* unfair, but Kern would have had a Park Adjusted LWTS of 86.5, the best season of the century for any starter.

On a lifetime basis, no reliever approaches Hoyt Wilhelm's adjusted LWTS of 287.2 (29.7 wins); in second place is John Hiller with 137.8 runs saved (14.4 wins). Wilhelm even ranks nineteenth among *all* pitchers in this century, leaving a trail of deities in his wake (Koufax, Walsh, Feller, Marichal, Roberts, Gomez, Lemon, et al.): See the Lifetime Pitching LWTS list of top one hundred. He may not make it to Cooperstown's Hall of Fame, but he's in ours.

----

[1] These men were among fifty managers, general managers, sportswriters, and sportscasters who responded to a survey conducted by James K. Skipper, Jr., and Donald Shoemaker, reported in the 1980 *Baseball Research Journal.*

[2] Holway refers to Pitching Averages as "the batters voting with their bats, a poll far more authoritative than the writers' Cy Young poll." He makes the distinction between value, as measured by runs allowed, and excellence, which he sees reflected in Pitching Averages. "Pitching Averages," he writes, "are as important as batting averages. No more, no less."

[3] For the exception to this rule, see the section on the intentional base on balls in Chapter 8.

[4] The figure of .15–.25 is the product of a study by Bill James written up for the *Baseball Research Journal* in 1977.

[5] For statistical records of relievers from 1876 on, see *The Baseball Encyclopedia* (Macmillan) and *The Relief Pitcher*, by John Thorn (Dutton, 1979).

[6] Examining the Millses' play-by-play data for 1969 and 1970, Pete discovered that the top relievers had a plus/minus swing on Win/Loss Points about double that of starters. The average of the top two relievers on each team gave swings of 50 percent higher than that of starters. What this means is that runs allowed (or saved) by the top firemen are worth 50–100 percent more—i.e., 5 runs saved by a top reliever could be the same as 10 runs for a starter. In 1970, for example, Pete Richert of Baltimore pitched 55 innings, in which he attained 862 Win Points and 606 Loss Points, or 26.7 Total Points per inning. In that same year, Jim Palmer's line read: 2,159 Win Points, 1,899 Loss Points, 305 innings—13.3 Total Points per inning.

11

MEASURING THE UNMEASURABLE

"There is nothing on earth anybody can do with fielding," Branch Rickey wrote in 1954, and until recently, few had tried to prove him wrong. Implicit in this remark, of course, was a thorough renunciation of the fielding percentage. The several deficiencies of this traditional measure were manifest almost since its inception: that one cannot commit an error on a ball one has not reached; that hometown scorers may look too kindly or too harshly upon the efforts of their lads, depending upon what seems to require protection, a pitcher's ERA or a fielder's errorless game streak; and that the statistic focused on failure rather than success.

Henry Chadwick had written in 1868: "The best player in a nine is he who makes the most good plays in a match, not the one who commits the fewest errors, and it is in the record of his good plays that we are to look for the most correct data for an estimate of his skill in the position he occupies." His gauntlet was picked up more than a hundred years later by Bill James, who in his discussions of Range Factor (total chances minus errors, divided by games) pointed out how absurd it had become, in a time when the best-fielding second baseman might commit 10 errors a season and the worst 20, to focus on this difference of 10 rather than on the difference of 250–300 in total chances which might separate the most agile second baseman from the exemplar of Lot's wife. Fielding percentage was a far better measure of ability in the 1860s, when one play in four produced an error, than now, when only two plays in a hundred are flubbed. By 1876, when major league play began, only one play in six was a miscue (not counting "battery errors" like passed balls and wild pitches, which were termed errors at the time), and the differences between the best fielders at each position and the worst as measured by fielding percentage looked like this:

Table XI, 1. *Fielding Percentages, 1876*

Position	Best	Worst	Diff.
1B	.964	.915	.049
2B	.910	.814	.096
SS	.932	.764	.168
3B	.867	.754	.113
OF	.923	.761	.162
C	.881	.736	.145
P	.951	.810	.141

Errors were so commonplace that two regular infielders committed an average of more than 1 each game, while a regular catcher committed more than 2 per game (incredibly, all three men played for the same woebegone team—the New York Mutuals). One hundred years later, the picture looked like this in the National League:

Table XI, 2. *Fielding Percentages (NL), 1976*

Position	Best	Worst	Diff.
1B	.998	.975	.023
2B	.988	.964	.024
SS	.986	.950	.036
3B	.969	.934	.035
OF	.994	.959	.035
C	.997	.978	.019
P	1.000	.853	.147

When the number of errors diminishes to such an extent that the most surehanded shortstop fields 98.6 percent of the balls he reaches and the stone-fingered one reaches only 3.6 percent less, fielding percentage may be measuring gradations so fine as to approach meaninglessness.

With fielding percentage, as with any absolute statistic, the figures require a complex historical understanding in order to compare performances across time. Putting aside for the moment the inadequacies of the measure, there is surely something to say on behalf of a stat that has endured since the 1860s, and if we wish to improve the accuracy or reasonableness of what it does measure, we must consider a relativist approach, as detailed in Chapter 6. In *The National Pastime* of 1983, Bill Deane employed Merritt Clifton's technique of normalizing to league leader rather than league average. Deane's tabulation of the average percentage leading the league at each position by decade in this century, and the all-time best as figured by his Relative Fielding Average, may be found on pages 115-16.

While the relativist approach is an interesting one that does succeed in leveling the fluctuations that arise from the conditions of a given chronological period, it cannot transform an inherently weak measure into a strong one. The weakness of the fielding percentage spills over into the primary measure of pitching effectiveness, the earned run average. A team whose fielders are surehanded but have limited range may offer a defensive profile of "good" fielding and "bad" pitching because only the pitcher suffers measurably for the fielders' shortcomings. To counter this injustice, James proposed a Defensive Efficiency Record, a measure of the percentage of all balls put into play that were converted into outs. This rating, however, suffers from a problem opposite to that of the ERA: Whereas the latter can mask an immobile, porous defense, the former can "protect" a lousy pitching staff that allows many hard-hit balls that no one could convert into outs. All attempts to rate team fielding as an element of overall defense run up against this truth: Pitching and fielding are linked in so many complex ways that no matter how many of the links one uncouples, some remain intact.

The other individual statistic in use, at least by baseball's front offices, before the advent of sophisticated statistical analysis in the 1970s, was Total Chances Per Game. David Neft included TC/G in the first edition of *The Baseball Encyclopedia* (1969) because he had read an interview in which Branch Rickey said he used the measure to evaluate fielding ability. So, it would seem, the statistic came into being somewhere between 1954, when Rickey wrote the remark which opens this chapter, and 1960. But Range Factor, which differs from TC/G only in that it does not include errors, was first formulated as "Fielding Average" in 1875 by Al Wright, then evidently forgotten.

Range Factor has been attacked for a variety of shortcomings: Neft cannot comprehend why a misplayed ball should not count as an indicator of range; Barry Codell has argued that errors need not be nonevents but rather should be *subtracted* from Total Chances (Neft would add them) before dividing by games. Others have argued that a fielder's number of chances may be dependent upon such variables as: (1) grass vs. turf; (2) a predominantly lefthanded or righthanded pitching staff—Yankee second basemen and Red Sox third basemen may get fewer chances for this reason; (3) the strikeouts registered by the pitching staff—more strikeouts mean fewer fielding plays and generally fewer ground balls; (4) the winning percentage on the road, where a poor team will more often than not record only 24 outs in the field; (5) home-park dimensions—Brooklyn Dodger outfielders averaged fewer chances because of cozy Ebbets Field; New York Giant left fielders and right fielders had fewer because of the crazily configured Polo Grounds; and (6) you name it. Still, James believed that Range Factor was a far better indicator of ability than anything else around, and he was right. If Larry Bowa in 1980 made 4.59 plays per game while committing 17 errors, and Garry Templeton made 5.86 plays per game while committing 29, why should Bowa be perceived as the better shortstop? Templeton booted 12 that Bowa didn't, but had he played in the same number of games as Bowa, he would have taken part in retiring 187 more batters. Bottom line: Templeton saves more *runs* than Bowa does, and thus is incontestably *better*.

James effectively countered most of the arguments against Range Factor by acknowledging some shortcomings—which stat has none?—and dismissing others with a convincing array of evidence. But he did not deal satisfactorily with the question of errors, any more than Neft or Codell did. Neft may have been right to consider a mishandled ball an indicator of range if not efficiency, but it violates one's sense of fair play to see a measure of an individual's ability *boosted* by a play which hurts his team. Codell's alternative, to subtract errors from total chances, implies that muffing a routine play may be balanced by successfully handling another, but this is not so, as explained in the paragraph below. And Range Factor treats the error as a nonevent, in the same league with the basehit, which does not count against a fielder's record—but this too is statistically invalid: Errors may be infrequent, but they are not insignificant.

An error hurts a team more than a routine putout or assist helps it, for it transforms into a hit (in effect) a batted ball which should have produced an out. The value of a hit is approximately .50 runs,[1] the value of an out approximately −.25. Because an error takes a −.25 situation and makes it a +.50, its cost to the defensive team is on average .75 runs, or the equivalent of *three* outs. Similarly, a fielder who makes a great play, a hit-saving play, has saved his team .75 runs, for he has transformed a +.50 situation into a −.25 one. An outfield error, because it so often produces more than one base for both batter and runners, costs about 1.10 runs; also, each error that allows an existing runner to take an extra base costs the team .25 runs. The upshot is that to balance an error on an individual's ledger, he must make one exceptional play, or three (for an infielder) or four to five (for an outfielder) routine plays. Dick Cramer did a study which determined that about 85 percent of all plays in the field are routine. This information suggests ways in which Range Factor might be improved, but it also provides a theoretical base for the Defensive Linear Weights formula, which we believe to be the best, although imperfect, measure of fielding ability.

Before proceeding to Defensive LWTS, let's look briefly at the three elements besides errors which go into it—putouts, assists, and double plays, weighted variously by position.

A double play is worth more than simply two outs (.50) in run value because it transforms a runner-on-base situation into two outs: The DP is worth .50 runs more than a lone out and 1.00 runs more than a single, so .75 is its averaged value. The DP is valuable, but it is also situation-dependent: The more baserunners the pitcher allows, the greater the opportunity for twin-killings. Also, breaking-ball pitchers tend to elicit more DPs than fastballers.

Putouts and assists? As a rule, putouts are the significant stat for outfielders and assists for infielders—even first basemen, for the overwhelming majority of their putouts, including double plays, are so routine as to be dropped from consideration in the LWTS formula. Outfield assists are of enormous defensive value, and their intimidating effect on runners contemplating an extra base must be reckoned, intuitively if not statistically. However, outfield assists have become four times less prevalent today than at the dawn of big-league play. In the nineteenth century, a season total of 20–30 assists was common for flychasers, and totals twice that high were registered. Because the ball was less resilient and outfielders had little fear of a ball being driven to the wall on a fly, they played so shallow that a line-drive "single" to right could result in the batter being thrown out at first base, and a liner over second base could produce, with a runner breaking from the bag, an unassisted double play by the center fielder. The top fifteen in lifetime outfield assists, excepting Sam Rice and Max Carey, played all or most of their games prior to 1920. The same holds true for lifetime outfield double plays. Do you imagine that all fifteen—or any—of these men had a finer throwing arm than Roberto Clemente? Of course, outfield assists are not necessarily the product of a great arm; we have no measure of extra bases given up, and a high assist total may reflect baserunners' disdain for an outfielder's arm.

For catchers, we subtract strikeouts from putouts, an obvious step, but still are left with a problem in that the remaining putouts are largely a function of the amount of foul territory in the catchers' home parks (the putout totals of other positions—3B, 1B, LF, RF, SS, 2B— are affected too) and high assist totals are usually accompanied by a high number of stolen bases allowed. A catcher's main defensive contribution is in calling the ballgame and keeping his pitcher in the proper frame of mind. Second in importance is his ability to throw— not only the would-be base thieves he intercepts, but also his demonstrated ability in the past which later keeps runners nailed to their bases. Who ran on Johnny Bench in the 1970s? His assist total led the NL in 1968 but never again. His reputation was sufficient to disrupt his opponents' offense. Catcher's assists are uniquely a product of fear, since the steal is an elective play; no matter how good a shortstop you are, you're not going to prevent the other team from hitting the ball to you. (Oddity: On June 14, 1870, the unbeaten Cincinnati Red Stockings led the Brooklyn Atlantics by two runs in the bottom of the eleventh inning; a righthanded Atlantics' batter, Bob Ferguson, turned around to bat lefty rather than hit in the direction of George Wright, the Reds' shortstop; Ferguson thus became the game's first switch-hitter.) The catcher is the only fielder whom opponents can deny the chance to strut his stuff. We have plugged catchers' fielding data into the Linear Weights formula but recognize that a better measure of their ability still lies out there on the horizon.[2]

The Defensive Runs (LWTS) formula is detailed in Chapter 4. In that chapter we presented a table of the top ten fielding performances since 1961; six times in that period a man saved his team 4 or more wins in a season above what an average fielder might have saved in his stead (three of these times the man was Pirate second baseman Bill Mazeroski). Here are the top defensive performances of the century, the twenty-two times a fielder surpassed the 4-win mark:

Table XI, 3. *Top Defensive Seasons, Post-1900, in Wins*

	Year	Player	Team	Position	Wins
1.	1914	Rabbit Maranville	BOS (N)	SS	6.3
2.	1908	Nap Lajoie	CLE	2B	5.8
3.	1963	Bill Mazeroski	PIT	2B	5.0
4.	1927	Frank Frisch	STL (N)	2B	5.0
5.	1910	Eddie Collins	PHI (A)	2B	5.0
6.	1928	Freddie Maguire	CHI (N)	2B	4.7
7.	1980	Ozzie Smith	SD	SS	4.5
8.	1907	Nap Lajoie	CLE	2B	4.5
9.	1908	Heinie Wagner	BOS (A)	SS	4.3

Table XI, 3. *Top Defensive Seasons, Post-1900, in Wins* continued

	Year	Player	Team	Position	Wins
10.	1966	Bill Mazeroski	PIT	2B	4.3
11.	1908	Bill Dahlen	BOS (N)	SS	4.3
12.	1908	Joe Tinker	CHI (N)	SS	4.2
13.	1910	Dave Shean	BOS (N)	2B	4.2
14.	1983	Ryne Sandberg	CHI (N)	2B	4.2
15.	1971	Graig Nettles	CLE	3B	4.2
16.	1908	George McBride	WAS	SS	4.2
17.	1933	Hughie Critz	NY (N)	2B	4.2
18.	1936	Dick Bartell	NY (N)	SS	4.2
19.	1962	Bill Mazeroski	PIT	2B	4.1
20.	1913	Buck Weaver	CHI (A)	SS	4.0
21.	1915	Buck Herzog	CIN	SS	4.0
22.	1914	Donie Bush	DET	SS	4.0

This list, which is a distillation of the top hundred single-seasons list in the rear, provokes some observations and a few timorously extended conclusions.

- No first baseman or catcher ever exceeded the average performance at his position by saving enough runs or wins to make the list of the top hundred season performances, and only one outfielder (Dave Parker in 1977) managed it. (However, the top hundred lifetime list includes many outfielders; first basemen Fred Tenney, George Sisler, and Vic Power; and catcher Bill Bergen.) The results for catcher must be viewed with some suspicion for reasons detailed above, but first base, left field, and right field—because the demands of the positions are less than those of the other five—have historically been occupied by a team's best hitters. The largest variations from the average fielding performance occur, as one would expect, at the positions requiring the greatest skill—shortstop and second base—which explains in part the dominance of the middle infielders in Table XI, 3 and in the Defensive LWTS tables at the back.
- Two men commonly thought to have made it into the Hall of Fame on a pass—Rabbit Maranville and Joe Tinker—emerge here as legitimate all-time greats at their position (as do HOFers Bobby Wallace, Dave Bancroft, and Lou Boudreau among the Lifetime Defensive Win Leaders). And Nap Lajoie, who is viewed today as an overlarge, out-of-position defensive liability who made his name with the bat, is likewise shown to have had exceptional range; on a lifetime basis, he saved more wins with his glove than anybody except Bill Mazeroski. Some others who

show up very well in these lists are George McBride, Lee Tannehill, Dick Bartell, Art Fletcher, and Sparky Adams; and among active players, Mike Schmidt, Ozzie Smith, Graig Nettles, Buddy Bell, and Manny Trillo.

Notable by their surprisingly indifferent showings in this measure are some men universally regarded as all-time greats at their positions: Going around the horn, Hal Chase, Bobby Richardson, Peewee Reese, Brooks Robinson, Paul Blair, and Johnny Bench. The last may be explained by the large intangible component of a catcher's defensive ability. As to the others: Chase was often trying to win bets rather than ballgames; Richardson's totals may have been hurt by the preponderance of lefthanders on the Yankees; Reese was renowned for his reliability more than his range; and Robinson and Blair, well, they won Gold Gloves year after year despite evidence in several seasons that others were doing the job more effectively (also, Robinson was active in a period that saw perhaps the best third-base play in history, by such men as Ron Santo, Clete Boyer, Graig Nettles, Aurelio Rodriguez, and Mike Schmidt). Robinson and Blair *looked* good while going about their business—which is the way to win Gold Gloves, alas. In 1971 Graig Nettles had the best fielding season by any third-sacker in this century, yet Robinson took fielding honors as usual. In 1982 Robin Yount won a Gold Glove with his bat.

The lack of consensus in past years on the meaning of the fielding percentage led sportswriters to throw up their hands and say, "Yes, we know Player B had a higher fielding percentage than Player A and accepted more chances per game, but everybody knows Player A is a great fielder—we've given him all those Gold Gloves already, haven't we?—so how can we give the Gold Glove to Player B?" The batting championship is not awarded to the batter who looks the best or who won it in the past, nor is the ERA title, but fielding is more deceptive: It's hard to become convinced that one must trust the numbers rather than one's eyes. In fact, the lesson is much the same as the one an aviator must learn to fly at night or in fog: Distrust your senses, have faith in the instruments. To look at them, who would have thought that Tom Foli covered more ground than Frank Taveras? Or Richie Ashburn more than Willie Mays? Or Bump Wills more than Charlie Gehringer? But it was so.

Fielding may be a far less important part of the game than batting or pitching, but the superlative fielder can save as many runs beyond the average as a front-rank pitcher or, less frequently, a top hitter. In a good number of the early years of this century, when peak performance diverged from the average to a greater extent than today, the Defensive Runs leader contributed more wins to his team than any batter, and a great fielder, like Rabbit Maranville in 1914, could be the game's top player.

Those days are not likely to return, but when we talk about the greatest players of all time in the final chapter of this book, we will do what no one else has done except in anecdotal fashion—evaluate fielding ability as a significant part of overall ability: For while fielding is only 6 percent of the game, it often is the difference between victory and defeat.

---

[1] Most errors place a man on first base who otherwise would have been retired, but several kinds of errors result in the batter taking two bases.
[2] A modification of the Defensive LWTS formula which we have employed for catchers of the nineteenth century, who took a fearful beating and thus played far fewer games at the position than their twentieth-century counterparts, is to award one additional win for each 100 games caught.

WHAT MAKES TEAMS WIN

Runs. A glib answer, perhaps, and one that will not come as a revelation to a reader of the previous eleven chapters; nonetheless, "runs" is the fundamental answer and the starting point for this discussion, as it is for the entire New Statistical movement. The difference between the runs a team scores and the runs it allows is the best predictor of won-lost percentage (see Chapter 4), but the year-end totals offer only the most basic guide to restructuring a team for the upcoming season: Score too few, then get yourself some hitters; allow too many, then buttress the pitching staff and/or the fielding. Such instruction is easy to give, but difficult to follow.

Our reason for writing this chapter, however, is to provide new, more detailed information about how teams win, information that will have more useful predictive value and that, if acted upon, will be likely to pay off in improved performance. Our studies of statistics previously unkept or uncollected reveal some surprising evidence of how home-park characteristics and home-road performance affect a team's chance of winning—and not in the ways one might suppose from a "reading" of The Book. We will also expand upon how, through use of the Linear Weights System, Pete has produced preseason predictions of the four divisional champions more accurate than those emanating from any other source: In 1971–83, his predictions have bettered those of 152 preseason magazine picks, compared to only 16 which have bettered his, and 3 ties. (*Sports Illustrated* is second-best, at 99-68-4; *Sport* is 16-50-1 since 1979.) His unique method is no longer secret and is now available to you.

The question posed in the title of this chapter has traditionally been understood to mean "Which is more important, offense or defense?" and, recently, "Power or speed?" Baseball pundits have answered with such remarks as "Pitching is 70 percent of baseball" (Connie Mack), or "If you ain't got a bullpen, you ain't got nothin'" (Yogi Berra), or "Give me pitching, defense, and three-run homers" (Earl Weaver), or "Power is out, speed is in" (George Steinbrenner, a presumptive quote). The notable exception among this group was Branch Rickey (yes, again), who in 1954 directed Allan Roth and a team of MIT mathematicians to test several old baseball ideas by the numbers. In the *Life* article previously cited, he wrote:

"Through the years I have felt, along with the best of baseball's old guard, that defense was infinitely more important than offense. Once again, I was faced by facts and forced to reverse my way of thinking. The figures show that offense has gradually taken over the game and has become more important in winning pennants than defense. For the last ten years in both major leagues, the ratio of importance for pennant winners was 54 percent for offense and 46 percent for defense, with pitching about 30 percent of the game. . . . Year by year the pendulum has swung back and forth between offense and defense. . . . But mathematical calculation shows offense clearly in command over the past decade [1945–54]."

Rickey acknowledged that such had not always been the case; that defense was dominant in the years before the lively ball arrived in 1920. How did Rickey come up with the figures to back these claims? He did not say, but we assume that he used his formula for identifying the constituent parts of offense and defense (from which he excluded fielding as being unmeasurable), and applied the formula to all pennant winners in this century. Then he and Roth presumably counted up the number of league champs who "won with offense" against those who "won with defense." He might have included fielding in this correlation study by working with runs scored for the offense and runs allowed for the defense, but his results would not have been substantially different.

A more sophisticated approach—a multivariant linear-regression analysis—was brought to bear on the subject by Arnold Soolman in an unpublished paper that received some distribution in 1970. Soolman's finding supported part of Rickey's contention—that hitting was becoming more important in the late 1940s to early 1950s—but contradicted his view that defense, by which he meant pitching, was diminishing in importance. In fact, the rise of pitching as a determinant of success had been dramatic throughout the century, while the role of batting was nearly a constant; the importance of fielding declined in a manner proportionate to the rise of pitching.

Soolman's study examined not solely the performance of pennant winners but the winning percentage for all 1,166 team-seasons to 1970. He correlated these winning percentages with pitching (as measured by ERA), batting (as measured by runs scored), and fielding (as measured by "UERA"—the average number of unearned runs allowed). Also, Soolman stratified the data—i.e., broke it into four segments corresponding to the generally accepted baseball periods of 1901–20, 1921–45, 1946–60, and 1961–70—to account for fluctuations caused by changes in equipment, style of play, scoring procedures, and so on.[1] The equation he established was:

Winning Percentage = K, (constant) + a (offense) + b (pitching) + c (fielding)

Soolman determined coefficients K a, b, and c through the analysis of the historical data; then he calculated the standard deviation of each variable as it modified these coefficients. His findings of the relative importance of each independent variable are summarized in the table below.

Table XII, 1. *Relative Importance by Period*

Period	Hitting	Pitching	Fielding
1901–20	45.1%	36.1%	18.8%
1921–45	46.2%	43.3%	10.5%
1946–60	48.0%	44.8%	7.2%
1961–70	46.2%	45.9%	7.9%

These figures closely resemble the theoretical breakdown we have cited in several places in the book, which was based upon logic rather than on detailed study: namely, that offense is 50 percent of the game, of which some portion, probably in the area of 1–2 percent, is not hitting; and that because 88 percent of all runs scored are earned, fielding comprises only 6 percent of defense and pitching 44 percent. It is interesting to note, by the way, that in the dead-ball era, when pitching was presumed (by Rickey, among others) to have dominated the game, success in that area of the game counted for only 36.1 percent of the pie; the team that could hit stood a much better chance to chalk up a high winning percentage. This finding is borne out by our study of the 203 pennant winners' rankings in pitching (as measured by ERA) and batting (as measured by runs scored) for all baseball history through the 1982 season, as summarized in Table XII, 2. (We will present this study once more, measuring pitching and batting by Linear Weights, when we discuss the influence of the home park on winning.)

Table XII, 2. *Pitching-Batting Profile of Pennant Winners, 1901–82*

	1876–1900	1901–19	1920–45	1946–60	1961–82	Total
P-1	14	16	27	17	15	89
B-1	23	23	29	14	17	106
Both	8	8	12	7	4	39
Neither	8	9	8	6	16	47
− Avg. P	3	3	0	0	2	8
− Avg. B	1	1	1	1	5	9

P-1 = Ranked first in pitching
B-1 = Ranked first in batting
Both = Ranked first in pitching and in batting
      (also included in P-1 and B-1)
Neither = Ranked first in neither pitching nor batting
− Avg. P = Below league average in ERA
− Avg. B = Below league average in batting average

Since 1920, fifty-nine pennant winners have had the league's best ERA, while sixty have scored the most runs. Pitching wins pennants? Only insofar as it is true that since 1920, seven teams have won pennants despite poor batters (a team batting average lower than the league average) and only two teams have won despite poor pitching (an ERA below the league average).

Another significant attempt to correlate winning with particular aspects of the game was George T. Wiley's analysis of how seventeen traditional baseball stats correlated with team winning percentage in the years 1920–59; his results were reported in "Computers in Baseball Analysis," in the 1976 *Baseball Research Journal*. What Wiley did was to rank each of the eight teams in each season as one to eight in the seventeen categories, then correlate the teams' won-lost standings with their standings in the categories. Next, he programmed the computer to formulate a predictor equation to determine the relative importance of each category in determining won-lost standing. Here are the correlations he obtained (the closer to 1.0, the better the "fit"):

Table XII, 3. *Correlation with Won-Lost Standings, 1920–59*

1.	Fewest Runs Allowed	.749
2.	ERA	.743
3.	Runs Scored	.737
4.	Slugging Percentage	.642
5.	Batting Average	.615
6.	Shutouts	.547
7.	Pitcher Strikeouts	.517
8.	Fielding Percentage	.498
9.	Fewest Errors	.472
10.	Saves	.453
11.	Complete Games	.420
12.	Homers	.419
13.	Fewest Bases on Balls	.368
14.	Doubles	.343
15.	Triples	.298
16.	Stolen Bases	.210
17.	Double Plays	.073

Wiley's study was tainted because the seventeen variables were not independent of each other but rather were interdependent: "fewest errors" interacts with fielding percentage; complete games tend to be a function of increased victories rather than the other way around; and slugging percentage is tied to batting average, homers, doubles, and triples. And of course it is not a surprise that runs allowed (with its interdependent variable, ERA) and runs scored have the best correlation with team standing—when combined, the correlation zooms to .901, and anything else one throws into the hopper raises the figure very slightly. What is interesting, however, is that such presumably important aspects of the game as stolen bases and pitchers' control had insignificant levels of correlation. One may argue that stolen bases did not show up well because the period of major-league play Wiley chose for analysis was one characterized by going for the big inning; however, in 1983 Bill James published a study which served to corroborate the belief of most New Statisticians that the stolen base just doesn't count for much: He found that since 1969 the average finish of the team stolen-base leader in each league has been lower than that of the leader in homers, slugging, batting average—even the leader in fewest walks. And, as you read in Chapter 4, the run value of the stolen base, .30, is set off by a value of − .60 for the failed attempt, making the requisite (break-even) point of success .667, which is approximately the actual rate of success since 1951.

Another area to discuss in determining what makes a team win is efficiency, making the most of what you've got. Rickey equated efficiency with clutch ability, which for a team meant scoring a comparatively high percentage of the men who reached base, and, for pitchers, stranding a high percentage of baserunners. This proposition is questionable; perhaps it merits the sort of detailed investigation we have devoted to other, no less dubious propositions; but we'll back off from this one in the conviction that clutch ability over a season or seasons is an illusion, and that in any event, the more batters who reach base, the more a team will score—that sheer quantity of baserunners is a better guide to a winning team than the rate at which they score.

Efficiency of a different sort may be represented by how close a team comes to the won-lost record predicted from its runs scored and runs allowed. It is revealing that the only two clubs which have consistently outperformed their projections—indicating that they got the most out of their runs scored and allowed—happen to be the best organizations in baseball, the Baltimore Orioles and the Los Angeles Dodgers. Over the six-year period 1976–81, the Orioles averaged 41 points better than expected; earlier, in 1954–63, the Dodgers averaged 27 points higher than expected. Chalk this up to the spirit of the organizations, to the leadership of Earl Weaver and Walter Alston, or to pure luck, but the fact remains: In this century, no other clubs have exceeded their predicted records to such an extent over time.

In 1982 Earl Weaver retired as manager of the Orioles, and for nearly two weeks, the game was without a resident genius; baseball writers were guru-less until the Cardinals defeated the Brewers in Game Seven of the World Series, upon which Whitey Herzog was installed as Weaver's successor. Deposed was Earl Weaver's "pitching, defense, and three-run homers" theory of winning baseball (which had always seemed to us unassailable—all he was asking for was offense and defense, and with that taken care of, he'd think of something if the need arose). Newly ascendant was "Whiteyball," with Herzog proclaimed the apostle of speed and defense, the mocker of the long ball, and the master sculptor of a team optimally fitted to its home park. Herzog's champions did steal more bases and hit fewer homers than any other club in the National League, and their pitching talent, as measured by LWTS, was the league's second best—but fitted to Busch Stadium? The Cards finished 46-35 at home, some 25 points above the historical home average, and 46-35 away, 111 points above the historical road average. Their record on natural grass was better than on the carpet. The Cards won the pennant because they were good enough to win anywhere—their road opponents outhomered them by only 46 to 40, and Card pitchers allowed 284 runs away, compared to 325 at home, a whopping difference. Indeed, because the

Cardinals scored only 13 more runs than their opponents at home, how they finished 46-35 there is something of a mystery, as is their pennant itself. The Cardinals outperformed the won-lost record that their LWTS predicted by 5.4 games. (Hmmm, maybe Herzog *is* a genius?) The Expos, who finished 6 games behind the Cards in the race, underperformed their LWTS prediction by 3.8 games—so you see who *should* have won the 1982 NL East. (More on the LWTS prediction method—as it would apply to *next* year—in the conclusion to this chapter.)

Herzog said, after winning the World Series, "You want to gear your team to what it can do best and capitalize on the possible home-field advantage, but the surest way to win is to keep the other team from scoring. We won with the same thing that the great teams in every sport win with—defense." Now, the Cards won with defense, but that's not the only way to go: We have already shown that offense takes the flag every bit as often. And he did not capitalize on the home-field advantage offered by Busch Stadium, which was 7 percent more friendly to hitters than the average home park. The 1982 Cards won the pennant not by "playing .500 on the road and fattening up at home," which is the axiomatic road to success, but by playing so well on the road that the difference between their home and road records was *smaller* than the league average, not greater—which happens to be the true, documentable path to success.

As discussed in Chapter 5, the average winning percentage of all home teams in this century is .543, while on the road the average is .457. Another way of expressing this is to say that the average team plays 86 percentage points better at home. The 166 pennant winners from 1901 to 1982 (including the two years of the Federal League) played only 76 points better at home—and National League winners had a margin of only 58. One would not have believed it, but it is so. Here are the results by era:

Table XII, 4. *Home-Away W-L Pct. Margins, by Period*

Home-Away Margin	1901–19		1920–45		1946–60		1961–82		TOTAL	
	Lg*	Pen†	Lg	Pen	Lg	Pen	Lg	Pen	Lg	Pen
National	71	39	90	79	85	55	79	55	82	58
American	109	120	89	97	75	75	75	75	88	92
Federal	82	135								
Combined	90	82	90	88	80	65	77	65	86	76

*Lg = League Average Margin
†Pen = Pennant Winners' Margin

How can we explain this phenomenon? Why do pennant winners not rely upon the home-park edge to the same extent as ordinary teams? Let's build a profile of the pennant winner, first looking at the ones that *did* play much better at home. Of the 166 champions in this century, 50 had a home-park margin of 125 points or more (45.3 percent above average). Breaking these totals down by era, and league, we get the following:

Table XII, 5. *Pennant Winners +.125 at Home*

	1901–19	1920–45	1946–60	1961–82	TOTAL
National	1	6	5	5	17
American	9	12	5	6	32
Federal	1				1

Yet of the 50 champions which performed substantially better at home, only 2 played in parks that benefited hitters or pitchers 10 percent more than average—that is, had a Park Factor of 110 or more, or 90 or less. Major league history offers many such parks—147, to be precise—yet only twice have they supplied pennant winners who could take advantage of them. All told, 19 times have such "extreme" parks been home to pennant winners, with 14 of these being hitters' parks (PF of 110 or more) and 5 being pitchers' parks (PF of 90 or less). Amazingly, the average home edge in winning percentage for the 14 teams from the hitters' parks was only .008, or one game over .500; indeed, 6 of these teams played better away from home. Not only did the 14 champions from the hitters' parks fail to take advantage of their circumstances, but 11 of them led their league in *pitching* (as measured by park-adjusted LWTS, the best measure of ability) and only one in batting. Similarly, of the 5 champions from pitchers' parks, 4 led in *batting* and none in pitching.

Rounding out this home-away profile (the basis of which is in Tables XII, 6–9) are the teams which won despite ordinary-to-poor home or away records. One might expect that a pennant winner, even if it did not perform much better at home than on the road, would at least play 10 percent better than the historical major league average both at home and on the road: to do that would mean a home mark of only .597 and a road mark of .503 (league averages being .543 and .457 respectively). And such is the case, although there are some surprises in this area, too. Of the 166 pennant winners, only 24 have had home or road performances less than 10 percent above average (only one of these has played at this lackluster level both at home *and* away—the 1973 Mets). One might expect these 24 cases to be predominantly below the road benchmark of .503, in line with the conventional wisdom, but only 7 teams (excluding the 1973 Mets) played below .503 on the road, while 16 (again excluding those Mets) won the flag despite home records of less than 10 percent above average. And what is more, of the latter group, 7 (sans those dreadful Mets) have cropped up in the last fifteen years!

Conclusions:

(1) The ability to take advantage of one's home park—that is, to an above-average degree—was never terribly important and is less so now than ever before. The ability to win away from home, however— again, to an above-average degree—*is* important, with less than 5 percent of all teams since 1901 being able to take a flag despite a road record of 10 percent or less above average. The team that plays well on the road is, virtually by definition, the team that has the best overall talent; such talent is not necessarily the kind required to pile up an impressive home-park record. Ask a Red Sox fan.

(2) Teams with home parks that favor hitters or pitchers to an extreme degree win the pennant about as often as other teams; 11.4 percent of extreme PF teams, 11.2 percent of all teams. In fact, of *all* teams with extreme PFs the percentage that win pennants is also about the same: 19 teams of 147, or 12.9 percent. Yet to win a pennant with an extreme Park Factor, a team must construct its talent to take maximum advantage of what its home park *hinders,* not what it helps.

(3) A team whose home park favors hitters to an extent 10 percent above average (PF 110) or more *cannot* have a won-lost record 10 percent above average at home *and* win a pennant—at least it has never been done by any of the 94 teams which have played in hitters' parks. So, for such a team to win a pennant, it must (a) have exceptional pitching and (b) win big on the road. The 14 pennant winners from hitters' parks have produced 11 league-leading pitching staffs and have played .627 on the road—*170 points,* or 37 percent, above average. Read well, Boston, Toronto, Pittsburgh, Chicago Cubs, Detroit, Seattle, Minnesota, and Atlanta.

(4) A team whose home park favors *pitchers* to an extent 10 percent above average (PF 90 or less) likewise must construct its team around its batting in order to win a pennant, but may not find it as necessary to excel on the road. The five teams that have won in such parks have averaged .672 at home and .532 on the road. However, the sampling here is much smaller than for the 110-PF parks, so conclusions might best be held in abeyance.

Here is the data upon which the preceding remarks are based:

Table XII, 6. *Home-Away Records, Pennant Winners 1901–82**
NATIONAL LEAGUE

Year	Team	Pct.	Home Pct.	Road Pct.	BPF
1901	PIT	647	652	643	102
1902	PIT	741	789	691	101
1903	PIT	650	657	643	109
1904	NY	693	683	704	109
1905	NY	686	720	654	102
1906	CHI	763	727	800	114
1907	CHI	704	740	671	111
1908	CHI	643	610	675	115
1909	PIT	724	727	720	105
1910	CHI	675	753	597	100
1911	NY	647	662	633	105
1912	NY	682	662	701	109
1913	NY	664	701	627	98
1914	BOS	614	671	558	107
1915	PHI	592	628	539	108
1916	BKN	610	649	571	102
1917	NY	636	641	632	97
1918	CHI	651	658	642	109
1919	CIN	686	732	638	102
1920	BKN	604	628	579	113
1921	NY	614	671	554	100
1922	NY	604	654	553	104
1923	NY	621	610	632	105
1924	NY	608	662	553	91
1925	PIT	621	675	566	99
1926	STL	578	610	545	109
1927	PIT	610	608	613	107
1928	STL	617	545	688	109
1929	CHI	645	675	613	102
1930	STL	597	688	506	102
1931	STL	656	692	618	110
1932	CHI	584	688	481	100
1933	NY	599	640	558	101
1934	STL	621	623	618	115
1935	CHI	649	727	571	98
1936	NY	597	667	526	98
1937	NY	625	667	584	99
1938	CHI	586	571	600	109
1939	CIN	630	688	568	105
1940	CIN	654	724	584	97
1941	BKN	649	675	623	101
1942	STL	688	779	597	103
1943	STL	682	734	627	105
1944	STL	682	711	654	102
1945	CHI	636	653	620	103

1946	STL	628	628	628	110
1947	BKN	610	675	545	103
1948	BOS	595	592	597	103
1949	BKN	630	623	636	106
1950	PHI	591	610	571	98
1951	NY	624	641	608	101
1952	BKN	627	577	680	106
1953	BKN	682	779	584	101
1954	NY	630	697	568	98
1955	BKN	641	727	553	105
1956	BKN	604	675	532	105
1957	MIL	617	584	649	96
1958	MIL	597	623	571	86
1959	LA	564	590	538	104
1960	PIT	617	675	558	100
1961	CIN	604	610	597	109
1962	SF	624	744	506	93
1963	LA	611	654	568	97
1964	STL	574	593	556	116
1965	LA	599	617	580	91
1966	LA	586	654	519	96
1967	STL	627	605	650	113
1968	STL	599	580	617	97
1969	NY	617	634	600	102
1970	CIN	630	704	556	100
1971	PIT	599	650	549	97
1972	CIN	617	553	679	98
1973	NY	509	531	488	99
1974	LA	630	642	617	95
1975	CIN	667	790	543	98
1976	CIN	630	605	654	108
1977	LA	605	630	580	100
1978	LA	586	667	506	102
1979	PIT	605	593	617	108
1980	PHI	562	605	519	109
1981	LA	573	589	556	100
1982	STL	568	568	568	107

AMERICAN LEAGUE

1901	CHI	610	600	515	96
1902	PHI	610	767	429	109
1903	BOS	659	710	609	116
1904	BOS	617	620	613	105
1905	PHI	622	685	560	107
1906	CHI	616	701	527	89
1907	DET	613	649	575	101
1908	DET	588	571	605	111
1909	DET	645	750	539	109
1910	PHI	680	750	608	103

WHAT MAKES TEAMS WIN ◇ 211

1911	PHI	669	730	610	93
1912	BOS	691	740	640	106
1913	PHI	627	658	597	93
1914	PHI	651	680	623	98
1915	BOS	669	733	605	96
1916	BOS	591	636	545	94
1917	CHI	649	727	571	93
1918	BOS	595	700	464	86
1919	CHI	629	686	571	108
1920	CLE	636	654	618	105
1921	NY	641	679	600	103
1922	NY	610	649	571	101
1923	NY	645	605	684	109
1924	WAS	597	610	584	97
1925	WAS	636	707	566	95
1926	NY	591	667	519	96
1927	NY	714	750	679	99
1928	NY	656	675	636	92
1929	PHI	693	780	610	106
1930	PHI	662	763	647	96
1931	PHI	704	800	610	104
1932	NY	695	805	584	93
1933	WAS	651	605	697	104
1934	DET	656	675	635	99
1935	DET	616	679	548	93
1936	NY	667	727	605	95
1937	NY	662	740	584	97
1938	NY	651	714	587	101
1939	NY	702	675	730	98
1940	DET	584	633	533	110
1941	NY	656	662	649	99
1942	NY	669	753	584	96
1943	NY	636	701	571	94
1944	STL	578	701	455	100
1945	DET	575	658	494	109
1946	BOS	675	792	558	108
1947	NY	630	714	545	91
1948	CLE	626	615	636	99
1949	NY	630	701	558	102
1950	NY	636	688	584	102
1951	NY	636	718	553	89
1952	NY	617	636	597	101
1953	NY	656	649	662	94
1954	CLE	721	766	675	103
1955	NY	623	675	571	94
1956	NY	630	636	623	98
1957	NY	636	623	649	97
1958	NY	597	571	623	112
1959	CHI	610	610	610	96

1960	NY	630	714	545	92
1961	NY	673	802	543	91
1962	NY	593	625	561	95
1963	NY	646	725	568	97
1964	NY	611	617	605	105
1965	MIN	630	630	630	105
1966	BAL	606	608	605	102
1967	BOS	568	605	531	117
1968	DET	636	691	580	106
1969	BAL	673	741	605	102
1970	BAL	667	728	605	95
1971	BAL	639	688	593	100
1972	OAK	600	623	577	100
1973	OAK	580	617	543	89
1974	OAK	556	605	506	98
1975	BOS	594	580	608	112
1976	NY	610	563	658	104
1977	NY	617	679	555	97
1978	NY	613	679	549	103
1979	BAL	642	696	588	96
1980	KC	589	605	593	99
1981	NY	551	627	482	97
1982	MIL	586	585	588	93

FEDERAL LEAGUE

1914	IND	575	697	454	110
1915	CHI	565	579	553	96

*Decimal points dropped for ease of expression and reading.

Now let's move on to the question of talent and its impact on winning, for talent may not be reflected accurately in either the team's standing or its run totals. Won-lost percentage correlates best with the differential between the runs scored and the runs allowed, but one or the other of these figures can be mightily distorted by home-park characteristics, as detailed above. The Red Sox may score more runs than anyone in the AL, or the Astros may allow fewer than anyone in the NL, but they may not be the best hitting or pitching teams in terms of *talent*. This will invariably show up in their road records or in their Park Adjusted Linear Weights, which can be calculated, for individuals as well as for teams, with more ease than obtaining the home-road record of all players back through history.

Remember Table XII, 2, which correlated pennant winners with their ranking in pitching and batting? That reflected a study of those teams' ERAs and runs scored. Below is Table XII, 7 with the same format, but the yardstick for pitching and batting talent is Park Adjusted Linear Weights. (The period 1876–1900 is not accounted for here because we do not yet have Park Factors and team LWTS in the database. In the team-stat section of the year-by-year analysis in the back of the book, the columns R/O and R/D reflect Offensive Runs [scored] above the league average, adjusted for games played, and Defensive Runs [allowed] below the league average, similarly adjusted.) Two categories have been added below—Ex-PF Pen (for pennant winners with extreme Park Factors) and Ex-PF Lg (for the league total of extreme Park Factor teams).

Table XII, 7. *Pitching, Batting Profile of Pennant Winners, 1901–82 (LWTS)*

	1901–19	1920–45	1946–60	1961–82	TOTAL
P-1	16	22	10	14	62
B-1	13	18	15	14	60
Both	1	2	2	1	6
Neither	12	14	7	17	50
– Avg. P	4	4	3	4	15
– Avg. B	6	8	2	6	22
Ex-PF Pen.	7	3	4	5	19
Ex-PF Lg.	36	34	16	61	147

Note that whereas in the table correlated with ERA and runs scored, 39 teams ranked first in both pitching and batting, here only 6 do. This implies that pennant winners which appeared to be well balanced, like the Cubs of 1906–08, were really quite unbalanced, with their high Park Factors masking both how extraordinary their pitching was and how ordinary their batting; to look at the run totals alone, one would get a very different impression. Taking a more recent example, the 1980 National League Championship pitted the Houston Astros, with the league's fewest runs allowed, against the Phillies, who had scored more than any team in the league but one. Pitching against hitting, a classic matchup, right? Yes—but not in the way you might think. Park Adjusted LWTS revealed the Phils to have by far the best pitching (and the seventh-best batting) and the Astros the third-best hitting (and second-best pitching).

The import of this is that team balance is not essential to winning: Lopsided teams are winning more frequently now than ever before. (Lopsided teams are defined as those with a ranking in one category, either pitching or batting, far lower than in the other, measured in this way to reflect the expanding number of teams: to 1961, a difference of five places—one in pitching and six in batting, or two in batting and seven in pitching, etc.; six places through 1968; eight places in the AL through 1976 and the NL to 1982; and nine places in the AL, 1977–82.)

Table XII, 8. *Lopsided Pennant Winners*

League	1901–19	1920–45	1946–60	1961–82	TOTAL
NL	1	3	0	5	9
AL	2	1	2	1	6

Table XII, 9 presents a year-by-year record of the talent levels of the pennant winners since 1901, as measured by Park Adjusted Linear Weights.

Table XII, 9. *Pitching-Batting Ranks of Pennant Winners*

NATIONAL LEAGUE

Year	Team	Pitching	Batting	Year	Team	Pitching	Batting
1901	PIT	2	2	1932	CHI	1	4
1902	PIT	2	1	1933	NY	1	5
1903	PIT	1	2	1934	STL	1	7
1904	NY	1	3	1935	CHI	2	1
1905	NY	2	1	1936	NY	1	4
1906	CHI	1	5	1937	NY	2	2
1907	CHI	1	6	1938	CHI	1	7
1908	CHI	1	5	1939	CIN	1	3
1909	PIT	2	1	1940	CIN	2	3
1910	CHI	2	2	1941	BKN	3	1
1911	NY	1	1	1942	STL	2	1
1912	NY	1	3	1943	STL	1	2
1913	NY	1	2	1944	STL	1	1
1914	BOS	2	5	1945	CHI	1	3
1915	PHI	1	5	1946	STL	1	5
1916	BKN	1	3	1947	BKN	2	3
1917	NY	3	2	1948	BOS	2	1
1918	CHI	1	3	1949	BKN	2	1
1919	CIN	2	2	1950	PHI	1	5
1920	BKN	1	7	1951	NY	1	2
1921	NY	4	1	1952	BKN	2	2
1922	NY	1	3	1953	BKN	3	1
1923	NY	3	4	1954	NY	1	2
1924	NY	5	1	1955	BKN	1	1
1925	PIT	3	1	1956	BKN	2	1
1926	STL	3	3	1957	MIL	3	1
1927	PIT	3	3	1958	MIL	3	1
1928	STL	1	4	1959	LA	1	5
1929	CHI	2	1	1960	PIT	3	3
1930	STL	2	3	1961	CIN	1	6
1931	STL	1	3	1962	SF	8	1

NATIONAL LEAGUE

Year	Team	Pitching	Batting	Year	Team	Pitching	Batting
1963	LA	3	4	1973	NY	3	9
1964	STL	1	9	1974	LA	3	1
1965	LA	3	5	1975	CIN	7	1
1966	LA	1	6	1976	CIN	5	1
1967	STL	1	7	1977	LA	1	3
1968	STL	2	4	1978	LA	1	1
1969	NY	2	2	1979	PIT	1	5
1970	CIN	4	2	1980	PHI	1	7
1971	PIT	7	1	1981	LA	2	6
1972	CIN	6	1	1982	STL	2	8

AMERICAN LEAGUE

Year	Team	Pitching	Batting	Year	Team	Pitching	Batting
1901	CHI	2	1	1933	WAS	2	3
1902	PHI	2	5	1934	DET	3	2
1903	BOS	1	5	1935	DET	5	1
1904	BOS	1	3	1936	NY	3	1
1905	PHI	1	3	1937	NY	2	1
1906	CHI	4	3	1938	NY	1	2
1907	DET	3	1	1939	NY	1	1
1908	DET	3	1	1940	DET	2	3
1909	DET	2	1	1941	NY	2	2
1910	PHI	1	2	1942	NY	2	1
1911	PHI	5	1	1943	NY	4	1
1912	BOS	1	3	1944	STL	3	5
1913	PHI	8	1	1945	DET	2	6
1914	PHI	7	1	1946	BOS	3	1
1915	BOS	2	2	1947	NY	6	1
1916	BOS	2	4	1948	CLE	1	2
1917	CHI	5	1	1949	NY	3	2
1918	BOS	5	1	1950	NY	2	2
1919	CHI	3	2	1951	NY	4	1
1920	CLE	1	2	1952	NY	2	2
1921	NY	1	3	1953	NY	3	2
1922	NY	2	3	1954	CLE	2	2
1923	NY	1	4	1955	NY	4	1
1924	WAS	1	5	1956	NY	2	1
1925	WAS	2	1	1957	NY	1	1
1926	NY	3	1	1958	NY	1	4
1927	NY	1	1	1959	CHI	1	5
1928	NY	6	1	1960	NY	6	1
1929	PHI	1	3	1961	NY	5	1
1930	PHI	5	2	1962	NY	6	1
1931	PHI	1	3	1963	NY	3	3
1932	NY	5	1	1964	NY	2	5

Year	Team	Pitching	Batting	Year	Team	Pitching	Batting
1965	MIN	2	2	1974	OAK	1	7
1966	BAL	4	1	1975	BOS	3	4
1967	BOS	2	5	1976	NY	1	3
1968	DET	3	2	1977	NY	5	1
1969	BAL	1	2	1978	NY	1	8
1970	BAL	3	1	1979	BAL	1	5
1971	BAL	4	2	1980	KC	6	3
1972	OAK	3	4	1981	NY	1	4
1973	OAK	5	1	1982	MIL	10	1

FEDERAL LEAGUE

Year	Team	Pitching	Batting
1914	IND	2	4
1915	CHI	3	2

Conclusions: To win a pennant, construct your team so that it excels in *either* pitching or batting (keeping in mind the recommendations based on Park Factor) rather than strive for balance. Of the 166 pennant winners, 1901–82, excluding the 6 which finished first in both pitching and batting (1911 Giants; Yankees of 1927, 1939, and 1957; Dodgers of 1955 and 1978), 110 ranked first in either pitching or batting talent, while 48 ranked first in neither. Of those 110, 55 finished fourth or worse in the category which was not their strength. Does it matter whether the team's pitching or batting is the aspect that excels? Since 1961, teams weighted strongly toward pitching have won five pennants in the NL, two in the AL; teams weighted strongly toward batting have won four NL flags, two in the AL. Combine these figures with the dramatically lower home-away winning margin in the NL, and draw your own conclusions.

And, last, how can we use the Linear Weights System to analyze the strengths and weaknesses of a team, to suggest how it might be improved, and to assess who figures to be the winner next year? We went into this somewhat at the end of Chapter 4, in analyzing the 1983 chances of the 1982 NL West champion Atlanta Braves. Here let's look at the New York Yankees of 1982 and 1983, and consider what they need to overtake Baltimore and Detroit in 1984.

In 1982 the Yankees finished fifth in the American League East, just one game out of a tie for the cellar spot. Their team was in disarray, and while their talent was too thin in all areas but the bullpen for them to have made a run at the pennant, it should have been good enough for a record of 82-80 (they finished 79-83). By adding Don Baylor and Steve Kemp, they figured to improve their offense, and Dale Murray figured to relieve the burden on Rich Gossage.

But Murray was no help, nor was Kemp; only Baylor of the new acquisitions was a big plus. The offense picked up dramatically, however, largely through efforts of holdovers Roy Smalley, Graig Nettles, Dave Winfield, and Butch Wynegar. But the offensive production of Wynegar (16 runs) was offset by that of Rick Cerone (-15); and playing Andre Robertson at shortstop did not improve the team defense by enough (if at all) to compensate for moving Smalley to the bench, or to another position which would necessitate the benching of a hitter like Lou Piniella or Oscar Gamble. The jury is still out on Don Mattingly, or should be—he doesn't draw walks or hit for power—and Omar Moreno will soon have the fans longing for Jerry Mumphrey. The answer for the Yankee offense looks to be to place Winfield in center, use a platoon alignment in left, and give Kemp another season in right to hit his stride—he is a proven quantity. Leave second base in Willie Randolph's capable hands and look to upgrade first base, but, failing that, go with Ken Griffey again as a stopgap.

Still, for all their problems, the 1983 Yanks were behind only the Brewers and Tigers in Batting Wins. Where they really needed help was on the mound, as only Gossage and Ron Guidry contributed so much as one win beyond average. Bob Shirley and Jay Howell by themselves negated (and then some) the work of Guidry and Gossage, and Matt Keough and Doyle Alexander were prodigiously awful in limited exposure. Dave Righetti will retain his spot in the rotation, but Shane Rawley or Ray Fontenot—one or the other—would be of more use to this jerry-built team (no lefthanded reliever, no righthanded starter—unless you believe that John Montefusco has, at the age of thirty-three, become an overnight sensation).

Bottom line: The Yanks need too many things to go right to win in 1984. Now if they could swing a major deal with Texas, which has the opposite problem from the Yanks (-11.4 Batting Wins, $+12.8$ Pitching Wins), they could help themselves, but the Yankees' best players are "mature," and Texas would be foolish to trade their live arms for New York's tired blood. No, the Yankees are not going to win the pennant by outpitching Baltimore . . . maybe they can improve their hitting? By trading a pitcher like Guidry or Righetti, perhaps, but the fans would scream. Any way you look at this bunch, they figure to decline in the standings unless they make some moves, and a pennant is out of the question barring divine intervention. How to go against the Linear Weights tea leaves? Move Rawley or Fontenot to the bullpen and Winfield to center; trade Cerone for whatever he will bring (this should have been done last year); let Butch Wynegar catch 130–140 games; play Smalley at short every day, and let Robertson sit; inject formaldehyde into the veins of Nettles and Piniella and hope to get another year out of each. Oh, and pick up a righthanded starter or two—Dave Stieb would be nice.

This is a team that is treading water and will have to struggle mightily not to go under.

This same approach may be used to evaluate the twenty-five other teams, for we give a complete statistical line—in the manner shown above for the Yankees—for all regulars and frequent substitutes who appeared in 1983. We could go through each team in this manner, but then this would be a different kind of book—and anyway, it will be more fun for you to try it yourself.

It's nice to win—it makes everybody feel good. But baseball management must keep in mind the bottom line when evaluating personnel moves and contract offers which seem to promise additional victories. By trading Neil Allen and Rick Ownbey to the Cards for Keith Hernandez, which meant the deportation of Dave Kingman to Siberia, the Mets figured to gain 2–3 net wins, with Hernandez worth 3 wins over Kingman for a full season and Ownbey and Allen worth about .5 wins together. In fact, they did gain 2 wins from Hernandez in his 95 games with New York, and Allen was worth − .2 wins to St. Louis for his 122 innings pitched with them. What was this gain worth to the Mets in attendance?

Pete has studied attendance figures for all clubs from 1969 through 1983 and found that each win beyond the previous year's total increased attendance by an average of about 25,000, or a 1.87 percent increase for each additional win. But the Mets' fans are the most starved for victory, and each additional win has meant an attendance gain at Shea Stadium of 56,000! Yankee fans are nearly as impressed by a winning team, averaging 35,000 for each extra victory, which is the high mark in the American League. The most indifferent (or least fickle) fans reside in Baltimore, where each Oriole victory boosts attendance by only 9,000, and Cleveland, where an Indian victory sells only 11,000 more tickets. (Baltimore's low figure is mainly the result of drawing just over 1 million in 1969 and 1970 with teams that won 108 and 109 games, then drawing 2 million with the 1983 team that won just 98 games.) Two thirds of the teams, however, are in the range of 20–36 thousand, with aberrant results for Seattle and Toronto because many fans were drawn to those parks in the first year of operation simply by novelty. (The strike-shortened season of 1981 was compensated for by multiplying wins and attendance by 1.5.) Here are the results of the study:

Table XII, 10. *Attendance Gain per Additional Victory*

NATIONAL LEAGUE		AMERICAN LEAGUE	
New York	56,000	New York	35,000
Chicago	36,000	Minnesota	24,000
Philadelphia	35,000	Chicago	24,000
Houston	34,000	California	24,000
Cincinnati	33,000	Boston	23,000
Atlanta	32,000	Kansas City	22,000
St. Louis	24,000	Milwaukee	20,000
San Diego	23,000	Oakland	18,000
Montreal	23,000	Detroit	15,000
Los Angeles	22,000	Texas	12,000
Pittsburgh	22,000	Cleveland	11,000
San Francisco	18,000	Baltimore	9,000
		Seattle	9,000*
		Toronto	3,000*

*Not statistically valid: There is too small a sample to date.

It appears that the Mets and Yanks have been correct to bid high for free agents and to sign their better players to generous contracts; the Orioles and Indians perhaps should exercise restraint in their spending, concentrating their efforts on their farm systems rather than anteing up for the free-agent game.

[1] In our own studies which follow, we close the dead-ball era at 1919 rather than 1920.

13

GREAT SINGLE-SEASON PERFORMANCES

In describing baseball, superlatives roll trippingly off the tongue; witness the current bankruptcy of the term "star," of which once there were few but now evidently there are so many that the absurd term "superstar" has come into vogue. "Great," another word that wore out its meaning, has been used so promiscuously for so long that it has come to denote anything out of the ordinary. Something "great" happens not only every season but also every game and, to listen to TV announcers, every inning. Pitchers make great pitches routinely. Fielders make great plays whenever they take more than two steps. Great ballgames occur several times a week. While it is surely shoveling sand against the tide to crusade for restoring "great" to greatness, we will confine ourselves in this chapter to a narrow and rigid use of the term, in all its archaic grandeur.

Amid salary negotiations, a player's agent might say his client had a great season if he batted .280 with 90 RBIs and 25 homers. By 1980s standards that is good indeed, but it is as far removed from greatness as it is from disgrace. A great season for purposes of this discussion is not simply a league-leading performance in a major category—in other words, even the 1 percent of all batters who may top their league in a given category in a given year is too lax a standard when looking over all the records of baseball since 1876. We will be looking for performances that are of such a magnitude that other seasons which were deemed great in their time, and won for their fashioners adulation and sometimes wealth, pale by comparison.

A few of these may be landmark seasons—like Babe Ruth's 1920 or Jim Konstanty's 1950 or Maury Wills's 1962—which led to changes in the fabric of the game itself, while others may be less influential, perhaps, but no less lofty: Rabbit Maranville in 1914, Walter Johnson in 1913, or Joe Morgan in 1975, for example. Our hope is that this chapter will spur you to peruse the tables—lifetime and seasonal highs, and annual summaries—where you will find a gold mine of new data that may well disclose aspects of the game that have remained hidden from us.

The average fan, whose memory or knowledge of baseball extends only spottily to the years before 1960, might recall some of these recent accomplishments as great ones: the .390 batting average by George Brett in 1980 and the .388 by Rod Carew in 1977; the triple crowns of Carl Yastrzemski in 1967 and Frank Robinson the year before; the dual home-run barrage of Mickey Mantle and Roger Maris in 1961; the pitching of Ron Guidry in 1978, Steve Carlton in 1972, Bob Gibson and Denny McLain in 1968; the base stealing of Lou Brock in 1974 and Rickey Henderson in 1982; the relief heroics of Mike Marshall in 1974 and Dan Quisenberry in 1983. In the years when these men recorded their deeds, all in their fields that had gone before dimmed in comparison—which may have been a good, if only temporary, counterforce to baseball's powerful undertow of nostalgia. And besides, the environment for achieving great records was far more propitious in earlier years because the average level of skill was lower, enabling stars like Hoss Radbourn and Babe Ruth, Rogers Hornsby and Grover Cleveland Alexander, to attain marks that seem unreal today—respectively, 60 wins; an .847 slugging percentage; a .424 BA; and 16 shutouts with tiny Baker Bowl as a home park.

Some "classic" accomplishments, like Bill Terry's .401 BA in 1930 or Nolan Ryan's 1.69 ERA in 1981, lose a good deal of their luster when recast into the new, more accurate measures introduced in this book, while others show to even better advantage, like Ty Cobb's .410 in 1912 (the twentieth century's best when normalized to league average and adjusted for home park) and Walter Johnson's 1913 season, when he went 36-7 with an ERA of 1.14 despite a Senator team that was under .500 without him and a home park that favored hitters to an extreme degree.

What follows is a series of tables delineating the top twenty performances since 1901 in traditional statistical terms, accompanied by a ranking of those performances in terms of the equivalent new stat—i.e., the one that measures accurately what the traditional stat is purported to measure.[1] We will then highlight the "surprise" great seasonal performances revealed by the New Statistics.

Our first inclination was to broaden the sampling of great seasons by employing somewhat generous cutoff points—a .370 BA rather than .400, 140 RBIs rather than 160, sub-2.00 ERAs rather than sub-1.50 ones—because all the great pitching stats posted in 1901–19 and the great hitting stats registered in 1920–41 were relegating recent players to the shadows; however, by lowering the fences in hopes that more modern players would clamber over, we instead generated a stampede of lesser-known oldtimers whose stats seemed largely a product of their era's dominant characteristic.

So, to begin:

Table XIII, 1. *Batting Average, Top Twenty Since 1901*

(Superscript numbers = Park Adjusted,
Normalized On Base Average rankings)

1.	1924	Rogers Hornsby, STL (N)	.424[5]
2.	1901	Nap Lajoie, CLE	.422[71]
3.	1922	George Sisler, STL (A)	.420
4.	1911	Ty Cobb, DET	.420
5.	1912	Ty Cobb, DET	.410[36]
6.	1911	Joe Jackson, CLE	.408[67]
7.	1920	George Sisler, STL (A)	.407
8.	1941	Ted Williams, BOS (A)	.406[1]
9.	1925	Rogers Hornsby, STL (N)	.403[34]
10.	1923	Harry Heilmann, DET	.403[59]
11.	1922	Ty Cobb, DET	.401
12.	1930	Bill Terry, NY (N)	.401
13.	1922	Rogers Hornsby, STL (N)	.401
14.	1929	Lefty O'Doul, PHI (N)	.398
15.	1927	Harry Heilmann, DET	.398
16.	1921	Rogers Hornsby, STL (N)	.397[46]
17.	1912	Joe Jackson, CLE	.395[64]
18.	1921	Harry Heilmann, DET	.394
19.	1923	Babe Ruth, NY (A)	.393[10]
20.	1925	Harry Heilmann, DET	.393
20.	1930	Babe Herman, BKN	.393

We compare batting average with On Base Average rather than with Linear Weights or On Base Plus Slugging because the intent of the BA is to measure the ability to reach base safely, not to produce runs or hit for extra bases. (If the base on balls had existed when the BA was invented in the 1860s, the statistic might well have taken the form of the OBA.) A walk is not as good as a hit, but it is more nearly so than any other offensive event (.33 runs for a walk, .46 for a single; a double, for example, is worth .81 runs). It is interesting to note that Ty Cobb's best season for batting average—his .420 in 1911—does not make the top hundred in NOBA, but five other Cobb seasons *do,* none of them a .400 season. Also, when George Sisler and Bill Terry passed the magic .400 mark, they did not "pass" enough to post impressive OBAs. One might correlate the top batting average marks with normalized slugging percentage, to see whether Sisler or Terry was turning up his nose at walks with the intent of maximizing his extra-base blows . . . nope, these were relatively soft batting averages, more like those of Mickey Rivers or Willie Wilson than like those of Ruth or Hornsby.

In the nineteenth century, the .400-plus batting averages numbered sixteen—of which twelve occurred in the years 1894–99, when pitchers were struggling with the adjustment to the new 60′6″ distance to home plate. The highest marks were Hugh Duffy's .438 in 1894, Tip O'Neill's .435 in 1887, Willie Keeler's .432 in 1897, Ross Barnes's .429 in 1876, and Jesse Burkett's .423 in 1895; both Burkett and Ed Delahanty topped the .400 mark three times, as Cobb and Hornsby were to do later. Of the men listed above, only Barnes and O'Neill are among the top twenty-five in NOPS for this era, and Delahanty joins Barnes and O'Neill in NSLG.

Bottom line: A .400 BA is gaudy, but unaccompanied by a high number of walks or extra-base hits, it is not a great performance.

Table XIII, 2. *Runs Batted In, Top Twenty Since 1901*

(Superscript numbers = Park Adjusted Linear Weights, ranked by Wins)

1.	1930	Hack Wilson, CHI (N)	191
2.	1931	Lou Gehrig, NY (A)	184[36]
3.	1937	Hank Greenberg, DET	183
4.	1938	Jimmie Foxx, BOS (A)	175[47]
5.	1927	Lou Gehrig, NY (A)	175[6]
6.	1930	Lou Gehrig, NY (A)	174[22]
7.	1921	Babe Ruth, NY (A)	171[1]
8.	1935	Hank Greenberg, DET (A)	170
9.	1930	Chuck Klein, PHI (N)	170
10.	1932	Jimmie Foxx, PHI (A)	169[32]
11.	1937	Joe DiMaggio, NY (A)	167
12.	1930	Al Simmons, PHI (A)	165
13.	1934	Lou Gehrig, NY (A)	165[18]
14.	1927	Babe Ruth, NY (A)	164[7]
15.	1931	Babe Ruth, NY (A)	163[17]
16.	1933	Jimmie Foxx, PHI (A)	163

17.	1936	Hal Trosky, CLE	162
18.	1929	Hack Wilson, CHI (N)	159
19.	1937	Lou Gehrig, NY (A)	159[56]
20.	1949	Ted Williams, BOS (A)	159
20.	1949	Vern Stephens, BOS (A)	159

The RBI is accepted by traditionalists as the best measure of run production and is thought to be the single most important batting stat by all major leaguers who bat in the middle of the order. But as we have shown in earlier chapters, the best measure of run production and thus value to the team, regardless of position in the batting order, is Linear Weights. And among the top twenty RBI seasons, the number that place in the top *hundred* in LWTS Wins is only ten—the same number as the BA-OBA correlation produced. It is hard to believe that Hack Wilson and Hank Greenberg, the top two RBI performers of all time on a seasonal basis, did not have even one top-hundred season in LWTS between them, yet it is true; their RBI totals were "sports," produced as much by circumstance as by superman efforts.

Runs batted in were not recorded continuously in the nineteenth century, and the deadened ball, the lack of acclaim for power hitting, and most important, the shorter schedules, combined to keep all players of the period below the cutoff point of Table XIII, 2—except for Sam Thompson, the most prolific RBI producer of all time. Not only did he drive in 166 runs in 1887 (in 127 games) and 165 in 1895 (in 119 games), but he also had the highest RBI-per-game ratio of anyone in history, for both season and career. Other significant RBI performances of the years before 1901 were by Cap Anson in 1880 and 1894, Ed Delahanty in 1893, Dave Orr in 1890, and Hugh Duffy in 1894. The correlation of these RBI highs with the nineteenth century's top twenty-five in LWTS is extremely poor, with only Thompson in 1895 and Duffy in 1894 making the cut.

Bottom line: RBIs correlate about as well with slugging percentage as with LWTS and are not indicative, in and of themselves, of run-producing ability. As mentioned early on, the RBI is situation-dependent to a high degree, with batting position, team ability, and the home park exerting a huge force on this overemphasized statistic.

As you can see, the correlation on p. 226 between the traditional measure of power hitting and the newer one, which measures that portion of slugging percentage which is *not* batting average (Isolated Power rises only by extra-base hits, not singles) is a significant one. That is because the top sluggers are all, with the exception of Mantle in 1956 and Williams in 1957, a product of the years before World War

Table XIII, 3. *Slugging Percentage, Top Twenty*

(Superscript numbers = Normalized,
Park Adjusted Isolated Power ranking)

1.	1920	Babe Ruth, NY (A)	.847[1]
2.	1921	Babe Ruth, NY (A)	.846[3]
3.	1927	Babe Ruth, NY (A)	.772[5]
4.	1927	Lou Gehrig, NY (A)	.765[6]
5.	1923	Babe Ruth, NY (A)	.764[9]
6.	1925	Rogers Hornsby, STL (N)	.756[16]
7.	1932	Jimmie Foxx, PHI (A)	.749[23]
8.	1924	Babe Ruth, NY (A)	.739[8]
9.	1926	Babe Ruth, NY (A)	.737[10]
10.	1941	Ted Williams, BOS	.735[49]
11.	1930	Babe Ruth, NY (A)	.732[17]
12.	1957	Ted Williams, BOS	.731[57]
13.	1930	Hack Wilson, CHI (N)	.723[85]
14.	1922	Rogers Hornsby, STL (N)	.722[27]
15.	1930	Lou Gehrig, NY (A)	.721[39]
16.	1928	Babe Ruth, NY (A)	.709[7]
17.	1930	Al Simmons, PHI (A)	.708[73]
18.	1934	Lou Gehrig, NY (A)	.706[14]
19.	1956	Mickey Mantle, NY (A)	.705[42]
20.	1938	Jimmie Foxx, BOS (A)	.704[46]

II, when power hitters customarily hit for average as well. Note that since 1962 no one has slugged .700, and only three times has anyone surpassed .650 (Willie McCovey in 1969, Henry Aaron in 1971, and George Brett in 1980). Interestingly, Ted Williams's 1941 season is not among the top fifty in ISO because his .406 BA contributed so heavily to his .735 SLG. Babe Ruth has a lock on the top spots because, unlike Williams or Foxx, he is not "hurt" by a high Park Factor or by normalizing—no one was hitting for power the way he was before 1930, except maybe Hornsby.

Ruth's 1920 season is the best any mortal has ever had. Not only did it provide the best slugging percentage and Isolated Power, it also ranks first in NOPS, second in Linear Weights (to Ruth's own 1921, when he played in ten more games), seventh in NOBA, first in home run percentage (11.8 homers for every 100 at bats), and seventh in home runs. Anyone who would disagree with calling this season the best ever would have to choose Ruth's 1921 or 1927. (Some diehards hold out for Williams's 1941, but they're wrong.) In 1920, moreover, Ruth's 54 home runs were more than fourteen of the other fifteen *teams* could muster (the year before, his last with the Red Sox, he had hit 29 of his own team's 33 homers).

There has never been another player like Ruth in terms of his impact on the game. Of the top eleven seasons in Isolated Power, Ruth has ten; of the top ten in LWTS, six; of the top fourteen in NOPS, nine. On a lifetime basis, where differences can be expected to narrow from the wide variances which characterize the top single-season feats, Ruth's NOPS is 9.5 percent higher than the second-best batter's, Williams; 32.6 percent better than the tenth-best, Joe DiMaggio; and 63.3 percent over the hundredth-best batter of this century, as measured by NOPS, Keith Hernandez (that's about the difference between Dale Murphy and Ron Hodges in the 1983 season).

No batter dominated the nineteenth century to anything like that extent, but Dan Brouthers very nearly had an impressive double: On a lifetime basis, he led all performers of the period in normalized slugging and was second to Billy Hamilton—by five thousandths—in normalized On Base Average; obviously his NOPS was tops, and his LWTS was too. Not much contest as far as we're concerned—Big Dan was the most dominant hitter before Ruth. There aren't too many other nineteenth-century sluggers worth mentioning—the ones to top .600 did so in 1894–96, excepting Tip O'Neill's .691 in 1887. The best of the lot (and one must remember they all were historical anomalies) were Ed Delahanty, Sam Thompson, Dave Orr, and Roger Connor; though Connor had no season which was among the top twenty-five in SLG, his lifetime mark was up there with that of the others, and he hit more career homers than anyone before Ruth.

We should, by all rights, move on to the last significant batting category, home runs—but what is there to say that hasn't been said? That they are desirable, that there is a price to pay in increased strikeouts, that those who hit them drive BMWs rather than Subarus? Only seventeen times in the history of the game has anyone hit 50 homers, and four of those times were by Ruth; Kiner, Mantle, Mays, and Foxx each managed it twice. All but two of the plus-50 men are in the Hall of Fame—Roger Maris will not make it, and George Foster will only if he takes a cue from Lazarus. There is no denying it: Home runs are glamor events which make for glamor stats; but once you know how many, there's not much else to know, except to compile home-away homer profiles of those who play in extreme parks. It is illuminating to learn, for example, that in 1983 Houston's Dickie Thon hit 16 of his 20 homers on the road and that even Jim Rice hit 23 of his 39 homers on the road. More of this type of homer analysis may be found in Chapter 5.

Statistically, home runs are as shallow as any "counting" stat—

doubles, triples, steals, wins, complete games, shutouts, pitcher strike-outs, etc.—although for many fans they record the most compelling events on the baseball field; in fact, it may be argued that "counting" statistics are not statistics at all but simply ledgerbook entries awaiting statistical analysis. Let's move on to the great pitching performances.

The earned run average has been viewed as the best index of a pitcher's ability ever since it became an official stat. Pitching for a poor offensive team would unfairly hurt one's won-lost record; pitching for a poor defensive team would unfairly boost one's runs allowed: ERA seemed the perfect solution, divorcing the pitcher's index of ability from the efforts of his teammates. Of course, it doesn't really do that because poor defense shows itself in ways more damaging to a pitcher's ERA than mere errors. Nonetheless, it is the best official pitching stat.

What to do about the utter domination of the top-twenty list by dead-ball hurlers? The first step one might consider would be to nor-malize by league average; this succeeds in adding, among post-1920 pitchers, Lefty Grove in 1931, Whitey Ford in 1958, Vida Blue in 1971, and Ron Guidry in 1978. However, a better measure of pitching effec-tiveness than NERA is Linear Weights, which has a built-in normaliz-ing factor and credits a pitcher for being effective over a greater number of innings—Harry Krause's 1.39 ERA over 213 innings in 1909 is not as fine an accomplishment as Walter Johnson's identical 1.39 over 368 innings in 1912 (disregarding for the moment the differ-ing league averages).

Fourteen of the top twenty ERAs are also among the top hundred in LWTS wins, but many are way down the list. Great performances pointed up by LWTS which do not seem exceptional from a glance at the ERA alone are: Dolf Luque in 1923; Dizzy Trout in 1944; Juan Marichal in 1965; Vida Blue and Wilbur Wood in 1971; Bert Blyleven in 1973; and more—see the tables.

Nineteenth-century ERAs, as computed by the ICI group for the 1969 *Baseball Encyclopedia,* are a statistical problem for a number of reasons cited in Chapter 10, the prime one being the great number of errors in the field and by the battery. A purely statistical problem is the changing schedule: In 1876 George Bradley, for example, threw 16 shutouts and won 45 games in a 70-game schedule—a sensational year, but how can we compare his 6.0 LWTS Wins to the 9.8 of Amos Rusie in 1894, when the schedule called for 132 games?

The early years produced baseball's all-time low ERA—Tim Keefe's 0.86 in 1880, and these other men would have qualified for the top twenty listed above: Denny Driscoll in 1882, Bradley in 1876 and

1880, Guy Hecker in 1882, and the great Hoss Radbourn in 1884. However, the marks of Keefe and Bradley were accomplished at a pitching distance of 45 feet and the others at a distance of 50 feet, and the rules for calling strikes and balls were radically different from modern practice.

Table XIII, 4. *Earned Run Average, Top Twenty Since 1901*

Superscript numbers = Linear Weights rankings, by Wins

1.	1914	Dutch Leonard, BOS (A)	0.96[93]
2.	1906	Three Finger Brown, CHI (N)	1.04[27]
3.	1968	Bob Gibson, STL	1.12[12]
4.	1913	Walter Johnson, WAS	1.14[1]
5.	1909	Christy Mathewson, NY (N)	1.14[46]
6.	1907	Jack Pfiester, CHI (N)	1.15
7.	1908	Addie Joss, CLE	1.16[59]
8.	1907	Carl Lundgren, CHI (N)	1.17
9.	1915	Pete Alexander, PHI (N)	1.22[3]
10.	1908	Cy Young, BOS (A)	1.26
11.	1910	Ed Walsh, CHI (A)	1.27
12.	1918	Walter Johnson, WAS	1.27[6]
13.	1905	Christy Mathewson, NY (N)	1.27[14]
14.	1910	Jack Coombs, PHI (A)	1.30[80]
15.	1909	Three Finger Brown, CHI (N)	1.31[47]
16.	1902	Jack Taylor, CHI (N)	1.33[33]
17.	1910	Walter Johnson, WAS	1.35[56]
18.	1912	Walter Johnson, WAS	1.39[2]
19.	1907	Three Finger Brown, CHI (N)	1.39
20.	1909	Harry Krause, PHI (A)	1.39

Won-lost percentage is linked to the quality of a pitcher's offensive and defensive support, so it is not a trustworthy measure of performance. But wins *are* important—that's the objective, after all—and so are losses. How to divorce a pitcher's ability to win from that of his team? Ted Oliver set out to answer the question back in 1940, when he started to work with the rating system we have modified slightly to become Wins Above Team (see Chapter 10). Note that the correlation between won-lost percentage and Wins Above Team is spotty, which is as one would expect because the two stats are measuring rather different things. An interesting example of their differences is that some stratospheric won-lost percentages, achieved with decent but not pennant-caliber clubs (Allen's, Fitzsimmons', etc.), don't make the top hundred, while a lower percentage like Denny McLain's in 1968, though achieved for a world champion, ranks eighteenth in Wins

Above Team. Why? Because McLain had 37 decisions while Allen, Fitzsimmons, Seaver, etc., all had fewer than 20. The Oliver method credited both efficiency *and* durability, in much the way that the LWTS System does.

Nineteenth-century won-lost percentages are not as impressive as those posted after 1901, largely because the average starter had so much additional exposure to defeat (in plain language, he pitched more often) that it was harder to maintain a winning rate of 80 percent. The only performance that would make the twentieth-century list is Fred Goldsmith's .875 for the Chicago White Stockings in 1880. In Wins Above Team, *eleven* pitchers top Steve Carlton's post-1901 high of 17.1, but this is a phenomenon tied to time as well as place, a time when one pitcher might start and complete anywhere from 50 to 100 percent of a team's games.

Table XIII, 5. *Won-Lost Percentage, Top Twenty*

Superscript numbers = Wins Above Team ranking

1.	1959	Roy Face, PIT	.947[23]
2.	1938	Johnny Allen, CLE	.938
3.	1978	Ron Guidry, NY (A)	.892[26]
4.	1940	Freddie Fitzsimmons, BKN	.889
5.	1931	Lefty Grove, PHI (A)	.886[70]
6.	1978	Bob Stanley, BOS	.882
7.	1951	Preacher Roe, BKN	.880[86]
8.	1912	Joe Wood, BOS (A)	.872[24]
9.	1907	Bill Donovan, DET	.862[41]
10.	1961	Whitey Ford, NY	.862
11.	1914	Chief Bender, PHI (A)	.850
12.	1930	Lefty Grove, PHI (A)	.848[83]
13.	1916	Tom Hughes, BOS (N)	.842
14.	1924	Emil Yde, PIT	.842
15.	1940	Schoolboy Rowe, DET	.842
16.	1954	Sandy Consuegra, CHI (A)	.842
17.	1961	Ralph Terry, NY	.842
18.	1963	Ron Perranoski, LA (N)	.842
19.	1934	Lefty Gomez, NY (A)	.839[48]
20.	1968	Denny McLain, DET	.838[18]

The correlation between saves and Relief Linear Weights is abysmal, with only two of the top fourteen relief seasons as measured by the former appearing among the top fifty in the latter. Is the new stat a dud? No, it simply is not situation-dependent in the way that saves and its cousin, Relief Points, are. A man designated as an ace—a Quisenberry, Gossage, or Sutter—will string together several scoreless outings, none more than 2 innings in length, but will ultimately absorb a pounding that will bring his ERA zooming up from near zero. Another man, used in a variety of circumstances, may pitch more innings at a greater rate of effectiveness (as measured by ERA) but may find himself off the Fireman of the Year charts because his manager hasn't used his talents in the payoff situations. Ah, but would he have been as effective if the game had hung in the balance? We'll never know; nor will the manager, unless he gets his shot—as so many current late-inning stars did when they were novices, from Fingers to Tekulve to Quisenberry.

Can you make the Relief Linear Weights top twenty without pitching as many innings as a spot starter, like Bob Stanley with the Red Sox of 1982? Yes . . . if one's ERA is extraordinary, like Bruce Sutter's 1.35 in 1977.

Nineteenth-century relief performances? That's a subject for *real* nuts.

Table XIII, 6. *Saves, Top Twenty*

Superscript numbers = Relief Linear Weights, ranked by Wins

1.	1983	Dan Quisenberry, KC	45[8]
2.	1973	John Hiller, DET	38[3]
3.	1979	Bruce Sutter, CHI (N)	37
4.	1972	Clay Carroll, CIN	37
5.	1978	Rollie Fingers, SD	37
6.	1982	Bruce Sutter, STL	36
7.	1972	Sparky Lyle, NY (A)	35
8.	1982	Dan Quisenberry, KC	35
9.	1977	Rollie Fingers, SD	35
10.	1970	Wayne Granger, CIN	35
11.	1970	Ron Perranoski, MIN	34
12.	1980	Dan Quisenberry, KC	33
13.	1980	Rich Gossage, NY (A)	33
14.	1983	Bob Stanley, BOS	33
15.	1979	Mike Marshall, MIN	32[6]
16.	1966	Jack Aker, KC	32
17.	1979	Kent Tekulve, PIT	31
18.	1977	Bill Campbell, MIN	31[14]
19.	1971	Ken Sanders, MIL	31[15]
20.	1969	Ron Perranoski, MIN	31
20.	1973	Mike Marshall, MON	31
20.	1965	Ted Abernathy, CHI (N)	31
20.	1977	Bruce Sutter, CHI (N)	31
20.	1978	Kent Tekulve, PIT	31

Table XIII, 7. *Stolen Bases, Top Twenty*

Superscript numbers = Base-Stealing LWTS, ranked in wins

1.	1982	Rickey Henderson, OAK	130[15]
2.	1974	Lou Brock, STL	118[10]
3.	1983	Rickey Henderson, OAK	108[2]
4.	1962	Maury Wills, LA (N)	104[1]
5.	1980	Rickey Henderson, OAK	100[16]
6.	1980	Ron LeFlore, MON	97[4]
7.	1915	Ty Cobb, DET	96
8.	1980	Omar Moreno, PIT	96[45]
9.	1965	Maury Wills, LA	94[41]
10.	1983	Tim Raines, MON	90[3]
11.	1912	Clyde Milan, WAS	88
12.	1911	Ty Cobb, DET	83
13.	1979	Willie Wilson, KC	83[6]
14.	1911	Bob Bescher, CIN	81
15.	1910	Eddie Collins, PHI (A)	81
16.	1980	Willie Wilson, KC	79[5]
17.	1980	Dave Collins, CIN	79[30]
18.	1979	Ron LeFlore, DET	78[2]
19.	1982	Tim Raines, MON	78[13]
20.	1975	Davey Lopes, LA	77[14]
20.	1979	Omar Moreno, PIT	77[36]
20.	1983	Rudy Law, CHI (A)	77[9]

The traditional measure of great baserunning performance is the stolen base total. However, a failed attempt is twice as damaging to the team as a success is helpful, for reasons explained in Chapter 4, so we can correlate the top twenty in steals with the top hundred in Stealing Linear Weights since 1951. (Full data is not continuously available before that date—note that not even fifty times since 1951 has anyone's base stealing produced even one extra win for his team; on a lifetime basis, only Joe Morgan and Lou Brock accounted for ten extra wins.)

The correlation here is good, as it should be: Why would a manager allow a player to attempt enough steals to make this list if he is not succeeding substantially more often than the historical average of two steals for three attempts? What is incredible is that Maury Wills in 1965 and Omar Moreno in 1980 were permitted free rein to so little effect: Moreno was caught stealing 33 times in 129 attempts that year, meaning that his frenzied baserunning netted 9 runs, less than a win; Wills was caught 31 times in 125 tries, a net gain of 9.6 runs.

For the years before 1901, not only do we lack caught-stealing data, but we also are uncertain to what degree the steals recorded in the box scores included miscellaneous extra bases taken by daring, such as scoring from second on an infield hit or even tagging up on a long fly ball. Suffice it to say that Harry Stovey has the all-time mark with 156 stolen bases in 1888 and is second to himself with his 143 of the previous year. This will provide Rickey Henderson with another mountain to climb should he feel the need of one.

Next and last of the traditional statistics we will plumb for greatness is fielding percentage, treated by position and rounded to three digits. This we compare with Defensive Linear Weights.

Table XIII, 8. *Fielding Percentage, Top Three at Each Position*

First Base

1.	1921	Stuffy McInnis, BOS (A)	.999
1.	1946	Frank McCormick, PHI (N)	.999
1.	1973	Jim Spencer, CAL-TEX	.999
1.	1968	Wes Parker, LA	.999
1.	1981	Steve Garvey, LA	.999

Second Base

1.	1970	Ken Boswell, NY (N)	.996
2.	1980	Rob Wilfong, MIN	.995
2.	1973	Bobby Grich, BAL	.995

Shortstop

1.	1979	Larry Bowa, PHI	.991
2.	1972	Eddie Brinkman, DET	.990
3.	1972	Larry Bowa, PHI	.987

Third Base

1.	1974	Don Money, MIL	.989
2.	1947	Hank Majeski, PHI (A)	.988
3.	1978	Aurelio Rodriguez, DET	.987

Outfield

1.		Held by 23 players	1.000

Catcher

1.	1946	Buddy Rosar, PHI (A)	1.000
1.	1958	Yogi Berra, NY (A)	1.000
2.	1950	Wes Westrum, NY (N)	.999

Pitcher

1.		Held by many	1.000

What are we to make of this? Surely the better defensive players at each position are listed in Table 9, not in Table 8. The normalizing factor built into the Defensive LWTS can produce a bias toward exceptional players from earlier periods, in which the average level of skill was lower; however, the question of average fielding skill 70–80

Table XIII, 9. *Defensive LWTS by Position (by wins)*

First Base

1.	1914	Chick Gandil, WAS	2.4
2.	1907	Jiggs Donahue, CHI (A)	2.4
3.	1905	Fred Tenney, BOS (A)	2.2

Second Base

1.	1908	Nap Lajoie, CLE	5.8
2.	1963	Bill Mazeroski, PIT	5.0
3.	1927	Frank Frisch, STL (N)	5.0

Shortstop

1.	1914	Rabbit Maranville BOS (N)	6.3
2.	1980	Ozzie Smith, SD	4.7
3.	1908	Heinie Wagner, BOS (A)	4.3

Third Base

1.	1971	Graig Nettles, CLE	4.2
2.	1937	Harlond Clift, STL (A)	3.8
3.	1982	Buddy Bell, TEX	3.7

Outfield

1.	1977	Dave Parker, PIT	3.1
2.	1930	Chuck Klein, PHI (N)	2.9
3.	1916	Max Carey, PIT	2.8

Catcher

1.	1910	Bill Bergen, BKN	2.4
2.	1915	Bill Rariden, NEW (F)	2.2
3.	1909	Bill Bergen, BKN	2.0

Pitcher

1.	1907	Ed Walsh, CHI (A)	2.3
2.	1911	Ed Walsh, CHI (A)	1.3
3.	1908	Ed Walsh, CHI (A)	1.3

years ago is muddied by the vastly different playing conditions and equipment.

Fielding stats from the nineteenth century are even harder to evaluate because for most of the years up to 1901 fielders wore no gloves, and even when gloves became common in the 1890s they were designed more to protect the hand than to aid in catching. Players from this period do not appear among the all-time leaders in fielding percentage or total chances, but on a chances-per-game basis they dominate, largely because the strikeout was an uncommon event. A preliminary application[2] of the Defensive LWTS formula to pre-1901 players reveals these men to have been the best fielders: second baseman Bid McPhee and shortstops Bill Dahlen, George Davis, Bobby Wallace, and Jack Glasscock.

The best way to review the truly great single-season performances—the ones which contributed or saved the most runs, and thus wins—is to flip through the tables at the rear. But here are some performances to look for in those tables, for they indicate how greatness can be present but unaccounted for (until now, that is):

- *Batting LWTS.* The domination by Ruth and Williams is incredible: Of the top twenty seasons, Ruth authored nine and Williams five. Mickey Mantle's 1957 was the tenth best ever, much better than his triple-crown year of 1956. And how about Norm Cash's twenty-third spot in 1961—better than Ty Cobb's 1911 when he hit .420?' Cash's accomplishment was buried amid all the Maris-Mantle hoopla that year. Or how about Dick Allen's coming in thirty-seventh with his MVP year of 1972? The only active players to record one of the top 100 seasons are Reggie Jackson, Rod Carew, and George Brett.
- *Pitching LWTS.* Dizzy Trout had the fourth-best year since 1901, better than Lefty Grove's 31-4 season of 1931? Yes. And Vida Blue, Wilbur Wood, Bert Blyleven, and Gaylord Perry, among other modern hurlers, each had a season better than any by Steve Carlton, Jim Palmer, or Tom Seaver. Aside from those mentioned above, active pitchers on the list of top hundred seasons are Ron Guidry, Steve Rogers, Mike Caldwell, and Phil Niekro (who is without a team as this book goes to press but is looking to pitch for somebody in 1984).
- *Relief LWTS.* Jim Kern stands at the top of the heap for his 1979 season, but the upper echelons are otherwise dotted with less than household names: Rod Scurry, Sid Monge, Ken Sanders. Relievers of the 1950s like Ellis Kinder, Joe Black, and Dick Hyde show up well here, though they are absent from the Relief Points lists—indicating that the latter is more a reflection of managerial trends and whims than quality moundwork.
- *Stealing LWTS.* Nothing to say here that hasn't been said already. A substantial stat measuring an insubstantial activity.
- *Defensive LWTS.* We didn't know that Harlond Clift was *that* good—did you? Or Dick Bartell? Or Joe Tinker? Or Ivan DeJesus? And Ozzie Smith has a good chance to surpass Bill Mazeroski as the best lifetime fielder.
- *Isolated Power.* They say that modern players don't hit the long ball the way they did in the 1930s—Ruth, Gehrig, Greenberg, Foxx, et al.—yet the top twenty seasons in the normalized version of this stat, a better measure of pure power than slugging

percentage, feature such names as Willie Stargell and Mike Schmidt.

- *On Base Average.* Norm Cash's 1961 was the thirteenth best ever in Normalized OBA, and Ken Singleton's 1977 places eighteenth. Gene Tenace, whose ability to draw a walk has been no secret the past few years, has no season among the top hundred, yet has the sixteenth-best mark on a lifetime basis. (Joe Morgan is twelfth, Mike Hargrove fourteenth.)

And there is more . . . more statistical categories, more surprise stars . . . and we have saved the best for last.

---

[1] A top hundred, in most instances, for each of the New Stats can be found at the back of the book. Also, pre-1901 performances are treated separately because Pete's database does not yet contain all the information necessary for the New Statistical scrutiny of those seasons.

[2] A full treatment of pre-1901 players and teams by the New Statistical measures is in the planning stage.

THE ULTIMATE BASEBALL STATISTIC

"Who is the best player (nonpitcher) in the major leagues?" That question was put to all 650 big-league performers by the New York *Times* over ten days in June of 1983, with the results reported in the issue of July 4, a neatly chosen date. Of all those who responded, 23 percent voted for Montreal's Andre Dawson; 18.9 supported Milwaukee's Robin Yount; and 15 percent went for Atlanta's Dale Murphy. Five other players received votes, but none were close to these three. In a sidebar survey of ten experts (people like Cedric Tallis, Hank Peters, Harry Dalton), six voted for Yount and the others for Dawson.

Alongside Joe Durso's story on the poll results (Steve Carlton was an overwhelming choice as best pitcher, Whitey Herzog an easy winner as best manager) was an account of the previous day's game at Philadelphia's Veterans Stadium, headed "Schmidt Hits 15th as Phils Beat Mets." And one month earlier, Mike Schmidt had been the choice of writer Glen Waggoner in an article for *Sport* magazine entitled "The Best Player in Baseball." To 1982 performance, Waggoner applied Tom Boswell's Total Average (which measures offense only), Bill James's Offensive and Defensive Won-Lost Records (a twenty-six-step process spelled out in the *Baseball Abstract*), and the "Grebey System" of rankings tabulated by the Elias Sports Bureau for the owners' use in categorizing free agents. (Dalton: "It's useful for propping up the fourth leg of an uneven table.") The eight players whom all three approaches identified as tops in their leagues at their positions in 1982 were: Eddie Murray, 1B; Bobby Grich and Joe Morgan, 2B; Robin Yount, SS; Schmidt, 3B; Pedro Guerrero and Dwight Evans, RF; and Gary Carter, C. Dawson, the players' choice, emerged as the NL's best center fielder by the Grebey and James systems, but trailed Murray in Total Average because he disdains the base on balls.

Paring the consensus stars to a final four of Yount, Carter, Dawson, and Schmidt, Waggoner went for the Phillie third baseman as the best "by a nose" because of his current Total Average (over 1.000, meaning more bases gained than outs lost) and his edge over Carter in Runs Created and RBIs; Yount had to prove 1982 wasn't a fluke, Dawson had to boost his On Base Average. Author Waggoner may also have been swayed by the comment of Expo manager Bill Virdon: "Excluding my own players, I pick Schmidt because he beats you offensively, defensively, can steal bases in key situations, and his instincts for the game are exceptional. He has a chance to be among the best of all time."

Actually, Schmidt has more than a chance—if he hung up his spikes tomorrow, he'd be the fifteenth-best player of all time, for reasons described below. As Pete said to *Sport* in connection with the Waggoner article, "He's had nine years in a row that were as good as Yount's one." Dawson the best player? Murphy? Yount? Cal Ripken, maybe? Perhaps someday, when Schmidt totters off to Valhalla like his 1983 teammates Rose and Morgan, but not now. And that's not our opinion—it's a matter of record.

The conventional definition of the complete ballplayer—probably Branch Rickey's—was restated last year by Dodger Vice-President Al Campanis: "There are five basic tools a player needs, in this order: He must hit, run, throw, field, and hit with power." To those who have read the previous chapters, Campanis's order will seem nonsensical, but never mind—the overall import is right. To express the diverse accomplishments of that rare player who is adept in all five areas, fans and baseball professionals have had to resort to a statistical line that looked like this: .280 BA, 14 stolen bases, 324 assists (presume the player is a third baseman—he is in fact Schmidt), .950 fielding percentage, and 35 home runs. (The "All American Average" mentioned in Chapter 2 was a 1910 attempt to present a total statistical profile in compact fashion, but it failed the test of logic.) Now, in the ultimate extension, the culmination of the Linear Weights System, we can express Schmidt's complete 1982 performance—and indeed anyone's—

in all five areas in a single portmanteau number—in Schmidt's case, 65.4. This number is *not* a rating, a house of cards built up by assigning so many "points" for finishing first in the league in slugging, so many for being third in homers, etc.; the number 65.4 states *how many runs* Schmidt contributed, through all his "tangible" efforts, beyond those of an average player. And now that we are comfortable with the Linear Weights concept, the formula is a breeze:

Overall Player Runs = Batting Runs + Base Stealing Runs + Defensive Runs − Average Defensive Skill at the Position

The first three factors require no further explanation (unless you scorned our advice in Chapter 1 and began reading here), but the fourth does. Because it is harder to play shortstop than left field, a left fielder who accounted for 10 Defensive Runs should not be regarded as having the same value to a team as a shortstop who also accounted for 10 Defensive Runs: Have the two men switch positions, and you'd soon see who made more of a defensive contribution. And because some positions—shortstop, catcher, second base, and third base—are more difficult to play than others, we see a relative scarcity of good hitters at these positions and an abundance at the others. Another way of looking at this situation is to say that the fielding demands of the skill positions are so great that a club will tolerate average or below average batwork from players at those positions.

This positional bias, or relative worth, may be expressed in terms of the average batting skill required at that position to hold down a major league spot: Historically middle infielders have presented, on average, the worst Batting LWTS, while left fielders have presented the best; their offensive production is an inverse measure of their comparative defensive worth. A league-average hitter, or worse, like Chicago's first baseman Mike Squires, for example, would have to save a mammoth number of runs with his glove to be of as much value as the poor-fielding, heavy-hitting stereotype of the position—which at first base neither Squires nor anyone else can do. Yet a terrific defensive middle infielder like Mark Belanger or a catcher like Bill Bergen was able to stay employed for many seasons despite abysmal batting stats.

Returning to the Overall Player Runs formula, the purpose of the final factor is to credit those men who play the most difficult positions and thus have more value to their teams. To determine the average defensive skill at the position, we simply take the average batting skill at the position and subtract it. This puts left fielders and shortstops, for example, on an equal footing. For example, let's say all NL outfielders last year accounted for 360 runs above average; dividing that figure by 3 (left field, center field, right field), we get 120 per outfield post. The positional adjustment (the factor for average defensive skill) for a left fielder who played in 162 games would be 162 × 120/1,944, where 1,944 is the number of games played in the league by all left fielders (or any other position, obviously). Thus an outfielder who played in all his team's games would have 10 runs subtracted from the sum of his Batting, Stealing, and Defensive LWTS. If all NL shortstops last year accounted for 84 runs *below* average, the adjustment for a shortstop would be figured in the same way, multiplying his games played by − 84/1,944. Thus a shortstop who played in 162 games would have an adjustment factor of 162 × − 84/1,944, or − 7 runs, meaning that 7 runs (minus a minus-7) would be *added* to his Overall mark.

The Overall Player LWTS is a powerful and elegant solution to such age-old questions as "Was Cobb a better all-around player than Ruth, and was Honus Wagner better than both of them?" (No, and no.) "Were Speaker or Joe DiMaggio, because of their defensive skills, better all-around than Musial?" (Speaker, yes; DiMaggio, no.) On a current level, those baseball writers who participated in the balloting for the 1983 AL MVP had to choose from Jim Rice, Wade Boggs, Eddie Murray, Cal Ripken, and Rickey Henderson, recognizing that Rice and Boggs played for a poor club in a hitters' park, Murray played for a world champion but at an easy position, while Ripken played the game's most difficult position, and Henderson had a base-stealing year even better than his record-setting 1982. Now these differences can be quantified in a uniform and clear fashion.

In 1978 the MVP question in the American League was whether to vote for Rice—who had 400 total bases (the first time for an American Leaguer since Joe DiMaggio in 1937), 46 homers, and 139 RBIs—or for Ron Guidry, whose 25-3 won-lost percentage was the all-time high for a starter with 20 wins, and whose ERA of 1.74 was less than half the league average. We can assess Rice's complete performance in all areas by the method detailed above—and we can also evaluate Guidry's complete performance, because a pitcher's worth to his team

Table XIV, 1. *Overall Player and Pitcher Wins—Season* (* = MVP Award)

Player Wins
Since 1901

	Wins
1. 1921 Babe Ruth, NY (A)	10.3
2. 1923 Babe Ruth, NY (A)*	9.6

			Wins
3.	1920	Babe Ruth, NY (A)	9.5
4.	1924	Babe Ruth, NY (A)	8.9
5.	1947	Ted Williams, BOS (A)	8.8
6.	1927	Babe Ruth, NY (A)	8.8
7.	1975	Joe Morgan, CIN*	8.7
8.	1942	Ted Williams, BOS (A)	8.6
9.	1920	Rogers Hornsby, STL (N)	8.5
10.	1946	Ted Williams, BOS (A)*	8.4

Since 1961

			Wins
7.	1975	Joe Morgan, CIN*	8.7
23.	1977	Mike Schmidt, PHI	7.8
24.	1966	Ron Santo, CHI (N)	7.8
32.	1974	Rod Carew, MIN	7.7
35.	1980	Mike Schmidt, PHI*	7.5
36.	1974	Mike Schmidt, PHI	7.5
37.	1961	Mickey Mantle, NY	7.5
44.	1968	Carl Yastrzemski, BOS	7.3
52.	1962	Frank Robinson, CIN	7.2
53.	1967	Carl Yastrzemski, BOS*	7.1
59.	1961	Norm Cash, DET	7.1

Pitcher Wins
Since 1901

			Wins
1.	1913	Walter Johnson, WAS*	9.5
2.	1944	Dizzy Trout, DET	9.3
3.	1912	Walter Johnson, WAS	8.6
4.	1915	Pete Alexander, PHI (N)	8.2
5.	1939	Bucky Walters, CIN	7.7
6.	1905	Christy Mathewson, NY (N)	7.7
7.	1918	Walter Johnson, WAS	7.6
8.	1945	Hal Newhouser, DET	7.4
9.	1912	Joe Wood, BOS (A)	7.3
10.	1904	Jack Chesbro, NY (A)	7.3

Since 1961

			Wins
13.	1978	Ron Guidry, NY (A)	6.8
14.	1968	Bob Gibson, STL*	6.8
22.	1973	Bert Blyleven, MIN	6.5
25.	1971	Wilbur Wood, CHI (A)	6.4
27.	1972	Gaylord Perry, CLE	6.4
28.	1965	Juan Marichal, SF	6.3
31.	1969	Bob Gibson, STL	6.3
32.	1972	Steve Carlton, PHI	6.3
33.	1971	Vida Blue, OAK	6.2
40.	1971	Tom Seaver, NY (N)	5.9

THE ULTIMATE BASEBALL STATISTIC ◇ 241

consists not only of his pitching but also his fielding and (except for AL hurlers since 1973) his hitting. If Rice contributed 5.5 Wins through all his efforts—5.4 batting, 0.7 fielding, −0.1 base stealing, with an adjustment for average outfield defensive skill of −0.5—and Guidry contributed 6.8 Wins—6.5 pitching, 0.2 fielding, and 0 (of course) batting—then we can see that Guidry was more valuable to the Yankees than Rice was to the Red Sox. The formula for the Pitchers Overall Runs is simply: Pitching Runs + Defensive Runs + Batting Runs. (It should be emphasized that we removed pitchers' batting stats from all league stats, so that everyday players are compared only with their peers and pitchers are compared only with other pitchers. This explains how a pitcher can have a low OBA of .280 and a SLG of .320, yet have a positive Batting LWTS figure.)

You will find listed in the tables the hundred-best Overall seasons for players and for pitchers. (Remember that the Overall formula includes no Base Stealing Runs before 1951.) Only one of the top twenty Player Overall seasons has occurred since expansion in 1961, and of the top twenty Pitcher Overall seasons, only one has come since 1945 (for the latter state of affairs, one may blame the rise of the reliever and the fall of pitchers' batting ability). We present in Table XIV, 1 the top ten player and pitcher seasons of the century, and the top ten since expansion, all ranked by LWTS Wins; the numbers alongside the post-expansion players refer to their rankings on the list of top performers in this century.

As with so many of baseball's statistics, new or old, Babe Ruth seems to have dropped to Earth from the planet Krypton. He had three seasons better than anyone else, basically via his bat alone—no matter what you read about his being an outstanding outfielder, it just wasn't so, for only in 1919, when he was slim and still a part-time pitcher, was his Overall Player LWTS boosted by 1 Defensive Win. In fact, there is a lack of well-rounded players in the top spots altogether (the most balanced season performance perhaps being Nap Lajoie's 1903, good for sixteenth place), for these are dominated by Ruth, Williams, and Hornsby—none better than average as fielders in their positions. Even Joe Morgan, the only modern player to crack the top twenty (at seventh place), let alone the top ten, tacked on only half a win for his fielding in 1975, and that was one of his better years. Despite the popular impression that Morgan was an outstanding fielder, his lifetime Defensive LWTS was, through 1983, −10.8 wins, the tenth-worst record among all 815 men in this century who have played in 1000 or more games! (Standing immediately above Morgan on the defensive charts is Harmon Killebrew; immediately below, Felix Millan and Greg Luzinski.)

That Morgan's 1975, the best season anyone has had since 1948, should be the only post-expansion one in the top twenty is not surprising, for the gap between average and peak performance has narrowed to such an extent that for a player to exceed the league norm by, well, Ruthian proportions he would have to do what Mike Schmidt nearly did in 1981: Although his best Overall Player mark was his 7.8 Wins in 1977, in the shortened year of the strike (1981) he accounted for 7.0 Overall Wins in 102 games—the sixtieth-best mark of this century; had he been given the chance to extend that level of performance to 162 games, he figured to notch 11.1 Overall Wins, which would have surpassed Ruth for the top season spot! Incidentally, seven Schmidt seasons appear among the top hundred since 1901; only Ruth with ten and Williams with nine have more (Hornsby also has seven).

On the top hundred player seasons, in only six did a man's Defensive Wins exceed his Batting Wins. In fact, the most Defensive Wins present on the list of 100 is 5.8, by Nap Lajoie in 1908, and the most Base Stealing Wins on the list is only 1.4, by Morgan in 1975.

The best hundred seasons for Pitchers Overall Runs presents basically the same profile, except for Ed Walsh's weird 1907 campaign, in which he accounted for half as many Wins Above Average with his glove—2.3—as he did with his arm. (However, pitchers' fielding records must be viewed with caution, inasmuch as a man's style of pitching—the fastball or the breaking ball—may influence his number of assists. In the late 1960s, for example, Giant sinkerballer Frank Linzy averaged three times the assists per nine innings of perennial Gold Glover Bob Gibson, a fastball pitcher.) Many pitchers on the list have been superior batters—Walter Johnson, Bob Lemon, Bob Gibson, and of course Babe Ruth—yet only one contributed 2 wins with his bat: Wes Ferrell in 1935, when he batted .347 and slugged .533, hitting 7 homers in his 150 at bats (including 32 at bats as a pinch-hitter). Ferrell's season at the plate, however, was not the best any pitcher has had—Guy Hecker of Louisville in 1886 actually had enough at bats to lead the league in batting while winning 27 games as a pitcher, and two years earlier he won 52 games while batting .296 and slugging .427.

This question may have occurred to you: What happens if we combine Babe Ruth's batting, fielding, and pitching records for the years 1918–19? Wouldn't these be the best years ever? No. In 1919, Ruth contributed 7.2 wins with his bat, 1.4 with his glove, and 0.9 as a pitcher. Ruth was only a part-time pitcher in 1918–19, as his everyday slugging value began to outweigh, for the Red Sox management, his considerable merits as a hurler. Ruth contributed 5.6 Overall Pitcher Wins for 1917, good for sixty-fourth place, but in no subsequent season did his pitching and batting talents combine to a total more incredible than his epic seasons of 1920, '21, '23, and '27.

It is interesting to note that of the top ten Overall Pitcher marks since 1961, none contains a batter contribution of 1 above average win—and only Bob Gibson in 1969 batted for more than *half* a win. The pitchers of today may be better than ever, but they must be specializing at a younger age than their predecessors, for the level of pitcher batting has sunk dramatically since World War II. Maybe the DH rule does make sense.

Also, just for fun, we have correlated the best Overall Player Performances with the Most Valuable Player Award. While it seems clear that the player who, through his batting, fielding, and base stealing, accounted for the most wins would be the most valuable to his team or any other, intangibles may have their proper place in the handing out of honors. Dale Murphy, for example, was not the best player in the NL in 1982, when he won the MVP (he was the twelfth best), but maybe he was an inspiration to his teammates.

The most interesting and unsettling application of the two Overall LWTS measures is to the careers of all those who have played in 1,000 or more games or pitched in at least 1,500 innings. (While some active players who have not yet played in so many games or pitched so many innings have compiled impressive Overall LWTS totals—Dawson, Ripken, Murray, Murphy, Valenzuela, Stieb, etc.—they have not yet stood the test of time, nor had their career totals diminished by the customarily weak seasons prior to retirement.) The tables at the rear supply full win data for the top 300 players and 200 pitchers since 1901—for players, Batting Wins, Defensive Wins, Base Stealing Wins (post-1950), Positional Adjustment for Average Defensive Skill, and Overall Wins; for pitchers, Pitching Wins, Batting Wins, Fielding Wins, and Overall Wins. One might think that any player meeting the 1,000-game or 1,500-inning standard would have been a better than average performer all told—if not in batting, then in fielding or base stealing. Yet of the 815 players who appeared in 1,000 or more games, *327* had *negative* Overall Player Wins, meaning that they were not even average performers, including some very well-known names (e.g., Lou Brock, Chris Chambliss, Bobby Richardson, Bill Buckner, and all three Alou brothers) and even two members of the Hall of Fame (Al Lopez and Lloyd Waner). Of the 433 ranked pitchers, 121 had negative Overall Pitcher Wins (e.g., Joe Niekro, Denny McLain, Jim Lonborg), and of these none was a Hall of Famer. It's harder to disguise inept pitching than inept batting or fielding, and as a rule they don't keep sending you out to the mound if you're getting your ears blasted off.

Here are the top ten players and pitchers of the century and the top ten who played most of their games since 1961.

Table XIV, 2. *Overall Player and Pitcher Wins—Lifetime*

Players

Since 1901	Wins	*Since 1961*	Wins
1. Babe Ruth	116.9	3. Hank Aaron	89.9
2. Ted Williams	96.9	5. Willie Mays	87.7
3. Hank Aaron	89.9	13. Frank Robinson	70.7
4. Ty Cobb	89.3	14. Joe Morgan	67.7
5. Willie Mays	87.7	15. Mike Schmidt	64.8
6. Nap Lajoie	85.3	20. Carl Yastrzemski	47.1
7. Rogers Hornsby	82.6	21. Al Kaline	46.8
8. Tris Speaker	81.0	24. Reggie Jackson	43.3
9. Eddie Collins	80.0	25. Rod Carew	42.8
10. Honus Wagner	79.3	27. Bobby Grich	42.3

Pitchers

Since 1901	Wins	*Since 1961*	Wins
1. Walter Johnson	73.3	6. Tom Seaver	45.0
2. Cy Young	69.7	7. Bob Gibson	44.8
3. Pete Alexander	60.5	10. Gaylord Perry	37.1
4. Christy Mathewson	51.6	11. Steve Carlton	37.1
5. Lefty Grove	50.9	12. Phil Niekro	36.7
6. Tom Seaver	45.0	16. Jim Palmer	35.4
7. Bob Gibson	44.8	19. Don Drysdale	33.3
8. Warren Spahn	41.9	23. Bert Blyleven	28.8
9. Ed Walsh	37.3	24. Fergy Jenkins	28.7
10. Gaylord Perry	37.1	29. Juan Marichal	27.4

Let's talk about the Players list first. Ruth is by far the class of the field, 20 Wins ahead of Williams. Yet Williams missed all of three seasons—1943–45—and nearly all, save for 43 games, of two others (1952–53) in service to his country. Had he played those years, would he have surpassed Ruth?

Some of the names one might have expected to find higher on the

Table XIV, 3. *Overall Player Wins Pre-1901*

1.	Bill Dahlen	44.6
2.	Cap Anson	42.2
3.	Dan Brouthers	41.2
4.	Ed Delahanty	40.7
5.	Roger Connor	39.2
6.	Bid McPhee	38.5
7.	George Davis	37.7
8.	King Kelly	37.4
9.	Billy Hamilton	35.4
10.	Buck Ewing	31.8

lifetime list, like Gehrig (16), DiMaggio (23), Dick Allen (42), or Rose (65), had only average or worse fielding records and/or had their totals reduced by substantial positional adjustments (Gehrig, Cobb, and Speaker each had their win totals reduced by more than 20, the only ones affected to such an extent). Conversely, it will come as a surprise or shock to some that Joe Morgan is the fourteenth-best player of the century (despite his dreadful fielding record) and that Ron Santo is thirty-sixth (Brooks Robinson is one hundred and eighth, some twenty-five spots below Darrell Evans and Ron Cey); George Brett is fifty-sixth, Graig Nettles sixty-third. The man to watch, however, is Mike Schmidt in fifteenth place: if he can continue to perform at current levels (5–6 Overall Wins per year) for another five or six years, he has a solid chance to overtake Williams for the number-two spot all-time.

Among the pitchers, the surprise is that despite the "poor" showing of modern-day stars in the single-season category, they do very well indeed on a lifetime basis, with Tom Seaver, Bob Gibson, and Gaylord Perry cracking the top ten and Steve Carlton, Phil Niekro, Jim Palmer, and Don Drysdale joining them in the top twenty. How, one might wonder, did Drysdale reach such heights while his more celebrated moundmate, Sandy Koufax, stands no better than sixtieth? Because Drysdale was a fine hitter and fielder, and Koufax was not, though he was much better strictly as a pitcher. But a pitcher may help his team win by saving runs with his arm or his glove, or by contributing them with his bat (and, to an inconsequential extent, his baserunning)—and a run is a run is a run, however it is saved or produced.

Nineteenth-century players are ghettoized in the list above because their seasons were shorter, the record keeping was spottier, and prior to 1893, the rules were significantly different. Above are the top ten players of the nineteenth century, and on page 247 the top pitchers.

Table XIV, 4. *Overall Pitcher Wins Pre-1901*

1.	Kid Nichols	47.1
2.	Tim Keefe	36.7
3.	Amos Rusie	35.0
4.	John Clarkson	33.1
5.	Hoss Radbourn	26.8
6.	Tony Mullane	22.4
7.	Clark Griffith	21.7
8.	Mickey Welch	19.6
9.	Jim McCormick	19.5
10.	Bob Caruthers	18.8

Illustrative of the problem just referred to is Cap Anson's total. Five years of his prime are not counted at all because, through a wrong-headed edict of the Special Baseball Committee in 1968, the National Association, which existed from 1871 through 1875, was denied major-league status. Then he played one year of a 70-game schedule, two at 60, three at 84, and one at 98 before playing his first 100-plus game schedule in 1884, his *fourteenth* year of big-time play. And even in his later years, he saw a schedule of as many as 154 games only once. So, although Anson played twenty-seven years in the top existing league, he appeared in only 2,276 games that counted for his major-league record. Reggie Jackson has played more games in seventeen years than Anson in twenty-seven: How can one compare their LWTS fairly? Perhaps by looking at nineteenth-century stars' OPS along with Batting LWTS, or their NERA along with Pitching LWTS, for these measure only efficiency, not longevity. (Pitchers of the period do not suffer as much as batters, for they worked in many more innings than would be common today.)

If these men—the top ten pitchers and players of each century, as measured by Overall LWTS—are the best to have played the game, why are some of them not in the Hall of Fame? Why are some of the forty-three pitchers now in the Hall of Fame not among the top hundred pitchers by the record? And why are some of the 97 players with plaques on the wall not among the top *200* overall?[1] Why does a Dick Allen get fourteen votes in the last election, or Ron Santo none?

Because the voters never heard of Linear Weights, for one thing; because life is not fair, for another. Anyway, the Hall of Fame *belongs* to the Clark Family Estates; if its executives please to let the Baseball Writers Association ballot for players of recent vintage, and if they please to appoint a Veterans Committee to select players of the remote past, that's fine. You know and we know that there are some clinkers on that wall in the Hall, that the old-boy network has been particularly effective in securing places for colleagues from the 1920s and 1930s; however, if we want to construct an Imaginary Museum, there is no one to stop us: They can have their Hall, and we can have ours. If members of the Baseball Writers Association or the Veterans Committee wish to join us, great. There are no hard feelings; we're not out to rip down any plaques.

---

[1] The ten Negro League stars enshrined in the Hall are not counted in this tabulation: Alas, their feats were recorded in fragmentary fashion.

15

RUMBLINGS IN THE PANTHEON

This chapter will disturb some people—friends and relatives of some Hall of Famers, proponents of a few longtime HOF bridesmaids, worshipers of some active stars whom they presume to have a lock on immortality, Cooperstown-style. Putting noses out of joint is not our aim, however; identifying excellence is—and to this purpose we apply the best measure of excellence yet devised, the Overall Player and Pitcher Rankings. We do not tailor this stat to any preconceived definition of greatness, nor do we "rate" players or pitchers by the stat: The number is what it is, a statement of the number of wins a player or pitcher accounted for over his career beyond what the average player might have contributed in his place (by definition, zero: credits and debits in balance). If your favorite player has contributed a surprisingly low Overall Player Win total, this is not our doing but his; write him about it. There may still be ample reason to admire the man on a personal and even a professional level, for he may excel in something we are not measuring—intestinal fortitude, clutch ability, or a peculiar knack in a narrow area of performance, like the ability to hit homers (Dave Kingman) or doubles (Hal McRae), or to strike out battalions of batters (Nolan Ryan).

The Overall Player Win total makes no differentiation between a run contributed by batting, fielding, or base stealing—the object of the game is to score runs while denying them to the opposition, and any run contributed or denied is of equal value (in the aggregate, of course; in any given game, similar events have different values according to their contexts). Because of the nature of the game, it is more difficult to amass a high LWTS Run/Win total with the glove or with the legs than it is with the bat, yet a high Batting Win total may shrink dramatically for the player—along with his value to his team—by poor fielding or reckless baserunning. In other words, a player who registers 50 Batting Wins but whose fielding is 5 Wins below average is not as good a player as another who contributes 40 Wins with his bat, 5 with his glove, and another 5 with his base stealing. Similarly, a pitcher who is such a mystery to the batters that he chalks up 24 career LWTS Wins does not give his team the full benefit of that artistry if he is 2.6 Wins below average at the bat and 2.5 below average with the glove (this happens to describe the career of Sandy Koufax). Another pitcher may be no more effective in his moundwork, also contributing 24 Wins over his career, but may have allowed his teammates to take better advantage of his pitching skill—that is, have a better shot at winning— if he saved 3 Wins with his glove and gained 6 more with his bat (this happens to describe the career of Don Drysdale).

We have applied the Overall Player Win measure to all 815 men who appeared in 1,000 or more games in this century, and to all players of substance prior to 1901; for pitchers, we have applied the measure to the 433 men who have pitched 1,500 or more innings since 1901, and to all pitchers of substance before that time. It needs to be said of the nearly 13,000 men who have played major-league ball since 1876 that less than 10 percent passed the criteria mentioned above, so the "worst" of these players had a career of more distinction than 90 percent of all those who played in the majors (excepting perhaps those few careers cut off in their prime, like Herb Score's or Ross Youngs'). Moreover, the 140 players who are enshrined in the Hall of Fame represent only a little over 1 percent of those who have played major league ball, and to be on the outside looking in hardly makes one a nondescript player.

How did those 140 men (as of 1983) manage to have their likenesses cast in bronze and their life's work memorialized? Some men were rewarded for consistently superior play over a long haul, despite few flashes of brilliance (Burleigh Grimes, Ted Lyons, Bobby Wallace). Others were elected for dazzling performance over the short term (Dizzy Dean, Sandy Koufax, Ralph Kiner). Some benefited from an association with a string of pennant winners (Earle Combs, George Kelly); others from anecdotal accomplishments handed down as lore

(Candy Cummings, King Kelly, Tommy McCarthy); still others from superior press agentry (the "trio of bear cubs . . . fleeter than birds, Tinker to Evers to Chance"). A few men rode an extraordinary single season into the Hall (Hugh Duffy, Hack Wilson), and a few who shall remain nameless have been elected/selected for reasons not entirely related to their accomplishments. All that's OK with us, mind you— we don't pay the rent on the place. But we are interested to know which players were really *the best*.

How do men of merit *not* get elected to the Hall of Fame? Moral lapses are what have kept Joe Jackson out, and perhaps Buck Weaver, Eddie Cicotte, and Mike Donlin, too; ethical/legal slips may in the future delay or negate the chances of Fergy Jenkins, Denny McLain, and more than a few active players. Contentious personalities (as perceived by baseball writers and officials) will not smooth the path to Cooperstown for Carl Mays, Dick Allen, or Cesar Cedeno. Lack of media attention combined with superior play during the war years of 1943–45 may be what has made forgotten men of Hal Newhouser and Dizzy Trout. No World Series showcase for their talents hampered Arky Vaughan and Wes Ferrell. And low totals in the glamor stats— homers or strikeouts—have hurt the candidacy of Richie Ashburn and will hurt Jim Kaat when he becomes eligible for election.

What gets a man in, what keeps him out—these are questions which do not impose upon our alternative vision of the pantheon, our hidden Hall of Fame. All that counts is performance—to what extent did the man's efforts translate into additional runs and therefore wins for his team? In the tables which follow this chapter, we list the top 300 players in Overall Lifetime Wins, and the top 200 pitchers, breaking down the Win totals into their components—for the former group, Batting Wins, Fielding Wins, Base Stealing Wins, and Positional Adjustment; for the latter group, Pitching Wins, Batting Wins, and Fielding Wins. The men who contribute the most Wins are the best players—the most productive, the most valuable—but our pantheon will not consist simply of the men who produced the 140 top Win totals: We will abide by some of the rules and practices of Cooperstown's pantheon so as not to take advantage of opportunities for identifying excellence that are not afforded to the Baseball Writers Association.

For example, the Board of Directors of the National Baseball Hall of Fame stipulates that no player's name will be placed on the ballot until five years have passed since his departure from active duty, except in the case of death. That makes sense to us, too: We believe in letting the heat of the moment pass into cool reflection; Carl Yastrzemski and Joe Morgan will not be in our Hall until they are eligible for Cooperstown's. The HOF distinguishes between men elected as players, those elected from the Negro Leagues, and those elected as managers, umpires, pioneers, and executives. We will abide by that, too, though with regrets in the case of the Negro League veterans, whose performances we are unable to measure. Also, a few men elected as managers were players of consequence—John McGraw, Clark Griffith, Al Lopez, Miller Huggins, to name a few—and their careers we will evaluate as players alone.

In addition, because it is easier to pile up LWTS Wins by playing every day than it is by playing one day in four or five,[1] we have decided to use the current composition of the Hall of Fame as a format for ours: Of the 140 men elected to the National Baseball Hall of Fame for their efforts on the field, 97 are everyday players and 43 are pitchers; our 140 will divide along the same lines. Furthermore, of these 140, 25 were active primarily in the nineteenth century; for reasons discussed at several junctures in this book, their records are best viewed and analyzed separately, so we will allot 25 places in our pantheon to men of this period—17 players, 8 pitchers, just as in Cooperstown. We will not attempt any further compartmentalization by era or by position. Nor will we advocate a different set of standards by which relievers, pinch-hitters, or designated hitters may find their way into our group, although we would certainly not frown upon the election of relief pitchers to the Cooperstown Hall.

So those are the guidelines. And these are the players and pitchers, the 140 best of all time whose careers ended before 1978. (For a fuller, rank-ordered statistical description of their careers, see the tables.) All those in our Hall and not in Cooperstown's are set in boldface . . . for interest's sake, not as a signal of indignation.

Table XV, 1. *The Best Players*

NAME	POS.	YEARS	SPAN	GAMES	OVERALL WINS
Aaron, Hank	OF	23	1954–76	3,298	89.8
Allen, Dick	1B, 3B	15	1963–77	1,749	35.9
Anson, Cap	1B	22	1876–97	2,276	42.2
Appling, Luke	SS	20	1930–50	2,422	37.9
Ashburn, Richie	OF	15	1948–62	2,189	25.6
Baker, Frank	3B	13	1908–22	1,575	38.1
Bancroft, Dave	SS	16	1915–30	1,913	33.4
Banks, Ernie	1B, SS	19	1953–71	2,528	24.0
Bartell, Dick	SS	18	1927–46	2,016	26.5
Berra, Yogi	C	19	1946–65	2,120	31.8

Boudreau, Lou	SS	15	1938–52	1,646	36.2
Brouthers, Dan	1B	19	1879–1904	1,673	41.2
Browning, Pete	OF	13	1882–94	1,179	27.4
Burkett, Jesse	OF	16	1890–1905	2,063	28.2
Cash, Norm	1B	17	1958–74	2,089	27.6
Childs, Cupid	2B	13	1888–1901	1,467	31.6
Clarke, Fred	OF	21	1894–1915	2,244	25.6
Clemente, Roberto	OF	18	1955–72	2,433	37.7
Clift, Harlond	3B	12	1934–45	1,582	22.2
Cobb, Ty	OF	24	1905–28	3,034	89.3
Cochrane, Mickey	C	13	1925–37	1,482	33.4
Colavito, Rocky	OF	14	1955–68	1,841	24.7
Collins, Eddie	2B	25	1906–30	2,826	80.0
Collins, Jimmy	3B	14	1895–1908	1,728	26.8
Connor, Roger	1B	18	1880–97	1,988	39.2
Crawford, Sam	OF	19	1899–1917	2,517	25.5
Cronin, Joe	SS	20	1926–45	2,124	42.3
Dahlen, Bill	SS	21	1891–1911	2,443	44.6
Davis, George	SS	21	1890–1909	2,377	37.7
Delahanty, Ed	OF	16	1888–1903	1,835	37.7
Dickey, Bill	C	17	1928–47	1,789	33.1
DiMaggio, Joe	OF	13	1936–51	1,736	44.5
Doerr, Bobby	2B	14	1937–51	1,865	42.5
Ewing, Buck	C	18	1880–97	1,315	31.8
Fletcher, Art	SS	13	1909–22	1,529	25.7
Flick, Elmer	OF	13	1898–1910	1,482	26.4
Fournier, Jack	1B	15	1912–27	1,530	24.5
Foxx, Jimmie	1B	20	1925–45	2,317	57.0
Frisch, Frank	2B	19	1919–37	2,311	35.6
Gehrig, Lou	1B	17	1923–39	2,164	64.8
Gehringer, Charlie	2B	19	1924–42	2,323	45.1
Glasscock, Jack	SS	17	1879–95	1,736	25.9
Gordon, Joe	2B	11	1938–50	1,566	29.5
Goslin, Goose	OF	18	1921–38	2,287	22.9
Greenberg, Hank	1B	13	1930–47	1,394	29.2
Groh, Heinie	3B	16	1912–27	1,676	27.3
Hack, Stan	3B	16	1932–47	1,938	22.8
Hamilton, Billy	OF	14	1888–1901	1,591	37.8
Hartnett, Gabby	C	20	1922–41	1,990	35.4
Heilmann, Harry	OF	17	1914–32	2,146	25.3
Herman, Billy	2B	15	1931–47	1,922	35.7
Hornsby, Rogers	2B	23	1915–37	2,259	82.6
Huggins, Miller	2B	13	1904–16	1,586	22.7
Jackson, Joe	OF	13	1908–20	1,331	38.6
Johnson, Bob	OF	13	1933–45	1,863	36.7
Kaline, Al	OF	22	1953–74	2,834	46.8
Keller, Charlie	OF	13	1939–52	1,170	24.5
Kelly, King	OF, C	16	1878–93	1,463	37.4
Killebrew, Harmon	1B, 3B	22	1954–75	2,435	31.0

NAME	POS.	YEARS	SPAN	GAMES	OVERALL WINS
Kiner, Ralph	OF	10	1946–55	1,472	26.1
Klein, Chuck	OF	17	1928–44	1,753	24.5
Lajoie, Nap	2B	21	1896–1916	2,481	85.3
Lombardi, Ernie	C	17	1931–47	1,853	23.7
McPhee, Bid	2B	18	1882–99	2,138	38.5
Magee, Sherry	OF	16	1904–19	2,087	29.6
Mantle, Mickey	OF	18	1951–68	2,401	70.9
Mathews, Eddie	3B	17	1952–68	2,391	48.1
Mays, Willie	OF	22	1951–73	2,992	87.7
Mazeroski, Bill	2B	17	1956–72	2,163	36.7
Medwick, Ducky	OF	17	1932–48	1,984	25.9
Mize, Johnny	1B	15	1936–53	1,884	36.1
Musial, Stan	OF, 1B	22	1941–63	3,026	76.1
Oliva, Tony	OF	15	1962–76	1,676	24.7
Ott, Mel	OF	22	1926–47	2,730	59.2
Pfeffer, Fred	2B	16	1882–97	1,670	25.7
Pratt, Del	2B	13	1912–24	1,835	25.7
Robinson, Brooks	3B	23	1955–77	2,896	22.5
Robinson, Frank	OF	21	1956–76	2,808	70.7
Robinson, Jackie	2B, 3B	10	1947–56	1,382	33.4
Ruth, Babe	OF	22	1914–35	2,503	116.9
Santo, Ron	3B	15	1960–74	2,243	37.7
Schang, Wally	C	19	1913–31	1,839	24.0
Sewell, Joe	SS	14	1920–33	1,903	37.7
Simmons, Al	OF	20	1924–44	2,215	25.9
Sisler, George	1B	15	1915–30	2,055	25.7
Snider, Duke	OF	18	1947–64	2,143	24.3
Speaker, Tris	OF	22	1907–28	2,789	81.0
Terry, Bill	1B	14	1923–36	1,721	26.0
Thompson, Sam	OF	15	1885–1906	1,410	26.7
Vaughan, Arky	SS	14	1932–48	1,817	42.1
Wagner, Honus	SS	21	1897–1917	2,789	79.4
Wallace, Bobby	SS	25	1894–1918	2,386	34.4
Waner, Paul	OF	20	1926–45	2,549	39.4
Wheat, Zack	OF	19	1909–27	2,410	23.7
Williams, Billy	OF	18	1959–76	2,488	27.9
Williams, Ted	OF	19	1939–60	2,292	96.9
Wynn, Jim	OF	15	1963–77	1,920	30.0

Table XV, 2. *The Forty-Three Best Pitchers*

NAME	YEARS	SPAN	GAMES	OVERALL WINS
Alexander, Pete	17	1911–30	703	60.5
Brecheen, Harry	12	1940–53	321	22.1
Bridges, Tommy	16	1930–46	424	24.7
Brown, Three Finger	14	1903–16	493	34.5

Clarkson, John	12	1882–84	531	33.1
Coveleski, Stan	14	1912–28	451	24.4
Drysdale, Don	14	1956–69	547	33.3
Faber, Red	20	1914–33	670	24.0
Feller, Bob	18	1936–56	570	24.3
Ferrell, Wes	15	1927–41	548	28.6
Fitzsimmons, Freddie	19	1925–43	513	23.4
Ford, Whitey	16	1950–67	500	36.5
Gibson, Bob	17	1959–75	596	44.8
Griffith, Clark	21	1891–1914	453	21.7
Grimes, Burleigh	19	1916–34	632	22.7
Grove, Lefty	17	1925–41	619	50.9
Hubbell, Carl	16	1928–43	535	36.1
Johnson, Walter	21	1907–27	933	73.3
Joss, Addie	9	1902–10	296	23.1
Keefe, Tim	14	1880–93	600	36.7
Lemon, Bob	15	1941–58	615	33.1
Luque, Dolf	20	1914–35	558	22.1
Lyons, Ted	21	1923–46	705	35.4
Marichal, Juan	16	1960–75	475	27.4
Mathewson, Christy	17	1900–16	640	51.6
Mays, Carl	15	1915–29	502	31.8
Mullane, Tony	13	1881–94	556	22.4
Newhouser, Hal	17	1939–55	492	36.3
Nichols, Kid	15	1890–1906	621	47.1
Plank, Eddie	17	1901–17	629	23.8
Radbourn, Hoss	12	1880–91	528	26.8
Roberts, Robin	19	1948–66	688	23.8
Rommel, Ed	13	1920–32	507	23.9
Ruffing, Red	22	1924–47	882	23.4
Rusie, Amos	10	1889–1901	462	35.0
Shocker, Urban	13	1916–28	412	22.5
Spahn, Warren	21	1942–65	783	41.3
Trout, Dizzy	15	1939–57	535	32.1
Walsh, Ed	14	1904–17	459	37.3
Walters, Bucky	19	1931–50	715	28.0
Welch, Mickey	13	1880–92	564	19.6
Wilhelm, Hoyt	21	1952–72	1,070	27.9
Young, Cy	22	1890–1911	906	69.7

That Babe Ruth should emerge as the best player hardly classifies as an upset—his 116.9 Overall Player Wins do not even include the 12.4 additional wins he contributed as a pitcher, mostly in 1915–17 with the Red Sox. In his eighteen years as primarily an outfielder, he *averaged* 6.5 Overall Wins a year; a 6.3 season would give a player a spot in the list of the top hundred seasons since 1901. And still . . . Ted Williams was 20 Wins behind him, a whopping margin indeed—but what if he had not missed five years at the height of his career to military service? In the years immediately before and after his hiatus of 1943–45, he averaged 8.5 Wins; in the years sandwiching his 1952–53 gap, he averaged 6.3. Add it up: Williams probably would have added 36.1 Wins (subtracting the 2.0 he gained after his return in '53) and surpassed The One and Only.

Pete Rose has taken enough knocks the last couple of years that he doesn't need more from this direction; his stats are Hall of Fame caliber any way you slice them. But if "catching" Cobb is his aim, forget it. Rose may get 4,191 or 4,192 hits, whichever will pass Cobb's total (we say 4,191 will do it; O.B. says 4,192), but as a complete player, Cobb ranks fourth and Rose sixty-fifth.

Some reevaluations of player ability resulting from the combination of factors which forms the Overall Player rankings: The poor-fielding Lou Gehrig slips badly from third in Batting LWTS to sixteenth Overall, as do for the same reason Harry Heilmann, Johnny Mize, and Duke Snider (!); meanwhile, Nap Lajoie moves to higher ground because of his fine glovework, and folks like Bobby Doerr, Bill Mazeroski, Dave Bancroft, Dick Bartell, and Art Fletcher establish themselves as stars of the first magnitude despite ordinary to downright awful bats (Mazeroski was −18 Batting Wins and +36.5 Defensive Wins, with a Positional Adjustment of +18.8). Another surprising ranking emanating from a high Defensive Wins total is Richie Ashburn's—only Max Carey and Tris Speaker saved more runs in the outfield than Ashburn (Mays and DiMaggio weren't even close). And Gabby Hartnett is the top-ranked catcher of the century, surpassing Dickey, Berra, Campanella, Cochrane, and Bench—a tribute to his underrated glove and to his longevity in a strenuous position. And among players of the nineteenth century, when defensive skill was so much more important than it has been in our time, five middle infielders not in Cooperstown emerge as some of the top all-round players—Bid McPhee, Cupid Childs, and Fred Pfeffer, all second basemen; and Bill Dahlen and Jack Glasscock, shortstops.

Other surprises? It could be said that every name in boldface is a surprise, inasmuch as the electors of Cooperstown have had their shot at all of them and passed them by. Of these, the top five in Overall Wins are Bobby Doerr, Arky Vaughan, Joe Jackson, Ron Santo, and Bill Mazeroski.

But doesn't the Overall LWTS measure reward longevity and bypass short-term stars? Yes and no. Yes, the more games played at an above-average level, the higher the win total; but no, short-term brilliance is not given short shrift—if the performance is brilliant, then even a brief stay at the top will offer totals that are Hall of Fame caliber. By Overall Wins, pitcher Amos Rusie excelled despite only nine active years of consequence; Jackie Robinson and Ralph Kiner make both pantheons despite ten-year careers, and Joe Gordon, Charlie Keller, Joe DiMaggio, Joe Jackson, and Frank Baker all compiled impressive Overall Wins in their brief tenures. In fact, if you look at the table at the rear listing the top 100 seasons in Overall Player Wins, you will find only six eligible players not represented in Cooperstown's Hall or ours (Cy Seymour, 1905; Snuffy Stirnweiss, 1945; Eddie Lake, 1945; Al Rosen, 1953; and two Federal League stars—Benny Kauff in 1915 and Duke Kenworthy the year before).

Before proceeding to discuss the forty-three top pitchers, let's have a look at the active or recently retired players whose Overall Player Wins exceed those of the last player to make our pantheon. In the years to come, these players will be added onto our lists, not supplanting anyone, just as is done in Cooperstown. And then let's see why some HOFers did not pass muster.

The proponents of the good old days will be dismayed to learn that fully eighteen active players and seven more who retired since 1980—not counting pitchers—have Win totals higher than a good many Hall of Famers. Table XV, 3 presents those twenty-five players.

Table XV, 3. *Active and Recently Retired Players with Pantheon Credentials*

NAME	YEARS	SPAN	GAMES	OVERALL WINS
Bench, Johnny	17	1967–83	2,158	31.2
Bonds, Bobby	14	1968–81	1,849	29.5
Brett, George	11	1973–83	1,358	31.3
Carew, Rod	17	1967–83	2,249	42.8
Cedeno, Cesar	14	1970–83	1,748	22.4
Cey, Ron	13	1971–83	1,640	25.0
Concepcion, Dave	14	1970–83	1,901	23.4
Evans, Darrell	15	1969–83	1,853	25.1
Fisk, Carlton	15	1969–83	1,447	22.4
Grich, Bobby	14	1970–83	1,650	42.7
Jackson, Reggie	17	1967–83	2,287	43.3
McCovey, Willie	22	1959–80	2,588	37.8
Minoso, Minnie	17	1949–80	1,835	23.0
Morgan, Joe	21	1963–83	2,533	67.7
Nettles, Graig	17	1967–83	2,121	27.8
Rice, Jim	10	1974–83	1,334	22.6
Rose, Pete	21	1963–83	3,250	27.3
Schmidt, Mike	12	1972–83	1,638	64.8
Simmons, Ted	16	1968–83	1,954	26.1
Smalley, Roy	9	1975–83	1,157	24.2
Smith, Reggie	17	1966–82	1,987	32.1
Stargell, Willie	21	1962–82	2,360	32.2
Staub, Rusty	21	1963–83	2,819	25.6
Winfield, Dave	11	1973–83	1,514	25.2
Yastrzemski, Carl	23	1961–83	3,308	47.1

Table XV, 4. *Hall of Fame Players Outside the LWTS Pantheon*

PLAYER	YEARS	SPAN	GAMES	OVERALL WINS
Averill, Earl	13	1929–41	1,669	20.6
Beckley, Jake	20	1888–1907	2,386	13.8
Bottomley, Jim	16	1922–37	1,991	0.9
Burkett, Jesse	16	1890–1905	2,070	18.4
Bresnahan, Roger	17	1897–1915	1,438	20.3
Campanella, Roy	10	1948–57	1,215	16.0
Carey, Max	20	1910–29	2,476	20.9
Chance, Frank	17	1898–1914	1,285	14.5
Combs, Earle	12	1924–35	1,455	11.9
Cuyler, Kiki	18	1921–38	1,879	18.0
Duffy, Hugh	17	1888–1906	1,737	12.3
Evers, Johnny	18	1902–29	1,784	16.3
Hafey, Chick	13	1924–37	1,283	11.3
Hooper, Harry	17	1909–25	2,308	2.4
Jackson, Travis	15	1922–36	1,656	21.0
Jennings, Hugh	17	1891–1918	1,285	10.6
Keeler, Willie	19	1892–1910	2,124	23.3
Kell, George	15	1943–57	1,795	9.1
Kelley, Joe	17	1891–1908	1,845	20.0
Kelly, George	16	1915–32	1,622	4.9
Lindstrom, Fred	13	1924–36	1,438	7.4
Manush, Heinie	17	1923–39	2,009	7.6
Maranville, Rabbit	23	1912–35	2,670	6.6
McCarthy, Tommy	13	1884–96	1,275	0.3
O'Rourke, Jim	19	1876–1904	1,774	15.2
Rice, Sam	20	1915–34	2,404	5.7
Robinson, Wilbert	17	1886–1902	1,371	5.7
Roush, Edd	18	1913–31	1,967	15.6
Schalk, Ray	18	1912–29	1,760	3.0
Tinker, Joe	15	1902–16	1,804	17.0
Traynor, Pie	17	1920–37	1,941	21.2
Waner, Lloyd	18	1927–45	1,993	−9.3
Ward, John	17	1878–94	1,825	14.5
Wilson, Hack	12	1923–34	1,348	18.1
Youngs, Ross	10	1917–26	1,211	12.6

And these active players all stand a solid chance of entering the Overall Player Win pantheon as well (in order of their Win totals through 1983): Buddy Bell, Robin Yount, George Foster, Ken Singleton, Dave Parker, Amos Otis, Keith Hernandez, and Gene Tenace.

Of the ninety-seven nonpitchers enshrined in upstate New York, thirty-five fall short of our list. It is interesting that of our top thirty players in this century, twenty-seven are also in Cooperstown; indeed, agreement holds up pretty well up to our forty-fourth spot (thirty-eight of the forty-four are in Cooperstown). After that, however, we come to a radical parting of the ways. Can a good case be made as to why any or all of the Hall of Famers listed in Table XV, 4 should be included in our list as well? Yes, to a degree. If short-term brilliance is the ticket for you, then these men rank among the top ninety-seven in OPS, which is not weighted for longevity: Hack Wilson, Earl Averill, Chick Hafey, Willie Keeler, and Frank Chance. And a special mention should be made of Max Carey, who would have easily entered our pantheon had we included his base-stealing exploits; although we had ample data (though not complete) to make a reasonable statistical projection for his Stealing LWTS Wins, we did not have numbers for so many of his contemporaries that we thought it unfair to give Carey the benefit of his expanded data. For Honus Wagner, who stole 722 bases, we have no caught-stealing information; for Ty Cobb, only spotty documentation: Both have a Stealing Wins total of 0.0.

Pitchers: Of the top thirty-five in Overall Wins in this century, excepting those not yet eligible for the Hall of Fame, thirteen are absent from Cooperstown. The great names are common to both lists—Johnson, Young, Mathewson, Alexander, Grove, Gibson, Spahn, Walsh; in fact, these are the top eight in Overall Wins, in descending order, with Walsh securing his rank in part through a prodigious (for pitchers) 7.8 Defensive Wins.

Another pitcher whose fielding vaulted him onto the list of all-time greats is Carl Mays, who notched 7.9 Fielding Wins, the most by any pitcher in history. Mays is remembered today, if at all, as a cantankerous, surly submarine pitcher who was not afraid to pitch inside; one pitch that got away killed Cleveland shortstop Ray Chapman in 1920 and barred the gates of Cooperstown to Mays, probably forever. Other top hurlers whose gloves helped their teams to more than 5 extra wins were Burleigh Grimes, Freddie Fitzsimmons, and Bob Lemon.

Those whose bats contributed more than 10 extra wins were Walter Johnson and Wes Ferrell (the only other pitchers to have accomplished this were Red Lucas and George Mullin). Among the nineteenth-century players who are in Cooperstown and on our list (we are in agreement here, except for Tony Mullane in place of Pud Galvin, and Clark Griffith, who was elected as an executive, for which he displayed no notable talent, rather than as a pitcher, for which he did), Nichols and Radbourn had good bats, and Nichols was a top fielder.

Unlike the situation that prevailed with the everyday players, there is not an endless flood of new pitching talent advancing upon the old. Seven active pitchers, listed below, along with one recently retired, all have stats good enough to crack the hidden Hall of Fame, but after that, the only ones close to the stats of Dolf Luque, the thirty-fifth-best Overall Pitcher since 1901 (and the last to make our "cut"), are Steve Rogers in forty-eighth place and Rick Reuschel in sixty-seventh; the next active pitcher with a chance to amass many extra wins is Ron Guidry, back in seventy-ninth place. The ones to watch are Dave Stieb and Bob Stanley, who have not yet pitched 1,500 innings but have amassed more Overall Wins than several current Hall of Famers listed in Table XV, 6.

Table XV, 5. *Active and Recently Retired Qualifiers*
for the Pitching Pantheon

PLAYER	YEARS	SPAN	GAMES	OVERALL WINS
Blyleven, Bert	14	1970–83	418	28.8
Carlton, Steve	19	1965–83	590	37.1
Jenkins, Fergy	19	1965–83	632	28.7
John, Tommy	20	1963–83	581	27.9
Niekro, Phil	20	1964–83	707	36.7
Palmer, Jim	18	1965–83	557	35.4
Perry, Gaylord	22	1962–83	757	37.1
Seaver, Tom	17	1967–83	546	45.0

Of the fifteen men listed in Table XV, 6, all Hall of Famers who contributed fewer Overall Wins than those in our pantheon, a case can be made for several "shooting stars" who burned briefly but brightly, on the basis of their NERAs. On the whole, in fact, the group of pitchers in Cooperstown presents fewer unfathomable choices, in new statistical terms or old, than their counterparts in the field. As mentioned earlier, quality shows more readily among pitchers than among players because responsibility for run prevention is less diffuse than is the responsibility for run scoring; similarly, long careers of unrelieved ineptness are more rare among pitchers than among other players.[2] These pitchers in Table XV, 6, while they missed the Overall Pitcher select group of thirty-five (forty-three in toto, eight pre-1901), did have top-thirty-five NERAs: Rube Waddell, Sandy Koufax, Dizzy Dean, Dazzy Vance, and Lefty Gomez.

Table XV, 6. *Hall of Fame Pitchers Outside the LWTS Pantheon*

PLAYER	YEARS	SPAN	GAMES	OVERALL WINS
Bender, Chief	16	1903–25	507	12.2
Chesbro, Jack	11	1899–1909	392	10.2
Dean, Dizzy	13	1930–47	325	19.2
Galvin, Pud	14	1879–92	697	13.7
Gomez, Lefty	14	1930–43	368	14.7
Haines, Jesse	19	1918–37	560	6.0

Hoyt, Waite	21	1918–38	674	14.0
Koufax, Sandy	12	1955–66	397	18.7
Marquard, Rube	18	1908–25	536	2.6
McGinnity, Joe	10	1899–1908	465	19.2
Pennock, Herb	22	1912–34	620	5.8
Rixey, Eppa	21	1912–35	694	21.8
Vance, Dazzy	16	1915–35	442	22.0
Waddell, Rube	13	1897–1910	407	20.4
Wynn, Early	23	1939–63	796	15.6

Will someone rush to buy us a vacant lot—Hoboken's Elysian Fields site would do nicely—and erect an impressive edifice to give tangible expression to our Imaginary Hall of Fame? No, and perhaps just as well, too; the hidden game of baseball needs no playing field, no museum but that of the mind. The players of seasons past continue to play in baseball's statistical world-within-a-world; even the Knickerbockers of 1845, experimenting with the shape of the game to come, reconstitute themselves for one who studies the scorebooks of their contests with the Gothams and Excelsiors.

The New Statisticians, too, are experimenting with the shape of the game to come—how it will be understood and, increasingly, how it will be played—for these new techniques are designed to be used. If players or pitchers can enjoy long careers, capped by a Hall of Fame selection, without contributing so much as one extra win for the effort of eighteen to twenty years, then there's still a lot to learn about the old ball game.

---

[1] For example, the fiftieth-ranked batter has 32.2 Overall Wins, while the fiftieth-ranked pitcher has only 20.7.

[2] Of the 815 players since 1901 who participated in 1,000 or more games, 39.9 percent were below average—i.e., had negative Overall Wins; of the 433 men who pitched 1,500 innings, 26.8 percent were below average overall. The lowest ranked players—Doc Cramer, Ken Reitz, and Pete Suder—were −27 in Overall Wins; the lowest ranked pitchers—Herm Wehmeier, Casey Patten, and Si Johnson—averaged −17 Overall Wins.

KEY TO SYMBOLS
USED IN THE TABLES

ADJ Adjusted for home-park effects

ALL Overall Wins, Player or Pitcher

AVG Batting average (pitchers excluded)

BAT Batting Wins (Linear Weights)

Batting Runs Batting Linear Weights, expressed in runs above or below the league average

BPF Batter Park Factor

BR Batting Runs without park adjustments (ADJ in adjacent column signifies Batting Runs with park adjustments)

BSR Base Stealing Runs (used in 1983 section only)

DEF Defensive Linear Weights Runs (used in 1983 section only)

DIFF For seasons before 1901, refers to the difference between a team's actual victories and the victories predicted by its runs scored versus runs allowed; for seasons from 1901 onward, refers to the difference between a team's actual victories and the victories predicted by its players' Batting Linear Weights and Pitching Linear Weights

ERA Earned run average

FLD Fielding Wins

G Games

GB Games behind

HR Home runs

IP Innings pitched

ISO Isolated Power

L Losses credited officially; for tables of nineteenth-century best seasons, L stands for the individual's league (National, American Association, Players, and Union Association)

LG League average for given season or career

LWT Linear Weight

M Middle-inning pitcher (used in 1983 section only)

NERA Normalized earned run average

NERA-A A two-column heading: the column at the left is the normalized earned run average without park adjustments; the column at the right is the normalized earned run average with park adjustments

NO Number

NOPS Normalized On Base Average Plus Slugging Percentage

NOPS-A A two-column heading: the column at the left is Normalized On Base Plus Slugging without park adjustments; the column at the right is Normalized On Base Plus Slugging with park adjustments

NSLG Normalized Slugging Percentage

OBA On Base Average

OPS On Base Plus Slugging

OR Opponents' Runs

PCT Percentage

PF Park Factor

POS In 1983 section, means position played; anywhere else it signifies the positional adjustment to the Overall Player Wins formula

PIT Pitching Wins, in Linear Weights

Pitching Runs Pitching Linear Weights, expressed in Runs

Pitching Wins Pitching Linear Weights, expressed in Wins

PPF Pitcher Park Factor

PR Pitching Runs (Linear Weights) without park adjustments; ADJ afterward signifies Pitching Runs with park adjustments

R In 1983 section only, stands for reliever; otherwise, signifies actual runs scored (not Linear Weights)

RBA Relative (Normalized) Batting Average

RBI Run batted in

Runs O/D Actual runs scored (not Linear Weights) above average on offense (O, the column at the left) and actual runs allowed below average on defense (D, the column at the right); pertains to nineteenth century only

RUN Running (Base Stealing) wins

R/W Runs Per Win

S Starter

SLG Slugging percentage

TL Team losses

TW Team Wins

W Wins credited officially, not Linear Weights

WINS Overall Wins, Player or Pitcher, for 1983 section only; otherwise, refers to Linear Weights Wins—when following batting data, signifies Batting Wins, and when following pitching data, signifies Pitching Wins

WINS O/D Wins (not Linear Weights) above average resulting from team offense (O) and team defense (D); pertains to nineteenth century only

NOTES

1. Pitcher batting has been removed from all league averages since 1901, which elevates the league averages and lowers the normalized statistics such as NERA, NOPS, NSLG, NOBA, RBA, and LWTS.

2. For the calculation of On Base Average and Linear Weights for nineteenth-century and dead-ball–era players, we have relied upon the hit-by-pitch research of John Tattersall, Alex Haas, and Pete Palmer for the years before the official adoption of the HBP (by the NL, in 1917; by the AL, in 1921). Newspaper research continues for the period 1897–1908; other years are incorporated in these tables (from 1884 in the American Association, 1887 in the National League—the first years in which a hit batsman was awarded first base).

3. Won-lost records for the American League in the years 1901–19 have been rectified through the research of Frank Williams, whose findings are reflected in the categories "Wins Above Team" and "Percentage of Team Wins."

4. Outfielders' Defensive LWTS figures are subject to some degree of error due to switching of fields within a game or season (Babe Ruth, for example,

was positioned in the field which required the lesser range—right field in Yankee Stadium, left field in most road parks). Also, short distances to the left- or right-field walls, as in the Polo Grounds or Ebbets Field, tend to depress putout totals.

5. From 1946 to date, individual pitcher strikeouts have been subtracted from the total number of potential outs when calculating Defensive LWTS for pitchers. For the years 1901–45, the strikeouts subtracted have been estimated on the basis of the strikeout-per-inning rate of each team's pitching staff; this tends to depress the Defensive LWTS of exceptional strikeout pitchers like Walter Johnson, Dazzy Vance, and Bob Feller in his years before 1946.

6. There is some variation—generally one or two tenths of a Win—between the Batting, Pitching, and Relief LWTS recorded in the tables for those categories and the Batting, Pitching, and Relief LWTS listed in the Overall Win categories. The reason for this variation is that the Overall figure represents the sum of the yearly values, each of which has been rounded to the nearest tenth of a Win, while the figures in the specific LWTS categories were not rounded until the final calculation.

7. One of the items touched upon in Chapters 14 and 15 bears elaboration: what Ted Williams' record would have been had he not missed nearly five years due to military service. We have projected the career totals of several other wartime-affected stars, in the same manner as we did for Williams, and came up with these adjustments: Bob Feller gains four years at 4.2 Wins each, less .5 Wins for his partial performance in 1945; adding this 16.3 Wins to his Overall total gives him 40.5, good for ninth place. Other pitchers to gain significant amounts were Ted Lyons (6.6 Wins) and Pete Alexander (5.6). Among the hitters, Hank Greenberg gains $4.25 \times 5 - 1.9 = 19.3$ (rounded), moving him all the way up to nineteenth place in Overall Player Wins. Joe DiMaggio gains 6.8, Joe Gordon 7.2, Bobby Doerr 5.0, Johnny Mize 9.8, Stan Musial 5.6, Enos Slaughter 10.8, and Arky Vaughan 3.9.

8. To be eligible for inclusion in any lifetime table, a pitcher must have worked 1,500 innings and a player 1,000 games (relief-pitching categories require only 750 innings).

9. In the 1984 Complete Player Data section, a position might be listed as 3*/21. The asterisk indicates that the man was the team's regular at third base, and played more than 100 games at the position. The figures after the slash indicate that he also played some second base and first base, but in fewer than ten games at each position.

	G	LWT	Adj	Wins	PF	R/W
1 Babe Ruth	2503	1322.5	1363.8	130.61	.969	10.44
2 Ted Williams	2292	1165.9	1130.0	112.81	1.031	10.02
3 Ty Cobb	3034	1032.6	1006.7	105.41	1.018	9.55
4 Stan Musial	3026	983.0	942.6	95.47	1.028	9.87
5 Hank Aaron	3298	877.7	899.2	93.63	.986	9.60
6 Lou Gehrig	2164	917.3	977.9	90.94	.952	10.75
7 Willie Mays	2992	826.6	848.2	87.52	.985	9.69
8 Mickey Mantle	2401	802.9	833.8	85.60	.972	9.74
9 Rogers Hornsby	2259	843.4	861.5	85.51	.984	10.07
10 Tris Speaker	2789	840.7	808.8	83.10	1.023	9.73
11 Frank Robinson	2808	773.4	764.3	79.89	1.007	9.57
12 Mel Ott	2730	767.2	773.8	76.55	.995	10.11
13 Jimmie Foxx	2317	802.7	788.2	72.97	1.011	10.80
14 Honus Wagner	2789	684.8	657.6	68.34	1.021	9.62
15 Eddie Collins	2826	603.9	616.2	63.98	.991	9.63
16 Nap Lajoie	2474	563.0	564.5	57.54	.999	9.81
17 Carl Yastrzemski	3308	617.3	537.1	56.98	1.052	9.42
18 Willie McCovey	2588	523.5	536.6	56.35	.988	9.52
19 Eddie Mathews	2391	480.1	525.2	53.23	.961	9.87
20 Harmon Killebrew	2435	532.2	497.1	52.47	1.033	9.47
21 Willie Stargell	2360	482.8	487.3	51.45	.995	9.47
22 Al Kaline	2834	513.1	489.5	51.26	1.019	9.55
23 Joe DiMaggio	1736	506.6	533.3	51.08	.972	10.44
24 Sam Crawford	2517	509.7	478.4	51.04	1.028	9.37
25 Johnny Mize	1884	519.9	504.1	50.87	1.019	9.91
26 Joe Morgan	2649	438.0	481.6	50.46	.964	9.54
27 Dick Allen	1749	469.5	468.8	50.16	1.001	9.35
28 Harry Heilmann	2146	516.8	517.0	49.81	1.000	10.38
29 Reggie Jackson	2430	443.4	467.5	48.60	.978	9.62
30 Mike Schmidt	1789	468.3	462.0	48.46	1.008	9.53
31 Paul Waner	2549	489.2	474.6	46.39	1.011	10.23
32 Joe Jackson	1331	451.9	444.8	46.21	1.011	9.63
33 Billy Williams	2488	463.0	404.0	42.85	1.052	9.43
34 Rod Carew	2342	429.9	414.3	42.77	1.014	9.69
35 Pete Rose	3371	420.3	394.8	42.05	1.016	9.39
36 Hank Greenberg	1394	468.4	438.1	41.56	1.039	10.54
37 Duke Snider	2143	441.4	394.6	39.40	1.049	10.01
38 Norm Cash	2089	389.9	369.6	38.77	1.024	9.53
39 Roberto Clemente	2433	353.5	358.4	38.07	.996	9.42
40 Reggie Smith	1987	379.2	353.9	37.41	1.030	9.46
41 Frank Howard	1895	324.0	346.4	36.99	.971	9.37
42 Ralph Kiner	1472	391.1	370.9	36.85	1.027	10.06
43 Bob Johnson	1863	366.1	381.7	36.81	.984	10.37
44 Elmer Flick	1493	348.2	360.5	35.96	.983	9.77
45 Fred Clarke	2244	405.0	368.9	35.89	1.030	9.98
46 Ken Singleton	2082	331.6	348.2	35.60	.983	9.78
47 Arky Vaughan	1817	361.2	356.1	35.49	1.006	10.03
48 Al Simmons	2215	398.7	377.7	35.16	1.017	10.74
49 Rusty Staub	2897	316.1	332.1	34.86	.987	9.53
50 Zach Wheat	2410	334.0	336.2	34.84	.998	9.65
51 Orlando Cepeda	2124	336.8	333.6	34.58	1.003	9.65
52 Sherry Magee	2087	330.4	319.6	34.47	1.012	9.27
53 Goose Goslin	2287	340.4	362.9	33.97	.982	10.68
54 Boog Powell	2042	304.6	301.7	32.03	1.004	9.42
55 Chuck Klein	1753	374.6	330.7	31.96	1.050	10.35
56 Joe Medwick	1984	353.5	318.4	31.78	1.037	10.02
57 Charlie Gehringer	2323	376.0	344.2	31.70	1.024	10.86
58 George Brett	1462	302.4	309.2	31.20	.991	9.91
59 Joe Torre	2209	297.6	293.0	30.90	1.005	9.48
60 Jim Wynn	1920	256.2	290.4	30.78	.960	9.43
61 Babe Herman	1552	304.1	321.1	30.71	.978	10.46
62 Dave Winfield	1655	250.8	297.3	30.59	.939	9.72
63 Bill Terry	1721	318.1	316.7	30.41	1.002	10.41
64 Charlie Keller	1170	286.5	299.6	30.27	.975	9.90
65 Ron Santo	2243	344.8	282.4	30.20	1.062	9.35
66 Hack Wilson	1348	305.3	316.8	30.05	.983	10.54
67 Greg Luzinski	1821	291.1	288.3	29.93	1.003	9.63
68 Rocky Colavito	1841	296.5	289.3	29.69	1.009	9.74
69 Willie Keeler	2125	374.5	340.7	29.48	1.028	10.69
70 Jack Fournier	1530	284.9	293.6	29.46	.987	9.97
71 Minnie Minoso	1835	298.9	290.7	29.34	1.009	9.91
72 Earl Averill	1668	333.9	318.9	29.28	1.016	10.89
73 Bobby Bonds	1849	274.4	282.0	29.08	.992	9.70
74 Dolph Camilli	1490	318.8	280.8	28.24	1.052	9.94
75 Enos Slaughter	2380	305.8	273.7	28.08	1.031	9.75
76 Joe Perez	2628	282.9	264.7	27.85	1.016	9.50
77 Larry Doby	1533	266.6	278.0	27.66	.985	10.05
78 Eddie Murray	1206	253.0	274.9	27.62	.964	9.95
79 Gavvy Cravath	1220	272.2	250.7	27.43	1.045	9.14
80 Johnny Bench	2158	263.9	256.6	26.96	1.008	9.52
81 George Sisler	2055	267.3	263.0	26.80	1.004	9.81
82 Ernie Banks	2528	287.3	260.2	26.71	1.024	9.74
83 Edd Roush	1967	231.0	254.8	26.40	.974	9.65
84 Tony Oliva	1676	272.5	245.5	26.27	1.037	9.34
85 Kiki Cuyler	1879	269.5	272.9	26.02	.997	10.49
86 Gene Tenace	1555	218.5	249.6	26.01	.948	9.60
87 Bobby Murcer	1908	225.8	244.6	25.90	.978	9.44
88 Jim Rice	1493	314.9	255.2	25.71	1.079	9.93
89 Rico Carty	1651	247.9	246.5	25.58	1.002	9.64
90 Frank Baker	1575	242.9	238.7	25.47	1.006	9.37
91 Al Oliver	2272	232.9	245.8	25.41	.988	9.67
92 Bob Elliott	1978	244.8	251.2	25.39	.993	9.90
93 Jeff Heath	1383	235.1	247.9	25.04	.980	9.90
94 Ken Williams	1397	277.9	259.9	24.72	1.026	10.51
95 Bob Watson	1832	212.8	238.5	24.71	.966	9.65
96 Jim Bottomley	1991	262.1	255.9	24.63	1.006	10.39
97 Keith Hernandez	1414	242.8	233.0	24.42	1.016	9.54
98 Wally Berger	1350	222.3	247.8	24.36	.963	10.17
99 Mickey Cochrane	1482	269.5	259.6	24.32	1.012	10.68
100 Bobby Grich	1766	232.8	234.3	24.18	.998	9.69

	G	LWT	Adj	Wins	PF	R/W		G	LWT	Adj	Wins	PF	R/W
101 Bill Dickey	1789	215.6	255.4	23.85	.957	10.71	151 Ron Fairly	2442	170.2	188.3	19.25	.980	9.78
102 Gabby Hartnett	1990	240.5	245.5	23.79	.994	10.32	152 Harry Hooper	2308	173.8	184.7	19.13	.991	9.65
103 Fred Lynn	1301	272.7	233.2	23.56	1.063	9.90	153 Mike Hargrove	1559	206.6	188.6	19.13	1.025	9.86
104 Hal McRae	1842	223.7	231.8	23.54	.990	9.85	154 Bill Madlock	1443	201.3	182.0	19.10	1.030	9.53
105 Yogi Berra	2120	208.4	234.6	23.47	.973	10.00	155 Don Baylor	1770	166.2	186.5	19.01	.976	9.81
106 Jackie Robinson	1382	269.3	235.0	23.44	1.051	10.03	156 Dwight Evans	1622	237.7	185.7	18.87	1.071	9.84
107 Topsy Hartsel	1355	228.8	220.6	23.33	1.013	9.45	157 Roger Maris	1463	173.7	184.0	18.85	.984	9.76
108 Roy Thomas	1470	229.2	228.1	23.29	1.001	9.79	158 Frank Chance	1286	197.5	176.2	18.83	1.039	9.36
109 Jose Cruz	1907	164.8	222.8	23.29	.926	9.57	159 Luke Appling	2422	195.9	191.6	18.73	1.003	10.23
110 Bill Nicholson	1677	220.6	223.0	23.14	.997	9.64	160 Jake Daubert	2014	159.1	174.9	18.61	.983	9.40
111 Roy White	1881	179.9	215.0	23.04	.957	9.33	161 Mickey Vernon	2409	154.2	179.2	18.45	.978	9.71
112 George Foster	1761	222.6	219.8	22.85	1.004	9.62	162 Lou Boudreau	1646	153.6	180.5	18.43	.966	9.79
113 Augie Galan	1742	215.9	222.6	22.70	.992	9.80	163 Richie Zisk	1453	174.3	178.7	18.37	.993	9.73
114 Cesar Cedeno	1858	176.9	214.8	22.59	.955	9.51	164 Richie Hebner	1825	167.1	176.4	18.37	.988	9.60
115 Earle Combs	1455	196.6	241.0	22.57	.948	10.68	165 Gary Matthews	1724	211.5	174.9	18.29	1.047	9.56
116 Cy Williams	2002	277.5	225.8	22.54	1.059	10.02	166 Richie Ashburn	2189	169.1	182.7	18.27	.988	10.00
117 Stan Hack	1938	230.6	219.9	22.51	1.011	9.77	167 Joe Judge	2171	156.7	186.1	18.19	.974	10.23
118 Bobby Veach	1821	220.9	214.3	22.45	1.008	9.55	168 Gil Hodges	2071	238.4	181.3	18.18	1.060	9.97
119 Jimmy Sheckard	2122	247.2	221.5	22.40	1.025	9.89	169 Ted Kluszewski	1718	190.1	183.0	18.17	1.009	10.08
120 Ron Cey	1786	198.5	213.1	22.35	.982	9.53	170 Heinie Groh	1676	154.6	162.6	18.13	.989	8.97
121 Rick Monday	1986	220.3	209.4	22.29	1.014	9.40	171 Roger Bresnahan	1446	177.9	170.7	18.07	1.012	9.44
122 Ted Simmons	2086	203.0	209.1	21.94	.994	9.53	172 John Titus	1402	173.6	169.6	18.04	1.007	9.40
123 Sid Gordon	1475	195.8	217.2	21.73	.968	10.00	173 Sam Rice	2404	140.1	184.9	18.03	.965	10.25
124 Ed Konetchy	2085	185.8	204.2	21.60	.980	9.45	174 Jack Clark	1044	158.7	167.3	17.61	.982	9.50
125 Roy Sievers	1887	197.9	211.4	21.50	.984	9.83	175 Willie Horton	2028	193.9	161.9	17.56	1.037	9.22
126 Dixie Walker	1905	219.5	210.8	21.49	1.010	9.81	176 Vern Stephens	1720	184.4	171.5	17.54	1.016	9.78
127 Larry Doyle	1766	211.2	201.8	21.48	1.012	9.40	177 George Stone	848	146.5	157.9	17.48	.969	9.03
128 Joe Cronin	2124	249.6	228.7	21.44	1.018	10.67	178 Ginger Beaumont	1463	181.6	170.2	17.41	1.016	9.78
129 Benny Kauff	859	192.7	193.5	20.92	.998	9.25	179 Ken Boyer	2034	207.1	170.8	17.37	1.039	9.83
130 Tommy Henrich	1284	202.7	214.5	20.84	.982	10.30	180 Wally Schang	1839	155.6	169.5	17.31	.982	9.79
131 Roy Cullenbine	1181	210.5	200.4	20.80	1.019	9.64	181 Pudge Fisk	1549	204.9	165.4	17.17	1.057	9.63
132 Riggs Stephenson	1310	207.7	214.0	20.64	.990	10.37	182 George Burns	1853	142.7	153.4	17.06	.988	8.99
133 Cecil Cooper	1545	197.8	203.9	20.60	.992	9.90	183 Harry Davis	1769	167.2	156.2	17.05	1.013	9.16
134 Ross Youngs	1211	201.1	198.1	20.44	1.005	9.69	184 Phil Cavaretta	2030	169.7	165.2	17.04	1.005	9.69
135 Hal Trosky	1347	201.7	222.3	20.40	.972	10.90	185 Jackie Jensen	1438	183.8	166.6	16.95	1.025	9.83
136 Darrell Evans	1984	206.9	194.3	20.32	1.014	9.56	186 Fred Tenney	1994	171.4	169.3	16.89	1.002	10.02
137 Sal Bando	2019	165.5	190.9	20.16	.972	9.47	187 Oscar Gamble	1514	160.4	167.4	16.86	.988	9.92
138 Dave Parker	1457	206.4	193.8	20.14	1.020	9.62	188 Rudy York	1603	223.3	171.4	16.82	1.066	10.19
139 Al Rosen	1044	187.7	200.8	20.06	.974	10.01	189 Jeff Burroughs	1603	171.5	163.2	16.78	1.012	9.72
140 Heinie Manush	2008	218.4	209.8	20.02	1.008	10.48	190 Bobby Doerr	1865	181.1	163.7	16.62	1.018	9.85
141 Joe Adcock	1959	157.6	195.9	20.00	.954	9.79	191 George Grantham	1444	184.7	173.9	16.58	1.015	10.49
142 Tony Lazzeri	1740	166.4	212.2	19.99	.952	10.61	192 Bob Nieman	1113	142.0	162.6	16.46	.954	9.88
143 Ernie Lombardi	1853	180.3	197.4	19.91	.977	9.91	193 Vic Wertz	1862	194.3	164.5	16.36	1.036	10.06
144 Lefty O'Doul	970	216.9	209.5	19.72	1.016	10.62	194 Amos Otis	1998	155.0	156.4	16.31	.998	9.59
145 Bob Allison	1541	211.3	184.9	19.48	1.041	9.49	195 Elbie Fletcher	1415	133.1	155.7	16.21	.965	9.61
146 John Mayberry	1620	194.8	186.4	19.39	1.012	9.61	196 Earl Torgeson	1668	138.6	160.8	16.03	.969	10.03
147 Gene Woodling	1796	185.4	192.0	19.38	.991	9.91	197 Eddie Yost	2109	154.1	161.3	15.98	.993	10.09
148 Chick Hafey	1283	212.3	201.6	19.36	1.017	10.41	198 Danny Murphy	1518	149.3	145.9	15.92	1.005	9.17
149 Andy Thornton	1285	201.9	191.1	19.35	1.018	9.88	199 Don Mincher	1400	144.1	146.9	15.70	.994	9.36
150 Steve Garvey	1988	157.1	184.7	19.31	.968	9.57	200 Del Ennis	1903	139.2	154.6	15.49	.983	9.98

	G	LWT	Adj	Wins	PF	R/W		G	LWT	Adj	Wins	PF	R/W
201 Chet Lemon	1195	154.4	154.7	15.49	.999	9.99	251 Dan Driessen	1531	121.2	117.1	12.17	1.007	9.62
202 Fielder Jones	1788	131.4	158.8	15.35	.971	10.34	252 Don Buford	1286	98.4	113.7	12.16	.972	9.35
203 Joe Gordon	1566	136.3	153.2	15.23	.979	10.06	253 Bob Meusel	1407	115.8	126.3	12.15	.986	10.40
204 Jim Gentile	936	140.9	149.9	15.18	.977	9.87	254 Andy Pafko	1852	115.1	122.1	12.14	.991	10.05
205 Rickey Henderson	791	121.3	150.3	15.17	.929	9.91	255 Bernie Carbo	1010	124.5	119.1	12.03	1.015	9.90
206 George Hendrick	1727	153.1	145.2	15.13	1.011	9.60	256 Frank McCormick	1534	114.5	117.2	12.00	.996	9.76
207 Dusty Baker	1845	152.2	144.4	15.11	1.010	9.55	257 Mike Epstein	907	100.3	109.5	11.93	.974	9.17
208 Billy Herman	1922	150.8	149.4	15.02	1.001	9.95	258 Heinie Zimmerman	1456	117.0	115.1	11.87	1.003	9.70
209 Ferris Fain	1151	169.2	147.9	14.98	1.037	9.88	259 John Grubb	1206	88.9	116.0	11.87	.943	9.77
210 Ben Chapman	1717	120.0	163.3	14.94	.956	10.93	260 Ripper Collins	1084	148.5	119.7	11.82	1.058	10.13
211 Gary Carter	1408	149.0	143.6	14.88	1.009	9.65	261 Casey Stengel	1277	118.3	111.4	11.78	1.013	9.46
212 Socks Seybold	974	150.6	145.7	14.87	1.011	9.79	262 Frankie Frisch	2311	158.3	123.8	11.76	1.028	10.52
213 Stan Spence	1112	123.7	141.7	14.87	.964	9.53	263 Jim Fregosi	1902	82.9	111.0	11.76	.964	9.44
214 Hank Sauer	1399	143.9	147.6	14.81	.994	9.97	264 Eddie Stanky	1259	113.5	116.1	11.71	.996	9.92
215 Tommy Holmes	1320	137.6	144.7	14.73	.989	9.82	265 Dave Kingman	1639	109.6	111.0	11.61	.998	9.57
216 Toby Harrah	1934	150.3	147.7	14.58	1.003	10.13	266 Joe Sewell	1903	116.0	120.3	11.50	.996	10.46
217 Ken Griffey	1412	144.5	138.7	14.50	1.009	9.57	267 George Selkirk	846	111.4	125.3	11.46	.969	10.93
218 Jason Thompson	1265	163.0	139.8	14.49	1.040	9.64	268 Lu Blue	1615	118.8	119.8	11.39	.999	10.51
219 Bill White	1673	176.8	138.2	14.38	1.053	9.61	269 Jim Delahanty	1185	103.8	106.9	11.35	.994	9.42
220 Jim Ray Hart	1125	139.3	133.7	14.34	1.013	9.32	270 Norm Siebern	1406	129.3	111.4	11.32	1.031	9.83
221 Cy Seymour	1528	156.1	132.5	14.29	1.033	9.28	271 Leon Wagner	1352	117.9	108.5	11.31	1.018	9.59
222 Sixto Lezcano	1219	134.2	139.8	14.24	.989	9.82	272 Buddy Bell	1827	101.1	111.4	11.28	.988	9.88
223 Felipe Alou	2082	101.0	130.8	14.10	.965	9.28	273 Bill Freehan	1774	119.7	99.7	11.11	1.028	8.98
224 Joe Harris	970	149.8	139.6	13.92	1.024	10.03	274 Buddy Myer	1923	88.8	117.2	10.92	.973	10.73
225 Johnny Bates	1154	115.8	130.9	13.86	.969	9.45	275 Al Smith	1517	105.8	109.1	10.86	.995	10.05
226 Harlond Clift	1582	173.0	151.4	13.83	1.025	10.94	276 Darrell Porter	1545	104.7	107.7	10.71	.996	10.06
227 Lee May	2071	116.9	126.0	13.83	.990	9.12	277 Harry Lumley	730	81.5	98.2	10.68	.944	9.20
228 Cliff Johnson	1156	125.9	132.2	13.43	.985	9.85	278 Whitey Kurowski	916	122.6	104.3	10.63	1.045	9.81
229 Dom DiMaggio	1399	139.8	133.9	13.31	1.008	10.06	279 Irish Meusel	1289	117.8	105.5	10.63	1.020	9.93
230 Elmer Valo	1806	146.5	132.6	13.30	1.020	9.97	280 Monte Irvin	764	101.0	104.8	10.56	.983	10.11
231 Miller Huggins	1586	120.8	125.2	13.28	.994	9.43	281 Pete Reiser	861	109.9	101.8	10.54	1.024	9.66
232 Graig Nettles	2245	123.9	125.0	13.18	.999	9.49	282 Doc Gessler	880	93.8	93.5	10.33	1.001	9.05
233 Bill Skowron	1658	104.1	129.7	12.99	.962	9.99	283 Smoky Burgess	1691	112.1	102.5	10.32	1.017	9.93
234 Max Carey	2476	127.0	126.5	12.97	1.000	9.76	284 Ray Boone	1373	83.9	102.1	10.25	.970	9.96
235 George Scott	2034	161.0	124.5	12.86	1.041	9.68	285 Brian Downing	1284	101.6	102.4	10.25	.998	9.99
236 Steve Kemp	1047	132.4	128.0	12.85	1.009	9.96	286 Bob Bailey	1931	106.0	100.4	10.22	1.007	9.83
237 Dale Murphy	1038	156.6	122.5	12.83	1.074	9.55	287 Doug DeCinces	1252	85.9	100.8	10.22	.974	9.86
238 Wally Moon	1457	141.7	127.1	12.80	1.023	9.93	288 Johnny Pesky	1270	105.3	101.8	10.21	1.005	9.97
239 Andre Dawson	1174	120.3	121.0	12.74	.999	9.50	289 Bob Horner	689	116.1	96.8	10.13	1.063	9.55
240 Larry Hisle	1197	130.6	123.7	12.71	1.013	9.73	290 Tommy Leach	2156	118.7	91.4	10.10	1.028	9.05
241 Pedro Guerrero	657	121.8	120.5	12.70	1.005	9.49	291 Ival Goodman	1107	89.8	100.1	10.07	.980	9.94
242 Steve Evans	978	111.4	119.6	12.61	.981	9.49	292 George Kell	1795	118.2	103.6	10.07	1.017	10.30
243 John Briggs	1366	116.2	117.8	12.58	.997	9.36	293 Gus Suhr	1435	106.0	102.0	10.06	1.006	10.13
244 Nick Etten	937	116.1	118.3	12.58	.995	9.41	294 Zeke Bonura	917	108.0	108.6	10.06	.999	10.80
245 Ben Oglivie	1550	125.8	124.2	12.57	1.002	9.88	295 Merv Rettenmund	1023	91.6	94.2	9.98	.992	9.44
246 Vada Pinson	2469	154.6	119.8	12.54	1.031	9.55	296 Tony Gonzalez	1559	95.3	93.9	9.97	1.002	9.42
247 Thurman Munson	1423	97.0	120.1	12.49	.964	9.61	297 Dick Wakefield	638	110.0	95.5	9.96	1.053	9.58
248 Roy Campanella	1215	158.8	126.6	12.48	1.057	10.14	298 Bobby Estalella	680	81.7	94.7	9.88	.955	9.59
249 Johnny Callison	1886	123.3	118.4	12.42	1.006	9.54	299 Amos Strunk	1509	68.1	90.8	9.81	.964	9.26
250 Joe Cunningham	1141	136.2	123.3	12.40	1.028	9.94	300 Patsy Dougherty	1233	98.5	91.6	9.80	1.013	9.34

	G	OBA	Norm	Adj	PF	LG		G	OBA	Norm	Adj	PF	LG
1 Ted Williams	2292	.483	138.6	136.5	1.031	.348	51 Eddie Murray	1206	.375	113.2	115.4	.964	.331
2 Babe Ruth	2503	.474	133.0	135.1	.969	.356	52 Augie Galan	1742	.390	114.6	115.1	.992	.341
3 Mickey Mantle	2401	.423	126.0	127.8	.972	.335	53 Gavvy Cravath	1220	.379	117.6	115.0	1.045	.322
4 Rogers Hornsby	2259	.434	126.7	127.8	.984	.342	54 Willie McCovey	2588	.377	114.3	115.0	.988	.330
5 Lou Gehrig	2164	.447	123.9	127.0	.952	.361	55 Johnny Mize	1884	.397	116.0	115.0	1.019	.342
6 Roy Thomas	1470	.407	127.1	127.0	1.001	.320	56 Fielder Jones	1788	.362	113.3	114.9	.971	.320
7 Ty Cobb	3034	.432	128.0	126.8	1.018	.338	57 Jack Fournier	1530	.392	114.0	114.7	.987	.344
8 Eddie Collins	2826	.424	124.0	124.6	.991	.342	58 Hank Aaron	3298	.377	113.9	114.7	.986	.331
9 Joe Jackson	1331	.423	125.2	124.5	1.011	.338	59 Earl Torgeson	1668	.387	112.7	114.6	.969	.343
10 Tris Speaker	2789	.427	124.4	123.0	1.023	.344	60 Jim Wynn	1920	.369	112.2	114.5	.960	.329
11 Topsy Hartsel	1355	.383	123.1	122.3	1.013	.311	61 Don Buford	1286	.364	112.7	114.3	.972	.323
12 Elmer Flick	1493	.381	120.9	122.0	.983	.315	62 Ralph Kiner	1472	.398	115.9	114.3	1.027	.344
13 Stan Musial	3026	.418	123.5	121.8	1.028	.339	63 Eddie Yost	2109	.395	113.9	114.3	.993	.347
14 Gene Tenace	1555	.391	118.2	121.4	.948	.331	64 Ron Hunt	1483	.369	113.0	114.3	.978	.327
15 Joe Morgan	2649	.395	119.2	121.3	.964	.332	65 Willie Randolph	1210	.373	112.8	114.2	.975	.331
16 Mel Ott	2730	.414	120.0	121.1	.995	.343	66 Minnie Minoso	1835	.391	114.7	114.2	1.009	.341
17 Rod Carew	2342	.397	121.2	120.4	1.014	.327	67 Bobby Grich	1766	.375	114.1	114.2	.998	.329
18 Mike Hargrove	1559	.401	121.7	120.2	1.025	.330	68 John Grubb	1206	.369	110.9	114.2	.943	.333
19 Eddie Stanky	1259	.410	119.9	120.1	.996	.342	69 Elmer Valo	1806	.399	115.2	114.0	1.020	.347
20 Charlie Keller	1170	.410	118.4	119.9	.975	.346	70 Floyd Robinson	1012	.367	111.9	114.0	.963	.328
21 Arky Vaughan	1817	.406	120.0	119.7	1.006	.338	71 Norm Cash	2089	.377	115.3	113.9	1.024	.327
22 Joe Cunningham	1141	.406	121.1	119.4	1.028	.335	72 Roy White	1881	.363	111.4	113.9	.957	.326
23 Ken Singleton	2082	.391	118.0	119.0	.983	.331	73 Carl Yastrzemski	3308	.382	116.7	113.8	1.052	.328
24 Jimmie Foxx	2317	.428	119.6	119.0	1.011	.358	74 Bill North	1169	.366	111.1	113.7	.956	.330
25 Frank Robinson	2808	.392	119.3	118.9	1.007	.329	75 Earle Combs	1455	.397	110.6	113.6	.948	.359
26 Ferris Fain	1151	.425	120.8	118.7	1.037	.352	76 Hack Wilson	1348	.395	112.6	113.6	.983	.351
27 Roy Cullenbine	1181	.408	119.7	118.5	1.019	.341	77 Fred Clarke	2244	.381	115.2	113.5	1.030	.331
28 Honus Wagner	2789	.388	119.5	118.3	1.021	.325	78 Johnny Bates	1154	.364	111.7	113.5	.969	.326
29 Miller Huggins	1586	.381	117.6	117.9	.994	.324	79 Hank Greenberg	1394	.412	115.7	113.5	1.039	.356
30 Richie Ashburn	2189	.397	116.8	117.5	.988	.340	80 Wally Schang	1839	.393	112.4	113.5	.982	.350
31 Willie Mays	2992	.387	116.4	117.3	.985	.332	81 Greg Gross	1357	.379	113.0	113.4	.993	.335
32 Keith Hernandez	1414	.394	118.1	117.2	1.016	.334	82 Harmon Killebrew	2435	.379	115.3	113.4	1.033	.328
33 Roger Bresnahan	1446	.380	117.8	117.1	1.012	.323	83 Pete Rose	3371	.377	114.3	113.4	1.016	.330
34 Mickey Cochrane	1010	.389	117.6	116.7	1.015	.331	84 Joe DiMaggio	1736	.398	111.8	113.3	.972	.356
35 Paul Waner	2549	.404	117.3	116.6	1.011	.345	85 Heinie Groh	1676	.373	112.6	113.2	.989	.331
36 Dick Allen	1749	.381	116.6	116.6	1.001	.326	86 George Brett	1462	.371	112.6	113.1	.991	.330
37 Jackie Robinson	1382	.410	119.4	116.5	1.051	.343	87 John Titus	1402	.367	113.5	113.1	1.007	.323
38 Frank Chance	1286	.380	118.7	116.4	1.039	.320	88 Rico Carty	1651	.372	113.1	113.0	1.002	.329
39 Nap Lajoie	2474	.375	116.4	116.4	.999	.323	89 Jose Cruz	1907	.362	108.8	113.0	.926	.333
40 Merv Rettenmund	1023	.383	115.8	116.3	.992	.331	90 Al Kaline	2834	.379	114.0	113.0	1.019	.332
41 Ross Youngs	1211	.399	116.5	116.2	1.005	.342	91 Gene Woodling	1796	.388	112.5	113.0	.991	.345
42 Max Bishop	1338	.423	117.8	116.1	1.028	.359	92 Luke Appling	2422	.399	112.8	112.7	1.003	.354
43 Mickey Cochrane	1482	.419	116.7	116.0	1.012	.359	93 Steve Braun	1361	.374	114.0	112.6	1.024	.328
44 Willie Keeler	2125	.384	117.5	115.9	1.028	.326	94 Johnny Pesky	1270	.394	112.9	112.6	1.005	.349
45 Riggs Stephenson	1310	.407	115.2	115.8	.990	.353	95 Bob Nieman	1113	.375	110.0	112.6	.954	.341
46 Eddie Mathews	2391	.378	113.4	115.7	.961	.333	96 Steve Kemp	1047	.374	113.0	112.5	1.009	.331
47 Harry Heilmann	2146	.410	115.6	115.6	1.000	.354	97 Fred Tenney	1994	.367	112.6	112.5	1.002	.326
48 Mike Schmidt	1789	.388	116.0	115.6	1.008	.334	98 Bill Terry	1721	.393	112.5	112.4	1.002	.349
49 Stan Hack	1938	.394	116.2	115.5	1.011	.339	99 Sid Gordon	1475	.377	110.4	112.2	.968	.342
50 Elbie Fletcher	1415	.384	113.5	115.5	.965	.338	100 Al Rosen	1044	.386	110.6	112.1	.974	.349

	G	SLG	Norm	Adj	PF	LG		G	SLG	Norm	Adj	PF	LG
1 Babe Ruth	2503	.690	170.7	173.4	.969	.404	51 Bob Johnson	1863	.506	125.3	126.3	.984	.404
2 Ted Williams	2292	.634	159.4	157.0	1.031	.398	52 Chick Hafey	1283	.526	127.3	126.2	1.017	.413
3 Lou Gehrig	2164	.632	150.9	154.6	.952	.419	53 Orlando Cepeda	2124	.499	126.3	126.1	1.003	.395
4 Rogers Hornsby	2259	.577	146.8	148.1	.984	.393	54 Frank Baker	1575	.442	126.4	126.1	1.006	.350
5 Joe Jackson	1331	.517	146.5	145.7	1.011	.353	55 Jack Fournier	1530	.483	125.2	126.0	.987	.386
6 Jimmie Foxx	2317	.609	146.1	145.3	1.011	.417	56 Earl Averill	1668	.534	126.5	125.5	1.016	.422
7 Joe DiMaggio	1736	.579	142.1	144.1	.972	.407	57 Norm Cash	2089	.488	126.7	125.2	1.024	.385
8 Mickey Mantle	2401	.557	141.9	144.0	.972	.392	58 George Foster	1761	.484	125.2	125.0	1.004	.387
9 Hank Greenberg	1394	.605	146.7	143.9	1.039	.412	59 Larry Doby	1533	.490	123.7	124.7	.985	.396
10 Johnny Mize	1884	.562	143.7	142.3	1.019	.391	60 Sherry Magee	2087	.427	124.9	124.2	1.012	.342
11 Hank Aaron	3298	.555	139.5	140.5	.986	.397	61 Tommy Henrich	1284	.491	122.9	124.0	.982	.400
12 Willie Mays	2992	.557	138.9	140.0	.985	.401	62 Fred Lynn	1301	.500	127.7	123.9	1.063	.391
13 Ty Cobb	3034	.512	141.1	139.9	1.018	.363	63 Billy Williams	2488	.492	127.0	123.8	1.052	.387
14 Dick Allen	1749	.534	139.7	139.6	1.001	.382	64 Dave Parker	1457	.485	124.9	123.7	1.020	.388
15 Mike Schmidt	1789	.535	138.2	137.6	1.008	.387	65 Jack Clark	1044	.477	122.5	123.6	.982	.390
16 Stan Musial	3026	.559	139.4	137.5	1.028	.401	66 Dolph Camilli	1490	.492	126.7	123.5	1.052	.388
17 Willie Stargell	2360	.529	137.0	137.4	.995	.386	67 Rocky Colavito	1841	.489	123.8	123.2	1.009	.395
18 Frank Robinson	2808	.537	136.5	136.0	1.007	.393	68 Tony Oliva	1676	.476	125.5	123.2	1.037	.380
19 Nap Lajoie	2474	.467	135.0	135.1	.999	.346	69 Andre Dawson	1174	.479	123.0	123.1	.999	.390
20 Gavvy Cravath	1220	.478	137.7	134.7	1.045	.347	70 Yogi Berra	2120	.482	121.4	123.0	.973	.397
21 Charlie Keller	1170	.518	132.8	134.5	.975	.390	71 Johnny Bench	2158	.476	123.4	123.0	1.008	.386
22 Mel Ott	2730	.533	134.1	134.4	.995	.398	72 Cecil Cooper	1545	.479	122.4	122.9	.992	.392
23 Tris Speaker	2789	.500	135.5	133.9	1.023	.369	73 Ernie Banks	2528	.500	124.2	122.7	1.024	.402
24 Honus Wagner	2789	.469	135.1	133.7	1.021	.347	74 Bob Nieman	1113	.474	119.7	122.6	.954	.396
25 Ralph Kiner	1472	.548	135.5	133.7	1.027	.404	75 Greg Luzinski	1821	.478	122.6	122.4	1.003	.390
26 Willie McCovey	2588	.515	132.0	132.8	.988	.390	76 Bobby Bonds	1849	.471	121.8	122.4	.992	.386
27 Sam Crawford	2517	.453	134.2	132.4	1.028	.337	77 Dick Stuart	1112	.489	121.0	122.0	.984	.404
28 Hack Wilson	1348	.545	131.1	132.2	.983	.415	78 Bill Nicholson	1677	.465	121.7	121.9	.997	.382
29 Frank Howard	1895	.499	130.0	131.9	.971	.384	79 Bill Terry	1721	.506	122.0	121.9	1.002	.415
30 Harry Heilmann	2146	.520	131.3	131.3	1.000	.397	80 Al Kaline	2834	.480	122.9	121.8	1.019	.390
31 Reggie Jackson	2430	.498	129.8	131.2	.978	.383	81 Hank Sauer	1399	.496	121.2	121.6	.994	.409
32 Elmer Flick	1493	.448	130.0	131.1	.983	.345	82 Roger Maris	1463	.476	120.5	121.5	.984	.395
33 Jeff Heath	1383	.509	129.6	130.9	.980	.392	83 Boog Powell	2042	.462	121.6	121.4	1.004	.380
34 Wally Berger	1350	.522	128.2	130.7	.963	.407	84 Goose Goslin	2287	.500	120.2	121.3	.982	.416
35 Babe Herman	1552	.532	128.9	130.3	.978	.413	85 Dale Murphy	1038	.486	125.5	121.2	1.074	.387
36 Eddie Murray	1206	.507	127.6	130.0	.964	.397	86 Joe Adcock	1959	.485	118.2	121.0	.954	.410
37 Chuck Klein	1753	.543	132.7	129.5	1.050	.409	87 Jim Ray Hart	1125	.467	121.4	120.6	1.013	.385
38 Harmon Killebrew	2435	.509	131.4	129.3	1.033	.387	88 Gus Zernial	1234	.486	122.5	120.6	1.032	.396
39 Ken Williams	1397	.530	130.9	129.2	1.026	.405	89 Gabby Hartnett	1990	.489	120.1	120.4	.994	.407
40 George Brett	1462	.500	128.4	129.0	.991	.389	90 Bob Meusel	1407	.497	119.5	120.3	.996	.416
41 Jim Rice	1493	.524	133.9	128.9	1.079	.391	91 Zach Wheat	2410	.450	120.1	120.2	.998	.375
42 Duke Snider	2143	.540	131.2	128.1	1.049	.411	92 Jim Bottomley	1991	.500	120.6	120.2	1.006	.415
43 Eddie Mathews	2391	.509	125.3	127.8	.961	.407	93 Rudy York	1603	.483	124.1	120.2	1.066	.389
44 Joe Medwick	1984	.505	130.2	127.8	1.037	.388	94 Ted Kluszewski	1718	.498	120.7	120.2	1.009	.412
45 Al Simmons	2215	.535	128.4	127.3	1.017	.417	95 Cliff Johnson	1156	.468	119.3	120.2	.985	.392
46 Hal Trosky	1347	.522	125.4	127.2	.972	.416	96 Lee May	2071	.459	119.4	120.0	.990	.385
47 Dave Winfield	1655	.482	123.2	127.1	.939	.391	97 Roy Sievers	1887	.475	119.0	120.0	.984	.399
48 Reggie Smith	1987	.489	128.7	126.9	1.030	.380	98 Fred Clarke	2244	.432	121.7	119.9	1.030	.355
49 Al Rosen	1044	.495	124.8	126.5	.974	.397	99 Heinie Zimmerman	1456	.420	120.0	119.9	1.003	.350
50 Dave Kingman	1639	.489	126.2	126.3	.998	.388	100 Ripper Collins	1084	.492	123.3	119.8	1.058	.399

	G	OPS	Norm	Adj	PF	LG		G	OPS	Norm	Adj	PF	LG
1 Babe Ruth	2503	1.163	203.7	210.2	.969	.760	51 Bob Nieman	1113	.849	129.7	135.9	.954	.737
2 Ted Williams	2292	1.116	198.0	192.0	1.031	.746	52 Arky Vaughan	1817	.859	136.4	135.6	1.006	.728
3 Lou Gehrig	2164	1.080	174.8	183.5	.952	.780	53 Sherry Magee	2087	.788	137.1	135.5	1.012	.664
4 Rogers Hornsby	2259	1.010	173.6	176.4	.984	.735	54 Chuck Klein	1753	.922	142.2	135.5	1.050	.755
5 Mickey Mantle	2401	.979	167.9	172.8	.972	.728	55 Al Kaline	2834	.859	137.0	134.5	1.019	.722
6 Joe Jackson	1331	.940	171.7	169.8	1.011	.691	56 Dolph Camilli	1490	.880	141.3	134.3	1.052	.727
7 Ty Cobb	3034	.945	169.1	166.1	1.018	.701	57 Bill Terry	1721	.899	134.5	134.3	1.002	.764
8 Jimmie Foxx	2317	1.038	165.7	163.9	1.011	.775	58 Frank Baker	1575	.805	135.0	134.2	1.006	.684
9 Stan Musial	3026	.977	162.9	158.5	1.028	.740	59 Jack Clark	1044	.841	131.5	134.0	.982	.723
10 Joe DiMaggio	1736	.977	153.8	158.2	.972	.764	60 Joe Morgan	2649	.823	129.1	133.9	.964	.720
11 Willie Mays	2992	.944	155.3	157.7	.985	.734	61 Earl Averill	1668	.928	135.9	133.8	1.016	.782
12 Johnny Mize	1884	.959	159.7	156.8	1.019	.733	62 Fred Lynn	1301	.878	142.2	133.8	1.063	.722
13 Hank Greenberg	1394	1.017	162.4	156.3	1.039	.768	63 Tommy Henrich	1284	.873	131.0	133.4	.982	.753
14 Tris Speaker	2789	.927	159.9	156.2	1.023	.713	64 Paul Waner	2549	.877	134.7	133.2	1.011	.748
15 Dick Allen	1749	.914	156.4	156.2	1.001	.708	65 Boog Powell	2042	.826	133.6	133.1	1.004	.705
16 Mel Ott	2730	.947	154.8	155.6	.995	.740	66 Fred Clarke	2244	.813	136.8	132.9	1.030	.686
17 Hank Aaron	3298	.932	153.4	155.6	.986	.729	67 Rod Carew	2342	.830	134.7	132.8	1.014	.709
18 Charlie Keller	1170	.928	151.3	155.1	.975	.736	68 Orlando Cepeda	2124	.852	133.2	132.7	1.003	.726
19 Frank Robinson	2808	.929	155.8	154.7	1.007	.722	69 Hal Trosky	1347	.892	128.8	132.5	.972	.775
20 Elmer Flick	1493	.829	150.9	153.5	.983	.660	70 Joe Medwick	1984	.867	137.3	132.3	1.037	.726
21 Mike Schmidt	1789	.923	154.2	153.0	1.008	.721	71 Greg Luzinski	1821	.844	132.8	132.3	1.003	.722
22 Nap Lajoie	2474	.842	151.4	151.6	.999	.668	72 Rico Carty	1651	.836	132.6	132.3	1.002	.718
23 Honus Wagner	2789	.857	154.6	151.5	1.021	.672	73 Rocky Colavito	1841	.851	133.3	132.2	1.009	.725
24 Gavvy Cravath	1220	.857	155.3	148.6	1.045	.670	74 Ken Singleton	2082	.827	129.9	132.2	.983	.721
25 Willie McCovey	2588	.892	146.2	148.1	.988	.720	75 Al Simmons	2215	.915	133.8	131.6	1.017	.777
26 Willie Stargell	2360	.892	147.2	147.9	.995	.715	76 Chick Hafey	1283	.898	133.8	131.6	1.017	.763
27 Ralph Kiner	1472	.946	151.3	147.3	1.027	.748	77 Jim Rice	1493	.881	141.9	131.4	1.079	.722
28 Harry Heilmann	2146	.930	146.9	146.9	1.000	.751	78 Topsy Hartsel	1355	.754	132.9	131.2	1.013	.649
29 Hack Wilson	1348	.940	143.8	146.2	.983	.766	79 Sid Gordon	1475	.844	126.9	131.1	.968	.742
30 Eddie Murray	1206	.882	140.9	146.2	.964	.729	80 Riggs Stephenson	1310	.880	129.8	131.1	.990	.766
31 Eddie Mathews	2391	.888	138.7	144.4	.961	.740	81 Roy Cullenbine	1181	.840	133.5	131.0	1.019	.720
32 Sam Crawford	2517	.814	147.3	143.4	1.028	.656	82 Keith Hernandez	1414	.841	133.1	131.0	1.016	.722
33 Frank Howard	1895	.853	139.0	143.1	.971	.709	83 Jim Wynn	1920	.805	125.8	131.0	.960	.713
34 George Brett	1462	.871	141.0	142.3	.991	.719	84 Billy Williams	2488	.856	137.8	131.0	1.052	.716
35 Reggie Jackson	2430	.856	139.1	142.2	.978	.711	85 Willie Keeler	2125	.801	134.6	130.9	1.028	.683
36 Eddie Collins	2826	.852	140.8	142.1	.991	.709	86 Bobby Bonds	1849	.827	129.3	130.4	.992	.718
37 Harmon Killebrew	2435	.887	146.7	142.0	1.033	.715	87 Tony Oliva	1676	.832	135.1	130.2	1.037	.705
38 Babe Herman	1552	.915	138.8	141.9	.978	.761	88 Frank Chance	1286	.774	135.2	130.2	1.039	.658
39 Jack Fournier	1530	.875	139.3	141.0	.987	.730	89 Bill Nicholson	1677	.830	129.5	130.0	.997	.721
40 Dave Winfield	1655	.842	131.4	139.9	.939	.724	90 Mickey Cochrane	1482	.897	131.4	129.8	1.012	.776
41 Jeff Heath	1383	.879	136.4	139.1	.980	.739	91 Goose Goslin	2287	.887	127.4	129.7	.982	.777
42 Al Rosen	1044	.882	135.5	139.1	.974	.746	92 Bob Watson	1832	.814	125.3	129.6	.966	.722
43 Norm Cash	2089	.865	142.0	138.7	1.024	.712	93 Carl Yastrzemski	3308	.844	136.4	129.6	1.052	.714
44 Bob Johnson	1863	.899	136.5	138.7	.984	.757	94 Cliff Johnson	1156	.828	127.7	129.6	.985	.724
45 Wally Berger	1350	.881	132.8	137.9	.963	.750	95 Roberto Clemente	2433	.837	128.9	129.5	.996	.728
46 Gene Tenace	1555	.819	130.3	137.4	.948	.713	96 Bill Dickey	1789	.868	123.9	129.5	.957	.773
47 Ken Williams	1397	.923	140.8	137.2	1.026	.763	97 Zach Wheat	2410	.817	129.1	129.3	.998	.712
48 Reggie Smith	1987	.859	141.1	137.0	1.030	.709	98 Minnie Minoso	1835	.851	130.5	129.3	1.009	.738
49 Duke Snider	2143	.921	143.7	137.0	1.049	.750	99 Andy Thornton	1285	.842	131.6	129.3	1.018	.726
50 Larry Doby	1533	.877	134.7	136.8	.985	.745	100 Earle Combs	1455	.859	122.0	128.7	.948	.774

	G	ISO	Norm	Adj	PF	LG		G	ISO	Norm	Adj	PF	LG
1 Babe Ruth	2503	.348	299.0	303.7	.969	.116	51 Jim Rice	1493	.222	172.6	166.1	1.079	.128
2 Lou Gehrig	2164	.292	226.5	232.1	.952	.129	52 Duke Snider	2143	.244	170.1	166.1	1.049	.144
3 Ted Williams	2292	.289	225.7	222.3	1.031	.128	53 Jack Fournier	1530	.170	164.9	165.9	.987	.103
4 Mike Schmidt	1789	.270	220.1	219.3	1.008	.122	54 Cliff Johnson	1156	.210	164.4	165.6	.985	.128
5 Hank Greenberg	1394	.292	223.0	218.7	1.039	.131	55 Al Rosen	1044	.210	163.0	165.1	.974	.129
6 Gavvy Cravath	1220	.191	222.1	217.3	1.045	.086	56 Hank Sauer	1399	.230	163.4	163.9	.994	.141
7 Jimmie Foxx	2317	.284	216.9	215.7	1.011	.131	57 Sherry Magee	2087	.135	164.6	163.6	1.012	.082
8 Johnny Mize	1884	.250	209.1	207.2	1.019	.120	58 Bobby Bonds	1849	.203	162.6	163.3	.992	.125
9 Rogers Hornsby	2259	.218	201.3	203.0	.984	.108	59 Tris Speaker	2789	.156	165.1	163.2	1.023	.094
10 Dave Kingman	1639	.251	202.7	202.9	.998	.124	60 Norm Cash	2089	.217	164.9	163.0	1.024	.131
11 Willie Stargell	2360	.247	198.9	199.4	.995	.124	61 Harry Heilmann	2146	.179	162.5	162.6	1.000	.110
12 Joe DiMaggio	1736	.254	196.1	198.9	.972	.130	62 Jack Clark	1044	.200	160.8	162.3	.982	.124
13 Mickey Mantle	2401	.259	195.1	197.9	.972	.133	63 Reggie Smith	1987	.202	164.5	162.1	1.030	.123
14 Dick Allen	1749	.242	197.5	197.4	1.001	.122	64 Larry Doby	1533	.207	160.9	162.1	.985	.129
15 Ralph Kiner	1472	.269	199.0	196.3	1.027	.135	65 Ripper Collins	1084	.196	166.8	162.1	1.058	.117
16 Mel Ott	2730	.229	195.6	196.0	.995	.117	66 Rocky Colavito	1841	.223	162.2	161.5	1.009	.137
17 Charlie Keller	1170	.231	193.1	195.5	.975	.120	67 Harry Davis	1769	.131	162.4	161.4	1.013	.081
18 Hack Wilson	1348	.238	193.0	194.6	.983	.123	68 Ernie Banks	2528	.225	163.3	161.4	1.024	.138
19 Willie McCovey	2588	.245	191.7	192.9	.988	.128	69 Earl Averill	1668	.216	162.2	160.9	1.016	.133
20 Harmon Killebrew	2435	.252	191.7	188.6	1.033	.132	70 Frank Baker	1575	.135	161.4	160.9	1.006	.084
21 Reggie Jackson	2430	.233	185.9	188.0	.978	.125	71 Eddie Murray	1206	.209	157.8	160.7	.964	.133
22 Willie Mays	2992	.256	186.5	187.9	.985	.137	72 Gabby Hartnett	1990	.192	160.1	160.6	.994	.120
23 Joe Jackson	1331	.162	188.8	187.8	1.011	.086	73 Dick Stuart	1112	.225	159.2	160.5	.984	.142
24 Hank Aaron	3298	.250	185.7	187.1	.986	.134	74 Joe Gordon	1566	.197	158.6	160.4	.979	.124
25 Wally Berger	1350	.221	182.7	186.1	.963	.121	75 Roger Maris	1463	.216	158.9	160.2	.984	.136
26 Dolph Camilli	1490	.215	190.0	185.3	1.052	.113	76 Greg Luzinski	1821	.202	159.9	159.6	1.003	.126
27 Sam Crawford	2517	.143	185.2	182.7	1.028	.077	77 Elmer Smith	1012	.161	161.3	159.3	1.025	.100
28 Ken Williams	1397	.211	184.8	182.4	1.026	.114	78 Ty Cobb	3034	.146	160.2	158.8	1.018	.091
29 Frank Robinson	2808	.243	181.4	180.8	1.007	.134	79 Andy Thornton	1285	.207	160.0	158.5	1.018	.130
30 Jeff Heath	1383	.216	178.2	180.0	.980	.121	80 Andre Dawson	1174	.197	158.4	158.5	.999	.125
31 Chuck Klein	1753	.223	181.7	177.3	1.050	.123	81 Gene Tenace	1555	.187	153.9	158.1	.948	.122
32 Frank Howard	1895	.225	174.4	176.9	.971	.129	82 Nap Lajoie	2474	.128	157.9	158.0	.999	.081
33 Gorman Thomas	1199	.224	174.3	174.2	1.000	.129	83 Joe Medwick	1984	.181	160.7	157.8	1.037	.113
34 Bill Nicholson	1677	.198	173.1	173.4	.997	.114	84 Al Simmons	2215	.201	158.2	156.8	1.017	.127
35 Eddie Mathews	2391	.238	174.7	172.1	.961	.141	85 Dave Winfield	1655	.193	152.0	156.8	.939	.127
36 Babe Herman	1552	.207	169.1	171.0	.978	.123	86 Roy Sievers	1887	.208	155.6	156.8	.984	.134
37 Rudy York	1603	.208	176.0	170.4	1.066	.118	87 Bob Meusel	1407	.187	155.7	156.8	.986	.120
38 Bob Johnson	1863	.210	168.6	170.0	.984	.124	88 Bob Allison	1541	.217	159.6	156.4	1.041	.136
39 Nate Colbert	1004	.207	165.7	169.3	.958	.125	89 Tony Armas	1000	.205	154.9	156.4	.981	.132
40 Cy Williams	2002	.178	174.2	169.2	1.059	.102	90 Tilly Walker	1418	.146	156.1	156.2	1.000	.094
41 Chick Hafey	1283	.209	170.5	169.1	1.017	.123	91 Lee May	2071	.192	155.3	156.1	.990	.124
42 Hal Trosky	1347	.219	166.7	169.1	.972	.131	92 Don Mincher	1400	.201	155.2	155.6	.994	.129
43 Honus Wagner	2789	.140	170.6	168.9	1.021	.082	93 Billy Williams	2488	.202	159.5	155.5	1.052	.127
44 Johnny Bench	2158	.208	169.5	168.8	1.008	.123	94 Jim Bottomley	1991	.191	155.3	154.8	1.006	.123
45 George Foster	1761	.207	169.0	168.7	1.004	.123	95 Fred Lynn	1301	.205	159.6	154.8	1.063	.128
46 Gus Zernial	1234	.221	171.1	168.5	1.032	.129	96 Vince DiMaggio	1110	.164	152.2	154.8	.967	.108
47 Tommy Henrich	1284	.209	166.6	168.1	.982	.126	97 Dwight Evans	1622	.202	160.0	154.5	1.071	.126
48 Stan Musial	3026	.228	170.4	168.1	1.028	.134	98 Boog Powell	2042	.196	154.5	154.2	1.004	.127
49 Elmer Flick	1493	.133	166.2	167.6	.983	.080	99 Jim Wynn	1920	.186	150.3	153.4	.960	.124
50 Dale Murphy	1038	.211	172.6	166.6	1.074	.122	100 Oscar Gamble	1514	.191	152.1	153.1	.988	.126

		G	Avg	Norm	Adj	PF	LG			G	Avg	Norm	Adj	PF	LG
1	Ty Cobb	3034	.366	134.7	133.6	1.018	.272	51	Bob Nieman	1113	.295	111.6	114.2	.954	.264
2	Joe Jackson	1331	.356	132.9	132.2	1.011	.268	52	Fred Clarke	2244	.315	115.9	114.2	1.030	.272
3	Nap Lajoie	2474	.339	128.0	128.1	.999	.265	53	Rico Carty	1651	.299	114.3	114.2	1.002	.262
4	Rogers Hornsby	2259	.358	126.1	127.1	.984	.284	54	Mickey Rivers	1467	.295	112.8	114.2	.975	.262
5	Rod Carew	2342	.330	128.0	127.1	1.014	.258	55	Al Kaline	2834	.297	115.2	114.2	1.019	.258
6	Ted Williams	2292	.344	127.8	125.9	1.031	.269	56	Johnny Pesky	1270	.307	114.4	114.1	1.005	.268
7	Willie Keeler	2125	.343	126.1	124.3	1.028	.272	57	Tommy Davis	1999	.294	113.3	114.0	.987	.259
8	Tris Speaker	2789	.344	125.3	.123.9	1.023	.275	58	Bob Watson	1832	.295	112.1	114.0	.966	.263
9	Honus Wagner	2789	.329	124.1	122.9	1.021	.265	59	Johnny Mize	1884	.312	114.9	113.8	1.019	.272
10	Eddie Collins	2826	.333	121.6	122.1	.991	.274	60	Sam Rice	2404	.322	111.8	113.8	.965	.288
11	Stan Musial	3026	.331	123.8	122.1	1.028	.267	61	George Kell	1795	.306	114.7	113.8	1.017	.267
12	Roberto Clemente	2433	.317	120.7	121.0	.996	.263	62	Heinie Manush	2008	.330	114.0	113.6	1.008	.289
13	Babe Ruth	2503	.342	118.8	120.7	.969	.288	63	Joe Torre	2209	.297	113.8	113.6	1.005	.261
14	Lou Gehrig	2164	.340	117.2	120.1	.952	.290	64	Tommy Holmes	1320	.302	112.9	113.6	.989	.267
15	George Brett	1462	.314	119.5	120.0	.991	.263	65	Cecil Travis	1328	.314	111.0	113.5	.956	.283
16	Elmer Flick	1493	.315	119.0	120.0	.983	.265	66	Ken Griffey	1412	.302	114.0	113.5	1.009	.265
17	Harry Heilmann	2146	.342	119.2	119.3	1.000	.284	67	Stuffy McInnis	2128	.307	111.6	113.2	.970	.276
18	Joe DiMaggio	1736	.325	116.8	118.5	.972	.278	68	Dave Parker	1457	.303	114.3	113.2	1.020	.265
19	George Sisler	2055	.340	118.7	118.5	1.004	.287	69	Babe Herman	1552	.324	111.9	113.1	.978	.290
20	Dale Mitchell	1127	.312	116.2	118.4	.962	.269	70	Jimmie Foxx	2317	.325	113.7	113.1	1.011	.286
21	Tony Oliva	1676	.304	120.4	118.2	1.037	.253	71	Gene Richards	1026	.290	109.0	112.9	.932	.266
22	Matty Alou	1667	.307	117.9	118.1	.997	.260	72	Frank Robinson	2808	.294	113.3	112.9	1.007	.260
23	Manny Mota	1536	.304	116.7	117.9	.981	.261	73	Jose Cruz	1907	.287	108.7	112.9	.926	.264
24	Sam Crawford	2517	.310	119.1	117.4	1.028	.260	74	Dave Winfield	1655	.289	109.3	112.8	.939	.265
25	Bill Terry	1721	.341	117.2	117.1	1.002	.291	75	Jackie Robinson	1382	.311	115.6	112.8	1.051	.269
26	Hank Aaron	3298	.305	115.9	116.7	.986	.263	76	Danny Cater	1289	.276	110.1	112.8	.953	.251
27	Paul Waner	2549	.333	117.3	116.6	1.011	.284	77	Orlando Cepeda	2124	.297	112.8	112.6	1.003	.263
28	Mickey Mantle	2401	.298	114.8	116.4	.972	.260	78	Floyd Robinson	1012	.283	110.5	112.6	.963	.256
29	Ginger Beaumont	1463	.311	117.4	116.4	1.016	.265	79	Cy Seymour	1528	.303	114.4	112.5	1.033	.265
30	Edd Roush	1967	.323	114.9	116.4	.974	.281	80	Minnie Minoso	1835	.298	113.0	112.5	1.009	.264
31	Al Oliver	2272	.305	115.6	116.4	.988	.264	81	Ernie Lombardi	1853	.306	111.2	112.5	.977	.275
32	Cecil Cooper	1545	.305	115.7	116.2	.992	.263	82	Keith Hernandez	1414	.300	113.3	112.4	1.016	.265
33	Bill Madlock	1443	.312	117.8	116.1	1.030	.265	83	Ross Youngs	1211	.322	112.7	112.4	1.005	.286
34	Riggs Stephenson	1310	.336	115.2	115.7	.990	.292	84	Heinie Zimmerman	1456	.295	112.6	112.4	1.003	.262
35	Pete Rose	3371	.305	116.5	115.6	1.016	.262	85	Dick Allen	1749	.292	112.4	112.4	1.001	.260
36	Joe Medwick	1984	.324	117.7	115.5	1.037	.275	86	Lou Piniella	1747	.291	112.2	112.2	.999	.259
37	Richie Ashburn	2189	.308	114.7	115.4	.988	.268	87	Bill Dickey	1789	.313	109.7	112.2	.957	.285
38	Willie Mays	2992	.302	114.2	115.1	.985	.264	88	Bobby Veach	1821	.310	112.6	112.2	1.008	.275
39	Frank Baker	1575	.307	115.5	115.1	1.006	.266	89	Hal McRae	1842	.293	111.6	112.2	.990	.262
40	Zach Wheat	2410	.317	114.9	115.0	.998	.276	90	Felipe Alou	2082	.286	110.1	112.1	.965	.260
41	Earle Combs	1455	.325	111.9	115.0	.948	.290	91	Greg Gross	1357	.295	111.6	112.0	.993	.264
42	Steve Garvey	1988	.299	113.1	115.0	.968	.264	92	Jake Daubert	2014	.303	111.0	112.0	.983	.273
43	Thurman Munson	1423	.292	112.7	114.8	.964	.264	93	Terry Puhl	1017	.283	106.8	112.0	.910	.265
44	Arky Vaughan	1817	.318	115.1	114.8	1.006	.276	94	Bob Fothergill	1106	.325	111.8	111.9	.998	.291
45	Taffy Wright	1029	.311	113.4	114.7	.977	.274	95	Sherry Magee	2087	.291	112.4	111.7	1.012	.259
46	Harvey Kuenn	1833	.303	114.4	114.7	.995	.265	96	Dixie Walker	1905	.306	112.1	111.6	1.010	.273
47	Eddie Murray	1206	.298	112.5	114.6	.964	.265	97	Lou Boudreau	1646	.295	109.6	111.5	.966	.269
48	Manny Sanguillen	1448	.296	112.9	114.5	.972	.262	98	Pete Runnels	1799	.291	110.4	111.5	.981	.263
49	Ralph Garr	1317	.306	116.7	114.4	1.040	.262	99	Jack Fournier	1530	.313	110.8	111.5	.987	.283
50	Al Simmons	2215	.334	115.3	114.4	1.017	.290	100	Luke Appling	2422	.310	111.7	111.5	1.003	.278

	G	All	Bat	Fld	Run	Pos			G	All	Bat	Fld	Run	Pos
1 Joe Morgan	2649	66.8	50.6	-11.6	11.6	16.2	26 Hank Aaron	3298	89.9	93.7	8.4	3.0	-15.2
2 Lou Brock	2616	-2.3	8.7	-8.5	10.3	-12.8	27 Frank Taveras	1150	-11.5	-16.5	-9.8	2.8	12.0
3 Davey Lopes	1482	9.9	4.4	-11.3	8.3	8.5	28 Rod Scott	690	-2.0	-6.7	-2.7	2.7	4.7
4 Bert Campaneris	2328	8.6	-13.3	-5.0	7.9	19.0	29 Mickey Rivers	1467	3.4	2.6	2.2	2.7	-4.1
5 Willie Wilson	970	8.0	1.5	1.8	7.7	-3.0	30 Andre Dawson	1174	18.2	12.7	7.4	2.7	-4.6
6 Luis Aparicio	2599	14.2	-25.5	10.6	7.2	21.9	31 Mickey Mantle	2401	70.9	85.6	-5.2	2.5	-12.0
7 Tim Raines	581	14.6	7.5	1.4	6.8	-1.1	32 Dave Collins	1132	-2.9	-1.4	-.8	2.5	-3.2
8 Rickey Henderson	791	25.9	15.1	6.5	6.5	-2.2	33 Paul Molitor	765	10.6	5.5	1.0	2.3	1.8
9 Cesar Cedeno	1858	22.7	22.8	2.6	5.9	-8.6	34 Rudy Law	537	-5.4	-2.5	-3.8	2.3	-1.4
10 Julio Cruz	984	8.7	-11.8	10.0	5.6	4.9	35 Mookie Wilson	584	.5	-2.4	2.7	2.2	-2.0
11 Tommy Harper	1810	.8	2.2	-1.7	5.5	-5.2	36 Bill North	1169	4.1	3.1	2.8	2.2	-4.0
12 Ron Le Flore	1099	3.2	1.7	.2	5.3	-4.0	37 Al Bumbry	1428	-.8	3.5	-2.5	2.2	-4.0
13 Maury Wills	1942	15.8	-12.5	7.8	5.2	15.3	38 Sandy Alomar	1481	-9.2	-20.9	.7	2.2	8.8
14 Amos Otis	1998	18.2	16.4	3.5	4.8	-6.5	39 Gene Richards	1026	4.1	6.3	-.8	2.1	-3.5
15 Omar Moreno	1206	-6.9	-13.2	6.0	4.5	-4.2	40 Garry Maddox	1638	3.2	.7	7.1	2.1	-6.7
16 Willie Davis	2429	1.4	4.8	4.2	4.3	-11.9	41 Don Baylor	1770	11.1	19.1	-3.1	2.1	-7.0
17 Willie Mays	2992	87.7	87.7	9.9	4.1	-14.0	42 Bump Wills	831	12.8	-2.4	8.8	2.0	4.4
18 George Case	1226	-4.5	-3.0	.6	4.0	-6.1	43 Willie Randolph	1210	19.9	6.2	5.2	2.0	6.5
19 Freddie Patek	1650	10.3	-15.1	4.6	3.9	16.9	44 Enzo Hernandez	714	-3.5	-12.7	-1.4	2.0	8.6
20 Bobby Bonds	1849	29.5	29.1	5.2	3.9	-8.7	45 Lyn Lary	1302	6.9	-3.8	.7	1.9	8.1
21 Ozzie Smith	1006	23.5	-13.2	22.3	3.5	10.9	46 Ken Griffey	1412	11.1	14.3	.6	1.9	-5.7
22 Dave Concepcion	2055	21.2	-10.2	4.8	3.4	23.2	47 Lonnie Smith	627	5.9	6.1	-.3	1.8	-1.7
23 Larry Bowa	2161	-9.0	-31.9	-6.7	3.3	26.3	48 Vada Pinson	2469	3.5	12.6	.3	1.8	-11.2
24 Miguel Dilone	722	-2.8	-4.2	-.5	3.1	-1.2	49 Bake McBride	1071	5.6	4.7	3.1	1.7	-3.9
25 Al Wiggins	389	-1.9	-1.1	-3.6	3.0	-.2	50 Larry Lintz	350	1.2	-2.9	.7	1.7	1.7

	G	All	Bat	Fld	Run	Pos			G	All	Bat	Fld	Run	Pos
1 Bill Mazeroski	2163	36.7	-18.0	36.5	-.6	18.8		51 Al Kaline	2834	46.8	51.0	9.2	-.0	-13.4
2 Nap Lajoie	2474	85.3	56.9	31.4	0.0	-3.0		52 Luke Appling	2422	38.0	18.7	9.2	-1.1	11.2
3 Mike Schmidt	1789	69.7	48.3	24.8	-.0	-3.4		53 Jimmy Sheckard	2122	20.2	21.6	9.1	0.0	-10.5
4 Ozzie Smith	1006	23.5	-13.2	22.3	3.5	10.9		54 Art Devlin	1313	18.4	7.2	9.1	0.0	2.1
5 Art Fletcher	1529	25.7	-2.1	20.4	0.0	7.4		55 Hughie Critz	1478	-6.6	-21.9	9.0	-.4	6.7
6 Buddy Bell	1827	28.2	11.3	19.6	-2.7	-.0		56 George Sisler	2055	26.1	27.1	8.9	.2	-10.1
7 Dave Bancroft	1913	32.6	-.2	19.5	0.0	13.3		57 Joe Sewell	1903	37.1	11.4	8.9	-.4	17.2
8 Joe Tinker	1804	18.7	-6.6	18.3	0.0	7.0		58 Bump Wills	831	12.8	-2.4	8.8	2.0	4.4
9 Clete Boyer	1725	8.8	-11.0	18.3	-.2	1.7		59 Honus Wagner	2789	79.3	67.6	8.8	0.0	2.9
10 Tris Speaker	2789	81.6	83.3	18.0	.5	-20.2		60 Miller Huggins	1586	22.7	13.4	8.7	0.0	.6
11 Max Carey	2476	21.6	13.1	17.8	.7	-10.0		61 Johnny Evers	1784	16.3	5.7	8.7	0.0	1.9
12 George McBride	1659	4.1	-24.7	17.5	0.0	11.3		62 Bill Bergen	947	-12.7	-30.0	8.7	0.0	8.6
13 Bobby Wallace	2380	34.4	9.6	17.1	0.0	7.7		63 Billy Herman	1922	35.8	15.1	8.6	0.0	12.1
14 Graig Nettles	2245	28.1	13.0	16.6	-1.2	-.3		64 Dick Groat	1929	10.5	-11.4	8.6	-1.2	14.5
15 Bobby Doerr	1865	40.3	16.7	16.6	-2.1	9.1		65 Burgess Whitehead	924	.1	-13.3	8.5	0.0	4.9
16 Richie Ashburn	2189	25.6	18.4	16.6	-.1	-9.3		66 Tim Foli	1677	-6.4	-31.4	8.4	-.9	17.5
17 Lee Tannehill	1089	9.5	-12.6	16.5	0.0	5.6		67 Hank Aaron	3298	89.9	93.7	8.4	3.0	-15.2
18 Manny Trillo	1373	9.7	-13.0	16.3	-1.4	7.8		68 Babe Pinelli	774	4.6	-7.2	8.2	-.8	4.4
19 Dick Bartell	2016	28.1	-1.3	16.3	.1	13.0		69 Tony Kubek	1092	3.7	-9.9	8.2	-.5	5.9
20 Mickey Doolan	1728	-.4	-24.9	16.2	0.0	8.3		70 Keith Hernandez	1414	23.9	24.3	8.2	-.4	-8.2
21 Rabbit Maranville	2670	7.1	-25.2	15.1	-.1	17.3		71 Doug DeCinces	1252	17.5	10.3	8.2	-.6	-.4
22 Aurelio Rodriguez	2017	-10.2	-24.6	14.5	-.8	.7		72 Carl Yastrzemski	3308	47.1	57.1	8.1	-2.2	-15.9
23 Brooks Robinson	2896	22.0	4.4	14.3	-.5	3.8		73 Travis Jackson	1656	21.4	-.2	8.1	.1	13.4
24 Bobby Knoop	1153	12.9	-9.0	14.3	-.6	8.2		74 Jim Wynn	1920	30.0	31.0	8.0	.6	-9.6
25 Frankie Frisch	2311	34.5	11.6	13.9	-.1	9.1		75 Vic Power	1627	-4.6	-5.4	8.0	-.7	-6.5
26 Ron Santo	2243	37.7	30.2	13.7	-1.6	-4.6		76 Nellie Fox	2367	13.6	-7.5	7.9	-2.6	15.8
27 Everett Scott	1654	-2.1	-29.3	12.9	-.3	14.6		77 Maury Wills	1942	15.8	-12.5	7.8	5.2	15.3
28 Red Schoendienst	2216	16.6	-7.4	12.3	-.8	12.5		78 Roy McMillan	2093	-4.9	-25.6	7.7	-1.1	14.1
29 Lou Boudreau	1646	34.7	18.5	12.2	-1.7	5.7		79 Frank Baker	1575	38.5	25.6	7.7	0.0	5.2
30 Gene Alley	1195	19.2	-6.4	12.0	-.1	13.7		80 Doug Rader	1465	7.7	3.4	7.6	-1.1	-2.2
31 Bobby Grich	1766	43.8	24.4	11.8	-1.6	9.2		81 Glenn Hubbard	770	5.2	-6.3	7.6	-1.0	4.9
32 Mark Belanger	2016	5.4	-24.4	11.7	.5	17.6		82 Dave Cash	1422	11.4	-3.8	7.5	-.8	8.5
33 Fred Tenney	1994	22.6	18.2	11.6	0.0	-7.2		83 Andre Dawson	1174	18.2	12.7	7.4	2.7	-4.6
34 Phil Rizzuto	1661	15.0	-4.5	11.6	1.2	6.7		84 Jackie Robinson	1382	33.4	23.4	7.3	1.2	1.5
35 Roy Smalley	1271	23.2	3.0	11.4	-1.0	9.8		85 Roger Peckinpaugh	2011	14.8	-11.4	7.2	.3	18.7
36 Darrell Evans	1984	25.6	20.4	11.4	-.5	-5.7		86 Garry Maddox	1638	3.2	.7	7.1	2.1	-6.7
37 Roberto Clemente	2433	37.7	38.2	11.2	-.1	-11.6		87 Marty Marion	1572	7.6	-14.3	7.0	-.1	15.0
38 Ski Melillo	1377	-11.7	-26.2	11.1	-1.3	4.7		88 Johnny Callison	1886	9.3	12.6	7.0	-.9	-9.4
39 Del Pratt	1835	25.7	9.7	11.0	0.0	5.0		89 Jerry Priddy	1296	10.3	-1.4	6.9	-1.4	6.2
40 Eddie Collins	2826	79.9	64.2	10.7	.3	4.7		90 Hal Lanier	1196	-9.0	-26.4	6.8	-.5	11.1
41 Luis Aparicio	2599	14.2	-25.5	10.6	7.2	21.9		91 Don Kessinger	2078	1.4	-29.4	6.8	-2.2	26.2
42 Rick Burleson	1191	11.8	-8.7	10.4	-1.6	11.7		92 Joe Gordon	1566	28.3	15.2	6.8	-.9	7.2
43 Gil McDougald	1336	21.1	7.3	10.2	-1.4	5.0		93 Curt Flood	1759	-4.0	-.9	6.8	-1.8	-8.1
44 Julio Cruz	984	8.7	-11.8	10.0	5.6	4.9		94 Fred Clarke	2244	25.6	35.6	6.8	0.0	-16.8
45 Willie Mays	2992	87.7	87.7	9.9	4.1	-14.0		95 Ken Boyer	2034	18.8	17.4	6.8	-1.6	-3.8
46 Johnny Logan	1503	13.3	-4.7	9.8	-.2	8.4		96 Sam West	1753	2.6	3.4	6.7	-1.6	-5.9
47 Tommy Leach	2156	18.6	10.4	9.7	0.0	-1.5		97 Bob Johnson	1863	35.8	36.7	6.7	-1.0	-6.6
48 Ron Hansen	1384	11.5	-4.5	9.6	-.6	7.0		98 Jimmy Barrett	703	9.1	6.2	6.7	0.0	-3.8
49 Billy Jurges	1816	7.7	-13.7	9.4	0.0	12.0		99 Roy Thomas	1470	21.4	23.3	6.6	0.0	-8.5
50 Hobe Ferris	1280	-.2	-11.1	9.3	0.0	1.6		100 Johnny Pesky	1270	17.2	10.3	6.6	-1.4	1.7

	G	All	Bat	Fld	Run	Pos			G	All	Bat	Fld	Run	Pos
1 Babe Ruth	2503	115.6	130.6	-2.0	-1.2	-11.8		51 Dave Bancroft	1913	32.6	-.2	19.5	0.0	13.3
2 Ted Williams	2292	96.5	112.8	-4.5	-.3	-11.5		52 Willie Stargell	2360	32.2	51.5	-6.9	-.4	-12.0
3 Hank Aaron	3298	89.9	93.7	8.4	3.0	-15.2		53 Reggie Smith	1987	32.1	37.4	4.0	-1.1	-8.2
4 Ty Cobb	3034	89.7	105.7	5.0	.2	-21.2		54 Yogi Berra	2120	31.8	23.4	2.3	-.8	6.9
5 Willie Mays	2992	87.7	87.7	9.9	4.1	-14.0		55 Harmon Killebrew	2435	31.3	52.3	-10.5	-.6	-9.9
6 Nap Lajoie	2474	85.3	56.9	31.4	0.0	-3.0		56 Johnny Bench	2158	31.2	26.8	-.5	-.7	5.6
7 Rogers Hornsby	2259	81.8	85.3	-6.0	0.0	2.5		57 Jim Wynn	1920	30.0	31.0	8.0	.6	-9.6
8 Tris Speaker	2789	81.6	83.3	18.0	.5	-20.2		58 Sherry Magee	2087	29.6	34.6	2.2	0.0	-7.2
9 Eddie Collins	2826	79.9	64.2	10.7	.3	4.7		59 Bobby Bonds	1849	29.5	29.1	5.2	3.9	-8.7
10 Honus Wagner	2789	79.3	67.6	8.8	0.0	2.9		60 Dave Winfield	1655	29.4	30.5	3.9	1.4	-6.4
11 Stan Musial	3026	76.1	95.6	-1.3	-.8	-17.4		61 Hank Greenberg	1394	29.3	41.6	1.4	.2	-13.9
12 Mickey Mantle	2401	70.9	85.6	-5.2	2.5	-12.0		62 Joe Gordon	1566	28.3	15.2	6.8	-.9	7.2
13 Frank Robinson	2808	70.7	79.7	2.9	1.5	-13.4		63 Buddy Bell	1827	28.2	11.3	19.6	-2.7	-.0
14 Mike Schmidt	1789	69.7	48.3	24.8	-.0	-3.4		64 Graig Nettles	2245	28.1	13.0	16.6	-1.2	-.3
15 Joe Morgan	2649	66.8	50.6	-11.6	11.6	16.2		65 Dick Bartell	2016	28.1	-1.3	16.3	.1	13.0
16 Lou Gehrig	2164	62.4	91.1	-4.9	-2.3	-21.5		66 Billy Williams	2488	27.9	42.9	-2.0	-.3	-12.7
17 Mel Ott	2730	59.9	76.3	-3.9	0.0	-12.5		67 Norm Cash	2089	27.6	38.4	3.5	-.5	-13.8
18 Jimmie Foxx	2317	54.9	73.0	2.1	-1.8	-18.4		68 Heinie Groh	1676	27.3	18.2	5.3	.0	3.8
19 Eddie Mathews	2391	48.1	53.4	-1.1	-.4	-3.8		69 Pete Rose	3371	27.2	41.9	-4.1	-3.6	-7.0
20 Carl Yastrzemski	3308	47.1	57.1	8.1	-2.2	-15.9		70 Joe Medwick	1984	26.5	31.7	3.7	0.0	-8.9
21 Al Kaline	2834	46.8	51.0	9.2	-.0	-14.4		71 Elmer Flick	1493	26.4	35.8	1.1	0.0	-10.5
22 Joe DiMaggio	1736	44.8	51.1	1.0	.3	-7.6		72 George Sisler	2055	26.1	27.1	8.9	.2	-10.1
23 Charlie Gehringer	2323	44.7	31.9	3.2	.6	9.0		73 Ralph Kiner	1472	26.0	36.8	-4.6	.2	-6.4
24 Bobby Grich	1766	43.8	24.4	11.8	-1.6	9.2		74 Rickey Henderson	791	25.9	15.1	6.5	6.5	-2.2
25 Rod Carew	2342	41.8	42.8	.3	-.4	-.9		75 Del Pratt	1835	25.7	9.7	11.0	0.0	5.0
26 Reggie Jackson	2430	41.7	48.5	1.6	.2	-8.6		76 Art Fletcher	1529	25.7	-2.1	20.4	0.0	7.4
27 Arky Vaughan	1817	41.4	35.3	-4.4	0.0	10.5		77 Darrell Evans	1984	25.6	20.4	11.4	-.5	-5.7
28 Bobby Doerr	1865	40.3	16.7	16.6	-2.1	9.1		78 Fred Clarke	2244	25.6	35.6	6.8	0.0	-16.8
29 Paul Waner	2549	39.6	46.6	6.1	0.0	-13.1		79 Richie Ashburn	2189	25.6	18.4	16.6	-.1	-9.3
30 Joe Cronin	2124	39.4	21.5	6.0	-1.7	13.6		80 Bill Terry	1721	25.5	30.4	6.0	-.2	-10.7
31 Joe Jackson	1331	38.6	46.2	2.3	0.0	-9.9		81 Rusty Staub	2897	25.5	34.6	3.1	-.6	-11.6
32 Frank Baker	1575	38.5	25.6	7.7	0.0	5.2		82 Sam Crawford	2517	25.5	50.7	-8.7	0.0	-16.5
33 Luke Appling	2422	38.0	18.7	9.2	-1.1	11.2		83 Al Simmons	2215	25.2	35.1	1.2	-.9	-10.2
34 Willie McCovey	2588	37.8	56.3	-3.2	-.6	-14.7		84 Robin Yount	1549	25.1	9.4	1.3	.8	13.6
35 Ron Santo	2243	37.7	30.2	13.7	-1.6	-4.6		85 Harry Heilmann	2146	24.9	49.8	-10.3	-1.3	-14.3
36 Roberto Clemente	2433	37.7	38.2	11.2	-.1	-11.6		86 Stan Hack	1938	24.8	22.4	-.9	0.0	3.3
37 Joe Sewell	1903	37.1	11.4	8.9	-.4	17.2		87 Tony Oliva	1676	24.7	26.4	5.9	-.8	-6.8
38 Bill Mazeroski	2163	36.7	-18.0	36.5	-.6	18.8		88 Rocky Colavito	1841	24.7	29.8	5.4	-1.1	-9.4
39 Johnny Mize	1884	36.0	51.0	-2.1	-.1	-12.8		89 Chuck Klein	1753	24.5	32.1	1.8	0.0	-9.4
40 Dick Allen	1749	35.9	50.2	-4.5	.9	-10.7		90 Charlie Keller	1170	24.5	30.3	.3	-.0	-6.1
41 Bob Johnson	1863	35.8	36.7	6.7	-1.0	-6.6		91 Duke Snider	2143	24.3	39.4	-6.4	-1.1	-7.6
42 Billy Herman	1922	35.8	15.1	8.6	0.0	12.1		92 Jack Fournier	1530	24.3	29.4	1.1	-.2	-6.0
43 Gabby Hartnett	1990	35.0	23.8	3.5	-.3	8.0		93 Ernie Banks	2528	24.0	26.6	-.6	-1.8	-.2
44 Lou Boudreau	1646	34.7	18.5	12.2	-1.7	5.7		94 Keith Hernandez	1414	23.9	24.3	8.2	-.4	-8.2
45 Frankie Frisch	2311	34.5	11.6	13.9	-.1	9.1		95 Zach Wheat	2410	23.8	34.9	-2.1	.1	-9.1
46 Bobby Wallace	2380	34.4	9.6	17.1	0.0	7.7		96 Wally Schang	1839	23.6	17.4	-2.7	-.1	9.0
47 Bill Dickey	1789	34.0	23.8	-.6	-.8	11.6		97 Ozzie Smith	1006	23.5	-13.2	22.3	3.5	10.9
48 Jackie Robinson	1382	33.4	23.4	7.3	1.2	1.5		98 Ernie Lombardi	1853	23.5	19.8	-2.9	0.0	6.6
49 Mickey Cochrane	1482	32.7	24.6	-.7	-.4	9.2		99 Goose Goslin	2287	23.5	34.1	1.3	.6	-12.5
50 George Brett	1462	32.7	31.3	1.9	-.3	-.2		100 Ted Simmons	2086	23.4	21.8	-2.4	-1.6	5.6

	G	All	Bat	Fld	Run	Pos		G	All	Bat	Fld	Run	Pos
101 Jim Rice	1493	23.3	25.7	2.6	-.4	-4.6	151 Babe Herman	1552	17.9	30.9	-4.0	.1	-9.1
102 Roy Smalley	1271	23.2	3.0	11.4	-1.0	9.8	152 Larry Doby	1533	17.9	27.7	-3.1	-.7	-6.0
103 Minnie Minoso	1835	23.0	29.2	4.1	-1.8	-8.5	153 Bill Freehan	1774	17.7	11.1	.2	-.6	7.0
104 Ron Cey	1786	23.0	22.2	5.2	-1.1	-3.3	154 Doug DeCinces	1252	17.5	10.3	8.2	-.6	-.4
105 Pudge Fisk	1549	22.8	17.2	-.2	.8	5.0	155 Dwight Evans	1622	17.4	19.1	4.4	-.8	-5.3
106 Miller Huggins	1586	22.7	13.4	8.7	0.0	.6	156 Johnny Pesky	1270	17.2	10.3	6.6	-1.4	1.7
107 Cesar Cedeno	1858	22.7	22.8	2.6	5.9	-8.6	157 Roy Cullenbine	1181	17.1	20.7	1.9	-.6	-4.9
108 Fred Tenney	1994	22.6	18.2	11.6	0.0	-7.2	158 Orlando Cepeda	2124	17.1	34.5	-4.1	-.4	-12.9
109 Harlond Clift	1582	22.4	13.6	6.2	-.7	3.3	159 Wally Berger	1350	17.0	24.5	-.3	0.0	-7.2
110 Eddie Stanky	1259	22.1	11.7	3.3	-.1	7.2	160 Tony Lazzeri	1740	16.7	19.9	-11.0	.4	7.4
111 Brooks Robinson	2896	22.0	4.4	14.3	-.5	3.8	161 Benny Kauff	859	16.7	21.0	.4	0.0	-4.7
112 Eddie Murray	1206	22.0	27.7	.8	.4	-6.9	162 Ben Chapman	1717	16.7	14.8	6.1	.3	-4.5
113 Joe Torre	2209	21.9	31.0	-6.7	-1.3	-1.1	163 Red Schoendienst	2216	16.6	-7.4	12.3	-.8	12.5
114 Ed Konetchy	2085	21.9	21.4	5.3	0.0	-4.8	164 Ferris Fain	1151	16.6	14.9	4.6	-.3	-2.6
115 George Foster	1761	21.8	23.0	6.2	-.3	-7.1	165 Enos Slaughter	2380	16.5	28.0	-2.8	-.1	-8.6
116 Gary Carter	1408	21.7	14.9	5.2	-1.0	2.6	166 Johnny Evers	1784	16.3	5.7	8.7	0.0	1.9
117 Max Carey	2476	21.6	13.1	17.8	.7	-10.0	167 Rudy York	1603	16.1	16.7	3.8	-.5	-3.9
118 Roy Thomas	1470	21.4	23.3	6.6	0.0	-8.5	168 Edd Roush	1967	16.1	26.3	-.8	-.5	-8.9
119 Travis Jackson	1656	21.4	-.2	8.1	.1	13.4	169 Lonny Frey	1535	16.0	3.9	4.1	.1	7.9
120 Dave Concepcion	2055	21.2	-10.2	4.8	3.4	23.2	170 Roy Campanella	1215	16.0	12.4	-.6	-.4	4.6
121 Gil McDougald	1336	21.1	7.3	10.2	-1.4	5.0	171 Maury Wills	1942	15.8	-12.5	7.8	5.2	15.3
122 Jim Fregosi	1902	20.9	11.5	1.0	-.2	8.6	172 Donie Bush	1945	15.8	-4.6	5.4	0.0	15.0
123 Roger Bresnahan	1446	20.4	18.0	-3.6	0.0	6.0	173 Jack Clark	1044	15.7	17.6	2.6	-.9	-3.6
124 Jimmy Sheckard	2122	20.2	21.6	9.1	0.0	-10.5	174 Cal Ripken	507	15.5	7.9	4.7	-.3	3.2
125 Fred Lynn	1301	20.1	23.5	1.2	-.8	-3.8	175 Jeff Heath	1383	15.4	25.0	-1.3	-1.5	-6.8
126 Ken Williams	1397	20.0	24.7	6.2	-.5	-10.4	176 Dolph Camilli	1490	15.4	28.3	.3	.1	-13.3
127 Roy White	1881	20.0	23.1	3.7	-.0	-6.8	177 Don Buford	1286	15.4	12.2	3.4	-.4	.2
128 Dave Parker	1457	20.0	20.2	6.3	-.6	-5.9	178 Thurman Munson	1423	15.2	12.4	.5	-1.6	3.9
129 Willie Randolph	1210	19.9	6.2	5.2	2.0	6.5	179 Tommy Henrich	1284	15.2	21.0	-.3	-.1	-5.4
130 Gavvy Cravath	1220	19.8	27.5	-3.4	0.0	-4.3	180 Sid Gordon	1475	15.2	21.8	-2.5	-.1	-4.0
131 Frank Howard	1895	19.7	36.7	-7.4	-.6	-9.0	181 Phil Rizzuto	1661	15.0	-4.5	11.6	1.2	6.7
132 Pie Traynor	1941	19.5	7.4	2.0	-.1	10.2	182 Roger Peckinpaugh	2011	14.8	-11.4	7.2	.3	18.7
133 Earl Averill	1668	19.5	29.2	-3.2	-1.5	-5.0	183 Toby Harrah	1934	14.8	14.6	-9.0	1.7	7.5
134 Gene Alley	1195	19.2	-6.4	12.0	-.1	13.7	184 Hal McRae	1842	14.7	23.5	-1.0	-1.5	-6.3
135 Gene Tenace	1555	18.9	26.1	-4.2	-1.4	-1.6	185 Cy Seymour	1528	14.6	15.6	4.2	0.0	-5.2
136 Ken Boyer	2034	18.8	17.4	6.8	-1.6	-3.8	186 Tim Raines	581	14.6	7.5	1.4	6.8	-1.1
137 Joe Tinker	1804	18.7	-6.6	18.3	0.0	7.0	187 Augie Galan	1742	14.6	22.6	-2.1	0.0	-5.9
138 Bob Elliott	1978	18.7	25.4	-4.0	0.0	-2.7	188 Bill Nicholson	1677	14.5	23.0	-.9	-.1	-7.5
139 Ray Chapman	1050	18.7	5.2	5.4	0.0	8.1	189 Frank Chance	1286	14.5	17.8	-.3	0.0	-3.0
140 Tommy Leach	2156	18.6	10.4	9.7	0.0	-1.5	190 Chet Lemon	1195	14.2	15.5	4.3	-2.3	-3.3
141 Ken Singleton	2082	18.5	35.7	-8.3	-1.7	-7.2	191 Luis Aparicio	2599	14.2	-25.5	10.6	7.2	21.9
142 Peewee Reese	2166	18.5	.2	1.1	.5	16.7	192 Andy Thornton	1285	14.0	19.2	2.1	-.7	-6.6
143 Art Devlin	1313	18.4	7.2	9.1	0.0	2.1	193 George Stone	848	14.0	17.3	1.2	0.0	-4.5
144 Kiki Cuyler	1879	18.4	26.1	1.8	.4	-9.9	194 Bobby Murcer	1908	14.0	25.8	-4.7	-.7	-6.4
145 Jose Cruz	1907	18.3	23.5	1.4	1.3	-7.9	195 Harry Davis	1769	14.0	16.7	-.1	0.0	-2.6
146 Amos Otis	1998	18.2	16.4	3.5	4.8	-6.5	196 Mike Hargrove	1559	13.8	19.0	4.1	-1.5	-7.8
147 Andre Dawson	1174	18.2	12.7	7.4	2.7	-4.6	197 Tony Cuccinello	1704	13.8	4.6	-.2	.1	9.3
148 Rico Carty	1651	18.1	25.3	-.2	-1.0	-6.0	198 Cy Williams	2002	13.6	22.5	-.6	0.0	-8.3
149 Hack Wilson	1348	18.0	30.2	-4.9	0.0	-7.3	199 Nellie Fox	2367	13.6	-7.5	7.9	-2.6	15.8
150 Vern Stephens	1720	18.0	17.4	-1.6	-.8	3.0	200 Boog Powell	2042	13.5	31.8	-5.2	-.7	-12.4

	G	All	Bat	Fld	Run	Pos		G	All	Bat	Fld	Run	Pos
201 Lou Whitaker	987	13.3	4.8	4.2	-.3	4.6	251 Willie Keeler	2125	10.6	28.6	-4.8	0.0	-13.2
202 John Romano	905	13.3	8.6	1.4	-.5	3.8	252 Monte Irvin	764	10.5	10.5	2.5	.3	-2.8
203 Johnny Logan	1503	13.3	-4.7	9.8	-.2	8.4	253 Dick Groat	1929	10.5	-11.4	8.6	-1.2	14.5
204 Willie Kamm	1693	13.3	-1.4	4.0	-.1	10.8	254 Joe Ferguson	1013	10.5	8.2	.4	-.1	2.0
205 Bill Bradley	1461	13.3	3.4	6.5	0.0	3.4	255 John Titus	1402	10.4	17.9	-2.3	0.0	-5.2
206 Solly Hemus	961	13.1	8.3	1.0	-.4	4.2	256 Snuffy Stirnweiss	1028	10.4	3.4	2.7	.8	3.5
207 Rick Ferrell	1884	13.1	-.6	2.0	-1.2	12.9	257 Bob Nieman	1113	10.4	16.3	-.3	-1.6	-4.0
208 Jake Daubert	2014	13.1	18.7	1.1	0.0	-6.7	258 Marty McManus	1831	10.4	1.0	1.4	-1.0	9.0
209 Buddy Myer	1923	13.0	10.7	-4.5	-1.3	8.1	259 Whitey Kurowski	916	10.4	10.6	-.5	0.0	.3
210 Topsy Hartsel	1355	13.0	23.1	-1.8	0.0	-8.3	260 Cliff Johnson	1156	10.4	13.3	.1	-.6	-2.4
211 Bobby Knoop	1153	12.9	-9.0	14.3	-.6	8.2	261 Jerry Priddy	1296	10.3	-1.4	6.9	-1.4	6.2
212 Heinie Zimmerman	1456	12.8	12.1	.0	0.0	.7	262 Freddie Patek	1650	10.3	-15.1	4.6	3.9	16.9
213 Bump Wills	831	12.8	-2.4	8.8	2.0	4.4	263 Lance Parrish	915	10.3	5.0	2.3	-.8	3.8
214 Pedro Guerrero	657	12.8	12.7	2.0	.2	-2.1	264 Tom Haller	1294	10.3	7.6	-.6	-1.6	4.9
215 Garry Templeton	1128	12.7	-3.7	4.6	-.6	12.4	265 Smoky Burgess	1691	10.3	10.7	-4.3	-.6	4.5
216 Ross Youngs	1211	12.6	20.5	-2.5	-.1	-5.3	266 Greg Luzinski	1821	10.2	30.0	-11.2	-.8	-7.8
217 Ron Hunt	1483	12.6	7.3	-1.9	-1.3	8.5	267 Jim Gentile	936	10.2	15.2	1.0	0.0	-6.0
218 Larry Gardner	1923	12.6	8.3	-2.7	0.0	7.0	268 Woody English	1261	10.2	-.5	.8	0.0	9.9
219 Walker Cooper	1473	12.6	8.9	-1.4	0.0	5.1	269 Al Rosen	1044	10.1	20.0	-6.4	-.8	-2.7
220 Ken Keltner	1526	12.2	7.0	3.2	-1.1	3.1	270 Claude Ritchey	1671	10.1	4.3	1.3	0.0	4.5
221 Dave Johnson	1435	12.2	6.7	-1.5	-.6	7.6	271 Tommy Holmes	1320	10.1	14.7	1.9	0.0	-6.5
222 Wade Boggs	415	12.2	8.2	4.0	-.1	.1	272 Oscar Gamble	1514	10.1	17.0	-1.3	-.9	-4.7
223 Richie Zisk	1453	12.1	18.4	.4	-.6	-6.1	273 Ray Boone	1373	10.1	10.3	-.4	-.5	.7
224 Roy Sievers	1887	12.1	21.5	-.2	-.7	-8.5	274 Max Bishop	1338	10.1	8.3	-.6	-1.4	3.8
225 Spud Davis	1458	12.1	6.2	-.0	0.0	5.9	275 Hank Sauer	1399	10.0	14.8	1.0	-.1	-5.7
226 Bob Allison	1541	12.0	19.5	-.3	-.6	-6.6	276 Andy Seminick	1304	9.9	4.1	1.5	-.3	4.6
227 Sixto Lezcano	1219	11.9	14.2	2.1	-.9	-3.5	277 Al Oliver	2272	9.9	25.4	-2.1	-1.6	-11.8
228 Bob Watson	1832	11.8	24.6	-2.0	-.8	-10.0	278 Bobby Avila	1300	9.9	3.5	.3	-.9	7.0
229 Dixie Walker	1905	11.8	21.4	-1.7	-.2	-7.7	279 Cecil Travis	1328	9.8	5.2	1.0	-1.4	5.0
230 Rick Burleson	1191	11.8	-8.7	10.4	-1.6	11.7	280 Babe Phelps	726	9.8	6.7	.9	0.0	2.2
231 Gil Hodges	2071	11.7	18.1	3.9	-.8	-9.5	281 Manny Trillo	1373	9.7	-13.0	16.3	-1.4	7.8
232 Tony Perez	2628	11.6	27.9	-4.3	-.5	-11.5	282 Lefty O'Doul	970	9.7	19.9	-4.9	0.0	-5.3
233 Dwayne Murphy	881	11.5	8.5	5.9	-.3	-2.6	283 Sherm Lollar	1752	9.7	4.2	-1.1	-.2	6.8
234 Ron Hansen	1384	11.5	-4.5	9.6	-.6	7.0	284 Sal Bando	2019	9.6	19.9	-10.5	-.3	.5
235 Darrell Porter	1545	11.4	10.6	-2.5	-1.1	4.4	285 Ed Bailey	1212	9.6	6.8	-1.8	-.6	5.2
236 Dom DiMaggio	1399	11.4	13.3	5.7	-.7	-6.9	286 Lee Tannehill	1089	9.5	-12.6	16.5	0.0	5.6
237 Dave Cash	1422	11.4	-3.8	7.5	-.8	8.5	287 Ryne Sandberg	483	9.5	1.0	6.0	1.4	1.1
238 Stan Spence	1112	11.3	14.9	2.1	-.8	-4.9	288 Bill Melton	1144	9.5	6.1	5.0	-.6	-1.0
239 Jackie Jensen	1438	11.3	16.8	.3	.9	-6.7	289 Hank Gowdy	1050	9.4	1.9	2.9	0.0	4.6
240 Riggs Stephenson	1310	11.2	20.4	-4.6	0.0	-4.6	290 Dick Dietz	646	9.4	9.5	-2.4	-.3	2.6
241 Larry Hisle	1197	11.2	12.7	3.0	.1	-4.6	291 George Grantham	1444	9.3	16.5	-5.8	.2	-1.6
242 Ken Griffey	1412	11.1	14.3	.6	1.9	-5.7	292 Johnny Callison	1886	9.3	12.6	7.0	-.9	-9.4
243 Don Baylor	1770	11.1	19.1	-3.1	2.1	-7.0	293 Gene Woodling	1796	9.2	19.3	-1.6	-1.9	-6.6
244 Davey Lopes	1570	11.0	5.2	-11.4	8.8	8.4	294 Alan Trammell	989	9.1	4.7	-3.6	-.4	8.4
245 Earle Combs	1455	11.0	22.7	-3.1	-1.0	-7.6	295 Hank Thompson	933	9.1	9.4	.4	-.3	-.4
246 Chick Hafey	1283	10.9	19.1	-1.1	-.3	-6.8	296 George Kell	1795	9.1	10.0	.3	-.8	-.4
247 Red Smith	1117	10.8	8.9	.2	0.0	1.7	297 Kid Elberfeld	1292	9.1	2.3	2.6	0.0	4.2
248 Cecil Cooper	1545	10.8	20.5	-1.2	-.4	-8.1	298 Vic Wertz	1862	9.0	16.6	-.2	-.8	-6.6
249 Roger Maris	1463	10.7	18.8	-1.0	0.0	-7.1	299 Rico Petrocelli	1553	9.0	7.1	-2.5	-1.0	5.4
250 Paul Molitor	765	10.6	5.5	1.0	2.3	1.8	300 Brian Downing	1284	9.0	10.3	-1.1	-.5	.3

		IP	LWT	Adj	Wins	PF	R/W
1	Cy Young	7357	749.7	780.1	72.3	1.010	10.79
2	Walter Johnson	5925	705.2	649.7	67.1	.974	9.68
3	Lefty Grove	3940	595.0	615.3	57.1	1.010	10.78
4	Pete Alexander	5189	482.9	533.2	55.8	1.026	9.56
5	Christy Mathewson	4778	417.6	408.2	42.3	.994	9.66
6	Tom Seaver	4367	392.7	402.4	42.3	1.006	9.52
7	Bob Gibson	3885	291.1	367.3	39.1	1.049	9.39
8	Gaylord Perry	5352	315.4	342.7	36.4	1.013	9.41
9	Carl Hubbell	3591	393.6	362.6	35.4	.980	10.24
10	Phil Niekro	4835	226.3	334.3	35.4	1.056	9.45
11	Steve Carlton	4787	296.4	324.9	34.2	1.015	9.49
12	Bert Blyleven	3421	257.6	325.0	33.8	1.048	9.61
13	Three Finger Brown	3171	293.5	313.0	33.5	1.019	9.33
14	Hal Newhouser	2993	256.3	325.8	33.4	1.055	9.76
15	Jim Palmer	3948	377.6	322.8	33.3	.966	9.70
16	Warren Spahn	5246	471.1	338.1	33.2	.941	10.18
17	Whitey Ford	3171	386.9	312.7	31.7	.945	9.87
18	Ted Lyons	4162	314.2	328.2	31.1	1.007	10.56
19	Hoyt Wilhelm	2253	309.9	287.2	29.7	.976	9.67
20	Ed Walsh	2965	310.4	260.4	28.4	.945	9.18
21	Bob Feller	3828	385.4	290.1	28.0	.946	10.37
22	Stan Coveleski	3082	257.9	275.1	26.5	1.014	10.37
23	Tommy Bridges	2827	256.6	281.3	26.3	1.018	10.68
24	Red Faber	4086	293.8	262.1	26.0	.982	10.09
25	Ferguson Jenkins	4498	160.0	245.4	25.7	1.047	9.55
26	Eddie Plank	4507	269.2	243.0	25.7	.982	9.46
27	Juan Marichal	3506	261.9	236.3	25.1	.982	9.41
28	Rube Waddell	2963	238.8	235.9	24.8	.997	9.51
29	Dazzy Vance	2967	280.7	263.5	24.8	.987	10.64
30	Don Drysdale	3432	266.2	231.4	24.6	.975	9.42
31	Dizzy Trout	2726	177.4	242.2	24.5	1.056	9.89
32	Sandy Koufax	2325	243.5	225.6	23.9	.981	9.44
33	Robin Roberts	4689	263.9	240.3	23.9	.988	10.07
34	Billy Pierce	3305	249.3	225.0	22.6	.983	9.95
35	Tommy John	4123	233.9	213.2	22.3	.988	9.58
36	Addie Joss	2327	214.3	199.0	21.6	.978	9.21
37	Eppa Rixey	4494	250.1	225.7	21.3	.987	10.59
38	Dutch Leonard	3220	267.5	218.4	21.2	.966	10.31
39	Urban Shocker	2681	209.7	216.9	21.0	1.006	10.34
40	Harry Brecheen	1905	193.0	205.0	20.8	1.015	9.87
41	Eddie Rommel	2557	179.5	215.1	20.5	1.030	10.51
42	Dizzy Dean	1966	184.4	207.1	20.4	1.027	10.16
43	Eddie Cicotte	3223	203.7	189.2	20.2	.986	9.37
44	Virgil Trucks	2684	141.8	195.7	19.7	1.047	9.95
45	Joe McGinnity	3441	185.3	212.5	19.5	1.023	10.18
46	Lefty Gomez	2503	322.4	210.2	19.4	.910	10.86
47	Jim Bunning	3759	178.8	185.3	19.1	1.004	9.72
48	Bucky Walters	3104	151.4	184.8	19.0	1.026	9.72
49	Jerry Koosman	3740	159.1	181.0	18.8	1.014	9.60
50	Ed Reulbach	2633	166.9	175.6	18.8	1.010	9.35

		IP	LWT	Adj	Wins	PF	R/W
51	Bob Lemon	2849	249.9	188.8	18.7	.952	10.09
52	Vic Willis	3997	169.8	182.3	18.6	1.009	9.78
53	Luis Tiant	3486	118.8	170.2	18.5	1.037	9.17
54	Hippo Vaughn	2731	123.8	163.2	18.5	1.045	8.82
55	Steve Rogers	2801	157.3	176.2	18.4	1.017	9.59
56	Lon Warneke	2781	186.9	179.8	18.2	.994	9.85
57	Carl Mays	3021	216.5	175.9	18.2	.966	9.67
58	Sam Leever	2661	173.0	188.1	17.8	1.017	9.93
59	Wes Ferrell	2623	147.1	191.5	17.6	1.034	10.91
60	Jack Quinn	3920	159.3	181.8	17.5	1.014	10.38
61	Dolf Luque	3221	245.4	180.3	17.3	.954	10.45
62	Wilbur Cooper	3482	168.9	162.9	17.0	.995	9.57
63	Waite Hoyt	3763	209.9	175.9	17.0	.980	10.35
64	Larry Jackson	3262	94.5	161.9	17.0	1.051	9.55
65	Thornton Lee	2331	169.0	176.6	16.8	1.007	10.53
66	Larry French	3152	177.9	162.9	16.3	.989	9.99
67	Andy Messersmith	2230	173.5	153.1	16.2	.977	9.45
68	Dave Stieb	1389	110.8	158.5	15.9	1.077	9.97
69	Joe Wood	1436	151.2	154.3	15.9	1.007	9.74
70	Freddie Fitzsimmons	3225	174.0	163.0	15.8	.992	10.30
71	Nolan Ryan	1722	155.1	153.0	15.3	.997	9.99
72	Bobby Shantz	1936	125.3	150.0	15.2	1.029	9.87
73	Claude Passeau	2718	118.6	146.6	15.2	1.025	9.66
74	Mel Harder	3426	190.3	162.7	15.0	.983	10.87
75	Don Sutton	4570	254.9	139.0	14.8	.937	9.39
76	Deacon Phillippe	2608	150.5	162.5	14.7	1.013	10.23
77	Babe Adams	2995	163.2	141.0	14.6	.979	9.65
78	John Hiller	1241	112.2	137.8	14.4	1.051	9.53
79	Mort Cooper	1843	132.4	135.3	14.3	1.004	9.46
80	Sal Maglie	3705	230.6	137.8	14.3	.938	9.67
81	Bob Shawkey	2938	150.7	141.7	14.0	.992	10.13
82	Dean Chance	2148	134.1	130.3	13.9	.995	9.40
83	Eddie Lopat	2439	185.5	140.4	13.8	.957	10.19
84	Noodles Hahn	1409	104.5	134.9	13.6	1.064	9.93
85	Max Lanier	1618	126.3	131.6	13.6	1.008	9.71
86	Mike Garcia	2176	174.7	135.3	13.5	.959	10.01
87	Bruce Sutter	888	107.9	129.0	13.4	1.059	9.59
88	Spud Chandler	1485	175.7	130.4	13.4	.930	9.71
89	Red Ruffing	4342	271.1	150.1	13.4	.943	11.18
90	Curt Simmons	3348	109.2	134.8	13.3	1.018	10.10
91	Wilbur Wood	2684	94.5	119.4	13.3	1.023	8.96
92	Ron Guidry	1768	165.0	132.3	13.3	.958	9.95
93	Jim Kaat	4529	75.2	124.7	13.3	1.027	9.39
94	Curt Davis	2324	95.9	131.3	13.3	1.036	9.89
95	Frank Tanana	2354	158.3	128.6	13.2	.971	9.77
96	Preacher Roe	1916	108.5	131.8	13.1	1.028	10.06
97	Murry Dickson	3053	81.5	129.5	13.1	1.036	9.90
98	Don Newcombe	2154	96.9	130.7	13.1	1.036	10.00
99	Jack Taylor	2617	132.2	130.8	13.0	.998	10.06
100	Kent Tekulve	1014	112.9	123.3	12.9	1.025	9.54

	IP	LWT	Adj	Wins	PF	R/W			IP	LWT	Adj	Wins	PF	R/W
101 John Candelaria	1800	109.6	124.3	12.9	1.020	9.66		151 Jack Pfiester	1059	66.2	82.3	9.3	1.053	8.88
102 Jesse Tannehill	2751	120.3	132.0	12.8	1.012	10.31		152 Bill Walker	1489	92.5	99.3	9.2	1.010	10.75
103 Rich Gossage	1338	134.1	127.5	12.7	.988	10.04		153 Gary Lavelle	981	89.7	88.7	9.2	.997	9.63
104 Milt Pappas	3187	87.2	123.5	12.6	1.028	9.79		154 Ewell Blackwell	1322	91.9	93.3	9.2	1.002	10.13
105 Sparky Lyle	1391	122.3	119.6	12.4	.995	9.63		155 Mike Marshall	1387	85.5	88.1	9.2	1.005	9.60
106 Nap Rucker	2375	124.7	116.9	12.4	.990	9.46		156 Bill Dinneen	3075	36.8	93.4	9.1	1.053	10.29
107 Jon Matlack	2363	147.9	117.1	12.3	.969	9.55		157 Jim Scott	1891	111.1	86.0	9.0	.958	9.50
108 Paul Derringer	3646	131.1	123.5	12.2	.995	10.09		158 Chief Bender	3017	106.0	85.0	9.0	.977	9.42
109 Burleigh Grimes	4181	118.6	127.6	12.1	1.005	10.54		159 Fred Hutchinson	1465	62.7	90.9	9.0	1.042	10.07
110 Rick Reuschel	2362	43.2	115.5	12.0	1.076	9.60		160 Fred Toney	2206	102.3	82.0	9.0	.973	9.09
111 Bob Stanley	1204	80.3	119.7	12.0	1.073	9.97		161 Don Mossi	1548	74.2	86.9	8.7	1.019	9.94
112 Sam McDowell	2492	88.4	110.4	11.9	1.023	9.25		162 Bobo Newsom	3762	68.7	91.8	8.7	1.013	10.58
113 Firpo Marberry	2066	155.5	125.9	11.8	.970	10.67		163 Clay Carroll	1353	93.6	84.8	8.7	.984	9.78
114 Slim Sallee	2819	124.3	111.7	11.7	.986	9.56		164 Jake Weimer	1473	91.4	83.3	8.5	.982	9.79
115 Rollie Fingers	1646	151.8	111.0	11.4	.939	9.71		165 Stu Miller	1694	106.9	80.4	8.5	.963	9.51
116 Howie Pollet	2105	81.1	109.4	11.3	1.031	9.66		166 Harvey Haddix	2236	75.8	84.7	8.4	1.009	10.05
117 Mel Parnell	1752	118.5	117.0	11.2	.998	10.42		167 Tug McGraw	1516	82.9	79.5	8.4	.994	9.47
118 Early Wynn	4566	171.0	111.0	11.0	.969	10.10		168 Allie Reynolds	2492	148.1	83.5	8.3	.939	10.01
119 Charlie Root	3197	147.4	112.6	10.9	.975	10.31		169 Sonny Siebert	2152	58.4	78.4	8.3	1.024	9.42
120 Bob Friend	3612	94.4	105.1	10.9	1.007	9.66		170 Jack Powell	4388	87.4	84.2	8.2	.998	10.22
121 Russ Ford	1487	77.6	105.0	10.9	1.054	9.66		171 Red Ames	3197	73.7	79.9	8.2	1.006	9.76
122 Johnny Antonelli	1992	141.1	108.8	10.9	.963	10.03		172 Lefty Leifield	1839	76.7	78.2	8.2	1.003	9.58
123 Bill Hands	1951	45.1	101.5	10.8	1.073	9.38		173 Jim Perry	3287	63.1	77.5	8.1	1.011	9.54
124 Jack Chesbro	2898	57.1	105.7	10.8	1.053	9.81		174 Joe Horlen	2003	79.2	72.1	8.1	.991	8.92
125 Johnny Rigney	1188	98.4	112.1	10.6	1.024	10.60		175 Joe Benz	1360	78.7	75.0	8.0	.992	9.33
126 Tex Hughson	1375	98.5	100.7	10.6	1.004	9.52		176 Tom Burgmeier	1258	73.2	80.4	8.0	1.014	10.05
127 Ellis Kinder	1481	101.8	106.7	10.4	1.007	10.22		177 Ron Perranoski	1176	99.9	78.0	8.0	.953	9.78
128 Dutch Leonard	2191	112.9	94.4	10.4	.976	9.06		178 Harry Coveleski	1248	54.6	71.3	7.9	1.043	9.02
129 Dennis Eckersley	2125	64.7	102.7	10.4	1.041	9.90		179 Mel Stottlemyre	2662	140.5	76.3	7.9	.937	9.65
130 Al Benton	1689	74.5	103.2	10.3	1.038	10.01		180 Don McMahon	1313	97.0	77.3	7.9	.963	9.81
131 Jesse Haines	3208	113.5	104.0	10.3	.993	10.10		181 Joe Dobson	2172	86.8	77.4	7.8	.990	9.87
132 Vida Blue	3056	159.9	95.8	10.3	.949	9.34		182 Bill Lee	2862	76.9	81.9	7.8	1.004	10.53
133 Frank Sullivan	1732	58.7	100.9	10.2	1.056	9.86		183 Dave Rozema	1007	66.6	77.5	7.7	1.024	10.01
134 Dan Quisenberry	635	106.9	101.0	10.1	.979	9.96		184 Fritz Ostermueller	2069	50.7	80.3	7.7	1.031	10.41
135 Jeff Pfeffer	2408	97.5	91.0	10.0	.992	9.07		185 Bob Ewing	2302	67.5	69.9	7.7	1.003	9.12
136 Burt Hooton	2528	94.1	95.3	10.0	1.001	9.57		186 Harry Howell	2567	61.3	64.3	7.6	1.004	8.43
137 Earl Moore	2776	93.0	99.5	9.9	1.007	10.02		187 Dave McNally	2730	75.3	71.1	7.6	.996	9.32
138 Ned Garver	2477	77.6	100.2	9.9	1.020	10.10		188 Hooks Wiltse	2111	77.0	70.6	7.6	.990	9.30
139 Al Brazle	1375	92.7	98.1	9.9	1.009	9.91		189 Hank Aguirre	1376	67.5	73.8	7.6	1.011	9.73
140 Orval Overall	1532	80.1	90.3	9.9	1.022	9.16		190 Jess Barnes	2571	74.6	69.9	7.6	.995	9.23
141 Ray Kremer	1955	90.3	103.1	9.8	1.014	10.47		191 Ken Raffensberger	2152	71.2	74.4	7.5	1.004	9.87
142 Claude Hendrix	2372	104.7	97.6	9.7	.991	10.03		192 Ed Siever	1508	63.7	76.9	7.5	1.026	10.23
143 Tiny Bonham	1553	111.0	95.4	9.7	.976	9.86		193 Mario Soto	1250	54.9	70.4	7.4	1.031	9.46
144 Jim Maloney	1849	64.3	92.7	9.6	1.040	9.62		194 Larry Jansen	1767	87.5	76.0	7.4	.985	10.23
145 Frank Lary	2161	94.1	96.8	9.6	1.003	10.15		195 Johnny Vander Meer	2104	63.1	71.2	7.4	1.009	9.59
146 Gary Nolan	1675	88.1	90.8	9.6	1.004	9.45		196 George Uhle	3120	59.8	78.7	7.4	1.013	10.66
147 Doc White	3041	132.9	87.2	9.6	.951	9.08		197 Johnny Allen	1951	138.7	83.0	7.3	.942	11.30
148 Bob Veale	1926	95.8	89.0	9.6	.991	9.30		198 Bob Rush	2409	86.9	71.9	7.3	.986	9.81
149 Hal Schumacher	2483	117.0	94.9	9.5	.979	10.02		199 Gene Garber	1218	48.3	69.8	7.3	1.043	9.55
150 Mike Cuellar	2807	123.3	88.2	9.4	.968	9.41		200 Schoolboy Rowe	2218	92.0	74.2	7.1	.983	10.44

		IP	ERA	NERA	Adj	PF	LG			IP	ERA	NERA	Adj	PF	LG
1	Lefty Grove	3940	3.06	144.4	146.0	1.010	4.42	51	Thornton Lee	2331	3.56	118.4	119.2	1.007	4.21
2	Walter Johnson	5925	2.17	149.4	145.5	.974	3.24	52	Warren Spahn	5246	3.08	126.2	118.8	.941	3.89
3	Hoyt Wilhelm	2253	2.52	149.0	145.4	.976	3.76	53	Dutch Leonard	3220	3.25	123.0	118.8	.966	4.00
4	Ed Walsh	2965	1.82	151.9	143.6	.945	2.76	54	Billy Pierce	3305	3.27	120.8	118.7	.983	3.95
5	Three Finger Brown	3171	2.06	140.5	143.2	1.019	2.89	55	Dean Chance	2148	2.92	119.2	118.7	.995	3.48
6	Addie Joss	2327	1.89	143.9	140.8	.978	2.72	56	Gaylord Perry	5352	3.10	117.1	118.6	1.013	3.63
7	Cy Young	7357	2.63	134.9	136.3	1.010	3.54	57	Bob Lemon	2849	3.23	124.4	118.4	.952	4.02
8	Pete Alexander	5189	2.56	132.7	136.1	1.026	3.40	58	Red Faber	4086	3.15	120.5	118.3	.982	3.80
9	Christy Mathewson	4778	2.13	136.9	136.1	.994	2.92	59	Nap Rucker	2375	2.43	119.5	118.3	.990	2.90
10	Rube Waddell	2963	2.16	133.6	133.2	.997	2.88	60	Lon Warneke	2781	3.18	119.0	118.3	.994	3.79
11	Harry Brecheen	1905	2.92	131.2	133.2	1.015	3.83	61	Tiny Bonham	1553	3.06	121.0	118.1	.976	3.70
12	Whitey Ford	3171	2.74	140.0	132.3	.945	3.84	62	Preacher Roe	1916	3.43	114.9	118.1	1.028	3.94
13	Hal Newhouser	2993	3.06	125.2	132.1	1.055	3.83	63	Steve Rogers	2801	3.14	116.1	118.0	1.017	3.64
14	Sandy Koufax	2325	2.76	134.1	131.6	.981	3.70	64	Carl Mays	3021	2.92	122.1	118.0	.966	3.56
15	Dizzy Dean	1966	3.04	127.8	131.2	1.027	3.88	65	Jim Scott	1891	2.30	123.0	117.8	.958	2.83
16	Carl Hubbell	3591	2.98	133.1	130.5	.980	3.96	66	Ed Siever	1508	2.60	114.6	117.7	1.026	2.98
17	Tom Seaver	4367	2.80	128.9	129.6	1.006	3.61	67	Mel Parnell	1752	3.50	117.4	117.2	.998	4.11
18	Bob Gibson	3885	2.91	123.1	129.2	1.049	3.59	68	Mike Garcia	2176	3.26	122.1	117.1	.959	3.99
19	Bert Blyleven	3421	3.00	122.6	128.5	1.048	3.68	69	Jack Taylor	2617	2.66	117.1	116.9	.998	3.12
20	Stan Coveleski	3082	2.89	126.1	127.8	1.014	3.64	70	Wes Ferrell	2623	4.04	112.5	116.3	1.034	4.54
21	Ed Reulbach	2633	2.28	125.0	126.3	1.010	2.85	71	Bucky Walters	3104	3.30	113.3	116.2	1.026	3.74
22	Jim Palmer	3948	2.86	130.1	125.8	.966	3.72	72	Eddie Lopat	2439	3.21	121.3	116.2	.957	3.89
23	Sam Leever	2661	2.47	123.7	125.7	1.017	3.06	73	Gary Nolan	1675	3.08	115.4	115.9	1.004	3.55
24	Sal Maglie	1722	3.15	125.7	125.4	.997	3.96	74	Vic Willis	3997	2.63	114.6	115.6	1.009	3.01
25	Tommy Bridges	2827	3.57	122.9	125.1	1.018	4.39	75	Dolf Luque	3221	3.24	121.1	115.5	.954	3.93
26	Dizzy Trout	2726	3.23	118.1	124.7	1.056	3.82	76	Lefty Leifield	1839	2.47	115.2	115.5	1.003	2.85
27	Dazzy Vance	2967	3.24	126.3	124.7	.987	4.09	77	Jesse Tannehill	2751	2.79	114.1	115.5	1.012	3.18
28	Max Lanier	1618	3.01	123.4	124.3	1.008	3.71	78	Babe Adams	2995	2.75	117.8	115.4	.979	3.24
29	Orval Overall	1532	2.22	121.2	123.9	1.022	2.69	79	Don Newcombe	2154	3.56	111.4	115.3	1.036	3.96
30	Urban Shocker	2681	3.17	122.2	123.0	1.006	3.88	80	Frank Tanana	2354	3.26	118.6	115.1	.971	3.86
31	Lefty Gomez	2503	3.34	134.7	122.6	.910	4.50	81	Firpo Marberry	2066	3.63	118.6	115.1	.970	4.31
32	Mort Cooper	1843	2.96	121.8	122.3	1.004	3.61	82	Tug McGraw	1516	3.13	115.7	115.1	.994	3.63
33	Eddie Cicotte	3223	2.37	124.0	122.3	.986	2.94	83	Al Benton	1689	3.66	110.9	115.0	1.038	4.05
34	Deacon Phillippe	2608	2.59	120.1	121.7	1.013	3.11	84	Jeff Tesreau	1679	2.43	119.6	115.0	.961	2.91
35	Hippo Vaughn	2731	2.48	116.4	121.6	1.045	2.89	85	Curt Davis	2324	3.42	110.8	114.9	1.036	3.79
36	Andy Messersmith	2230	2.86	124.5	121.6	.977	3.56	86	Don Mossi	1548	3.43	112.6	114.7	1.019	3.86
37	Rollie Fingers	1646	2.83	129.3	121.4	.939	3.66	87	Johnny Antonelli	1992	3.34	119.1	114.7	.963	3.98
38	Eddie Rommel	2557	3.54	117.8	121.4	1.030	4.17	88	Ferguson Jenkins	4498	3.34	109.6	114.7	1.047	3.66
39	Ron Guidry	1768	3.16	126.6	121.3	.958	4.00	89	Claude Passeau	2718	3.32	111.8	114.6	1.025	3.71
40	Juan Marichal	3506	2.89	123.3	121.0	.982	3.56	90	Tommy John	4123	3.19	116.0	114.6	.988	3.70
41	Joe McGinnity	3441	2.65	118.3	121.0	1.023	3.14	91	Frank Sullivan	1732	3.60	108.5	114.6	1.056	3.90
42	Bob Feller	3828	3.25	127.8	121.0	.946	4.16	92	Wilbur Cooper	3482	2.89	115.1	114.6	.995	3.33
43	Eddie Plank	4507	2.34	122.9	120.7	.982	2.88	93	Eppa Rixey	4494	3.15	115.9	114.4	.987	3.65
44	Bobby Shantz	1936	3.38	117.3	120.7	1.029	3.96	94	Jim Maloney	1849	3.19	109.8	114.2	1.040	3.50
45	Don Drysdale	3432	2.95	123.7	120.6	.975	3.65	95	Bob Shawkey	2938	3.09	115.0	114.1	.992	3.55
46	Steve Carlton	4787	3.04	118.4	120.1	1.015	3.59	96	Dutch Leonard	2191	2.76	116.8	114.0	.976	3.22
47	John Candelaria	1800	3.10	117.7	120.0	1.020	3.65	97	Jon Matlack	2363	3.18	117.7	114.0	.969	3.74
48	Virgil Trucks	4835	3.19	113.2	119.5	1.056	3.61	98	Claude Hendrix	2372	2.65	115.0	114.0	.991	3.05
49	Virgil Trucks	2684	3.38	114.0	119.4	1.047	3.86	99	Bill Hands	1951	3.35	106.2	114.0	1.073	3.56
50	Ted Lyons	4162	3.67	118.5	119.4	1.007	4.35	100	Slim Sallee	2819	2.56	115.5	113.9	.986	2.96

		W	L	Pct	Team	Wins			W	L	Pct	Team	Wins
1	Cy Young	510	313	.620	.495	102.3	51	Firpo Marberry	148	88	.627	.541	20.3
2	Walter Johnson	417	279	.599	.460	96.6	52	Tommy John	255	197	.564	.519	20.2
3	Pete Alexander	373	208	.642	.498	83.5	53	George Uhle	200	166	.546	.491	20.2
4	Tom Seaver	288	181	.614	.480	63.0	54	Herb Pennock	240	162	.597	.547	20.0
5	Christy Mathewson	372	189	.663	.560	58.0	55	Camilo Pascual	174	170	.506	.449	19.5
6	Lefty Grove	300	141	.680	.561	52.6	56	Sal Maglie	119	62	.657	.552	19.2
7	Steve Carlton	313	207	.602	.515	44.9	57	Luis Tiant	229	172	.571	.523	19.1
8	Warren Spahn	363	245	.597	.526	43.5	58	Tommy Bridges	194	138	.584	.528	18.8
9	Ted Lyons	260	230	.531	.442	43.3	59	Jeff Pfeffer	158	112	.585	.516	18.7
10	Phil Niekro	284	238	.544	.463	42.3	60	Larry Dierker	139	123	.531	.460	18.5
11	Juan Marichal	243	142	.631	.534	37.2	61	Wild Bill Donovan	185	133	.582	.524	18.5
12	Dazzy Vance	197	140	.585	.477	36.3	62	Art Nehf	184	120	.605	.545	18.4
13	Wes Ferrell	193	128	.601	.489	36.0	63	Sam Leever	194	100	.660	.598	18.3
14	Whitey Ford	236	106	.690	.587	35.1	64	Earl Whitehill	218	185	.541	.496	18.2
15	Jesse Tannehill	186	110	.628	.512	34.5	65	Hal Newhouser	207	150	.580	.529	18.2
16	Bob Feller	266	162	.621	.541	34.3	66	Virgil Trucks	177	135	.567	.509	18.2
17	Robin Roberts	286	245	.539	.476	33.3	67	Jim Bunning	224	184	.549	.504	18.2
18	Eddie Rommel	171	119	.590	.478	32.5	68	Ray Kremer	143	85	.627	.549	17.9
19	Eddie Plank	326	193	.628	.566	32.1	69	Babe Adams	194	140	.581	.527	17.9
20	Carl Hubbell	253	154	.622	.544	31.8	70	Rube Waddell	194	141	.579	.527	17.6
21	Bob Gibson	251	174	.591	.516	31.7	71	Lon Warneke	192	121	.613	.558	17.5
22	Ed Walsh	195	126	.607	.510	31.2	72	Chief Bender	212	128	.624	.573	17.1
23	Urban Shocker	187	117	.615	.513	31.1	73	Lee Meadows	188	180	.511	.465	17.1
24	Nap Rucker	134	134	.500	.387	30.4	74	Stan Coveleski	215	142	.602	.555	17.0
25	Ferguson Jenkins	284	226	.557	.500	29.0	75	Andy Messersmith	130	99	.568	.494	16.8
26	Addie Joss	160	97	.623	.514	28.0	76	Freddie Fitzsimmons	217	146	.598	.552	16.8
27	Red Lucas	157	135	.538	.442	28.0	77	Hooks Dauss	222	182	.550	.509	16.5
28	Red Faber	254	213	.544	.484	27.8	78	Ned Garver	129	157	.451	.393	16.5
29	Sandy Koufax	165	87	.655	.544	27.8	79	Frank Tanana	135	130	.509	.448	16.4
30	Dutch Leonard	191	181	.513	.439	27.7	80	Jim Hunter	224	166	.574	.533	16.2
31	Joe McGinnity	247	144	.632	.563	26.8	81	Bobby Shantz	119	99	.546	.472	16.2
32	Rip Sewell	143	97	.596	.484	26.7	82	Vida Blue	191	143	.572	.524	16.1
33	Schoolboy Rowe	158	101	.610	.508	26.4	83	Fred Frankhouse	106	97	.522	.443	16.1
34	Casey Patten	105	128	.451	.343	25.2	84	Joe Niekro	193	167	.536	.491	16.1
35	Ron Guidry	132	62	.680	.551	25.0	85	Don Newcombe	149	90	.623	.557	15.8
36	Jim Palmer	268	152	.638	.579	24.8	86	Wilbur Wood	164	156	.513	.463	15.8
37	Dizzy Dean	150	83	.644	.537	24.8	87	Allie Reynolds	182	107	.630	.575	15.7
38	Claude Passeau	162	150	.519	.440	24.7	88	Carl Mays	207	126	.622	.575	15.6
39	Gaylord Perry	314	265	.542	.501	24.2	89	Sam McDowell	141	134	.513	.457	15.4
40	Hippo Vaughn	178	137	.565	.490	23.7	90	Denny McLain	131	91	.590	.521	15.3
41	Three Finger Brown	239	130	.648	.586	22.9	91	Fred Toney	139	102	.577	.515	15.0
42	J.R. Richard	107	71	.601	.474	22.6	92	Vic Raschi	132	66	.667	.591	14.9
43	Slim Sallee	173	143	.547	.477	22.2	93	Mel Stottlemyre	164	139	.541	.492	14.9
44	Wilbur Cooper	216	178	.548	.494	21.4	94	Van Mungo	120	115	.511	.448	14.7
45	Jack Chesbro	197	132	.599	.534	21.2	95	John Candelaria	122	80	.604	.531	14.6
46	Burleigh Grimes	270	212	.560	.516	21.1	96	Milt Pappas	209	164	.560	.521	14.6
47	Jim Maloney	134	84	.615	.518	21.0	97	George Suggs	99	91	.521	.446	14.3
48	Bucky Walters	198	160	.553	.495	20.7	98	Vern Law	162	147	.524	.479	14.1
49	Johnny Allen	142	75	.654	.561	20.4	99	Don Sutton	280	218	.562	.534	13.9
50	Nolan Ryan	231	206	.529	.482	20.4	100	Rick Reuschel	139	131	.515	.464	13.8

	IP	ERA	NERA	Adj	PF	LG		IP	ERA	NERA	Adj	PF	LG
1 Bruce Sutter	888	2.54	143.0	151.4	1.059	3.64	26 Stu Miller	1694	3.24	117.5	113.2	.963	3.81
2 Hoyt Wilhelm	2253	2.52	149.0	145.4	.976	3.76	27 Clem Labine	1079	3.63	109.3	112.6	1.031	3.96
3 Kent Tekulve	1014	2.64	138.0	141.5	1.025	3.64	28 Al McBean	1072	3.13	112.0	112.5	1.004	3.51
4 John Hiller	1241	2.84	128.7	135.2	1.051	3.65	29 Hugh Casey	941	3.44	109.1	112.2	1.029	3.76
5 Rich Gossage	1338	2.86	131.6	130.0	.988	3.76	30 Elias Sosa	918	3.32	111.2	112.1	1.009	3.69
6 Gary Lavelle	981	2.82	129.2	128.9	.997	3.64	31 Bill Campbell	1099	3.49	107.4	111.8	1.041	3.75
7 Sparky Lyle	1391	2.88	127.5	126.9	.995	3.67	32 Jim Konstanty	946	3.46	115.0	111.5	.970	3.98
8 Frank Linzy	817	2.85	122.6	123.9	1.011	3.50	33 Terry Forster	1005	3.28	110.1	111.3	1.011	3.61
9 Bob Locker	878	2.76	124.0	123.2	.994	3.42	34 Jim Brosnan	832	3.54	109.6	111.0	1.012	3.88
10 Rollie Fingers	1646	2.83	129.3	121.4	.939	3.66	35 Mace Brown	1075	3.47	110.3	110.4	1.001	3.82
11 Ron Perranoski	1176	2.79	127.4	121.4	.953	3.55	36 Dick Hall	1259	3.32	111.9	110.2	.985	3.71
12 Jim Kern	755	3.11	124.3	121.3	.976	3.87	37 Jim Brewer	1041	3.07	116.7	109.9	.942	3.58
13 Bill Henry	914	3.26	115.6	119.9	1.037	3.77	38 Roy Face	1375	3.48	109.0	109.9	1.008	3.79
14 Rick Camp	814	3.28	110.4	119.3	1.081	3.63	39 Al Worthington	1245	3.39	108.7	109.0	1.003	3.68
15 Clay Carroll	1353	2.94	121.2	119.2	.984	3.56	40 Joe Heving	1039	3.90	109.0	108.5	.995	4.25
16 Ted Wilks	913	3.26	116.6	118.7	1.018	3.81	41 Lindy McDaniel	2140	3.45	106.8	107.8	1.009	3.69
17 Mike Marshall	1387	3.14	117.7	118.2	1.005	3.70	42 Tom Hall	854	3.27	109.5	107.6	.983	3.58
18 Don McMahon	1313	2.95	122.5	117.9	.963	3.62	43 Fritz Dorish	836	3.82	104.1	107.3	1.030	3.98
19 Tom Burgmeier	1258	3.23	116.2	117.8	1.014	3.76	44 Dave LaRoche	1049	3.53	106.5	106.6	1.001	3.76
20 Gene Garber	1218	3.31	110.8	115.6	1.043	3.67	45 Bob Miller	1552	3.37	106.7	106.6	.999	3.59
21 Marv Grissom	809	3.42	115.2	115.1	.999	3.93	46 Don Elston	755	3.70	103.5	106.4	1.028	3.82
22 Ed Roebuck	789	3.35	112.6	115.1	1.022	3.78	47 Ted Abernathy	1146	3.46	102.5	106.3	1.037	3.55
23 Tug McGraw	1516	3.13	115.7	115.1	.994	3.63	48 Tom Morgan	1025	3.60	107.9	105.8	.980	3.89
24 Johnny Murphy	1044	3.50	124.0	114.5	.923	4.34	49 Grant Jackson	1359	3.46	105.1	105.2	1.001	3.64
25 Darold Knowles	1091	3.12	114.1	114.4	1.003	3.56	50 Pete Richert	1164	3.19	108.1	104.9	.971	3.45

	IP	All	Pit	Bat	Fld	Rel		IP	All	Pit	Bat	Fld	Rel
1 Rollie Fingers	1646	12.2	11.5	.2	.5	765	26 Tom Burgmeier	1258	10.9	8.1	.3	2.5	309
2 Hoyt Wilhelm	2253	27.9	29.7	-2.0	.2	600	27 Dick Radatz	694	3.6	4.6	-.4	-.6	305
3 Sparky Lyle	1391	12.2	12.4	.4	-.6	598	28 Jack Aker	747	1.7	.7	-.3	1.3	295
4 Bruce Sutter	888	13.4	13.5	-.4	.3	576	29 Frank Linzy	817	9.0	6.6	.1	2.3	289
5 Rich Gossage	1338	12.0	12.6	-.1	-.5	569	30 Tippy Martinez	745	6.5	5.5	0.0	1.0	288
6 Roy Face	1375	5.0	5.2	-.5	.3	496	31 Al Hrabosky	721	4.2	5.1	.1	-1.0	287
7 Lindy McDaniel	2140	8.8	6.9	-.5	2.4	494	32 Terry Forster	1005	7.3	4.4	1.4	1.5	282
8 Tug McGraw	1516	9.0	8.4	.7	-.1	470	33 Fred Gladding	600	1.7	2.6	-.6	-.3	280
9 Mike Marshall	1387	11.8	9.3	.5	2.0	463	34 Al Worthington	1245	4.6	4.5	-.6	.7	278
10 Ron Perranoski	1176	8.0	8.1	-.3	.2	442	35 Bob Stanley	1204	14.1	11.9	0.0	2.2	277
11 Don McMahon	1313	7.9	7.9	-.3	.3	420	36 Ron Kline	2078	-2.4	.9	-2.9	-.4	277
12 Dan Quisenberry	635	11.8	10.2	0.0	1.6	406	37 Ron Reed	2477	5.4	6.1	-.4	-.3	273
13 Gene Garber	1218	8.9	7.2	-.0	1.7	405	38 Clem Labine	1079	5.0	5.3	-1.0	.7	273
14 Stu Miller	1694	10.3	8.3	-.2	2.2	399	39 Bob Locker	878	8.3	6.8	-.3	1.8	265
15 Kent Tekulve	1014	14.2	13.0	-.6	1.8	395	40 Ellis Kinder	1481	7.2	10.4	-1.6	-1.6	262
16 Clay Carroll	1353	10.1	8.6	-.6	2.1	394	41 Pedro Borbon	1026	.5	.7	.3	-.5	261
17 Dave Giusti	1718	-2.7	-5.0	2.0	.3	363	42 Phil Regan	1373	-2.8	-2.5	.2	-.5	260
18 Ted Abernathy	1146	5.9	3.1	-.3	3.1	362	43 Greg Minton	682	4.7	3.7	.1	.9	255
19 Darold Knowles	1091	7.8	5.9	0.0	1.9	343	44 Ron Davis	569	2.8	3.1	0.0	-.3	252
20 Jim Brewer	1041	3.5	3.8	.2	-.5	339	45 Wayne Granger	640	2.8	2.0	-.1	.9	251
21 John Hiller	1241	13.1	14.5	-.5	-.9	337	46 Grant Jackson	1359	1.6	2.9	-.2	-1.1	244
22 Gary Lavelle	981	9.0	9.2	-.6	.4	336	47 John Wyatt	687	1.9	2.8	-.6	-.3	243
23 Bill Campbell	1099	5.8	5.2	0.0	.6	327	48 Joe Hoerner	563	3.5	4.5	-.1	-.9	242
24 Dave LaRoche	1049	3.3	2.7	.5	.1	326	49 Turk Farrell	1704	.4	2.3	-.1	-1.8	239
25 Johnny Murphy	1044	6.2	5.6	-.3	.9	318	50 Aurelio Lopez	708	3.3	3.7	0.0	-.4	237

		IP	All	Pit	Bat	Fld			IP	All	Pit	Bat	Fld
1	Walter Johnson	5925	73.5	66.6	11.1	-4.2	51	Jack Quinn	3920	20.4	17.5	-.7	3.6
2	Cy Young	7357	69.7	72.6	.8	-3.7	52	Don Newcombe	2154	20.2	13.1	7.8	-.7
3	Pete Alexander	5189	60.4	55.2	2.4	2.8	53	Bobby Shantz	1936	20.1	15.0	1.0	4.1
4	Christy Mathewson	4778	53.4	41.9	5.3	6.2	54	Dizzy Dean	1966	19.7	20.4	1.2	-1.9
5	Lefty Grove	3940	52.0	57.1	-3.5	-1.6	55	Ed Reulbach	2633	19.3	18.4	-1.4	2.3
6	Tom Seaver	4367	46.1	42.5	2.8	.8	56	Joe Wood	1436	19.2	15.7	2.9	.6
7	Bob Gibson	3885	44.8	38.9	6.7	-.8	57	Joe McGinnity	3441	19.2	19.5	-.9	.6
8	Warren Spahn	5246	41.9	33.0	8.0	.9	58	Larry Jackson	3262	19.1	17.1	-.4	2.4
9	Phil Niekro	4835	38.0	35.3	-.4	3.1	59	Murry Dickson	3053	19.0	13.1	2.5	3.4
10	Steve Carlton	4787	37.6	34.3	4.9	-1.6	60	Steve Rogers	2801	18.8	18.3	-.9	1.4
11	Ed Walsh	2965	37.3	28.4	1.3	7.6	61	Sandy Koufax	2325	18.7	23.9	-2.6	-2.6
12	Gaylord Perry	5352	37.1	36.3	-1.8	2.6	62	Jim Kaat	4529	18.2	13.3	4.4	.5
13	Hal Newhouser	2993	36.7	33.4	.9	2.4	63	Wilbur Cooper	3482	18.2	17.0	4.2	-3.0
14	Carl Hubbell	3591	36.7	35.2	-.9	2.4	64	Vic Willis	3997	18.0	17.1	-1.6	2.5
15	Whitey Ford	3171	36.5	31.8	2.0	2.7	65	Andy Messersmith	2230	17.9	16.0	1.9	-.0
16	Ted Lyons	4162	35.0	30.9	2.9	1.2	66	Spud Chandler	1485	17.6	13.5	1.6	2.5
17	Three Finger Brown	3171	34.5	33.4	1.3	-.2	67	Hippo Vaughn	2731	17.4	18.3	-.4	-.5
18	Jim Palmer	3948	34.4	33.2	.5	.7	68	Dave Stieb	1389	17.4	15.9	0.0	1.5
19	Don Drysdale	3432	33.3	24.4	5.9	3.0	69	Claude Passeau	2718	17.3	15.2	1.2	.9
20	Bob Lemon	2849	33.1	18.7	8.3	6.1	70	Luis Tiant	3486	17.2	18.4	.5	-1.7
21	Bert Blyleven	3421	32.6	33.8	-1.4	.2	71	Jesse Tannehill	2751	17.2	11.9	4.3	1.0
22	Dizzy Trout	2726	32.1	24.6	3.2	4.3	72	Curt Davis	2324	17.1	13.2	1.0	2.9
23	Carl Mays	3021	31.9	18.0	6.0	7.9	73	Virgil Trucks	2684	16.9	19.6	-2.2	-.5
24	Ferguson Jenkins	4498	28.7	25.9	1.8	1.0	74	Jack Taylor	2617	16.7	13.8	2.3	.6
25	Wes Ferrell	2623	28.6	17.7	10.8	.1	75	Eddie Lopat	2439	16.6	13.6	2.6	.4
26	Hoyt Wilhelm	2253	27.9	29.7	-2.0	.2	76	Red Lucas	2543	16.5	6.9	10.8	-1.2
27	Bucky Walters	3104	27.8	18.7	6.3	2.8	77	Rick Reuschel	2362	16.3	12.0	.8	3.5
28	Tommy John	4123	27.5	22.3	.1	5.1	78	Claude Hendrix	2372	16.3	9.6	5.1	1.6
29	Juan Marichal	3506	27.4	25.0	1.2	1.2	79	Doc White	3041	16.2	9.5	3.9	2.8
30	Tommy Bridges	2827	25.3	26.3	-.5	-.5	80	Thornton Lee	2331	16.0	16.8	-.0	-.8
31	Bob Feller	3828	25.1	28.0	-.7	-2.2	81	Jerry Koosman	3740	16.0	18.9	-2.4	-.5
32	Red Faber	4086	24.2	25.7	-3.0	1.5	82	Early Wynn	4566	15.6	11.0	6.7	-2.1
33	Stan Coveleski	3082	24.2	26.5	-3.6	1.3	83	Larry French	3152	15.5	16.3	-.1	-.7
34	Robin Roberts	4689	23.8	23.8	3.3	-3.3	84	George Mullin	3686	15.3	2.1	10.6	2.6
35	Eddie Plank	4507	23.8	25.7	1.7	-3.6	85	Sam Leever	2661	15.2	17.8	-.7	-1.9
36	Eddie Rommel	2557	23.6	20.3	-.8	4.1	86	Lefty Gomez	2503	15.2	19.5	-2.8	-1.5
37	Addie Joss	2327	23.1	21.7	-1.4	2.8	87	Harry Howell	2567	14.8	7.6	4.0	3.2
38	Freddie Fitzsimmons	3225	23.1	15.8	1.6	5.7	88	Jim Bunning	3759	14.8	19.0	-.0	-4.2
39	Dazzy Vance	2967	22.9	24.5	-1.7	.1	89	George Uhle	3120	14.6	7.2	8.8	-1.4
40	Red Ruffing	4342	22.7	13.5	13.0	-3.8	90	Hal Schumacher	2483	14.6	9.6	2.2	2.8
41	Urban Shocker	2681	22.5	20.9	2.8	-1.2	91	Schoolboy Rowe	2218	14.6	7.0	7.6	.0
42	Burleigh Grimes	4181	22.3	12.0	5.0	5.3	92	Mel Harder	3426	14.3	15.0	-2.4	1.7
43	Harry Brecheen	1905	22.2	20.8	1.0	.4	93	Kent Tekulve	1014	14.2	13.0	-.6	1.8
44	Eddie Cicotte	3223	21.9	20.2	.3	1.4	94	Mel Stottlemyre	2662	14.2	8.0	2.3	3.9
45	Dolf Luque	3221	21.8	17.0	3.6	1.2	95	Fred Hutchinson	1465	14.2	8.9	3.8	1.5
46	Dutch Leonard	3220	21.6	21.2	-1.9	2.3	96	Waite Hoyt	3763	14.2	16.7	-2.0	-.5
47	Eppa Rixey	4494	21.4	21.0	-1.4	1.8	97	Bob Stanley	1204	14.1	11.9	0.0	2.2
48	Billy Pierce	3305	20.8	22.4	.0	-1.6	98	Ned Garver	2477	14.0	9.9	3.2	.9
49	Lon Warneke	2781	20.5	18.1	2.7	-.3	99	Babe Adams	2995	13.9	14.4	2.2	-2.7
50	Rube Waddell	2963	20.4	24.2	-1.6	-2.2	100	Frank Tanana	2354	13.7	13.2	0.0	.5

		IP	All	Pit	Bat	Fld			IP	All	Pit	Bat	Fld
101	Deacon Phillippe	2608	13.6	14.7	-.0	-1.1	151	Firpo Marberry	2066	9.8	11.8	-.9	-1.1
102	Max Lanier	1618	13.6	13.5	-.3	.4	152	Johnny Rigney	1188	9.7	10.6	-.8	-.1
103	Wilbur Wood	2684	13.5	13.5	-1.5	1.5	153	Jim Perry	3287	9.5	8.2	2.4	-1.1
104	Mike Garcia	2176	13.5	13.4	-.2	.3	154	Van Mungo	2111	9.4	6.7	1.9	.8
105	Bruce Sutter	888	13.4	13.5	-.4	.3	155	Lefty Leifield	1839	9.4	8.1	.6	.7
106	Sal Maglie	1722	13.3	15.1	-1.9	.1	156	Harry Coveleski	1248	9.4	7.8	-.2	1.8
107	Ron Guidry	1768	13.3	13.3	0.0	.0	157	Fernando Valenzuela	1013	9.3	6.9	1.0	1.4
108	John Candelaria	1800	13.2	12.8	1.4	-1.0	158	Jess Barnes	2571	9.3	7.5	-.1	1.9
109	John Hiller	1241	13.1	14.5	-.5	-.9	159	Joe Horlen	2003	9.2	8.0	-1.4	2.6
110	Mort Cooper	1843	13.1	14.4	.5	-1.8	160	Dennis Eckersley	2125	9.2	10.2	-.3	-.7
111	Bob Shawkey	2938	13.0	13.8	.1	-.9	161	Al Brazle	1375	9.2	9.8	-1.1	.5
112	Babe Ruth	1221	12.4	7.1	4.7	.6	162	Preacher Roe	1916	9.1	13.0	-3.2	-.7
113	Johnny Antonelli	1992	12.3	10.7	1.5	.1	163	Fritz Ostermueller	2069	9.1	7.7	1.9	-.5
114	Jim Tobin	1899	12.2	4.5	5.8	1.9	164	Don Mossi	1548	9.0	8.8	.3	-.1
115	Don Sutton	4570	12.2	14.8	-.7	-1.9	165	Tug McGraw	1516	9.0	8.4	.7	-.1
116	Sparky Lyle	1391	12.2	12.4	.4	-.6	166	Frank Linzy	817	9.0	6.6	.1	2.3
117	Rollie Fingers	1646	12.2	11.5	.2	.5	167	Gary Lavelle	981	9.0	9.2	-.6	.4
118	Chief Bender	3017	12.2	9.0	3.8	-.6	168	Gene Garber	1218	8.9	7.2	-.0	1.7
119	Nolan Ryan	3705	12.1	14.4	-.8	-1.5	169	Frank Sullivan	1732	8.8	10.1	-1.7	.4
120	Noodles Hahn	1409	12.1	13.6	0.0	-1.5	170	Lindy McDaniel	2140	8.8	6.9	-.5	2.4
121	Howie Pollet	2105	12.0	11.3	.9	-.2	171	Burt Hooton	2528	8.8	10.1	-1.1	-.2
122	Rich Gossage	1338	12.0	12.6	-.1	-.5	172	Slim Sallee	2819	8.6	11.5	-.8	-2.1
123	Jon Matlack	2363	11.9	12.1	.2	-.4	173	Tex Hughson	1375	8.6	10.7	-1.7	-.4
124	Dan Quisenberry	635	11.8	10.2	0.0	1.6	174	Bill Hands	1951	8.5	10.8	-2.3	-.0
125	Mike Marshall	1387	11.8	9.3	.5	2.0	175	Red Ames	3197	8.5	8.2	-2.4	2.7
126	Harvey Haddix	2236	11.8	8.2	3.5	.1	176	John Denny	1747	8.4	5.4	.2	2.8
127	Claude Osteen	3459	11.7	5.7	3.9	2.1	177	Charlie Root	3197	8.3	10.9	.4	-3.0
128	Jim Maloney	1849	11.6	9.7	3.0	-1.1	178	Jeff Pfeffer	2408	8.3	9.9	.5	-2.1
129	Russ Ford	1487	11.3	10.7	1.1	-.5	179	Gary Nolan	1675	8.3	9.6	.2	-1.5
130	Curt Simmons	3348	11.2	13.2	.2	-2.2	180	Bob Locker	878	8.3	6.8	-.3	1.8
131	Nap Rucker	2375	11.2	12.2	.2	-1.2	181	Paul Derringer	3646	8.3	12.1	-2.1	-1.7
132	Sonny Siebert	2152	11.1	8.5	1.9	.7	182	Doc Crandall	1548	8.3	2.1	5.7	.5
133	Milt Pappas	3187	11.1	12.8	-1.4	-.3	183	Lefty Tyler	2228	8.2	2.8	3.5	1.9
134	Mel Parnell	1752	11.0	11.3	-.3	-.0	184	Jeff Tesreau	1679	8.2	6.8	1.6	-.2
135	Art Nehf	2708	11.0	6.1	3.1	1.8	185	Bob Rush	2409	8.2	7.5	.4	.3
136	Tom Burgmeier	1258	10.9	8.1	.3	2.5	186	Clint Brown	1485	8.1	5.9	.5	1.7
137	Camilo Pascual	2930	10.8	6.0	3.1	1.7	187	Dave Rozema	1007	8.0	7.7	0.0	.3
138	Dean Chance	2148	10.8	14.0	-4.0	.8	188	Ron Perranoski	1176	8.0	8.1	-.3	.2
139	Hooks Dauss	3391	10.7	3.9	3.1	3.7	189	Vida Blue	3056	8.0	10.3	-.4	-1.9
140	Sam McDowell	2492	10.6	11.9	-.8	-.5	190	Rip Sewell	2119	7.9	5.7	1.7	.5
141	Sherry Smith	2053	10.5	4.9	2.5	3.1	191	Don McMahon	1313	7.9	7.9	-.3	.3
142	Frank Lary	2161	10.4	9.6	.6	.2	192	Tom Zachary	3128	7.8	5.9	2.6	-.7
143	Ewell Blackwell	1322	10.4	9.3	-.3	1.4	193	Rick Rhoden	1650	7.8	4.2	3.3	.3
144	Stu Miller	1694	10.3	8.3	-.2	2.2	194	Darold Knowles	1091	7.8	5.9	0.0	1.9
145	Jake Weimer	1473	10.2	8.6	1.3	.3	195	Mike Cuellar	2807	7.8	9.5	-1.3	-.4
146	Jack Chesbro	2898	10.2	10.9	.4	-1.1	196	Joe Benz	1360	7.8	8.0	-2.3	2.1
147	Clay Carroll	1353	10.1	8.6	-.6	2.1	197	Bill Lee	2862	7.7	7.5	-1.4	1.6
148	Hooks Wiltse	2111	10.0	7.6	2.1	.3	198	Larry Jansen	1767	7.7	7.4	-.6	.9
149	Gary Peters	2081	9.9	4.0	6.0	-.1	199	Dick Rudolph	2048	7.6	4.4	1.3	1.9
150	Orval Overall	1532	9.9	9.9	.8	-.8	200	Johnny Sain	2125	7.5	3.9	3.8	-.2

Batter	LWTS		NOPS		OBA		SLG		NOBA		NSLG		Wins	
Cap Anson	594.1	2	153	7	.394		.446		1.263	6	1.274		53.9	2
Dan Brouthers	671.6	1	179	1	.423	3	.519	1	1.333	2	1.463	1	60.1	1
Pete Browning	378.0	8	166	2	.403	8	.466	5	1.290	4	1.373	3	33.3	9
Jesse Burkett	559.1	5	148		.412	5	.446		1.248	8	1.232		51.2	5
Oyster Burns	210.2		138		.368		.446		1.119		1.260		18.6	
Clarence Childs	258.8		128		.414	4	.389		1.216		1.063		22.6	
Roger Connor	590.0	3	161	3	.397	9	.485	4	1.250	7	1.360	6	53.0	3
Ed Delahanty	575.1	4	158	4	.409	6	.505	3	1.211		1.366	4	51.4	4
Hugh Duffy	317.0		134		.384		.448	10	1.126		1.208		27.1	
Fred Dunlap	153.2		136		.339		.406		1.163		1.202		14.0	
Buck Ewing	230.3		138		.351		.455	9	1.103		1.274	10	20.7	
George Gore	317.7		147		.386		.411		1.276	5	1.191		28.9	
Billy Hamilton	513.1	6	151	9	.454	2	.432		1.338	1	1.176		45.1	6
Mike Kelly	341.3		150	10	.367		.438		1.219	10	1.276	9	31.3	
Denny Lyons	264.3		145		.406	7	.443		1.208		1.241		22.9	
John McGraw	288.3		143		.456	1	.410		1.333	3	1.094		25.8	
Tip O'Neill	291.0		156	5	.393		.458	8	1.228	9	1.335	7	25.4	
Jim O'Rourke	331.0		140		.355		.422		1.172		1.228		30.0	
John Reilly	172.4		136		.325		.437		1.061		1.303	8	15.5	
Hardy Richardson	247.2		139		.344		.436		1.132		1.255		22.3	
Jimmy Ryan	347.3	10	134		.372		.444		1.121		1.218		31.4	10
Elmer Smith	215.2		131		.396	10	.434		1.155		1.158		18.7	
Harry Stovey	373.8	9	153	8	.360		.461	7	1.168		1.362	5	33.5	8
Sam Thompson	399.3	7	154	6	.384		.506	2	1.154		1.385	2	34.7	7
Mike Tiernan	332.7		143		.391		.464	6	1.163		1.269		29.6	

Pitcher	LWTS		NERA		WAT		Wins	
Tommy Bond	82.1		116		5.3		7.2	
Ted Breitenstein	94.1		109		26.4		7.4	
Charlie Buffington	159.0		119		33.3	10	14.2	
Bob Caruthers	214.0	8	129	6	23.6		18.8	10
John Clarkson	368.3	4	129	7	18.8		33.1	4
George Cuppy	198.3		124		13.9		17.1	
Jim Galvin	156.1		111		70.6	2	13.7	
Clark Griffith	239.5	7	125		46.8	5	21.7	7
Guy Hecker	152.4		130	5	34.3	9	14.1	
Tim Keefe	400.2	3	135	3	13.1		36.7	2
Silver King	205.7		129	9	2.6		17.9	
Jim McCormick	212.3	10	123		71.8	1	19.6	9
Sadie McMahon	167.7		103		27.9		14.2	
Ed Morris	176.2		127	10	33.3		16.4	
Tony Mullane	248.4	6	122		43.5	6	22.4	6
Kid Nichols	531.8	1	136	2	29.0		47.1	1
Charlie Radbourn	298.5	5	134	4	49.9	4	26.8	5
Amos Rusie	415.7	2	137	1	35.0	8	35.0	3
Monte Ward	136.8		127		6.8		12.9	
Mickey Welch	212.9	9	119		35.1	7	19.6	8
Gus Weyhing	63.3		108		24.9		5.5	
Will White	198.2		129	8	62.7	3	18.2	

		Year	G	LWT	Adj	Wins	PF	R/W
1	Babe Ruth	1921	152	119.3	116.7	10.91	1.028	10.70
2	Babe Ruth	1920	142	113.2	111.7	10.82	1.019	10.33
3	Babe Ruth	1923	152	119.1	110.7	10.71	1.095	10.34
4	Ted Williams	1941	143	102.0	101.4	9.82	1.009	10.32
5	Babe Ruth	1926	152	97.4	100.6	9.74	.961	10.33
6	Lou Gehrig	1927	155	100.8	101.5	9.64	.992	10.53
7	Babe Ruth	1927	151	100.7	101.4	9.63	.992	10.53
8	Rogers Hornsby	1924	143	94.1	96.9	9.62	.963	10.08
9	Babe Ruth	1924	153	100.7	100.9	9.52	.998	10.59
10	Mickey Mantle	1957	144	88.8	90.9	9.39	.970	9.68
11	Ted Williams	1942	150	92.6	91.5	9.37	1.014	9.76
12	Ted Williams	1947	156	91.1	90.0	9.33	1.014	9.65
13	Ted Williams	1946	150	94.2	88.5	9.28	1.078	9.54
14	Stan Musial	1948	155	90.2	89.8	8.97	1.005	10.01
15	Ted Williams	1957	132	89.9	85.1	8.79	1.079	9.68
16	Babe Ruth	1928	154	83.9	90.9	8.78	.919	10.35
17	Babe Ruth	1931	145	91.5	94.1	8.76	.971	10.74
18	Lou Gehrig	1934	154	85.6	93.6	8.71	.912	10.75
19	Babe Ruth	1930	145	89.5	96.3	8.71	.927	11.06
20	Ted Williams	1949	155	88.8	89.4	8.70	.994	10.28
21	Jimmie Foxx	1933	149	82.8	91.3	8.64	.901	10.56
22	Lou Gehrig	1930	154	88.4	95.5	8.63	.927	11.06
23	Norm Cash	1961	159	86.1	85.8	8.51	1.003	10.09
24	Rogers Hornsby	1925	138	87.1	89.0	8.33	.976	10.68
25	Rogers Hornsby	1922	154	90.0	88.1	8.32	1.020	10.60
26	Mickey Mantle	1956	150	83.2	84.7	8.29	.982	10.22
27	Mickey Mantle	1961	153	76.3	83.4	8.26	.908	10.09
28	Lou Gehrig	1928	154	76.0	82.8	8.00	.919	10.35
29	Rogers Hornsby	1921	154	74.4	80.9	7.98	.921	10.13
30	Willie McCovey	1969	149	76.1	75.8	7.96	1.004	9.53
31	Ty Cobb	1917	152	74.8	71.4	7.94	1.051	8.99
32	Jimmie Foxx	1932	154	96.7	85.7	7.93	1.117	10.81
33	Ted Williams	1948	137	75.7	80.3	7.78	.941	10.32
34	Lou Gehrig	1936	155	82.2	87.7	7.75	.946	11.32
35	Frank Robinson	1966	155	73.6	72.0	7.73	1.023	9.31
36	Lou Gehrig	1931	155	79.9	82.7	7.70	.971	10.74
37	Dick Allen	1972	148	66.0	67.3	7.65	.978	8.80
38	Nap Lajoie	1910	159	68.2	66.6	7.40	1.024	9.01
39	Ty Cobb	1911	146	78.5	75.2	7.39	1.040	10.18
40	Ted Williams	1954	117	70.9	71.1	7.35	.997	9.68
41	Frank Robinson	1962	162	66.6	73.2	7.30	.921	10.02
42	Al Rosen	1953	155	63.3	72.7	7.28	.883	9.98
43	Cy Seymour	1905	149	65.8	69.3	7.21	.951	9.62
44	Stan Musial	1951	152	70.1	71.9	7.21	.977	9.98
45	Babe Ruth	1919	130	66.5	68.7	7.19	.963	9.56
46	Ty Cobb	1912	140	67.9	72.1	7.18	.943	10.03
47	Jimmie Foxx	1938	149	78.2	79.2	7.16	.989	11.07
48	Tris Speaker	1916	151	64.6	64.5	7.14	1.002	9.03
49	Carl Yastrzemski	1967	161	76.4	64.6	7.14	1.174	9.06
50	Ty Cobb	1910	140	67.9	64.0	7.11	1.066	9.01
51	Rogers Hornsby	1928	140	72.5	72.8	7.11	.996	10.24
52	Babe Ruth	1932	133	71.2	76.8	7.10	.930	10.81
53	Joe Jackson	1912	152	70.3	71.2	7.10	.989	10.03
54	Joe Jackson	1911	147	71.8	72.0	7.07	.998	10.18
55	Harry Heilmann	1923	144	70.7	72.5	7.01	.977	10.34
56	Lou Gehrig	1937	157	73.3	76.3	7.01	.968	10.90
57	Ty Cobb	1915	156	71.6	65.8	6.98	1.077	9.42
58	Lou Gehrig	1932	156	68.6	75.2	6.96	.930	10.81
59	Honus Wagner	1908	151	65.2	59.9	6.95	1.089	8.61
60	Harmon Killebrew	1969	162	65.6	65.9	6.91	.996	9.54
61	Carl Yastrzemski	1970	161	71.7	66.6	6.90	1.067	9.64
62	Joe Medwick	1937	156	68.8	68.8	6.83	1.000	10.09
63	Reggie Jackson	1969	152	62.0	65.1	6.82	.958	9.54
64	Rod Carew	1977	155	67.0	68.7	6.82	.980	10.07
65	Carl Yastrzemski	1968	157	57.5	59.2	6.79	.973	8.72
66	Stan Musial	1946	156	70.9	63.8	6.77	1.096	9.42
67	Tris Speaker	1912	153	72.8	67.8	6.76	1.062	10.03
68	Hank Aaron	1959	154	62.8	67.1	6.74	.947	9.94
69	Rogers Hornsby	1920	149	62.6	62.7	6.71	.999	9.34
70	Harmon Killebrew	1967	163	62.3	60.7	6.70	1.024	9.06
71	Nap Lajoie	1904	140	61.5	59.8	6.69	1.029	8.94
72	Willie Stargell	1973	148	57.8	64.2	6.68	.904	9.60
73	Joe Jackson	1913	148	65.5	62.4	6.65	1.046	9.39
74	Willie McCovey	1970	152	62.0	66.5	6.63	.940	10.04
75	Stan Musial	1943	157	65.4	62.0	6.62	1.047	9.36
76	Johnny Mize	1939	153	68.7	66.2	6.62	1.032	10.00
77	Rogers Hornsby	1929	156	74.1	72.5	6.59	1.017	11.00
78	Lou Gehrig	1933	152	59.5	69.5	6.58	.887	10.56
79	Ty Cobb	1909	156	63.3	57.9	6.57	1.088	8.81
80	Ralph Kiner	1949	152	70.0	66.2	6.57	1.047	10.09
81	George Brett	1980	117	64.8	65.6	6.55	.987	10.02
82	Ralph Kiner	1951	151	70.8	65.2	6.53	1.071	9.98
83	Jimmie Foxx	1935	147	67.2	69.6	6.52	.972	10.68
84	George Sisler	1920	154	73.2	67.2	6.51	1.069	10.33
85	Tris Speaker	1923	150	70.9	67.3	6.51	1.042	10.34
86	Joe Torre	1971	161	62.5	60.5	6.49	1.026	9.33
87	Willie Mays	1963	157	55.8	59.7	6.48	.943	9.22
88	Willie Stargell	1971	141	58.7	60.3	6.47	.974	9.33
89	Hank Aaron	1963	161	63.0	59.6	6.47	1.046	9.22
90	Willie Mays	1957	152	60.5	63.3	6.44	.964	9.84
91	Frank Howard	1969	161	56.8	61.4	6.43	.940	9.54
92	Stan Musial	1949	157	72.4	64.8	6.42	1.088	10.09
93	Nap Lajoie	1901	131	71.6	71.0	6.41	1.007	11.07
94	Mel Ott	1938	150	61.9	63.7	6.41	.976	9.94
95	Hank Aaron	1971	139	65.2	59.8	6.41	1.092	9.33
96	Mickey Mantle	1955	147	59.0	63.2	6.33	.944	9.98
97	Stan Musial	1953	157	62.7	65.5	6.31	.968	10.38
98	Willie Mays	1954	151	61.8	63.6	6.30	.976	10.10
99	Joe DiMaggio	1941	139	64.3	64.8	6.28	.993	10.32
100	George Stone	1906	154	57.1	56.9	6.26	1.003	9.09

	Year	G	OBA	Norm	Adj	PF	LG			Year	G	OBA	Norm	Adj	PF	LG
1 Ted Williams	1941	143	.551	156.5	155.8	1.009	.352	51 Jim Wynn	1969	149	.440	132.5	136.3	.945	.332
2 Mickey Mantle	1957	144	.515	152.3	154.6	.970	.338	52 Babe Ruth	1928	154	.461	130.5	136.1	.919	.354
3 Ted Williams	1954	117	.516	150.2	150.4	.997	.344	53 Babe Ruth	1919	130	.456	133.5	136.0	.963	.342
4 Ted Williams	1957	132	.528	156.2	150.3	1.079	.338	54 Eddie Collins	1913	148	.441	131.2	135.9	.932	.336
5 Rogers Hornsby	1924	143	.507	146.9	149.7	.963	.345	55 Augie Galan	1947	124	.449	128.5	135.7	.896	.349
6 Mickey Mantle	1962	123	.488	144.5	148.5	.947	.338	56 Ted Williams	1949	155	.490	135.2	135.6	.994	.363
7 Babe Ruth	1920	142	.530	149.0	147.6	1.019	.356	57 Ted Williams	1956	136	.479	136.2	135.4	1.011	.352
8 Ted Williams	1942	150	.499	146.8	145.8	1.014	.340	58 Babe Ruth	1927	151	.487	134.8	135.3	.992	.361
9 Babe Ruth	1926	152	.516	142.6	145.5	.961	.362	59 Harry Heilmann	1923	144	.481	133.6	135.2	.977	.360
10 Babe Ruth	1923	152	.545	151.3	144.6	1.095	.360	60 Al Kaline	1967	131	.415	132.1	135.2	.955	.314
11 Ted Williams	1947	156	.499	145.6	144.6	1.014	.343	61 Honus Wagner	1904	132	.419	135.5	135.1	1.005	.309
12 Ted Williams	1948	137	.497	138.7	143.0	.941	.358	62 Nap Lajoie	1904	140	.405	137.0	135.0	1.029	.296
13 Norm Cash	1961	159	.488	142.8	142.6	1.003	.342	63 Jesse Burkett	1901	142	.432	133.5	135.0	.977	.324
14 Babe Ruth	1930	145	.493	137.3	142.6	.927	.359	64 Joe Jackson	1912	152	.458	134.2	134.9	.989	.341
15 Babe Ruth	1932	133	.489	137.4	142.5	.930	.356	65 Sherry Magee	1910	154	.445	131.9	134.9	.956	.337
16 Tris Speaker	1916	151	.470	142.1	142.0	1.002	.331	66 Eddie Collins	1911	132	.451	129.8	134.8	.927	.347
17 Babe Ruth	1931	145	.495	139.9	142.0	.971	.354	67 Joe Jackson	1911	147	.468	134.7	134.8	.998	.347
18 Ken Singleton	1977	152	.442	132.7	141.8	.876	.333	68 Dick Allen	1972	148	.422	133.2	134.7	.978	.317
19 Ted Williams	1946	150	.497	147.1	141.6	1.078	.338	69 Bobby Murcer	1971	146	.429	130.2	134.6	.936	.329
20 Eddie Collins	1915	155	.460	138.1	141.4	.954	.333	70 Lou Gehrig	1934	154	.465	128.5	134.6	.912	.362
21 Rogers Hornsby	1928	140	.498	140.4	140.7	.996	.355	71 Nap Lajoie	1901	131	.451	135.0	134.5	1.007	.334
22 Carl Yastrzemski	1968	157	.429	138.7	140.6	.973	.309	72 Jimmie Foxx	1933	149	.449	127.7	134.5	.901	.352
23 Ty Cobb	1915	156	.486	145.9	140.5	1.077	.333	73 Willie McCovey	1970	152	.446	130.3	134.4	.940	.342
24 Joe Morgan	1975	146	.471	139.0	140.2	.983	.339	74 Miller Huggins	1913	121	.432	129.4	134.2	.929	.334
25 Ty Cobb	1913	122	.467	139.1	139.9	.988	.336	75 Fred Snodgrass	1910	123	.440	130.4	134.2	.943	.337
26 Ty Cobb	1910	140	.456	144.4	139.9	1.066	.316	76 George Stone	1906	154	.411	134.4	134.2	1.003	.306
27 Roy Thomas	1903	130	.450	134.9	139.6	.934	.334	77 Eddie Stanky	1950	152	.460	132.7	134.2	.978	.346
28 Babe Ruth	1924	153	.513	139.4	139.5	.998	.368	78 Wade Boggs	1983	153	.449	135.9	134.1	1.027	.330
29 Nap Lajoie	1910	159	.445	140.8	139.2	1.024	.316	79 Frank Chance	1905	118	.430	134.5	134.0	1.008	.319
30 Rogers Hornsby	1923	107	.459	130.2	138.9	.878	.352	80 Mickey Mantle	1956	150	.467	132.6	133.9	.982	.352
31 George Brett	1980	117	.461	137.8	138.7	.987	.334	81 Al Kaline	1971	133	.421	127.8	133.8	.912	.329
32 Mickey Mantle	1961	153	.452	132.2	138.7	.908	.342	82 Joe Jackson	1913	148	.460	136.9	133.8	1.046	.336
33 Eddie Collins	1914	152	.452	136.9	138.3	.980	.330	83 Roy Thomas	1901	129	.428	132.4	133.8	.979	.324
34 Rogers Hornsby	1925	138	.489	136.6	138.3	.976	.358	84 Ross Youngs	1924	133	.441	127.6	133.7	.912	.345
35 Babe Ruth	1921	152	.512	140.2	138.2	1.028	.365	85 Willie Mays	1971	136	.429	130.6	133.7	.955	.328
36 Ty Cobb	1912	140	.458	134.2	138.2	.943	.341	86 Joe Cunningham	1959	144	.456	134.6	133.7	1.014	.338
37 Lou Gehrig	1928	154	.467	132.2	137.9	.919	.354	87 Ty Cobb	1918	111	.440	132.7	133.6	.986	.331
38 Mike Hargrove	1981	94	.432	133.5	137.9	.937	.323	88 Max Bishop	1933	117	.446	126.7	133.4	.901	.352
39 Ted Williams	1958	129	.462	138.8	137.8	1.015	.333	89 Ted Williams	1951	148	.464	133.0	133.4	.994	.349
40 Willie McCovey	1969	149	.458	138.0	137.7	1.004	.332	90 Babe Ruth	1933	137	.442	125.5	133.3	.887	.352
41 Benny Kauff	1915	136	.440	135.5	137.5	.971	.325	91 Topsy Hartsel	1907	143	.405	133.3	133.2	1.001	.304
42 Roy Thomas	1904	139	.411	133.0	137.5	.936	.309	92 Rickey Henderson	1981	108	.411	127.1	133.1	.912	.323
43 Mickey Cochrane	1933	130	.459	130.5	137.4	.901	.352	93 Babe Ruth	1918	95	.410	123.6	133.1	.862	.331
44 Rod Carew	1977	155	.452	135.8	137.1	.980	.333	94 Stan Musial	1951	152	.449	131.5	133.1	.977	.341
45 Cy Seymour	1905	149	.427	133.7	137.1	.951	.319	95 Ty Cobb	1917	152	.444	136.3	132.9	1.051	.326
46 Rogers Hornsby	1921	154	.458	131.6	137.1	.921	.348	96 Fred Tenney	1902	134	.404	128.4	132.9	.934	.315
47 Arky Vaughan	1935	137	.491	144.0	137.1	1.104	.341	97 Stan Musial	1952	154	.432	129.2	132.9	.945	.334
48 Rod Carew	1974	153	.435	133.6	136.8	.955	.325	98 Rickey Henderson	1980	158	.422	126.3	132.7	.906	.334
49 Lou Gehrig	1930	154	.473	131.7	136.8	.927	.359	99 Richie Ashburn	1955	140	.449	132.6	132.6	.999	.339
50 Ed Delahanty	1902	123	.449	133.9	136.3	.965	.335	100 Tris Speaker	1920	150	.483	135.7	132.6	1.048	.356

	Player	Year	G	SLG	Norm	Adj	PF	LG
1	Babe Ruth	1920	142	.847	211.7	209.7	1.019	.400
2	Babe Ruth	1921	152	.846	201.2	198.5	1.028	.421
3	Babe Ruth	1927	151	.772	187.7	188.5	.992	.411
4	Lou Gehrig	1927	155	.765	186.1	186.8	.992	.411
5	Babe Ruth	1926	152	.737	181.4	185.0	.961	.406
6	Jimmie Foxx	1933	149	.703	173.8	183.1	.901	.405
7	Babe Ruth	1923	152	.764	190.7	182.2	1.095	.401
8	Babe Ruth	1924	153	.739	180.9	181.1	.998	.409
9	Babe Ruth	1919	130	.657	177.3	180.7	.963	.371
10	Babe Ruth	1928	154	.709	172.9	180.4	.919	.410
11	Babe Ruth	1918	95	.555	167.4	180.3	.862	.332
12	Ted Williams	1941	143	.735	180.9	180.1	1.009	.406
13	Rogers Hornsby	1925	138	.756	176.4	178.5	.976	.429
14	Lou Gehrig	1934	154	.706	170.2	178.2	.912	.415
15	Ted Williams	1957	132	.731	184.1	177.3	1.079	.397
16	Stan Musial	1948	155	.702	176.4	176.0	1.005	.398
17	Rogers Hornsby	1924	143	.696	172.4	175.6	.963	.404
18	Mickey Mantle	1961	153	.687	167.3	175.5	.908	.411
19	Babe Ruth	1930	145	.732	168.6	175.2	.927	.434
20	Mickey Mantle	1956	150	.705	172.9	174.5	.982	.408
21	Willie Stargell	1973	148	.646	165.4	173.9	.904	.390
22	Babe Ruth	1931	145	.700	171.0	173.6	.971	.410
23	Ted Williams	1942	150	.648	174.1	172.9	1.014	.372
24	Lou Gehrig	1930	154	.721	166.2	172.6	.927	.434
25	Dick Allen	1972	148	.603	169.4	171.3	.978	.356
26	Rogers Hornsby	1922	154	.722	172.4	170.7	1.020	.419
27	Willie McCovey	1969	149	.656	170.6	170.3	1.004	.384
28	Mickey Mantle	1957	144	.665	167.4	170.0	.970	.397
29	Ted Williams	1946	156	.667	176.0	169.5	1.078	.379
30	Ty Cobb	1917	152	.571	173.3	169.1	1.051	.330
31	Jimmie Foxx	1932	154	.749	178.6	169.0	1.117	.419
32	Ty Cobb	1912	140	.586	163.6	168.4	.943	.358
33	Hank Aaron	1971	139	.669	175.9	168.3	1.092	.380
34	Babe Ruth	1929	135	.697	165.5	168.2	.969	.421
35	Nap Lajoie	1901	131	.643	168.5	167.9	1.007	.382
36	Willie Stargell	1971	141	.628	165.2	167.4	.974	.380
37	George Brett	1980	117	.664	166.3	167.4	.987	.399
38	Ted Williams	1947	156	.634	168.1	166.9	1.014	.377
39	Cy Seymour	1905	149	.559	162.1	166.2	.951	.345
40	Al Simmons	1930	138	.708	163.1	166.0	.965	.434
41	Frank Howard	1968	158	.552	156.7	165.9	.892	.352
42	Ted Williams	1949	155	.650	165.3	165.8	.994	.393
43	Ty Cobb	1910	140	.551	170.9	165.6	1.066	.323
44	Ty Cobb	1911	146	.621	168.1	164.9	1.040	.369
45	Nap Lajoie	1904	140	.552	167.2	164.8	1.029	.330
46	Lou Gehrig	1928	154	.648	158.0	164.8	.919	.410
47	Frank Robinson	1966	155	.637	166.5	164.6	1.023	.383
48	Al Rosen	1953	155	.613	154.4	164.3	.883	.397
49	Lou Gehrig	1931	155	.662	161.7	164.1	.971	.410
50	Rogers Hornsby	1923	107	.627	153.7	164.1	.878	.408
51	Mike Schmidt	1981	102	.644	171.3	164.0	1.090	.376
52	Ted Williams	1954	117	.635	163.6	163.9	.997	.388
53	Jimmie Foxx	1938	149	.704	162.9	163.8	.989	.432
54	Babe Ruth	1932	133	.661	157.6	163.5	.930	.419
55	Honus Wagner	1908	151	.542	170.6	163.5	1.089	.318
56	Lou Gehrig	1936	155	.696	158.7	163.1	.946	.439
57	Joe Jackson	1912	152	.579	161.5	162.4	.989	.358
58	Babe Ruth	1922	110	.672	163.4	162.3	1.014	.411
59	Rogers Hornsby	1921	154	.639	155.6	162.1	.921	.410
60	Reggie Jackson	1969	152	.608	158.6	162.0	.958	.384
61	Joe Medwick	1937	156	.641	161.6	161.6	1.000	.397
62	Norm Cash	1961	159	.662	161.1	160.9	1.003	.411
63	Ted Williams	1948	137	.615	155.9	160.7	.941	.394
64	Joe DiMaggio	1939	120	.671	158.4	160.5	.975	.424
65	Joe Jackson	1911	147	.590	159.8	159.9	.998	.369
66	Rocky Colavito	1958	143	.620	155.9	159.5	.955	.397
67	Harry Heilmann	1923	144	.632	157.6	159.4	.977	.401
68	Willie Mays	1954	151	.667	157.5	159.4	.976	.424
69	Johnny Mize	1940	155	.636	162.6	159.4	1.040	.391
70	Joe DiMaggio	1941	139	.643	158.4	159.0	.993	.406
71	Lou Gehrig	1933	152	.605	149.6	158.8	.887	.405
72	Ralph Kiner	1949	152	.658	162.5	158.8	1.047	.405
73	Frank Robinson	1962	162	.624	152.3	158.7	.921	.410
74	Mickey Mantle	1955	147	.611	154.1	158.6	.944	.397
75	Roger Maris	1961	161	.620	151.1	158.5	.908	.411
76	Joe DiMaggio	1937	151	.673	155.9	158.4	.968	.432
77	Frank Robinson	1967	129	.576	158.0	158.3	.997	.365
78	Hank Greenberg	1938	155	.683	158.1	157.9	1.002	.432
79	Bobby Murcer	1972	153	.537	150.8	157.9	.913	.356
80	Willie Mays	1963	157	.582	153.3	157.8	.943	.380
81	Nap Lajoie	1910	159	.514	159.5	157.6	1.024	.323
82	Carl Yastrzemski	1967	161	.622	170.5	157.4	1.174	.365
83	Jimmie Foxx	1934	150	.653	157.3	157.3	1.001	.415
84	Dick Allen	1966	141	.632	158.4	157.3	1.015	.399
85	Ed Delahanty	1902	123	.590	154.4	157.1	.965	.382
86	Hank Greenberg	1935	152	.628	151.3	157.0	.928	.415
87	Willie McCovey	1968	148	.545	153.4	157.0	.955	.355
88	Ty Cobb	1918	111	.515	155.4	156.5	.986	.332
89	Jim Gentile	1961	148	.646	157.4	156.4	1.012	.411
90	Andre Dawson	1981	103	.553	147.1	156.4	.885	.376
91	Hank Aaron	1959	154	.636	152.2	156.4	.947	.418
92	Jimmie Foxx	1939	124	.694	163.8	156.2	1.100	.424
93	Honus Wagner	1904	132	.520	156.5	156.2	1.005	.332
94	Willie Mays	1965	157	.645	165.4	155.9	1.125	.390
95	Hank Aaron	1969	147	.607	157.9	155.8	1.027	.384
96	Willie Mays	1955	152	.659	156.7	155.6	1.015	.420
97	Willie Mays	1962	162	.615	150.2	155.4	.934	.410
98	Jimmie Foxx	1935	147	.636	153.0	155.2	.972	.415
99	Harry Lumley	1906	133	.477	148.2	155.1	.913	.322
100	Joe Jackson	1913	148	.551	158.6	155.1	1.046	.347

	Year	G	OPS	Norm	Adj	PF	LG		Year	G	OPS	Norm	Adj	PF	LG
1 Babe Ruth	1920	142	1.378	260.7	255.8	1.019	.756	51 Nap Lajoie	1910	159	.960	200.3	195.6	1.024	.639
2 Ted Williams	1941	143	1.286	237.5	235.4	1.009	.758	52 Ty Cobb	1913	122	1.002	193.0	195.4	.988	.684
3 Babe Ruth	1921	152	1.358	241.4	234.8	1.028	.786	53 Ed Delahanty	1902	123	1.038	188.3	195.1	.965	.717
4 Babe Ruth	1926	152	1.253	224.0	233.1	.961	.768	54 Joe Jackson	1911	147	1.058	194.4	194.8	.998	.717
5 Rogers Hornsby	1924	143	1.203	219.3	227.7	.963	.749	55 Ty Cobb	1911	146	1.088	202.4	194.7	1.040	.717
6 Mickey Mantle	1957	144	1.179	219.7	226.5	.970	.735	56 Rogers Hornsby	1928	140	1.130	193.5	194.3	.996	.767
7 Babe Ruth	1927	151	1.259	222.5	224.3	.992	.772	57 Lou Gehrig	1931	155	1.108	187.7	193.4	.971	.763
8 Jimmie Foxx	1933	149	1.153	201.5	223.6	.901	.757	58 Frank Robinson	1966	155	1.052	197.3	192.8	1.023	.700
9 Ted Williams	1957	132	1.259	240.3	222.7	1.079	.735	59 Willie Stargell	1971	141	1.029	187.4	192.4	.974	.708
10 Babe Ruth	1930	145	1.225	205.9	222.1	.927	.793	60 Frank Robinson	1962	162	1.048	177.2	192.4	.921	.749
11 Babe Ruth	1918	95	.965	191.0	221.6	.862	.663	61 Willie McCovey	1970	152	1.058	180.4	191.9	.940	.750
12 Babe Ruth	1928	154	1.170	203.4	221.3	.919	.764	62 Lou Gehrig	1933	152	1.030	170.1	191.8	.887	.757
13 Babe Ruth	1923	152	1.309	242.0	221.0	1.095	.761	63 Babe Ruth	1929	135	1.128	185.6	191.5	.969	.780
14 Babe Ruth	1924	153	1.252	220.3	220.7	.998	.776	64 Tris Speaker	1916	151	.972	191.9	191.5	1.002	.666
15 Mickey Mantle	1961	153	1.138	199.4	219.6	.908	.752	65 Honus Wagner	1904	132	.939	192.0	191.1	1.005	.642
16 Lou Gehrig	1927	155	1.240	217.4	219.1	.992	.772	66 Babe Ruth	1933	137	1.023	169.3	190.8	.887	.757
17 Babe Ruth	1919	130	1.114	210.8	218.9	.963	.713	67 Ty Cobb	1918	111	.955	188.1	190.8	.986	.663
18 Rogers Hornsby	1925	138	1.245	213.0	218.2	.976	.787	68 Jimmie Foxx	1938	149	1.166	188.1	190.2	.989	.801
19 Ted Williams	1942	150	1.147	221.0	217.9	1.014	.712	69 Reggie Jackson	1969	152	1.019	181.7	189.7	.958	.717
20 Lou Gehrig	1934	154	1.172	198.7	217.9	.912	.777	70 Honus Wagner	1908	151	.952	206.3	189.5	1.089	.620
21 Babe Ruth	1931	145	1.195	210.9	217.2	.971	.763	71 Al Kaline	1967	131	.957	180.6	189.1	.955	.679
22 Ted Williams	1954	117	1.151	213.8	214.5	.997	.732	72 Mickey Mantle	1955	147	1.044	178.3	188.8	.944	.745
23 Lou Gehrig	1930	154	1.194	197.9	213.5	.927	.793	73 Jimmie Foxx	1932	154	1.218	210.5	188.4	1.117	.775
24 Ted Williams	1947	137	1.133	213.7	210.7	1.014	.720	74 Frank Robinson	1967	129	.984	187.1	188.3	.997	.679
25 Babe Ruth	1932	133	1.150	195.0	209.7	.930	.775	75 Lou Gehrig	1932	156	1.072	174.9	188.0	.930	.775
26 Ty Cobb	1912	140	1.043	197.7	209.7	.943	.699	76 Harry Lumley	1906	133	.863	171.3	187.6	.913	.635
27 Rogers Hornsby	1923	107	1.086	183.9	209.5	.878	.760	77 Stan Musial	1951	152	1.063	183.1	187.4	.977	.747
28 Mickey Mantle	1956	150	1.172	205.6	209.4	.982	.760	78 Al Simmons	1930	138	1.130	180.7	187.3	.965	.793
29 Willie McCovey	1969	149	1.114	208.6	207.8	1.004	.716	79 Mike Schmidt	1981	102	1.083	204.0	187.1	1.090	.707
30 Dick Allen	1972	148	1.025	202.6	207.1	.978	.673	80 Joe DiMaggio	1939	120	1.119	182.3	187.0	.975	.785
31 Ted Williams	1946	150	1.164	223.1	207.0	1.078	.717	81 Frank Howard	1968	158	.892	166.8	187.0	.892	.661
32 Lou Gehrig	1928	154	1.115	190.2	206.9	.919	.764	82 Joe Jackson	1913	148	1.011	195.5	186.9	1.046	.684
33 Ted Williams	1948	137	1.112	194.6	206.8	.941	.753	83 George Stone	1906	154	.912	187.4	186.9	1.003	.633
34 George Brett	1980	117	1.124	204.1	206.8	.987	.733	84 Rocky Colavito	1958	143	1.027	178.2	186.6	.955	.730
35 Stan Musial	1948	155	1.152	207.6	206.5	1.005	.741	85 Jimmie Foxx	1935	147	1.096	181.0	186.2	.972	.775
36 Cy Seymour	1905	149	.987	195.8	205.9	.951	.665	86 Benny Kauff	1915	136	.949	180.5	185.9	.971	.676
37 Norm Cash	1961	159	1.150	204.0	203.4	1.003	.752	87 Sherry Magee	1910	154	.952	177.4	185.5	.956	.686
38 Rogers Hornsby	1921	154	1.097	187.1	203.2	.921	.759	88 Bobby Murcer	1971	146	.972	173.6	185.5	.936	.708
39 Willie Stargell	1973	148	1.041	183.7	203.2	.904	.724	89 Hank Aaron	1971	139	1.082	201.9	184.9	1.092	.708
40 Mickey Mantle	1962	123	1.093	192.1	202.8	.947	.748	90 Willie Mays	1963	157	.966	174.0	184.6	.943	.698
41 Nap Lajoie	1901	131	1.094	203.5	202.1	1.007	.716	91 Joe DiMaggio	1941	139	1.083	183.3	184.5	.993	.758
42 Ty Cobb	1910	140	1.008	215.4	202.0	1.066	.639	92 Willie McCovey	1968	148	.928	176.1	184.4	.955	.668
43 Ted Williams	1949	155	1.141	200.5	201.7	.994	.756	93 Lou Gehrig	1937	157	1.116	178.5	184.4	.968	.797
44 Al Rosen	1953	155	1.034	176.3	199.7	.883	.743	94 Carl Yastrzemski	1968	157	.924	179.3	184.3	.973	.661
45 Ty Cobb	1917	152	1.016	209.6	199.4	1.051	.656	95 Mel Ott	1938	150	1.024	179.5	183.9	.976	.729
46 Nap Lajoie	1904	140	.957	204.1	198.4	1.029	.626	96 Jack Fournier	1915	126	.920	175.2	183.6	.954	.668
47 Joe Jackson	1912	154	1.036	195.8	197.9	.989	.699	97 Tris Speaker	1912	153	1.031	194.4	183.1	1.062	.699
48 Lou Gehrig	1936	155	1.174	186.2	196.9	.946	.813	98 Ted Williams	1958	129	1.046	185.8	183.0	1.015	.730
49 Rogers Hornsby	1922	154	1.181	200.1	196.2	1.020	.778	99 Rogers Hornsby	1920	149	.990	182.7	182.8	.999	.697
50 Harry Heilmann	1923	144	1.113	191.2	195.7	.977	.761	100 Joe Medwick	1937	156	1.056	182.8	182.8	1.000	.739

#		Year	G	ISO	Norm	Adj	PF	LG		#		Year	G	ISO	Norm	Adj	PF	LG
1	Babe Ruth	1920	142	.472	433.5	429.4	1.019	.109		51	Roger Maris	1961	161	.351	239.0	250.8	.908	.147
2	Babe Ruth	1918	95	.256	359.0	386.7	.862	.071		52	Willie McCovey	1968	148	.252	244.5	250.1	.955	.103
3	Babe Ruth	1921	152	.469	387.3	382.0	1.028	.121		53	Gavvy Cravath	1915	150	.224	258.6	249.4	1.075	.087
4	Babe Ruth	1919	130	.336	354.6	361.3	.963	.095		54	Dolph Camilli	1941	149	.270	248.8	247.8	1.008	.109
5	Babe Ruth	1927	151	.417	352.3	353.7	.992	.118		55	Babe Ruth	1932	133	.319	238.4	247.2	.930	.134
6	Lou Gehrig	1927	155	.392	331.5	332.8	.992	.118		56	Dave Kingman	1976	123	.268	240.0	247.0	.944	.112
7	Babe Ruth	1928	154	.386	318.0	331.7	.919	.121		57	Ted Williams	1957	132	.343	256.0	246.4	1.079	.134
8	Babe Ruth	1924	153	.361	325.8	326.1	.998	.111		58	Harry Davis	1906	145	.167	233.3	246.4	.897	.072
9	Babe Ruth	1923	152	.372	335.4	320.5	1.095	.111		59	Ty Cobb	1917	152	.189	252.3	246.1	1.051	.075
10	Babe Ruth	1926	152	.366	313.9	320.2	.961	.116		60	Ken Williams	1922	153	.296	249.7	245.8	1.032	.118
11	Babe Ruth	1922	110	.357	301.6	299.5	1.014	.118		61	Dave Kingman	1979	145	.325	249.6	243.8	1.048	.130
12	Jimmie Foxx	1933	149	.347	280.3	295.3	.901	.124		62	Hank Greenberg	1935	152	.300	234.8	243.8	.928	.128
13	Willie Stargell	1973	148	.347	271.4	285.4	.904	.128		63	Rogers Hornsby	1924	143	.272	239.2	243.7	.963	.114
14	Lou Gehrig	1934	154	.344	270.5	283.3	.912	.127		64	Ralph Kiner	1947	152	.326	246.1	242.5	1.030	.132
15	Willie Stargell	1971	141	.333	276.0	279.6	.974	.121		65	Hank Aaron	1969	147	.307	245.7	242.4	1.027	.125
16	Rogers Hornsby	1925	138	.353	275.5	278.8	.976	.128		66	Mike Schmidt	1979	160	.311	238.4	241.6	.973	.130
17	Babe Ruth	1930	145	.373	268.3	278.7	.927	.139		67	Tim Jordan	1906	129	.160	230.5	241.2	.913	.069
18	Babe Ruth	1929	135	.353	272.2	276.5	.969	.130		68	Mickey Mantle	1955	147	.306	234.3	241.2	.944	.130
19	Mike Schmidt	1981	102	.328	288.3	276.2	1.090	.114		69	Babe Ruth	1933	137	.281	226.8	240.8	.887	.124
20	Hank Greenberg	1946	142	.327	283.1	273.8	1.069	.115		70	Hack Wilson	1927	146	.261	237.6	240.3	.978	.110
21	Ted Williams	1942	150	.291	274.4	272.5	1.014	.106		71	Willie Mays	1964	157	.311	243.9	239.9	1.034	.128
22	Dick Allen	1972	148	.294	269.3	272.3	.978	.109		72	Tommy Leach	1902	135	.148	240.8	239.8	1.008	.061
23	Jimmie Foxx	1932	154	.385	287.0	271.5	1.117	.134		73	Al Simmons	1930	138	.327	235.3	239.5	.965	.139
24	Hank Aaron	1971	139	.341	283.2	271.0	1.092	.121		74	Harry Lumley	1907	127	.159	231.9	239.2	.940	.068
25	Ted Williams	1946	150	.325	281.3	271.0	1.078	.115		75	Rocky Colavito	1958	143	.317	233.3	238.7	.955	.136
26	Babe Ruth	1931	145	.328	266.8	270.7	.971	.123		76	Harmon Killebrew	1969	162	.308	238.1	238.6	.996	.129
27	Rogers Hornsby	1922	154	.321	272.6	269.9	1.020	.118		77	Ken Williams	1923	147	.267	240.7	238.5	1.018	.111
28	Mike Schmidt	1980	150	.338	280.3	268.7	1.088	.120		78	Jim Bottomley	1928	149	.304	248.0	238.1	1.085	.122
29	Charlie Keller	1943	141	.254	261.1	268.7	.944	.097		79	Ted Williams	1954	117	.290	236.6	236.9	.997	.123
30	Willie McCovey	1969	149	.336	268.8	268.3	1.004	.125		80	Mel Ott	1944	120	.256	235.7	236.8	.991	.108
31	Lou Gehrig	1931	155	.321	261.7	265.6	.971	.123		81	Gavvy Cravath	1914	149	.200	229.4	236.7	.939	.087
32	Johnny Mize	1940	155	.321	269.5	264.2	1.040	.119		82	Willie McCovey	1970	152	.323	229.4	236.7	.940	.141
33	Mickey Mantle	1961	153	.370	251.8	264.2	.908	.147		83	Mel Ott	1938	150	.271	233.8	236.6	.976	.116
34	Reggie Jackson	1969	152	.333	257.6	263.2	.958	.129		84	Mike Schmidt	1975	158	.274	232.3	236.5	.965	.118
35	Rudy York	1943	155	.256	262.9	260.2	1.021	.097		85	Hack Wilson	1930	155	.368	241.8	236.0	1.050	.152
36	Hank Greenberg	1938	155	.369	259.3	259.0	1.002	.142		86	Frank Robinson	1966	155	.321	238.6	235.9	1.023	.135
37	Wally Berger	1933	137	.254	251.2	258.0	.948	.101		87	Lou Gehrig	1928	154	.274	225.6	235.4	.919	.121
38	Frank Howard	1968	158	.278	242.9	257.1	.892	.114		88	Johnny Mize	1947	154	.312	236.0	235.3	1.006	.132
39	Lou Gehrig	1930	154	.343	246.6	256.2	.927	.139		89	Andre Dawson	1981	103	.251	221.1	235.0	.885	.114
40	Honus Wagner	1908	151	.188	264.8	253.7	1.089	.071		90	Harmon Killebrew	1967	163	.289	237.8	235.0	1.024	.121
41	Ralph Kiner	1949	152	.348	259.3	253.4	1.047	.134		91	Sam Crawford	1901	131	.194	230.6	234.8	.965	.084
42	Mickey Mantle	1956	150	.353	250.8	253.1	.982	.141		92	Willie Mays	1965	157	.328	248.4	234.2	1.125	.132
43	Dick Allen	1968	152	.257	249.1	253.1	.969	.103		93	Bill Nicholson	1944	156	.258	237.6	234.0	1.031	.108
44	Ted Williams	1947	156	.292	253.9	252.1	1.014	.115		94	Bobby Murcer	1972	153	.244	223.5	233.9	.913	.109
45	Stan Musial	1948	155	.326	252.5	251.9	1.005	.129		95	Willie McCovey	1963	152	.285	227.0	233.8	.943	.126
46	Jimmie Foxx	1938	149	.356	250.2	251.6	.989	.142		96	Joe DiMaggio	1937	151	.327	230.0	233.7	.968	.142
47	Ted Williams	1949	155	.307	250.7	251.4	.994	.123		97	Hack Wilson	1928	145	.275	224.5	233.4	.925	.122
48	Jimmie Foxx	1934	150	.319	251.2	251.0	1.001	.127		98	Dick Allen	1966	141	.315	234.9	233.2	1.015	.134
49	Ted Williams	1941	143	.329	252.2	251.0	1.009	.130		99	Frank Schulte	1911	154	.234	230.5	232.8	.980	.102
50	Lou Gehrig	1936	155	.342	244.1	251.0	.946	.140		100	Jim Gentile	1961	148	.344	234.0	232.7	1.012	.147

		Year	G	Avg	Norm	Adj	PF	LG			Year	G	Avg	Norm	Adj	PF	LG
1	Ty Cobb	1912	140	.410	150.7	155.2	.943	.272	51	Dixie Walker	1944	147	.357	133.2	136.1	.959	.268
2	Nap Lajoie	1910	159	.384	153.7	151.9	1.024	.250	52	Nap Lajoie	1903	125	.344	130.9	136.0	.926	.263
3	Tris Speaker	1916	151	.386	150.6	150.5	1.002	.257	53	Nap Lajoie	1912	117	.368	135.2	136.0	.989	.272
4	Nap Lajoie	1901	131	.426	150.0	149.5	1.007	.284	54	Rogers Hornsby	1925	138	.403	134.1	135.7	.976	.300
5	Rogers Hornsby	1924	143	.424	146.1	148.9	.963	.290	55	Eddie Collins	1914	152	.344	134.2	135.5	.980	.256
6	Ty Cobb	1910	140	.383	153.4	148.6	1.066	.250	56	Bobby Veach	1919	139	.355	128.6	135.5	.900	.276
7	Ty Cobb	1913	122	.390	147.5	148.4	.988	.265	57	Dave Winfield	1984	141	.340	129.1	135.5	.908	.264
8	Nap Lajoie	1904	140	.376	150.1	147.9	1.029	.251	58	Honus Wagner	1904	132	.349	135.6	135.3	1.005	.257
9	Ty Cobb	1918	111	.382	146.8	147.9	.986	.260	59	George Sisler	1920	154	.407	139.8	135.2	1.069	.291
10	Rod Carew	1977	155	.388	145.7	147.2	.980	.266	60	Ed Delahanty	1902	123	.376	132.8	135.2	.965	.283
11	Ty Cobb	1919	124	.384	139.2	146.7	.900	.276	61	Tommy Davis	1962	163	.346	128.1	135.1	.899	.270
12	Ted Williams	1941	143	.406	147.2	146.6	1.009	.276	62	Rico Carty	1970	136	.366	137.1	135.1	1.030	.267
13	Ty Cobb	1911	146	.420	149.4	146.5	1.040	.281	63	Lou Gehrig	1928	154	.374	129.5	135.1	.919	.289
14	Ty Cobb	1917	152	.383	150.1	146.4	1.051	.255	64	Eddie Collins	1911	132	.365	130.0	135.0	.927	.281
15	Cy Seymour	1905	149	.377	142.4	146.1	.951	.265	65	Jim Bottomley	1923	134	.371	126.4	134.9	.878	.293
16	Joe Jackson	1912	152	.395	145.1	145.9	.989	.272	66	Hank Aaron	1959	154	.355	131.3	134.9	.947	.270
17	George Brett	1980	117	.390	144.8	145.8	.987	.269	67	Eddie Collins	1913	148	.345	130.2	134.9	.932	.265
18	Joe Jackson	1911	147	.408	145.2	145.4	.998	.281	68	Joe Jackson	1920	146	.382	131.3	134.9	.947	.291
19	Rod Carew	1974	153	.364	140.9	144.2	.955	.258	69	Joe DiMaggio	1939	120	.381	133.0	134.7	.975	.286
20	Ty Cobb	1909	156	.377	149.3	143.1	1.088	.253	70	Harry Walker	1947	140	.363	132.4	134.4	.969	.274
21	George Sisler	1917	135	.353	138.3	142.4	.943	.255	71	Harry Lumley	1906	133	.324	128.4	134.4	.913	.253
22	Ted Williams	1957	132	.388	147.5	142.0	1.079	.263	72	Mike Donlin	1908	155	.334	135.3	134.4	1.014	.247
23	George Sisler	1922	142	.420	143.3	141.0	1.032	.293	73	Mickey Vernon	1946	148	.353	133.8	134.2	.993	.264
24	Roberto Clemente	1967	147	.357	138.5	140.9	.966	.258	74	Stan Musial	1951	152	.355	132.6	134.2	.977	.267
25	Mickey Mantle	1957	144	.365	138.8	140.9	.970	.263	75	Bobby Murcer	1971	146	.331	129.8	134.1	.936	.255
26	Harry Heilmann	1923	144	.403	138.8	140.5	.977	.290	76	Honus Wagner	1907	142	.350	138.7	134.1	1.069	.252
27	Rogers Hornsby	1923	107	.384	131.1	139.9	.878	.293	77	Rod Carew	1975	143	.359	139.1	134.1	1.076	.258
28	George Stone	1906	154	.358	140.0	139.8	1.003	.256	78	Wade Boggs	1983	153	.361	135.9	134.1	1.027	.266
29	Ty Cobb	1916	145	.371	144.5	139.7	1.071	.257	79	Honus Wagner	1906	142	.339	134.3	134.0	1.004	.253
30	Stan Musial	1948	155	.376	140.0	139.6	1.005	.269	80	Rogers Hornsby	1920	149	.370	133.9	134.0	.999	.276
31	Ty Cobb	1915	156	.369	144.8	139.5	1.077	.255	81	Joe Medwick	1937	156	.374	133.8	133.8	1.000	.280
32	Ted Williams	1948	137	.369	135.3	139.5	.941	.273	82	Rogers Hornsby	1928	140	.387	133.4	133.7	.996	.290
33	Rogers Hornsby	1921	154	.397	133.2	138.8	.921	.298	83	Ginger Beaumont	1902	130	.357	134.1	133.6	1.008	.266
34	Tris Speaker	1910	141	.340	136.1	138.8	.962	.250	84	Jimmie Foxx	1933	149	.356	126.8	133.6	.901	.281
35	Ty Cobb	1922	137	.401	136.9	138.7	.974	.293	85	Honus Wagner	1905	147	.363	137.2	133.5	1.056	.265
36	Edd Roush	1917	136	.341	132.9	138.5	.922	.256	86	Roberto Clemente	1969	138	.345	133.1	133.4	.996	.259
37	Jesse Burkett	1901	142	.376	136.6	138.2	.977	.275	87	Matty Alou	1967	139	.338	131.1	133.4	.966	.258
38	Joe Torre	1971	161	.363	139.7	137.9	1.026	.260	88	Lou Gehrig	1930	154	.379	128.4	133.3	.927	.295
39	Joe Jackson	1913	148	.373	141.0	137.9	1.046	.265	89	Al Simmons	1931	128	.390	136.0	133.3	1.041	.287
40	Honus Wagner	1908	151	.354	143.4	137.4	1.089	.247	90	Eddie Collins	1915	155	.332	130.2	133.3	.954	.255
41	Rusty Staub	1967	149	.333	129.2	137.2	.887	.258	91	Ted Williams	1942	150	.356	134.1	133.2	1.014	.266
42	Ty Cobb	1907	150	.350	137.9	137.1	1.012	.254	92	Mickey Mantle	1956	150	.353	132.0	133.2	.982	.267
43	Nap Lajoie	1906	152	.355	139.0	136.9	1.031	.256	93	Roberto Clemente	1971	132	.341	131.3	133.1	.974	.260
44	Don Mattingly	1984	153	.343	130.2	136.6	.908	.264	94	Mike Donlin	1905	150	.356	134.7	133.0	1.025	.265
45	Tris Speaker	1913	141	.363	137.4	136.6	1.012	.265	95	Stan Musial	1946	156	.365	139.3	133.0	1.096	.262
46	Norm Cash	1961	159	.361	136.8	136.6	1.003	.264	96	Hal Chase	1916	142	.339	133.5	133.0	1.008	.254
47	Tris Speaker	1912	153	.383	140.6	136.4	1.062	.272	97	George Brett	1976	159	.333	130.2	132.8	.961	.256
48	Tony Gwynn	1984	158	.351	133.5	136.3	.960	.263	98	Alex Johnson	1970	156	.329	127.5	132.7	.924	.258
49	Pete Rose	1969	156	.348	134.1	136.2	.969	.259	99	Harry Heilmann	1921	149	.394	131.4	132.6	.982	.300
50	George Burns	1918	130	.352	135.3	136.2	.988	.260	100	Al Rosen	1953	155	.336	124.5	132.5	.883	.270

SINGLE-SEASON LEADERS ◇ 291

		Year	G	All	Bat	Fld	Run	Pos			Year	G	All	Bat	Fld	Run	Pos
1	Maury Wills	1962	165	4.5	.4	.2	2.3	1.6	26	Rod Scott	1980	154	-.4	-2.1	-.7	1.2	1.2
2	Rickey Henderson	1983	145	5.7	3.0	.8	2.1	-.2	27	Amos Otis	1971	147	3.7	1.8	1.4	1.2	-.7
3	Tim Raines	1983	156	5.6	2.5	1.4	1.9	-.2	28	Joe Morgan	1973	157	6.1	4.6	-.3	1.2	.6
4	Ron Le Flore	1980	139	.9	-.1	-.2	1.9	-.7	29	Davey Lopes	1978	151	3.9	1.3	.4	1.2	1.0
5	Willie Wilson	1980	161	3.2	1.2	.6	1.8	-.4	30	Tommy Harper	1969	148	.7	-.6	-.1	1.2	.2
6	Willie Wilson	1979	154	2.7	.2	1.4	1.7	-.6	31	Julio Cruz	1978	147	2.0	-2.5	2.4	1.2	.9
7	Tim Raines	1984	160	5.1	3.9	.1	1.7	-.6	32	Dave Collins	1980	144	-.2	.1	-.8	1.2	-.7
8	Davey Lopes	1975	155	1.1	1.0	-2.3	1.7	.7	33	Lou Brock	1966	156	.6	.2	-.2	1.2	-.6
9	Tim Raines	1981	88	3.4	2.2	-.1	1.6	-.3	34	Bobby Bonds	1969	158	2.4	2.5	-.3	1.2	-1.0
10	Rudy Law	1983	141	.3	-.1	-1.0	1.6	-.2	35	Jim Wynn	1965	157	5.3	3.8	1.1	1.1	-.7
11	Lou Brock	1974	153	1.3	.7	-.2	1.6	-.8	36	Willie Wilson	1984	128	1.4	.5	.2	1.1	-.4
12	Joe Morgan	1975	146	8.7	6.1	.5	1.5	.6	37	Frank Taveras	1976	144	1.7	-1.5	.7	1.1	1.4
13	Ron Le Flore	1979	148	2.0	1.2	-0.0	1.5	-.7	38	Joe Morgan	1974	149	6.2	4.9	-.6	1.1	.8
14	Tim Raines	1982	156	.6	-.8	0.0	1.4	0.0	39	Omar Moreno	1979	162	1.0	-1.2	1.7	1.1	-.6
15	Davey Lopes	1976	117	.8	-.4	-.6	1.4	.4	40	Davey Lopes	1979	153	2.0	2.7	-3.1	1.1	1.3
16	Rickey Henderson	1982	149	3.3	1.8	.7	1.4	-.6	41	Larry Lintz	1974	113	1.4	-1.2	.7	1.1	.8
17	Rickey Henderson	1980	158	6.6	3.6	1.9	1.4	-.3	42	Ron Le Flore	1978	155	1.6	.7	.2	1.1	-.4
18	Bert Campaneris	1969	135	-.1	-2.0	-.3	1.4	.8	43	George Case	1943	141	.6	.3	-.3	1.1	-.5
19	Willie Wilson	1983	137	-1.2	-1.1	-1.1	1.3	-.3	44	Luis Aparicio	1960	153	4.1	-1.4	3.2	1.1	1.2
20	Al Wiggins	1983	144	0.0	-0.0	-.7	1.3	-.6	45	Maury Wills	1965	158	5.0	-.4	2.3	1.0	2.1
21	Juan Samuel	1984	160	.1	-0.0	-1.7	1.3	.5	46	Frank Taveras	1977	147	-.2	-2.1	-1.0	1.0	1.9
22	Mickey Rivers	1975	155	.9	-.1	.1	1.3	-.4	47	Snuffy Stirnweiss	1944	154	7.0	3.4	2.0	1.0	.6
23	Jerry Mumphrey	1980	160	1.3	1.0	-.2	1.3	-.8	48	Gene Richards	1977	146	1.3	.9	0.0	1.0	-.6
24	Joe Morgan	1976	141	5.6	6.0	-2.4	1.3	.7	49	Mitchell Page	1977	145	4.9	4.2	.3	1.0	-.6
25	Lou Brock	1968	159	2.0	2.2	-.5	1.3	-1.0	50	Omar Moreno	1980	162	-.5	-2.5	1.8	1.0	-.8

		Year	G	All	Bat	Fld	Run	Pos			Year	G	All	Bat	Fld	Run	Pos
1	Rabbit Maranville	1914	156	5.4	-1.5	6.3	0.0	.6	51	Art Fletcher	1917	151	3.8	.2	3.4	0.0	.2
2	Nap Lajoie	1908	157	7.6	2.2	5.8	0.0	-.4	52	Johnny Evers	1907	151	2.6	-1.1	3.4	0.0	.3
3	Bill Mazeroski	1963	142	4.6	-1.6	5.0	.1	1.1	53	George Cutshaw	1914	153	2.4	-.9	3.4	0.0	-.1
4	Eddie Collins	1910	153	7.3	3.5	5.0	0.0	-1.2	54	Garry Templeton	1980	118	5.2	.5	3.3	0.0	1.4
5	Frankie Frisch	1927	153	6.6	2.0	4.8	0.0	-.2	55	Brooks Robinson	1967	158	5.5	2.0	3.3	-.2	.4
6	Freddie Maguire	1928	140	3.1	-1.7	4.7	0.0	.1	56	Del Pratt	1916	158	5.4	1.6	3.3	0.0	.5
7	Ozzie Smith	1980	158	5.5	-1.9	4.5	.9	2.0	57	Leo Cardenas	1969	160	4.7	.8	3.3	-.2	.8
8	Nap Lajoie	1907	137	7.1	2.6	4.5	0.0	0.0	58	Ed Brinkman	1970	158	2.7	-1.8	3.3	-.3	1.5
9	Heinie Wagner	1908	153	3.9	-.9	4.3	0.0	.5	59	Pep Young	1938	149	3.7	-.6	3.2	0.0	1.1
10	Bill Mazeroski	1966	162	4.6	-.9	4.3	-.1	1.3	60	Lee Tannehill	1906	116	2.1	-1.9	3.2	0.0	.8
11	Bill Dahlen	1908	144	3.4	-.6	4.3	0.0	-.3	61	Ozzie Smith	1984	124	5.4	-0.0	3.2	.7	1.5
12	Hughie Critz	1933	133	2.8	-2.6	4.3	0.0	1.1	62	Roy Smalley	1979	162	5.6	.3	3.2	-.1	2.2
13	Joe Tinker	1908	157	4.4	.5	4.2	0.0	-.3	63	George McBride	1910	154	3.2	-.3	3.2	0.0	.3
14	Dave Shean	1910	150	1.6	-2.9	4.2	0.0	.3	64	Bill Mazeroski	1965	130	2.7	-1.3	3.2	0.0	.8
15	Ryne Sandberg	1983	158	4.4	-1.0	4.2	.5	.7	65	Clete Boyer	1961	148	2.9	-1.1	3.2	-.1	.9
16	Graig Nettles	1971	158	5.1	1.5	4.2	-0.0	-.6	66	Luis Aparicio	1969	156	4.0	-.4	3.2	.5	.7
17	George McBride	1908	155	4.2	-.4	4.2	0.0	.4	67	Luis Aparicio	1968	155	3.4	-1.0	3.2	-.2	1.4
18	Dick Bartell	1936	145	5.7	.6	4.2	0.0	.9	68	Luis Aparicio	1960	153	4.1	-1.4	3.2	1.1	1.2
19	Bill Mazeroski	1962	159	4.7	-.8	4.1	-.2	1.6	69	Freddie Patek	1972	136	3.1	-2.3	3.1	.6	1.7
20	Buck Weaver	1913	151	4.8	-.6	4.0	0.0	1.4	70	Dave Parker	1977	159	6.2	4.3	3.1	-.6	-.6
21	Buck Herzog	1915	155	3.6	-.8	4.0	0.0	.4	71	Ski Melillo	1931	151	3.2	-.5	3.1	-.4	1.0
22	Donie Bush	1914	157	6.2	.5	4.0	0.0	1.7	72	Bill Mazeroski	1961	152	3.5	-1.7	3.1	0.0	2.1
23	Cal Ripken	1984	162	9.4	4.1	3.9	0.0	1.4	73	Billy Herman	1933	153	4.4	.1	3.1	0.0	1.2
24	Rabbit Maranville	1919	131	4.9	.5	3.9	0.0	.5	74	Mickey Doolan	1910	148	3.5	-.2	3.1	0.0	.6
25	Everett Scott	1921	154	2.4	-3.4	3.8	0.0	2.0	75	Art Devlin	1906	148	5.9	3.4	3.1	0.0	-.6
26	Bobby Knoop	1964	162	3.6	-2.1	3.8	-0.0	1.9	76	Ray Chapman	1917	156	5.4	1.6	3.1	0.0	.7
27	Nap Lajoie	1903	125	8.3	4.4	3.7	0.0	.2	77	Dave Cash	1974	162	4.9	.6	3.1	.1	1.1
28	Ivan DeJesus	1977	155	3.7	-2.3	3.7	0.0	2.3	78	Rick Burleson	1980	155	2.9	-1.2	3.1	-.4	1.4
29	Harlond Clift	1937	155	7.2	3.3	3.7	-.1	.3	79	Dave Bancroft	1916	142	3.1	-.8	3.1	0.0	.8
30	Buddy Bell	1982	148	5.6	2.3	3.7	-.1	-.3	80	Lee Tannehill	1911	141	1.9	-2.0	3.0	0.0	.9
31	Ossie Vitt	1916	153	2.6	-1.8	3.6	0.0	.8	81	Lee Tannehill	1904	153	2.1	-1.5	3.0	0.0	.6
32	Manny Trillo	1978	152	2.4	-1.9	3.6	-.4	1.1	82	Ozzie Smith	1981	110	1.8	-2.6	3.0	-.1	1.5
33	Manny Trillo	1977	152	3.2	-1.2	3.6	-.2	1.0	83	Red Schoendienst	1954	148	4.2	.3	3.0	0.0	.9
34	Ozzie Smith	1982	140	4.0	-1.3	3.6	.5	1.2	84	Jerry Priddy	1950	157	3.9	.1	3.0	-.3	1.1
35	Tommy Leach	1904	146	4.3	.1	3.6	0.0	.6	85	Roger Peckinpaugh	1918	122	2.6	-1.3	3.0	0.0	.9
36	Clete Boyer	1962	158	4.1	.4	3.6	-0.0	.1	86	Roy McMillan	1956	150	3.4	-.7	3.0	-.1	1.2
37	Dick Bartell	1937	128	6.5	1.8	3.6	0.0	1.1	87	Art Fletcher	1918	124	2.8	-.6	3.0	0.0	.4
38	Joe Tinker	1911	144	4.0	-0.0	3.5	0.0	.5	88	Bobby Doerr	1946	151	5.2	1.1	3.0	-.2	1.3
39	Ozzie Smith	1978	159	4.0	-1.4	3.5	.5	1.4	89	Bucky Dent	1979	141	2.5	-2.2	3.0	0.0	1.7
40	Del Pratt	1919	140	5.0	.2	3.5	0.0	1.3	90	Bucky Dent	1975	157	3.0	-1.6	3.0	-.2	1.8
41	Freddie Patek	1973	135	3.6	-1.9	3.5	.2	1.8	91	Lou Boudreau	1943	152	5.8	3.1	3.0	-.3	.0
42	George McBride	1912	152	2.3	-2.6	3.5	0.0	1.4	92	Mark Belanger	1977	144	2.8	-1.8	3.0	-0.0	1.6
43	Bill Mazeroski	1964	162	4.3	-.8	3.5	-0.0	1.6	93	Mark Belanger	1975	152	2.7	-1.9	3.0	.2	1.4
44	Rabbit Maranville	1916	155	3.8	-.7	3.5	0.0	1.0	94	Dave Bancroft	1920	108	4.2	.5	3.0	0.0	.7
45	Miller Huggins	1905	149	6.5	2.1	3.5	0.0	.9	95	Dave Bancroft	1917	127	2.6	-.6	3.0	0.0	.2
46	Art Fletcher	1915	149	3.0	-.9	3.5	0.0	.4	96	Bobby Wallace	1905	156	5.2	1.8	2.9	0.0	.5
47	Johnny Evers	1904	152	3.4	-.5	3.5	0.0	.4	97	Garry Templeton	1978	155	3.7	-1.1	2.9	.4	1.5
48	Horace Clarke	1968	148	.7	-3.7	3.5	.2	.7	98	Lee Tannehill	1905	142	1.7	-1.8	2.9	0.0	.6
49	Zoilo Versalles	1962	160	1.9	-2.4	3.4	-.2	1.1	99	Mike Schmidt	1977	154	7.8	5.0	2.9	-0.0	-.1
50	Ron Santo	1967	161	6.7	4.7	3.4	-.3	-1.1	100	Phil Rizzuto	1942	144	4.4	.4	2.9	.3	.8

		Year	G	All	Bat	Fld	Run	Pos			Year	G	All	Bat	Fld	Run	Pos
1	Babe Ruth	1921	152	10.3	10.9	.7	0.0	-1.3	51	Frank Robinson	1962	162	7.2	7.3	.9	0.0	-1.0
2	Babe Ruth	1923	152	9.6	10.7	.4	0.0	-1.5	52	Rogers Hornsby	1929	156	7.2	6.6	.1	0.0	.5
3	Babe Ruth	1920	142	9.5	10.8	-.1	0.0	-1.2	53	Joe Cronin	1930	154	7.2	3.4	2.3	-.1	1.6
4	Cal Ripken	1984	162	9.4	4.1	3.9	0.0	1.4	54	Harlond Clift	1937	155	7.2	3.3	3.7	-.1	.3
5	Babe Ruth	1927	151	9.0	9.6	.2	.2	-1.0	55	Carl Yastrzemski	1967	161	7.1	7.1	.8	-.2	-.6
6	Babe Ruth	1924	153	8.9	9.5	.6	0.0	-1.2	56	Cy Seymour	1905	149	7.1	7.2	1.0	0.0	-1.1
7	Ted Williams	1947	156	8.8	9.3	.3	-.1	-.7	57	Nap Lajoie	1907	137	7.1	2.6	4.5	0.0	0.0
8	Joe Morgan	1975	146	8.7	6.1	.5	1.5	.6	58	Joe Jackson	1912	152	7.1	7.1	.9	0.0	-.9
9	Lou Gehrig	1927	155	8.7	9.6	-.6	.3	-.6	59	Norm Cash	1961	159	7.1	8.5	.4	0.0	-1.8
10	Ted Williams	1942	150	8.6	9.4	.5	-0.0	-1.3	60	Honus Wagner	1909	137	7.0	5.0	1.9	0.0	.1
11	Rogers Hornsby	1920	149	8.5	6.7	1.4	0.0	.4	61	Snuffy Stirnweiss	1944	154	7.0	3.4	2.0	1.0	.6
12	Ted Williams	1946	150	8.4	9.3	0.0	0.0	-.9	62	Mike Schmidt	1981	102	7.0	4.9	2.4	.1	-.4
13	Ted Williams	1949	155	8.3	8.7	.1	-0.0	-.5	63	Lou Boudreau	1944	150	7.0	3.4	2.5	.2	.9
14	Nap Lajoie	1903	125	8.3	4.4	3.7	0.0	.2	64	Robin Yount	1982	156	6.9	5.6	-.3	.2	1.4
15	Ted Williams	1941	143	8.2	9.8	-.5	-.2	-.9	65	Ted Williams	1957	132	6.9	8.8	-1.0	-.1	-.8
16	Babe Ruth	1926	152	8.2	9.7	-.2	-.2	-1.1	66	Snuffy Stirnweiss	1945	152	6.9	4.0	2.7	-0.0	.2
17	Stan Musial	1948	155	8.2	9.0	.4	0.0	-1.2	67	Mike Schmidt	1982	148	6.9	5.4	1.9	0.0	-.4
18	Mickey Mantle	1957	144	8.1	9.4	-.7	.3	-.9	68	Willie Mays	1954	151	6.9	6.3	1.3	-.1	-.6
19	Rogers Hornsby	1924	143	8.1	9.6	-.7	0.0	-.8	69	Nap Lajoie	1904	140	6.9	6.7	.2	0.0	.0
20	George Sisler	1920	154	8.0	6.5	1.5	0.0	0.0	70	Ty Cobb	1911	146	6.9	7.4	.7	0.0	-1.2
21	Babe Ruth	1919	130	8.0	7.2	1.4	0.0	-.6	71	George Brett	1980	117	6.9	6.5	.3	.1	.0
22	Jimmie Foxx	1933	149	7.9	8.6	.5	-.1	-1.1	72	Nap Lajoie	1906	152	6.8	4.7	2.3	0.0	-.2
23	Mike Schmidt	1977	154	7.8	5.0	2.9	-0.0	-.1	73	Reggie Jackson	1969	152	6.8	6.8	.4	.1	-.5
24	Ron Santo	1966	155	7.8	5.4	2.8	-.2	-.2	74	Joe Jackson	1911	147	6.8	7.1	1.0	0.0	-1.3
25	Willie Mays	1955	152	7.8	6.1	1.6	.5	-.4	75	Rogers Hornsby	1917	145	6.8	4.3	2.3	0.0	.2
26	Mickey Mantle	1956	150	7.8	8.3	.1	.2	-.8	76	Frank Baker	1913	149	6.8	5.4	1.1	0.0	.3
27	Rogers Hornsby	1922	154	7.8	8.3	-.3	0.0	-.2	77	Honus Wagner	1912	145	6.7	3.1	2.2	0.0	1.4
28	Nap Lajoie	1910	159	7.7	7.4	1.3	0.0	-1.0	78	Mike Schmidt	1975	158	6.7	3.8	2.7	.2	-.0
29	Nap Lajoie	1901	131	7.7	6.4	1.7	0.0	-.4	79	Ron Santo	1967	161	6.7	4.7	3.4	-.3	-1.1
30	Rogers Hornsby	1921	154	7.7	8.0	-.4	0.0	.1	80	Ryne Sandberg	1984	156	6.7	3.2	2.4	.6	.5
31	Rod Carew	1974	153	7.7	5.0	1.8	.2	.7	81	Babe Ruth	1931	145	6.7	8.8	-1.1	-.1	-.9
32	Babe Ruth	1930	145	7.6	8.7	-.3	-.3	-.5	82	Eddie Lake	1945	133	6.7	3.4	2.9	-.2	.6
33	Jackie Robinson	1951	153	7.6	4.7	2.1	.3	.5	83	Tris Speaker	1916	151	6.6	7.1	.5	0.0	-1.0
34	Nap Lajoie	1908	157	7.6	2.2	5.8	0.0	-.4	84	Mike Schmidt	1976	160	6.6	4.3	2.6	-.1	-.2
35	Tris Speaker	1914	158	7.5	6.0	2.6	0.0	-1.1	85	Frank Robinson	1966	155	6.6	7.7	-.3	-.1	-.7
36	Mike Schmidt	1980	150	7.5	5.3	2.2	.1	-.1	86	Willie Mays	1958	152	6.6	5.6	1.0	.6	-.6
37	Mike Schmidt	1974	162	7.5	5.2	2.7	-0.0	-.4	87	Rickey Henderson	1980	158	6.6	3.6	1.9	1.4	-.3
38	Mickey Mantle	1961	153	7.5	8.3	-.4	.3	-.7	88	Charlie Gehringer	1936	154	6.6	4.2	1.4	.1	.9
39	Honus Wagner	1906	142	7.4	4.7	1.9	0.0	.8	89	Frankie Frisch	1927	153	6.6	2.0	4.8	0.0	-.2
40	Tris Speaker	1912	153	7.4	6.8	1.7	0.0	-1.1	90	Arky Vaughan	1938	148	6.5	3.5	1.9	0.0	1.1
41	Chuck Klein	1930	156	7.4	5.6	2.9	0.0	-1.1	91	Willie Mays	1957	152	6.5	6.4	.6	0.0	-.5
42	Eddie Collins	1915	155	7.4	5.9	1.7	0.0	-.2	92	Duke Kenworthy	1914	146	6.5	3.8	2.5	0.0	.2
43	Carl Yastrzemski	1968	157	7.3	6.8	1.3	0.0	-.8	93	Miller Huggins	1905	149	6.5	2.1	3.5	0.0	.9
44	Honus Wagner	1905	147	7.3	5.0	2.0	0.0	.3	94	Lou Gehrig	1930	154	6.5	8.6	.1	-.4	-1.8
45	Tris Speaker	1913	141	7.3	5.9	2.4	0.0	-1.0	95	Eddie Collins	1913	148	6.5	5.4	1.6	0.0	-.5
46	Stan Musial	1951	152	7.3	7.2	.8	-.2	-.5	96	Dick Bartell	1937	128	6.5	1.8	3.6	0.0	1.1
47	Eddie Collins	1910	153	7.3	3.5	5.0	0.0	-1.2	97	Hank Aaron	1959	154	6.5	6.7	-.2	.2	-.2
48	Ty Cobb	1917	152	7.3	7.9	.6	0.0	-1.2	98	Honus Wagner	1908	151	6.4	7.0	-.3	0.0	-.3
49	Ted Williams	1948	137	7.2	7.8	-.1	.1	-.6	99	Willie Mays	1964	157	6.4	5.7	.9	.3	-.5
50	Babe Ruth	1928	154	7.2	8.8	-.5	-.2	-.9	100	Benny Kauff	1915	136	6.4	5.7	1.4	0.0	-.7

		Year	IP	LWT	Adj	Wins	PF	R/W			Year	IP	LWT	Adj	Wins	PF	R/W
1	Walter Johnson	1913	346	68.6	81.1	8.6	1.111	9.39	51	Claude Hendrix	1914	362	60.9	51.9	5.4	.930	9.62
2	Walter Johnson	1912	368	79.9	78.8	7.9	.992	10.03	52	Lefty Gomez	1937	278	70.7	58.8	5.4	.916	10.90
3	Pete Alexander	1915	376	64.0	69.8	7.7	1.051	9.01	53	Virgil Trucks	1949	275	42.3	55.3	5.4	1.102	10.28
4	Dizzy Trout	1944	352	51.2	71.6	7.5	1.152	9.52	54	Tom Seaver	1971	286	54.3	50.0	5.4	.961	9.33
5	Lefty Grove	1931	289	74.7	75.4	7.0	1.005	10.74	55	Dazzy Vance	1928	280	59.0	54.8	5.4	.966	10.24
6	Walter Johnson	1918	325	53.9	62.0	6.9	1.081	8.96	56	Walter Johnson	1910	374	48.7	47.7	5.3	.991	9.01
7	Vida Blue	1971	312	57.2	64.3	6.9	1.059	9.30	57	Whitey Ford	1958	219	42.7	51.0	5.3	1.090	9.67
8	Hal Newhouser	1945	313	54.0	63.3	6.8	1.079	9.34	58	Buck Newsom	1940	264	45.6	55.7	5.3	1.079	10.59
9	Wilbur Wood	1971	334	57.7	62.8	6.8	1.040	9.30	59	Addie Joss	1908	325	44.2	46.1	5.3	1.022	8.78
10	Dizzy Dean	1934	312	48.9	68.6	6.7	1.140	10.27	60	Lefty Grove	1930	291	68.5	57.9	5.2	.930	11.06
11	Bert Blyleven	1973	325	47.0	65.1	6.7	1.131	9.78	61	Carl Hubbell	1934	313	61.3	53.5	5.2	.945	10.27
12	Bob Gibson	1968	305	63.2	57.9	6.6	.948	8.72	62	Robin Roberts	1953	347	59.3	54.1	5.2	.969	10.38
13	Ron Guidry	1978	274	62.0	63.9	6.6	1.016	9.71	63	Johnny Rigney	1940	281	39.9	55.2	5.2	1.112	10.59
14	Christy Mathewson	1905	339	64.7	61.4	6.4	.971	9.62	64	Thornton Lee	1941	300	59.2	53.7	5.2	.960	10.32
15	Jack Chesbro	1904	455	39.3	57.0	6.4	1.135	8.94	65	Rube Waddell	1905	329	42.8	47.3	5.2	1.047	9.10
16	Cy Young	1902	385	60.7	67.1	6.4	1.042	10.54	66	Dick Ellsworth	1963	291	38.3	47.8	5.2	1.089	9.22
17	Dolf Luque	1923	322	74.2	66.2	6.4	.944	10.41	67	Juan Marichal	1969	300	49.9	49.3	5.2	.995	9.53
18	Juan Marichal	1965	295	46.0	59.9	6.3	1.120	9.46	68	Rube Waddell	1902	276	46.5	54.3	5.2	1.072	10.54
19	Hal Newhouser	1944	312	41.9	60.0	6.3	1.152	9.52	69	Tom Seaver	1973	290	51.3	49.5	5.2	.985	9.60
20	Cy Young	1901	371	84.0	69.3	6.3	.903	11.07	70	Mike Caldwell	1978	293	46.0	49.8	5.1	1.031	9.71
21	Sandy Koufax	1966	323	67.4	59.7	6.3	.940	9.54	71	Phil Niekro	1974	302	41.8	49.2	5.1	1.061	9.61
22	Ed Walsh	1912	393	52.2	62.7	6.2	1.072	10.03	72	Russ Ford	1911	281	33.4	52.0	5.1	1.179	10.18
23	Lefty Grove	1935	273	53.1	66.5	6.2	1.099	10.68	73	Bob Feller	1939	297	58.4	55.4	5.1	.980	10.85
24	Lefty Grove	1932	292	53.3	66.7	6.2	1.092	10.81	74	Jim Palmer	1973	296	46.7	49.9	5.1	1.025	9.78
25	Dazzy Vance	1924	309	58.8	61.8	6.1	1.023	10.08	75	Dean Chance	1964	278	61.0	48.3	5.1	.887	9.51
26	Bucky Walters	1939	319	57.8	61.2	6.1	1.024	10.00	76	Red Faber	1922	352	48.1	52.2	5.1	1.026	10.31
27	Three Finger Brown	1906	277	49.0	54.8	6.1	1.072	9.01	77	Carl Hubbell	1936	304	57.7	51.9	5.1	.957	10.25
28	Gaylord Perry	1972	343	44.0	53.3	6.1	1.079	8.80	78	Juan Marichal	1966	307	47.0	48.1	5.0	1.009	9.54
29	Walter Johnson	1919	290	55.9	57.2	6.0	1.013	9.56	79	Luis Tiant	1968	258	39.4	43.9	5.0	1.053	8.72
30	Steve Carlton	1972	346	56.9	55.6	6.0	.990	9.30	80	Jack Coombs	1910	353	47.8	45.3	5.0	.975	9.01
31	Joe McGinnity	1904	408	50.7	55.9	5.9	1.042	9.45	81	Christy Mathewson	1911	307	48.1	50.2	5.0	1.018	9.98
32	Hal Newhouser	1946	293	51.0	56.4	5.9	1.047	9.54	82	Bert Blyleven	1977	235	35.3	50.6	5.0	1.143	10.07
33	Jack Taylor	1902	325	52.3	55.9	5.9	1.036	9.48	83	Christy Mathewson	1912	310	44.5	50.7	5.0	1.053	10.19
34	Lefty Grove	1926	258	43.3	60.5	5.9	1.150	10.33	84	Cy Blanton	1935	254	40.5	51.2	5.0	1.094	10.29
35	Red Faber	1921	331	66.5	62.5	5.8	.975	10.70	85	Pete Alexander	1920	363	49.4	46.2	5.0	.975	9.34
36	Carl Hubbell	1933	309	57.6	55.2	5.8	.979	9.45	86	Randy Jones	1975	285	44.0	47.3	4.9	1.028	9.58
37	Bob Gibson	1969	314	49.5	54.7	5.7	1.041	9.53	87	Ferguson Jenkins	1970	313	23.0	49.5	4.9	1.188	10.04
38	Joe Wood	1912	344	54.9	57.2	5.7	1.018	10.03	88	Ed Reulbach	1905	292	51.1	47.3	4.9	.961	9.62
39	Eddie Cicotte	1919	307	47.9	53.8	5.6	1.053	9.56	89	Pete Alexander	1917	388	36.9	43.5	4.9	1.056	8.86
40	Steve Carlton	1980	304	42.9	53.0	5.6	1.083	9.46	90	Cy Young	1903	342	33.1	47.2	4.9	1.126	9.63
41	Warren Spahn	1953	266	64.7	57.8	5.6	.946	10.38	91	Lefty Grove	1939	191	44.0	52.9	4.9	1.090	10.85
42	Steve Rogers	1982	277	37.1	52.9	5.6	1.143	9.52	92	Ted Lyons	1927	308	44.7	51.2	4.9	1.046	10.53
43	Frank Sullivan	1955	260	30.5	55.1	5.5	1.215	9.98	93	Dutch Leonard	1914	225	44.4	44.0	4.9	.994	9.05
44	Lefty Grove	1936	253	62.6	62.5	5.5	.999	11.32	94	Howie Pollet	1946	266	38.9	45.8	4.9	1.068	9.42
45	Dazzy Vance	1930	259	68.1	62.4	5.5	.960	11.33	95	Warren Spahn	1947	290	56.1	49.3	4.8	.948	10.16
46	Christy Mathewson	1909	275	44.4	49.6	5.5	1.065	9.03	96	Mort Cooper	1942	279	47.7	45.2	4.8	.975	9.32
47	Three Finger Brown	1909	343	49.1	49.3	5.5	1.002	9.03	97	Jim Palmer	1975	323	61.0	47.1	4.8	.898	9.79
48	Cy Falkenberg	1914	377	41.2	52.4	5.4	1.083	9.62	98	Bob Gibson	1970	294	30.4	48.2	4.8	1.134	10.04
49	Bobby Shantz	1952	280	37.3	52.3	5.4	1.131	9.64	99	Denny McLain	1968	336	38.2	41.8	4.8	1.032	8.72
50	Walter Johnson	1915	337	52.4	51.0	5.4	.988	9.42	100	Wes Ferrell	1930	297	44.6	52.9	4.8	1.054	11.06

SINGLE-SEASON LEADERS ◇ 295

	Name	Year	IP	ERA	NERA	Adj	PF	LG
1	Walter Johnson	1913	346	1.14	256.9	285.4	1.111	2.93
2	Dutch Leonard	1914	225	.96	285.2	283.5	.994	2.74
3	Three Finger Brown	1906	277	1.04	253.0	271.2	1.072	2.63
4	Bob Gibson	1968	305	1.12	266.6	252.7	.948	2.99
5	Christy Mathewson	1909	275	1.15	226.1	240.8	1.065	2.60
6	Walter Johnson	1912	368	1.39	240.8	238.9	.992	3.35
7	Pete Alexander	1915	376	1.22	225.5	237.1	1.051	2.75
8	Walter Johnson	1918	325	1.27	217.9	235.5	1.081	2.77
9	Jack Pfiester	1907	195	1.15	214.3	229.8	1.072	2.46
10	Christy Mathewson	1905	339	1.27	235.5	228.7	.971	2.99
11	Carl Lundgren	1907	207	1.17	210.7	225.8	1.072	2.46
12	Ron Guidry	1978	274	1.74	217.2	220.6	1.016	3.78
13	Walter Johnson	1919	290	1.49	216.3	219.1	1.013	3.22
14	Jack Taylor	1902	325	1.33	208.9	216.4	1.036	2.78
15	Lefty Grove	1931	289	2.06	212.7	213.7	1.005	4.38
16	Addie Joss	1908	325	1.16	205.8	210.4	1.022	2.39
17	Whitey Ford	1958	219	2.01	187.5	204.4	1.090	3.77
18	Cy Young	1901	371	1.63	224.7	202.9	.903	3.66
19	Ed Reulbach	1905	292	1.42	210.7	202.4	.961	2.99
20	Vida Blue	1971	312	1.82	190.5	201.8	1.059	3.47
21	Monty Stratton	1937	165	2.40	192.5	201.4	1.046	4.62
22	Hal Newhouser	1945	313	1.81	185.9	200.6	1.079	3.36
23	Three Finger Brown	1909	343	1.31	198.4	198.8	1.002	2.60
24	Lefty Grove	1939	191	2.54	181.9	198.2	1.090	4.62
25	Carl Hubbell	1933	309	1.66	201.1	196.9	.979	3.34
26	Luis Tiant	1968	258	1.60	186.2	196.0	1.053	2.98
27	Sandy Koufax	1966	323	1.73	208.5	196.0	.940	3.61
28	Dolf Luque	1923	322	1.93	207.4	195.7	.944	4.00
29	Billy Pierce	1955	206	1.97	201.1	195.5	.972	3.96
30	Dean Chance	1964	278	1.65	219.7	194.9	.887	3.63
31	Warren Spahn	1953	266	2.10	204.1	193.1	.946	4.29
32	Cy Young	1908	299	1.26	189.5	192.0	1.013	2.39
33	Sandy Koufax	1964	223	1.74	203.2	191.6	.943	3.54
34	Three Finger Brown	1907	233	1.39	177.3	190.1	1.072	2.46
35	Tom Seaver	1971	286	1.76	197.2	189.5	.961	3.47
36	Hal Newhouser	1946	293	1.94	180.6	189.1	1.047	3.50
37	Ed Siever	1902	188	1.91	186.9	188.9	1.011	3.57
38	Jack Coombs	1910	353	1.30	193.7	188.9	.975	2.52
39	Wilbur Wood	1971	334	1.91	181.5	188.8	1.040	3.47
40	Hank Aguirre	1962	216	2.21	179.7	188.3	1.048	3.97
41	Walter Johnson	1915	337	1.55	190.2	187.9	.988	2.95
42	Rube Waddell	1905	329	1.48	178.9	187.3	1.047	2.65
43	Harry Krause	1909	213	1.39	178.3	186.8	1.048	2.48
44	Fred Toney	1915	223	1.57	175.3	186.7	1.065	2.75
45	Rube Waddell	1902	276	2.05	174.1	186.6	1.072	3.57
46	Eddie Cicotte	1919	307	1.82	177.1	186.5	1.053	3.22
47	Dizzy Trout	1944	352	2.12	161.8	186.4	1.152	3.43
48	Vean Gregg	1911	245	1.80	185.7	185.9	1.001	3.34
49	Juan Marichal	1965	295	2.14	165.4	185.2	1.120	3.54
50	Walter Johnson	1910	374	1.35	186.6	184.9	.991	2.52
51	Phil Niekro	1967	207	1.87	180.5	184.9	1.024	3.38
52	Dazzy Vance	1928	280	2.09	190.7	184.3	.966	3.99
53	Lefty Grove	1926	258	2.51	160.2	184.2	1.150	4.02
54	Orval Overall	1909	285	1.42	183.1	183.4	1.002	2.60
55	Spud Chandler	1943	253	1.64	201.0	183.3	.912	3.30
56	Jeff Tesreau	1912	243	1.96	174.0	183.2	1.053	3.41
57	Dazzy Vance	1924	309	2.16	179.0	183.1	1.023	3.87
58	Dazzy Vance	1930	259	2.61	190.5	182.9	.960	4.97
59	Joe Wood	1915	157	1.49	197.8	182.8	.924	2.95
60	Mort Cooper	1942	279	1.77	187.2	182.5	.975	3.31
61	Lefty Gomez	1937	278	2.33	198.3	181.7	.916	4.62
62	Lefty Grove	1935	273	2.70	165.0	181.3	1.099	4.46
63	Jack Pfiester	1906	242	1.56	168.7	180.8	1.072	2.63
64	Ed Walsh	1910	370	1.26	199.9	180.5	.903	2.52
65	Russ Ford	1914	247	1.82	176.1	179.8	1.021	3.20
66	Al Benton	1945	192	2.02	166.6	179.7	1.079	3.36
67	Lefty Grove	1936	253	2.81	179.3	179.1	.999	5.04
68	Max Lanier	1943	213	1.90	177.6	179.0	1.008	3.37
69	Eddie Cicotte	1913	268	1.58	185.4	178.7	.964	2.93
70	Joe Wood	1912	344	1.91	175.2	178.4	1.018	3.35
71	Three Finger Brown	1908	312	1.47	159.5	178.1	1.116	2.35
72	Hal Newhouser	1944	312	2.22	154.5	178.0	1.152	3.43
73	Joe McGinnity	1904	408	1.61	169.5	176.6	1.042	2.73
74	Claude Hendrix	1914	362	1.69	189.6	176.4	.930	3.20
75	Mike Garcia	1949	176	2.35	178.6	175.9	.985	4.20
76	Bill Doak	1914	256	1.72	162.1	175.7	1.084	2.79
77	Bucky Walters	1939	319	2.29	171.1	175.2	1.024	3.92
78	Whitey Ford	1964	245	2.13	170.2	175.1	1.029	3.63
79	Dizzy Dean	1934	312	2.65	153.3	174.8	1.140	4.06
80	Fred Anderson	1917	162	1.44	188.3	174.6	.927	2.71
81	Harry Brecheen	1948	233	2.24	176.7	174.5	.988	3.96
82	Christy Mathewson	1911	307	1.99	171.0	174.1	1.018	3.40
83	Hoyt Wilhelm	1959	226	2.19	176.3	174.0	.987	3.86
84	Tom Seaver	1973	290	2.08	176.4	173.8	.985	3.67
85	Howie Pollet	1946	266	2.10	162.6	173.7	1.068	3.41
86	Russ Ford	1911	281	2.27	147.2	173.6	1.179	3.34
87	Sam McDowell	1968	269	1.81	164.6	173.3	1.053	2.98
88	Joe Horlen	1964	211	1.88	192.8	173.2	.898	3.63
89	Willie Mitchell	1913	217	1.74	168.3	173.0	1.028	2.93
90	Cy Young	1902	385	2.15	166.0	173.0	1.042	3.57
91	Allie Reynolds	1952	244	2.07	177.5	172.9	.974	3.67
92	Steve Carlton	1972	346	1.98	174.6	172.8	.990	3.46
93	Steve Carlton	1969	236	2.17	165.8	172.6	1.041	3.60
94	Ted Lyons	1942	180	2.10	174.1	172.6	.991	3.66
95	Gaylord Perry	1972	343	1.92	159.9	172.6	1.079	3.07
96	Lefty Grove	1932	292	2.84	157.7	172.2	1.092	4.48
97	Steve Rogers	1982	277	2.40	150.4	171.9	1.143	3.61
98	Bob Gibson	1969	314	2.18	165.1	171.8	1.041	3.60
99	Bert Blyleven	1973	325	2.52	151.7	171.6	1.131	3.82
100	Stan Coveleski	1917	298	1.81	147.1	171.5	1.166	2.66

		Year	Clb	L	W	L	TW	TL	Wins			Year	Clb	L	W	L	TW	TL	Wins
1	Steve Carlton	1972	PHI	N	27	10	59	97	17.1	51	Juan Marichal	1966	SF	N	25	6	93	68	8.8
2	Walter Johnson	1913	WAS	A	36	7	90	64	15.1	52	Irv Young	1905	BOS	N	20	21	51	103	8.8
3	Eddie Rommel	1922	PHI	A	27	13	65	89	13.7	53	Slim Sallee	1913	STL	N	18	15	51	99	8.7
4	Jack Chesbro	1904	NY	A	41	12	92	59	13.4	54	Carl Hubbell	1936	NY	N	26	6	92	62	8.7
5	Ed Walsh	1908	CHI	A	40	15	88	64	12.8	55	Urban Shocker	1921	STL	A	27	12	81	73	8.7
6	Walter Johnson	1911	WAS	A	25	13	64	90	12.2	56	Jim Maloney	1963	CIN	N	23	7	86	76	8.7
7	Red Faber	1921	CHI	A	25	15	62	92	12.0	57	Bob Feller	1941	CLE	A	25	13	75	79	8.6
8	Cy Young	1901	BOS	A	33	10	79	57	11.7	58	General Crowder	1928	STL	A	21	5	82	72	8.6
9	Ned Garver	1951	STL	A	20	12	52	102	11.6	59	Walter Johnson	1912	WAS	A	33	12	91	61	8.6
10	Cy Young	1902	BOS	A	32	11	77	60	11.4	60	Eddie Cicotte	1919	CHI	A	29	7	88	52	8.6
11	Bob Gibson	1970	STL	N	23	7	76	86	11.0	61	Mario Soto	1984	CIN	N	18	7	70	92	8.5
12	Bob Feller	1946	CLE	A	26	15	68	86	10.8	62	Lefty Stewart	1930	STL	A	20	12	64	90	8.5
13	Robin Roberts	1952	PHI	N	28	7	87	67	10.6	63	Walter Johnson	1919	WAS	A	20	14	56	84	8.5
14	Dazzy Vance	1925	BRO	N	22	9	68	85	10.3	64	Gaylord Perry	1978	SD	N	21	6	84	78	8.4
15	Bobby Shantz	1952	PHI	A	24	7	79	75	10.1	65	Dave Ferriss	1945	BOS	A	21	10	71	83	8.4
16	Dazzy Vance	1924	BRO	N	28	6	92	62	9.9	66	Sandy Koufax	1964	LA	N	19	5	80	82	8.4
17	Dutch Leonard	1939	WAS	A	20	8	65	87	9.8	67	Paul Derringer	1935	CIN	N	22	13	68	85	8.4
18	Denny McLain	1968	DET	A	31	6	103	59	9.7	68	Phil Collins	1930	PHI	N	16	11	52	102	8.3
19	Hal Newhouser	1944	DET	A	29	9	88	66	9.7	69	Jeff Tesreau	1914	NY	N	26	10	84	70	8.3
20	Larry Jackson	1964	CHI	N	24	11	76	86	9.7	70	Don Newcombe	1956	BRO	N	27	7	93	61	8.3
21	Ewell Blackwell	1947	CIN	N	22	8	73	81	9.7	71	Bert Blyleven	1984	CLE	A	19	7	75	87	8.3
22	Pete Alexander	1920	CHI	N	27	14	75	79	9.6	72	Lefty Grove	1931	PHI	A	31	4	107	45	8.3
23	Roy Face	1959	PIT	N	18	1	78	76	9.6	73	Sandy Koufax	1963	LA	N	25	5	99	63	8.2
24	Joe Wood	1912	BOS	A	34	5	105	47	9.5	74	Dolf Luque	1923	CIN	N	27	8	91	63	8.2
25	Noodles Hahn	1901	CIN	N	22	19	52	87	9.4	75	Christy Mathewson	1901	NY	N	20	17	52	85	8.2
26	Ron Guidry	1978	NY	A	25	3	100	63	9.4	76	Russ Ford	1914	BUF	F	21	6	80	71	8.2
27	Wilbur Cooper	1917	PIT	N	17	11	51	103	9.4	77	Jack Taylor	1902	CHI	N	23	11	68	69	8.1
28	Christy Mathewson	1908	NY	N	37	11	98	56	9.4	78	Ed Morris	1928	BOS	A	19	15	57	96	8.1
29	Pete Alexander	1914	PHI	N	27	15	74	80	9.4	79	Burleigh Grimes	1918	BRO	N	19	9	57	69	8.1
30	Ted Lyons	1930	CHI	A	22	15	62	92	9.4	80	Elmer Knetzer	1914	PIT	F	20	12	64	86	8.1
31	Claude Hendrix	1914	CHI	F	29	10	87	67	9.3	81	Casey Patten	1906	WAS	A	19	16	55	95	8.0
32	Russ Ford	1910	NY	A	26	6	88	63	9.3	82	Christy Mathewson	1909	NY	N	25	6	92	61	8.0
33	Dizzy Dean	1934	STL	N	30	7	95	58	9.3	83	Nolan Ryan	1974	CAL	A	22	16	68	94	7.9
34	Lefty Grove	1933	PHI	A	24	8	79	72	9.2	84	Bob Hooper	1950	PHI	A	15	10	52	102	7.8
35	Pete Alexander	1915	PHI	N	31	10	90	62	9.2	85	Lefty Grove	1930	PHI	A	28	5	102	52	7.8
36	Walter Johnson	1910	WAS	A	25	17	66	85	9.2	86	Joe Bush	1916	PHI	A	15	24	36	117	7.8
37	Pete Alexander	1911	PHI	N	28	13	79	73	9.2	87	Tom Seaver	1975	NY	N	22	9	82	80	7.8
38	Bobo Newsom	1938	STL	A	20	16	55	97	9.1	88	Preacher Roe	1951	BRO	N	22	3	97	60	7.8
39	Warren Spahn	1963	MIL	N	23	7	84	78	9.1	89	Ray Scarborough	1948	WAS	A	15	8	56	97	7.7
40	Vean Gregg	1911	CLE	A	23	7	80	73	9.1	90	Jeff Pfeffer	1914	BRO	N	23	12	75	79	7.7
41	Wild Bill Donovan	1907	DET	A	25	4	92	58	8.9	91	Addie Joss	1907	CLE	A	27	11	85	67	7.7
42	Al Mamaux	1915	PIT	N	21	8	73	81	8.9	92	Jesse Tannehill	1905	BOS	A	22	9	78	74	7.7
43	Jimmy Ring	1923	PHI	N	18	16	50	104	8.9	93	Tex Hughson	1944	BOS	A	18	5	77	77	7.6
44	Eddie Plank	1912	PHI	A	26	6	90	62	8.9	94	Bob Porterfield	1953	WAS	A	22	10	76	76	7.6
45	Juan Marichal	1968	SF	N	26	9	88	74	8.9	95	Al Mamaux	1916	PIT	N	21	15	65	89	7.6
46	Cy Young	1907	BOS	A	21	15	59	90	8.9	96	Mel Stottlemyre	1965	NY	A	20	9	77	85	7.6
47	Juan Marichal	1963	SF	N	25	8	88	74	8.9	97	Sandy Koufax	1966	LA	N	27	9	95	67	7.6
48	Lefty Gomez	1934	NY	A	26	5	94	60	8.9	98	Mike Norris	1980	OAK	A	22	9	83	79	7.6
49	Pete Alexander	1916	PHI	N	33	12	91	62	8.8	99	Fred Toney	1915	CIN	N	17	6	71	83	7.5
50	Larry Jansen	1947	NY	N	21	5	81	73	8.8	100	Pete Alexander	1917	PHI	N	30	13	87	65	7.5

	Year	Clb	L	W	L	TW	TL	Pct			Year	Clb	L	W	L	TW	TL	Pct
1 Steve Carlton	1972	PHI	N	27	10	59	97	45.8	51 Ed Morris	1928	BOS	A	19	15	57	96	33.3
2 Ed Walsh	1908	CHI	A	40	15	88	64	45.5	52 Al Mattern	1909	BOS	N	15	21	45	108	33.3
3 Jack Chesbro	1904	NY	A	41	12	92	59	44.6	53 Claude Hendrix	1914	CHI	F	29	10	87	67	33.3
4 Noodles Hahn	1901	CIN	N	22	19	52	87	42.3	54 Burleigh Grimes	1918	BRO	N	19	9	57	69	33.3
5 Cy Young	1901	BOS	A	33	10	79	57	41.8	55 Bob Feller	1941	CLE	A	25	13	75	79	33.3
6 Joe Bush	1916	PHI	A	15	24	36	117	41.7	56 Murry Dickson	1952	PIT	N	14	21	42	112	33.3
7 Cy Young	1902	BOS	A	32	11	77	60	41.6	57 Wilbur Cooper	1917	PIT	N	17	11	51	103	33.3
8 Eddie Rommel	1922	PHI	A	27	13	65	89	41.5	58 Joe McGinnity	1904	NY	N	35	8	106	47	33.0
9 Red Faber	1921	CHI	A	25	15	62	92	40.3	59 Hal Newhouser	1944	DET	A	29	9	88	66	33.0
10 Walter Johnson	1913	WAS	A	36	7	90	64	40.0	60 Eddie Cicotte	1919	CHI	A	29	7	88	52	33.0
11 Irv Young	1905	BOS	N	20	21	51	103	39.2	61 Walter Johnson	1916	WAS	A	25	20	76	77	32.9
12 Walter Johnson	1911	WAS	A	25	13	64	90	39.1	62 Noodles Hahn	1902	CIN	N	23	12	70	70	32.9
13 Elmer Myers	1916	PHI	A	14	23	36	117	38.9	63 Howard Ehmke	1923	BOS	A	20	17	61	91	32.8
14 Scott Perry	1918	PHI	A	20	19	52	76	38.5	64 Vic Willis	1904	BOS	N	18	25	55	98	32.7
15 Christy Mathewson	1901	NY	N	20	17	52	85	38.5	65 Irv Young	1906	BOS	N	16	25	49	102	32.7
16 Ned Garver	1951	STL	A	20	12	52	102	38.5	66 Tom Seaton	1914	BRO	F	25	14	77	77	32.5
17 Joe McGinnity	1901	BAL	A	26	20	68	65	38.2	67 Joe Wood	1912	BOS	A	34	5	105	47	32.4
18 Bob Feller	1946	CLE	A	26	15	68	86	38.2	68 Dazzy Vance	1925	BRO	N	22	9	68	85	32.4
19 Walter Johnson	1910	WAS	A	25	17	66	85	37.9	69 Nolan Ryan	1974	CAL	A	22	16	68	94	32.4
20 Christy Mathewson	1908	NY	N	37	11	98	56	37.8	70 Paul Derringer	1935	CIN	N	22	13	68	85	32.4
21 Vic Willis	1902	BOS	N	27	19	73	64	37.0	71 Willie Sudhoff	1903	STL	A	21	15	65	74	32.3
22 Togie Pittinger	1902	BOS	N	27	15	73	64	37.0	72 Al Mamaux	1916	PIT	N	21	15	65	89	32.3
23 Joe McGinnity	1903	NY	N	31	20	84	55	36.9	73 Robin Roberts	1952	PHI	N	28	7	87	67	32.2
24 Casey Patten	1904	WAS	A	14	23	38	113	36.8	74 Eddie Plank	1904	PHI	A	26	16	81	70	32.1
25 Pete Alexander	1914	PHI	N	27	15	74	80	36.5	75 Nap Rucker	1908	BRO	N	17	19	53	101	32.1
26 Buck Newsom	1938	STL	A	20	16	55	97	36.4	76 Frank Smith	1909	CHI	A	25	17	78	74	32.1
27 Walter Johnson	1912	WAS	A	33	12	91	61	36.3	77 Wes Ferrell	1935	BOS	A	25	14	78	75	32.1
28 Pete Alexander	1916	PHI	N	33	12	91	62	36.3	78 Walter Johnson	1918	WAS	A	23	13	72	56	31.9
29 Jimmy Ring	1923	PHI	N	18	16	50	104	36.0	79 Phil Niekro	1979	ATL	N	21	20	66	94	31.8
30 Pete Alexander	1920	CHI	N	27	14	75	79	36.0	80 Addie Joss	1907	CLE	A	27	11	85	67	31.8
31 Christy Mathewson	1903	NY	N	30	13	84	55	35.7	81 Walter Johnson	1915	WAS	A	27	13	85	68	31.8
32 Walter Johnson	1919	WAS	A	20	14	56	84	35.7	82 George Uhle	1923	CLE	A	26	16	82	71	31.7
33 Cy Young	1907	BOS	A	21	15	59	90	35.5	83 Wild Bill Donovan	1901	BRO	N	25	15	79	57	31.6
34 Ted Lyons	1930	CHI	A	22	15	62	92	35.5	84 Jim Bagby	1920	CLE	A	31	12	98	56	31.6
35 Pete Alexander	1911	PHI	N	28	13	79	73	35.4	85 Larry Jackson	1964	CHI	N	24	11	76	86	31.6
36 Slim Sallee	1913	STL	N	18	15	51	99	35.3	86 Dizzy Dean	1934	STL	N	30	7	95	58	31.6
37 Ed Walsh	1911	CHI	A	27	18	77	74	35.1	87 Henry Schmidt	1903	BRO	N	22	13	70	66	31.4
38 Ed Walsh	1912	CHI	A	27	17	78	76	34.6	88 Ted Lyons	1927	CHI	A	22	14	70	83	31.4
39 Dummy Taylor	1901	NY	N	18	27	52	85	34.6	89 Bill Voiselle	1944	NY	N	21	16	67	87	31.3
40 Walter Johnson	1914	WAS	A	28	18	81	73	34.6	90 Jake Weimer	1906	CIN	N	20	14	64	87	31.3
41 Casey Patten	1906	WAS	A	19	16	55	95	34.5	91 Lefty Stewart	1930	STL	A	20	12	64	90	31.3
42 Pete Alexander	1917	PHI	N	30	13	87	65	34.5	92 Elmer Knetzer	1914	PIT	F	20	12	64	86	31.3
43 Pete Alexander	1915	PHI	N	31	10	90	62	34.4	93 Murry Dickson	1951	PIT	N	20	16	64	90	31.3
44 Nap Rucker	1911	BRO	N	22	18	64	86	34.4	94 Wilbur Wood	1973	CHI	A	24	20	77	85	31.2
45 Curt Davis	1934	PHI	N	19	17	56	93	33.9	95 Al Orth	1902	WAS	A	19	18	61	75	31.1
46 Jack Taylor	1902	CHI	N	23	11	68	69	33.8	96 Christy Mathewson	1904	NY	N	33	12	106	47	31.1
47 Roscoe Miller	1901	DET	A	25	13	74	61	33.8	97 Hippo Vaughn	1917	CHI	N	23	13	74	80	31.1
48 Urban Shocker	1921	STL	A	27	12	81	73	33.3	98 Walter Johnson	1917	WAS	A	23	16	74	79	31.1
49 William Reidy	1901	MIL	A	16	20	48	89	33.3	99 Doc White	1907	CHI	A	27	13	87	64	31.0
50 Gaylord Perry	1972	CLE	A	24	16	72	84	33.3	100 Nap Rucker	1912	BRO	N	18	21	58	95	31.0

		Year	IP	ERA	NERA	Adj	PF	LG			Year	IP	ERA	NERA	Adj	PF	LG
1	Rollie Fingers	1981	78	1.04	352.2	346.9	.985	3.66	26	Ken Sanders	1970	92	1.76	211.3	210.4	.996	3.72
2	Bruce Sutter	1977	107	1.35	289.9	327.3	1.129	3.91	27	Sparky Lyle	1974	114	1.66	218.4	210.1	.962	3.63
3	Ted Abernathy	1967	106	1.27	265.8	283.9	1.068	3.38	28	Rich Gossage	1975	142	1.84	205.9	209.6	1.018	3.79
4	Frank Linzy	1965	82	1.43	247.5	277.2	1.120	3.54	29	Phil Regan	1966	117	1.62	222.6	209.3	.940	3.61
5	Jim Kern	1979	143	1.57	269.5	267.3	.992	4.23	30	Bob Miller	1971	99	1.64	211.6	209.2	.989	3.47
6	John Hiller	1973	125	1.44	265.5	267.1	1.006	3.82	31	Dick Hall	1964	88	1.84	197.0	208.2	1.057	3.63
7	Tug McGraw	1980	92	1.47	245.5	265.9	1.083	3.61	32	Mike Marshall	1972	116	1.78	194.2	207.8	1.070	3.46
8	Ken Tatum	1969	86	1.36	266.7	261.9	.982	3.63	33	Dan Quisenberry	1983	139	1.9;	210.1	206.7	.984	4.08
9	Bob Apodaca	1975	85	1.48	245.5	244.2	.995	3.63	34	Doug Corbett	1980	136	1.99	203.2	206.2	1.015	4.04
10	Jesse Orosco	1983	110	1.47	247.4	243.7	.985	3.64	35	Gary Lavelle	1977	118	2.06	190.0	205.8	1.083	3.91
11	Marv Grissom	1956	81	1.56	241.8	243.0	1.005	3.77	36	Rick Camp	1981	76	1.78	196.2	203.0	1.035	3.49
12	Rich Gossage	1977	133	1.62	241.6	237.3	.982	3.91	37	Doug Corbett	1984	85	2.12	188.6	202.5	1.074	4.00
13	Ellis Kinder	1953	107	1.85	216.0	237.1	1.098	4.00	38	Jumbo Brown	1938	90	1.80	210.5	202.3	.961	3.79
14	Kent Tekulve	1983	99	1.64	221.8	235.8	1.063	3.64	39	Bob Reynolds	1973	111	1.95	190.0	200.9	1.025	3.82
15	Rod Scurry	1982	104	1.74	207.4	233.3	1.125	3.61	40	Tom House	1974	103	1.92	189.0	200.6	1.061	3.63
16	Hoyt Wilhelm	1967	89	1.31	246.6	232.3	.942	3.23	41	Jeff Reardon	1982	109	2.06	175.2	200.2	1.143	3.61
17	Frank Linzy	1967	96	1.50	225.1	228.2	1.014	3.38	42	Johnny Murphy	1941	77	1.99	208.4	199.7	.958	4.15
18	Al Hrabosky	1975	97	1.67	217.5	226.7	1.042	3.63	43	Al Holland	1980	82	1.76	205.0	199.3	.972	3.61
19	Bruce Sutter	1984	123	1.54	233.3	225.1	.965	3.59	44	Moe Drabowsky	1967	95	1.61	200.7	197.9	.986	3.23
20	Dan Quisenberry	1981	62	1.74	210.5	224.2	1.065	3.66	45	Bill Caudill	1982	96	2.34	174.4	197.5	1.132	4.08
21	Tom Burgmeier	1980	99	2.00	202.1	222.0	1.098	4.04	46	Ryne Duren	1959	77	1.87	206.5	197.2	.955	3.86
22	Dick Hyde	1958	103	1.75	215.4	221.4	1.028	3.77	47	Tug McGraw	1971	111	1.70	204.1	196.2	.961	3.47
23	Roy Face	1962	91	1.88	209.6	218.0	1.040	3.94	48	Hersh Freeman	1955	92	2.15	187.8	195.5	1.041	4.04
24	Bob Lee	1964	137	1.51	240.1	212.9	.887	3.63	49	Tom Burgmeier	1971	88	1.74	199.3	195.3	.980	3.47
25	Lee Smith	1983	103	1.65	220.4	211.9	.961	3.64	50	Tom Niedenfuer	1983	95	1.90	191.4	195.1	1.019	3.64

		Year	IP	All	Pit	Bat	Fld	Rel			Year	IP	All	Pit	Bat	Fld	Rel
1	Dan Quisenberry	1984	129	2.1	1.8	0.0	.3	97	25	Kent Tekulve	1979	134	2.0	1.9	-0.0	.1	74
2	Dan Quisenberry	1983	139	3.3	3.2	0.0	.1	97	26	Dick Radatz	1963	132	2.2	2.6	-.2	-.2	74
3	Bruce Sutter	1984	123	2.8	2.8	-.1	.1	93	27	Ron Perranoski	1970	111	1.2	1.5	-.2	-.1	74
4	John Hiller	1973	125	3.4	3.4	0.0	0.0	91	28	Lindy McDaniel	1960	116	2.6	2.4	.1	.1	74
5	Dan Quisenberry	1980	128	1.5	1.2	0.0	.3	83	29	Bruce Sutter	1977	107	3.8	3.7	-0.0	.1	73
6	Sparky Lyle	1972	108	1.1	1.2	.1	-.2	83	30	Sparky Lyle	1977	137	2.5	2.6	0.0	-.1	73
7	Bill Caudill	1984	96	.8	.9	0.0	-.1	83	31	Rollie Fingers	1978	107	.9	.9	-0.0	0.0	73
8	Luis Arroyo	1961	119	1.8	1.7	.2	-.1	83	32	Bob Stanley	1983	145	2.1	2.2	0.0	-.1	72
9	Bruce Sutter	1982	102	1.1	1.1	-.0	0.0	82	33	Joe Page	1949	135	1.9	2.2	-.1	-.2	72
10	Clay Carroll	1972	96	1.4	1.2	-0.0	.2	82	34	Kent Tekulve	1978	135	2.3	2.1	-.1	.3	71
11	Dick Radatz	1964	157	2.0	2.2	-0.0	-.2	81	35	Ron Perranoski	1963	129	2.3	2.3	-0.0	0.0	71
12	Dan Quisenberry	1982	137	2.4	1.8	0.0	.6	81	36	Lindy McDaniel	1970	112	1.8	1.7	0.0	.1	71
13	Bruce Sutter	1979	101	2.1	2.0	.1	-.0	80	37	Eddie Fisher	1965	165	1.6	1.4	0.0	.2	71
14	Mike Marshall	1973	179	2.7	2.2	.1	.4	79	38	Ron Perranoski	1969	120	2.0	2.0	-.1	.1	70
15	Jim Kern	1979	143	4.0	4.1	0.0	-.1	79	39	Mike Marshall	1979	143	3.4	3.2	0.0	.2	70
16	Willie Hernandez	1984	140	2.7	2.8	0.0	-.1	79	40	Hoyt Wilhelm	1964	131	1.8	1.9	-0.0	-.1	69
17	Bill Campbell	1977	140	2.9	2.8	0.0	.1	79	41	Minnie Rojas	1967	122	.5	.8	-.1	-.2	69
18	Lee Smith	1984	101	.2	.3	-.1	.0	77	42	Phil Regan	1966	117	2.5	2.4	0.0	.1	69
19	Wayne Granger	1970	85	1.4	1.3	-.1	.2	77	43	Stu Miller	1965	119	2.3	2.3	-.1	.1	69
20	Rollie Fingers	1977	132	.7	.9	-.2	.0	77	44	Jim Konstanty	1950	152	1.8	2.1	-.2	-.1	69
21	Jesse Orosco	1984	87	1.0	.9	.1	.0	76	45	Rawly Eastwick	1976	108	1.7	2.1	-.2	-.2	69
22	Greg Minton	1982	123	2.2	2.1	0.0	.1	76	46	Bill Campbell	1976	168	.9	.9	0.0	0.0	69
23	Rich Gossage	1980	99	1.7	1.8	0.0	-.1	76	47	Ellis Kinder	1953	107	3.3	3.0	.3	-.0	68
24	Jack Aker	1966	113	1.9	1.6	-.1	.4	76									

		Year	IP	All	Pit	Bat	Fld				Year	IP	All	Pit	Bat	Fld
1	Walter Johnson	1913	346	9.5	8.7	1.1	-.3	51	Thornton Lee	1941	300	5.6	5.2	.5	-.1
2	Dizzy Trout	1944	352	9.3	7.5	1.1	.7	52	Red Faber	1921	331	5.6	5.8	-.3	.1
3	Walter Johnson	1912	368	8.6	7.8	.9	-.1	53	Pete Alexander	1920	363	5.6	4.9	.5	.2
4	Pete Alexander	1915	376	8.2	7.7	-.1	.6	54	Cy Young	1903	342	5.5	4.9	1.1	-.5
5	Bucky Walters	1939	319	7.7	6.1	1.2	.4	55	Rube Waddell	1902	276	5.5	5.2	.6	-.3
6	Christy Mathewson	1905	339	7.7	6.4	.8	.5	56	Dazzy Vance	1928	280	5.5	5.3	.2	-.0
7	Walter Johnson	1918	325	7.6	6.9	1.0	-.3	57	Steve Rogers	1982	277	5.5	5.6	-.1	.0
8	Hal Newhouser	1945	313	7.4	6.8	.4	.2	58	Carl Mays	1921	337	5.5	4.2	1.0	.3
9	Joe Wood	1912	344	7.3	5.7	1.0	.6	59	Juan Marichal	1969	300	5.5	5.2	-.1	.4
10	Jack Chesbro	1904	455	7.3	6.4	.5	.4	60	Bob Lemon	1949	280	5.5	3.5	1.4	.6
11	Ed Walsh	1912	393	7.1	6.2	.4	.5	61	Carl Hubbell	1934	313	5.5	5.2	-.1	.4
12	Ron Guidry	1978	274	6.8	6.6	0.0	.2	62	Carl Hubbell	1932	284	5.5	4.6	.3	.6
13	Bob Gibson	1968	305	6.8	6.6	.5	-.3	63	Pete Alexander	1917	388	5.5	4.9	.4	.2
14	Ed Walsh	1907	422	6.7	4.6	-.2	2.3	64	Pete Alexander	1916	389	5.5	4.4	1.0	.1
15	Jack Taylor	1902	325	6.7	5.9	.4	.4	65	Ed Walsh	1908	464	5.4	3.8	.3	1.3
16	Christy Mathewson	1909	275	6.7	5.5	.7	.5	66	Tom Seaver	1973	290	5.4	5.2	.3	-.1
17	Wes Ferrell	1935	322	6.7	4.6	2.0	.1	67	Carl Mays	1917	289	5.4	4.0	.5	.9
18	Carl Hubbell	1933	309	6.6	5.8	0.0	.8	68	Ned Garver	1950	260	5.4	4.7	.5	.2
19	Dizzy Dean	1934	312	6.6	6.7	.2	-.3	69	Whitey Ford	1958	219	5.4	5.3	0.0	.1
20	Hal Newhouser	1944	312	6.5	6.3	.2	-.0	70	Bob Feller	1939	297	5.4	5.1	.4	-.1
21	Dolf Luque	1923	322	6.5	6.3	.2	-.0	71	Steve Carlton	1980	304	5.4	5.6	-.1	-.1
22	Lefty Grove	1931	289	6.5	7.0	-.2	-.3	72	Babe Ruth	1917	326	5.3	3.4	1.7	.2
23	Bert Blyleven	1973	325	6.5	6.7	0.0	-.2	73	Robin Roberts	1953	347	5.3	5.2	.2	-.1
24	Cy Young	1901	371	6.4	6.3	.1	-.0	74	Phil Niekro	1978	334	5.3	4.6	.4	.3
25	Wilbur Wood	1971	334	6.4	6.8	-.5	.1	75	Christy Mathewson	1912	310	5.3	4.9	.4	.0
26	Gaylord Perry	1972	343	6.4	6.1	0.0	.3	76	Christy Mathewson	1908	391	5.3	4.2	.1	1.0
27	Claude Hendrix	1914	362	6.4	5.4	.5	.5	77	Ted Lyons	1927	308	5.3	4.8	.4	.1
28	Juan Marichal	1965	295	6.3	6.3	-0.0	0.0	78	Sandy Koufax	1966	323	5.3	6.3	-.6	-.4
29	Walter Johnson	1915	337	6.3	5.4	.8	.1	79	Randy Jones	1975	285	5.3	4.9	-.1	.5
30	Bob Gibson	1969	314	6.3	5.7	.7	-.1	80	Lefty Grove	1936	253	5.3	5.5	-.2	.0
31	Steve Carlton	1972	346	6.3	6.0	.5	-.2	81	Three Finger Brown	1909	343	5.3	5.4	-0.0	-.1
32	Walter Johnson	1919	290	6.2	6.0	.2	-.0	82	Ed Walsh	1911	369	5.2	4.1	-.2	1.3
33	Three Finger Brown	1906	277	6.2	6.1	0.0	.1	83	Ferguson Jenkins	1971	325	5.2	3.7	1.4	.1
34	Vida Blue	1971	312	6.2	6.9	-.3	-.4	84	Lefty Grove	1930	291	5.2	5.2	.1	-.1
35	Warren Spahn	1953	266	6.1	5.6	.4	.1	85	Wes Ferrell	1930	297	5.2	4.8	.8	-.4
36	Cy Young	1902	385	6.0	6.4	.2	-.6	86	Cy Falkenberg	1914	377	5.2	5.4	-.3	.1
37	Ed Walsh	1910	370	6.0	4.6	.5	.9	87	Eddie Cicotte	1919	307	5.2	5.6	-0.0	-.4
38	Joe McGinnity	1904	408	6.0	5.9	-.3	.4	88	Spud Chandler	1943	253	5.2	4.2	.8	.2
39	Tom Seaver	1971	286	5.9	5.4	.4	.1	89	Mike Caldwell	1978	293	5.2	5.1	0.0	.1
40	Hal Newhouser	1946	293	5.9	5.9	-.2	.2	90	Rick Reuschel	1977	252	5.1	4.6	.2	.3
41	Bob Lemon	1948	294	5.9	3.9	1.2	.8	91	Phil Niekro	1974	302	5.1	5.1	-0.0	0.0
42	Burleigh Grimes	1920	304	5.9	4.6	1.0	.3	92	Dolf Luque	1925	291	5.1	4.3	.5	.3
43	Bobby Shantz	1952	280	5.8	5.4	-.1	.3	93	Walter Johnson	1911	323	5.1	4.6	.4	.1
44	Lefty Grove	1935	273	5.8	6.2	-.6	.2	94	Carl Hubbell	1936	304	5.1	5.1	0.0	0.0
45	Bob Gibson	1970	294	5.8	4.8	1.1	-.1	95	Dick Ellsworth	1963	291	5.1	5.2	-.3	.2
46	Dazzy Vance	1924	309	5.7	6.1	-.3	-.1	96	Don Drysdale	1960	269	5.1	4.5	.1	.5
47	Juan Marichal	1966	307	5.7	5.0	.7	-.0	97	Bert Blyleven	1977	235	5.1	5.0	0.0	.1
48	Addie Joss	1908	325	5.7	5.3	.2	.2	98	Dizzy Trout	1946	276	5.0	4.2	.3	.5
49	Walter Johnson	1914	372	5.7	4.5	1.0	.2	99	Frank Sullivan	1955	260	5.0	5.5	-.6	.1
50	Lefty Grove	1932	292	5.7	6.2	-.1	-.4	100	Johnny Rigney	1940	281	5.0	5.2	-.1	-.1

Batting Runs

		Year	L	Club	Value
1	Tip O'Neill	1887	A	STL	88.5
2	Hugh Duffy	1894	N	BOS	79.8
3	Fred Dunlap	1884	U	STL	73.1
4	Ed Delahanty	1899	N	PHI	72.4
5	Ed Delahanty	1895	N	PHI	68.8
6	Dan Brouthers	1886	N	DET	68.4
7	Ed Delahanty	1896	N	PHI	67.8
8	Mike Kelly	1886	N	CHI	66.6
9	Joe Kelley	1894	N	BAL	63.5
10	Billy Hamilton	1894	N	PHI	63.3
11	Pete Browning	1887	A	LOU	63.2
12	Cap Anson	1886	N	CHI	60.8
13	Dan Brouthers	1892	N	BRO	60.3
14	Dan Brouthers	1891	A	BOS	59.8
15	Sam Thompson	1895	N	PHI	59.7
16	Jesse Burkett	1896	N	CLE	58.0
17	Joe Kelley	1896	N	BAL	58.0
18	Jesse Burkett	1895	N	CLE	56.9
19	Jesse Burkett	1899	N	STL	56.6
20	John McGraw	1899	N	BAL	56.6
21	Ed Delahanty	1893	N	PHI	56.3
22	Honus Wagner	1900	N	PIT	56.2
23	Willie Keeler	1897	N	BAL	56.1
24	Elmer Flick	1900	N	PHI	54.3
25	Roger Connor	1892	N	PHI	54.1

Normalized OPS

		Year	L	Club	Value
1	Fred Dunlap	1884	U	STL	261
2	Ross Barnes	1876	N	CHI	250
3	Tip O'Neill	1887	A	STL	234
4	Pete Browning	1882	A	LOU	222
5	Dan Brouthers	1886	N	DET	218
6	Mike Kelly	1886	N	CHI	217
7	Dan Brouthers	1885	N	BUF	213
8	George Hall	1876	N	PHI	208
9	Roger Connor	1885	N	NY	207
10	George Shaffer	1884	U	STL	205
11	Dan Brouthers	1882	N	BUF	204
12	Cap Anson	1886	N	CHI	203
13	Cap Anson	1881	N	CHI	203
14	Deacon White	1877	N	BOS	201
15	John Reilly	1884	A	CIN	201
16	Harry Stovey	1884	A	ATH	200
17	Mike Kelly	1884	N	CHI	198
18	Dan Brouthers	1884	N	BUF	197
19	Ed Delahanty	1899	N	PHI	197
20	Ed Delahanty	1896	N	PHI	196
21	Pete Browning	1885	A	LOU	196
22	Ed Swartwood	1882	A	PIT	196
23	Dave Orr	1884	A	MET	196
24	Dan Brouthers	1883	N	BUF	196
25	George Gore	1880	N	CHI	194

On Base Average

		Year	L	Club	Value
1	John McGraw	1899	N	BAL	.535
2	Billy Hamilton	1894	N	PHI	.523
3	Hugh Duffy	1894	N	BOS	.506
4	Joe Kelley	1894	N	BAL	.502
5	Ed Delahanty	1895	N	PHI	.500
6	Bill Joyce	1894	N	WAS	.496
7	Billy Hamilton	1895	N	PHI	.490
8	Billy Hamilton	1893	N	PHI	.490
9	Tip O'Neill	1887	A	STL	.490
10	Jesse Burkett	1895	N	CLE	.486
11	Fred Carroll	1889	N	PIT	.486
12	Mike Kelly	1886	N	CHI	.483
13	Ed Delahanty	1894	N	PHI	.478
14	Billy Hamilton	1898	N	BOS	.478
15	Billy Hamilton	1896	N	BOS	.477
16	Clarence Childs	1894	N	CLE	.475
17	Hughie Jennings	1896	N	BAL	.472
18	Ed Delahanty	1896	N	PHI	.472
19	Dan Brouthers	1891	A	BOS	.471
20	Joe Kelley	1896	N	BAL	.469
21	Clarence Childs	1896	N	CLE	.467
22	Dan Brouthers	1890	P	BOS	.466
23	Pete Browning	1887	A	LOU	.464
24	Bob Caruthers	1887	A	STL	.463
25	Clarence Childs	1893	N	CLE	.463

Slugging Percentage

		Year	L	Club	Value
1	Tip O'Neill	1887	A	STL	.691
2	Hugh Duffy	1894	N	BOS	.690
3	Sam Thompson	1894	N	PHI	.686
4	Sam Thompson	1895	N	PHI	.654
5	Bill Joyce	1894	N	WAS	.648
6	Ed Delahanty	1896	N	PHI	.631
7	Fred Dunlap	1884	U	STL	.621
8	Ed Delahanty	1895	N	PHI	.617
9	Joe Kelley	1894	N	BAL	.602
10	Ross Barnes	1876	N	CHI	.590
11	Ed Delahanty	1894	N	PHI	.585
12	Ed Delahanty	1893	N	PHI	.583
13	Ed Delahanty	1899	N	PHI	.582
14	Bill Lange	1895	N	CHI	.582
15	Dan Brouthers	1886	N	DET	.581
16	Jake Stenzel	1894	N	PIT	.580
17	Honus Wagner	1900	N	PIT	.573
18	Dan Brouthers	1883	N	BUF	.572
19	Sam Thompson	1887	N	DET	.571
20	Nap Lajoie	1897	N	PHI	.569
21	Bill Dahlen	1894	N	CHI	.566
22	Buck Freeman	1899	N	WAS	.563
23	Dan Brouthers	1884	N	BUF	.563
24	Dan Brouthers	1887	N	DET	.562
25	Dan Brouthers	1894	N	BAL	.560

Normalized On Base

		Year	L	Club	Value
1	Ross Barnes	1876	N	CHI	1.666
2	Fred Dunlap	1884	U	STL	1.645
3	Mike Kelly	1886	N	CHI	1.610
4	John McGraw	1899	N	BAL	1.599
5	Pete Browning	1882	A	LOU	1.586
6	Roger Connor	1885	N	NY	1.533
7	Cap Anson	1881	N	CHI	1.523
8	George Gore	1880	N	CHI	1.494
9	Dan Brouthers	1886	N	DET	1.483
10	Billy Hamilton	1898	N	BOS	1.469
11	George Shaffer	1884	U	STL	1.464
12	Fred Carroll	1889	N	PIT	1.456
13	Tip O'Neill	1887	A	STL	1.454
14	George Gore	1886	N	CHI	1.447
15	Dan Brouthers	1882	N	BUF	1.444
16	Cap Anson	1886	N	CHI	1.443
17	Mike Kelly	1884	N	CHI	1.442
18	Dan Brouthers	1885	N	BUF	1.439
19	George Gore	1885	N	CHI	1.427
20	Cap Anson	1882	N	CHI	1.421
21	John McGraw	1898	N	BAL	1.411
22	Jim O'Rourke	1877	N	BOS	1.408
23	George Gore	1884	N	CHI	1.407
24	Cap Anson	1888	N	CHI	1.402
25	Deacon White	1877	N	BOS	1.401

Normalized Slugging

		Year	L	Club	Value
1	Fred Dunlap	1884	U	STL	1.967
2	Tip O'Neill	1887	A	STL	1.882
3	Ross Barnes	1876	N	CHI	1.838
4	Dan Brouthers	1886	N	DET	1.698
5	George Hall	1876	N	PHI	1.697
6	John Reilly	1884	A	CIN	1.696
7	Dan Brouthers	1885	N	BUF	1.686
8	Harry Stovey	1884	A	ATH	1.676
9	Dave Orr	1884	A	MET	1.659
10	Dan Brouthers	1884	N	BUF	1.655
11	Dave Orr	1885	A	MET	1.655
12	Pete Browning	1882	A	LOU	1.636
13	Sam Thompson	1895	N	PHI	1.636
14	Dave Orr	1886	A	MET	1.632
15	Ed Delahanty	1896	N	PHI	1.631
16	Ned Williamson	1884	N	CHI	1.629
17	Pete Browning	1885	A	LOU	1.616
18	Deacon White	1877	N	BOS	1.613
19	Henry Larkin	1885	A	ATH	1.602
20	Dan Brouthers	1881	N	BUF	1.600
21	Ed Swartwood	1882	A	PIT	1.598
22	Cap Anson	1884	N	CHI	1.597
23	Dan Brouthers	1882	N	BUF	1.595
24	John Reilly	1888	A	CIN	1.590
25	Cap Anson	1886	N	CHI	1.590

Batting Wins

		Year	L	Club	Value
1	Tip O'Neill	1887	A	STL	7.3
2	Ed Delahanty	1899	N	PHI	6.7
3	Fred Dunlap	1884	U	STL	6.5
4	Dan Brouthers	1886	N	DET	6.3
5	Hugh Duffy	1894	N	BOS	6.2
6	Mike Kelly	1886	N	CHI	6.2
7	Ed Delahanty	1896	N	PHI	5.9
8	Ed Delahanty	1895	N	PHI	5.7
9	Dan Brouthers	1892	N	BRO	5.7
10	Cap Anson	1886	N	CHI	5.6
11	Jesse Burkett	1899	N	STL	5.2
12	John McGraw	1899	N	BAL	5.2
13	Dan Brouthers	1891	A	BOS	5.2
14	Pete Browning	1887	A	LOU	5.2
15	Jimmy Ryan	1888	N	CHI	5.2
16	Honus Wagner	1900	N	PIT	5.2
17	Cap Anson	1888	N	CHI	5.2
18	Roger Connor	1885	N	NY	5.1
19	Roger Connor	1892	N	PHI	5.1
20	Elmer Flick	1900	N	PHI	5.0
21	Jesse Burkett	1896	N	CLE	5.0
22	Joe Kelley	1896	N	BAL	5.0
23	Jimmy Williams	1899	N	PIT	5.0
24	Joe Kelley	1894	N	BAL	5.0
25	Dan Brouthers	1888	N	DET	5.0

Wins Above Team

		Year	L	Club	Value
1	Will White	1879	N	CIN	43.0
2	Jim McCormick	1880	N	CLE	31.7
3	Jim Devlin	1876	N	LOU	30.0
4	Jim Galvin	1883	N	BUF	25.5
5	Tony Mullane	1884	A	TOL	23.2
6	Bobby Mathews	1876	N	MUT	21.0
7	Guy Hecker	1884	A	LOU	20.0
8	Sadie McMahon	1890	A	ATH	19.4
9	Charlie Radbourn	1883	N	PRO	18.8
10	Terry Larkin	1878	N	CHI	18.0
11	Jim Galvin	1884	N	BUF	17.5
12	Matt Kilroy	1887	A	BAL	17.2
13	Charlie Radbourn	1884	N	PRO	16.8
14	Ed Morris	1885	A	PIT	16.7
15	Frank Mountain	1883	A	COL	16.7
16	Bill Sweeney	1884	U	BAL	15.0
17	Henry Porter	1885	A	BRO	14.4
18	Charlie Buffinton	1884	N	BOS	14.0
19	Lee Richmond	1881	N	WOR	13.5
20	Mickey Welch	1884	N	NY	12.5
21	Frank Killen	1892	N	WAS	12.4
22	Toad Ramsey	1886	A	LOU	12.4
23	Will White	1882	A	CIN	12.1
24	Tommy Bond	1879	N	BOS	12.0
25	Bobby Mathews	1885	A	ATH	11.9

Pitching Runs

		Year	L	Club	Value
1	Amos Rusie	1894	N	NY	125.3
2	Charlie Radbourn	1884	N	PRO	120.7
3	Guy Hecker	1884	A	LOU	107.4
4	Silver King	1888	A	STL	91.8
5	John Clarkson	1889	N	BOS	88.9
6	Matt Kilroy	1887	A	BAL	80.5
7	Pink Hawley	1895	N	PIT	78.9
8	Silver King	1890	P	CHI	78.9
9	Will White	1883	A	CIN	77.6
10	Amos Rusie	1893	N	NY	76.6
11	Charlie Radbourn	1883	N	PRO	75.8
12	Dave Foutz	1886	A	STL	75.0
13	Jouett Meekin	1894	N	NY	73.6
14	Billy Rhines	1890	N	CIN	73.5
15	Scott Stratton	1890	A	LOU	71.8
16	Jim Galvin	1884	N	BUF	70.0
17	George Bradley	1876	N	STL	68.8
18	Kid Nichols	1897	N	BOS	68.3
19	Cy Young	1892	N	CLE	68.0
20	Elmer Smith	1887	A	CIN	67.5
21	John Clarkson	1885	N	CHI	67.1
22	Kid Nichols	1890	N	BOS	65.9
23	Toad Ramsey	1886	A	LOU	65.4
24	Amos Rusie	1890	N	NY	63.4
25	Mickey Welch	1885	N	NY	63.4

Normalized ERA

		Year	L	Club	Value
1	Tim Keefe	1880	N	TRO	276
2	Denny Driscoll	1882	A	PIT	225
3	Charlie Radbourn	1884	N	PRO	216
4	Guy Hecker	1882	A	LOU	209
5	Al Maul	1895	N	WAS	195
6	Jim McCormick	1884	U	CIN	194
7	Charlie Sweeney	1884	N	PRO	192
8	Clark Griffith	1898	N	CHI	191
9	Amos Rusie	1894	N	NY	191
10	George Bradley	1876	N	STL	188
11	Silver King	1888	A	STL	186
12	Billy Rhines	1890	N	CIN	185
13	Guy Hecker	1884	A	LOU	180
14	Harry McCormick	1882	A	CIN	179
15	Tim Keefe	1885	N	NY	178
16	Billy Rhines	1896	N	CIN	178
17	Billy Taylor	1884	U	STL	177
18	Will White	1882	A	CIN	177
19	George Bradley	1880	N	PRO	172
20	Al Maul	1898	N	BAL	171
21	Henry Boyle	1884	U	STL	171
22	Jack Stivetts	1889	A	STL	171
23	Jim McCormick	1883	N	CLE	170
24	Cy Young	1892	N	CLE	170
25	Mickey Welch	1885	N	NY	170

Percent Of Team Wins

		Year	L	Club	Value
1	Will White	1879	N	CIN	1.000
2	Bobby Mathews	1876	N	MUT	1.000
3	Jim Devlin	1877	N	LOU	1.000
4	Jim Devlin	1876	N	LOU	1.000
5	George Bradley	1876	N	STL	1.000
6	Tommy Bond	1878	N	BOS	.976
7	Terry Larkin	1878	N	CHI	.967
8	Jim McCormick	1880	N	CLE	.957
9	Tommy Bond	1877	N	BOS	.952
10	Terry Larkin	1877	N	HAR	.935
11	Al Spalding	1876	N	CHI	.885
12	Jim Galvin	1883	N	BUF	.885
13	Will White	1880	N	CIN	.857
14	Jim McCormick	1882	N	CLE	.857
15	Mickey Welch	1880	N	TRO	.829
16	Charlie Radbourn	1883	N	PRO	.828
17	Harry McCormick	1879	N	SYR	.818
18	Jim Whitney	1881	N	BOS	.816
19	Frank Mountain	1883	A	COL	.813
20	Will White	1878	N	CIN	.811
21	Tony Mullane	1884	A	TOL	.804
22	Jim Galvin	1879	N	BUF	.804
23	Sam Weaver	1878	N	MIL	.800
24	Lee Richmond	1880	N	WOR	.800
25	Monte Ward	1879	N	PRO	.797

Pitching Wins

		Year	L	Club	Value
1	Charlie Radbourn	1884	N	PRO	10.9
2	Guy Hecker	1884	A	LOU	10.0
3	Amos Rusie	1894	N	NY	9.8
4	Silver King	1888	A	STL	8.5
5	John Clarkson	1889	N	BOS	7.8
6	Will White	1883	A	CIN	6.9
7	Charlie Radbourn	1883	N	PRO	6.7
8	Dave Foutz	1886	A	STL	6.7
9	Matt Kilroy	1887	A	BAL	6.7
0	Billy Rhines	1890	N	CIN	6.6
11	Pink Hawley	1895	N	PIT	6.5
12	Scott Stratton	1890	A	LOU	6.4
13	John Clarkson	1885	N	CHI	6.4
14	Cy Young	1892	N	CLE	6.4
15	Silver King	1890	P	CHI	6.4
16	Amos Rusie	1893	N	NY	6.3
17	Jim Galvin	1884	N	BUF	6.3
18	Mickey Welch	1885	N	NY	6.0
19	Kid Nichols	1898	N	BOS	6.0
20	George Bradley	1876	N	STL	6.0
21	Kid Nichols	1897	N	BOS	6.0
22	Clark Griffith	1898	N	CHI	5.9
23	Kid Nichols	1890	N	BOS	5.9
24	Will White	1882	A	CIN	5.8
25	Toad Ramsey	1886	A	LOU	5.8

SINGLE-SEASON LEADERS ◇ 303

How to Read a Nineteenth Century Team Line

Club	G	W	L	Pct	GB	R	OR	Runs O/D		Wins O/D		Diff
Boston	61	42	18	.700	0.0	419	263	73	83	6.5	7.4	-1.9
Louisville	61	35	25	.583	7.0	339	288	-6	58	-.6	5.1	.5
Hartford	60	31	27	.534	10.0	341	311	1	29	.1	2.6	-.7
St. Louis	60	28	32	.467	14.0	284	318	-55	22	-5.0	2.0	1.0
Chicago	60	26	33	.441	15.5	366	375	26	-34	2.3	-3.1	-2.7
Cincinnati	58	15	42	.263	25.5	291	485	-37	-155	-3.4	-13.9	3.8

Boston won the 1877 pennant by seven games over Louisville, which collapsed in the late going in part because four of its players were beholden to gamblers. The first six columns speak for themselves, but the seventh and eighth—runs above average on offense and runs below average on defense—are the key to understanding a team's record. Boston, by scoring 73 runs more than the average for the league that year and by allowing 83 below the league average, was clearly the class of the six-team circuit. Using the formula for converting runs into wins (detailed in Chapter 4), we see that Boston's run totals figured to lead to 6.5 wins thanks to its offense and another 7.4 wins on account of its pitching and fielding. Thus their predicted won-lost record was 6.5 + 7.4 beyond the average in their 60-game schedule. So, rather than finishing 30-30, defined as average, Boston figured to finish 44-16 (30 + 6.5 + 7.4). In fact, the pennant winners finished 42-18, two games off their prediction; the column headed "DIFF" carries the precise mathematical differential, although of course all fractions must be rounded to whole wins.

How to Read a Twentieth Century Team Line

East	W	L	R	OR	Avg	OBA	SLG	BPF	NOPS-A	BR	Adj	Wins	ERA	PPF	NERA-A	PR	Adj	Wins	Diff
NY	100	62	632	541	.242	.313	.351	102	94/ 93	-51	-60	-6.4	2.99	100	121/121	100	102	10.7	14.7
CHI	92	70	720	611	.253	.326	.384	106	108/102	37	0	-.0	3.34	105	108/113	41	68	7.1	3.9
PIT	88	74	725	652	.277	.336	.398	100	115/115	84	87	9.1	3.61	99	100/ 98	-0	-8	-1.0	-1.1
STL	87	75	595	540	.253	.318	.359	105	98/ 94	-24	-55	-5.9	2.94	104	122/127	106	130	13.7	-1.8
PHI	63	99	645	745	.241	.314	.372	100	101/101	-11	-13	-1.4	4.17	102	86/ 88	-90	-80	-8.5	-8.1
MON	52	110	582	791	.240	.312	.359	104	96/ 92	-39	-66	-7.1	4.33	107	83/ 89	-115	-74	-7.9	-14.0

The New York Mets won a miracle pennant in 1969, coming from ninth place the previous year to capture the NL East, the League Championship Series, and finally the World Series. How did they do it? With mirrors, in part, but they had a good club too, good enough to be in a pennant race to the end if not to coast to a flag. The first four columns are self-explanatory. The fifth, AVG, reveals the Mets to have had a poor-hitting team, a view ratified by the next two columns, On Base Average and slugging percentage. Their home park, Shea Stadium, was marginally beneficial to hitters that year, as stated in the BPF (Batter Park Factor) of 102, or 2 percent above average. Having seen the puny totals in the OBA and SLG columns, one will not be surprised by the division-low Normalized OPS (94 unadjusted for Shea, 93 adjusted). The Mets' batters contributed 51 runs below the league average on an unadjusted basis and 60 below average when their park factor is figured in; in 1969 this performance should have driven the Mets' record 6.4 wins below the .500 mark (81-81). Their pitching and fielding were excellent, however; they were second only to the Cardinals in ERA, Linear Weights Runs (100 unadjusted for Shea, 102 adjusted), and Wins (10.7). So: Their offense cost them 6.4 wins, while their defense contributed 10.7—on the basis of their LWTS, the Mets projected to a 1969 record of 85 wins (81 − 6.4 + 10.7), not 100. On paper, the division winner should have been neither the Mets nor their perceived rivals, the Cubs, but the third-place Pittsburgh Pirates, who figured to win 89 games to the Cubs' 88 and the Mets' 85. The "DIFF" column tells one the extent to which a team outperformed or underperformed its talent.

Batting Runs			Normalized OPS			Pitching Runs			Normalized ERA		
Ross Barnes	CHI	50.3	Ross Barnes	CHI	250	George Bradley	STL	68.8	George Bradley	STL	188
George Hall	PHI	29.4	George Hall	PHI	208	Jim Devlin	LOU	51.8	Jim Devlin	LOU	148
Cap Anson	CHI	23.6	Cap Anson	CHI	174	Al Spalding	CHI	32.9	Candy Cummings	HAR	138
On Base Average			**Slugging Percentage**			**Percent of Team Wins**			**Wins Above Team**		
Ross Barnes	CHI	.462	Ross Barnes	CHI	.590	Bobby Mathews	MUT	1.000	Jim Devlin	LOU	30.0
George Hall	PHI	.384	George Hall	PHI	.545	Jim Devlin	LOU	1.000	Bobby Mathews	MUT	21.0
Cap Anson	CHI	.380	Lip Pike	STL	.472	George Bradley	STL	1.000	Lon Knight	ATH	5.3

Club	G	W	L	Pct	GB	R	OR	Runs O/D		Wins O/D		Diff
Chicago	66	52	14	.788	0.0	624	257	235	132	20.5	11.5	-13.1
St. Louis	64	45	19	.703	6.0	386	229	9	148	.8	13.0	-.7
Hartford	69	47	21	.691	6.0	429	261	22	146	1.9	12.7	-1.7
Boston	70	39	31	.557	15.0	471	450	58	-36	5.1	-3.3	2.2
Louisville	69	30	36	.455	22.0	280	344	-126	63	-11.1	5.5	2.6
Mutuals	57	21	35	.375	26.0	260	412	-75	-75	-6.6	-6.6	6.3
Athletics	60	14	45	.237	34.5	378	534	24	-179	2.1	-15.7	-1.9
Cincinnati	65	9	56	.138	42.5	238	579	-144	-195	-12.7	-17.1	6.3

Batting Runs			Normalized OPS			Pitching Runs			Normalized ERA		
Deacon White	BOS	29.0	Deacon White	BOS	201	Tommy Bond	BOS	40.5	Tommy Bond	BOS	133
Jim O'Rourke	BOS	22.0	Jim O'Rourke	BOS	173	Terry Larkin	HAR	37.3	Terry Larkin	HAR	131
Cal McVey	CHI	19.7	John Cassidy	HAR	169	Jim Devlin	LOU	34.8	Jim Devlin	LOU	125
On Base Average			**Slugging Percentage**			**Percent of Team Wins**			**Wins Above Team**		
Jim O'Rourke	BOS	.407	Deacon White	BOS	.545	Jim Devlin	LOU	1.000	Bobby Mitchell	CIN	3.8
Deacon White	BOS	.405	Charley Jones	CIN	.471	Tommy Bond	BOS	.952	Tommy Bond	BOS	2.0
Cal McVey	CHI	.387	John Cassidy	HAR	.458	Terry Larkin	HAR	.935	Terry Larkin	HAR	2.0

Club	G	W	L	Pct	GB	R	OR	Runs O/D		Wins O/D		Diff
Boston	61	42	18	.700	0.0	419	263	73	83	6.5	7.4	-1.9
Louisville	61	35	25	.583	7.0	339	288	-6	58	-.6	5.1	.5
Hartford	60	31	27	.534	10.0	341	311	1	29	.1	2.6	-.7
St. Louis	60	28	32	.467	14.0	284	318	-55	22	-5.0	2.0	1.0
Chicago	60	26	33	.441	15.5	366	375	26	-34	2.3	-3.1	-2.7
Cincinnati	58	15	42	.263	25.5	291	485	-37	-155	-3.4	-13.9	3.8

Batting Runs			Normalized OPS			Pitching Runs			Normalized ERA		
Paul Hines	PRO	21.0	Paul Hines	PRO	183	Monte Ward	PRO	29.3	Monte Ward	PRO	152
George Shaffer	IND	20.6	George Shaffer	IND	175	Will White	CIN	26.5	Jim McCormick	IND	136
Joe Start	CHI	18.3	Joe Start	CHI	165	Sam Weaver	MIL	14.9	Will White	CIN	128
On Base Average			**Slugging Percentage**			**Percent of Team Wins**			**Wins Above Team**		
Bob Ferguson	CHI	.375	Paul Hines	PRO	.486	Tommy Bond	BOS	.976	Terry Larkin	CHI	18.0
Cap Anson	CHI	.372	Tom York	PRO	.465	Terry Larkin	CHI	.967	Monte Ward	PRO	6.6
George Shaffer	IND	.369	George Shaffer	IND	.455	Will White	CIN	.811	Sam Weaver	MIL	4.4

Club	G	W	L	Pct	GB	R	OR	Runs O/D		Wins O/D		Diff
Boston	60	41	19	.683	0.0	298	241	-11	69	-1.2	6.5	5.7
Cincinnati	61	37	23	.617	4.0	333	281	17	35	1.6	3.2	2.2
Providence	62	33	27	.550	8.0	353	337	32	-15	3.0	-1.5	1.5
Chicago	61	30	30	.500	11.0	371	331	55	-14	5.2	-1.4	-3.7
Indianapolis	63	24	36	.400	17.0	293	328	-32	-1	-3.1	-.2	-2.7
Milwaukee	61	15	45	.250	26.0	256	386	-59	-69	-5.6	-6.6	-2.9

NATIONAL LEAGUE   1879

Batting Runs			Normalized OPS			Pitching Runs			Normalized ERA		
Charley Jones	BOS	34.3	Charley Jones	BOS	191	Will White	CIN	38.5	Tommy Bond	BOS	128
Paul Hines	PRO	34.2	John O'Rourke	BOS	190	Tommy Bond	BOS	33.3	Will White	CIN	126
Mike Kelly	CIN	29.3	Mike Kelly	CIN	184	Monte Ward	PRO	22.8	Monte Ward	PRO	116
On Base Average			**Slugging Percentage**			**Percent of Team Wins**			**Wins Above Team**		
Jim O'Rourke	PRO	.371	John O'Rourke	BOS	.521	Will White	CIN	1.000	Will White	CIN	43.0
Paul Hines	PRO	.369	Charley Jones	BOS	.510	Harry McCormick	SYR	.818	Tommy Bond	BOS	12.0
Charley Jones	BOS	.367	Mike Kelly	CIN	.493	Jim Galvin	BUF	.804	Harry McCormick	SYR	7.3

Club	G	W	L	Pct	GB	R	OR	Runs O/D		Wins O/D		Diff
Providence	85	59	25	.702	0.0	612	355	161	96	14.8	8.9	-6.7
Boston	84	54	30	.643	5.0	562	348	116	98	10.7	9.0	-7.7
Buffalo	79	46	32	.590	10.0	394	365	-24	54	-2.3	5.0	4.3
Chicago	83	46	33	.582	10.5	437	411	-3	30	-.3	2.7	4.1
Cincinnati	81	43	37	.538	14.0	485	464	55	-33	5.1	-3.1	1.1
Cleveland	82	27	55	.329	31.0	322	461	-112	-25	-10.4	-2.4	-1.2
Syracuse	71	22	48	.314	30.0	276	462	-100	-84	-9.3	-7.8	4.1
Troy	77	19	56	.253	35.5	321	543	-87	-133	-8.1	-12.3	1.9

NATIONAL LEAGUE   1880

Batting Runs			Normalized OPS			Pitching Runs			Normalized ERA		
George Gore	CHI	30.9	George Gore	CHI	194	Monte Ward	PRO	41.7	Tim Keefe	TRO	276
Roger Connor	TRO	26.0	Roger Connor	TRO	177	Jim McCormick	CLE	38.0	George Bradley	PRO	172
Abner Dalrymple	CHI	25.2	Abner Dalrymple	CHI	168	Larry Corcoran	CHI	25.0	Monte Ward	PRO	136
On Base Average			**Slugging Percentage**			**Percent of Team Wins**			**Wins Above Team**		
George Gore	CHI	.399	George Gore	CHI	.463	Jim McCormick	CLE	.957	Jim McCormick	CLE	31.7
Cap Anson	CHI	.362	Roger Connor	TRO	.459	Will White	CIN	.857	Mickey Welch	TRO	10.4
Roger Connor	TRO	.357	Abner Dalrymple	CHI	.458	Mickey Welch	TRO	.829	Jim Galvin	BUF	9.4

Club	G	W	L	Pct	GB	R	OR	Runs O/D		Wins O/D		Diff
Chicago	86	67	17	.798	0.0	538	317	134	87	13.2	8.5	3.4
Providence	87	52	32	.619	15.0	419	299	11	109	1.1	10.7	-1.8
Cleveland	85	47	37	.560	20.0	387	337	-11	62	-1.2	6.1	.1
Troy	83	41	42	.494	25.5	392	438	3	-48	.2	-4.8	4.0
Worcester	85	40	43	.482	26.5	412	370	13	29	1.3	2.8	-5.6
Boston	86	40	44	.476	27.0	416	456	12	-51	1.2	-5.1	1.9
Buffalo	85	24	58	.293	42.0	331	502	-67	-102	-6.6	-10.1	-.3
Cincinnati	83	21	59	.263	44.0	296	472	-92	-82	-9.2	-8.1	-1.8

NATIONAL LEAGUE   1881

Batting Runs			Normalized OPS			Pitching Runs			Normalized ERA		
Cap Anson	CHI	39.8	Cap Anson	CHI	203	George Derby	DET	31.4	Stump Weidman	DET	154
Dan Brouthers	BUF	24.9	Dan Brouthers	BUF	184	Monte Ward	PRO	23.5	Monte Ward	PRO	130
Fred Dunlap	CLE	21.1	Charlie Bennett	DET	159	Jim Galvin	BUF	21.1	George Derby	DET	126
On Base Average			**Slugging Percentage**			**Percent of Team Wins**			**Wins Above Team**		
Cap Anson	CHI	.442	Dan Brouthers	BUF	.541	Jim Whitney	BOS	.816	Lee Richmond	WOR	13.5
Tom York	PRO	.362	Cap Anson	CHI	.510	Lee Richmond	WOR	.781	Charlie Radbourn	PRO	8.5
Dan Brouthers	BUF	.361	Charlie Bennett	DET	.478	Jim McCormick	CLE	.722	Jim Whitney	BOS	7.4

Club	G	W	L	Pct	GB	R	OR	Runs O/D		Wins O/D		Diff
Chicago	84	56	28	.667	0.0	550	379	122	49	11.4	4.6	-2.1
Providence	85	47	37	.560	9.0	447	426	14	7	1.3	.7	3.0
Buffalo	83	45	38	.542	10.5	440	447	17	-23	1.6	-2.2	4.2
Detroit	84	41	43	.488	15.0	439	429	11	-0	1.0	-.1	-1.9
Troy	85	39	45	.464	17.0	399	429	-33	4	-3.2	.4	-.2
Boston	83	38	45	.458	17.5	349	410	-73	13	-7.0	1.2	2.2
Cleveland	85	36	48	.429	20.0	392	414	-40	19	-3.9	1.8	-3.9
Worcester	83	32	50	.390	23.0	410	492	-12	-68	-1.2	-6.5	-1.3

NATIONAL LEAGUE 1882

Batting Runs			Normalized OPS			Pitching Runs			Normalized ERA		
Dan Brouthers	BUF	39.7	Dan Brouthers	BUF	204	Charlie Radbourn	PRO	42.1	Larry Corcoran	CHI	148
Cap Anson	CHI	33.3	Cap Anson	CHI	188	Larry Corcoran	CHI	37.2	Charlie Radbourn	PRO	138
Roger Connor	TRO	30.2	Jim Whitney	BOS	185	Jim McCormick	CLE	34.4	Jim McCormick	CLE	122
On Base Average			**Slugging Percentage**			**Percent of Team Wins**			**Wins Above Team**		
Dan Brouthers	BUF	.403	Dan Brouthers	BUF	.547	Jim McCormick	CLE	.857	Larry Corcoran	CHI	11.3
Cap Anson	CHI	.397	Roger Connor	TRO	.530	Lee Richmond	WOR	.778	Lee Richmond	WOR	8.9
Jim Whitney	BOS	.382	Jim Whitney	BOS	.510	Charlie Radbourn	PRO	.635	Stump Weidman	DET	4.9

Club	G	W	L	Pct	GB	R	OR	Runs O/D		Wins O/D		Diff
Chicago	84	55	29	.655	0.0	604	353	150	101	13.6	9.2	-9.9
Providence	84	52	32	.619	3.0	463	356	9	98	.8	9.0	.2
Buffalo	84	45	39	.536	10.0	500	461	46	-6	4.2	-.6	-.6
Boston	85	45	39	.536	10.0	472	414	12	46	1.1	4.2	-2.3
Cleveland	84	42	40	.512	12.0	402	411	-51	43	-4.8	4.0	1.8
Detroit	86	42	41	.506	12.5	407	488	-57	-22	-5.3	-2.1	7.9
Troy	85	35	48	.422	19.5	430	522	-29	-61	-2.7	-5.7	1.9
Worcester	84	18	66	.214	37.0	379	652	-74	-197	-6.9	-18.0	.9

AMERICAN ASSOCIATION 1882

Batting Runs			Normalized OPS			Pitching Runs			Normalized ERA		
Pete Browning	LOU	36.2	Pete Browning	LOU	222	Will White	CIN	62.9	Denny Driscoll	PIT	225
Ed Swartwood	PIT	31.4	Ed Swartwood	PIT	196	Tony Mullane	LOU	42.9	Guy Hecker	LOU	209
Hick Carpenter	CIN	23.2	Hick Carpenter	CIN	168	Denny Driscoll	PIT	33.7	Harry McCormick	CIN	179
On Base Average			**Slugging Percentage**			**Percent of Team Wins**			**Wins Above Team**		
Pete Browning	LOU	.430	Pete Browning	LOU	.510	Will White	CIN	.727	Will White	CIN	12.1
Ed Swartwood	PIT	.370	Ed Swartwood	PIT	.498	Tony Mullane	LOU	.714	George McGinnis	STL	11.7
Hick Carpenter	CIN	.360	Billy Taylor	PIT	.455	George McGinnis	STL	.676	Sam Weaver	ATH	7.9

Club	G	W	L	Pct	GB	R	OR	Runs O/D		Wins O/D		Diff
Cincinnati	80	55	25	.688	0.0	489	268	72	149	6.7	13.8	-5.5
Athletics	75	41	34	.547	11.5	406	389	15	2	1.4	.2	1.9
Louisville	80	42	38	.525	13.0	443	352	26	65	2.4	6.0	-6.5
Pittsburgh	79	39	39	.500	15.0	428	418	16	-5	1.5	-.6	-.9
St. Louis	80	37	43	.463	18.0	399	496	-17	-78	-1.6	-7.4	6.0
Baltimore	74	19	54	.260	32.5	273	515	-111	-129	-10.5	-12.0	5.0

NATIONAL LEAGUE 1883

Batting Runs			Normalized OPS			Pitching Runs			Normalized ERA		
Dan Brouthers	BUF	46.1	Dan Brouthers	BUF	196	Charlie Radbourn	PRO	75.8	Jim McCormick	CLE	170
Roger Connor	NY	36.3	Roger Connor	NY	177	Jim Whitney	BOS	50.8	Charlie Radbourn	PRO	153
John Morrill	BOS	29.3	John Morrill	BOS	164	Jim McCormick	CLE	49.0	Jim Whitney	BOS	140
On Base Average			**Slugging Percentage**			**Percent of Team Wins**			**Wins Above Team**		
Dan Brouthers	BUF	.397	Dan Brouthers	BUF	.572	Jim Galvin	BUF	.885	Jim Galvin	BUF	25.5
Roger Connor	NY	.394	John Morrill	BOS	.525	Charlie Radbourn	PRO	.828	Charlie Radbourn	PRO	18.8
George Gore	CHI	.377	Roger Connor	NY	.506	John Coleman	PHI	.706	Jim McCormick	CLE	9.1

Club	G	W	L	Pct	GB	R	OR	Runs O/D		Wins O/D		Diff
Boston	98	63	35	.643	0.0	669	456	103	110	9.1	9.7	-4.8
Chicago	98	59	39	.602	4.0	679	540	113	26	9.9	2.3	-2.3
Providence	98	58	40	.592	5.0	636	436	70	130	6.2	11.5	-8.6
Cleveland	100	55	42	.567	7.5	476	443	-101	135	-9.0	11.9	3.6
Buffalo	98	52	45	.536	10.5	614	576	48	-9	4.2	-.9	.1
New York	98	46	50	.479	16.0	530	577	-35	-10	-3.2	-.9	2.1
Detroit	101	40	58	.408	23.0	524	650	-59	-65	-5.3	-5.9	2.1
Philadelphia	99	17	81	.173	46.0	437	887	-134	-314	-11.9	-27.8	7.7

AMERICAN ASSOCIATION 1883

Batting Runs

Ed Swartwood	PIT	36.3
Harry Stovey	ATH	32.8
Pete Browning	LOU	27.9

On Base Average

Ed Swartwood	PIT	.391
Pete Browning	LOU	.378
Jim Clinton	BAL	.357

Normalized OPS

Ed Swartwood	PIT	182
Harry Stovey	ATH	173
Pete Browning	LOU	172

Slugging Percentage

Harry Stovey	ATH	.504
John Reilly	CIN	.485
Ed Swartwood	PIT	.475

Pitching Runs

Will White	CIN	77.6
Tim Keefe	MET	61.2
Tony Mullane	STL	56.9

Percent of Team Wins

Frank Mountain	COL	.813
Tim Keefe	MET	.759
Will White	CIN	.705

Normalized ERA

Will White	CIN	158
Tony Mullane	STL	151
Ren Deagle	CIN	143

Wins Above Team

Frank Mountain	COL	16.7
Tim Keefe	MET	9.4
Denny Driscoll	PIT	9.4

Club	G	W	L	Pct	GB	R	OR	Runs O/D		Wins O/D		Diff
Athletics	98	66	32	.673	0.0	720	547	159	14	14.1	1.2	1.7
St. Louis	98	65	33	.663	1.0	549	409	-11	152	-1.1	13.5	3.6
Cincinnati	98	61	37	.622	5.0	662	413	101	148	9.0	13.1	-10.1
Metropolitans	97	54	42	.563	11.0	498	405	-56	150	-5.1	13.3	-2.2
Louisville	98	52	45	.536	13.5	564	562	3	-0	.3	-.1	3.3
Columbus	97	32	65	.330	33.5	476	659	-78	-103	-7.0	-9.2	-.3
Pittsburgh	98	31	67	.316	35.0	525	728	-35	-166	-3.2	-14.8	-.0
Baltimore	96	28	68	.292	37.0	471	742	-78	-191	-7.0	-17.1	4.0

NATIONAL LEAGUE 1884

Batting Runs

Mike Kelly	CHI	50.4
Cap Anson	CHI	46.3
Dan Brouthers	BUF	42.7

On Base Average

Mike Kelly	CHI	.414
George Gore	CHI	.404
Jim O'Rourke	BUF	.392

Normalized OPS

Mike Kelly	CHI	198
Dan Brouthers	BUF	197
Cap Anson	CHI	190

Slugging Percentage

Dan Brouthers	BUF	.563
Ned Williamson	CHI	.554
Cap Anson	CHI	.543

Pitching Runs

Charlie Radbourn	PRO	120.7
Jim Galvin	BUF	70.0
Charlie Buffington	BOS	54.1

Percent of Team Wins

Jim Galvin	BUF	.719
Charlie Radbourn	PRO	.714
Charlie Buffington	BOS	.658

Normalized ERA

Charlie Radbourn	PRO	216
Charlie Sweeney	PRO	192
Charlie Getzein	DET	153

Wins Above Team

Jim Galvin	BUF	17.5
Charlie Radbourn	PRO	16.8
Charlie Buffington	BOS	14.0

Club	G	W	L	Pct	GB	R	OR	Runs O/D		Wins O/D		Diff
Providence	114	84	28	.750	0.0	665	388	37	240	3.3	21.7	3.0
Boston	116	73	38	.658	10.5	684	468	44	172	4.0	15.5	-2.0
Buffalo	114	64	47	.577	19.5	700	626	72	2	6.5	.2	1.8
New York	116	62	50	.554	22.0	693	623	53	17	4.8	1.5	-.3
Chicago	112	62	50	.554	22.0	834	647	217	-29	19.6	-2.7	-10.9
Philadelphia	113	39	73	.348	45.0	549	824	-73	-200	-6.7	-18.2	7.8
Cleveland	113	35	77	.313	49.0	458	716	-164	-92	-14.9	-8.4	2.3
Detroit	114	28	84	.250	56.0	445	736	-182	-107	-16.6	-9.7	-1.7

AMERICAN ASSOCIATION 1884

Batting Runs

Harry Stovey	ATH	46.5
John Reilly	CIN	46.2
Dave Orr	MET	43.8

On Base Average

Charley Jones	CIN	.376
Jack Nelson	MET	.375
Harry Stovey	ATH	.368

Normalized OPS

John Reilly	CIN	201
Harry Stovey	ATH	200
Dave Orr	MET	196

Slugging Percentage

John Reilly	CIN	.551
Harry Stovey	ATH	.545
Dave Orr	MET	.539

Pitching Runs

Guy Hecker	LOU	107.4
Tim Keefe	MET	51.9
Ed Morris	COL	50.6

Percent of Team Wins

Tony Mullane	TOL	.804
Guy Hecker	LOU	.765
Larry McKeon	IND	.621

Normalized ERA

Guy Hecker	LOU	180
Ed Morris	COL	149
Dave Foutz	STL	149

Wins Above Team

Tony Mullane	TOL	23.2
Guy Hecker	LOU	20.0
Ed Morris	COL	7.0

Club	G	W	L	Pct	GB	R	OR	Runs O/D		Wins O/D		Diff
Metropolitans	112	75	32	.701	0.0	734	423	149	162	13.8	15.1	-7.4
Columbus	110	69	39	.639	6.5	585	459	10	116	.9	10.8	3.3
Louisville	110	68	40	.630	7.5	573	425	-1	150	-.2	13.9	.3
St. Louis	110	67	40	.626	8.0	658	539	83	36	7.7	3.3	2.5
Cincinnati	112	68	41	.624	8.0	754	512	169	73	15.6	6.8	-9.0
Baltimore	108	63	43	.594	11.5	636	515	71	50	6.6	4.6	-1.2
Athletics	108	61	46	.570	14.0	700	546	135	19	12.6	1.7	-6.8
Toledo	110	46	58	.442	27.5	463	571	-111	4	-10.4	.4	4.0
Brooklyn	109	40	64	.385	33.5	476	644	-93	-73	-8.7	-6.9	3.6
Pittsburgh	110	30	78	.278	45.5	406	725	-168	-149	-15.7	-13.9	5.6
Indianapolis	110	29	78	.271	46.0	462	755	-112	-179	-10.5	-16.7	2.7
Washington	109	24	81	.229	50.0	442	775	-127	-204	-11.9	-19.0	2.4

UNION ASSOCIATION   1884

Batting Runs			Normalized OPS			Pitching Runs			Normalized ERA		
Fred Dunlap	STL	73.1	Fred Dunlap	STL	261	Hugh Daily	C-P	57.7	Jim McCormick	CIN	194
George Shaffer	STL	49.8	George Shaffer	STL	205	Dupee Shaw	BOS	42.5	Billy Taylor	STL	177
Harry Moore	WAS	29.6	Jack Gleason	STL	170	Billy Taylor	STL	38.0	Henry Boyle	STL	171
On Base Average			Slugging Percentage			Percent of Team Wins			Wins Above Team		
Fred Dunlap	STL	.448	Fred Dunlap	STL	.621	Bill Sweeney	BAL	.690	Bill Sweeney	BAL	15.0
George Shaffer	STL	.398	George Shaffer	STL	.501	Ed Bakely	PHI	.667	Jim McCormick	CIN	6.8
Harry Moore	WAS	.363	Dick Burns	CIN	.457	Hugh Daily	C-P	.659	Hugh Daily	C-P	6.6

Club	G	W	L	Pct	GB	R	OR	Runs O/D		Wins O/D		Diff
St. Louis	114	94	19	.832	0.0	887	429	244	214	21.8	19.1	-3.4
Milwaukee	12	8	4	.667	35.5	53	34	-14	34	-1.3	3.0	.3
Cincinnati	105	69	36	.657	21.0	703	466	111	126	9.9	11.2	-4.7
Baltimore	106	58	47	.552	32.0	662	627	65	-29	5.8	-2.6	2.4
Boston	111	58	51	.532	34.0	636	558	10	68	.9	6.0	-3.5
Chi-Pit	93	41	50	.451	42.0	438	482	-85	42	-7.7	3.8	-.6
Washington	114	47	65	.420	46.5	572	679	-70	-35	-6.3	-3.3	.6
Philadelphia	67	21	46	.313	50.0	414	545	36	-166	3.2	-15.0	-.8
St. Paul	9	2	6	.250	39.5	24	57	-26	-5	-2.4	-.6	.9
Altoona	25	6	19	.240	44.0	90	216	-50	-74	-4.5	-6.7	4.8
Kansas City	82	16	63	.203	61.0	311	618	-150	-155	-13.5	-13.9	3.9
Wilmington	18	2	16	.111	44.5	35	114	-65	-12	-5.9	-1.1	.1

NATIONAL LEAGUE   1885

Batting Runs			Normalized OPS			Pitching Runs			Normalized ERA		
Roger Connor	NY	53.5	Dan Brouthers	BUF	213	John Clarkson	CHI	67.1	Tim Keefe	NY	178
Dan Brouthers	BUF	48.5	Roger Connor	NY	207	Mickey Welch	NY	63.4	Mickey Welch	NY	170
George Gore	CHI	42.6	George Gore	CHI	184	Tim Keefe	NY	54.8	John Clarkson	CHI	152
On Base Average			Slugging Percentage			Percent of Team Wins			Wins Above Team		
Roger Connor	NY	.435	Dan Brouthers	BUF	.543	John Clarkson	CHI	.609	Charlie Radbourn	PRO	7.9
Dan Brouthers	BUF	.408	Roger Connor	NY	.495	Charlie Radbourn	PRO	.528	Mickey Welch	NY	4.4
George Gore	CHI	.405	Buck Ewing	NY	.471	Mickey Welch	NY	.518	Charlie Ferguson	PHI	4.4

Club	G	W	L	Pct	GB	R	OR	Runs O/D		Wins O/D		Diff
Chicago	113	87	25	.777	0.0	834	470	274	90	26.2	8.5	-3.7
New York	112	85	27	.759	2.0	691	370	136	185	13.0	17.6	-1.6
Philadelphia	111	56	54	.509	30.0	513	511	-36	39	-3.5	3.7	.8
Providence	110	53	57	.482	33.0	442	531	-102	14	-9.8	1.3	6.5
Boston	113	46	66	.411	41.0	528	589	-31	-28	-3.0	-2.8	-4.2
Detroit	108	41	67	.380	44.0	514	582	-20	-46	-2.0	-4.5	-6.5
Buffalo	112	38	74	.339	49.0	495	761	-59	-205	-5.7	-19.7	7.4
St. Louis	111	36	72	.333	49.0	390	593	-159	-42	-15.2	-4.1	1.4

AMERICAN ASSOCIATION   1885

Batting Runs			Normalized OPS			Pitching Runs			Normalized ERA		
Pete Browning	LOU	49.1	Pete Browning	LOU	196	Bob Caruthers	STL	62.7	Bob Caruthers	STL	157
Henry Larkin	ATH	42.6	Dave Orr	MET	188	Ed Morris	PIT	57.5	Guy Hecker	LOU	149
Harry Stovey	ATH	40.2	Henry Larkin	ATH	188	Guy Hecker	LOU	56.5	Ed Morris	PIT	138
On Base Average			Slugging Percentage			Percent of Team Wins			Wins Above Team		
Pete Browning	LOU	.393	Dave Orr	MET	.543	Ed Morris	PIT	.696	Ed Morris	PIT	16.7
Henry Larkin	ATH	.373	Pete Browning	LOU	.530	Henry Porter	BRO	.623	Henry Porter	BRO	14.4
Harry Stovey	ATH	.371	Henry Larkin	ATH	.525	Hardie Henderson	BAL	.610	Bobby Mathews	ATH	11.9

Club	G	W	L	Pct	GB	R	OR	Runs O/D		Wins O/D		Diff
St. Louis	112	79	33	.705	0.0	677	461	62	154	5.6	13.9	3.4
Cincinnati	112	63	49	.563	16.0	642	575	27	40	2.5	3.6	.9
Pittsburgh	111	56	55	.505	22.5	548	539	-60	70	-5.6	6.4	-.3
Athletics	113	55	57	.491	24.0	764	691	144	-70	13.0	-6.4	-7.6
Louisville	112	53	59	.473	26.0	564	599	-50	16	-4.6	1.4	.2
Brooklyn	112	53	59	.473	26.0	624	650	9	-34	.8	-3.2	-.6
Metropolitans	108	44	64	.407	33.0	526	688	-66	-94	-6.1	-8.6	4.7
Baltimore	110	41	68	.376	36.5	541	683	-62	-78	-5.7	-7.2	-.6

NATIONAL LEAGUE 1886

Batting Runs			Normalized OPS			Pitching Runs			Normalized ERA		
Dan Brouthers	DET	68.4	Dan Brouthers	DET	218	Charlie Ferguson	PHI	58.5	Charlie Ferguson	PHI	167
Mike Kelly	CHI	66.6	Mike Kelly	CHI	217	Lady Baldwin	DET	57.9	Jocko Flynn	CHI	148
Cap Anson	CHI	60.8	Cap Anson	CHI	203	Tim Keefe	NY	46.8	Henry Boyle	STL	148

On Base Average			Slugging Percentage			Percent of Team Wins			Wins Above Team		
Mike Kelly	CHI	.483	Dan Brouthers	DET	.581	Tim Keefe	NY	.547	Charlie Ferguson	PHI	8.7
Dan Brouthers	DET	.445	Cap Anson	CHI	.544	Lady Baldwin	DET	.483	Lady Baldwin	DET	5.6
George Gore	CHI	.434	Roger Connor	NY	.540	Charlie Radbourn	BOS	.482	Tim Keefe	NY	5.2

Club	G	W	L	Pct	GB	R	OR	Runs O/D		Wins O/D		Diff
Chicago	126	90	34	.726	0.0	900	555	236	109	21.8	10.1	-3.9
Detroit	126	87	36	.707	2.5	829	538	165	126	15.3	11.6	-1.4
New York	124	75	44	.630	12.5	692	558	39	95	3.6	8.8	3.1
Philadelphia	119	71	43	.623	14.0	621	498	-5	129	-.6	11.9	2.6
Boston	118	56	61	.479	30.5	657	661	35	-38	3.3	-3.6	-2.1
St. Louis	126	43	79	.352	46.0	547	712	-116	-47	-10.8	-4.4	-2.8
Kansas City	123	30	91	.248	58.5	494	872	-153	-223	-14.2	-20.7	4.4
Washington	122	28	92	.233	60.0	445	791	-197	-147	-18.3	-13.7	-.0

AMERICAN ASSOCIATION 1886

Batting Runs			Normalized OPS			Pitching Runs			Normalized ERA		
Dave Orr	MET	47.7	Dave Orr	MET	182	Dave Foutz	STL	75.0	Dave Foutz	STL	164
Henry Larkin	ATH	43.2	Henry Larkin	ATH	167	Toad Ramsey	LOU	65.4	Bob Caruthers	STL	149
Bob Caruthers	STL	42.4	Pete Browning	LOU	164	Ed Morris	PIT	61.7	Toad Ramsey	LOU	141

On Base Average			Slugging Percentage			Percent of Team Wins			Wins Above Team		
Henry Larkin	ATH	.390	Dave Orr	MET	.527	Matt Kilroy	BAL	.604	Toad Ramsey	LOU	12.4
Pete Browning	LOU	.389	Henry Larkin	ATH	.450	Toad Ramsey	LOU	.576	Matt Kilroy	BAL	11.4
Tip O'Neill	STL	.385	Pete Browning	LOU	.441	Ed Morris	PIT	.513	Ed Morris	PIT	9.7

Club	G	W	L	Pct	GB	R	OR	Runs O/D		Wins O/D		Diff
St. Louis	139	93	46	.669	0.0	944	592	155	197	13.8	17.6	-7.8
Pittsburgh	140	80	57	.584	12.0	810	647	15	148	1.3	13.2	-3.0
Brooklyn	141	76	61	.555	16.0	832	832	31	-30	2.8	-2.8	7.5
Louisville	138	66	70	.485	25.5	833	805	49	-20	4.4	-1.9	-4.5
Cincinnati	141	65	73	.471	27.5	883	865	82	-63	7.3	-5.7	-5.6
Athletics	139	63	72	.467	28.0	772	942	-16	-152	-1.6	-13.6	10.6
Metropolitans	137	53	82	.393	38.0	628	766	-149	12	-13.4	1.1	-2.2
Baltimore	139	48	83	.366	41.0	625	878	-163	-88	-14.6	-7.9	5.0

NATIONAL LEAGUE 1887

Batting Runs			Normalized OPS			Pitching Runs			Normalized ERA		
Sam Thompson	DET	53.9	Dan Brouthers	DET	178	John Clarkson	CHI	56.4	Dan Casey	PHI	142
Dan Brouthers	DET	53.2	Sam Thompson	DET	177	Dan Casey	PHI	51.6	Pete Conway	DET	140
Cap Anson	CHI	41.8	Cap Anson	CHI	165	Tim Keefe	NY	50.6	Charlie Ferguson	PHI	135

On Base Average			Slugging Percentage			Percent of Team Wins			Wins Above Team		
Dan Brouthers	DET	.426	Sam Thompson	DET	.571	John Clarkson	CHI	.535	Jim Whitney	WAS	11.1
Cap Anson	CHI	.422	Dan Brouthers	DET	.562	Jim Whitney	WAS	.522	Jim Galvin	PIT	10.4
Sam Thompson	DET	.416	Roger Connor	NY	.541	Tim Keefe	NY	.515	Tim Keefe	NY	9.2

Club	G	W	L	Pct	GB	R	OR	Runs O/D		Wins O/D		Diff
Detroit	127	79	45	.637	0.0	969	714	197	58	16.9	5.0	-4.9
Philadelphia	128	75	48	.610	3.5	901	702	122	77	10.5	6.6	-3.6
Chicago	127	71	50	.587	6.5	813	716	41	56	3.5	4.9	2.2
New York	129	68	55	.553	10.5	816	723	31	62	2.7	5.3	-1.5
Boston	127	61	60	.504	16.5	831	792	59	-19	5.0	-1.7	-2.9
Pittsburgh	125	55	69	.444	24.0	621	750	-138	10	-12.0	.9	4.1
Washington	126	46	76	.377	32.0	601	818	-164	-51	-14.2	-4.4	3.7
Indianapolis	127	37	89	.294	43.0	628	965	-143	-192	-12.4	-16.6	3.0

AMERICAN ASSOCIATION 1887

Batting Runs			Normalized OPS			Pitching Runs			Normalized ERA		
Tip O'Neill	STL	88.5	Tip O'Neill	STL	234	Matt Kilroy	BAL	80.5	Elmer Smith	CIN	146
Pete Browning	LOU	63.2	Pete Browning	LOU	187	Elmer Smith	CIN	67.5	Matt Kilroy	BAL	140
Denny Lyons	ATH	49.3	Bob Caruthers	STL	187	Toad Ramsey	LOU	54.2	Tony Mullane	CIN	133

On Base Average			Slugging Percentage			Percent of Team Wins			Wins Above Team		
Tip O'Neill	STL	.490	Tip O'Neill	STL	.691	Matt Kilroy	BAL	.597	Matt Kilroy	BAL	17.2
Pete Browning	LOU	.464	Bob Caruthers	STL	.547	Toad Ramsey	LOU	.487	Elmer Smith	CIN	5.5
Bob Caruthers	STL	.463	Pete Browning	LOU	.547	Elmer Smith	CIN	.420	Tony Mullane	CIN	3.4

Club	G	W	L	Pct	GB	R	OR	Runs O/D		Wins O/D		Diff
St. Louis	138	95	40	.704	0.0	1131	761	223	147	18.5	12.1	-3.1
Cincinnati	136	81	54	.600	14.0	892	745	-1	149	-.2	12.4	1.3
Baltimore	141	77	58	.570	18.0	975	861	48	66	3.9	5.5	.1
Louisville	139	76	60	.559	19.5	956	854	42	60	3.5	5.0	-.4
Athletics	137	64	69	.481	30.0	893	890	-7	11	-.7	.9	-2.7
Brooklyn	138	60	74	.448	34.5	904	918	-3	-9	-.3	-.9	-5.8
Metropolitans	138	44	89	.331	50.0	754	1093	-153	-184	-12.7	-15.3	5.5
Cleveland	133	39	92	.298	54.0	729	1112	-145	-236	-12.0	-19.6	5.2

NATIONAL LEAGUE 1888

Batting Runs			Normalized OPS			Pitching Runs			Normalized ERA		
Jimmy Ryan	CHI	52.6	Cap Anson	CHI	194	Tim Keefe	NY	52.6	Tim Keefe	NY	163
Cap Anson	CHI	52.3	Jimmy Ryan	CHI	191	Charlie Buffington	PHI	42.5	Ben Sanders	PHI	149
Dan Brouthers	DET	49.8	Roger Connor	NY	184	Charlie Buffington	PHI	40.9	Charlie Buffington	PHI	148

On Base Average			Slugging Percentage			Percent of Team Wins			Wins Above Team		
Cap Anson	CHI	.400	Jimmy Ryan	CHI	.515	John Clarkson	BOS	.471	Pete Conway	DET	10.8
Dan Brouthers	DET	.399	Cap Anson	CHI	.499	Pete Conway	DET	.441	John Clarkson	BOS	8.8
Roger Connor	NY	.389	Roger Connor	NY	.480	Ed Morris	PIT	.439	Tim Keefe	NY	7.2

Club	G	W	L	Pct	GB	R	OR	Runs O/D		Wins O/D		Diff
New York	137	84	47	.641	0.0	659	479	35	145	3.5	14.4	.6
Chicago	135	77	58	.570	9.0	734	659	119	-43	11.8	-4.4	2.0
Philadelphia	131	69	61	.531	14.5	535	509	-61	88	-6.1	8.7	1.4
Boston	137	70	64	.522	15.5	669	619	45	5	4.5	.5	-2.0
Detroit	134	68	63	.519	16.0	721	629	111	-18	11.0	-1.9	-6.6
Pittsburgh	138	66	68	.493	19.5	534	580	-94	49	-9.4	4.8	3.6
Indianapolis	136	50	85	.370	36.0	603	731	-15	-111	-1.6	-11.1	-4.8
Washington	136	48	86	.358	37.5	482	731	-136	-111	-13.7	-11.1	5.8

AMERICAN ASSOCIATION 1888

Batting Runs			Normalized OPS			Pitching Runs			Normalized ERA		
John Reilly	CIN	43.8	John Reilly	CIN	181	Silver King	STL	91.8	Silver King	STL	186
Tip O'Neill	STL	41.7	Tip O'Neill	STL	173	Ed Seward	ATH	60.0	Ed Seward	ATH	152
Harry Stovey	ATH	39.9	Harry Stovey	ATH	169	Mickey Hughes	BRO	37.1	Adonis Terry	BRO	150

On Base Average			Slugging Percentage			Percent of Team Wins			Wins Above Team		
Yank Robinson	STL	.400	John Reilly	CIN	.501	Ed Bakely	CLE	.500	Ed Bakely	CLE	5.4
Tip O'Neill	STL	.390	Harry Stovey	ATH	.460	Silver King	STL	.489	Lee Viau	CIN	3.6
Pete Browning	LOU	.380	Tip O'Neill	STL	.446	Ed Seward	ATH	.432	Ed Seward	ATH	3.6

Club	G	W	L	Pct	GB	R	OR	Runs O/D		Wins O/D		Diff
St. Louis	137	92	43	.681	0.0	789	501	78	210	7.2	19.6	-2.3
Brooklyn	143	88	52	.629	6.5	758	584	15	159	1.4	14.8	1.8
Athletics	136	81	52	.609	10.0	827	594	121	112	11.2	10.4	-7.2
Cincinnati	137	80	54	.597	11.5	745	628	34	83	3.1	7.8	2.1
Baltimore	137	57	80	.416	36.0	653	779	-57	-67	-5.4	-6.3	.2
Cleveland	135	50	82	.379	40.5	651	839	-49	-137	-4.7	-12.8	1.5
Louisville	139	48	87	.356	44.0	689	870	-32	-147	-3.0	-13.8	-2.7
Kansas City	132	43	89	.326	47.5	579	896	-105	-210	-9.9	-19.6	6.5

Batting Runs
Player	Team	
Dan Brouthers	BOS	53.2
Mike Tiernan	NY	51.7
Roger Connor	NY	50.2

On Base Average
Player	Team	
Fred Carroll	PIT	.486
Dan Brouthers	BOS	.462
Mike Tiernan	NY	.447

Normalized OPS
Player	Team	
Fred Carroll	PIT	180
Dan Brouthers	BOS	180
Roger Connor	NY	175

Slugging Percentage
Player	Team	
Roger Connor	NY	.528
Dan Brouthers	BOS	.507
Mike Tiernan	NY	.501

Pitching Runs
Player	Team	
John Clarkson	BOS	88.9
Mickey Welch	NY	41.7
Ed Bakely	CLE	35.8

Percent of Team Wins
Player	Team	
John Clarkson	BOS	.590
Charlie Buffington	PHI	.444
Alex Ferson	WAS	.415

Normalized ERA
Player	Team	
John Clarkson	BOS	147
Ed Bakely	CLE	136
Mickey Welch	NY	133

Wins Above Team
Player	Team	
John Clarkson	BOS	10.5
Charlie Buffington	PHI	9.4
Alex Ferson	WAS	7.9

Club	G	W	L	Pct	GB	R	OR	Runs O/D		Wins O/D		Diff
New York	131	83	43	.659	0.0	935	708	170	57	14.9	5.0	.1
Boston	133	83	45	.648	1.0	826	626	49	151	4.3	13.2	1.4
Chicago	136	67	65	.508	19.0	867	814	73	-19	6.4	-1.7	-3.7
Philadelphia	130	63	64	.496	20.5	742	748	-16	11	-1.5	1.0	.0
Pittsburgh	134	61	71	.462	25.0	726	801	-56	-17	-5.0	-1.6	1.6
Cleveland	136	61	72	.459	25.5	656	720	-137	74	-12.1	6.5	.1
Indianapolis	135	59	75	.440	28.0	819	894	30	-104	2.7	-9.3	-1.4
Washington	127	41	83	.331	41.0	632	892	-109	-149	-9.6	-13.2	1.8

AMERICAN ASSOCIATION 1889

Batting Runs
Player	Team	
Tommy Tucker	BAL	50.9
Harry Stovey	ATH	45.8
Tip O'Neill	STL	42.8

On Base Average
Player	Team	
Tommy Tucker	BAL	.450
Henry Larkin	ATH	.428
Denny Lyons	ATH	.426

Normalized OPS
Player	Team	
Tommy Tucker	BAL	172
Harry Stovey	ATH	166
Tip O'Neill	STL	161

Slugging Percentage
Player	Team	
Harry Stovey	ATH	.525
Bug Holliday	CIN	.497
Tommy Tucker	BAL	.484

Pitching Runs
Player	Team	
Jesse Duryea	CIN	57.0
Matt Kilroy	BAL	52.9
Gus Weyhing	ATH	44.4

Percent of Team Wins
Player	Team	
Mark Baldwin	COL	.450
Bob Caruthers	BRO	.430
Jesse Duryea	CIN	.421

Normalized ERA
Player	Team	
Jack Stivetts	STL	171
Jesse Duryea	CIN	150
Matt Kilroy	BAL	135

Wins Above Team
Player	Team	
Bob Caruthers	BRO	8.6
Jesse Duryea	CIN	6.5
Jim Conway	KC	5.2

Club	G	W	L	Pct	GB	R	OR	Runs O/D		Wins O/D		Diff
Brooklyn	140	93	44	.679	0.0	995	706	146	143	12.5	12.3	-.4
St. Louis	141	90	45	.667	2.0	957	680	102	175	8.7	15.1	-1.4
Athletics	138	75	58	.564	16.0	880	787	43	50	3.7	4.3	.5
Cincinnati	141	76	63	.547	18.0	897	769	42	86	3.6	7.4	-4.5
Baltimore	139	70	65	.519	22.0	791	795	-51	48	-4.5	4.2	2.8
Columbus	140	60	78	.435	33.5	779	924	-69	-74	-6.1	-6.4	3.5
Kansas City	139	55	82	.401	38.0	852	1031	9	-187	.7	-16.2	1.9
Louisville	140	27	111	.196	66.5	632	1091	-216	-241	-18.7	-20.8	-2.5

NATIONAL LEAGUE 1890

Batting Runs
Player	Team	
Mike Tiernan	NY	41.0
Cap Anson	CHI	40.4
Billy Hamilton	PHI	34.3

On Base Average
Player	Team	
Cap Anson	CHI	.443
Billy Hamilton	PHI	.430
George Pinckney	BRO	.411

Normalized OPS
Player	Team	
Mike Tiernan	NY	162
Cap Anson	CHI	158
George Pinckney	BRO	151

Slugging Percentage
Player	Team	
Mike Tiernan	NY	.495
John Reilly	CIN	.472
Oyster Burns	BRO	.464

Pitching Runs
Player	Team	
Billy Rhines	CIN	73.5
Kid Nichols	BOS	65.9
Amos Rusie	NY	63.4

Percent of Team Wins
Player	Team	
Bill Hutchison	CHI	.500
Ed Beatin	CLE	.500
Kid Gleason	PHI	.487

Normalized ERA
Player	Team	
Billy Rhines	CIN	185
Kid Nichols	BOS	163
Tony Mullane	CIN	161

Wins Above Team
Player	Team	
Kid Gleason	PHI	9.4
Ed Beatin	CLE	7.2
Cy Young	CLE	4.5

Club	G	W	L	Pct	GB	R	OR	Runs O/D		Wins O/D		Diff
Brooklyn	129	86	43	.667	0.0	884	621	164	99	14.8	8.9	-2.1
Chicago	139	84	53	.613	6.0	847	695	72	80	6.4	7.2	1.8
Philadelphia	133	78	54	.591	9.5	827	707	85	35	7.6	3.1	1.2
Cincinnati	134	77	55	.583	10.5	753	633	5	115	.5	10.3	.2
Boston	134	76	57	.571	12.0	763	593	15	155	1.4	13.9	-5.8
New York	135	63	68	.481	24.0	713	698	-39	55	-3.6	5.0	-3.8
Cleveland	136	44	88	.333	43.5	630	832	-128	-72	-11.6	-6.6	-3.9
Pittsburgh	138	23	113	.169	66.5	597	1235	-172	-464	-15.5	-41.8	12.3

AMERICAN ASSOCIATION 1890

Batting Runs			Normalized OPS			Pitching Runs			Normalized ERA		
Tommy McCarthy	STL	48.4	Clarence Childs	SYR	177	Scott Stratton	LOU	71.8	Scott Stratton	LOU	164
Clarence Childs	SYR	47.5	Chicken Wolf	LOU	172	Phil Ehret	LOU	53.1	Phil Ehret	LOU	153
Chicken Wolf	LOU	45.7	Tommy McCarthy	STL	171	John Healy	TOL	41.9	Frank Knauss	COL	137
On Base Average			**Slugging Percentage**			**Percent of Team Wins**			**Wins Above Team**		
Ed Swartwood	TOL	.442	Clarence Childs	SYR	.481	Sadie McMahon	ATH	.537	Sadie McMahon	ATH	19.4
Clarence Childs	SYR	.434	Chicken Wolf	LOU	.479	Bob Barr	ROC	.444	Hank Gastright	COL	6.0
Tommy McCarthy	STL	.429	Tommy McCarthy	STL	.467	Scott Stratton	LOU	.386	Bob Barr	ROC	3.4

Club	G	W	L	Pct	GB	R	OR	Runs O/D		Wins O/D		Diff
Louisville	136	88	44	.667	0.0	819	588	58	173	5.2	15.5	1.3
Columbus	140	79	55	.590	10.0	831	617	48	166	4.3	14.9	-7.2
St. Louis	139	78	58	.574	12.0	870	736	92	42	8.3	3.7	-2.0
Toledo	134	68	64	.515	20.0	739	689	-10	61	-1.0	5.4	-2.5
Rochester	133	63	63	.500	22.0	709	711	-34	33	-3.1	3.0	.2
Syracuse	128	55	72	.433	30.5	698	831	-17	-114	-1.6	-10.3	3.4
Athletics	138	54	78	.409	34.0	702	945	-58	-183	-5.3	-16.5	9.8
Baltimore	134	41	92	.308	47.5	674	925	-75	-174	-6.8	-15.7	-3.0

PLAYERS LEAGUE 1890

Batting Runs			Normalized OPS			Pitching Runs			Normalized ERA		
Roger Connor	NY	49.9	Roger Connor	NY	171	Silver King	CHI	78.9	Silver King	CHI	157
Pete Browning	CLE	48.3	Pete Browning	CLE	167	Mark Baldwin	CHI	51.2	Henry Staley	PIT	131
Dan Brouthers	BOS	41.3	Dave Orr	BRO	160	Henry Staley	PIT	43.1	Charlie Radbourn	BOS	128
On Base Average			**Slugging Percentage**			**Percent of Team Wins**			**Wins Above Team**		
Dan Brouthers	BOS	.466	Roger Connor	NY	.541	Mark Baldwin	CHI	.453	Phil Knell	PHI	6.5
Pete Browning	CLE	.459	Jake Beckley	PIT	.541	Silver King	CHI	.400	Gus Weyhing	BRO	5.4
Roger Connor	NY	.450	Dave Orr	BRO	.537	Gus Weyhing	BRO	.395	Mark Baldwin	CHI	3.9

Club	G	W	L	Pct	GB	R	OR	Runs O/D		Wins O/D		Diff
Boston	130	81	48	.628	0.0	992	767	98	127	7.9	10.3	-1.7
Brooklyn	133	76	56	.576	6.5	964	893	49	22	4.0	1.8	4.3
New York	132	74	57	.565	8.0	1018	876	110	32	8.9	2.6	-3.0
Chicago	138	75	62	.547	10.0	886	770	-62	179	-5.1	14.5	-2.9
Philadelphia	132	68	63	.519	14.0	942	855	34	53	2.7	4.3	-4.5
Pittsburgh	128	60	68	.469	20.5	835	892	-45	-10	-3.7	-.9	.6
Cleveland	131	55	75	.423	26.5	849	1027	-51	-125	-4.2	-10.2	4.4
Buffalo	134	36	96	.273	46.5	793	1199	-128	-276	-10.4	-22.4	2.8

NATIONAL LEAGUE 1891

Batting Runs			Normalized OPS			Pitching Runs			Normalized ERA		
Billy Hamilton	PHI	47.5	Mike Tiernan	NY	165	Kid Nichols	BOS	45.0	John Ewing	NY	147
Mike Tiernan	NY	42.4	Billy Hamilton	PHI	162	Amos Rusie	NY	43.9	Kid Nichols	BOS	140
Harry Stovey	BOS	38.8	Harry Stovey	BOS	160	Bill Hutchison	CHI	33.0	Henry Staley	BOS	134
On Base Average			**Slugging Percentage**			**Percent of Team Wins**			**Wins Above Team**		
Billy Hamilton	PHI	.453	Mike Tiernan	NY	.500	Bill Hutchison	CHI	.537	Bill Hutchison	CHI	10.8
Roger Connor	NY	.399	Harry Stovey	BOS	.498	Amos Rusie	NY	.465	Amos Rusie	NY	7.5
Clarence Childs	CLE	.395	Bug Holliday	CIN	.473	Cy Young	CLE	.415	John Ewing	NY	6.9

Club	G	W	L	Pct	GB	R	OR	Runs O/D		Wins O/D		Diff
Boston	140	87	51	.630	0.0	847	658	72	117	6.5	10.6	1.0
Chicago	137	82	53	.607	3.5	832	730	73	29	6.6	2.6	5.3
New York	136	71	61	.538	13.0	754	711	1	42	.1	3.8	1.1
Philadelphia	138	68	69	.496	18.5	756	773	-7	-8	-.7	-.8	1.0
Cleveland	141	65	74	.468	22.5	835	888	54	-106	4.9	-9.7	.3
Brooklyn	137	61	76	.445	25.5	765	820	6	-60	.6	-5.5	-2.5
Cincinnati	138	56	81	.409	30.5	646	790	-117	-25	-10.7	-2.3	.5
Pittsburgh	137	55	80	.407	30.5	679	744	-79	15	-7.2	1.3	-6.6

AMERICAN ASSOCIATION 1891

Batting Runs			Normalized OPS			Pitching Runs			Normalized ERA		
Dan Brouthers	BOS	59.8	Dan Brouthers	BOS	188	George Haddock	BOS	51.9	Ed Crane	CIN	152
Denny Lyons	STL	41.9	Jocko Milligan	ATH	165	Sadie McMahon	BAL	50.9	George Haddock	BOS	149
Tom Brown	BOS	38.8	Denny Lyons	STL	164	Charlie Buffington	BOS	47.3	Charlie Buffington	BOS	146
On Base Average			Slugging Percentage			Percent of Team Wins			Wins Above Team		
Dan Brouthers	BOS	.471	Dan Brouthers	BOS	.512	Sadie McMahon	BAL	.486	Frank Foreman	WAS	7.8
Denny Lyons	STL	.445	Jocko Milligan	ATH	.505	Phil Knell	COL	.459	Gus Weyhing	ATH	6.7
Dummy Hoy	STL	.424	Duke Farrell	BOS	.474	Gus Weyhing	ATH	.425	Sadie McMahon	BAL	6.3

Club	G	W	L	Pct	GB	R	OR	Runs O/D		Wins O/D		Diff
Boston	139	93	42	.689	0.0	1028	676	213	139	18.7	12.2	-5.3
St. Louis	141	86	52	.623	8.5	976	753	149	74	13.1	6.5	-2.5
Baltimore	139	72	63	.533	21.0	850	798	35	17	3.1	1.5	-.1
Athletics	143	73	66	.525	22.0	819	794	-19	45	-1.7	3.9	1.3
Cincinnati	138	64	72	.471	29.5	777	801	-31	8	-2.8	.7	-1.9
Columbus	138	61	76	.445	33.0	702	777	-106	32	-9.4	2.8	-.9
Louisville	141	55	84	.396	40.0	713	890	-113	-62	-10.0	-5.5	1.0
Washington	139	43	92	.319	50.0	691	1067	-123	-251	-10.9	-22.1	8.4

NATIONAL LEAGUE 1892

Batting Runs			Normalized OPS			Pitching Runs			Normalized ERA		
Dan Brouthers	BRO	60.3	Dan Brouthers	BRO	183	Cy Young	CLE	68.0	Cy Young	CLE	170
Roger Connor	PHI	54.1	Roger Connor	PHI	174	Bill Hutchison	CHI	37.6	Tim Keefe	PHI	139
Clarence Childs	CLE	48.9	Ed Delahanty	PHI	165	John Clarkson	CLE	34.6	John Clarkson	CLE	132
On Base Average			Slugging Percentage			Percent of Team Wins			Wins Above Team		
Clarence Childs	CLE	.443	Ed Delahanty	PHI	.495	Bill Hutchison	CHI	.529	Frank Killen	WAS	12.4
Dan Brouthers	BRO	.432	Dan Brouthers	BRO	.480	Frank Killen	WAS	.500	Cy Young	CLE	8.9
Billy Hamilton	PHI	.423	Roger Connor	PHI	.463	Amos Rusie	NY	.437	Adonis Terry	PIT	5.3

Club	G	W	L	Pct	GB	R	OR	Runs O/D		Wins O/D		Diff
Boston	152	102	48	.680	0.0	862	649	86	127	8.1	11.9	7.0
Cleveland	153	93	56	.624	8.5	855	613	74	168	7.0	15.7	-4.2
Brooklyn	158	95	59	.617	9.0	935	733	129	73	12.1	6.9	-1.0
Philadelphia	155	87	66	.569	16.5	860	690	69	101	6.5	9.5	-5.5
Cincinnati	155	82	68	.547	20.0	766	731	-24	60	-2.3	5.6	3.7
Pittsburgh	155	80	73	.523	23.5	802	796	11	-4	1.0	-.5	2.9
Chicago	147	70	76	.479	30.0	635	735	-114	15	-10.8	1.4	6.4
New York	153	71	80	.470	31.5	811	826	30	-44	2.9	-4.3	-3.1
Louisville	154	63	89	.414	40.0	649	804	-136	-17	-12.8	-1.7	1.6
Washington	151	58	93	.384	44.5	731	869	-38	-98	-3.7	-9.3	-4.5
St. Louis	155	56	94	.373	46.0	703	922	-87	-130	-8.2	-12.3	1.6
Baltimore	152	46	101	.313	54.5	779	1020	3	-243	.3	-23.0	-4.9

NATIONAL LEAGUE 1893

Batting Runs			Normalized OPS			Pitching Runs			Normalized ERA		
Ed Delahanty	PHI	56.3	Billy Hamilton	PHI	176	Amos Rusie	NY	76.6	Ted Breitenstein	STL	147
Sam Thompson	PHI	47.0	Ed Delahanty	PHI	173	Ted Breitenstein	STL	63.0	Amos Rusie	NY	144
Jesse Burkett	CLE	46.8	George Davis	NY	161	Cy Young	CLE	61.1	Cy Young	CLE	139
On Base Average			Slugging Percentage			Percent of Team Wins			Wins Above Team		
Billy Hamilton	PHI	.490	Ed Delahanty	PHI	.583	Amos Rusie	NY	.485	Cy Young	CLE	9.0
Clarence Childs	CLE	.463	George Davis	NY	.554	Cy Young	CLE	.466	Amos Rusie	NY	8.8
Jesse Burkett	CLE	.459	Sam Thompson	PHI	.530	Frank Killen	PIT	.432	Frank Killen	PIT	6.8

Club	G	W	L	Pct	GB	R	OR	Runs O/D		Wins O/D		Diff
Boston	131	86	43	.667	0.0	1008	795	147	66	12.2	5.4	3.9
Pittsburgh	131	81	48	.628	5.0	970	766	109	95	9.0	7.8	-.4
Cleveland	129	73	55	.570	12.5	976	839	128	9	10.6	.7	-2.3
Philadelphia	133	72	57	.558	14.0	1011	841	137	33	11.4	2.7	-6.6
New York	136	68	64	.515	19.5	941	845	47	49	3.9	4.0	-5.9
Cincinnati	131	65	63	.508	20.5	759	814	-101	47	-8.4	3.9	5.6
Brooklyn	130	65	63	.508	20.5	775	845	-78	9	-6.5	.8	6.8
Baltimore	130	60	70	.462	26.5	820	893	-33	-38	-2.8	-3.2	1.0
Chicago	128	56	71	.441	29.0	829	874	-11	-32	-1.0	-2.7	-3.8
St. Louis	135	57	75	.432	30.5	745	829	-141	58	-11.7	4.8	-2.0
Louisville	126	50	75	.400	34.0	759	942	-68	-113	-5.7	-9.4	2.6
Washington	130	40	89	.310	46.0	722	1032	-131	-177	-10.9	-14.7	1.2

NATIONAL LEAGUE   1894

Batting Runs			Normalized OPS			Pitching Runs			Normalized ERA		
Hugh Duffy	BOS	79.8	Hugh Duffy	BOS	192	Amos Rusie	NY	125.3	Amos Rusie	NY	191
Joe Kelley	BAL	63.5	Bill Joyce	WAS	180	Jouett Meekin	NY	73.6	Jouett Meekin	NY	144
Billy Hamilton	PHI	63.3	Sam Thompson	PHI	179	Cy Young	CLE	62.7	Win Mercer	WAS	141
On Base Average			**Slugging Percentage**			**Percent of Team Wins**			**Wins Above Team**		
Billy Hamilton	PHI	.523	Hugh Duffy	BOS	.690	Ted Breitenstein	STL	.482	Ted Breitenstein	STL	9.3
Hugh Duffy	BOS	.506	Sam Thompson	PHI	.686	Amos Rusie	NY	.409	Clark Griffith	CHI	8.0
Joe Kelley	BAL	.502	Bill Joyce	WAS	.648	Kid Nichols	BOS	.386	Jouett Meekin	NY	7.3

Club	G	W	L	Pct	GB	R	OR	Runs O/D		Wins O/D		Diff
Baltimore	129	89	39	.695	0.0	1171	820	222	129	17.4	10.1	-2.5
New York	137	88	44	.667	3.0	940	789	-67	219	-5.3	17.1	10.2
Boston	133	83	49	.629	8.0	1222	1002	244	-23	19.1	-1.9	-.2
Philadelphia	131	71	57	.555	18.0	1143	966	179	-1	14.0	-.2	-6.8
Brooklyn	134	70	61	.534	20.5	1021	1007	35	-20	2.8	-1.7	3.4
Cleveland	130	68	61	.527	21.5	932	896	-23	60	-1.9	4.7	.7
Pittsburgh	132	65	65	.500	25.0	955	972	-15	-0	-1.2	-.1	1.3
Chicago	135	57	75	.432	34.0	1041	1066	48	-72	3.8	-5.7	-7.0
St. Louis	133	56	76	.424	35.0	771	954	-206	24	-16.2	1.9	4.3
Cincinnati	132	55	75	.423	35.0	910	1085	-60	-113	-4.8	-8.9	3.7
Washington	132	45	87	.341	46.0	882	1122	-88	-150	-7.0	-11.8	-2.2
Louisville	130	36	94	.277	54.0	692	1001	-263	-44	-20.7	-3.5	-4.8

NATIONAL LEAGUE   1895

Batting Runs			Normalized OPS			Pitching Runs			Normalized ERA		
Ed Delahanty	PHI	68.8	Ed Delahanty	PHI	192	Pink Hawley	PIT	78.9	Al Maul	WAS	195
Sam Thompson	PHI	59.7	Sam Thompson	PHI	183	Cy Young	CLE	62.5	Dad Clarke	NY	153
Jesse Burkett	CLE	56.9	Bill Lange	CHI	172	Kid Nichols	BOS	57.8	Pink Hawley	PIT	150
On Base Average			**Slugging Percentage**			**Percent of Team Wins**			**Wins Above Team**		
Ed Delahanty	PHI	.500	Sam Thompson	PHI	.654	Ted Breitenstein	STL	.487	Cy Young	CLE	9.1
Billy Hamilton	PHI	.490	Ed Delahanty	PHI	.617	Pink Hawley	PIT	.437	Bill Hoffer	BAL	8.7
Jesse Burkett	CLE	.486	Bill Lange	CHI	.582	Cy Young	CLE	.417	Ted Breitenstein	STL	7.0

Club	G	W	L	Pct	GB	R	OR	Runs O/D		Wins O/D		Diff
Baltimore	132	87	43	.669	0.0	1009	647	140	222	11.5	18.4	-7.9
Cleveland	131	84	46	.646	3.0	917	721	54	142	4.5	11.7	2.8
Philadelphia	133	78	53	.595	9.5	1068	957	192	-80	15.9	-6.7	3.3
Chicago	133	72	58	.554	15.0	867	854	-8	22	-.7	1.8	5.9
Brooklyn	133	71	60	.542	16.5	867	834	-8	42	-.7	3.5	2.8
Boston	132	71	60	.542	16.5	907	826	38	43	3.1	3.6	-1.2
Pittsburgh	134	71	61	.538	17.0	811	787	-71	96	-5.9	7.9	3.0
Cincinnati	132	66	64	.508	21.0	903	854	34	15	2.8	1.3	-3.1
New York	132	66	65	.504	21.5	854	834	-14	35	-1.3	2.9	-1.2
Washington	132	43	85	.336	43.0	837	1049	-31	-179	-2.7	-14.8	-3.5
St. Louis	135	39	92	.298	48.5	747	1032	-141	-142	-11.7	-11.8	-2.9
Louisville	133	35	96	.267	52.5	698	1090	-177	-213	-14.7	-17.7	1.9

NATIONAL LEAGUE   1896

Batting Runs			Normalized OPS			Pitching Runs			Normalized ERA		
Ed Delahanty	PHI	67.8	Ed Delahanty	PHI	196	Kid Nichols	BOS	63.2	Billy Rhines	CIN	178
Jesse Burkett	CLE	58.0	Joe Kelley	BAL	173	Cy Young	CLE	51.5	Kid Nichols	BOS	154
Joe Kelley	BAL	58.0	Jesse Burkett	CLE	170	George Cuppy	CLE	48.5	George Cuppy	CLE	139
On Base Average			**Slugging Percentage**			**Percent of Team Wins**			**Wins Above Team**		
Billy Hamilton	BOS	.477	Ed Delahanty	PHI	.631	Frank Killen	PIT	.455	Jouett Meekin	NY	9.3
Hughie Jennings	BAL	.472	Bill Dahlen	CHI	.553	Ted Breitenstein	STL	.450	Win Mercer	WAS	8.9
Ed Delahanty	PHI	.472	Tom McCreery	LOU	.546	Win Mercer	WAS	.431	Frank Killen	PIT	8.7

Club	G	W	L	Pct	GB	R	OR	Runs O/D		Wins O/D		Diff
Baltimore	132	90	39	.698	0.0	995	662	199	134	17.2	11.6	-3.3
Cleveland	135	80	48	.625	9.5	840	651	26	163	2.2	14.1	-.3
Cincinnati	128	77	50	.606	12.0	784	620	12	152	1.0	13.1	-.7
Boston	132	74	57	.565	17.0	860	761	64	35	5.5	3.1	-.1
Chicago	132	71	57	.555	18.5	815	799	19	-2	1.6	-.2	5.6
Pittsburgh	131	66	63	.512	24.0	787	741	-2	49	-.3	4.3	-2.5
New York	133	64	67	.489	27.0	829	821	27	-18	2.3	-1.6	-2.2
Philadelphia	131	62	68	.477	28.5	890	891	100	-100	8.6	-8.7	-2.9
Washington	133	58	73	.443	33.0	818	920	16	-117	1.4	-10.2	1.3
Brooklyn	133	58	73	.443	33.0	692	764	-109	38	-9.5	3.3	-1.3
St. Louis	131	40	90	.308	50.5	593	929	-196	-138	-17.0	-12.0	4.0
Louisville	133	38	93	.290	53.0	653	997	-148	-194	-12.9	-16.8	2.2

NATIONAL LEAGUE 1897

Batting Runs			Normalized OPS			Pitching Runs			Normalized ERA		
Willie Keeler	BAL	56.1	Willie Keeler	BAL	173	Kid Nichols	BOS	68.3	Amos Rusie	NY	170
Ed Delahanty	PHI	49.4	Ed Delahanty	PHI	167	Amos Rusie	NY	63.3	Kid Nichols	BOS	163
Fred Clarke	PIT	46.1	Fred Clarke	PIT	165	Joe Corbett	BAL	41.7	Jerry Nops	BAL	153
On Base Average			**Slugging Percentage**			**Percent of Team Wins**			**Wins Above Team**		
Jesse Burkett	CLE	.462	Nap Lajoie	PHI	.569	Clark Griffith	CHI	.356	Amos Rusie	NY	5.5
John McGraw	BAL	.461	Willie Keeler	BAL	.544	Red Donahue	STL	.345	Clark Griffith	CHI	5.1
Willie Keeler	BAL	.457	Ed Delahanty	PHI	.538	Amos Rusie	NY	.337	Jack Dunn	BRO	4.1

Club	G	W	L	Pct	GB	R	OR	Runs O/D		Wins O/D		Diff
Boston	135	93	39	.705	0.0	1025	665	231	129	20.2	11.3	-4.5
Baltimore	136	90	40	.692	2.0	964	674	164	126	14.3	11.0	-.4
New York	137	83	48	.634	9.5	896	695	90	111	7.9	9.7	-.1
Cincinnati	134	76	56	.576	17.0	763	705	-25	84	-2.2	7.3	4.9
Cleveland	132	69	62	.527	23.5	773	680	-3	97	-.3	8.5	-4.6
Washington	135	61	71	.462	32.0	781	793	-12	1	-1.2	.1	-4.0
Brooklyn	136	61	71	.462	32.0	802	845	2	-44	.2	-3.9	-1.2
Pittsburgh	135	60	71	.458	32.5	676	835	-117	-40	-10.4	-3.6	8.4
Chicago	138	59	73	.447	34.0	832	894	20	-81	1.7	-7.2	-1.6
Philadelphia	134	55	77	.417	38.0	752	792	-36	-2	-3.2	-.3	-7.5
Louisville	134	52	78	.400	40.0	669	860	-119	-70	-10.5	-6.3	3.7
St. Louis	132	29	102	.221	63.5	588	1083	-188	-305	-16.5	-26.8	6.8

NATIONAL LEAGUE 1898

Batting Runs			Normalized OPS			Pitching Runs			Normalized ERA		
Billy Hamilton	BOS	46.7	Billy Hamilton	BOS	177	Kid Nichols	BOS	63.4	Clark Griffith	CHI	191
John McGraw	BAL	44.3	Ed Delahanty	PHI	159	Clark Griffith	CHI	62.3	Al Maul	BAL	171
Ed Delahanty	PHI	42.2	Elmer Flick	PHI	156	Doc McJames	BAL	51.5	Kid Nichols	BOS	169
On Base Average			**Slugging Percentage**			**Percent of Team Wins**			**Wins Above Team**		
Billy Hamilton	BOS	.478	John Anderson	WAS	.494	Bert Cunningham	LOU	.400	Bert Cunningham	LOU	11.3
John McGraw	BAL	.459	Jimmy Collins	BOS	.479	Jesse Tannehill	STL	.385	Jesse Tannehill	PIT	8.8
Elmer Smith	CIN	.423	Nap Lajoie	PHI	.461	Jesse Tannehill	PIT	.347	Clark Griffith	CHI	6.1

Club	G	W	L	Pct	GB	R	OR	Run O/D		Wins O/D		Diff
Boston	152	102	47	.685	0.0	872	614	119	139	11.3	13.3	2.9
Baltimore	154	96	53	.644	6.0	933	623	170	140	16.2	13.4	-8.0
Cincinnati	157	92	60	.605	11.5	831	740	53	38	5.0	3.6	7.3
Chicago	152	85	65	.567	17.5	828	679	75	74	7.1	7.1	-4.2
Cleveland	156	81	68	.544	21.0	730	683	-42	90	-4.1	8.6	2.0
Philadelphia	150	78	71	.523	24.0	823	784	80	-40	7.6	-3.9	-.2
New York	157	77	73	.513	25.5	837	800	59	-21	5.6	-2.1	-1.5
Pittsburgh	152	72	76	.486	29.5	634	694	-118	59	-11.4	5.7	3.7
Louisville	154	70	81	.464	33.0	728	833	-34	-69	-3.4	-6.6	4.5
Brooklyn	149	54	91	.372	46.0	638	811	-99	-72	-9.6	-6.9	-2.0
Washington	155	51	101	.336	52.5	704	939	-63	-170	-6.1	-16.3	-2.6
St. Louis	154	39	111	.260	63.5	571	929	-191	-165	-18.3	-15.8	-1.9

NATIONAL LEAGUE 1899

Batting Runs			Normalized OPS			Pitching Runs			Normalized ERA		
Ed Delahanty	PHI	72.4	Ed Delahanty	PHI	197	Cy Young	STL	52.1	Vic Willis	BOS	154
Jesse Burkett	STL	56.6	John McGraw	BAL	182	Vic Willis	BOS	51.5	Cy Young	STL	149
John McGraw	BAL	56.6	Jesse Burkett	STL	174	Joe McGinnity	BAL	47.6	Joe McGinnity	BAL	144
On Base Average			**Slugging Percentage**			**Percent of Team Wins**			**Wins Above Team**		
John McGraw	BAL	.535	Ed Delahanty	PHI	.582	Joe McGinnity	BAL	.326	Noodles Hahn	CIN	7.4
Jesse Burkett	STL	.461	Buck Freeman	WAS	.563	Jesse Tannehill	PIT	.316	Vic Willis	BOS	6.7
Ed Delahanty	PHI	.461	Jimmy Williams	PIT	.532	Gus Weyhing	WAS	.315	Jim Hughes	BRO	6.2

Club	G	W	L	Pct	GB	R	OR	Runs O/D		Wins O/D		Diff
Brooklyn	150	101	47	.682	0.0	892	658	106	128	9.8	11.8	5.3
Boston	153	95	57	.625	8.0	859	645	58	156	5.3	14.5	-.8
Philadelphia	154	94	58	.618	9.0	916	744	109	63	10.1	5.8	2.1
Baltimore	152	86	62	.581	15.0	827	691	31	105	2.9	9.8	-.6
St. Louis	155	84	67	.556	18.5	819	739	7	73	.7	6.8	1.1
Cincinnati	156	83	67	.553	19.0	856	770	39	47	3.6	4.4	.0
Pittsburgh	154	76	73	.510	25.5	835	766	28	41	2.6	3.8	-4.9
Chicago	152	75	73	.507	26.0	812	763	16	33	1.5	3.1	-3.5
Louisville	155	75	77	.493	28.0	827	775	15	37	1.4	3.4	-5.8
New York	152	60	90	.400	42.0	734	863	-61	-66	-5.8	-6.2	-3.0
Washington	155	54	98	.355	49.0	743	983	-68	-170	-6.4	-15.9	.2
Cleveland	154	20	134	.130	84.0	529	1252	-277	-444	-25.7	-41.3	10.0

NATIONAL LEAGUE 1900

Batting Runs				Normalized OPS				Pitching Runs				Normalized ERA		
Honus Wagner	PIT	56.2		Honus Wagner	PIT	185		Ned Garvin	CHI	35.0		Rube Waddell	PIT	156
Elmer Flick	PHI	54.3		Elmer Flick	PHI	178		Rube Waddell	PIT	30.7		Ned Garvin	CHI	153
Jesse Burkett	STL	43.9		Jesse Burkett	STL	159		Joe McGinnity	BRO	28.6		Jack W. Taylor	CHI	145
On Base Average				**Slugging Percentage**				**Percent of Team Wins**				**Wins Above Team**		
Billy Hamilton	BOS	.447		Honus Wagner	PIT	.573		Joe McGinnity	BRO	.341		Joe McGinnity	BRO	8.6
Roy Thomas	PHI	.438		Elmer Flick	PHI	.545		Bill Carrick	NY	.317		Jesse Tannehill	PIT	6.4
Jesse Burkett	STL	.427		Nap Lajoie	PHI	.510		Bill Dinneen	BOS	.303		Bill Dinneen	BOS	5.0

Club	G	W	L	Pct	GB	R	OR	Runs O/D			Wins O/D		Diff
Brooklyn	142	82	54	.603	0.0	816	722	76	18	7.1	1.7		5.3
Pittsburgh	140	79	60	.568	4.5	733	612	3	118	.3	10.9		-1.7
Philadelphia	141	75	63	.543	8.0	810	791	75	-55	7.0	-5.2		4.2
Boston	142	66	72	.478	17.0	778	739	38	1	3.5	.1		-6.6
St. Louis	142	65	75	.464	19.0	743	747	3	-6	.3	-.7		-4.6
Chicago	146	65	75	.464	19.0	635	751	-125	10	-11.7	.9		5.8
Cincinnati	144	62	77	.446	21.5	702	745	-47	5	-4.5	.5		-3.5
New York	141	60	78	.435	23.0	713	823	-21	-87	-2.0	-8.2		1.2

AMERICAN LEAGUE 1901

Batting Runs				Park Adjusted				Pitching Runs				Park Adjusted		
Nap Lajoie	PHI	71.6		Nap Lajoie	PHI	71.0		Cy Young	BOS	84.0		Cy Young	BOS	69.3
Buck Freeman	BOS	34.6		Buck Freeman	BOS	39.3		Clark Griffith	CHI	29.6		Roscoe Miller	DET	38.1
Mike Donlin	BAL	30.4		Jimmy Collins	BOS	33.7		Nixey Callahan	CHI	29.5		Joe Yeager	DET	30.6
Normalized OPS				**Park Adjusted**				**Normalized ERA**				**Park Adjusted**		
Nap Lajoie	PHI	203		Nap Lajoie	PHI	202		Cy Young	BOS	225		Cy Young	BOS	203
Buck Freeman	BOS	154		Buck Freeman	BOS	164		Nixey Callahan	CHI	151		Joe Yeager	DET	153
Socks Seybold	PHI	148		Jimmy Collins	BOS	149		Joe Yeager	DET	140		Nixey Callahan	CHI	139
On Base Average				**Slugging Percentage**				**Percent of Team Wins**				**Wins Above Team**		
Nap Lajoie	PHI	.451		Nap Lajoie	PHI	.643		Cy Young	BOS	.418		Cy Young	BOS	11.7
Fielder Jones	CHI	.407		Buck Freeman	BOS	.520		Joe McGinnity	BAL	.382		Casey Patten	WAS	6.6
Mike Donlin	BAL	.406		Socks Seybold	PHI	.503		William Reidy	MIL	.333		Clark Griffith	CHI	6.6
Isolated Power				**Players Overall**				**Pitchers Overall**				**Defensive Runs**		
Nap Lajoie	PHI	.217		Nap Lajoie	PHI	85.0		Cy Young	BOS	70.9		Kid Elberfeld	DET	22.8
Buck Freeman	BOS	.182		Jimmy Collins	BOS	45.0		Roscoe Miller	DET	42.4		Billy Clingman	WAS	19.8
Jimmy Williams	BAL	.178		Kid Elberfeld	DET	41.9		Nixey Callahan	CHI	41.8		Nap Lajoie	PHI	18.5

Club	W	L	R	OR	Avg	OBA	SLG	BPF	NOPS-A	BR	Adj	Wins	ERA	PPF	NERA-A	PR	Adj	Wins	Diff
CHI	83	53	819	631	.276	.342	.370	96	106/110	37	69	6.2	2.98	92	123/113	93	53	4.8	4.0
BOS	79	57	759	608	.278	.324	.381	94	104/111	10	56	5.0	3.04	90	120/109	84	36	3.2	2.7
DET	74	61	741	694	.279	.333	.370	109	103/ 95	14	-50	-4.6	3.30	109	111/121	48	91	8.2	2.8
PHI	74	62	805	761	.289	.330	.395	101	110/109	42	37	3.4	4.00	100	92/ 92	-44	-44	-4.0	6.7
BAL	68	65	760	750	.294	.346	.397	104	115/111	79	51	4.6	3.73	104	98/102	-7	11	1.0	-4.2
WAS	61	73	678	767	.269	.319	.364	103	98/ 95	-23	-45	-4.1	4.09	105	89/ 94	-56	-33	-3.0	1.2
CLE	55	82	663	827	.271	.306	.348	100	89/ 89	-79	-82	-7.5	4.12	103	89/ 92	-59	-44	-4.0	-2.0
MIL	48	89	641	828	.261	.308	.345	94	89/ 95	-78	-32	-3.0	4.06	96	90/ 87	-53	-71	-6.5	-11.1
			733		.277	.326	.371						3.66						

NATIONAL LEAGUE 1901

Batting Runs				Park Adjusted				Pitching Runs				Park Adjusted		
Jesse Burkett	STL	57.4		Jesse Burkett	STL	59.3		Deacon Phillippe	PIT	36.2		Vic Willis	BOS	44.8
Ed Delahanty	PHI	53.7		Ed Delahanty	PHI	55.3		Christy Mathewson	NY	34.0		Deacon Phillippe	PIT	33.2
Jimmy Sheckard	BRO	49.9		Jimmy Sheckard	BRO	47.3		Al Orth	PHI	33.1		Jesse Tannehill	PIT	29.4
Normalized OPS				**Park Adjusted**				**Normalized ERA**				**Park Adjusted**		
Ed Delahanty	PHI	178		Ed Delahanty	PHI	181		Jesse Tannehill	PIT	152		Vic Willis	BOS	156
Jesse Burkett	STL	175		Jesse Burkett	STL	179		Deacon Phillippe	PIT	150		Jesse Tannehill	PIT	148
Jimmy Sheckard	BRO	174		Jimmy Sheckard	BRO	168		Al Orth	PHI	147		Deacon Phillippe	PIT	145
On Base Average				**Slugging Percentage**				**Percent of Team Wins**				**Wins Above Team**		
Jesse Burkett	STL	.432		Jimmy Sheckard	BRO	.534		Noodles Hahn	CIN	.423		Noodles Hahn	CIN	9.4
Roy Thomas	PHI	.428		Ed Delahanty	PHI	.528		Christy Mathewson	NY	.385		Christy Mathewson	NY	8.2
Ed Delahanty	PHI	.423		Sam Crawford	CIN	.524		Dummy Taylor	NY	.346		Jack Harper	STL	4.7
Isolated Power				**Players Overall**				**Pitchers Overall**				**Defensive Runs**		
Sam Crawford	CIN	.194		Bobby Wallace	STL	53.6		Vic Willis	BOS	43.0		Bobby Wallace	STL	24.9
Jimmy Sheckard	BRO	.181		Jesse Burkett	STL	46.5		Deacon Phillippe	PIT	40.3		George Davis	NY	15.8
Ed Delahanty	PHI	.173		George Davis	NY	44.3		Al Orth	PHI	36.6		Cupid Childs	CHI	11.0

Club	W	L	R	OR	Avg	OBA	SLG	BPF	NOPS-A	BR	Adj	Wins	ERA	PPF	NERA-A	PR	Adj	Wins	Diff
PIT	90	49	776	534	.286	.338	.378	102	117/115	98	85	8.4	2.58	97	129/125	102	90	8.8	3.4
PHI	83	57	668	543	.266	.326	.346	98	104/106	26	39	3.8	2.86	95	116/110	63	40	4.0	5.2
BRO	79	57	744	600	.287	.330	.384	103	118/114	89	66	6.5	3.14	101	106/107	24	28	2.7	1.8
STL	76	64	792	689	.284	.326	.381	98	115/118	75	90	8.9	3.68	96	90/ 86	-50	-70	-7.0	4.1
BOS	69	69	531	556	.249	.294	.310	110	83/ 76	-97	-159	-15.6	2.90	111	114/127	59	109	10.7	4.9
CHI	53	86	578	699	.258	.303	.326	102	91/ 89	-55	-66	-6.5	3.33	104	100/104	-1	17	1.7	-11.6
NY	52	85	544	755	.253	.297	.318	92	87/ 94	-79	-30	-3.0	3.87	90	86/ 82	-74	-93	-9.2	-4.3
CIN	52	87	561	818	.251	.297	.338	96	93/ 96	-52	-29	-2.9	4.18	101	79/ 81	-120	-113	-11.2	-3.4
			649		.267	.314	.348						3.32						

AMERICAN LEAGUE 1902

Batting Runs			Park Adjusted			Pitching Runs			Park Adjusted		
Ed Delahanty	WAS	56.9	Ed Delahanty	WAS	59.4	Cy Young	BOS	60.7	Cy Young	BOS	67.1
Bill Bradley	CLE	30.9	Bill Bradley	CLE	34.4	Rube Waddell	PHI	46.5	Rube Waddell	PHI	54.3
Socks Seybold	PHI	28.0	Jimmy Williams	BAL	22.3	Ed Siever	DET	34.6	Ed Siever	DET	35.4
Normalized OPS			**Park Adjusted**			**Normalized ERA**			**Park Adjusted**		
Ed Delahanty	WAS	188	Ed Delahanty	WAS	195	Ed Siever	DET	186	Ed Siever	DET	188
Bill Bradley	CLE	145	Bill Bradley	CLE	152	Rube Waddell	PHI	174	Rube Waddell	PHI	186
Socks Seybold	PHI	142	Jimmy Williams	BAL	136	Cy Young	BOS	166	Cy Young	BOS	173
On Base Average			**Slugging Percentage**			**Percent of Team Wins**			**Wins Above Team**		
Ed Delahanty	WAS	.449	Ed Delahanty	WAS	.590	Cy Young	BOS	.416	Cy Young	BOS	11.4
Patsy Dougherty	BOS	.400	Bill Bradley	CLE	.515	Al Orth	WAS	.311	Rube Waddell	PHI	6.6
Jimmy Barrett	DET	.391	Socks Seybold	PHI	.506	Rube Waddell	PHI	.289	Joe McGinnity	BAL	5.5
Isolated Power			**Players Overall**			**Pitchers Overall**			**Defensive Runs**		
Ed Delahanty	WAS	.214	Ed Delahanty	WAS	55.7	Cy Young	BOS	55.7	Hobe Ferris	BOS	27.5
Buck Freeman	BOS	.193	Bill Bradley	CLE	42.9	Rube Waddell	PHI	57.9	Monte Cross	PHI	14.7
Socks Seybold	PHI	.190	George Davis	CHI	25.5	Red Donahue	STL	31.3	Bill Bradley	CLE	12.9

Club	W	L	R	OR	Avg	OBA	SLG	BPF	NOPS-A	BR	Adj	Wins	ERA	PPF	NERA-A	PR	Adj	Wins	Diff
PHI	83	53	775	636	.287	.335	.389	109	111/102	51	-7	-.8	3.29	107	109/116	38	73	6.9	8.8
STL	78	58	619	607	.265	.318	.353	104	95/ 91	-36	-64	-6.2	3.34	104	107/112	32	53	5.0	11.1
BOS	77	60	664	600	.278	.317	.383	105	104/ 99	1	-32	-3.2	3.02	104	118/123	75	96	9.1	2.6
CHI	74	60	675	602	.268	.328	.335	92	92/100	-40	14	1.3	3.41	89	105/ 93	21	-30	-2.9	8.6
CLE	69	67	686	667	.289	.332	.389	95	110/115	43	75	7.1	3.28	94	109/103	39	11	1.1	-7.2
WAS	61	75	707	790	.283	.329	.396	96	111/115	47	71	6.7	4.36	98	82/ 80	-105	-115	-11.0	-2.7
DET	52	83	566	657	.251	.305	.320	100	81/ 81	-111	-107	-10.3	3.56	101	100/101	1	6	.6	-5.8
BAL	50	88	715	848	.277	.335	.385	102	109/107	47	34	3.3	4.31	105	83/ 87	-99	-76	-7.3	-14.9
			676		.275	.325	.369						3.57						

NATIONAL LEAGUE 1902

Batting Runs			Park Adjusted			Pitching Runs			Park Adjusted		
Sam Crawford	CIN	39.9	Honus Wagner	PIT	36.6	Jack Taylor	CHI	52.3	Jack Taylor	CHI	55.9
Honus Wagner	PIT	37.1	Ginger Beaumont	PIT	33.8	Noodles Hahn	CIN	36.1	Noodles Hahn	CIN	44.5
Ginger Beaumont	PIT	34.3	Sam Crawford	CIN	33.7	Vic Willis	BOS	26.5	Christy Mathewson	NY	24.0
Normalized OPS			**Park Adjusted**			**Normalized ERA**			**Park Adjusted**		
Sam Crawford	CIN	163	Honus Wagner	PIT	161	Jack Taylor	CHI	209	Jack Taylor	CHI	216
Honus Wagner	PIT	162	Fred Clarke	PIT	158	Noodles Hahn	CIN	157	Noodles Hahn	CIN	171
Fred Clarke	PIT	159	Ginger Beaumont	PIT	154	Jesse Tannehill	PIT	143	Carl Lundgren	CHI	146
On Base Average			**Slugging Percentage**			**Percent of Team Wins**			**Wins Above Team**		
Roy Thomas	PHI	.412	Honus Wagner	PIT	.463	Vic Willis	BOS	.370	Jack Taylor	CHI	8.1
Fred Tenney	BOS	.404	Sam Crawford	CIN	.461	Togie Pittinger	BOS	.370	Noodles Hahn	CIN	7.3
Ginger Beaumont	PIT	.400	Fred Clarke	PIT	.449	Jack Taylor	CHI	.338	Togie Pittinger	BOS	6.7
Isolated Power			**Players Overall**			**Pitchers Overall**			**Defensive Runs**		
Tommy Leach	PIT	.148	Honus Wagner	PIT	41.9	Jack Taylor	CHI	63.5	John Farrell	STL	22.2
Honus Wagner	PIT	.133	Fred Tenney	BOS	39.6	Noodles Hahn	CIN	42.2	Bobby Lowe	CHI	20.4
Fred Clarke	PIT	.133	Tommy Leach	PIT	37.5	Jesse Tannehill	PIT	28.0	Germany Long	BOS	19.8

Club	W	L	R	OR	Avg	OBA	SLG	BPF	NOPS-A	BR	Adj	Wins	ERA	PPF	NERA-A	PR	Adj	Wins	Diff
PIT	103	36	775	440	.286	.336	.374	101	128/127	147	142	15.0	2.30	93	121/113	67	41	4.3	14.2
BRO	75	63	564	519	.256	.302	.319	90	99/110	-7	47	4.9	2.69	88	103/ 91	12	-34	-3.7	4.8
BOS	73	64	572	516	.249	.307	.305	93	96/103	-14	22	2.4	2.61	91	106/ 97	24	-8	-1.0	3.1
CIN	70	70	633	566	.282	.323	.362	109	120/109	97	44	4.7	2.67	109	104/113	14	47	4.9	-9.6
CHI	68	69	530	501	.250	.301	.298	104	92/ 88	-39	-62	-6.7	2.21	104	126/130	81	95	10.0	-3.8
STL	56	78	517	695	.258	.298	.304	95	93/ 97	-38	-13	-1.4	3.48	100	80/ 79	-94	-96	-10.3	.7
PHI	56	81	484	649	.247	.301	.293	107	90/ 84	-45	-81	-8.6	3.50	111	79/ 88	-96	-55	-5.9	2.1
NY	48	88	401	590	.238	.277	.290	100	82/ 82	-96	-95	-10.1	2.81	104	99/103	-4	11	1.1	-11.0
				560	.259	.306	.319						2.78						

AMERICAN LEAGUE 1903

Batting Runs			Park Adjusted			Pitching Runs			Park Adjusted		
Nap Lajoie	CLE	37.8	Nap Lajoie	CLE	42.2	Earl Moore	CLE	33.3	Cy Young	BOS	47.2
Sam Crawford	DET	35.3	Sam Crawford	DET	36.8	Cy Young	BOS	33.1	Bill Dinneen	BOS	36.4
Bill Bradley	CLE	31.3	Bill Bradley	CLE	36.1	Doc White	CHI	27.3	Earl Moore	CLE	25.7
Normalized OPS			**Park Adjusted**			**Normalized ERA**			**Park Adjusted**		
Nap Lajoie	CLE	168	Nap Lajoie	CLE	181	Earl Moore	CLE	169	Cy Young	BOS	160
Topsy Hartsel	PHI	162	Bill Bradley	CLE	164	Cy Young	BOS	142	Earl Moore	CLE	153
Sam Crawford	DET	156	Sam Crawford	DET	160	Bill Bernhard	CLE	140	Bill Dinneen	BOS	149
On Base Average			**Slugging Percentage**			**Percent of Team Wins**			**Wins Above Team**		
Jimmy Barrett	DET	.401	Nap Lajoie	CLE	.518	Willie Sudhoff	STL	.323	Willie Sudhoff	STL	5.6
Topsy Hartsel	PHI	.391	Bill Bradley	CLE	.496	Cy Young	BOS	.308	Cy Young	BOS	4.9
Billy Lush	DET	.377	Buck Freeman	BOS	.496	Eddie Plank	PHI	.307	Earl Moore	CLE	4.5
Isolated Power			**Players Overall**			**Pitchers Overall**			**Defensive Runs**		
Buck Freeman	BOS	.208	Nap Lajoie	CLE	80.0	Cy Young	BOS	52.6	Nap Lajoie	CLE	35.5
Bill Bradley	CLE	.183	Bill Bradley	CLE	46.1	George Mullin	DET	34.0	Jimmy Barrett	DET	20.7
Nap Lajoie	CLE	.173	Jimmy Barrett	DET	42.4	Bill Dinneen	BOS	30.9	Bobby Wallace	STL	17.0

Club	W	L	R	OR	Avg	OBA	SLG	BPF	NOPS-A	BR	Adj	Wins	ERA	PPF	NERA-A	PR	Adj	Wins	Diff
BOS	91	47	708	504	.272	.308	.392	116	121/105	93	-0	-.1	2.56	113	115/130	54	106	11.0	11.1
PHI	75	60	597	519	.264	.304	.362	103	110/107	38	19	2.0	2.98	102	99/101	-2	4	.4	5.1
CLE	77	63	639	579	.265	.303	.373	93	113/122	51	93	9.7	2.65	91	111/101	41	3	.3	-3.0
NY	72	62	579	573	.249	.300	.330	103	98/ 95	-13	-31	-3.4	3.08	103	96/ 99	-16	-2	-.3	8.7
DET	65	71	567	539	.268	.312	.351	98	109/111	37	50	5.2	2.75	97	107/104	26	13	1.4	-9.6
STL	65	74	500	525	.244	.286	.317	97	90/ 92	-61	-45	-4.8	2.77	97	107/104	25	14	1.5	-1.2
CHI	60	77	516	613	.247	.296	.314	88	91/104	-44	21	2.2	3.02	89	98/ 87	-8	-52	-5.5	-5.2
WAS	43	94	437	691	.231	.272	.311	104	83/ 80	-97	-119	-12.5	3.82	110	77/ 85	-117	-75	-7.9	-5.1
			568		.255	.298	.344						2.95						

NATIONAL LEAGUE 1903

Batting Runs			Park Adjusted			Pitching Runs			Park Adjusted		
Mike Donlin	CIN	43.2	Jimmy Sheckard	BRO	39.0	Christy Mathewson	NY	41.0	Joe McGinnity	NY	49.4
Honus Wagner	PIT	41.5	Frank Chance	CHI	36.4	Joe McGinnity	NY	40.7	Christy Mathewson	NY	48.3
Jimmy Sheckard	BRO	39.6	Roy Thomas	PHI	35.9	Sam Leever	PIT	38.2	Sam Leever	PIT	44.0
Normalized OPS			**Park Adjusted**			**Normalized ERA**			**Park Adjusted**		
Fred Clarke	PIT	169	Frank Chance	CHI	160	Sam Leever	PIT	159	Sam Leever	PIT	168
Mike Donlin	CIN	168	Fred Clarke	PIT	156	Christy Mathewson	NY	145	Christy Mathewson	NY	153
Roger Bresnahan	NY	167	Jimmy Sheckard	BRO	155	Jake Weimer	CHI	142	Noodles Hahn	CIN	143
On Base Average			**Slugging Percentage**			**Percent of Team Wins**			**Wins Above Team**		
Roy Thomas	PHI	.450	Fred Clarke	PIT	.532	Joe McGinnity	NY	.369	Christy Mathewson	NY	5.8
Roger Bresnahan	NY	.435	Honus Wagner	PIT	.518	Christy Mathewson	NY	.357	Sam Leever	PIT	5.4
Frank Chance	CHI	.428	Mike Donlin	CIN	.516	Henry Schmidt	BRO	.314	Henry Schmidt	BRO	5.4
Isolated Power			**Players Overall**			**Pitchers Overall**			**Defensive Runs**		
Fred Clarke	PIT	.180	Honus Wagner	PIT	56.3	Christy Mathewson	NY	50.4	Ed Gremminger	BOS	23.7
Harry Steinfeldt	CIN	.169	Jimmy Sheckard	BRO	49.5	Joe McGinnity	NY	42.4	Jimmy Sheckard	BRO	18.6
Mike Donlin	CIN	.165	Roy Thomas	PHI	37.3	Sam Leever	PIT	39.1	Bill Dahlen	BRO	17.2

Club	W	L	R	OR	Avg	OBA	SLG	BPF	NOPS-A	BR	Adj	Wins	ERA	PPF	NERA-A	PR	Adj	Wins	Diff
PIT	91	49	793	613	.287	.335	.393	109	117/108	90	31	2.9	2.91	106	112/119	50	75	7.2	10.9
NY	84	55	729	567	.272	.326	.344	108	100/ 92	-3	-57	-5.5	2.95	105	111/117	45	70	6.7	13.3
CHI	82	56	695	599	.275	.334	.347	94	103/110	19	60	5.8	2.77	92	118/108	69	31	3.0	4.3
CIN	74	65	765	656	.288	.343	.390	112	119/106	102	22	2.1	3.07	110	106/117	27	73	7.0	-4.6
BRO	70	66	667	682	.265	.341	.339	101	102/101	26	20	1.9	3.44	101	95/ 96	-22	-17	-1.7	1.8
BOS	58	80	578	699	.245	.304	.318	94	85/ 91	-84	-44	-4.3	3.34	96	98/ 94	-8	-28	-2.7	-4.0
PHI	49	86	617	738	.268	.317	.341	93	96/103	-28	15	1.4	3.97	95	82/ 79	-92	-113	-10.9	-9.0
STL	43	94	505	795	.251	.293	.313	97	81/ 83	-118	-98	-9.4	3.76	102	87/ 89	-63	-54	-5.2	-10.8
			669		.269	.324	.349						3.27						

AMERICAN LEAGUE 1904

Batting Runs			Park Adjusted			Pitching Runs			Park Adjusted		
Nap Lajoie	CLE	61.3	Nap Lajoie	CLE	59.6	Rube Waddell	PHI	41.5	Jack Chesbro	NY	57.0
Elmer Flick	CLE	37.5	Elmer Flick	CLE	35.6	Jack Chesbro	NY	39.3	Rube Waddell	PHI	35.5
Chick Stahl	BOS	32.2	Harry Davis	PHI	29.6	Cy Young	BOS	26.6	Cy Young	BOS	28.6
Normalized OPS			**Park Adjusted**			**Normalized ERA**			**Park Adjusted**		
Nap Lajoie	CLE	204	Nap Lajoie	CLE	198	Addie Joss	CLE	163	Jack Chesbro	NY	162
Elmer Flick	CLE	158	Elmer Flick	CLE	154	Rube Waddell	PHI	160	Addie Joss	CLE	162
Willie Keeler	NY	153	Danny Murphy	PHI	145	Doc White	CHI	146	Rube Waddell	PHI	151
On Base Average			**Slugging Percentage**			**Percent of Team Wins**			**Wins Above Team**		
Nap Lajoie	CLE	.405	Nap Lajoie	CLE	.552	Jack Chesbro	NY	.446	Jack Chesbro	NY	13.4
Willie Keeler	NY	.382	Elmer Flick	CLE	.449	Casey Patten	WAS	.368	Casey Patten	WAS	6.2
Elmer Flick	CLE	.362	Danny Murphy	PHI	.440	Eddie Plank	PHI	.321	Fred Glade	STL	5.0
Isolated Power			**Players Overall**			**Pitchers Overall**			**Defensive Runs**		
lap Lajoie	CLE	.175	Nap Lajoie	CLE	61.7	Jack Chesbro	NY	65.4	Lee Tannehill	CHI	26.6
anny Murphy	PHI	.153	Danny Murphy	PHI	38.8	Frank Owen	CHI	30.0	Jimmy Barrett	DET	23.7
:lmer Flick	CLE	.143	Bill Bradley	CLE	33.5	George Mullin	DET	29.1	Bobby Wallace	STL	17.2

Club	W	L	R	OR	Avg	OBA	SLG	BPF	NOPS-A	BR	Adj	Wins	ERA	PPF	NERA-A	PR	Adj	Wins	Diff
BOS	95	59	608	466	.247	.294	.340	105	109/104	41	15	1.6	2.12	102	123/125	75	82	9.2	7.2
NY	92	59	598	526	.259	.301	.347	114	114/100	66	-12	-1.4	2.57	113	101/115	4	58	6.5	11.5
CHI	89	65	600	482	.242	.294	.294	99	101/102	6	12	1.3	2.30	96	113/108	45	28	3.2	7.5
CLE	86	65	647	482	.260	.302	.354	103	117/113	78	62	7.0	2.22	99	117/116	56	53	6.0	-2.4
PHI	81	70	557	503	.249	.292	.336	96	107/112	30	51	5.7	2.35	95	111/105	38	17	1.9	-2.0
STL	65	87	481	604	.239	.284	.294	97	90/ 93	-54	-36	-4.1	2.83	100	92/ 91	-35	-37	-4.3	-2.6
DET	62	90	505	627	.231	.278	.292	98	88/ 90	-69	-56	-6.3	2.77	100	94/ 94	-26	-25	-2.9	-4.8
WAS	38	113	437	743	.227	.267	.288	92	83/ 90	-96	-54	-6.2	3.62	99	72/ 71	-154	-158	-17.8	-13.6
			554		.244	.289	.321						2.60						

NATIONAL LEAGUE 1904

Batting Runs			Park Adjusted			Pitching Runs			Park Adjusted		
Honus Wagner	PIT	52.7	Honus Wagner	PIT	52.4	Joe McGinnity	NY	50.7	Joe McGinnity	NY	55.9
Roy Thomas	PHI	28.3	Roy Thomas	PHI	32.5	Christy Mathewson	NY	28.6	Christy Mathewson	NY	33.3
Cy Seymour	CIN	25.5	Jake Beckley	STL	24.6	Jake Weimer	CHI	28.1	Noodles Hahn	CIN	30.2
Normalized OPS			**Park Adjusted**			**Normalized ERA**			**Park Adjusted**		
Honus Wagner	PIT	192	Honus Wagner	PIT	191	Joe McGinnity	NY	169	Joe McGinnity	NY	177
Frank Chance	CHI	146	Frank Chance	CHI	147	Ned Garvin	BRO	162	Ned Garvin	BRO	163
Roger Bresnahan	NY	145	Roy Thomas	PHI	146	Three Finger Brown	CHI	146	Noodles Hahn	CIN	144
On Base Average			**Slugging Percentage**			**Percent of Team Wins**			**Wins Above Team**		
Honus Wagner	PIT	.419	Honus Wagner	PIT	.520	Joe McGinnity	NY	.330	Joe McGinnity	NY	7.0
Roy Thomas	PHI	.411	Cy Seymour	CIN	.439	Vic Willis	BOS	.327	Kid Nichols	STL	5.9
Miller Huggins	CIN	.375	Frank Chance	CHI	.430	Christy Mathewson	NY	.311	Jack Harper	CIN	5.8
Isolated Power			**Players Overall**			**Pitchers Overall**			**Defensive Runs**		
Honus Wagner	PIT	.171	Honus Wagner	PIT	46.8	Joe McGinnity	NY	57.0	Tommy Leach	PIT	33.8
Harry Lumley	BRO	.149	Tommy Leach	PIT	40.6	Christy Mathewson	NY	42.9	Johnny Evers	CHI	33.4
Dave Brain	STL	.141	Roy Thomas	PHI	39.2	Patsy Flaherty	PIT	29.1	Bill Dahlen	NY	23.6

Club	W	L	R	OR	Avg	OBA	SLG	BPF	NOPS-A	BR	Adj	Wins	ERA	PPF	NERA-A	PR	Adj	Wins	Diff
NY	106	47	744	476	.262	.319	.344	109	114/104	78	20	2.2	2.17	104	126/131	87	104	11.0	16.3
CHI	93	60	599	517	.248	.289	.315	100	95/ 95	-37	-35	-3.8	2.30	98	118/116	65	56	5.9	14.4
CIN	88	65	695	547	.255	.307	.338	111	108/ 97	43	-24	-2.7	2.35	109	116/126	58	94	10.0	4.2
PIT	87	66	675	592	.258	.311	.338	101	109/109	49	46	4.9	2.89	99	94/ 93	-23	-28	-3.1	8.7
STL	75	79	602	595	.253	.300	.327	100	102/103	7	9	1.0	2.64	100	104/103	14	12	1.3	-4.2
BRO	56	97	497	614	.232	.292	.295	98	89/ 91	-54	-42	-4.5	2.70	100	101/101	4	5	.5	-16.5
BOS	55	98	491	749	.237	.281	.300	95	87/ 92	-75	-43	-4.7	3.43	100	79/ 79	-105	-105	-11.2	-5.6
PHI	52	100	571	784	.248	.300	.316	94	99/105	-7	31	3.2	3.39	98	81/ 79	-97	-107	-11.4	-15.8
			609		.249	.300	.322						2.73						

AMERICAN LEAGUE 1905

Batting Runs			Park Adjusted			Pitching Runs			Park Adjusted		
Elmer Flick	CLE	37.5	Elmer Flick	CLE	37.5	Rube Waddell	PHI	42.8	Rube Waddell	PHI	47.3
Topsy Hartsel	PHI	34.6	Sam Crawford	DET	34.4	Cy Young	BOS	29.4	Cy Young	BOS	30.8
Sam Crawford	DET	31.7	George Stone	STL	34.0	Nick Altrock	CHI	26.9	Andy Coakley	PHI	26.5
Normalized OPS			**Park Adjusted**			**Normalized ERA**			**Park Adjusted**		
Elmer Flick	CLE	167	Elmer Flick	CLE	167	Rube Waddell	PHI	179	Rube Waddell	PHI	188
Sam Crawford	DET	150	Sam Crawford	DET	157	Doc White	CHI	150	Andy Coakley	PHI	151
Topsy Hartsel	PHI	144	George Stone	STL	155	Cy Young	BOS	145	Cy Young	PHI	147
On Base Average			**Slugging Percentage**			**Percent of Team Wins**			**Wins Above Team**		
Topsy Hartsel	PHI	.410	Elmer Flick	CLE	.462	Rube Waddell	PHI	.293	Jesse Tannehill	BOS	7.7
Elmer Flick	CLE	.374	Sam Crawford	DET	.430	Ed Killian	DET	.291	Rube Waddell	PHI	5.3
Sam Crawford	DET	.354	Harry Davis	PHI	.422	Jesse Tannehill	BOS	.282	Addie Joss	CLE	5.3
Isolated Power			**Players Overall**			**Pitchers Overall**			**Defensive Runs**		
Elmer Flick	CLE	.154	Bobby Wallace	STL	47.5	Rube Waddell	PHI	43.6	Bobby Wallace	STL	26.5
Hobe Ferris	BOS	.141	George Davis	CHI	33.8	Cy Young	BOS	26.4	Bobby Wallace	STL	26.3
Harry Davis	PHI	.138	Sam Crawford	DET	33.0	Addie Joss	CLE	25.0	Joseph Cassidy	WAS	19.4

Club	W	L	R	OR	Avg	OBA	SLG	BPF	NOPS-A	BR	Adj	Wins	ERA	PPF	NERA-A	PR	Adj	Wins	Diff
PHI	92	56	623	492	.254	.305	.338	107	113/105	65	24	2.6	2.19	105	121/126	70	89	9.8	5.6
CHI	92	60	612	451	.237	.297	.304	97	99/102	-1	18	1.9	1.99	93	133/124	105	74	8.2	5.9
DET	79	74	512	602	.243	.296	.311	96	101/105	4	27	2.9	2.83	98	94/ 91	-26	-36	-4.1	3.7
BOS	78	74	579	564	.234	.301	.311	102	102/100	18	8	.8	2.84	102	93/ 95	-28	-22	-2.5	3.7
CLE	76	78	567	587	.255	.294	.334	100	108/108	31	31	3.4	2.84	101	93/ 94	-29	-26	-3.0	-1.4
NY	71	78	586	622	.248	.299	.319	100	104/105	20	21	2.4	2.93	101	90/ 91	-41	-38	-4.3	-1.5
WAS	64	87	559	623	.224	.267	.302	109	88/ 81	-70	-123	-13.6	2.87	111	92/103	-32	12	1.4	.7
STL	54	99	511	608	.232	.282	.289	90	88/ 98	-63	-7	-.9	2.74	92	97/ 89	-13	-46	-5.2	-16.4
			569		.241	.293	.314						2.65						

NATIONAL LEAGUE 1905

Batting Runs			Park Adjusted			Pitching Runs			Park Adjusted		
Cy Seymour	CIN	65.8	Cy Seymour	CIN	69.3	Christy Mathewson	NY	64.7	Christy Mathewson	NY	61.4
Mike Donlin	NY	52.5	Mike Donlin	NY	50.6	Ed Reulbach	CHI	51.1	Ed Reulbach	CHI	47.3
Honus Wagner	PIT	51.7	Honus Wagner	PIT	47.9	Deacon Phillippe	PIT	24.7	Tully Sparks	PHI	30.8
Normalized OPS			**Park Adjusted**			**Normalized ERA**			**Park Adjusted**		
Cy Seymour	CIN	196	Cy Seymour	CIN	206	Christy Mathewson	NY	235	Christy Mathewson	NY	228
Honus Wagner	PIT	178	Honus Wagner	PIT	169	Ed Reulbach	CHI	211	Ed Reulbach	CHI	203
Mike Donlin	NY	172	Mike Donlin	NY	168	Bob Wicker	CHI	148	Tully Sparks	PHI	149
On Base Average			**Slugging Percentage**			**Percent of Team Wins**			**Wins Above Team**		
Frank Chance	CHI	.430	Cy Seymour	CIN	.559	Irv Young	BOS	.392	Irv Young	BOS	8.8
Cy Seymour	CIN	.427	Honus Wagner	PIT	.505	Christy Mathewson	NY	.295	Doc Scanlan	BRO	7.0
Honus Wagner	PIT	.420	Mike Donlin	NY	.495	Doc Scanlan	BRO	.292	Sam Leever	PIT	5.2
Isolated Power			**Players Overall**			**Pitchers Overall**			**Defensive Runs**		
Cy Seymour	CIN	.182	Honus Wagner	PIT	70.0	Christy Mathewson	NY	74.5	Miller Huggins	CIN	33.3
Honus Wagner	PIT	.142	Cy Seymour	CIN	68.2	Ed Reulbach	CHI	40.0	Honus Wagner	PIT	19.5
Mike Donlin	NY	.139	Miller Huggins	CIN	62.3	Tully Sparks	PHI	21.6	Tommy Corcoran	CIN	18.3

Club	W	L	R	OR	Avg	OBA	SLG	BPF	NOPS-A	BR	Adj	Wins	ERA	PPF	NERA-A	PR	Adj	Wins	Diff
NY	105	48	778	505	.273	.340	.368	102	122/119	131	115	11.9	2.39	97	125/121	91	78	8.1	8.4
PIT	96	57	692	570	.266	.316	.350	106	108/103	43	7	.8	2.86	103	104/108	20	36	3.7	15.0
CHI	92	61	667	442	.245	.305	.314	101	94/ 93	-31	-37	-3.9	2.04	96	147/141	149	130	13.6	5.8
PHI	83	69	708	602	.260	.313	.336	110	103/ 94	16	-48	-5.1	2.81	109	106/115	28	68	7.0	5.1
CIN	79	74	735	698	.269	.325	.354	95	113/119	74	105	11.0	3.01	94	99/ 93	-2	-29	-3.1	-5.4
STL	58	96	535	734	.248	.301	.321	95	95/ 99	-32	-2	-.3	3.59	99	83/ 83	-88	-92	-9.7	-9.0
BOS	51	103	468	731	.234	.277	.293	98	78/ 79	-134	-124	-13.0	3.52	103	85/ 88	-80	-64	-6.8	-6.2
BRO	48	104	506	807	.246	.292	.317	95	90/ 96	-62	-28	-3.0	3.76	100	80/ 80	-113	-112	-11.7	-13.2
			636		.255	.309	.332						2.99						

AMERICAN LEAGUE   1906

Batting Runs			Park Adjusted			Pitching Runs			Park Adjusted		
George Stone	STL	57.1	George Stone	STL	56.9	Barney Pelty	STL	32.0	Otto Hess	CLE	31.1
Nap Lajoie	CLE	44.5	Nap Lajoie	CLE	42.5	Otto Hess	CLE	31.8	Barney Pelty	STL	31.0
Elmer Flick	CLE	37.6	Harry Davis	PHI	39.5	Dusty Rhoads	CLE	31.1	Dusty Rhoads	CLE	30.4
Normalized OPS			**Park Adjusted**			**Normalized ERA**			**Park Adjusted**		
George Stone	STL	187	George Stone	STL	187	Doc White	CHI	177	Barney Pelty	STL	167
Nap Lajoie	CLE	168	Harry Davis	PHI	173	Barney Pelty	STL	170	Addie Joss	CLE	155
Harry Davis	PHI	155	Nap Lajoie	CLE	163	Addie Joss	CLE	156	Doc White	CHI	151
On Base Average			**Slugging Percentage**			**Percent of Team Wins**			**Wins Above Team**		
George Stone	STL	.411	George Stone	STL	.501	Casey Patten	WAS	.345	Casey Patten	WAS	8.0
Nap Lajoie	CLE	.386	Nap Lajoie	CLE	.465	Al Orth	NY	.300	Eddie Plank	PHI	6.7
Elmer Flick	CLE	.366	Harry Davis	PHI	.459	George Mullin	DET	.292	Jesse Tannehill	BOS	6.4
Isolated Power			**Players Overall**			**Pitchers Overall**			**Defensive Runs**		
Harry Davis	PHI	.167	Nap Lajoie	CLE	61.5	Barney Pelty	STL	33.5	Lee Tannehill	CHI	28.7
George Stone	STL	.143	George Stone	STL	52.1	Al Orth	NY	33.2	Terry Turner	CLE	22.8
Charlie Hickman	WAS	.137	Terry Turner	CLE	36.5	Addie Joss	CLE	33.1	Nap Lajoie	CLE	21.3

Club	W	L	R	OR	Avg	OBA	SLG	BPF	NOPS-A	BR	Adj	Wins	ERA	PPF	NERA-A	PR	Adj	Wins	Diff
CHI	93	58	570	460	.230	.295	.286	89	89/ 99	-48	11	1.2	2.13	86	126/108	85	25	2.8	13.6
NY	90	61	644	543	.266	.311	.339	115	111/ 97	58	-26	-3.0	2.78	114	97/111	-12	45	5.0	12.5
CLE	89	64	663	482	.279	.320	.356	103	121/117	112	93	10.2	2.09	99	129/128	94	91	10.0	-7.8
PHI	78	67	561	542	.247	.302	.330	90	106/118	29	85	9.4	2.60	88	103/ 91	13	-34	-3.8	-.1
STL	76	73	558	498	.247	.298	.312	100	99/ 98	-6	-8	-.9	2.23	99	121/119	70	65	7.2	-4.7
DET	71	78	518	599	.242	.290	.306	107	94/ 88	-32	-68	-7.6	3.06	109	88/ 96	-54	-17	-2.0	6.2
WAS	55	95	518	664	.238	.282	.309	91	93/101	-44	1	.2	3.25	94	83/ 78	-82	-105	-11.6	-8.5
BOS	49	105	462	706	.237	.278	.304	105	90/ 85	-63	-93	-10.3	3.41	111	79/ 88	-110	-64	-7.2	-10.5
				562		.249	.297	.318						2.69					

NATIONAL LEAGUE   1906

Batting Runs			Park Adjusted			Pitching Runs			Park Adjusted		
Honus Wagner	PIT	43.0	Harry Lumley	BRO	43.0	Three Finger Brown	CHI	49.0	Three Finger Brown	CHI	54.8
Harry Lumley	BRO	38.3	Honus Wagner	PIT	42.8	Vic Willis	PIT	32.1	Jack Pfiester	CHI	33.8
Frank Chance	CHI	36.7	Art Devlin	NY	30.4	Jack Pfiester	CHI	28.7	Vic Willis	PIT	29.1
Normalized OPS			**Park Adjusted**			**Normalized ERA**			**Park Adjusted**		
Honus Wagner	PIT	172	Harry Lumley	BRO	188	Three Finger Brown	CHI	253	Three Finger Brown	CHI	271
Harry Lumley	BRO	171	Honus Wagner	PIT	172	Jack Pfiester	CHI	168	Jack Pfiester	CHI	181
Frank Chance	CHI	163	Tim Jordan	BRO	156	Ed Reulbach	CHI	159	Ed Reulbach	CHI	171
On Base Average			**Slugging Percentage**			**Percent of Team Wins**			**Wins Above Team**		
Frank Chance	CHI	.406	Harry Lumley	BRO	.477	Irv Young	BOS	.327	Jake Weimer	CIN	7.2
Honus Wagner	PIT	.406	Honus Wagner	PIT	.459	Jake Weimer	CIN	.313	Doc Scanlan	BRO	5.7
Roger Bresnahan	NY	.401	Harry Steinfeldt	CHI	.430	Joe McGinnity	NY	.281	Sam Leever	PIT	5.4
Isolated Power			**Players Overall**			**Pitchers Overall**			**Defensive Runs**		
Tim Jordan	BRO	.160	Honus Wagner	PIT	66.5	Three Finger Brown	CHI	56.2	Art Devlin	NY	28.1
Harry Lumley	BRO	.153	Art Devlin	NY	53.1	Vic Willis	PIT	33.4	Billy Gilbert	NY	20.0
Sherry Magee	PHI	.124	Harry Lumley	BRO	38.2	Jake Weimer	CIN	30.5	Dave Brain	BOS	19.8

Club	W	L	R	OR	Avg	OBA	SLG	BPF	NOPS-A	BR	Adj	Wins	ERA	PPF	NERA-A	PR	Adj	Wins	Diff
CHI	116	36	705	381	.262	.323	.339	114	116/101	85	6	.7	1.76	107	149/160	133	162	18.0	21.3
NY	96	56	625	510	.255	.334	.321	97	113/117	84	100	11.1	2.49	94	106/ 99	21	-1	-.3	9.2
PIT	93	60	623	470	.261	.318	.327	100	110/110	57	54	6.0	2.21	97	119/115	63	50	5.6	4.9
PHI	71	82	528	564	.241	.302	.307	95	99/104	-5	19	2.1	2.58	96	102/ 98	8	-7	-.9	-6.7
BRO	66	86	496	625	.236	.292	.308	91	96/105	-27	19	2.1	3.13	94	84/ 79	-73	-97	-10.9	-1.2
CIN	64	87	533	582	.238	.294	.304	110	95/ 86	-30	-88	-9.9	2.69	113	98/110	-7	42	4.6	-6.2
STL	52	98	470	607	.235	.286	.296	100	89/ 89	-62	-62	-7.0	3.05	103	86/ 89	-61	-49	-5.6	-10.5
BOS	49	102	408	649	.226	.279	.281	97	82/ 85	-96	-81	-9.1	3.18	102	83/ 85	-79	-69	-7.8	-9.6
				549		.244	.304	.310						2.63					

AMERICAN LEAGUE   1907

Batting Runs			Park Adjusted			Pitching Runs			Park Adjusted		
Ty Cobb	DET	44.2	Ty Cobb	DET	43.4	Ed Walsh	CHI	44.1	Ed Walsh	CHI	41.6
Sam Crawford	DET	39.5	Elmer Flick	CLE	39.4	Addie Joss	CLE	26.7	Ed Killian	DET	24.5
George Stone	STL	35.1	Sam Crawford	DET	38.7	Ed Killian	DET	26.6	Cy Young	BOS	23.0

Normalized OPS			Park Adjusted			Normalized ERA			Park Adjusted		
Ty Cobb	DET	171	Ty Cobb	DET	169	Ed Walsh	CHI	159	Ed Walsh	CHI	155
Sam Crawford	DET	165	Elmer Flick	CLE	168	Ed Killian	DET	143	Ed Killian	DET	140
Elmer Flick	CLE	153	Sam Crawford	DET	163	Addie Joss	CLE	139	Harry Howell	STL	132

On Base Average			Slugging Percentage			Percent of Team Wins			Wins Above Team		
Topsy Hartsel	PHI	.405	Ty Cobb	DET	.468	Cy Young	BOS	.356	Wild Bill Donovan	DET	8.9
George Stone	STL	.382	Sam Crawford	DET	.460	Addie Joss	CLE	.318	Cy Young	BOS	8.9
Elmer Flick	CLE	.375	Elmer Flick	CLE	.412	Doc White	CHI	.310	Addie Joss	CLE	7.7

Isolated Power			Players Overall			Pitchers Overall			Defensive Runs		
Sam Crawford	DET	.137	Nap Lajoie	CLE	64.1	Ed Walsh	CHI	60.4	Nap Lajoie	CLE	41.1
Harry Davis	PHI	.132	Sam Crawford	DET	47.7	Harry Howell	STL	35.0	Jiggs Donahue	CHI	21.9
Ty Cobb	DET	.117	Ty Cobb	DET	45.6	Ed Killian	DET	34.1	Ed Walsh	CHI	20.8

Club	W	L	R	OR	Avg	OBA	SLG	BPF	NOPS-A	BR	Adj	Wins	ERA	PPF	NERA-A	PR	Adj	Wins	Diff
DET	92	58	694	532	.266	.308	.335	101	113/111	63	57	6.2	2.33	98	109/106	32	23	2.5	8.2
PHI	88	57	582	511	.254	.307	.329	100	111/111	56	55	6.1	2.34	98	108/107	30	24	2.6	6.8
CHI	87	64	588	474	.238	.296	.283	101	91/ 91	-36	-39	-4.5	2.22	98	114/112	50	41	4.6	11.4
CLE	85	67	530	525	.241	.288	.310	91	98/108	-14	35	3.9	2.26	90	112/101	43	4	.4	4.7
NY	70	78	605	665	.249	.292	.314	113	101/ 89	-2	-73	-8.1	3.03	115	84/ 97	-71	-14	-1.7	5.8
STL	69	83	542	555	.253	.303	.313	101	103/103	20	16	1.7	2.61	101	97/ 98	-10	-6	-.7	-8.0
BOS	59	90	464	558	.234	.276	.292	100	88/ 88	-70	-72	-8.1	2.45	102	104/106	14	23	2.5	-10.0
WAS	49	102	506	691	.243	.297	.299	93	97/104	-11	25	2.8	3.11	97	82/ 80	-84	-94	-10.5	-18.8
			564		.247	.296	.310						2.54						

NATIONAL LEAGUE   1907

Batting Runs			Park Adjusted			Pitching Runs			Park Adjusted		
Honus Wagner	PIT	50.4	Honus Wagner	PIT	46.5	Carl Lundgren	CHI	29.7	Carl Lundgren	CHI	33.8
Sherry Magee	PHI	38.6	Sherry Magee	PHI	34.5	Jack Pfiester	CHI	28.4	Three Finger Brown	CHI	32.4
Ginger Beaumont	BOS	30.2	Ginger Beaumont	BOS	32.7	Three Finger Brown	CHI	27.8	Jack Pfiester	CHI	32.2

Normalized OPS			Park Adjusted			Normalized ERA			Park Adjusted		
Honus Wagner	PIT	189	Honus Wagner	PIT	177	Jack Pfiester	CHI	214	Jack Pfiester	CHI	229
Sherry Magee	PHI	168	Sherry Magee	PHI	156	Carl Lundgren	CHI	210	Carl Lundgren	CHI	225
Ginger Beaumont	BOS	149	Ginger Beaumont	BOS	155	Three Finger Brown	CHI	177	Three Finger Brown	CHI	190

On Base Average			Slugging Percentage			Percent of Team Wins			Wins Above Team		
Honus Wagner	PIT	.403	Honus Wagner	PIT	.513	Ed Karger	STL	.294	Tully Sparks	PHI	6.4
Sherry Magee	PHI	.392	Sherry Magee	PHI	.455	Christy Mathewson	NY	.293	Christy Mathewson	NY	6.2
Fred Clarke	PIT	.374	Harry Lumley	BRO	.425	Stoney McGlynn	STL	.275	Ed Karger	STL	4.6

Isolated Power			Players Overall			Pitchers Overall			Defensive Runs		
Honus Wagner	PIT	.163	Honus Wagner	PIT	49.8	Three Finger Brown	CHI	34.4	Johnny Evers	CHI	29.5
Harry Lumley	BRO	.159	Dave Brain	BOS	40.6	Orval Overall	CHI	32.0	Whitey Alperman	BRO	23.3
Dave Brain	BOS	.141	Sherry Magee	PHI	38.8	Carl Lundgren	CHI	29.9	Dave Brain	BOS	22.2

Club	W	L	R	OR	Avg	OBA	SLG	BPF	NOPS-A	BR	Adj	Wins	ERA	PPF	NERA-A	PR	Adj	Wins	Diff
CHI	107	45	572	390	.250	.311	.311	111	104/ 94	25	-32	-3.7	1.73	107	142/153	112	139	15.8	18.9
PIT	91	63	634	510	.254	.319	.324	107	111/104	62	25	2.8	2.30	105	107/112	25	42	4.8	6.4
PHI	83	64	512	476	.236	.299	.305	108	98/ 91	-9	-47	-5.4	2.43	107	101/109	5	30	3.4	11.5
NY	82	71	574	510	.251	.322	.317	103	110/107	61	47	5.3	2.45	101	101/102	2	7	.8	-.6
BRO	65	83	446	522	.232	.281	.298	94	90/ 96	-59	-28	-3.3	2.38	95	104/ 99	13	-4	-.6	-5.2
CIN	66	87	526	519	.247	.299	.318	90	102/114	7	60	6.9	2.41	89	102/ 91	8	-33	-3.8	-13.5
BOS	58	90	502	652	.243	.301	.309	96	100/104	0	21	2.4	3.34	100	74/ 73	-129	-131	-15.0	-3.5
STL	52	101	419	606	.232	.277	.288	95	85/ 89	-84	-60	-7.0	2.69	100	91/ 91	-34	-35	-4.1	-13.4
			523		.243	.301	.309						2.46						

AMERICAN LEAGUE 1908

Batting Runs			Park Adjusted			Pitching Runs			Park Adjusted		
Ty Cobb	DET	43.6	Ty Cobb	DET	36.7	Ed Walsh	CHI	50.0	Addie Joss	CLE	46.1
Sam Crawford	DET	39.1	Sam Crawford	DET	32.0	Addie Joss	CLE	44.2	Cy Young	BOS	38.3
Matty McIntyre	DET	36.0	Doc Gessler	BOS	29.9	Cy Young	BOS	37.3	Ed Walsh	CHI	33.0
Normalized OPS			**Park Adjusted**			**Normalized ERA**			**Park Adjusted**		
Ty Cobb	DET	174	Doc Gessler	BOS	160	Addie Joss	CLE	205	Addie Joss	CLE	210
Sam Crawford	DET	165	Ty Cobb	DET	156	Cy Young	BOS	189	Cy Young	BOS	191
Doc Gessler	BOS	164	Sam Crawford	DET	148	Ed Walsh	CHI	169	Ed Summers	DET	159
On Base Average			**Slugging Percentage**			**Percent of Team Wins**			**Wins Above Team**		
Matty McIntyre	DET	.385	Ty Cobb	DET	.475	Ed Walsh	CHI	.455	Ed Walsh	CHI	12.8
Doc Gessler	BOS	.381	Sam Crawford	DET	.457	Cy Young	BOS	.280	Cy Young	BOS	6.8
Topsy Hartsel	PHI	.371	Doc Gessler	BOS	.423	Jack Chesbro	NY	.275	Addie Joss	CLE	4.6
Isolated Power			**Players Overall**			**Pitchers Overall**			**Defensive Runs**		
Ty Cobb	DET	.151	Nap Lajoie	CLE	66.5	Addie Joss	CLE	49.9	Nap Lajoie	CLE	51.2
Sam Crawford	DET	.146	George McBride	WAS	37.2	Ed Walsh	CHI	47.4	Heinie Wagner	BOS	37.7
Claude Rossman	DET	.124	Heinie Wagner	BOS	34.2	Cy Young	BOS	32.7	George McBride	WAS	36.7

Club	W	L	R	OR	Avg	OBA	SLG	BPF	NOPS-A	BR	Adj	Wins	ERA	PPF	NERA-A	PR	Adj	Wins	Diff
DET	90	63	647	547	.263	.307	.347	111	121/109	104	43	4.9	2.40	109	99/109	-1	32	3.7	5.0
CLE	90	64	568	457	.239	.290	.309	105	103/ 98	13	-11	-1.4	2.02	102	118/121	59	67	7.6	6.8
CHI	88	64	537	470	.224	.290	.271	89	90/100	-40	17	2.0	2.22	86	107/ 93	26	-25	-3.0	13.0
STL	83	69	544	483	.245	.292	.310	102	104/102	18	6	.7	2.15	101	111/112	37	40	4.6	1.7
BOS	75	79	564	513	.245	.286	.312	102	103/100	6	-5	-.7	2.27	101	105/106	17	22	2.5	-3.8
PHI	68	85	486	562	.223	.276	.292	100	92/ 92	-41	-43	-5.0	2.56	102	93/ 95	-26	-17	-2.1	-1.4
WAS	67	85	479	539	.235	.287	.296	93	97/105	-12	25	2.8	2.34	93	102/ 95	7	-16	-1.9	-9.9
NY	51	103	459	713	.236	.277	.291	98	92/ 94	-43	-31	-3.7	3.16	104	75/ 78	-116	-102	-11.8	-10.6
				536		.239	.288	.304									2.39		

NATIONAL LEAGUE 1908

Batting Runs			Park Adjusted			Pitching Runs			Park Adjusted		
Honus Wagner	PIT	65.2	Honus Wagner	PIT	59.9	Christy Mathewson	NY	39.9	Three Finger Brown	CHI	39.7
Mike Donlin	NY	36.6	Mike Donlin	NY	35.8	George McQuillan	PHI	32.8	Christy Mathewson	NY	36.5
Roger Bresnahan	NY	26.5	Sherry Magee	PHI	27.1	Three Finger Brown	CHI	30.3	George McQuillan	PHI	28.1
Normalized OPS			**Park Adjusted**			**Normalized ERA**			**Park Adjusted**		
Honus Wagner	PIT	206	Honus Wagner	PIT	189	Christy Mathewson	NY	164	Three Finger Brown	CHI	178
Mike Donlin	NY	161	Mike Donlin	NY	159	Three Finger Brown	CHI	159	Howie Camnitz	PIT	160
Johnny Evers	CHI	149	Sherry Magee	PHI	151	George McQuillan	PHI	154	Christy Mathewson	NY	159
On Base Average			**Slugging Percentage**			**Percent of Team Wins**			**Wins Above Team**		
Honus Wagner	PIT	.410	Honus Wagner	PIT	.542	Christy Mathewson	NY	.378	Christy Mathewson	NY	9.4
Johnny Evers	CHI	.396	Mike Donlin	NY	.452	Nap Rucker	BRO	.321	Three Finger Brown	CHI	6.1
Roger Bresnahan	NY	.395	Sherry Magee	PHI	.417	Bugs Raymond	STL	.306	Nap Rucker	BRO	6.0
Isolated Power			**Players Overall**			**Pitchers Overall**			**Defensive Runs**		
Honus Wagner	PIT	.188	Honus Wagner	PIT	54.7	Christy Mathewson	NY	45.3	Bill Dahlen	BOS	36.6
Sherry Magee	PHI	.134	Joe Tinker	CHI	38.2	Three Finger Brown	CHI	38.9	Joe Tinker	CHI	36.3
Tim Jordan	BRO	.124	Mike Donlin	NY	29.8	George McQuillan	PHI	25.4	Art Devlin	NY	17.2

Club	W	L	R	OR	Avg	OBA	SLG	BPF	NOPS-A	BR	Adj	Wins	ERA	PPF	NERA-A	PR	Adj	Wins	Diff
CHI	99	55	624	461	.249	.306	.321	115	110/ 95	51	-27	-3.2	2.14	112	110/122	33	76	8.8	16.4
PIT	98	56	585	469	.247	.304	.332	109	113/103	65	18	2.1	2.12	106	111/117	35	57	6.6	12.2
NY	98	56	652	456	.267	.333	.333	101	122/121	128	121	14.1	2.14	97	110/106	32	20	2.3	4.6
PHI	83	71	504	445	.244	.291	.316	97	103/107	10	26	3.1	2.10	95	112/106	38	20	2.3	.6
CIN	73	81	489	544	.227	.282	.294	98	92/ 94	-39	-30	-3.6	2.37	99	99/ 98	-3	-6	-.8	.4
BOS	63	91	537	622	.239	.296	.293	100	97/ 97	-13	-11	-1.4	2.79	102	84/ 86	-69	-62	-7.3	-5.3
BRO	53	101	377	516	.213	.262	.277	94	80/ 85	-105	-74	-8.7	2.47	97	95/ 92	-18	-29	-3.5	-11.8
STL	49	105	371	626	.223	.265	.283	91	83/ 92	-94	-46	-5.5	2.64	96	89/ 86	-44	-57	-6.7	-15.8
				517		.239	.293	.306									2.35		

AMERICAN LEAGUE 1909

Batting Runs			Park Adjusted			Pitching Runs			Park Adjusted		
Ty Cobb	DET	63.3	Ty Cobb	DET	57.9	Frank Smith	CHI	27.1	Harry Krause	PHI	28.2
Eddie Collins	PHI	48.9	Eddie Collins	PHI	43.2	Ed Walsh	CHI	27.1	Eddie Plank	PHI	27.1
Sam Crawford	DET	36.7	Sam Crawford	DET	31.2	Harry Krause	PHI	25.4	Chief Bender	PHI	25.9
Normalized OPS			**Park Adjusted**			**Normalized ERA**			**Park Adjusted**		
Ty Cobb	DET	199	Ty Cobb	DET	183	Harry Krause	PHI	177	Harry Krause	PHI	186
Eddie Collins	PHI	173	Eddie Collins	PHI	159	Ed Walsh	CHI	175	Ed Walsh	CHI	167
Sam Crawford	DET	158	Nap Lajoie	CLE	148	Chief Bender	PHI	149	Chief Bender	PHI	156
On Base Average			**Slugging Percentage**			**Percent of Team Wins**			**Wins Above Team**		
Ty Cobb	DET	.431	Ty Cobb	DET	.517	Frank Smith	CHI	.321	George Mullin	DET	6.8
Eddie Collins	PHI	.416	Sam Crawford	DET	.452	Walter Johnson	WAS	.310	Frank Smith	CHI	4.8
Donie Bush	DET	.380	Eddie Collins	PHI	.449	George Mullin	DET	.296	Cy Young	CLE	4.1
Isolated Power			**Players Overall**			**Pitchers Overall**			**Defensive Runs**		
Frank Baker	PHI	.142	Eddie Collins	PHI	54.6	Frank Smith	CHI	37.3	Nap Lajoie	CLE	24.5
Jake Stahl	BOS	.140	Ty Cobb	DET	49.9	Ed Walsh	CHI	34.2	Fred Parent	CHI	17.6
Ty Cobb	DET	.140	Nap Lajoie	CLE	47.5	Eddie Plank	PHI	29.7	Tris Speaker	BOS	17.5

Club	W	L	R	OR	Avg	OBA	SLG	BPF	NOPS-A	BR	Adj	Wins	ERA	PPF	NERA-A	PR	Adj	Wins	Diff
DET	98	54	666	493	.267	.325	.342	109	118/109	98	50	5.7	2.26	106	109/115	33	54	6.2	10.1
PHI	95	58	605	408	.256	.321	.343	109	117/108	91	43	4.9	1.92	105	129/135	85	103	11.7	1.8
BOS	88	63	597	550	.263	.321	.333	114	114/100	73	0	.0	2.59	114	95/108	-18	33	3.7	8.8
CHI	78	74	492	463	.221	.292	.275	97	85/88	-71	-52	-6.1	2.04	95	121/115	68	50	5.7	2.4
NY	74	77	590	587	.248	.313	.311	95	104/109	25	51	5.8	2.68	95	92/87	-30	-49	-5.7	-1.6
CLE	71	82	493	532	.241	.288	.313	105	97/92	-27	-53	-6.1	2.38	106	104/110	13	36	4.1	-3.5
STL	61	89	441	575	.232	.287	.279	88	85/97	-78	-15	-1.9	2.88	90	86/77	-61	-98	-11.2	-.9
WAS	42	110	380	656	.223	.276	.275	90	80/89	-108	-54	-6.3	3.03	95	81/78	-86	-103	-11.8	-15.9
			533		.244	.303	.309						2.47						

NATIONAL LEAGUE 1909

Batting Runs			Park Adjusted			Pitching Runs			Park Adjusted		
Honus Wagner	PIT	48.1	Honus Wagner	PIT	45.3	Three Finger Brown	CHI	48.7	Three Finger Brown	CHI	49.3
Mike Mitchell	CIN	30.3	Mike Mitchell	CIN	33.5	Christy Mathewson	NY	44.1	Christy Mathewson	NY	48.9
Larry Doyle	NY	26.2	Doc Hoblitzel	CIN	27.4	Orval Overall	CHI	37.0	Orval Overall	CHI	37.2
Normalized OPS			**Park Adjusted**			**Normalized ERA**			**Park Adjusted**		
Honus Wagner	PIT	182	Honus Wagner	PIT	173	Christy Mathewson	NY	226	Christy Mathewson	NY	241
Mike Mitchell	CIN	150	Mike Mitchell	CIN	159	Three Finger Brown	CHI	197	Three Finger Brown	CHI	198
Doc Hoblitzel	CIN	142	Doc Hoblitzel	CIN	150	Orval Overall	CHI	182	Orval Overall	CHI	183
On Base Average			**Slugging Percentage**			**Percent of Team Wins**			**Wins Above Team**		
Honus Wagner	PIT	.420	Honus Wagner	PIT	.489	Al Mattern	BOS	.333	Christy Mathewson	NY	8.0
Al Bridwell	NY	.386	Mike Mitchell	CIN	.430	George Bell	BRO	.291	George Bell	BRO	6.1
Fred Clarke	PIT	.384	Larry Doyle	NY	.419	Fred Beebe	STL	.278	Al Mattern	BOS	5.8
Isolated Power			**Players Overall**			**Pitchers Overall**			**Defensive Runs**		
Honus Wagner	PIT	.149	Honus Wagner	PIT	63.5	Christy Mathewson	NY	60.8	Mickey Doolan	PHI	26.0
Sherry Magee	PHI	.128	Fred Clarke	PIT	32.5	Three Finger Brown	CHI	47.7	Dick Egan	CIN	25.7
Mike Mitchell	CIN	.120	Dick Egan	CIN	30.3	Orval Overall	CHI	43.0	Joe Tinker	CHI	19.0

Club	W	L	R	OR	Avg	OBA	SLG	BPF	NOPS-A	BR	Adj	Wins	ERA	PPF	NERA-A	PR	Adj	Wins	Diff
PIT	110	42	699	447	.260	.327	.353	105	118/113	100	72	8.0	2.07	99	125/124	82	79	8.7	17.2
CHI	104	49	635	390	.245	.308	.323	105	102/97	7	-22	-2.6	1.75	100	148/148	131	132	14.6	15.4
NY	92	61	623	546	.255	.330	.329	108	111/103	72	26	2.9	2.28	106	114/121	50	77	8.6	4.0
CIN	77	76	606	599	.250	.319	.323	95	106/112	35	66	7.3	2.52	94	103/97	11	-12	-1.4	-5.4
PHI	74	79	516	518	.244	.303	.309	105	96/91	-24	-54	-6.2	2.44	106	106/112	24	47	5.2	-1.5
BRO	55	98	444	627	.229	.279	.297	94	85/90	-96	-61	-6.9	3.10	97	84/81	-78	-89	-10.0	-4.7
STL	54	98	583	731	.243	.326	.303	99	102/103	24	32	3.5	3.41	102	76/78	-125	-116	-13.0	-12.5
BOS	45	108	435	683	.223	.285	.274	92	79/86	-115	-72	-8.1	3.20	97	81/79	-91	-102	-11.4	-12.0
			568		.244	.310	.314						2.59						

Batting Runs
Nap Lajoie	CLE	68.2
Ty Cobb	DET	67.9
Tris Speaker	BOS	44.9

Park Adjusted
Nap Lajoie	CLE	66.6
Ty Cobb	DET	64.0
Tris Speaker	BOS	47.3

Pitching Runs
Ed Walsh	CHI	51.5
Walter Johnson	WAS	48.6
Jack Coombs	PHI	47.8

Park Adjusted
Walter Johnson	WAS	47.7
Jack Coombs	PHI	45.3
Ed Walsh	CHI	41.5

Normalized OPS
Ty Cobb	DET	215
Nap Lajoie	CLE	200
Tris Speaker	BOS	173

Park Adjusted
Ty Cobb	DET	202
Nap Lajoie	CLE	196
Tris Speaker	BOS	180

Normalized ERA
Ed Walsh	CHI	199
Jack Coombs	PHI	194
Walter Johnson	WAS	187

Park Adjusted
Jack Coombs	PHI	189
Walter Johnson	WAS	185
Ed Walsh	CHI	180

On Base Average
Ty Cobb	DET	.456
Nap Lajoie	CLE	.445
Tris Speaker	BOS	.404

Slugging Percentage
Ty Cobb	DET	.551
Nap Lajoie	CLE	.514
Tris Speaker	BOS	.468

Percent of Team Wins
Walter Johnson	WAS	.379
Jack Coombs	PHI	.304
Russ Ford	NY	.295

Wins Above Team
Russ Ford	NY	9.3
Walter Johnson	WAS	9.2
Jack Coombs	PHI	5.2

Isolated Power
Ty Cobb	DET	.168
Jake Stahl	BOS	.153
Danny Murphy	PHI	.136

Players Overall
Nap Lajoie	CLE	69.0
Eddie Collins	PHI	65.7
Ty Cobb	DET	55.2

Pitchers Overall
Ed Walsh	CHI	54.1
Walter Johnson	WAS	43.9
Jack Coombs	PHI	42.2

Defensive Runs
Eddie Collins	PHI	44.7
George McBride	WAS	29.1
Donie Bush	DET	18.2

Club	W	L	R	OR	Avg	OBA	SLG	BPF	NOPS-A	BR	Adj	Wins	ERA	PPF	NERA-A	PR	Adj	Wins	Diff
PHI	102	48	673	441	.266	.326	.355	103	119/116	104	89	9.9	1.79	98	141/137	116	106	11.7	5.4
NY	88	63	626	557	.248	.320	.322	106	107/101	44	8	.9	2.62	105	96/101	-16	4	.5	11.2
DET	86	68	679	582	.261	.329	.344	107	117/110	95	57	6.4	2.82	105	89/ 94	-46	-27	-3.1	5.7
BOS	81	72	638	564	.259	.323	.351	96	117/122	94	116	12.9	2.45	94	103/ 97	10	-11	-1.4	-7.0
CLE	71	81	548	657	.244	.297	.308	102	95/ 93	-36	-50	-5.7	2.90	105	87/ 91	-40	-40	-4.5	5.2
CHI	68	85	457	479	.211	.275	.261	91	73/ 80	-149	-99	-11.1	2.03	90	124/112	78	39	4.4	-1.8
WAS	66	85	501	550	.236	.309	.288	98	92/ 94	-31	-22	-2.5	2.46	99	103/102	10	6	.7	-7.6
STL	47	107	451	743	.218	.281	.274	96	79/ 82	-116	-93	-10.5	3.09	102	81/ 83	-87	-79	-8.9	-10.7
			572		.243	.308	.313						2.52						

Batting Runs
Sherry Magee	PHI	54.9
Solly Hofman	CHI	32.4
Fred Snodgrass	NY	31.9

Park Adjusted
Sherry Magee	PHI	58.0
Fred Snodgrass	NY	35.0
Ed Konetchy	STL	34.3

Pitching Runs
Christy Mathewson	NY	39.9
Three Finger Brown	CHI	38.2
King Cole	CHI	32.7

Park Adjusted
Three Finger Brown	CHI	33.6
Christy Mathewson	NY	30.1
Babe Adams	PIT	29.9

Normalized OPS
Sherry Magee	PHI	177
Fred Snodgrass	NY	154
Solly Hofman	CHI	153

Park Adjusted
Sherry Magee	PHI	186
Fred Snodgrass	NY	164
Ed Konetchy	STL	154

Normalized ERA
King Cole	CHI	168
Three Finger Brown	CHI	163
Christy Mathewson	NY	160

Park Adjusted
King Cole	CHI	160
Three Finger Brown	CHI	155
Babe Adams	PIT	149

On Base Average
Sherry Magee	PHI	.445
Fred Snodgrass	NY	.440
Johnny Evers	CHI	.413

Slugging Percentage
Sherry Magee	PHI	.507
Solly Hofman	CHI	.461
Frank Schulte	CHI	.460

Percent of Team Wins
Al Mattern	BOS	.302
Christy Mathewson	NY	.297
Earl Moore	PHI	.282

Wins Above Team
Christy Mathewson	NY	7.5
Deacon Phillippe	PIT	5.6
George Suggs	CIN	5.6

Isolated Power
Sherry Magee	PHI	.175
Frank Schulte	CHI	.159
Fred Merkle	NY	.148

Players Overall
Sherry Magee	PHI	46.1
Ed Konetchy	STL	40.4
Mickey Doolan	PHI	33.1

Pitchers Overall
Christy Mathewson	NY	44.0
Three Finger Brown	CHI	35.1
King Cole	CHI	29.7

Defensive Runs
Dave Shean	BOS	39.7
Mickey Doolan	PHI	29.1
Otto Knabe	PHI	26.0

Club	W	L	R	OR	Avg	OBA	SLG	BPF	NOPS-A	BR	Adj	Wins	ERA	PPF	NERA-A	PR	Adj	Wins	Diff
CHI	104	50	712	499	.268	.344	.362	100	112/112	75	76	8.0	2.51	95	121/115	79	58	6.1	12.9
NY	91	63	715	567	.275	.354	.366	94	116/124	107	143	15.0	2.68	91	113/103	54	11	1.1	-2.1
PIT	86	67	655	576	.266	.328	.360	111	107/ 96	31	-38	-4.1	2.83	110	107/118	29	77	8.1	5.4
PHI	78	75	674	639	.255	.327	.338	96	100/104	-2	25	2.7	3.06	95	99/ 94	-4	-29	-3.2	2.0
CIN	75	79	620	684	.259	.332	.333	102	100/ 97	0	-13	-1.5	3.09	104	98/102	-8	8	.9	-1.4
BRO	64	90	497	623	.229	.294	.305	94	80/ 85	-131	-94	-10.0	3.07	96	99/ 95	-6	-25	-2.7	-.3
STL	63	90	639	718	.248	.345	.319	91	100/110	20	78	8.2	3.78	92	80/ 73	-110	-148	-15.7	-6.1
BOS	53	100	495	701	.246	.301	.317	114	86/ 76	-97	-181	-19.2	3.22	119	94/112	-29	58	6.1	-10.4
			626		.256	.328	.337						3.03						

Batting Runs			Park Adjusted			Pitching Runs			Park Adjusted		
Ty Cobb	DET	78.5	Ty Cobb	DET	75.2	Walter Johnson	WAS	51.8	Russ Ford	NY	51.9
Joe Jackson	CLE	71.8	Joe Jackson	CLE	72.0	Ed Walsh	CHI	45.9	Walter Johnson	WAS	47.1
Sam Crawford	DET	52.4	Sam Crawford	DET	49.2	Vean Gregg	CLE	41.9	Vean Gregg	CLE	42.0
Normalized OPS			**Park Adjusted**			**Normalized ERA**			**Park Adjusted**		
Ty Cobb	DET	202	Joe Jackson	CLE	195	Vean Gregg	CLE	185	Vean Gregg	CLE	186
Joe Jackson	CLE	194	Ty Cobb	DET	195	Walter Johnson	WAS	176	Russ Ford	NY	173
Sam Crawford	DET	168	Eddie Collins	PHI	173	Joe Wood	BOS	165	Walter Johnson	WAS	169
On Base Average			**Slugging Percentage**			**Percent of Team Wins**			**Wins Above Team**		
Joe Jackson	CLE	.468	Ty Cobb	DET	.621	Walter Johnson	WAS	.391	Walter Johnson	WAS	12.2
Ty Cobb	DET	.467	Joe Jackson	CLE	.590	Ed Walsh	CHI	.351	Vean Gregg	CLE	9.1
Eddie Collins	PHI	.451	Sam Crawford	DET	.526	Joe Wood	BOS	.295	Russ Ford	NY	7.0
Isolated Power			**Players Overall**			**Pitchers Overall**			**Defensive Runs**		
Ty Cobb	DET	.201	Ty Cobb	DET	70.4	Ed Walsh	CHI	52.7	Lee Tannehill	CHI	30.1
Joe Jackson	CLE	.182	Joe Jackson	CLE	69.6	Walter Johnson	WAS	52.3	George McBride	WAS	29.8
Frank Baker	PHI	.171	Eddie Collins	PHI	55.9	Joe Wood	BOS	48.6	Pepper Austin	STL	19.0

Club	W	L	R	OR	Avg	OBA	SLG	BPF	NOPS-A	BR	Adj	Wins	ERA	PPF	NERA-A	PR	Adj	Wins	Diff
PHI	101	50	861	601	.296	.357	.397	93	117/127	108	160	15.7	3.01	87	111/ 97	51	-14	-1.4	11.2
DET	89	65	831	776	.292	.355	.388	104	114/110	92	62	6.1	3.73	104	89/ 93	-60	-41	-4.2	10.0
CLE	80	73	691	712	.262	.333	.369	100	102/102	1	3	.3	3.36	100	99/ 99	-2	-2	-.3	3.5
CHI	77	74	719	624	.269	.325	.350	99	94/ 95	-45	-36	-3.7	2.96	97	113/109	58	42	4.1	1.1
BOS	78	75	680	643	.274	.350	.362	101	105/104	37	31	3.0	2.74	100	122/122	89	91	8.9	-10.5
NY	76	76	684	724	.272	.344	.362	116	103/ 89	24	-90	-8.9	3.54	118	94/111	-29	61	6.0	3.0
WAS	64	90	625	766	.258	.330	.329	94	87/ 93	-74	-32	-3.3	3.52	96	95/ 91	-26	-45	-4.6	-5.2
STL	45	107	567	812	.239	.307	.311	94	78/ 83	-141	-101	-10.0	3.86	98	87/ 85	-76	-84	-8.4	-12.6
			707		.273	.338	.358						3.34						

NATIONAL LEAGUE 1911

Batting Runs			Park Adjusted			Pitching Runs			Park Adjusted		
Frank Schulte	CHI	40.6	Honus Wagner	PIT	42.7	Christy Mathewson	NY	47.7	Christy Mathewson	NY	49.7
Larry Doyle	NY	39.3	Frank Schulte	CHI	42.1	Babe Adams	PIT	34.4	Pete Alexander	PHI	45.1
Honus Wagner	PIT	38.8	Larry Doyle	NY	35.3	Pete Alexander	PHI	33.3	Earl Moore	PHI	36.0
Normalized OPS			**Park Adjusted**			**Normalized ERA**			**Park Adjusted**		
Honus Wagner	PIT	160	Honus Wagner	PIT	170	Christy Mathewson	NY	170	Christy Mathewson	NY	173
Larry Doyle	NY	158	Frank Schulte	CHI	159	Lew Richie	CHI	147	Pete Alexander	PHI	143
Frank Schulte	CHI	156	Larry Doyle	NY	149	Babe Adams	PIT	145	Earl Moore	PHI	140
On Base Average			**Slugging Percentage**			**Percent of Team Wins**			**Wins Above Team**		
Jimmy Sheckard	CHI	.434	Frank Schulte	CHI	.534	Pete Alexander	PHI	.354	Pete Alexander	PHI	9.2
Honus Wagner	PIT	.423	Larry Doyle	NY	.527	Nap Rucker	BRO	.344	Nap Rucker	BRO	6.7
Johnny Bates	CIN	.415	Honus Wagner	PIT	.507	Bob Harmon	STL	.307	Rube Marquard	NY	4.9
Isolated Power			**Players Overall**			**Pitchers Overall**			**Defensive Runs**		
Frank Schulte	CHI	.234	Jimmy Sheckard	CHI	48.5	Christy Mathewson	NY	42.0	Joe Tinker	CHI	35.0
Larry Doyle	NY	.217	Joe Tinker	CHI	40.2	Pete Alexander	PHI	41.5	Jimmy Sheckard	CHI	20.5
Sherry Magee	PHI	.196	Honus Wagner	PIT	36.9	Rube Marquard	NY	37.4	Miller Huggins	STL	20.4

Club	W	L	R	OR	Avg	OBA	SLG	BPF	NOPS-A	BR	Adj	Wins	ERA	PPF	NERA-A	PR	Adj	Wins	Diff
NY	99	54	756	542	.279	.358	.390	105	117·111	109	72	7.2	2.69	102	126·128	107	116	11.7	3.7
CHI	92	62	757	607	.260	.341	.374	98	107·109	43	57	5.7	2.89	95	117·111	78	51	5.2	4.1
PIT	85	69	744	557	.262	.336	.371	94	105·112	27	68	6.8	2.83	90	120·107	85	32	3.2	-2.1
PHI	79	73	658	669	.259	.328	.360	108	100· 93	-9	-62	-6.3	3.30	109	103·112	14	59	5.9	3.4
STL	75	74	671	745	.252	.337	.340	102	96· 94	-16	-32	-3.4	3.68	104	92· 96	-44	-24	-2.5	6.4
CIN	70	83	682	706	.261	.337	.346	90	98·109	-9	59	6.0	3.26	90	104· 94	21	-31	-3.2	-9.2
BRO	64	86	539	659	.237	.301	.311	99	77· 79	-151	-141	-14.3	3.40	101	100·100	-0	2	.2	3.0
BOS	44	107	699	1021	.267	.340	.355	101	101·100	10	2	.2	5.08	107	67· 72	-257	-219	-22.1	-9.6
			688		.260	.335	.356						3.39						

AMERICAN LEAGUE 1912

Batting Runs			Park Adjusted			Pitching Runs			Park Adjusted		
Tris Speaker	BOS	72.8	Ty Cobb	DET	72.1	Walter Johnson	WAS	79.6	Walter Johnson	WAS	78.5
Joe Jackson	CLE	70.3	Joe Jackson	CLE	71.2	Joe Wood	BOS	54.7	Ed Walsh	CHI	62.4
Ty Cobb	DET	67.9	Tris Speaker	BOS	67.8	Ed Walsh	CHI	51.9	Joe Wood	BOS	57.0
Normalized OPS			**Park Adjusted**			**Normalized ERA**			**Park Adjusted**		
Ty Cobb	DET	198	Ty Cobb	DET	210	Walter Johnson	WAS	240	Walter Johnson	WAS	238
Joe Jackson	CLE	196	Joe Jackson	CLE	198	Joe Wood	BOS	175	Joe Wood	BOS	178
Tris Speaker	BOS	194	Tris Speaker	BOS	183	Ed Walsh	CHI	155	Ed Walsh	CHI	166
On Base Average			**Slugging Percentage**			**Percent of Team Wins**			**Wins Above Team**		
Tris Speaker	BOS	.464	Ty Cobb	DET	.586	Walter Johnson	WAS	.363	Joe Wood	BOS	9.5
Joe Jackson	CLE	.458	Joe Jackson	CLE	.579	Ed Walsh	CHI	.346	Eddie Plank	PHI	8.9
Ty Cobb	DET	.458	Tris Speaker	BOS	.567	Joe Wood	BOS	.324	Walter Johnson	WAS	8.6
Isolated Power			**Players Overall**			**Pitchers Overall**			**Defensive Runs**		
Frank Baker	PHI	.194	Tris Speaker	BOS	74.4	Walter Johnson	WAS	86.0	George McBride	WAS	35.0
Tris Speaker	BOS	.184	Joe Jackson	CLE	71.1	Joe Wood	BOS	73.1	Donie Bush	DET	26.8
Joe Jackson	CLE	.184	Eddie Collins	PHI	61.0	Ed Walsh	CHI	71.6	Morrie Rath	CHI	22.5

Club	W	L	R	OR	Avg	OBA	SLG	BPF	NOPS-A	BR	Adj	Wins	ERA	PPF	NERA-A	PR	Adj	Wins	Diff
BOS	105	47	799	544	.277	.355	.380	106	116/109	104	61	6.1	2.76	102	121/123	89	98	9.8	13.1
WAS	91	61	698	581	.256	.324	.341	101	96/ 94	-31	-40	-4.1	2.69	99	124/123	99	95	9.5	9.6
PHI	90	62	779	658	.282	.349	.377	102	114/111	83	70	7.0	3.33	100	100/100	2	1	.1	6.9
CHI	78	76	638	646	.255	.317	.329	107	90/ 84	-67	-114	-11.5	3.06	107	109/117	45	83	8.3	4.3
CLE	75	78	676	680	.273	.333	.352	99	101/102	3	11	1.1	3.30	99	101/100	6	1	.1	-2.7
DET	69	84	720	777	.268	.343	.349	94	103/110	27	67	6.6	3.80	95	88/ 84	-69	-93	-9.4	-4.8
STL	53	101	552	764	.248	.315	.320	97	87/ 89	-85	-63	-6.4	3.71	100	90/ 90	-55	-53	-5.4	-12.2
NY	50	102	630	842	.259	.329	.334	96	95/ 99	-31	-3	-.4	4.12	100	81/ 81	-115	-115	-11.6	-13.9
			687		.265	.333	.348						3.34						

NATIONAL LEAGUE 1912

Batting Runs			Park Adjusted			Pitching Runs			Park Adjusted		
Heinie Zimmerman	CHI	49.8	Heinie Zimmerman	CHI	47.6	Christy Mathewson	NY	44.0	Christy Mathewson	NY	50.2
Honus Wagner	PIT	32.1	Honus Wagner	PIT	31.3	Nap Rucker	BRO	39.5	Jeff Tesreau	NY	43.6
Bill Sweeney	BOS	30.9	Bill Sweeney	BOS	28.8	Jeff Tesreau	NY	38.7	Nap Rucker	BRO	37.7
Normalized OPS			**Park Adjusted**			**Normalized ERA**			**Park Adjusted**		
Heinie Zimmerman	CHI	169	Heinie Zimmerman	CHI	164	Jeff Tesreau	NY	173	Jeff Tesreau	NY	182
Honus Wagner	PIT	143	Honus Wagner	PIT	141	Christy Mathewson	NY	160	Christy Mathewson	NY	169
Johnny Evers	CHI	139	Ed Konetchy	STL	136	Nap Rucker	BRO	154	Nap Rucker	BRO	152
On Base Average			**Slugging Percentage**			**Percent of Team Wins**			**Wins Above Team**		
Johnny Evers	CHI	.431	Heinie Zimmerman	CHI	.571	Nap Rucker	BRO	.310	Larry Cheney	CHI	5.5
Miller Huggins	STL	.422	Chief Wilson	PIT	.513	Bob Harmon	STL	.286	Claude Hendrix	PIT	4.7
Dode Paskert	PHI	.420	Honus Wagner	PIT	.496	Larry Cheney	CHI	.286	Nap Rucker	BRO	4.3
Isolated Power			**Players Overall**			**Pitchers Overall**			**Defensive Runs**		
Chief Wilson	PIT	.213	Honus Wagner	PIT	68.4	Christy Mathewson	NY	54.0	Bill Sweeney	BOS	28.1
Heinie Zimmerman	CHI	.199	Bill Sweeney	BOS	51.3	Claude Hendrix	PIT	41.4	Joe Tinker	CHI	25.6
Gavvy Cravath	PHI	.186	Heinie Zimmerman	CHI	43.5	Nap Rucker	BRO	41.3	Honus Wagner	PIT	21.9

Club	W	L	R	OR	Avg	OBA	SLG	BPF	NOPS-A	BR	Adj	Wins	ERA	PPF	NERA-A	PR	Adj	Wins	Diff
NY	103	48	823	571	.286	.360	.395	109	114/104	89	23	2.3	2.58	105	132/139	124	152	14.9	10.3
PIT	93	58	751	565	.284	.340	.398	101	109/108	44	37	3.6	2.85	98	119/116	84	71	7.0	6.9
CHI	91	59	756	668	.277	.354	.387	103	110/106	63	41	4.1	3.42	102	99/101	-3	5	.5	11.5
CIN	75	78	656	722	.256	.323	.339	90	87/ 97	-89	-16	-1.7	3.42	90	99/ 89	-2	-55	-5.5	-5.7
PHI	73	79	670	688	.267	.332	.367	103	98/ 95	-22	-45	-4.5	3.25	104	105/108	22	41	4.0	-2.6
STL	63	90	659	830	.268	.340	.352	96	95/100	-26	2	.2	3.85	99	88/ 87	-67	-73	-7.3	-6.4
BRO	58	95	651	754	.268	.336	.358	97	96/ 99	-27	-4	-.5	3.64	98	93/ 92	-36	-44	-4.4	-13.6
BOS	52	101	693	861	.273	.335	.361	103	97/ 94	-27	-46	-4.6	4.17	106	81/ 86	-118	-87	-8.6	-11.3
			707		.272	.340	.369						3.40						

AMERICAN LEAGUE 1913

Batting Runs			Park Adjusted			Pitching Runs			Park Adjusted		
Joe Jackson	CLE	65.5	Joe Jackson	CLE	62.4	Walter Johnson	WAS	68.7	Walter Johnson	WAS	81.2
Tris Speaker	BOS	55.9	Tris Speaker	BOS	55.1	Eddie Cicotte	CHI	40.3	Eddie Cicotte	CHI	37.2
Ty Cobb	DET	51.8	Ty Cobb	DET	52.5	Reb Russell	CHI	36.0	Reb Russell	CHI	32.2
Normalized OPS			**Park Adjusted**			**Normalized ERA**			**Park Adjusted**		
Joe Jackson	CLE	195	Ty Cobb	DET	195	Walter Johnson	WAS	256	Walter Johnson	WAS	285
Ty Cobb	DET	193	Joe Jackson	CLE	187	Eddie Cicotte	CHI	186	Eddie Cicotte	CHI	179
Tris Speaker	BOS	184	Tris Speaker	BOS	182	Willie Mitchell	CLE	168	Willie Mitchell	CLE	173
On Base Average			**Slugging Percentage**			**Percent of Team Wins**			**Wins Above Team**		
Ty Cobb	DET	.467	Joe Jackson	CLE	.551	Walter Johnson	WAS	.400	Walter Johnson	WAS	15.1
Joe Jackson	CLE	.460	Ty Cobb	DET	.535	Reb Russell	CHI	.282	Ray Collins	BOS	5.8
Eddie Collins	PHI	.441	Tris Speaker	BOS	.533	Cy Falkenberg	CLE	.267	Cy Falkenberg	CLE	5.5
Isolated Power			**Players Overall**			**Pitchers Overall**			**Defensive Runs**		
Joe Jackson	CLE	.178	Tris Speaker	BOS	68.9	Walter Johnson	WAS	89.5	Buck Weaver	CHI	37.1
Sam Crawford	DET	.172	Frank Baker	PHI	64.3	Eddie Cicotte	CHI	40.9	Tris Speaker	BOS	22.3
Tris Speaker	BOS	.169	Eddie Collins	PHI	60.9	Joe Boehling	WAS	31.2	Nap Lajoie	CLE	17.6

Club	W	L	R	OR	Avg	OBA	SLG	BPF	NOPS-A	BR	Adj	Wins	ERA	PPF	NERA-A	PR	Adj	Wins	Diff
PHI	96	57	794	592	.280	.356	.376	93	122/130	135	177	18.8	3.19	88	92 81	-38	-89	-9.6	10.2
WAS	90	64	596	561	.252	.317	.327	111	95/ 85	-33	-103	-11.0	2.73	111	107 119	31	81	8.7	15.4
CLE	86	66	633	536	.268	.331	.348	105	105/101	29	1	.1	2.52	103	117/120	64	77	8.2	1.7
BOS	79	71	631	610	.269	.336	.365	101	112/111	67	60	6.4	2.95	101	99/100	-2	1	.1	-2.5
CHI	78	74	488	498	.236	.299	.310	97	85/ 87	-94	-75	-8.1	2.33	96	126 121	91	75	8.0	2.1
DET	66	87	624	716	.265	.336	.355	99	109/111	56	63	6.7	3.41	101	86/ 87	-72	-68	-7.3	-9.9
NY	57	94	529	668	.237	.320	.293	100	85/ 85	-74	-75	-8.1	3.28	103	89/ 92	-50	-36	-3.9	-6.4
STL	57	96	528	642	.237	.305	.312	96	87/ 91	-81	-55	-6.0	3.06	98	96 94	-19	-29	-3.2	-10.3
			603		.256	.325	.336						2.93						

NATIONAL LEAGUE 1913

Batting Runs			Park Adjusted			Pitching Runs			Park Adjusted		
Gavvy Cravath	PHI	50.5	Gavvy Cravath	PHI	39.7	Christy Mathewson	NY	38.7	Tom Seaton	PHI	39.6
Vic Saier	CHI	27.5	Vic Saier	CHI	32.9	Babe Adams	PIT	36.5	Christy Mathewson	NY	32.3
Heinie Zimmerman	CHI	26.3	Heinie Zimmerman	CHI	30.9	Jeff Tesreau	NY	32.2	Pete Alexander	PHI	31.0
Normalized OPS			**Park Adjusted**			**Normalized ERA**			**Park Adjusted**		
Gavvy Cravath	PHI	176	Heinie Zimmerman	CHI	160	Christy Mathewson	NY	155	Ad Brennan	PHI	155
Heinie Zimmerman	CHI	147	Vic Saier	CHI	154	Babe Adams	PIT	149	Christy Mathewson	NY	146
Vic Saier	CHI	142	Gavvy Cravath	PHI	152	Jeff Tesreau	NY	147	Tom Seaton	PHI	143
On Base Average			**Slugging Percentage**			**Percent of Team Wins**			**Wins Above Team**		
Miller Huggins	STL	.432	Gavvy Cravath	PHI	.568	Slim Sallee	STL	.360	Slim Sallee	STL	8.9
Gavvy Cravath	PHI	.407	Heinie Zimmerman	CHI	.490	Tom Seaton	PHI	.307	Babe Adams	PIT	6.0
Jake Daubert	BRO	.405	Vic Saier	CHI	.480	Babe Adams	PIT	.269	Tom Seaton	PHI	5.8
Isolated Power			**Players Overall**			**Pitchers Overall**			**Defensive Runs**		
Gavvy Cravath	PHI	.227	Joe Tinker	CIN	32.5	Christy Mathewson	NY	37.1	Johnny Evers	CHI	23.6
Vic Saier	CHI	.191	Heinie Zimmerman	CHI	32.5	Babe Adams	PIT	36.2	George Cutshaw	BRO	22.3
Heinie Zimmerman	CHI	.177	Johnny Evers	CHI	31.5	Tom Seaton	PHI	34.4	Max Carey	PIT	22.3

Club	W	L	R	OR	Avg	OBA	SLG	BPF	NOPS-A	BR	Adj	Wins	ERA	PPF	NERA-A	PR	Adj	Wins	Diff
NY	101	51	684	515	.273	.338	.361	98	107 109	43	58	6.0	2.43	94	131 124	120	91	9.4	9.5
PHI	88	63	693	636	.265	.318	.382	116	108 93	29	-77	-8.2	3.15	116	101 117	7	89	9.3	11.4
CHI	88	65	720	625	.257	.335	.369	92	108 118	52	104	10.8	3.12	89	102 91	11	-41	-4.3	5.0
PIT	78	71	673	585	.263	.319	.356	95	100 105	-12	17	1.8	2.90	93	110 103	46	12	1.3	.4
BOS	69	82	641	690	.256	.326	.335	96	95 99	-25	-2	-.3	3.19	97	100 97	0	-11	-1.3	-4.9
BRO	65	84	595	613	.270	.321	.363	110	103 93	4	-58	-6.2	3.12	111	102 113	11	64	6.6	-10.0
CIN	64	89	607	717	.261	.325	.347	103	99 96	-9	-29	-3.2	3.46	106	92 98	-39	-12	-1.3	-8.0
STL	51	99	523	755	.247	.315	.316	93	86 93	-78	-34	-3.6	4.24	97	75 73	-155	-168	-17.6	-2.8
			642		.262	.325	.354						3.20						

Batting Runs / Park Adjusted / Pitching Runs / Park Adjusted

Batting Runs			Park Adjusted			Pitching Runs			Park Adjusted		
Tris Speaker	BOS	54.9	Tris Speaker	BOS	54.3	Dutch Leonard	BOS	44.4	Dutch Leonard	BOS	44.0
Eddie Collins	PHI	51.6	Eddie Collins	PHI	52.9	Walter Johnson	WAS	42.1	Walter Johnson	WAS	41.1
Ty Cobb	DET	42.3	Ty Cobb	DET	41.1	Rube Foster	BOS	24.5	Rube Foster	BOS	24.1

Normalized OPS			Park Adjusted			Normalized ERA			Park Adjusted		
Tris Speaker	BOS	178	Tris Speaker	BOS	176	Dutch Leonard	BOS	285	Dutch Leonard	BOS	283
Eddie Collins	PHI	172	Eddie Collins	PHI	175	Rube Foster	BOS	161	Rube Foster	BOS	160
Sam Crawford	DET	162	Sam Crawford	DET	157	Walter Johnson	WAS	159	Walter Johnson	WAS	158

On Base Average			Slugging Percentage			Percent of Team Wins			Wins Above Team		
Eddie Collins	PHI	.452	Tris Speaker	BOS	.503	Walter Johnson	WAS	.346	Ray Caldwell	NY	6.2
Tris Speaker	BOS	.423	Sam Crawford	DET	.483	Harry Coveleski	DET	.275	Dutch Leonard	BOS	5.6
Joe Jackson	CLE	.399	Joe Jackson	CLE	.464	Carl Weilman	STL	.254	Harry Coveleski	DET	5.4

Isolated Power			Players Overall			Pitchers Overall			Defensive Runs		
Sam Crawford	DET	.168	Tris Speaker	BOS	67.6	Walter Johnson	WAS	51.2	Donie Bush	DET	36.2
Tris Speaker	BOS	.165	Donie Bush	DET	55.8	Dutch Leonard	BOS	38.1	Chick Gandil	WAS	22.0
Tilly Walker	STL	.143	Eddie Collins	PHI	54.7	Eddie Cicotte	CHI	27.4	Tilly Walker	STL	21.7

Club	W	L	R	OR	Avg	OBA	SLG	BPF	NOPS-A	BR	Adj	Wins	ERA	PPF	NERA-A	PR	Adj	Wins	Diff
PHI	99	53	749	529	.272	.348	.352	98	118/120	113	124	13.7	2.78	93	99/ 92	-5	-35	-4.0	13.3
BOS	91	62	588	511	.250	.320	.338	101	105/104	25	20	2.2	2.37	99	116/115	59	56	6.2	6.2
WAS	81	73	572	519	.244	.313	.321	100	97/ 97	-16	-18	-2.2	2.54	99	108/107	31	27	3.0	3.2
DET	80	73	615	618	.258	.337	.344	103	112/109	76	59	6.5	2.86	103	96/ 99	-18	-5	-.6	-2.4
STL	71	82	523	614	.243	.306	.319	96	95/ 99	-38	-15	-1.8	2.85	98	96/ 94	-16	-26	-3.0	-.7
NY	70	84	538	550	.229	.315	.287	97	87/ 90	-59	-44	-4.9	2.82	97	97/ 95	-11	-23	-2.6	.6
CHI	70	84	487	560	.239	.302	.311	100	91/ 91	-57	-57	-6.4	2.48	102	110/112	40	46	5.1	-5.7
CLE	51	102	538	709	.245	.310	.312	103	94/ 92	-38	-53	-6.0	3.21	107	85/ 91	-73	-44	-5.0	-14.5
			576		.248	.319	.323						2.74						

NATIONAL LEAGUE   1914

Batting Runs			Park Adjusted			Pitching Runs			Park Adjusted		
Gavvy Cravath	PHI	42.7	Gavvy Cravath	PHI	46.6	Bill James	BOS	32.7	Bill James	BOS	38.1
Sherry Magee	PHI	40.7	Sherry Magee	PHI	44.6	Bill Doak	STL	30.2	Bill Doak	STL	36.9
George Burns	NY	34.0	George Burns	NY	32.4	Jeff Pfeffer	BRO	28.5	Slim Sallee	STL	28.6

Normalized OPS			Park Adjusted			Normalized ERA			Park Adjusted		
Gavvy Cravath	PHI	167	Gavvy Cravath	PHI	178	Bill Doak	STL	162	Bill Doak	STL	175
Sherry Magee	PHI	164	Sherry Magee	PHI	174	Bill James	BOS	147	Bill James	BOS	154
Casey Stengel	BRO	146	Beals Becker	PHI	151	Jeff Pfeffer	BRO	141	Slim Sallee	STL	143

On Base Average			Slugging Percentage			Percent of Team Wins			Wins Above Team		
Casey Stengel	BRO	.404	Sherry Magee	PHI	.509	Pete Alexander	PHI	.365	Pete Alexander	PHI	9.4
George Burns	NY	.403	Gavvy Cravath	PHI	.499	Jeff Tesreau	NY	.310	Jeff Tesreau	NY	8.3
Gavvy Cravath	PHI	.402	Zach Wheat	BRO	.452	Jeff Pfeffer	BRO	.307	Jeff Pfeffer	BRO	7.7

Isolated Power			Players Overall			Pitchers Overall			Defensive Runs		
Gavvy Cravath	PHI	.200	Rabbit Maranville	BOS	50.3	Bill James	BOS	39.7	Rabbit Maranville	BOS	58.5
Sherry Magee	PHI	.195	Gavvy Cravath	PHI	41.7	Bill Doak	STL	37.2	George Cutshaw	BRO	31.9
Vic Saier	CHI	.175	Sherry Magee	PHI	41.6	Slim Sallee	STL	28.4	Buck Herzog	CIN	21.4

Club	W	L	R	OR	Avg	OBA	SLG	BPF	NOPS-A	BR	Adj	Wins	ERA	PPF	NERA-A	PR	Adj	Wins	Diff
BOS	94	59	657	548	.251	.323	.335	107	102/ 96	17	-25	-2.8	2.74	105	102/107	7	29	3.2	17.1
NY	84	70	672	576	.265	.330	.348	102	109/106	53	39	4.2	2.95	100	95/ 95	-24	-22	-2.5	5.3
STL	81	72	558	540	.248	.314	.333	108	99/ 91	-8	-57	-6.2	2.38	108	117/127	63	100	10.8	-.1
CHI	78	76	605	638	.243	.317	.337	95	101/106	7	36	3.9	2.71	95	103/ 98	12	-8	-1.0	-2.0
BRO	75	79	622	618	.269	.323	.355	100	109/109	43	45	4.9	2.82	100	99/ 98	-4	-6	-.7	-6.1
PHI	74	80	651	687	.263	.329	.361	94	112/120	69	106	11.4	3.06	94	91/ 86	-41	-66	-7.2	-7.2
PIT	69	85	503	540	.233	.295	.303	91	84/ 91	-100	-49	-5.4	2.69	91	103/ 94	15	-23	-2.5	-.0
CIN	60	94	530	651	.236	.305	.300	105	86/ 82	-78	-104	-11.3	2.94	107	95/102	-23	8	.8	-6.5
			600		.251	.317	.334						2.78						

AMERICAN LEAGUE 1915

Batting Runs			Park Adjusted			Pitching Runs			Park Adjusted		
Ty Cobb	DET	71.6	Ty Cobb	DET	65.8	Walter Johnson	WAS	52.2	Walter Johnson	WAS	50.9
Eddie Collins	CHI	51.9	Eddie Collins	CHI	55.2	Ernie Shore	BOS	35.8	Ernie Shore	BOS	29.6
Jack Fournier	CHI	40.9	Jack Fournier	CHI	43.4	Jim Scott	CHI	29.8	Guy Morton	CLE	28.6
Normalized OPS			**Park Adjusted**			**Normalized ERA**			**Park Adjusted**		
Ty Cobb	DET	191	Jack Fournier	CHI	184	Joe Wood	BOS	197	Walter Johnson	WAS	188
Jack Fournier	CHI	175	Ty Cobb	DET	177	Walter Johnson	WAS	190	Joe Wood	BOS	182
Eddie Collins	CHI	168	Eddie Collins	CHI	176	Ernie Shore	BOS	179	Ernie Shore	BOS	166
On Base Average			**Slugging Percentage**			**Percent of Team Wins**			**Wins Above Team**		
Ty Cobb	DET	.486	Jack Fournier	CHI	.491	Walter Johnson	WAS	.318	Walter Johnson	WAS	6.5
Eddie Collins	CHI	.460	Ty Cobb	DET	.487	Carl Weilman	STL	.286	Ray Fisher	NY	6.0
Jack Fournier	CHI	.429	Sam Crawford	DET	.436	Guy Morton	CLE	.281	Guy Morton	CLE	5.5
Isolated Power			**Players Overall**			**Pitchers Overall**			**Defensive Runs**		
Jack Fournier	CHI	.168	Eddie Collins	CHI	69.5	Walter Johnson	WAS	59.3	Luke Boone	NY	19.6
Sam Crawford	DET	.137	Ty Cobb	DET	51.4	Ernie Shore	BOS	30.4	Ossie Vitt	DET	19.1
Amos Strunk	PHI	.130	Tris Speaker	BOS	40.5	Hooks Dauss	DET	28.3	Jack Lapp	PHI	15.9

Club	W	L	R	OR	Avg	OBA	SLG	BPF	NOPS-A	BR	Adj	Wins	ERA	PPF	NERA-A	PR	Adj	Wins	Diff
BOS	101	50	668	499	.260	.336	.339	96	108/112	46	68	7.2	2.39	92	123/114	86	51	5.4	12.8
DET	100	54	778	597	.268	.357	.358	108	120/111	131	82	8.7	2.86	105	103/108	13	35	3.7	10.6
CHI	93	61	717	509	.258	.345	.348	95	113/119	83	111	11.8	2.43	91	121/110	80	37	3.9	.3
WAS	85	68	569	491	.244	.312	.312	101	92/ 91	-51	-54	-5.9	2.31	99	128/126	99	93	9.9	4.5
NY	69	83	584	588	.233	.317	.305	98	91/ 93	-45	-36	-4.0	3.08	98	95/ 94	-21	-28	-3.1	.0
STL	63	91	521	679	.246	.315	.315	97	93/ 96	-43	-25	-2.7	3.07	100	96/ 96	-19	-20	-2.2	-9.1
CLE	57	95	539	670	.240	.312	.317	106	93/ 88	-42	-80	-8.6	3.13	109	94/103	-28	12	1.3	-11.7
PHI	43	109	545	888	.237	.304	.311	96	89/ 93	-73	-47	-5.1	4.33	103	68/ 70	-206	-193	-20.7	-7.3
				615		.248	.325	.326						2.94					

NATIONAL LEAGUE 1915

Batting Runs			Park Adjusted			Pitching Runs			Park Adjusted		
Gavvy Cravath	PHI	46.8	Gavvy Cravath	PHI	42.1	Pete Alexander	PHI	63.7	Pete Alexander	PHI	69.5
Fred Luderus	PHI	30.3	Larry Doyle	NY	32.5	Fred Toney	CIN	29.0	Fred Toney	CIN	33.4
Bill Hinchman	PIT	29.7	Bill Hinchman	PIT	31.7	Jeff Pfeffer	BRO	21.0	Tom Hughes	BOS	19.7
Normalized OPS			**Park Adjusted**			**Normalized ERA**			**Park Adjusted**		
Gavvy Cravath	PHI	172	Gavvy Cravath	PHI	160	Pete Alexander	PHI	225	Pete Alexander	PHI	236
Fred Luderus	PHI	151	Larry Doyle	NY	154	Fred Toney	CIN	174	Fred Toney	CIN	186
Bill Hinchman	PIT	144	Bill Hinchman	PIT	148	Al Mamaux	PIT	135	Tom Hughes	BOS	130
On Base Average			**Slugging Percentage**			**Percent of Team Wins**			**Wins Above Team**		
Gavvy Cravath	PHI	.393	Gavvy Cravath	PHI	.510	Pete Alexander	PHI	.344	Pete Alexander	PHI	9.2
Fred Luderus	PHI	.376	Fred Luderus	PHI	.457	Al Mamaux	PIT	.288	Al Mamaux	PIT	8.9
Bill Hinchman	PIT	.368	Tom Long	STL	.446	Jeff Tesreau	NY	.275	Fred Toney	CIN	7.5
Isolated Power			**Players Overall**			**Pitchers Overall**			**Defensive Runs**		
Gavvy Cravath	PHI	.224	Gavvy Cravath	PHI	39.2	Pete Alexander	PHI	73.9	Buck Herzog	CIN	36.4
Vic Saier	CHI	.181	Buck Herzog	CIN	32.7	Fred Toney	CIN	29.0	Art Fletcher	NY	31.4
Tom Long	STL	.152	Heinie Groh	CIN	32.0	Erskine Mayer	PHI	21.3	George Cutshaw	BRO	20.5

Club	W	L	R	OR	Avg	OBA	SLG	BPF	NOPS-A	BR	Adj	Wins	ERA	PPF	NERA-A	PR	Adj	Wins	Diff
PHI	90	62	589	463	.247	.316	.340	108	106/ 98	31	-9	-1.1	2.17	105	126/133	87	109	12.1	3.0
BOS	83	69	582	545	.240	.321	.319	101	100/ 99	15	9	1.0	2.58	100	107/107	26	28	3.1	2.9
BRO	80	72	536	560	.248	.295	.317	97	92/ 95	-55	-37	-4.2	2.65	97	104/100	14	1	.1	8.1
CHI	73	80	570	620	.244	.303	.342	103	102/ 99	4	-15	-1.8	3.11	105	88/ 93	-56	-35	-4.0	2.3
PIT	73	81	557	520	.246	.309	.334	97	102/105	8	25	2.8	2.59	96	106/101	23	5	.6	-7.4
STL	72	81	590	601	.254	.320	.333	100	105/105	30	32	3.5	2.89	100	95/ 95	-21	-21	-2.4	-5.6
CIN	71	83	516	585	.253	.308	.331	105	100/ 96	-4	-31	-3.6	2.84	106	97/103	-14	13	1.5	-3.9
NY	69	83	582	628	.251	.300	.329	91	97/106	-26	22	2.4	3.11	92	88/ 81	-56	-90	-10.1	.7
				565		.248	.309	.331						2.74					

FEDERAL LEAGUE 1914

Batting Runs			Park Adjusted			Pitching Runs			Park Adjusted		
Benny Kauff	IND	59.1	Benny Kauff	IND	51.6	Claude Hendrix	CHI	60.9	Cy Falkenberg	IND	52.4
Steve Evans	BRO	47.6	Steve Evans	BRO	47.7	Cy Falkenberg	IND	41.2	Claude Hendrix	CHI	51.9
Ed Lennox	PIT	33.9	Ed Lennox	PIT	37.3	Russ Ford	BUF	38.0	Russ Ford	BUF	39.8

Normalized OPS			Park Adjusted			Normalized ERA			Park Adjusted		
Benny Kauff	IND	178	Steve Evans	BRO	174	Claude Hendrix	CHI	190	Russ Ford	BUF	180
Steve Evans	BRO	174	Ed Lennox	PIT	168	Russ Ford	BUF	176	Claude Hendrix	CHI	176
Ed Lennox	PIT	158	Benny Kauff	IND	162	Doc Watson	STL	157	Cy Falkenberg	IND	156

On Base Average			Slugging Percentage			Percent of Team Wins			Wins Above Team		
Benny Kauff	IND	.440	Steve Evans	BRO	.556	Claude Hendrix	CHI	.333	Claude Hendrix	CHI	9.3
Ed Lennox	PIT	.409	Benny Kauff	IND	.534	Russ Ford	BRO	.325	Russ Ford	BUF	8.2
Steve Evans	BRO	.406	Duke Kenworthy	KC	.525	Elmer Knetzer	PIT	.313	Elmer Knetzer	PIT	8.1

Isolated Power			Players Overall			Pitchers Overall			Defensive Runs		
Steve Evans	BRO	.208	Duke Kenworthy	KC	62.4	Claude Hendrix	CHI	61.9	Mickey Doolan	BAL	28.0
Duke Kenworthy	KC	.207	Benny Kauff	IND	48.1	Cy Falkenberg	IND	49.7	Duke Kenworthy	KC	24.1
Ed Lennox	PIT	.181	Steve Evans	BRO	42.8	Russ Ford	BUF	37.8	Bill McKechnie	IND	23.0

Club	W	L	R	OR	Avg	OBA	SLG	BPF	NOPS-A	BR	Adj	Wins	ERA	PPF	NERA-A	PR	Adj	Wins	Diff
IND	88	65	762	622	.285	.344	.383	110	116/105	94	29	3.0	3.06	108	105/113	23	64	6.7	1.8
CHI	87	67	621	517	.258	.326	.352	95	101/106	6	35	3.6	2.44	93	131/122	121	85	8.9	-2.5
BAL	84	70	645	628	.268	.332	.357	96	104/108	24	47	4.9	3.13	96	102/98	12	-8	-.9	3.1
BUF	80	71	620	602	.250	.308	.336	102	91/89	-61	-76	-8.0	3.16	102	101/104	7	17	1.8	10.7
BRO	77	77	662	677	.269	.321	.368	100	105/105	15	16	1.7	3.33	100	96/96	-19	-17	-1.9	.3
KC	67	84	644	683	.267	.320	.364	95	103/109	6	41	4.2	3.40	95	94/89	-29	-54	-5.7	-7.0
PIT	64	86	605	698	.262	.316	.352	94	98/104	-19	17	1.8	3.56	95	90/86	-54	-76	-8.0	-4.8
STL	62	89	565	697	.247	.315	.326	108	90/83	-61	-113	-11.9	3.59	111	89/100	-57	-1	-.2	-1.4
			641		.263	.323	.355						3.20						

FEDERAL LEAGUE 1915

Batting Runs			Park Adjusted			Pitching Runs			Park Adjusted		
Benny Kauff	BRO	51.6	Benny Kauff	BRO	53.4	Dave Davenport	STL	36.4	Eddie Plank	STL	40.3
Ed Konetchy	PIT	31.7	Dutch Zwilling	CHI	28.4	Earl Moseley	NEW	33.3	Dave Davenport	STL	31.0
Dutch Zwilling	CHI	25.7	Ed Konetchy	PIT	28.2	Eddie Plank	STL	28.3	Earl Moseley	NEW	29.6

Normalized OPS			Park Adjusted			Normalized ERA			Park Adjusted		
Benny Kauff	BRO	180	Benny Kauff	BRO	186	Earl Moseley	NEW	158	Earl Moseley	NEW	152
Ed Konetchy	PIT	148	Dutch Zwilling	CHI	144	Eddie Plank	STL	146	Eddie Plank	STL	150
Dutch Zwilling	CHI	138	Ed Konetchy	PIT	141	Three Finger Brown	CHI	145	Dave Davenport	STL	142

On Base Average			Slugging Percentage			Percent of Team Wins			Wins Above Team		
Benny Kauff	BRO	.440	Benny Kauff	BRO	.509	George McConnell	CHI	.291	George McConnell	CHI	6.8
Ward Miller	STL	.395	Ed Konetchy	PIT	.483	Albert Schulz	BUF	.284	Ed Reulbach	NEW	5.9
Babe Borton	STL	.388	Hal Chase	BUF	.471	Nick Cullop	KC	.272	Nick Cullop	KC	5.8

Isolated Power			Players Overall			Pitchers Overall			Defensive Runs		
Hal Chase	BUF	.180	Benny Kauff	BRO	59.1	Doc Crandall	STL	33.7	Mickey Doolan	CHI	24.9
Ed Konetchy	PIT	.168	Claude Cooper	BRO	36.4	Eddie Plank	STL	32.5	Bill Rariden	NEW	20.5
Benny Kauff	BRO	.168	Bill Rariden	NEW	36.1	George McConnell	CHI	30.1	Ernie Johnson	STL	20.2

Club	W	L	R	OR	Avg	OBA	SLG	BPF	NOPS-A	BR	Adj	Wins	ERA	PPF	NERA-A	PR	Adj	Wins	Diff
CHI	86	66	640	538	.257	.316	.352	96	104/109	19	44	4.7	2.64	93	115/107	61	29	3.2	2.1
STL	87	67	634	527	.261	.336	.345	105	108/103	56	25	2.7	2.73	103	111/114	48	62	6.7	.7
PIT	86	67	592	524	.262	.322	.341	105	102/97	13	-17	-2.0	2.79	104	109/113	37	56	6.0	5.5
KC	81	72	547	551	.244	.297	.329	93	91/98	-61	-19	-2.2	2.82	92	107/99	32	-3	-.4	7.1
NEW	80	72	585	562	.252	.311	.334	97	97/101	-21	-0	-.1	2.61	96	116/112	67	47	5.1	-.9
BUF	74	78	574	634	.249	.306	.338	100	97/97	-26	-27	-3.0	3.38	102	90/91	-52	-45	-5.0	6.0
BRO	70	82	647	673	.268	.331	.360	97	111/114	62	80	8.6	3.37	98	90/88	-50	-61	-6.6	-8.0
BAL	47	107	550	760	.244	.308	.325	106	93/88	-39	-77	-8.4	3.95	111	77/85	-139	-86	-9.4	-12.2
			596		.255	.316	.340						3.03						

AMERICAN LEAGUE 1916

Batting Runs			Park Adjusted			Pitching Runs			Park Adjusted		
Tris Speaker	CLE	64.6	Tris Speaker	CLE	64.5	Babe Ruth	BOS	38.7	Harry Coveleski	DET	36.8
Ty Cobb	DET	57.5	Ty Cobb	DET	53.0	Walter Johnson	WAS	38.4	Walter Johnson	WAS	32.0
Joe Jackson	CHI	44.9	Joe Jackson	CHI	40.2	Harry Coveleski	DET	30.7	Babe Ruth	BOS	30.5
Normalized OPS			**Park Adjusted**			**Normalized ERA**			**Park Adjusted**		
Tris Speaker	CLE	192	Tris Speaker	CLE	192	Babe Ruth	BOS	161	Eddie Cicotte	CHI	167
Ty Cobb	DET	184	Ty Cobb	DET	171	Eddie Cicotte	CHI	159	Harry Coveleski	DET	152
Joe Jackson	CHI	167	Joe Jackson	CHI	155	Walter Johnson	WAS	149	Babe Ruth	BOS	148
On Base Average			**Slugging Percentage**			**Percent of Team Wins**			**Wins Above Team**		
Tris Speaker	CLE	.470	Tris Speaker	CLE	.502	Joe Bush	PHI	.417	Joe Bush	PHI	7.8
Ty Cobb	DET	.452	Joe Jackson	CHI	.495	Elmer Myers	PHI	.389	Elmer Myers	PHI	7.0
Eddie Collins	CHI	.405	Ty Cobb	DET	.493	Walter Johnson	WAS	.329	Bob Shawkey	NY	5.7
Isolated Power			**Players Overall**			**Pitchers Overall**			**Defensive Runs**		
Wally Pipp	NY	.154	Tris Speaker	CLE	59.3	Babe Ruth	BOS	44.4	Ossie Vitt	DET	32.4
Joe Jackson	CHI	.154	Ty Cobb	DET	53.1	Harry Coveleski	DET	42.0	Del Pratt	STL	29.5
Jack Graney	CLE	.143	Del Pratt	STL	48.4	Walter Johnson	WAS	35.3	Doc Lavan	STL	24.0

Club	W	L	R	OR	Avg	OBA	SLG	BPF	NOPS-A	BR	Adj	Wins	ERA	PPF	NERA-A	PR	Adj	Wins	Diff
BOS	91	63	550	480	.248	.317	.318	94	97/103	-19	12	1.4	2.47	92	114/105	55	20	2.2	10.5
CHI	89	65	601	497	.251	.319	.339	107	104/ 97	18	-23	-2.6	2.37	105	119/126	72	96	10.7	4.0
DET	87	67	670	595	.264	.337	.350	107	113/106	78	37	4.1	2.97	106	95/101	-22	4	.4	5.5
NY	80	74	577	561	.246	.318	.326	101	100/ 99	-3	-8	-1.0	2.77	101	102/103	8	11	1.2	2.8
STL	79	75	588	545	.245	.331	.307	94	98/105	2	39	4.4	2.58	92	109/101	39	3	.3	-2.7
CLE	77	77	630	602	.250	.324	.331	100	103/103	17	16	1.8	2.98	100	95/ 94	-24	-26	-3.0	1.2
WAS	76	77	536	543	.242	.320	.306	95	94/ 99	-29	-0	-.1	2.66	95	106/100	26	1	.1	-.5
PHI	36	117	447	776	.242	.303	.313	100	91/ 91	-60	-61	-6.9	3.84	108	73/ 79	-151	-118	-13.2	-20.4
			575		.249	.321	.324						2.82						

NATIONAL LEAGUE 1916

Batting Runs			Park Adjusted			Pitching Runs			Park Adjusted		
Zach Wheat	BRO	34.5	Zach Wheat	BRO	33.2	Pete Alexander	PHI	45.9	Pete Alexander	PHI	38.8
Bill Hinchman	PIT	31.1	Rogers Hornsby	STL	31.8	Ferdie Schupp	NY	26.6	Hippo Vaughn	CHI	28.1
Hal Chase	CIN	30.2	Gavvy Cravath	PHI	30.6	Jeff Pfeffer	BRO	25.5	Jeff Pfeffer	BRO	25.0
Normalized OPS			**Park Adjusted**			**Normalized ERA**			**Park Adjusted**		
Cy Williams	CHI	154	Rogers Hornsby	STL	160	Pete Alexander	PHI	169	Rube Marquard	BRO	164
Zach Wheat	BRO	153	Gavvy Cravath	PHI	157	Rube Marquard	BRO	165	Pete Alexander	PHI	158
Hal Chase	CIN	152	Hal Chase	CIN	150	Eppa Rixey	PHI	141	Wilbur Cooper	PIT	146
On Base Average			**Slugging Percentage**			**Percent of Team Wins**			**Wins Above Team**		
Gavvy Cravath	PHI	.379	Zach Wheat	BRO	.461	Pete Alexander	PHI	.363	Pete Alexander	PHI	8.8
Bill Hinchman	PIT	.378	Hal Chase	CIN	.459	Al Mamaux	PIT	.323	Al Mamaux	PIT	7.6
Cy Williams	CHI	.372	Cy Williams	CHI	.459	Jeff Pfeffer	BRO	.266	Tom Hughes	BOS	5.6
Isolated Power			**Players Overall**			**Pitchers Overall**			**Defensive Runs**		
Cy Williams	CHI	.180	Art Fletcher	NY	42.7	Pete Alexander	PHI	48.2	Rabbit Maranville	BOS	30.8
Gavvy Cravath	PHI	.156	Zach Wheat	BRO	36.6	Jeff Pfeffer	BRO	28.4	Dave Bancroft	PHI	26.7
Zach Wheat	BRO	.150	Heinie Groh	CIN	35.4	Hippo Vaughn	CHI	23.9	Max Carey	PIT	24.2

Club	W	L	R	OR	Avg	OBA	SLG	BPF	NOPS-A	BR	Adj	Wins	ERA	PPF	NERA-A	PR	Adj	Wins	Diff
BRO	94	60	585	471	.261	.313	.345	102	109/107	46	35	4.0	2.12	100	123/123	78	76	8.7	4.3
PHI	91	62	581	489	.250	.310	.341	96	107/111	34	53	6.1	2.36	94	111/104	40	14	1.6	6.8
BOS	89	63	542	453	.233	.299	.307	98	92/ 94	-37	-27	-3.2	2.19	96	119/114	66	49	5.6	10.7
NY	86	66	597	504	.253	.307	.343	94	107/113	31	63	7.2	2.60	91	101/ 92	2	-33	-3.8	6.7
CHI	67	86	520	541	.239	.298	.325	116	98/ 84	-17	-102	-11.7	2.65	117	99/116	-5	65	7.4	-5.2
PIT	65	89	484	586	.240	.298	.316	101	95/ 93	-31	-39	-4.6	2.76	104	95/ 98	-23	-6	-.8	-6.6
STL	60	93	476	629	.243	.295	.318	93	95/102	-34	1	.1	3.14	96	83/ 80	-79	-93	-10.8	-5.8
CIN	60	93	505	617	.254	.307	.331	101	103/102	13	8	.9	3.10	104	84/ 87	-75	-61	-7.0	-10.4
			536		.247	.303	.328						2.61						

AMERICAN LEAGUE   1917

Batting Runs			Park Adjusted			Pitching Runs			Park Adjusted		
Ty Cobb	DET	74.8	Ty Cobb	DET	71.4	Eddie Cicotte	CHI	43.7	Stan Coveleski	CLE	42.8
Tris Speaker	CLE	51.2	Tris Speaker	CLE	41.2	Carl Mays	BOS	29.5	Jim Bagby	CLE	40.8
Bobby Veach	DET	40.0	George Sisler	STL	37.3	Stan Coveleski	CLE	28.2	Carl Mays	BOS	35.7

Normalized OPS			Park Adjusted			Normalized ERA			Park Adjusted		
Ty Cobb	DET	210	Ty Cobb	DET	199	Eddie Cicotte	CHI	174	Stan Coveleski	CLE	171
Tris Speaker	CLE	180	George Sisler	STL	166	Carl Mays	BOS	153	Carl Mays	BOS	164
Bobby Veach	DET	159	Joe Jackson	CHI	157	Stan Coveleski	CLE	147	Jim Bagby	CLE	158

On Base Average			Slugging Percentage			Percent of Team Wins			Wins Above Team		
Ty Cobb	DET	.444	Ty Cobb	DET	.571	Walter Johnson	WAS	.311	Dave Davenport	STL	5.7
Tris Speaker	CLE	.432	Tris Speaker	CLE	.486	Dave Davenport	STL	.298	Walter Johnson	WAS	5.6
Bobby Veach	DET	.393	Bobby Veach	DET	.457	Eddie Cicotte	CHI	.280	Carl Mays	BOS	4.6

Isolated Power			Players Overall			Pitchers Overall			Defensive Runs		
Ty Cobb	DET	.189	Ty Cobb	DET	66.1	Carl Mays	BOS	48.7	Ray Chapman	CLE	28.3
Bobby Veach	DET	.138	Ray Chapman	CLE	48.7	Babe Ruth	BOS	47.5	Bill Wambsganss	CLE	23.3
Wally Pipp	NY	.136	Joe Jackson	CHI	35.4	Jim Bagby	CLE	40.3	Everett Scott	BOS	20.7

Club	W	L	R	OR	Avg	OBA	SLG	BPF	NOPS-A	BR	Adj	Wins	ERA	PPF	NERA-A	PR	Adj	Wins	Diff
CHI	100	54	656	464	.253	.329	.326	93	105/114	36	78	8.7	2.16	87	123/108	80	27	3.0	11.3
BOS	90	62	555	454	.246	.314	.319	109	99/ 90	-9	-60	-6.7	2.20	107	121/130	73	104	11.5	9.2
CLE	88	66	584	543	.245	.324	.322	116	103/ 88	21	-71	-8.1	2.52	117	106/123	22	91	10.1	8.9
DET	78	75	639	577	.259	.328	.344	105	111/105	61	32	3.6	2.56	104	104/108	16	32	3.6	-5.6
WAS	74	79	543	566	.241	.313	.304	90	94/104	-33	25	2.8	2.77	89	96/ 86	-16	-61	-6.9	1.6
NY	71	82	524	558	.239	.310	.308	107	94/ 88	-33	-73	-8.2	2.66	108	100/108	0	34	3.8	-1.1
STL	57	97	510	687	.246	.305	.315	94	95/100	-38	-6	-.7	3.20	98	83/ 82	-82	-89	-10.1	-9.2
PHI	55	98	529	691	.254	.316	.323	91	100/110	-0	49	5.4	3.27	94	82/ 77	-90	-114	-12.7	-14.2
			568		.248	.318	.320						2.66						

NATIONAL LEAGUE   1917

Batting Runs			Park Adjusted			Pitching Runs			Park Adjusted		
Rogers Hornsby	STL	39.6	Heinie Groh	CIN	38.7	Pete Alexander	PHI	36.6	Pete Alexander	PHI	43.1
Gavvy Cravath	PHI	34.6	Rogers Hornsby	STL	38.3	Hippo Vaughn	CHI	23.0	Hippo Vaughn	CHI	28.7
Heinie Groh	CIN	33.5	Edd Roush	CIN	36.6	Ferdie Schupp	NY	22.7	Fred Anderson	NY	19.1

Normalized OPS			Park Adjusted			Normalized ERA			Park Adjusted		
Rogers Hornsby	STL	165	Edd Roush	CIN	168	Fred Anderson	NY	187	Fred Anderson	NY	174
Gavvy Cravath	PHI	157	Rogers Hornsby	STL	162	Pete Alexander	PHI	146	Pete Alexander	PHI	154
Edd Roush	CIN	155	Heinie Groh	CIN	156	Pol Perritt	NY	144	Hippo Vaughn	CHI	143

On Base Average			Slugging Percentage			Percent of Team Wins			Wins Above Team		
Heinie Groh	CIN	.385	Rogers Hornsby	STL	.484	Pete Alexander	PHI	.345	Wilbur Cooper	PIT	9.4
Rogers Hornsby	STL	.385	Gavvy Cravath	PHI	.473	Wilbur Cooper	PIT	.333	Pete Alexander	PHI	7.5
George Burns	NY	.380	Edd Roush	CIN	.454	Hippo Vaughn	CHI	.311	Hippo Vaughn	CHI	7.4

Isolated Power			Players Overall			Pitchers Overall			Defensive Runs		
Gavvy Cravath	PHI	.193	Rogers Hornsby	STL	60.0	Pete Alexander	PHI	48.6	Art Fletcher	NY	29.8
Rogers Hornsby	STL	.157	Heinie Groh	CIN	44.5	Hippo Vaughn	CHI	29.8	Dave Bancroft	PHI	26.8
Dave Robertson	NY	.132	Max Carey	PIT	39.5	Eppa Rixey	PHI	20.2	Rabbit Maranville	BOS	24.0

Club	W	L	R	OR	Avg	OBA	SLG	BPF	NOPS-A	BR	Adj	Wins	ERA	PPF	NERA-A	PR	Adj	Wins	Diff
NY	98	56	635	457	.261	.317	.343	97	109/113	50	67	7.6	2.27	93	119/110	69	38	4.2	9.2
PHI	87	65	578	500	.248	.310	.339	107	106/ 99	29	-8	-1.0	2.46	106	110/116	37	61	6.9	5.1
STL	82	70	531	567	.250	.303	.333	102	102/ 99	2	-9	-1.1	3.04	103	89/ 92	-51	-38	-4.4	11.5
CIN	78	76	601	611	.264	.309	.354	92	110/120	48	91	10.3	2.70	92	100/ 92	1	-31	-3.7	-5.7
CHI	74	80	552	567	.239	.299	.313	106	94/ 89	-34	-66	-7.5	2.62	106	103/110	14	41	4.6	-.1
BOS	72	81	536	552	.246	.309	.320	98	99/101	-1	11	1.2	2.77	98	98/ 96	-9	-18	-2.1	-3.6
BRO	70	81	511	559	.247	.296	.322	100	96/ 96	-31	-32	-3.7	2.78	101	97/ 98	-11	-6	-.8	-1.0
PIT	51	103	464	595	.238	.298	.298	99	89/ 90	-60	-53	-6.1	3.01	102	90/ 91	-47	-40	-4.6	-15.2
			551		.249	.305	.328						2.70						

AMERICAN LEAGUE 1918

Batting Runs

Ty Cobb	DET	44.0
Babe Ruth	BOS	34.8
George Burns	PHI	32.8

Normalized OPS

Babe Ruth	BOS	191
Ty Cobb	DET	188
George Burns	PHI	159

On Base Average

Ty Cobb	DET	.440
Babe Ruth	BOS	.410
Eddie Collins	CHI	.407

Isolated Power

Babe Ruth	BOS	.256
Ty Cobb	DET	.133
Tilly Walker	PHI	.128

Park Adjusted

Ty Cobb	DET	44.7
Babe Ruth	BOS	40.0
George Burns	PHI	33.5

Park Adjusted

Babe Ruth	BOS	222
Ty Cobb	DET	191
Harry Hooper	BOS	162

Slugging Percentage

Babe Ruth	BOS	.555
Ty Cobb	DET	.515
George Burns	PHI	.467

Players Overall

Ty Cobb	DET	36.0
Babe Ruth	BOS	35.0
Frank Baker	NY	31.5

Pitching Runs

Walter Johnson	WAS	53.8
Stan Coveleski	CLE	32.5
Scott Perry	PHI	29.0

Normalized ERA

Walter Johnson	WAS	217
Stan Coveleski	CLE	152
Allen Sothoron	STL	143

Percent of Team Wins

Scott Perry	PHI	.385
Walter Johnson	WAS	.319
Stan Coveleski	CLE	.301

Pitchers Overall

Walter Johnson	WAS	68.5
Stan Coveleski	CLE	34.8
Scott Perry	PHI	26.8

Park Adjusted

Walter Johnson	WAS	61.9
Stan Coveleski	CLE	37.4
Scott Perry	PHI	31.3

Park Adjusted

Walter Johnson	WAS	235
Stan Coveleski	CLE	159
Allen Sothoron	STL	150

Wins Above Team

Scott Perry	PHI	6.0
Bernie Boland	DET	4.4
Sam Jones	BOS	4.2

Defensive Runs

Roger Peckinpaugh	NY	27.1
Everett Scott	BOS	20.4
Joe Gedeon	STL	19.3

Club	W	L	R	OR	Avg	OBA	SLG	BPF	NOPS-A	BR	Adj	Wins	ERA	PPF	NERA-A	PR	Adj	Wins	Diff
BOS	75	51	474	380	.249	.322	.327	86	101/117	5	66	7.4	2.30	82	120/ 99	57	-3	-.4	5.0
CLE	73	54	504	447	.260	.344	.341	106	112/106	68	39	4.4	2.64	105	105/110	17	35	3.9	1.3
WAS	72	56	461	412	.256	.318	.316	109	96/ 88	-20	-64	-7.2	2.14	108	129/140	85	115	12.9	2.3
NY	60	63	493	475	.257	.331	.320	97	101/104	2	14	1.6	3.00	97	92/ 89	-29	-41	-4.6	1.6
STL	58	64	426	448	.259	.331	.320	104	102/ 97	11	-7	-.9	2.75	105	100/106	1	19	2.1	-4.2
CHI	57	67	457	446	.256	.322	.321	101	99/ 98	-4	-8	-1.0	2.68	101	103/104	10	12	1.3	-5.3
DET	55	71	476	557	.249	.325	.318	99	99/100	-1	5	.5	3.40	101	81/ 82	-81	-78	-8.8	.3
PHI	52	76	412	538	.243	.303	.308	99	89/ 90	-58	-53	-6.0	3.22	102	86/ 88	-57	-49	-5.5	-.5
			463		.254	.323	.323						2.76						

NATIONAL LEAGUE 1918

Batting Runs

Heinie Groh	CIN	26.4
Edd Roush	CIN	24.3
Charlie Hollocher	CHI	23.0

Normalized OPS

Edd Roush	CIN	151
Heinie Groh	CIN	141
Jake Daubert	BRO	140

On Base Average

Heinie Groh	CIN	.395
Charlie Hollocher	CHI	.379
Red Smith	BOS	.373

Isolated Power

Gavvy Cravath	PHI	.143
Albert Wickland	BOS	.136
Rogers Hornsby	STL	.135

Park Adjusted

Heinie Groh	CIN	30.0
Edd Roush	CIN	27.2
Sherry Magee	CIN	22.0

Park Adjusted

Edd Roush	CIN	161
Heinie Groh	CIN	151
Sherry Magee	CIN	149

Slugging Percentage

Edd Roush	CIN	.455
Jake Daubert	BRO	.429
Rogers Hornsby	STL	.416

Players Overall

Heinie Groh	CIN	34.8
Lee Magee	CIN	25.0
Art Fletcher	NY	24.9

Pitching Runs

Hippo Vaughn	CHI	33.1
Lefty Tyler	CHI	22.6
Wilbur Cooper	PIT	19.8

Normalized ERA

Hippo Vaughn	CHI	159
Lefty Tyler	CHI	138
Wilbur Cooper	PIT	131

Percent of Team Wins

Burleigh Grimes	BRO	.333
Wilbur Cooper	PIT	.292
Art Nehf	BOS	.283

Pitchers Overall

Hippo Vaughn	CHI	40.4
Lefty Tyler	CHI	31.2
Wilbur Cooper	PIT	24.4

Park Adjusted

Hippo Vaughn	CHI	37.7
Lefty Tyler	CHI	26.9
Wilbur Cooper	PIT	23.7

Park Adjusted

Hippo Vaughn	CHI	167
Lefty Tyler	CHI	145
Phil Douglas	CHI	137

Wins Above Team

Burleigh Grimes	BRO	8.1
Gene Packard	STL	3.1
Claude Hendrix	CHI	3.1

Defensive Runs

Art Fletcher	NY	27.3
Dave Bancroft	PHI	24.5
Max Carey	PIT	15.4

Club	W	L	R	OR	Avg	OBA	SLG	BPF	NOPS-A	BR	Adj	Wins	ERA	PPF	NERA-A	PR	Adj	Wins	Diff
CHI	84	45	538	393	.265	.325	.342	109	109/100	47	5	.6	2.18	105	127/133	78	97	10.8	8.1
NY	71	53	480	415	.260	.310	.330	100	101/101	1	0	-.0	2.64	98	105/103	15	10	1.1	7.9
CIN	68	60	530	496	.278	.330	.366	94	119/127	86	115	12.8	3.00	92	92/ 85	-29	-56	-6.3	-2.5
PIT	65	60	466	412	.248	.315	.321	106	99/ 94	1	-24	-2.8	2.48	105	111/117	36	52	5.8	-.5
BRO	57	69	360	463	.250	.291	.315	97	90/ 93	-52	-39	-4.5	2.81	99	98/ 98	-5	-7	-.9	-.6
PHI	55	68	430	507	.244	.305	.313	109	94/ 86	-28	-69	-7.8	3.15	112	88/ 98	-48	-7	-.8	2.1
BOS	53	71	424	469	.244	.307	.307	97	92/ 96	-31	-15	-1.8	2.90	98	95/ 93	-16	-24	-2.8	-4.4
STL	51	78	454	527	.244	.301	.325	95	97/102	-20	5	.5	2.97	96	93/ 90	-26	-40	-4.6	-9.5
			460		.254	.311	.328						2.76						

AMERICAN LEAGUE 1919

Batting Runs
Babe Ruth BOS 66.5
Joe Jackson CHI 41.7
Ty Cobb DET 41.6

Park Adjusted
Babe Ruth BOS 68.7
Ty Cobb DET 47.6
Bobby Veach DET 44.1

Pitching Runs
Walter Johnson WAS 55.8
Eddie Cicotte CHI 47.9
Allen Sothoron STL 30.6

Park Adjusted
Walter Johnson WAS 57.2
Eddie Cicotte CHI 53.7
Claude Williams CHI 24.9

Normalized OPS
Babe Ruth BOS 211
Ty Cobb DET 164
Joe Jackson CHI 160

Park Adjusted
Babe Ruth BOS 219
Ty Cobb DET 183
Bobby Veach DET 173

Normalized ERA
Walter Johnson WAS 216
Eddie Cicotte CHI 177
Carl Weilman STL 156

Park Adjusted
Walter Johnson WAS 219
Eddie Cicotte CHI 187
Carl Weilman STL 145

On Base Average
Babe Ruth BOS .456
Ty Cobb DET .429
Joe Jackson CHI .422

Slugging Percentage
Babe Ruth BOS .657
George Sisler STL .530
Bobby Veach DET .519

Percent of Team Wins
Walter Johnson WAS .357
Eddie Cicotte CHI .330
Jim Shaw WAS .304

Wins Above Team
Eddie Cicotte CHI 8.6
Walter Johnson WAS 8.6
Herb Pennock BOS 5.4

Isolated Power
Babe Ruth BOS .336
George Sisler STL .178
Bobby Veach DET .164

Players Overall
Babe Ruth BOS 76.2
Roger Peckinpaugh NY 50.5
Del Pratt NY 47.9

Pitchers Overall
Walter Johnson WAS 59.0
Eddie Cicotte CHI 50.0
Stan Coveleski CLE 30.5

Defensive Runs
Del Pratt NY 33.1
Roger Peckinpaugh NY 25.9
Happy Felsch CHI 20.0

Club	W	L	R	OR	Avg	OBA	SLG	BPF	NOPS-A	BR	Adj	Wins	ERA	PPF	NERA-A	PR	Adj	Wins	Diff
CHI	88	52	667	534	.287	.351	.380	108	112/104	69	25	2.6	3.04	105	106/112	26	50	5.3	10.1
CLE	84	55	636	537	.278	.354	.381	107	113/106	78	40	4.2	2.94	105	109/115	39	61	6.4	4.0
NY	80	59	578	506	.267	.326	.356	102	98/96	-19	-32	-3.5	2.82	101	114/115	57	61	6.3	7.6
DET	80	60	618	578	.283	.346	.381	90	11/123	60	118	12.3	3.30	88	98/86	-10	-63	-6.7	4.4
STL	67	72	533	567	.264	.326	.355	93	97/105	-21	18	1.9	3.13	93	103/96	12	-19	-2.1	-2.4
BOS	66	71	564	552	.261	.336	.344	96	97/101	-10	10	1.1	3.30	96	98/93	-10	-28	-3.1	-.5
WAS	56	84	533	570	.260	.325	.339	101	92/92	-46	-49	-5.2	3.01	101	107/109	30	36	3.8	-12.6
PHI	36	104	457	742	.244	.300	.334	105	84/80	-107	-135	-14.2	4.26	112	76/85	-142	-90	-9.5	-10.3
			573		.268	.333	.359						3.22						

NATIONAL LEAGUE 1919

Batting Runs
Gavvy Cravath PHI 32.2
George Burns NY 30.0
Rogers Hornsby STL 28.5

Park Adjusted
Rogers Hornsby STL 32.3
Gavvy Cravath PHI 31.2
George Burns NY 31.2

Pitching Runs
Hippo Vaughn CHI 38.2
Pete Alexander CHI 31.0
Dutch Ruether CIN 29.5

Park Adjusted
Hippo Vaughn CHI 38.3
Pete Alexander CHI 31.0
Dutch Ruether CIN 28.1

Normalized OPS
Heinie Groh CIN 147
Rogers Hornsby STL 145
Edd Roush CIN 143

Park Adjusted
Rogers Hornsby STL 155
Heinie Groh CIN 144
George Burns NY 143

Normalized ERA
Pete Alexander CHI 169
Hippo Vaughn CHI 163
Dutch Ruether CIN 160

Park Adjusted
Pete Alexander CHI 169
Hippo Vaughn CHI 163
Dutch Ruether CIN 157

On Base Average
George Burns NY .396
Heinie Groh CIN .392
Rogers Hornsby STL .384

Slugging Percentage
Hi Myers BRO .436
Larry Doyle NY .433
Heinie Groh CIN .431

Percent of Team Wins
Jess Barnes NY .287
Hippo Vaughn CHI .280
Wilbur Cooper PIT .268

Wins Above Team
Jess Barnes NY 5.1
Red Causey BOS 4.5
Babe Adams PIT 4.0

Isolated Power
Benjamin Kauff NY .145
Larry Doyle NY .144
Hi Myers BRO .129

Players Overall
Rabbit Maranville BOS 43.8
Rogers Hornsby STL 40.6
Heinie Groh CIN 32.1

Pitchers Overall
Pete Alexander CHI 35.2
Hippo Vaughn CHI 34.9
Dutch Ruether CIN 30.5

Defensive Runs
Rabbit Maranville BOS 34.7
Art Fletcher NY 25.1
Morrie Rath CIN 15.6

Club	W	L	R	OR	Avg	OBA	SLG	BPF	NOPS-A	BR	Adj	Wins	ERA	PPF	NERA-A	PR	Adj	Wins	Diff
CIN	96	44	577	401	.263	.327	.342	102	107/105	43	30	3.3	2.23	98	130/128	96	88	9.8	12.8
NY	87	53	605	470	.269	.322	.366	98	113/115	64	73	8.1	2.70	95	108/102	29	7	.8	8.1
CHI	75	65	454	407	.256	.308	.332	101	98/97	-13	-20	-2.3	2.21	100	131/132	98	98	10.9	-3.6
PIT	71	68	472	466	.249	.306	.325	100	95/95	-24	-24	-2.8	2.88	100	101/101	4	3	.3	4.0
BRO	69	71	525	513	.263	.304	.340	97	99/102	-12	1	.1	2.73	97	106/103	25	11	1.2	-2.3
BOS	57	82	465	563	.253	.311	.324	101	97/96	-16	-20	-2.3	3.17	103	92/95	-35	-22	-2.6	-7.6
STL	54	83	463	552	.256	.305	.326	93	95/102	-27	5	.5	3.23	95	90/86	-42	-62	-7.0	-8.1
PHI	47	90	510	699	.251	.303	.342	104	100/96	-8	-27	-3.1	4.14	109	70/77	-171	-133	-15.0	-3.4
			509		.258	.311	.337						2.91						

Batting Runs				Park Adjusted				Pitching Runs				Park Adjusted		
Babe Ruth	NY	113.2		Babe Ruth	NY	111.7		Stan Coveleski	CLE	45.6		Stan Coveleski	CLE	47.3
George Sisler	STL	73.2		George Sisler	STL	67.2		Bob Shawkey	NY	39.8		Bob Shawkey	NY	38.0
Tris Speaker	CLE	65.7		Joe Jackson	CHI	62.6		Jim Bagby	CLE	34.1		Urban Shocker	STL	36.8
Normalized OPS				**Park Adjusted**				**Normalized ERA**				**Park Adjusted**		
Babe Ruth	NY	261		Babe Ruth	NY	256		Bob Shawkey	NY	155		Stan Coveleski	CLE	154
George Sisler	STL	184		Joe Jackson	CHI	182		Stan Coveleski	CLE	152		Bob Shawkey	NY	152
Tris Speaker	CLE	176		George Sisler	STL	172		Urban Shocker	STL	140		Urban Shocker	STL	150
On Base Average				**Slugging Percentage**				**Percent of Team Wins**				**Wins Above Team**		
Babe Ruth	NY	.530		Babe Ruth	NY	.847		Jim Bagby	CLE	.316		Urban Shocker	STL	6.3
Tris Speaker	CLE	.483		George Sisler	STL	.632		Carl Mays	NY	.274		Jim Bagby	CLE	5.0
George Sisler	STL	.449		Joe Jackson	CHI	.589		Urban Shocker	STL	.263		Carl Mays	NY	4.2
Isolated Power				**Players Overall**				**Pitchers Overall**				**Defensive Runs**		
Babe Ruth	NY	.472		Babe Ruth	NY	97.7		Stan Coveleski	CLE	51.9		Everett Scott	BOS	28.1
George Sisler	STL	.225		George Sisler	STL	82.8		Urban Shocker	STL	38.3		Wally Gerber	STL	17.2
Joe Jackson	CHI	.207		Eddie Collins	CHI	60.6		Jim Bagby	CLE	35.2		Del Pratt	NY	16.9

Club	W	L	R	OR	Avg	OBA	SLG	BPF	NOPS-A	BR	Adj	Wins	ERA	PPF	NERA-A	PR	Adj	Wins	Diff
CLE	98	56	857	642	.303	.376	.417	105	117/112	124	88	8.5	3.41	101	111/112	58	65	6.3	6.2
CHI	96	58	794	665	.295	.357	.402	95	108/114	50	89	8.6	3.59	92	106/ 97	31	-15	-1.5	11.9
NY	95	59	838	629	.280	.349	.426	102	113/111	68	54	5.2	3.31	98	114/113	72	63	6.1	6.6
STL	76	77	797	766	.308	.363	.419	107	114/107	91	40	3.9	4.04	107	94/100	-37	3	.2	-4.6
BOS	72	81	650	698	.269	.342	.351	92	89/ 97	-69	-12	-1.3	3.82	92	99/ 91	-3	-50	-4.9	1.7
WAS	68	84	723	802	.290	.351	.386	97	102/105	6	29	2.8	4.17	98	91/ 89	-57	-67	-6.6	-4.2
DET	61	93	652	833	.270	.334	.358	106	90/ 84	-78	-123	-12.1	4.04	110	94/103	-39	18	1.8	-5.7
Ph.	48	106	558	834	.252	.304	.337	97	76/ 78	-189	-168	-16.4	3.92	102	97/ 98	-20	-10	-1.1	-11.6
			734		.283	.347	.387						3.79						

Batting Runs				Park Adjusted				Pitching Runs				Park Adjusted		
Rogers Hornsby	STL	62.6		Rogers Hornsby	STL	62.7		Pete Alexander	CHI	49.3		Pete Alexander	CHI	46.2
Ross Youngs	NY	47.1		Ross Youngs	NY	43.6		Burleigh Grimes	BRO	30.8		Burleigh Grimes	BRO	43.1
Cy Williams	PHI	32.1		Edd Roush	CIN	35.9		Babe Adams	PIT	28.5		Babe Adams	PIT	28.4
Normalized OPS				**Park Adjusted**				**Normalized ERA**				**Park Adjusted**		
Rogers Hornsby	STL	183		Rogers Hornsby	STL	183		Pete Alexander	CHI	164		Pete Alexander	CHI	160
Ross Youngs	NY	159		Edd Roush	CIN	154		Babe Adams	PIT	145		Burleigh Grimes	BRO	157
Cy Williams	PHI	146		Ross Youngs	NY	152		Burleigh Grimes	BRO	141		Babe Adams	PIT	145
On Base Average				**Slugging Percentage**				**Percent of Team Wins**				**Wins Above Team**		
Rogers Hornsby	STL	.431		Rogers Hornsby	STL	.559		Pete Alexander	CHI	.360		Pete Alexander	CHI	9.6
Ross Youngs	NY	.427		Cy Williams	PHI	.497		Wilbur Cooper	PIT	.304		Bill Doak	STL	5.6
Edd Roush	CIN	.386		Ross Youngs	NY	.477		Bill Doak	STL	.267		Wilbur Cooper	PIT	5.3
Isolated Power				**Players Overall**				**Pitchers Overall**				**Defensive Runs**		
Rogers Hornsby	STL	.188		Rogers Hornsby	STL	79.3		Burleigh Grimes	BRO	55.3		Mickey O'Neil	BOS	15.3
Cy Williams	PHI	.171		Edd Roush	CIN	39.1		Pete Alexander	CHI	52.2		Rabbit Maranville	BOS	14.5
Irish Meusel	PHI	.164		Ross Youngs	NY	35.3		Sherry Smith	BRO	31.5		Peter Kilduff	BRO	14.0

Club	W	L	R	OR	Avg	OBA	SLG	BPF	NOPS-A	BR	Adj	Wins	ERA	PPF	NERA-A	PR	Adj	Wins	Diff
BRO	93	61	660	528	.277	.324	.367	113	105/ 92	19	-64	-6.9	2.62	112	119/133	81	139	14.8	8.1
NY	86	68	682	543	.269	.327	.363	105	104/ 99	23	-7	-.8	2.81	102	112/114	51	62	6.6	3.2
CIN	82	71	639	569	.277	.332	.349	91	101/112	11	66	7.1	2.90	89	108/ 96	36	-18	-2.0	.4
PIT	79	75	530	552	.257	.310	.332	100	90/ 90	-66	-64	-6.9	2.89	100	108/108	38	38	4.1	4.9
STL	75	79	675	682	.289	.337	.385	100	114/114	83	83	8.9	3.43	100	91/ 91	-46	-46	-5.0	-6.0
CHI	75	79	619	635	.264	.326	.354	97	101/104	7	22	2.4	3.27	98	96/ 94	-20	-32	-3.5	-.9
BOS	62	90	523	670	.260	.315	.339	93	93/100	-44	-1	-.3	3.54	95	89/ 84	-61	-84	-9.1	-4.7
PHI	62	91	565	714	.263	.305	.364	105	98/ 94	-29	-58	-6.3	3.63	109	86/ 94	-76	-35	-3.8	-4.3
			812		.270	.322	.357						3.13						

Batting Runs
Babe Ruth	NY	119.3
Harry Heilmann	DET	58.5
Ty Cobb	DET	50.4

Park Adjusted
Babe Ruth	NY	116.7
Harry Heilmann	DET	60.1
Ty Cobb	DET	51.7

Pitching Runs
Red Faber	CHI	66.4
Carl Mays	NY	46.2
George Mogridge	WAS	40.9

Park Adjusted
Red Faber	CHI	62.4
Carl Mays	NY	44.8
Sam Jones	BOS	36.7

Normalized OPS
Babe Ruth	NY	241
Harry Heilmann	DET	166
Ty Cobb	DET	166

Park Adjusted
Babe Ruth	NY	235
Harry Heilmann	DET	169
Ty Cobb	DET	169

Normalized ERA
Red Faber	CHI	173
George Mogridge	WAS	143
Carl Mays	NY	141

Park Adjusted
Red Faber	CHI	169
Carl Mays	NY	139
George Mogridge	WAS	138

On Base Average
Babe Ruth	NY	.512
Ty Cobb	DET	.452
Harry Heilmann	DET	.444

Slugging Percentage
Babe Ruth	NY	.846
Harry Heilmann	DET	.606
Ty Cobb	DET	.596

Percent of Team Wins
Red Faber	CHI	.403
Urban Shocker	STL	.333
Sam Jones	BOS	.307

Wins Above Team
Red Faber	CHI	12.0
Urban Shocker	STL	8.7
Dickie Kerr	CHI	5.9

Isolated Power
Babe Ruth	NY	.469
Bob Meusel	NY	.241
Elmer Smith	CLE	.218

Players Overall
Babe Ruth	NY	110.6
Ty Cobb	DET	60.5
Eddie Collins	CHI	53.5

Pitchers Overall
Red Faber	CHI	60.0
Carl Mays	NY	59.0
Urban Shocker	STL	41.5

Defensive Runs
Everett Scott	BOS	40.8
Eddie Collins	CHI	31.3
Jimmy Dykes	PHI	29.5

Club	W	L	R	OR	Avg	OBA	SLG	BPF	NOPS-A	BR	Adj	Wins	ERA	PPF	NERA-A	PR	Adj	Wins	Diff
NY	98	55	948	708	.300	.374	.464	103	122/118	138	116	10.9	3.81	99	112/111	71	65	6.1	4.6
CLE	94	60	925	712	.308	.383	.431	98	115/117	114	130	12.1	3.90	94	110/104	58	21	2.0	2.9
STL	81	73	835	845	.303	.356	.423	104	106/102	23	-6	-.6	4.61	104	93/ 97	-50	-22	-2.2	6.8
WAS	80	73	704	738	.278	.342	.384	97	92/ 95	-77	-50	-4.8	3.97	97	108/104	48	26	2.4	5.9
BOS	75	79	668	696	.276	.334	.360	101	83/ 82	-134	-140	-13.2	3.99	101	107/109	45	52	4.8	6.4
DET	71	82	883	852	.316	.385	.433	98	116/118	121	136	12.7	4.40	98	97/ 95	-18	-33	-3.2	-15.0
CHI	62	92	683	858	.284	.344	.380	95	91/ 96	-79	-41	-3.9	4.93	98	87/ 85	-99	-115	-10.8	-.3
PHI	53	100	657	894	.274	.329	.390	104	90/ 86	-103	-137	-12.9	4.61	108	93/101	-50	5	.5	-11.0
				788	.292	.356	.408						4.28						

Batting Runs
Rogers Hornsby	STL	74.4
Jack Fournier	STL	34.5
Austin McHenry	STL	33.5

Park Adjusted
Rogers Hornsby	STL	80.9
Jack Fournier	STL	40.7
Austin McHenry	STL	39.5

Pitching Runs
Eppa Rixey	CIN	33.4
Burleigh Grimes	BRO	31.9
Bill Doak	STL	27.8

Park Adjusted
Burleigh Grimes	BRO	37.6
Eppa Rixey	CIN	30.4
Whitey Glazner	PIT	28.4

Normalized OPS
Rogers Hornsby	STL	187
Austin McHenry	STL	142
Jack Fournier	STL	141

Park Adjusted
Rogers Hornsby	STL	203
Austin McHenry	STL	154
Jack Fournier	STL	153

Normalized ERA
Bill Doak	STL	146
Babe Adams	PIT	143
Whitey Glazner	PIT	137

Park Adjusted
Babe Adams	PIT	146
Burleigh Grimes	BRO	140
Whitey Glazner	PIT	139

On Base Average
Rogers Hornsby	STL	.458
Ross Youngs	NY	.411
Jack Fournier	STL	.409

Slugging Percentage
Rogers Hornsby	STL	.639
Austin McHenry	STL	.531
George Kelly	NY	.528

Percent of Team Wins
Burleigh Grimes	BRO	.286
Eppa Rixey	CIN	.271
Joe Oeschger	BOS	.253

Wins Above Team
Burleigh Grimes	BRO	5.5
Pete Alexander	CHI	4.0
Bill Doak	STL	3.5

Isolated Power
Rogers Hornsby	STL	.242
George Kelly	NY	.220
Austin McHenry	STL	.181

Players Overall
Rogers Hornsby	STL	77.8
Dave Bancroft	NY	59.2
Frankie Frisch	NY	39.5

Pitchers Overall
Burleigh Grimes	BRO	42.6
Pete Alexander	CHI	28.9
Clarence Mitchell	BRO	28.8

Defensive Runs
Dave Bancroft	NY	26.2
Doc Lavan	STL	19.6
Carson Bigbee	PIT	19.0

Club	W	L	R	OR	Avg	OBA	SLG	BPF	NOPS-A	BR	Adj	Wins	ERA	PPF	NERA-A	PR	Adj	Wins	Diff
NY	94	59	840	637	.298	.359	.421	100	115/115	94	96	9.4	3.55	96	106/102	35	12	1.2	6.8
PIT	90	63	692	595	.285	.330	.387	104	98/ 94	-36	-63	-6.3	3.17	102	119/122	97	109	10.8	9.0
STL	87	66	809	681	.308	.358	.437	92	119/129	114	169	16.7	3.63	89	104/ 93	23	-38	-3.8	-2.4
BOS	79	74	721	697	.290	.339	.400	91	103/114	6	70	6.9	3.90	90	97/ 87	-17	-77	-7.7	3.3
BRO	77	75	667	681	.280	.325	.386	104	96/ 92	-51	-79	-7.9	3.70	105	102/107	13	38	3.8	5.1
CIN	70	83	618	649	.278	.333	.370	97	93/ 96	-55	-37	-3.7	3.46	98	109/107	49	35	3.4	-6.2
CHI	64	89	668	773	.292	.339	.378	111	97/ 88	-30	-105	-10.5	4.39	113	86/ 97	-92	-17	-1.8	-.3
PHI	51	103	617	919	.284	.324	.397	102	99/ 97	-36	-47	-4.7	4.48	107	84/ 90	-104	-63	-6.3	-15.0
				704	.289	.338	.397						3.78						

AMERICAN LEAGUE 1922

Batting Runs			Park Adjusted			Pitching Runs			Park Adjusted		
George Sisler	STL	64.4	George Sisler	STL	61.8	Red Faber	CHI	47.7	Red Faber	CHI	51.8
Ken Williams	STL	56.1	Ty Cobb	DET	54.3	Urban Shocker	STL	40.9	Urban Shocker	STL	40.3
Tris Speaker	CLE	52.8	Ken Williams	STL	53.4	Bob Shawkey	NY	37.4	Bob Shawkey	NY	36.0
Normalized OPS			**Park Adjusted**			**Normalized ERA**			**Park Adjusted**		
Babe Ruth	NY	185	Babe Ruth	NY	182	Red Faber	CHI	143	Red Faber	CHI	147
Tris Speaker	CLE	180	Tris Speaker	CLE	175	Herman Pillette	DET	142	Rasty Wright	STL	137
George Sisler	STL	175	Ty Cobb	DET	171	Bob Shawkey	NY	139	Bob Shawkey	NY	137
On Base Average			**Slugging Percentage**			**Percent of Team Wins**			**Wins Above Team**		
Tris Speaker	CLE	.474	Babe Ruth	NY	.672	Eddie Rommel	PHI	.415	Eddie Rommel	PHI	13.7
George Sisler	STL	.467	Ken Williams	STL	.627	George Uhle	CLE	.282	Joe Bush	NY	7.5
Ty Cobb	DET	.462	Tris Speaker	CLE	.606	Joe Bush	NY	.277	George Mogridge	WAS	5.1
Isolated Power			**Players Overall**			**Pitchers Overall**			**Defensive Runs**		
Babe Ruth	NY	.357	George Sisler	STL	60.4	Red Faber	CHI	50.2	Bucky Harris	WAS	25.6
Ken Williams	STL	.296	Ken Williams	STL	54.8	Urban Shocker	STL	37.7	Everett Scott	NY	19.3
Tilly Walker	PHI	.265	Tris Speaker	CLE	47.6	Eddie Rommel	PHI	35.4	Ray Schalk	CHI	15.3

Club	W	L	R	OR	Avg	OBA	SLG	BPF	NOPS-A	BR	Adj	Wins	ERA	PPF	NERA-A	PR	Adj	Wins	Diff
NY	94	60	758	618	.287	.353	.411	101	107/105	35	25	2.4	3.39	99	119/118	99	93	9.1	5.5
STL	93	61	867	643	.310	.369	.453	103	123/119	147	123	11.9	3.38	100	119/119	101	99	9.6	-5.5
DET	79	75	828	791	.306	.373	.416	97	113/116	95	114	11.1	4.27	97	94/ 91	-35	-56	-5.5	-3.6
CLE	78	76	817	768	.292	.364	.398	102	106/103	44	25	2.4	4.60	104	88/ 91	-86	-63	-6.3	4.8
CHI	77	77	691	691	.278	.343	.374	102	94/ 92	-49	-68	-6.7	3.94	103	102/105	15	31	3.0	3.6
WAS	69	85	650	706	.268	.334	.367	89	89/101	-84	-4	-.5	3.81	89	106/ 94	34	-33	-3.3	-4.2
PHI	65	89	830	705	.270	.331	.402	105	99/ 94	-36	-75	-7.4	4.59	108	88/ 95	-84	-33	-3.3	-1.3
BOS	61	93	598	769	.263	.317	.356	98	82/ 83	-147	-136	-13.3	4.30	101	94/ 95	-40	-33	-3.3	.5
			733		.285	.348	.397						4.03						

NATIONAL LEAGUE 1922

Batting Runs			Park Adjusted			Pitching Runs			Park Adjusted		
Rogers Hornsby	STL	90.0	Rogers Hornsby	STL	88.1	Wilbur Cooper	PIT	30.3	Wilbur Cooper	PIT	37.9
Oscar Grimes	CHI	47.1	Oscar Grimes	CHI	47.7	Pete Donohue	CIN	26.1	Johnny Morrison	PIT	28.6
Cy Williams	PHI	27.5	Zach Wheat	BRO	32.0	Phil Douglas	NY	25.9	Phil Douglas	NY	26.9
Normalized OPS			**Park Adjusted**			**Normalized ERA**			**Park Adjusted**		
Rogers Hornsby	STL	200	Rogers Hornsby	STL	196	Phil Douglas	NY	156	Phil Douglas	NY	159
Oscar Grimes	CHI	160	Oscar Grimes	CHI	161	Rosy Ryan	NY	137	Rosy Ryan	NY	138
Cy Williams	PHI	132	Zach Wheat	BRO	141	Pete Donohue	CIN	131	Lefty Weinert	PHI	137
On Base Average			**Slugging Percentage**			**Percent of Team Wins**			**Wins Above Team**		
Rogers Hornsby	STL	.459	Rogers Hornsby	STL	.722	Eppa Rixey	CIN	.291	Dutch Ruether	BRO	6.0
Oscar Grimes	CHI	.442	Oscar Grimes	CHI	.572	Dutch Ruether	BRO	.276	Eppa Rixey	CIN	5.0
Bob O'Farrell	CHI	.437	Cotton Tierney	PIT	.515	Wilbur Cooper	PIT	.271	Dazzy Vance	BRO	4.0
Isolated Power			**Players Overall**			**Pitchers Overall**			**Defensive Runs**		
Rogers Hornsby	STL	.321	Rogers Hornsby	STL	82.9	Wilbur Cooper	PIT	43.9	Frank Parkinson	PHI	23.8
Oscar Grimes	CHI	.218	Dave Bancroft	NY	42.9	Art Nehf	NY	29.2	Dave Bancroft	NY	23.5
Cy Williams	PHI	.205	Oscar Grimes	CHI	42.7	Phil Douglas	NY	27.3	Babe Pinelli	CIN	22.2

Club	W	L	R	OR	Avg	OBA	SLG	BPF	NOPS-A	BR	Adj	Wins	ERA	PPF	NERA-A	PR	Adj	Wins	Diff
NY	93	61	852	658	.305	.363	.428	104	112/108	77	44	4.1	3.45	101	119/120	100	109	10.3	1.6
CIN	86	68	766	677	.296	.353	.401	93	103/110	8	60	5.6	3.53	91	116/106	87	30	2.8	.5
STL	85	69	863	819	.301	.357	.444	102	116/114	91	75	7.1	4.44	102	92/ 94	-51	-41	-4.0	4.9
PIT	85	69	865	736	.308	.360	.419	107	109/102	56	-1	-.2	3.98	106	103/109	18	54	5.1	3.1
CHI	80	74	771	808	.293	.359	.390	99	101/102	6	13	1.2	4.34	100	94/ 94	-36	-38	-3.6	5.4
BRO	76	78	743	754	.290	.335	.392	91	95/105	-57	11	1.1	4.05	90	101/ 91	8	-52	-5.0	3.0
PHI	57	96	738	920	.282	.341	.415	110	103/ 94	-2	-77	-7.4	4.64	114	88/100	-82	2	.2	-12.3
BOS	53	100	596	822	.263	.317	.341	94	76/ 81	-178	-134	-12.7	4.37	97	94/ 91	-40	-57	-5.4	-5.3
			774		.292	.348	.404						4.10						

Batting Runs			Park Adjusted			Pitching Runs			Park Adjusted		
Babe Ruth	NY	119.1	Babe Ruth	NY	110.7	Stan Coveleski	CLE	30.9	Stan Coveleski	CLE	32.9
Tris Speaker	CLE	70.9	Harry Heilmann	DET	72.5	Elam Vangilder	STL	28.8	Waite Hoyt	NY	32.6
Harry Heilmann	DET	70.7	Tris Speaker	CLE	67.3	Waite Hoyt	NY	25.7	Elam Vangilder	STL	31.8
Normalized OPS			**Park Adjusted**			**Normalized ERA**			**Park Adjusted**		
Babe Ruth	NY	242	Babe Ruth	NY	221	Stan Coveleski	CLE	144	Stan Coveleski	CLE	147
Harry Heilmann	DET	191	Harry Heilmann	DET	196	Waite Hoyt	NY	132	Waite Hoyt	NY	141
Tris Speaker	CLE	183	Tris Speaker	CLE	175	Allan Russell	WAS	131	Herb Pennock	NY	135
On Base Average			**Slugging Percentage**			**Percent of Team Wins**			**Wins Above Team**		
Babe Ruth	NY	.545	Babe Ruth	NY	.764	Howard Ehmke	BOS	.328	Howard Ehmke	BOS	6.8
Harry Heilmann	DET	.481	Harry Heilmann	DET	.632	George Uhle	CLE	.317	Urban Shocker	STL	5.6
Tris Speaker	CLE	.469	Ken Williams	STL	.623	Urban Shocker	STL	.270	George Uhle	CLE	4.8
Isolated Power			**Players Overall**			**Pitchers Overall**			**Defensive Runs**		
Babe Ruth	NY	.372	Babe Ruth	NY	99.3	Eddie Rommel	PHI	31.8	Rube Lutzke	CLE	25.3
Ken Williams	STL	.267	Joe Sewell	CLE	60.3	Joe Bush	NY	31.7	Roger Peckinpaugh	WAS	23.8
Tris Speaker	CLE	.230	Harry Heilmann	DET	57.9	Waite Hoyt	NY	29.3	Aaron Ward	NY	18.3

Club	W	L	R	OR	Avg	OBA	SLG	BPF	NOPS-A	BR	Adj	Wins	ERA	PPF	NERA-A	PR	Adj	Wins	Diff
NY	98	54	823	622	.291	.357	.423	109	112/103	72	2	.1	3.62	106	110/117	55	95	9.2	12.7
DET	83	71	831	741	.300	.377	.401	98	111/114	91	108	10.5	4.09	96	97/93	-16	-40	-4.0	-.5
CLE	82	71	888	746	.301	.381	.421	104	118/113	134	103	9.9	3.91	102	102/104	10	23	2.2	-6.6
WAS	75	78	720	747	.274	.346	.367	92	94/101	-46	10	1.0	3.98	92	100/92	0	-47	-4.6	2.1
STL	74	78	688	720	.282	.340	.398	102	101/99	-12	-25	-2.5	3.93	102	101/104	8	23	2.2	-1.7
PHI	69	83	661	761	.271	.334	.370	98	92/93	-71	-59	-5.8	4.08	100	97/97	-15	-15	-1.6	.3
CHI	69	85	692	741	.279	.350	.374	100	97/97	-23	-23	-2.4	4.03	101	99/100	-7	-2	-.2	-5.4
BOS	61	91	584	809	.262	.318	.351	98	82/83	-141	-129	-12.6	4.20	102	95/97	-32	-19	-2.0	-.4
			736		.283	.351	.388						3.98						

Batting Runs			Park Adjusted			Pitching Runs			Park Adjusted		
Rogers Hornsby	STL	52.1	Rogers Hornsby	STL	59.6	Dolf Luque	CIN	74.0	Dolf Luque	CIN	66.0
Jack Fournier	BRO	43.7	Jim Bottomley	STL	48.1	Eppa Rixey	CIN	41.2	Eppa Rixey	CIN	33.5
Jim Bottomley	STL	39.2	Jack Fournier	BRO	43.4	Pete Alexander	CHI	27.4	Jimmy Ring	PHI	27.5
Normalized OPS			**Park Adjusted**			**Normalized ERA**			**Park Adjusted**		
Rogers Hornsby	STL	184	Rogers Hornsby	STL	209	Dolf Luque	CIN	207	Dolf Luque	CIN	196
Jack Fournier	BRO	161	Jim Bottomley	STL	173	Eppa Rixey	CIN	143	Eppa Rixey	CIN	135
Jim Bottomley	STL	152	Jack Fournier	BRO	160	Vic Keen	CHI	133	Vic Keen	CHI	131
On Base Average			**Slugging Percentage**			**Percent of Team Wins**			**Wins Above Team**		
Rogers Hornsby	STL	.459	Rogers Hornsby	STL	.627	Jimmy Ring	PHI	.360	Jimmy Ring	PHI	8.9
Jim Bottomley	STL	.425	Jack Fournier	BRO	.588	Dolf Luque	CIN	.297	Dolf Luque	CIN	8.2
Ross Youngs	NY	.412	Cy Williams	PHI	.576	Johnny Morrison	PIT	.287	Pete Alexander	CHI	4.7
Isolated Power			**Players Overall**			**Pitchers Overall**			**Defensive Runs**		
Cy Williams	PHI	.282	Rogers Hornsby	STL	53.0	Dolf Luque	CIN	67.7	Sam Bohne	CIN	19.7
Rogers Hornsby	STL	.243	Pie Traynor	PIT	42.0	Eppa Rixey	CIN	30.8	James Johnston	BRO	17.5
Jack Fournier	BRO	.237	Jack Fournier	BRO	40.2	Pete Alexander	CHI	29.4	Dave Bancroft	NY	16.0

Club	W	L	R	OR	Avg	OBA	SLG	BPF	NOPS-A	BR	Adj	Wins	ERA	PPF	NERA-A	PR	Adj	Wins	Diff
NY	95	58	854	679	.295	.356	.415	105	112/107	72	37	3.5	3.90	102	102/104	15	26	2.5	12.5
CIN	91	63	708	629	.285	.344	.392	96	101/106	-0	27	2.6	3.22	94	124/117	121	86	8.3	3.1
PIT	87	67	786	696	.295	.347	.404	93	106/113	26	76	7.3	3.87	91	103/94	19	-33	-3.3	6.0
CHI	83	71	756	704	.288	.348	.406	99	107/107	33	39	3.7	3.82	98	105/103	27	16	1.5	.8
STL	79	74	746	732	.286	.343	.398	88	103/117	5	98	9.5	3.87	87	103/90	20	-62	-6.0	-.9
BRO	76	78	753	741	.285	.340	.387	101	99/99	-19	-23	-2.3	3.74	100	107/107	39	41	4.0	-2.6
BOS	54	100	636	798	.273	.331	.353	104	87/83	-98	-128	-12.4	4.22	107	95/101	-34	8	.7	-11.4
PHI	50	104	748	1008	.278	.333	.401	112	101/90	-14	-106	-10.3	5.34	117	75/88	-204	-98	-9.5	-7.2
			748		.286	.343	.395						4.00						

AMERICAN LEAGUE 1924

Batting Runs			Park Adjusted			Pitching Runs			Park Adjusted		
Babe Ruth	NY	100.7	Babe Ruth	NY	100.9	Walter Johnson	WAS	46.8	Herb Pennock	NY	41.3
Harry Heilmann	DET	40.4	Harry Heilmann	DET	40.1	Herb Pennock	NY	44.5	Walter Johnson	WAS	39.6
Goose Goslin	WAS	35.5	Goose Goslin	WAS	37.9	Sherry Smith	CLE	33.7	Stan Baumgartner	PHI	31.0
Normalized OPS			**Park Adjusted**			**Normalized ERA**			**Park Adjusted**		
Babe Ruth	NY	220	Babe Ruth	NY	221	Walter Johnson	WAS	156	Stan Baumgartner	PHI	153
Harry Heilmann	DET	147	Tris Speaker	CLE	149	Tom Zachary	WAS	154	Walter Johnson	WAS	147
Ken Williams	STL	146	Harry Heilmann	DET	146	Herb Pennock	NY	149	Herb Pennock	NY	146
On Base Average			**Slugging Percentage**			**Percent of Team Wins**			**Wins Above Team**		
Babe Ruth	NY	.513	Babe Ruth	NY	.739	Sloppy Thurston	CHI	.303	Sloppy Thurston	CHI	6.9
Eddie Collins	CHI	.441	Harry Heilmann	DET	.533	Joe Shaute	CLE	.299	Walter Johnson	WAS	6.3
Tris Speaker	CLE	.432	Ken Williams	STL	.533	Howard Ehmke	BOS	.284	Joe Shaute	CLE	5.0
Isolated Power			**Players Overall**			**Pitchers Overall**			**Defensive Runs**		
Babe Ruth	NY	.361	Babe Ruth	NY	94.5	Walter Johnson	WAS	43.3	Joe Sewell	CLE	21.1
Joe Hauser	PHI	.228	Joe Sewell	CLE	38.0	Herb Pennock	NY	38.0	Rube Lutzke	CLE	18.4
Baby Doll Jacobson	STL	.211	Harry Heilmann	DET	32.1	Tom Zachary	WAS	33.4	Bill Wambsganss	BOS	17.7

Club	W	L	R	OR	Avg	OBA	SLG	BPF	NOPS-A	BR	Adj	Wins	ERA	PPF	NERA-A	PR	Adj	Wins	Diff
WAS	92	62	755	613	.294	.364	.388	97	100/103	1	23	2.1	3.35	95	126/120	136	100	9.5	3.4
NY	89	63	798	667	.291	.353	.427	100	108/108	35	36	3.4	3.86	98	110/107	56	41	3.9	5.7
DET	86	68	849	796	.298	.374	.404	100	107/107	57	55	5.2	4.20	100	101/100	6	2	.2	3.6
STL	74	78	769	809	.295	.356	.408	108	103/ 96	8	-50	-4.8	4.58	109	92/101	-52	6	.6	2.3
PHI	71	81	685	778	.281	.336	.389	103	93/ 91	-72	-93	-8.9	4.39	105	97/101	-22	6	.5	3.3
CLE	67	86	755	814	.298	.362	.401	95	103/108	17	52	4.9	4.40	96	96/ 92	-24	-49	-4.7	-9.7
BOS	67	87	737	806	.277	.355	.373	100	94/ 94	-44	-43	-4.1	4.35	101	97/ 98	-18	-12	-1.2	-4.7
CHI	66	87	793	858	.289	.366	.383	96	99/103	0	28	2.7	4.76	97	89/ 87	-78	-96	-9.1	-4.0
			768		.290	.359	.397						4.23						

NATIONAL LEAGUE 1924

Batting Runs			Park Adjusted			Pitching Runs			Park Adjusted		
Rogers Hornsby	STL	94.1	Rogers Hornsby	STL	96.9	Dazzy Vance	BRO	58.7	Dazzy Vance	BRO	61.7
Zach Wheat	BRO	48.2	Ross Youngs	NY	51.8	Eppa Rixey	CIN	29.2	Eppa Rixey	CIN	31.1
Jack Fournier	BRO	48.1	Zach Wheat	BRO	46.0	Hugh McQuillan	NY	24.0	Emil Yde	PIT	23.1
Normalized OPS			**Park Adjusted**			**Normalized ERA**			**Park Adjusted**		
Rogers Hornsby	STL	219	Rogers Hornsby	STL	228	Dazzy Vance	BRO	179	Dazzy Vance	BRO	183
Zach Wheat	BRO	160	Ross Youngs	NY	172	Hugh McQuillan	NY	144	Eppa Rixey	CIN	143
Jack Fournier	BRO	157	Zach Wheat	BRO	156	Eppa Rixey	CIN	140	Rube Benton	CIN	143
On Base Average			**Slugging Percentage**			**Percent of Team Wins**			**Wins Above Team**		
Rogers Hornsby	STL	.507	Rogers Hornsby	STL	.696	Dazzy Vance	BRO	.304	Dazzy Vance	BRO	9.9
Ross Youngs	NY	.441	Cy Williams	PHI	.552	Jess Barnes	BOS	.283	Emil Yde	PIT	5.5
Jack Fournier	BRO	.428	Zach Wheat	BRO	.549	Carl Mays	CIN	.241	Carl Mays	CIN	5.3
Isolated Power			**Players Overall**			**Pitchers Overall**			**Defensive Runs**		
Rogers Hornsby	STL	.272	Rogers Hornsby	STL	81.2	Dazzy Vance	BRO	57.0	Babe Pinelli	CIN	27.3
Cy Williams	PHI	.224	Frankie Frisch	NY	47.5	Carl Mays	CIN	34.5	Frankie Frisch	NY	23.3
George Kelly	NY	.207	Zach Wheat	BRO	47.1	Eppa Rixey	CIN	31.9	Jigger Statz	CHI	13.2

Club	W	L	R	OR	Avg	OBA	SLG	BPF	NOPS-A	BR	Adj	Wins	ERA	PPF	NERA-A	PR	Adj	Wins	Diff
NY	93	60	857	641	.300	.358	.432	91	120/131	124	187	18.6	3.61	87	107/ 93	39	-40	-4.1	2.0
BRO	92	62	717	675	.287	.345	.391	103	104/101	19	-0	-.1	3.63	102	106/109	35	49	4.8	10.3
PIT	90	63	724	588	.287	.336	.399	103	104/101	11	-12	-1.3	3.27	101	118/119	92	98	9.7	5.1
CIN	83	70	649	579	.290	.337	.397	103	104/101	8	-12	-1.3	3.12	102	124/126	115	126	12.5	-4.7
CHI	81	72	698	699	.276	.340	.378	99	99/101	-9	-0	-.1	3.83	98	101/ 99	5	-3	-.4	4.9
STL	65	89	740	750	.290	.341	.411	96	109/113	42	68	6.7	4.14	96	93/ 90	-41	-63	-6.4	-12.4
PHI	55	96	676	849	.275	.328	.397	111	102/ 92	-12	-87	-8.8	4.87	115	79/ 91	-150	-64	-6.5	-5.3
BOS	53	100	520	800	.256	.306	.327	94	75/ 80	-180	-137	-13.7	4.46	98	87/ 85	-91	-100	-10.0	.2
			698		.283	.337	.392						3.86						

AMERICAN LEAGUE 1925

Batting Runs			Park Adjusted			Pitching Runs			Park Adjusted		
Harry Heilmann	DET	52.4	Harry Heilmann	DET	53.4	Herb Pennock	NY	44.1	Herb Pennock	NY	37.1
Al Simmons	PHI	50.1	Al Simmons	PHI	49.1	Stan Coveleski	WAS	41.6	Ted Blankenship	CHI	34.7
Tris Speaker	CLE	47.0	Ty Cobb	DET	46.1	Ted Blankenship	CHI	35.2	Ted Lyons	CHI	32.8
Normalized OPS			**Park Adjusted**			**Normalized ERA**			**Park Adjusted**		
Ty Cobb	DET	169	Ty Cobb	DET	171	Stan Coveleski	WAS	155	Ted Blankenship	CHI	145
Tris Speaker	CLE	167	Harry Heilmann	DET	161	Herb Pennock	NY	149	Stan Coveleski	WAS	142
Harry Heilmann	DET	159	Tris Speaker	CLE	158	Ted Blankenship	CHI	145	Jake Miller	CLE	142
On Base Average			**Slugging Percentage**			**Percent of Team Wins**			**Wins Above Team**		
Tris Speaker	CLE	.479	Ty Cobb	DET	.598	Ted Lyons	CHI	.266	Ted Lyons	CHI	5.8
Ty Cobb	DET	.468	Al Simmons	PHI	.596	Ted Wingfield	BOS	.255	Ted Blankenship	CHI	5.0
Eddie Collins	CHI	.461	Tris Speaker	CLE	.578	Eddie Rommel	PHI	.239	Stan Coveleski	WAS	4.9
Isolated Power			**Players Overall**			**Pitchers Overall**			**Defensive Runs**		
Bob Meusel	NY	.252	Joe Sewell	CLE	51.4	Walter Johnson	WAS	37.0	Joe Sewell	CLE	19.9
Lou Gehrig	NY	.236	Tris Speaker	CLE	37.1	Herb Pennock	NY	31.7	Frank O'Rourke	DET	14.6
Ty Cobb	DET	.219	Al Wingo	DET	33.8	Ted Lyons	CHI	31.6	Sam Rice	WAS	11.0

Club	W	L	R	OR	Avg	OBA	SLG	BPF	NOPS-A	BR	Adj	Wins	ERA	PPF	NERA-A	PR	Adj	Wins	Diff
WAS	96	55	829	670	.304	.374	.412	95	106/111	44	81	7.5	3.67	92	120/110	109	56	5.2	7.8
PHI	88	64	831	713	.307	.364	.434	101	110/109	52	43	4.0	3.87	99	113/113	80	74	6.9	1.1
STL	82	71	900	906	.298	.360	.440	108	110/103	53	-8	-.8	4.92	109	89/ 97	-81	-24	-2.3	8.6
DET	81	73	903	829	.302	.380	.413	99	108/109	64	73	6.8	4.60	98	95/ 93	-32	-47	-4.5	1.7
CHI	79	75	811	770	.284	.369	.385	100	98/ 97	-7	-9	-1.0	4.28	100	103/102	18	15	1.4	1.6
CLE	70	84	782	817	.297	.361	.399	106	99/ 94	-12	-61	-5.8	4.49	107	98/105	-15	32	3.0	-4.2
NY	69	85	706	774	.275	.336	.411	94	97/102	-59	-14	-1.4	4.33	95	101/ 96	9	-25	-2.5	-4.2
BOS	47	105	639	922	.266	.336	.365	96	84/ 87	-132	-100	-9.3	4.96	100	88/ 89	-84	-82	-7.7	-12.0
			800		.292	.360	.408						4.39						

NATIONAL LEAGUE 1925

Batting Runs			Park Adjusted			Pitching Runs			Park Adjusted		
Rogers Hornsby	STL	87.1	Rogers Hornsby	STL	89.0	Dolf Luque	CIN	53.0	Dolf Luque	CIN	46.3
Kiki Cuyler	PIT	53.1	Jack Fournier	BRO	54.7	Eppa Rixey	CIN	44.1	Eppa Rixey	CIN	37.5
Jack Fournier	BRO	50.1	Kiki Cuyler	PIT	53.7	Pete Donohue	CIN	39.7	Pete Donohue	CIN	32.9
Normalized OPS			**Park Adjusted**			**Normalized ERA**			**Park Adjusted**		
Rogers Hornsby	STL	213	Rogers Hornsby	STL	218	Dolf Luque	CIN	162	Dolf Luque	CIN	155
Kiki Cuyler	PIT	158	Jack Fournier	BRO	166	Eppa Rixey	CIN	148	Eppa Rixey	CIN	141
Jack Fournier	BRO	157	Kiki Cuyler	PIT	159	Pete Donohue	CIN	139	Larry Benton	BOS	138
On Base Average			**Slugging Percentage**			**Percent of Team Wins**			**Wins Above Team**		
Rogers Hornsby	STL	.489	Rogers Hornsby	STL	.756	Dazzy Vance	BRO	.324	Dazzy Vance	BRO	10.3
Jack Fournier	BRO	.446	Kiki Cuyler	PIT	.598	Eppa Rixey	CIN	.263	Eppa Rixey	CIN	5.4
Ray Blades	STL	.423	Jim Bottomley	STL	.578	Pete Donohue	CIN	.263	Bill Sherdel	STL	5.1
Isolated Power			**Players Overall**			**Pitchers Overall**			**Defensive Runs**		
Rogers Hornsby	STL	.353	Rogers Hornsby	STL	70.9	Dolf Luque	CIN	54.7	Sparky Adams	CHI	24.9
Kiki Cuyler	PIT	.241	Kiki Cuyler	PIT	54.9	Pete Donohue	CIN	37.8	Pie Traynor	PIT	23.3
Irish Meusel	NY	.221	Pie Traynor	PIT	46.7	Jack Scott	NY	35.6	Babe Pinelli	CIN	21.0

Club	W	L	R	OR	Avg	OBA	SLG	BPF	NOPS-A	BR	Adj	Wins	ERA	PPF	NERA-A	PR	Adj	Wins	Diff
PIT	95	58	912	715	.307	.369	.449	99	118/119	116	121	11.3	3.88	96	110/106	59	34	3.2	4.0
NY	86	66	736	702	.283	.337	.415	100	101/101	-25	-24	-2.4	3.94	99	108/107	49	44	4.1	8.3
CIN	80	73	690	643	.285	.339	.387	96	93/ 96	-65	-39	-3.7	3.38	95	126/120	135	104	9.7	-2.5
STL	77	76	828	764	.299	.356	.445	98	114/117	72	91	8.5	4.36	96	98/ 94	-13	-36	-3.5	-4.5
BOS	70	83	708	802	.292	.345	.390	99	95/ 97	-47	-36	-3.5	4.39	100	97/ 97	-18	-17	-1.7	-1.3
PHI	68	85	812	930	.295	.354	.425	115	108/ 93	35	-85	-8.0	5.03	119	85/101	-113	6	.5	-1.0
BRO	68	85	786	866	.296	.351	.406	95	101/107	-5	36	3.4	4.77	96	90/ 86	-74	-102	-9.7	-2.2
CHI	68	86	723	773	.275	.329	.397	96	94/ 97	-76	-49	-4.7	4.41	97	97/ 94	-21	-40	-3.9	-.4
			774		.292	.348	.414						4.27						

AMERICAN LEAGUE 1926

Batting Runs				Park Adjusted				Pitching Runs				Park Adjusted		
Babe Ruth	NY	97.4		Babe Ruth	NY	100.6		Lefty Grove	PHI	43.2		Lefty Grove	PHI	60.4
Lou Gehrig	NY	44.2		Lou Gehrig	NY	47.5		George Uhle	CLE	42.0		Eddie Rommel	PHI	37.4
Goose Goslin	WAS	42.6		Goose Goslin	WAS	42.2		Ted Lyons	CHI	31.8		George Uhle	CLE	35.3
Normalized OPS				**Park Adjusted**				**Normalized ERA**				**Park Adjusted**		
Babe Ruth	NY	224		Babe Ruth	NY	233		Lefty Grove	PHI	160		Lefty Grove	PHI	184
Heinie Manush	DET	155		Lou Gehrig	NY	157		George Uhle	CLE	142		Eddie Rommel	PHI	150
Harry Heilmann	DET	154		Heinie Manush	DET	152		Ted Lyons	CHI	133		Jack Quinn	PHI	136
On Base Average				**Slugging Percentage**				**Percent of Team Wins**				**Wins Above Team**		
Babe Ruth	NY	.516		Babe Ruth	NY	.737		George Uhle	CLE	.307		George Uhle	CLE	7.0
Harry Heilmann	DET	.445		Al Simmons	PHI	.566		Herb Pennock	NY	.253		Joe Pate	PHI	4.3
Max Bishop	PHI	.431		Heinie Manush	DET	.564		Ted Wingfield	BOS	.239		Herb Pennock	NY	3.7
Isolated Power				**Players Overall**				**Pitchers Overall**				**Defensive Runs**		
Babe Ruth	NY	.366		Babe Ruth	NY	51.6		Lefty Grove	PHI	51.6		Topper Rigney	BOS	15.3
Lou Gehrig	NY	.236		Willie Kamm	CHI	40.5		George Uhle	CLE	38.0		Max Bishop	PHI	14.7
Al Simmons	PHI	.224		Goose Goslin	WAS	37.4		Eddie Rommel	PHI	36.0		Willie Kamm	CHI	14.2

Club	W	L	R	OR	Avg	OBA	SLG	BPF	NOPS-A	BR	Adj	Wins	ERA	PPF	NERA-A	PR	Adj	Wins	Diff
NY	91	63	847	713	.289	.370	.438	96	119/124	125	154	14.9	3.86	93	104/ 97	24	-15	-1.6	.7
CLE	88	66	738	612	.289	.349	.386	98	99/101	-17	-1	-.2	3.40	95	118/113	94	66	6.3	4.8
PHI	83	67	677	570	.270	.341	.384	116	96/ 83	-35	-146	-14.3	3.00	115	134/154	153	243	23.5	-1.3
WAS	81	69	802	761	.292	.364	.401	100	107/107	49	46	4.4	4.34	100	93/ 92	-4	-48	-4.7	6.3
CHI	81	72	730	665	.289	.360	.390	91	103/114	22	90	8.7	3.74	89	107/ 95	43	-25	-2.6	-1.6
DET	79	75	793	830	.291	.366	.398	102	101/ 99	51	36	3.5	4.41	103	91/ 94	-60	-43	-4.3	2.7
STL	62	92	682	845	.276	.335	.395	97	98/101	-36	-12	-1.2	4.66	99	86/ 86	-97	-102	-9.9	-3.8
BOS	46	107	562	835	.255	.321	.342	102	80/ 78	-154	-165	-16.1	4.72	106	85/ 90	-106	-67	-6.6	-7.8
			729		.281	.351	.392						4.02						

NATIONAL LEAGUE 1926

Batting Runs				Park Adjusted				Pitching Runs				Park Adjusted		
Paul Waner	PIT	39.5		Hack Wilson	CHI	36.2		Ray Kremer	PIT	31.1		Ray Kremer	PIT	42.5
Hack Wilson	CHI	39.2		Paul Waner	PIT	30.5		Jesse Petty	BRO	30.2		Charlie Root	CHI	33.3
Les Bell	STL	31.5		Cy Williams	PHI	28.3		Charlie Root	CHI	30.1		Jesse Petty	BRO	32.9
Normalized OPS				**Park Adjusted**				**Normalized ERA**				**Park Adjusted**		
Hack Wilson	CHI	152		Hack Wilson	CHI	146		Ray Kremer	PIT	146		Ray Kremer	PIT	163
Paul Waner	PIT	152		Paul Waner	PIT	135		Charlie Root	CHI	135		Charlie Root	CHI	139
Les Bell	STL	140		Dave Bancroft	BOS	134		Jesse Petty	BRO	135		Jesse Petty	BRO	138
On Base Average				**Slugging Percentage**				**Percent of Team Wins**				**Wins Above Team**		
Paul Waner	PIT	.413		Hack Wilson	CHI	.538		Hal Carlson	PHI	.293		Hal Carlson	PHI	7.3
Ray Blades	STL	.409		Paul Waner	PIT	.528		Jesse Petty	BRO	.239		Ray Kremer	PIT	6.9
Hack Wilson	CHI	.406		Les Bell	STL	.518		Lee Meadows	PIT	.238		Flint Rhem	STL	5.3
								Ray Kremer	PIT	.238				
Isolated Power				**Players Overall**				**Pitchers Overall**				**Defensive Runs**		
Hack Wilson	CHI	.217		Paul Waner	PIT	30.6		Ray Kremer	PIT	41.0		Hughie Critz	CIN	25.9
Jim Bottomley	STL	.207		Sparky Adams	CHI	30.2		Charlie Root	CHI	26.8		Sparky Adams	CHI	24.9
Les Bell	STL	.193		Hack Wilson	CHI	29.3		Hal Carlson	PHI	25.4		Bernie Friberg	PHI	20.5

Club	W	L	R	OR	Avg	OBA	SLG	BPF	NOPS-A	BR	Adj	Wins	ERA	PPF	NERA-A	PR	Adj	Wins	Diff
STL	89	65	817	678	.286	.348	.415	109	113/103	76	8	.8	3.67	107	104/112	24	67	6.7	4.5
CIN	87	67	747	651	.290	.349	.400	91	109/119	52	115	11.3	3.42	88	112/ 99	63	-6	-.7	-.6
PIT	84	69	769	689	.285	.343	.396	112	106/ 95	28	-56	-5.7	3.66	112	104/116	24	92	9.1	4.0
CHI	82	72	682	602	.278	.338	.390	104	103/ 99	7	-20	-2.1	3.26	103	117/120	86	102	10.1	-3.0
NY	74	77	663	668	.278	.325	.384	100	98/ 98	-36	-33	-3.3	3.76	100	102/101	10	8	.8	1.1
BRO	71	82	623	705	.263	.328	.359	101	91/ 91	-64	-69	-6.9	3.82	102	100/102	0	13	1.3	.1
BOS	66	86	624	719	.277	.335	.350	83	90/108	-65	50	5.0	4.01	83	95/ 79	-28	-125	-12.5	-2.5
PHI	58	93	687	900	.281	.337	.390	103	103/100	5	-17	-1.8	5.03	108	76/ 82	-178	-133	-13.3	-2.5
			702		.280	.338	.386						3.82						

AMERICAN LEAGUE 1927

Batting Runs			Park Adjusted			Pitching Runs			Park Adjusted		
Lou Gehrig	NY	100.8	Lou Gehrig	NY	101.5	Ted Lyons	CHI	44.5	Ted Lyons	CHI	51.0
Babe Ruth	NY	100.7	Babe Ruth	NY	101.4	Cy Moore	NY	43.9	Tommy Thomas	CHI	46.0
Harry Heilmann	DET	62.3	Harry Heilmann	DET	54.7	Waite Hoyt	NY	42.6	Cy Moore	NY	36.7
Normalized OPS			**Park Adjusted**			**Normalized ERA**			**Park Adjusted**		
Babe Ruth	NY	223	Babe Ruth	NY	224	Cy Moore	NY	181	Cy Moore	NY	168
Lou Gehrig	NY	217	Lou Gehrig	NY	219	Waite Hoyt	NY	157	Ted Lyons	CHI	153
Harry Heilmann	DET	181	Harry Heilmann	DET	165	Ted Lyons	CHI	146	Waite Hoyt	NY	145
On Base Average			**Slugging Percentage**			**Percent of Team Wins**			**Wins Above Team**		
Babe Ruth	NY	.487	Babe Ruth	NY	.772	Ted Lyons	CHI	.314	Ted Lyons	CHI	7.2
Harry Heilmann	DET	.475	Lou Gehrig	NY	.765	Slim Harriss	BOS	.275	Willis Hudlin	CLE	6.3
Lou Gehrig	NY	.474	Harry Heilmann	DET	.616	Willis Hudlin	CLE	.273	Tommy Thomas	CHI	3.9
Isolated Power			**Players Overall**			**Pitchers Overall**			**Defensive Runs**		
Babe Ruth	NY	.417	Babe Ruth	NY	92.5	Ted Lyons	CHI	55.8	Bibb Falk	CHI	21.9
Lou Gehrig	NY	.392	Lou Gehrig	NY	88.6	Cy Moore	NY	38.1	Mark Koenig	NY	16.3
Harry Heilmann	DET	.218	Tony Lazzeri	NY	39.4	Tommy Thomas	CHI	37.0	Charlie Gehringer	DET	16.1

Club	W	L	R	OR	Avg	OBA	SLG	BPF	NOPS-A	BR	Adj	Wins	ERA	PPF	NERA-A	PR	Adj	Wins	Diff
NY	110	44	975	599	.307	.384	.488	99	135/136	235	242	23.0	3.20	93	129/120	145	98	9.3	.7
PHI	91	63	841	726	.304	.373	.415	97	111/115	82	105	9.9	3.97	95	104/ 99	26	-5	-.6	4.7
WAS	85	69	782	730	.287	.351	.385	92	97/106	-25	37	3.5	3.97	90	104/ 94	26	-38	-3.7	8.1
DET	82	71	845	805	.289	.363	.409	110	107/ 97	47	-30	-2.9	4.14	110	100/110	0	65	6.2	2.2
CHI	70	83	662	708	.278	.344	.378	104	94/ 90	-53	-81	-7.8	3.91	105	106/111	35	64	6.0	-4.8
CLE	66	87	668	766	.283	.337	.379	100	92/ 93	-70	-67	-6.4	4.27	101	97/ 98	-19	-12	-1.3	-2.8
STL	59	94	724	904	.277	.339	.381	101	93/ 93	-62	-68	-6.5	4.95	104	84/ 87	-121	-95	-9.1	-1.9
BOS	51	103	597	856	.259	.320	.357	97	82/ 84	-149	-125	-12.0	4.72	101	88/ 88	-88	-82	-7.9	-6.1
			762		.286	.352	.399						4.14						

NATIONAL LEAGUE 1927

Batting Runs			Park Adjusted			Pitching Runs			Park Adjusted		
Rogers Hornsby	NY	62.5	Rogers Hornsby	NY	60.9	Pete Alexander	STL	41.5	Pete Alexander	STL	46.4
Paul Waner	PIT	54.4	Paul Waner	PIT	48.2	Jesse Haines	STL	39.8	Jesse Haines	STL	45.3
Hack Wilson	CHI	45.7	Hack Wilson	CHI	47.4	Dazzy Vance	BRO	36.7	Dazzy Vance	BRO	43.2
Normalized OPS			**Park Adjusted**			**Normalized ERA**			**Park Adjusted**		
Rogers Hornsby	NY	175	Rogers Hornsby	NY	172	Ray Kremer	PIT	158	Ray Kremer	PIT	166
Paul Waner	PIT	161	Hack Wilson	CHI	163	Pete Alexander	STL	155	Pete Alexander	STL	162
Hack Wilson	CHI	160	Paul Waner	PIT	150	Dazzy Vance	BRO	145	Dazzy Vance	BRO	153
On Base Average			**Slugging Percentage**			**Percent of Team Wins**			**Wins Above Team**		
Rogers Hornsby	NY	.448	Rogers Hornsby	NY	.586	Charlie Root	CHI	.306	Red Lucas	CIN	4.7
Paul Waner	PIT	.437	Hack Wilson	CHI	.579	Jesse Haines	STL	.261	Jesse Haines	STL	4.6
George Harper	NY	.435	Paul Waner	PIT	.543	Dazzy Vance	BRO	.246	Charlie Root	CHI	4.4
Isolated Power			**Players Overall**			**Pitchers Overall**			**Defensive Runs**		
Hack Wilson	CHI	.261	Frankie Frisch	STL	68.3	Pete Alexander	STL	49.9	Frankie Frisch	STL	50.3
Cy Williams	PHI	.228	Rogers Hornsby	NY	62.4	Jesse Haines	STL	44.0	Travis Jackson	NY	23.0
Rogers Hornsby	NY	.225	Travis Jackson	NY	49.0	Dazzy Vance	BRO	38.4	Bernie Friberg	PHI	16.4

Club	W	L	R	OR	Avg	OBA	SLG	BPF	NOPS-A	BR	Adj	Wins	ERA	PPF	NERA-A	PR	Adj	Wins	Diff
PIT	94	60	817	659	.305	.361	.412	107	115·107	99	45	4.5	3.65	105	107·112	40	70	6.9	5.6
STL	92	61	754	665	.278	.343	.408	105	109·104	47	9	.9	3.57	104	109·114	51	76	7.5	7.1
NY	92	62	817	720	.297	.356	.427	102	118·116	113	98	9.7	3.96	100	99· 99	-7	-4	-.5	5.8
CHI	85	68	750	661	.284	.346	.400	98	108·110	43	59	5.8	3.65	96	107·103	40	16	1.6	1.1
CIN	75	78	643	653	.278	.332	.367	98	94· 96	-46	-32	-3.3	3.54	98	111·108	57	45	4.5	-2.7
BRO	65	88	541	619	.253	.306	.342	104	80· 77	-151	-178	-17.7	3.36	105	117·123	85	118	11.6	-5.4
BOS	60	94	651	771	.279	.326	.363	95	91· 96	-72	-37	-3.8	4.22	97	93· 90	-47	-65	-6.6	-6.7
PHI	51	103	678	903	.280	.337	.370	93	97·104	-29	20	1.9	5.36	97	73· 71	-217	-236	-23.4	-4.5
			706		.282	.339	.386						3.91						

AMERICAN LEAGUE 1928

Batting Runs
Babe Ruth	NY	83.9
Lou Gehrig	NY	76.0
Heinie Manush	STL	50.3

Park Adjusted
Babe Ruth	NY	90.9
Lou Gehrig	NY	82.8
Heinie Manush	STL	47.0

Pitching Runs
Lefty Grove	PHI	42.6
Garland Braxton	WAS	36.8
Herb Pennock	NY	34.7

Park Adjusted
Garland Braxton	WAS	41.1
Lefty Grove	PHI	36.5
Sam Jones	WAS	34.4

Normalized OPS
Babe Ruth	NY	203
Lou Gehrig	NY	190
Goose Goslin	WAS	175

Park Adjusted
Babe Ruth	NY	221
Lou Gehrig	NY	207
Goose Goslin	WAS	168

Normalized ERA
Garland Braxton	WAS	160
Herb Pennock	NY	158
Lefty Grove	PHI	157

Park Adjusted
Garland Braxton	WAS	167
Lefty Grove	PHI	149
Sam Jones	WAS	148

On Base Average
Lou Gehrig	NY	.467
Babe Ruth	NY	.461
Goose Goslin	WAS	.443

Slugging Percentage
Babe Ruth	NY	.709
Lou Gehrig	NY	.648
Goose Goslin	WAS	.614

Percent of Team Wins
Ed Morris	BOS	.333
General Crowder	STL	.256
Lefty Grove	PHI	.245

Wins Above Team
General Crowder	STL	8.6
Ed Morris	BOS	8.1
Sam Jones	WAS	6.3

Isolated Power
Babe Ruth	NY	.386
Lou Gehrig	NY	.274
Goose Goslin	WAS	.235

Players Overall
Babe Ruth	NY	76.1
Lou Gehrig	NY	64.0
Goose Goslin	WAS	44.3

Pitchers Overall
Sam Jones	WAS	39.5
Garland Braxton	WAS	36.6
Lefty Grove	PHI	33.5

Defensive Runs
Joe Sewell	CLE	19.4
Fred Schulte	STL	11.0
William Regan	BOS	10.9

Club	W	L	R	OR	Avg	OBA	SLG	BPF	NOPS-A	BR	Adj	Wins	ERA	PPF	NERA-A	PR	Adj	Wins	Diff
NY	101	53	894	685	.295	.365	.450	92	123/134	143	204	19.8	3.73	88	108/ 95	46	-29	-2.9	7.1
PHI	98	55	829	615	.294	.363	.436	99	118/120	113	122	11.8	3.36	95	120/114	103	71	6.9	2.8
STL	82	72	772	742	.274	.346	.393	104	101/ 98	1	-27	-2.7	4.17	104	97/100	-19	2	.2	7.5
WAS	75	79	718	705	.284	.347	.393	104	102/ 97	2	-30	-3.0	3.88	104	104/109	25	52	5.1	-4.1
CHI	72	82	656	725	.270	.333	.357	101	88/ 87	-91	-100	-9.8	3.99	102	101/104	8	23	2.3	2.5
DET	68	86	744	804	.279	.340	.401	98	102/105	-1	14	1.4	4.32	99	93/ 92	-42	-49	-4.9	-5.5
CLE	62	92	674	830	.285	.336	.382	105	96/ 92	-45	-78	-7.7	4.47	108	90/ 97	-65	-18	-1.8	-5.5
BOS	57	96	589	770	.264	.319	.361	98	85/ 87	-117	-102	-9.9	4.40	101	92/ 93	-53	-48	-4.7	-4.9
			735		.281	.344	.397						4.04						

NATIONAL LEAGUE 1928

Batting Runs
Rogers Hornsby	BOS	72.5
Paul Waner	PIT	53.4
Jim Bottomley	STL	52.5

Park Adjusted
Rogers Hornsby	BOS	72.8
Paul Waner	PIT	52.7
Hack Wilson	CHI	47.7

Pitching Runs
Dazzy Vance	BRO	58.9
Larry Benton	NY	43.2
Sheriff Blake	CHI	40.6

Park Adjusted
Dazzy Vance	BRO	54.7
Larry Benton	NY	39.8
Bill Sherdel	STL	37.5

Normalized OPS
Rogers Hornsby	BOS	194
Jim Bottomley	STL	165
Paul Waner	PIT	158

Park Adjusted
Rogers Hornsby	BOS	194
Hack Wilson	CHI	169
Paul Waner	PIT	157

Normalized ERA
Dazzy Vance	BRO	191
Sheriff Blake	CHI	162
Art Nehf	CHI	151

Park Adjusted
Dazzy Vance	BRO	184
Bill Sherdel	STL	147
Sheriff Blake	CHI	145

On Base Average
Rogers Hornsby	BOS	.498
Paul Waner	PIT	.446
George Grantham	PIT	.408

Slugging Percentage
Rogers Hornsby	BOS	.632
Jim Bottomley	STL	.628
Chick Hafey	STL	.604

Percent of Team Wins
Burleigh Grimes	PIT	.294
Dazzy Vance	BRO	.286
Larry Benton	NY	.269

Wins Above Team
Dazzy Vance	BRO	7.5
Larry Benton	NY	5.7
Burleigh Grimes	PIT	4.3

Isolated Power
Jim Bottomley	STL	.304
Hack Wilson	CHI	.275
Chick Hafey	STL	.267

Players Overall
Rogers Hornsby	BOS	51.0
Paul Waner	PIT	49.6
Gabby Hartnett	CHI	44.7

Pitchers Overall
Dazzy Vance	BRO	56.5
Burleigh Grimes	PIT	50.9
Bill Sherdel	STL	36.2

Defensive Runs
Freddie Maguire	CHI	47.8
Taylor Douthit	STL	23.5
Travis Jackson	NY	20.1

Club	W	L	R	OR	Avg	OBA	SLG	BPF	NOPS-A	BR	Adj	Wins	ERA	PPF	NERA-A	PR	Adj	Wins	Diff
STL	95	59	807	636	.281	.353	.425	109	113/104	73	10	1.0	3.38	106	118/125	96	131	12.8	4.2
NY	93	61	807	653	.293	.349	.430	100	113/113	68	67	6.5	3.67	98	108/106	48	33	3.2	6.3
CHI	91	63	714	615	.278	.345	.402	93	104/112	11	65	6.4	3.40	90	117/105	90	28	2.8	4.8
PIT	85	67	837	704	.309	.364	.421	101	114/113	89	83	8.2	3.94	99	101/100	7	-1	-.2	1.0
CIN	78	74	648	686	.280	.333	.368	94	91/ 96	-75	-35	-3.5	3.94	94	101/ 95	6	-27	-2.7	8.2
BRO	77	76	665	640	.266	.340	.374	97	94/ 97	-45	-26	-2.6	3.25	97	122/118	113	92	9.0	-5.9
BOS	50	103	631	878	.275	.335	.367	100	91/ 92	-69	-67	-6.6	4.83	104	82/ 86	-128	-103	-10.2	-9.7
PHI	43	109	660	957	.267	.333	.382	104	95/ 92	-48	-77	-7.6	5.52	110	72/ 79	-230	-172	-16.9	-8.5
			721		.281	.344	.397						3.98						

AMERICAN LEAGUE 1929

Batting Runs			Park Adjusted			Pitching Runs			Park Adjusted		
Jimmie Foxx	PHI	63.2	Babe Ruth	NY	64.3	Lefty Grove	PHI	43.6	Lefty Grove	PHI	45.3
Babe Ruth	NY	61.9	Jimmie Foxx	PHI	58.5	Firpo Marberry	WAS	32.8	Firpo Marberry	WAS	35.3
Lou Gehrig	NY	51.3	Lou Gehrig	NY	54.1	Tommy Thomas	CHI	30.5	Willis Hudlin	CLE	31.2
Normalized OPS			**Park Adjusted**			**Normalized ERA**			**Park Adjusted**		
Babe Ruth	NY	186	Babe Ruth	NY	192	Lefty Grove	PHI	151	Lefty Grove	PHI	153
Jimmie Foxx	PHI	177	Jimmie Foxx	PHI	168	Firpo Marberry	WAS	139	Firpo Marberry	WAS	141
Al Simmons	PHI	163	Lou Gehrig	NY	164	Tommy Thomas	CHI	133	George Earnshaw	PHI	131
On Base Average			**Slugging Percentage**			**Percent of Team Wins**			**Wins Above Team**		
Jimmie Foxx	PHI	.463	Babe Ruth	NY	.697	Firpo Marberry	WAS	.268	Firpo Marberry	WAS	5.7
Lou Gehrig	NY	.431	Al Simmons	PHI	.642	Wes Ferrell	CLE	.259	Wes Ferrell	CLE	5.6
Tony Lazzeri	NY	.430	Jimmie Foxx	PHI	.625	Ed Morris	BOS	.241	Tom Zachary	NY	5.6
Isolated Power			**Players Overall**			**Pitchers Overall**			**Defensive Runs**		
Babe Ruth	NY	.353	Al Simmons	PHI	56.2	Lefty Grove	PHI	41.1	Roy Johnson	DET	19.2
Lou Gehrig	NY	.282	Tony Lazzeri	NY	48.3	Willis Hudlin	CLE	35.3	Ski Melillo	STL	19.1
Al Simmons	PHI	.277	Babe Ruth	NY	45.2	Firpo Marberry	WAS	34.1	Ray Gardner	CLE	18.7

Club	W	L	R	OR	Avg	OBA	SLG	BPF	NOPS-A	BR	Adj	Wins	ERA	PPF	NERA-A	PR	Adj	Wins	Diff
PHI	104	46	901	615	.296	.365	.450	106	119/112	112	69	6.5	3.43	101	124/125	122	130	12.3	10.2
NY	88	66	899	775	.296	.364	.451	97	119/123	114	139	13.1	4.19	95	101/96	8	-26	-2.5	.4
CLE	81	71	717	736	.293	.351	.415	102	105/103	17	1	.1	4.05	102	105/107	28	44	4.1	.7
STL	79	73	733	713	.276	.352	.381	100	96/95	-32	-35	-3.4	4.08	100	104/104	25	25	2.4	4.1
WAS	71	81	730	776	.276	.347	.375	101	93/92	-56	-65	-6.2	4.34	102	98/100	-14	-1	-.2	1.4
DET	70	84	926	928	.299	.360	.453	98	118/121	108	128	12.1	4.96	98	86/84	-110	-125	-11.9	-7.2
CHI	59	93	627	792	.268	.325	.363	95	84/88	-133	-99	-9.4	4.42	98	96/94	-26	-39	-3.8	-3.8
BOS	58	96	605	803	.267	.325	.365	101	84/84	-127	-134	-12.7	4.43	104	96/100	-28	-1	-.2	-6.1
			767		.284	.349	.407						4.24						

NATIONAL LEAGUE 1929

Batting Runs			Park Adjusted			Pitching Runs			Park Adjusted		
Rogers Hornsby	CHI	74.1	Rogers Hornsby	CHI	72.5	Burleigh Grimes	PIT	40.9	Burleigh Grimes	PIT	40.4
Lefty O'Doul	PHI	68.5	Lefty O'Doul	PHI	59.1	Charlie Root	CHI	37.3	Carl Hubbell	NY	36.2
Mel Ott	NY	58.4	Babe Herman	BRO	55.4	Pat Malone	CHI	33.7	Bill Walker	NY	36.1
Normalized OPS			**Park Adjusted**			**Normalized ERA**			**Park Adjusted**		
Rogers Hornsby	CHI	178	Rogers Hornsby	CHI	175	Bill Walker	NY	153	Bill Walker	NY	159
Lefty O'Doul	PHI	167	Babe Herman	BRO	167	Burleigh Grimes	PIT	150	Burleigh Grimes	PIT	150
Mel Ott	NY	165	Mel Ott	NY	154	Charlie Root	CHI	136	Charlie Root	CHI	133
On Base Average			**Slugging Percentage**			**Percent of Team Wins**			**Wins Above Team**		
Lefty O'Doul	PHI	.465	Rogers Hornsby	CHI	.679	Red Lucas	CIN	.288	Red Lucas	CIN	7.2
Rogers Hornsby	CHI	.459	Chuck Klein	PHI	.657	Watty Clark	BRO	.229	Johnny Morrison	BRO	4.4
Mel Ott	NY	.449	Mel Ott	NY	.635	Pat Malone	CHI	.224	Burleigh Grimes	PIT	3.8
Isolated Power			**Players Overall**			**Pitchers Overall**			**Defensive Runs**		
Mel Ott	NY	.306	Rogers Hornsby	CHI	78.2	Burleigh Grimes	PIT	49.0	Pinky Whitney	PHI	24.9
Chuck Klein	PHI	.302	Mel Ott	NY	54.1	Red Lucas	CIN	35.3	Travis Jackson	NY	18.3
Rogers Hornsby	CHI	.299	Pinky Whitney	PHI	43.3	Carl Hubbell	NY	33.5	Woody English	CHI	17.7

Club	W	L	R	OR	Avg	OBA	SLG	BPF	NOPS-A	BR	Adj	Wins	ERA	PPF	NERA-A	PR	Adj	Wins	Diff
CHI	98	54	982	758	.303	.373	.452	102	114/112	92	78	7.1	4.15	98	113/111	86	74	6.7	8.2
PIT	88	65	904	780	.303	.364	.430	101	106/105	28	16	1.5	4.35	100	108/108	54	52	4.7	5.3
NY	84	67	897	709	.296	.358	.436	107	106/99	18	-38	-3.6	3.97	104	119/124	113	144	13.1	-1.0
STL	78	74	831	806	.293	.354	.438	98	106/108	12	30	2.8	4.66	97	101/98	7	-11	-1.1	.4
PHI	71	82	897	1032	.309	.377	.467	109	119/109	125	46	4.2	6.13	112	77/86	-212	-125	-11.5	1.8
BRO	70	83	755	888	.291	.355	.427	93	103/110	-0	52	4.8	4.92	95	96/91	-31	-67	-6.2	-5.0
CIN	66	88	686	760	.281	.336	.379	96	85/88	-132	-104	-9.5	4.41	97	107/104	45	24	2.2	-3.7
BOS	56	98	657	876	.280	.335	.375	93	84/90	-141	-88	-8.1	5.12	96	92/88	-61	-88	-8.1	-4.8
				826	.294	.357	.426						4.71						

AMERICAN LEAGUE 1930

Batting Runs			Park Adjusted			Pitching Runs			Park Adjusted		
Babe Ruth	NY	89.5	Babe Ruth	NY	96.3	Lefty Grove	PHI	68.3	Lefty Grove	PHI	57.8
Lou Gehrig	NY	88.4	Lou Gehrig	NY	95.5	Wes Ferrell	CLE	44.4	Wes Ferrell	CLE	52.7
Al Simmons	PHI	64.2	Al Simmons	PHI	67.1	Lefty Stewart	STL	36.0	Lefty Stewart	STL	51.0
Normalized OPS			**Park Adjusted**			**Normalized ERA**			**Park Adjusted**		
Babe Ruth	NY	206	Babe Ruth	NY	222	Lefty Grove	PHI	183	Lefty Grove	PHI	170
Lou Gehrig	NY	198	Lou Gehrig	NY	213	Wes Ferrell	CLE	141	Lefty Stewart	STL	149
Al Simmons	PHI	181	Al Simmons	PHI	187	Lefty Stewart	STL	135	Wes Ferrell	CLE	148
On Base Average			**Slugging Percentage**			**Percent of Team Wins**			**Wins Above Team**		
Babe Ruth	NY	.493	Babe Ruth	NY	.732	Ted Lyons	CHI	.355	Ted Lyons	CHI	9.4
Lou Gehrig	NY	.473	Lou Gehrig	NY	.721	Lefty Stewart	STL	.313	Lefty Stewart	STL	8.5
Jimmie Foxx	PHI	.429	Al Simmons	PHI	.708	Wes Ferrell	CLE	.309	Lefty Grove	PHI	7.8
Isolated Power			**Players Overall**			**Pitchers Overall**			**Defensive Runs**		
Babe Ruth	NY	.373	Joe Cronin	WAS	87.4	Lefty Grove	PHI	57.4	Joe Cronin	WAS	32.4
Lou Gehrig	NY	.343	Babe Ruth	NY	86.6	Wes Ferrell	CLE	57.0	Ski Melillo	STL	21.3
Al Simmons	PHI	.327	Lou Gehrig	NY	77.4	Lefty Stewart	STL	53.1	Johnny Hodapp	CLE	18.3

Club	W	L	R	OR	Avg	OBA	SLG	BPF	NOPS-A	BR	Adj	Wins	ERA	PPF	NERA-A	PR	Adj	Wins	Diff
PHI	102	52	951	751	.294	.369	.452	96	116/121	102	132	11.9	4.28	93	109/101	56	7	.6	12.5
WAS	94	60	892	689	.302	.369	.425	96	109/113	57	90	8.2	3.97	92	117/108	104	49	4.4	4.4
NY	86	68	1062	898	.309	.384	.488	93	130/141	207	270	24.4	4.88	90	95/86	-34	-106	-9.7	-5.7
CLE	81	73	890	915	.304	.363	.431	105	109/105	51	13	1.2	4.88	105	95/100	-34	3	.3	2.5
DET	75	79	783	833	.284	.344	.421	103	102/99	-17	-42	-3.9	4.70	104	99/103	-7	20	1.8	.1
STL	64	90	751	886	.268	.333	.391	108	91/84	-94	-160	-14.6	5.07	111	92/101	-64	11	1.0	.6
CHI	62	92	729	884	.276	.329	.392	105	90/86	-107	-146	-13.3	4.71	107	99/106	-8	43	3.9	-5.6
BOS	52	102	612	814	.263	.312	.364	95	78/82	-194	-150	-13.6	4.70	97	99/96	-6	-28	-2.6	-8.7
				834	.288	.351	.421						4.65						

NATIONAL LEAGUE 1930

Batting Runs			Park Adjusted			Pitching Runs			Park Adjusted		
Hack Wilson	CHI	75.4	Babe Herman	BRO	70.4	Dazzy Vance	BRO	68.1	Dazzy Vance	BRO	62.3
Babe Herman	BRO	68.7	Hack Wilson	CHI	70.3	Pat Malone	CHI	31.2	Pat Malone	CHI	36.7
Chuck Klein	PHI	67.3	Chuck Klein	PHI	63.8	Carl Hubbell	NY	29.7	Carl Hubbell	NY	26.2
Normalized OPS			**Park Adjusted**			**Normalized ERA**			**Park Adjusted**		
Hack Wilson	CHI	179	Babe Herman	BRO	172	Dazzy Vance	BRO	191	Dazzy Vance	BRO	183
Babe Herman	BRO	169	Hack Wilson	CHI	170	Carl Hubbell	NY	129	Pat Malone	CHI	131
Chuck Klein	PHI	166	Chuck Klein	PHI	161	Bill Walker	NY	126	Carl Hubbell	NY	125
On Base Average			**Slugging Percentage**			**Percent of Team Wins**			**Wins Above Team**		
Mel Ott	NY	.458	Hack Wilson	CHI	.723	Phil Collins	PHI	.308	Phil Collins	PHI	8.3
Babe Herman	BRO	.455	Chuck Klein	PHI	.687	Ray Kremer	PIT	.250	Freddie Fitzsimmons	NY	5.2
Hack Wilson	CHI	.454	Babe Herman	BRO	.678	Red Lucas	CIN	.237	Erv Brame	PIT	4.8
Isolated Power			**Players Overall**			**Pitchers Overall**			**Defensive Runs**		
Hack Wilson	CHI	.368	Chuck Klein	PHI	83.3	Dazzy Vance	BRO	55.3	Chuck Klein	PHI	32.6
Chick Hafey	STL	.316	Bill Terry	NY	63.2	Pat Malone	CHI	37.8	Frankie Frisch	STL	23.7
Wally Berger	BOS	.305	Hack Wilson	CHI	61.6	Freddie Fitzsimmons	NY	25.5	Pinky Whitney	PHI	17.5

Club	W	L	R	OR	Avg	OBA	SLG	BPF	NOPS-A	BR	Adj	Wins	ERA	PPF	NERA-A	PR	Adj	Wins	Diff
STL	92	62	1004	784	.314	.372	.471	102	114/112	72	58	5.2	4.40	98	113/111	88	76	6.7	3.1
CHI	90	64	998	870	.309	.378	.481	105	118/112	110	64	5.7	4.80	104	104/107	27	54	4.8	2.5
NY	87	67	959	814	.319	.369	.473	100	114/114	67	70	6.2	4.61	97	108/105	54	35	3.1	.7
BRO	86	68	871	738	.304	.364	.454	98	107/109	21	36	3.2	4.03	96	123/118	144	114	10.0	-4.2
PIT	80	74	891	928	.303	.365	.449	99	106/107	15	20	1.8	5.24	100	95/95	-40	-40	-3.6	4.8
BOS	70	84	693	835	.281	.326	.393	100	82/82	-185	-180	-16.0	4.91	101	101/103	9	19	1.7	7.3
CIN	59	95	665	857	.281	.339	.400	91	87/95	-134	-59	-5.3	5.08	93	98/91	-15	-67	-6.0	-6.7
PHI	52	102	944	1199	.315	.367	.458	103	109/106	37	7	.6	6.71	108	74/80	-264	-206	-18.2	-7.4
				878	.303	.360	.448						4.97						

Batting Runs			Park Adjusted			Pitching Runs			Park Adjusted		
Babe Ruth	NY	91.5	Babe Ruth	NY	94.1	Lefty Grove	PHI	74.6	Lefty Grove	PHI	75.4
Lou Gehrig	NY	79.9	Lou Gehrig	NY	82.7	Lefty Gomez	NY	47.3	Lefty Gomez	NY	37.8
Al Simmons	PHI	59.6	Al Simmons	PHI	56.5	Lloyd Brown	WAS	34.0	George Uhle	DET	29.3
Normalized OPS			**Park Adjusted**			**Normalized ERA**			**Park Adjusted**		
Babe Ruth	NY	211	Babe Ruth	NY	217	Lefty Grove	PHI	213	Lefty Grove	PHI	214
Lou Gehrig	NY	188	Lou Gehrig	NY	193	Lefty Gomez	NY	167	Lefty Gomez	NY	153
Al Simmons	PHI	182	Al Simmons	PHI	175	Bump Hadley	WAS	144	George Uhle	DET	139
On Base Average			**Slugging Percentage**			**Percent of Team Wins**			**Wins Above Team**		
Babe Ruth	NY	.495	Babe Ruth	NY	.700	Lefty Grove	PHI	.290	Lefty Grove	PHI	8.3
Ed Morgan	CLE	.451	Lou Gehrig	NY	.662	Wes Ferrell	CLE	.282	Wes Ferrell	CLE	6.1
Lou Gehrig	NY	.446	Al Simmons	PHI	.641	Danny MacFayden	BOS	.258	Danny MacFayden	BOS	5.6
Isolated Power			**Players Overall**			**Pitchers Overall**			**Defensive Runs**		
Babe Ruth	NY	.328	Babe Ruth	NY	72.7	Lefty Grove	PHI	69.9	Ski Melillo	STL	34.7
Lou Gehrig	NY	.321	Lou Gehrig	NY	56.9	Wes Ferrell	CLE	46.0	Jack Burns	STL	18.5
Jimmie Foxx	PHI	.276	Joe Cronin	WAS	55.5	George Uhle	DET	33.5	Roy Johnson	DET	17.2

Club	W	L	R	OR	Avg	OBA	SLG	BPF	NOPS-A	BR	Adj	Wins	ERA	PPF	NERA-A	PR	Adj	Wins	Diff
PHI	107	45	858	626	.287	.355	.434	104	116/111	93	60	5.6	3.47	101	126/127	138	142	13.2	12.2
NY	94	59	1067	760	.297	.383	.457	97	130/134	220	245	22.8	4.20	92	104/ 96	28	-26	-2.5	-2.8
WAS	92	62	843	691	.286	.346	.400	98	104/106	11	31	2.9	3.76	95	116/110	96	59	5.5	6.6
CLE	78	76	885	833	.296	.364	.420	104	114/110	92	63	5.9	4.63	103	95/ 98	-37	-15	-1.5	-3.4
STL	63	91	722	870	.271	.333	.389	102	97/ 96	-39	-52	-4.9	4.76	104	92/ 96	-57	-29	-2.8	-6.3
BOS	62	90	625	800	.262	.315	.349	93	81/ 87	-152	-100	-9.4	4.60	95	95/ 91	-32	-62	-5.9	1.3
DET	61	93	651	836	.268	.330	.371	107	91/ 85	-76	-134	-12.6	4.59	111	95/106	-32	41	3.8	-7.2
CHI	56	97	704	939	.260	.323	.343	95	81/ 85	-144	-106	-10.0	5.04	99	87/ 86	-101	-110	-10.3	-.2
				794	.279	.344	.396						4.38						

Batting Runs			Park Adjusted			Pitching Runs			Park Adjusted		
Chuck Klein	PHI	48.9	Chuck Klein	PHI	45.0	Bill Walker	NY	42.6	Bill Walker	NY	37.0
Bill Terry	NY	39.0	Bill Terry	NY	40.7	Carl Hubbell	NY	33.5	Ed Brandt	BOS	28.8
Chick Hafey	STL	36.2	Mel Ott	NY	36.5	Heinie Meine	PIT	27.9	Heinie Meine	PIT	28.7
Normalized OPS			**Park Adjusted**			**Normalized ERA**			**Park Adjusted**		
Chuck Klein	PHI	161	Chuck Klein	PHI	153	Bill Walker	NY	171	Bill Walker	NY	162
Chick Hafey	STL	159	Mel Ott	NY	153	Carl Hubbell	NY	146	Carl Hubbell	NY	138
Mel Ott	NY	149	Bill Terry	NY	150	Ed Brandt	BOS	133	Syl Johnson	STL	137
On Base Average			**Slugging Percentage**			**Percent of Team Wins**			**Wins Above Team**		
Chick Hafey	STL	.404	Chuck Klein	PHI	.584	Jumbo Elliott	PHI	.288	Ed Brandt	BOS	7.3
Kiki Cuyler	CHI	.404	Chick Hafey	STL	.569	Ed Brandt	BOS	.281	Jumbo Elliott	PHI	6.2
Paul Waner	PIT	.404	Mel Ott	NY	.545	Heinie Meine	PIT	.253	Red Lucas	CIN	4.6
Isolated Power			**Players Overall**			**Pitchers Overall**			**Defensive Runs**		
Mel Ott	NY	.254	Rogers Hornsby	CHI	39.5	Ed Brandt	BOS	35.4	Paul Waner	PIT	20.5
Chuck Klein	PHI	.247	Paul Waner	PIT	37.9	Freddie Fitzsimmons	NY	33.5	Frankie Frisch	STL	14.4
Buzz Arlett	PHI	.225	Bill Terry	NY	37.0	Carl Hubbell	NY	28.5	Lloyd Waner	PIT	12.5

Club	W	L	R	OR	Avg	OBA	SLG	BPF	NOPS-A	BR	Adj	Wins	ERA	PPF	NERA-A	PR	Adj	Wins	Diff
STL	101	53	815	614	.286	.342	.411	110	112/102	61	-5	-.6	3.45	107	112/119	64	103	10.2	14.4
NY	87	65	768	599	.289	.340	.416	98	112/115	61	76	7.6	3.30	95	117/110	84	52	5.2	-1.8
CHI	84	70	828	710	.289	.360	.423	95	120/126	131	166	16.5	3.98	93	97/ 90	-17	-61	-6.2	-3.3
BRO	79	73	681	673	.276	.331	.390	101	102/102	-2	-7	-.8	3.85	100	100/101	3	5	.5	3.2
PIT	75	79	636	691	.266	.330	.360	100	93/ 93	-53	-51	-5.2	3.65	101	106/106	33	36	3.6	-.4
PHI	66	88	684	828	.279	.336	.400	105	107/101	27	-7	-.8	4.58	108	84/ 91	-107	-58	-5.9	-4.3
BOS	64	90	533	680	.258	.309	.341	100	82/ 82	-136	-135	-13.5	3.90	102	99/101	-5	8	.8	-.3
CIN	58	96	592	742	.269	.323	.352	94	89/ 95	-84	-40	-4.1	4.22	96	92/ 88	-52	-76	-7.6	-7.3
				692	.277	.334	.387						3.86						

AMERICAN LEAGUE 1932

Batting Runs			Park Adjusted			Pitching Runs			Park Adjusted		
Jimmie Foxx	PHI	96.7	Jimmie Foxx	PHI	85.7	Lefty Grove	PHI	53.3	Lefty Grove	PHI	86.7
Babe Ruth	NY	71.2	Babe Ruth	NY	76.8	General Crowder	WAS	41.7	Wes Ferrell	CLE	40.9
Lou Gehrig	NY	68.6	Lou Gehrig	NY	75.2	Red Ruffing	NY	39.9	General Crowder	WAS	34.6
Normalized OPS			**Park Adjusted**			**Normalized ERA**			**Park Adjusted**		
Jimmie Foxx	PHI	210	Babe Ruth	NY	210	Lefty Grove	PHI	158	Lefty Grove	PHI	172
Babe Ruth	NY	195	Jimmie Foxx	PHI	188	Red Ruffing	NY	145	Tommy Bridges	DET	138
Lou Gehrig	NY	175	Lou Gehrig	NY	188	Ted Lyons	CHI	137	Wes Ferrell	CLE	135
On Base Average			**Slugging Percentage**			**Percent of Team Wins**			**Wins Above Team**		
Babe Ruth	NY	.489	Jimmie Foxx	PHI	.749	General Crowder	WAS	.280	Bob Kline	BOS	5.1
Jimmie Foxx	PHI	.469	Babe Ruth	NY	.661	Lefty Grove	PHI	.266	Lefty Grove	PHI	4.7
Lou Gehrig	NY	.451	Lou Gehrig	NY	.621	Wes Ferrell	CLE	.264	Monte Weaver	WAS	3.4
Isolated Power			**Players Overall**			**Pitchers Overall**			**Defensive Runs**		
Jimmie Foxx	PHI	.385	Jimmie Foxx	PHI	67.0	Lefty Grove	PHI	61.8	Rabbit Warstler	BOS	21.7
Babe Ruth	NY	.319	Babe Ruth	NY	63.3	Wes Ferrell	CLE	45.3	Earl Averill	CLE	16.2
Lou Gehrig	NY	.272	Lou Gehrig	NY	52.1	Mel Harder	CLE	35.6	Sam West	WAS	14.2

Club	W	L	R	OR	Avg	OBA	SLG	BPF	NOPS-A	BR	Adj	Wins	ERA	PPF	NERA-A	PR	Adj	Wins	Diff
NY	107	47	1002	724	.286	.376	.454	93	124/134	172	231	21.4	3.98	88	112/ 99	78	-7	-.7	9.4
PHI	94	60	981	752	.290	.367	.457	112	123/110	150	52	4.8	4.45	109	101/110	4	67	6.2	5.9
WAS	93	61	840	716	.284	.347	.408	98	104/106	8	25	2.3	4.16	96	108/103	50	19	1.8	11.9
CLE	87	65	845	747	.285	.357	.413	111	108/ 97	44	-45	-4.2	4.11	110	109/120	56	126	11.6	3.6
DET	76	75	799	787	.273	.335	.401	103	99/ 96	-34	-61	-5.7	4.30	103	104/108	27	50	4.6	1.6
STL	63	91	736	898	.276	.340	.389	100	97/ 97	-42	-40	-3.8	5.01	102	89/ 92	-80	-63	-6.0	-4.2
CHI	49	102	667	897	.267	.327	.359	90	85/ 95	-123	-43	-4.0	4.83	93	93/ 86	-51	-101	-9.4	-13.0
BOS	43	111	566	915	.251	.314	.351	98	79/ 81	-170	-152	-14.1	5.01	103	89/ 92	-80	-60	-5.7	-14.2
			805		.277	.346	.404						4.48						

NATIONAL LEAGUE 1932

Batting Runs			Park Adjusted			Pitching Runs			Park Adjusted		
Chuck Klein	PHI	68.2	Chuck Klein	PHI	62.4	Lon Warneke	CHI	46.4	Carl Hubbell	NY	46.9
Mel Ott	NY	59.8	Mel Ott	NY	57.0	Carl Hubbell	NY	43.4	Lon Warneke	CHI	44.7
Lefty O'Doul	BRO	51.2	Lefty O'Doul	BRO	54.5	Red Lucas	CIN	27.9	Red Lucas	CIN	30.1
Normalized OPS			**Park Adjusted**			**Normalized ERA**			**Park Adjusted**		
Chuck Klein	PHI	177	Lefty O'Doul	BRO	167	Lon Warneke	CHI	164	Lon Warneke	CHI	161
Mel Ott	NY	172	Mel Ott	NY	166	Carl Hubbell	NY	155	Carl Hubbell	NY	159
Lefty O'Doul	BRO	160	Chuck Klein	PHI	166	Huck Betts	BOS	139	Steve Swetonic	PIT	137
On Base Average			**Slugging Percentage**			**Percent of Team Wins**			**Wins Above Team**		
Mel Ott	NY	.424	Chuck Klein	PHI	.646	Carl Hubbell	NY	.250	Lon Warneke	CHI	6.9
Lefty O'Doul	BRO	.423	Mel Ott	NY	.601	Dizzy Dean	STL	.250	Carl Hubbell	NY	5.5
Don Hurst	PHI	.412	Bill Terry	NY	.580	Watty Clark	BRO	.247	Bob Brown	BOS	4.1
Isolated Power			**Players Overall**			**Pitchers Overall**			**Defensive Runs**		
Chuck Klein	PHI	.298	Chuck Klein	PHI	57.1	Carl Hubbell	NY	55.5	Billy Jurges	CHI	26.5
Mel Ott	NY	.283	Bill Terry	NY	47.7	Lon Warneke	CHI	44.1	Tony Cuccinello	BRO	20.1
Hack Wilson	BRO	.241	Mel Ott	NY	44.7	Red Lucas	CIN	43.3	Billy Herman	CHI	17.9

Club	W	L	R	OR	Avg	OBA	SLG	BPF	NOPS-A	BR	Adj	Wins	ERA	PPF	NERA-A	PR	Adj	Wins	Diff
CHI	90	64	720	633	.278	.330	.392	100	103/103	0	-1	-.2	3.44	99	113/111	68	59	5.9	7.3
PIT	86	68	701	711	.285	.333	.395	99	105/105	10	14	1.4	3.75	100	104/103	20	18	1.8	5.9
BRO	81	73	752	747	.283	.334	.419	96	112/117	54	84	8.3	4.27	95	91/ 87	-59	-87	-8.7	4.4
PHI	78	76	844	796	.292	.348	.442	107	123/115	132	83	8.2	4.47	107	87/ 93	-90	-49	-4.9	-2.2
BOS	77	77	649	655	.265	.311	.366	94	90/ 96	-93	-49	-4.9	3.52	93	110/103	56	15	1.5	3.4
STL	72	82	684	717	.269	.324	.385	101	99/ 99	-26	-29	-3.0	3.97	101	98/ 99	-14	-7	-.8	-1.2
NY	72	82	755	706	.276	.322	.406	103	105/102	1	-23	-2.4	3.83	103	101/104	8	25	2.4	-5.1
CIN	60	94	575	715	.263	.320	.362	100	91/ 92	-75	-72	-7.2	3.79	102	102/104	14	26	2.5	-12.3
			710		.276	.328	.396						3.88						

AMERICAN LEAGUE 1933

Batting Runs			Park Adjusted			Pitching Runs			Park Adjusted		
Jimmie Foxx	PHI	82.8	Jimmie Foxx	PHI	91.3	Mel Harder	CLE	37.4	Mel Harder	CLE	39.1
Lou Gehrig	NY	59.5	Lou Gehrig	NY	69.5	Lefty Grove	PHI	32.9	Bump Hadley	STL	38.5
Babe Ruth	NY	49.6	Babe Ruth	NY	58.0	Tommy Bridges	DET	30.9	Tommy Bridges	DET	35.1

Normalized OPS			Park Adjusted			Normalized ERA			Park Adjusted		
Jimmie Foxx	PHI	201	Jimmie Foxx	PHI	224	Mel Harder	CLE	145	Mel Harder	CLE	147
Lou Gehrig	NY	170	Lou Gehrig	NY	192	Tommy Bridges	DET	139	Tommy Bridges	DET	144
Babe Ruth	NY	169	Babe Ruth	NY	191	Lefty Gomez	NY	135	Firpo Marberry	DET	135

On Base Average			Slugging Percentage			Percent of Team Wins			Wins Above Team		
Mickey Cochrane	PHI	.459	Jimmie Foxx	PHI	.703	Lefty Grove	PHI	.304	Lefty Grove	PHI	9.2
Jimmie Foxx	PHI	.449	Lou Gehrig	NY	.605	Bump Hadley	STL	.273	Bump Hadley	STL	3.5
Max Bishop	PHI	.446	Babe Ruth	NY	.582	George Blaeholder	STL	.273	George Blaeholder	STL	3.4

Isolated Power			Players Overall			Pitchers Overall			Defensive Runs		
Jimmie Foxx	PHI	.347	Jimmie Foxx	PHI	84.3	Mel Harder	CLE	46.4	Jimmie Foxx	DET	19.0
Babe Ruth	NY	.281	Mickey Cochrane	PHI	54.8	Tommy Bridges	DET	37.2	Jackie Hayes	CHI	13.5
Lou Gehrig	NY	.272	Lou Gehrig	NY	53.4	Monte Pearson	CLE	30.2	Ski Melillo	STL	13.2

Club	W	L	R	OR	Avg	OBA	SLG	BPF	NOPS-A	BR	Adj	Wins	ERA	PPF	NERA-A	PR	Adj	Wins	Diff
WAS	99	53	850	665	.287	.352	.402	104	108/103	47	13	1.3	3.82	101	112/114	71	80	7.6	14.1
NY	91	59	927	768	.284	.369	.440	89	123/139	155	243	23.0	4.36	85	98/ 84	-11	-107	-10.2	3.3
PHI	79	72	875	853	.285	.362	.440	90	121/135	136	212	20.1	4.82	89	89/ 79	-78	-148	-14.1	-2.5
CLE	75	76	654	669	.260	.320	.360	101	87/ 86	-104	-113	-10.8	3.71	101	115/117	86	95	9.0	1.3
DET	75	79	722	733	.269	.329	.380	103	95/ 92	-52	-78	-7.5	3.96	104	108/112	50	76	7.2	-1.7
CHI	67	83	683	814	.272	.342	.360	101	93/ 92	-48	-54	-5.2	4.45	103	96/ 99	-24	-4	-.4	-2.3
BOS	63	86	700	758	.270	.339	.377	102	97/ 96	-29	-40	-3.8	4.35	103	98/101	-9	7	.6	-8.3
STL	55	96	669	820	.253	.322	.360	113	88/ 77	-101	-202	-19.2	4.82	117	89/104	-80	29	2.7	-4.0
			760		.273	.342	.390						4.28						

NATIONAL LEAGUE 1933

Batting Runs			Park Adjusted			Pitching Runs			Park Adjusted		
Chuck Klein	PHI	69.1	Chuck Klein	PHI	58.1	Carl Hubbell	NY	57.5	Carl Hubbell	NY	55.1
Wally Berger	BOS	40.5	Wally Berger	BOS	43.6	Lon Warneke	CHI	42.3	Hal Schumacher	NY	31.9
Arky Vaughan	PIT	35.2	Arky Vaughan	PIT	40.8	Hal Schumacher	NY	33.9	Lon Warneke	CHI	31.0

Normalized OPS			Park Adjusted			Normalized ERA			Park Adjusted		
Chuck Klein	PHI	190	Wally Berger	BOS	172	Carl Hubbell	NY	201	Carl Hubbell	NY	197
Wally Berger	BOS	163	Chuck Klein	PHI	164	Lon Warneke	CHI	166	Hal Schumacher	NY	152
Spud Davis	PHI	147	Arky Vaughan	PIT	159	Hal Schumacher	NY	155	Lon Warneke	CHI	149

On Base Average			Slugging Percentage			Percent of Team Wins			Wins Above Team		
Chuck Klein	PHI	.422	Chuck Klein	PHI	.602	Carl Hubbell	NY	.253	Ben Cantwell	BOS	4.8
Spud Davis	PHI	.395	Wally Berger	BOS	.566	Van Mungo	BRO	.246	Van Mungo	BRO	3.5
Arky Vaughan	PIT	.388	Babe Herman	CHI	.502	Dizzy Dean	STL	.244	Eppa Rixey	CIN	2.7

Isolated Power			Players Overall			Pitchers Overall			Defensive Runs		
Wally Berger	BOS	.254	Chuck Klein	PHI	55.3	Carl Hubbell	NY	62.0	Hughie Critz	NY	40.0
Chuck Klein	PHI	.234	Arky Vaughan	PIT	42.2	Lon Warneke	CHI	44.0	Billy Herman	CHI	28.7
Babe Herman	CHI	.213	Billy Herman	CHI	41.5	Hal Schumacher	NY	35.9	Billy Jurges	CHI	21.6

Club	W	L	R	OR	Avg	OBA	SLG	BPF	NOPS-A	BR	Adj	Wins	ERA	PPF	NERA-A	PR	Adj	Wins	Diff
NY	91	61	636	515	.263	.312	.361	101	100/ 99	-13	-17	-1.9	2.71	98	123/121	98	87	9.2	7.7
PIT	87	67	667	619	.285	.333	.383	92	113/123	71	122	12.9	3.27	90	102/ 92	10	-39	-4.3	1.4
CHI	86	68	646	536	.271	.325	.380	93	109/118	47	92	9.8	2.92	89	114/10	63	10	1.0	-1.8
BOS	83	71	552	531	.252	.299	.345	95	91/ 96	-69	-38	-4.1	2.96	94	113/106	58	26	2.7	7.4
STL	82	71	687	609	.276	.329	.378	105	110/105	53	25	2.6	3.37	103	99/102	-5	11	1.1	1.7
BRO	65	88	617	695	.263	.316	.359	99	100/101	-7	-2	-.3	3.73	101	89/ 90	-60	-55	-5.9	-5.3
PHI	60	92	607	760	.274	.326	.369	116	106/ 92	32	-62	-6.6	4.34	120	77/ 92	-149	-49	-5.3	-4.1
CIN	58	94	496	643	.246	.298	.320	98	83/ 85	-110	-95	-10.2	3.42	101	97/ 98	-12	-10	-1.1	-6.7
			614		.266	.317	.362						3.33						

AMERICAN LEAGUE 1934

Batting Runs			Park Adjusted			Pitching Runs			Park Adjusted		
Lou Gehrig	NY	85.6	Lou Gehrig	NY	93.6	Lefty Gomez	NY	67.9	Lefty Gomez	NY	50.0
Jimmie Foxx	PHI	66.4	Jimmie Foxx	PHI	66.3	Mel Harder	CLE	53.4	Mel Harder	CLE	49.1
Charlie Gehringer	DET	47.8	Charlie Gehringer	DET	48.8	Johnny Murphy	NY	31.9	Fritz Ostermueller	BOS	28.0
Normalized OPS			**Park Adjusted**			**Normalized ERA**			**Park Adjusted**		
Lou Gehrig	NY	199	Lou Gehrig	NY	218	Lefty Gomez	NY	193	Lefty Gomez	NY	168
Jimmie Foxx	PHI	181	Jimmie Foxx	PHI	181	Mel Harder	CLE	172	Mel Harder	CLE	166
Hank Greenberg	DET	156	Babe Ruth	NY	168	Johnny Murphy	NY	144	Bobby Burke	WAS	145
On Base Average			**Slugging Percentage**			**Percent of Team Wins**			**Wins Above Team**		
Lou Gehrig	NY	.465	Lou Gehrig	NY	.706	Lefty Gomez	NY	.277	Lefty Gomez	NY	8.9
Charlie Gehringer	DET	.450	Jimmie Foxx	PHI	.653	George Earnshaw	CHI	.264	George Earnshaw	CHI	6.3
Jimmie Foxx	PHI	.449	Hank Greenberg	DET	.600	Buck Newsom	STL	.239	Wes Ferrell	BOS	5.1
Isolated Power			**Players Overall**			**Pitchers Overall**			**Defensive Runs**		
Lou Gehrig	NY	.344	Lou Gehrig	NY	68.0	Mel Harder	CLE	49.8	Odell Hale	CLE	24.1
Jimmie Foxx	PHI	.319	Charlie Gehringer	DET	66.9	Lefty Gomez	NY	44.7	Billy Werber	BOS	23.6
Hal Trosky	CLE	.269	Earl Averill	CLE	50.6	Schoolboy Rowe	DET	34.6	Joe Cronin	WAS	15.3

Club	W	L	R	OR	Avg	OBA	SLG	BPF	NOPS-A	BR	Adj	Wins	ERA	PPF	NERA-A	PR	Adj	Wins	Diff
DET	101	53	958	708	.300	.376	.424	99	115/116	111	120	11.2	4.06	95	111/105	67	31	2.9	10.0
NY	94	60	842	669	.278	.364	.419	91	111/121	70	141	13.1	3.75	87	120/105	114	26	2.4	1.4
CLE	85	69	814	763	.287	.352	.423	98	109/111	43	61	5.7	4.28	97	105/101	32	9	.9	1.4
BOS	76	76	820	775	.274	.350	.383	106	97/ 92	-27	-73	-6.9	4.32	106	104/110	27	65	6.1	.8
PHI	68	82	764	838	.280	.343	.425	100	107/107	20	19	1.8	5.01	101	90/ 91	-75	-66	-6.2	-2.5
STL	67	85	674	800	.268	.335	.373	104	90/ 86	-84	-118	-11.1	4.49	107	100/107	1	46	4.2	-2.2
WAS	66	86	729	806	.278	.348	.382	102	96/ 94	-34	-54	-5.1	4.67	104	96/100	-26	-0	-.1	-4.9
CHI	53	99	704	946	.263	.336	.363	99	88/ 89	-96	-88	-8.3	5.41	103	83/ 86	-136	-116	-10.9	-3.8
				788	.279	.351	.399						4.50						

NATIONAL LEAGUE 1934

Batting Runs			Park Adjusted			Pitching Runs			Park Adjusted		
Mel Ott	NY	55.2	Mel Ott	NY	56.8	Carl Hubbell	NY	61.3	Dizzy Dean	STL	68.5
Ripper Collins	STL	53.3	Paul Waner	PIT	47.4	Dizzy Dean	STL	48.8	Carl Hubbell	NY	53.5
Paul Waner	PIT	50.2	Arky Vaughan	PIT	41.6	Curt Davis	PHI	33.7	Curt Davis	PHI	43.6
Normalized OPS			**Park Adjusted**			**Normalized ERA**			**Park Adjusted**		
Mel Ott	NY	166	Mel Ott	NY	169	Carl Hubbell	NY	177	Dizzy Dean	STL	174
Ripper Collins	STL	165	Len Koenecke	BRO	157	Dizzy Dean	STL	153	Carl Hubbell	NY	167
Paul Waner	PIT	157	Wally Berger	BOS	155	Waite Hoyt	PIT	139	Bill Walker	STL	149
On Base Average			**Slugging Percentage**			**Percent of Team Wins**			**Wins Above Team**		
Arky Vaughan	PIT	.431	Ripper Collins	STL	.615	Curt Davis	PHI	.339	Dizzy Dean	STL	9.3
Paul Waner	PIT	.429	Mel Ott	NY	.591	Dizzy Dean	STL	.316	Curt Davis	PHI	7.2
Mel Ott	NY	.415	Wally Berger	BOS	.546	Paul Derringer	CIN	.288	Waite Hoyt	PIT	5.4
Isolated Power			**Players Overall**			**Pitchers Overall**			**Defensive Runs**		
Ripper Collins	STL	.282	Arky Vaughan	PIT	46.7	Dizzy Dean	STL	67.8	Hughie Critz	NY	28.0
Mel Ott	NY	.265	Paul Waner	PIT	44.1	Carl Hubbell	NY	56.1	Blondy Ryan	NY	15.0
Wally Berger	BOS	.249	Mel Ott	NY	39.0	Curt Davis	PHI	51.7	John Vergez	NY	14.6

Club	W	L	R	OR	Avg	OBA	SLG	BPF	NOPS-A	BR	Adj	Wins	ERA	PPF	NERA-A	PR	Adj	Wins	Diff
STL	95	58	799	656	.288	.337	.425	115	113/ 98	63	-48	-4.8	3.68	114	110/126	58	146	14.2	9.0
NY	93	60	760	583	.275	.329	.405	98	105/107	8	23	2.2	3.19	95	127/120	132	98	9.6	4.7
CHI	86	65	705	639	.279	.330	.402	100	104/105	5	8	.8	3.76	98	108/106	46	35	3.4	6.3
BOS	78	73	683	714	.272	.323	.378	89	96/108	-51	30	2.9	4.11	88	99/ 87	-7	-80	-7.9	7.5
PIT	74	76	735	713	.287	.344	.398	103	107/103	34	10	.9	4.20	103	97/100	-19	1	.1	-2.0
BRO	71	81	748	795	.281	.350	.396	92	108/117	51	108	10.5	4.48	93	91/ 84	-62	-108	-10.6	-4.9
PHI	56	93	675	794	.284	.338	.384	105	101/ 96	-2	-38	-3.8	4.76	108	85/ 92	-100	-53	-5.2	-9.4
CIN	52	99	590	801	.266	.311	.364	100	88/ 88	-104	-107	-10.5	4.36	104	93/ 97	-44	-18	-1.8	-11.1
				712	.279	.333	.394						4.06						

AMERICAN LEAGUE 1935

Batting Runs			Park Adjusted			Pitching Runs			Park Adjusted		
Jimmie Foxx	PHI	67.2	Jimmie Foxx	PHI	69.6	Lefty Grove	BOS	53.1	Lefty Grove	BOS	66.5
Lou Gehrig	NY	61.6	Hank Greenberg	DET	64.1	Mel Harder	CLE	37.0	Wes Ferrell	BOS	49.1
Hank Greenberg	DET	57.5	Lou Gehrig	NY	62.5	Lefty Gomez	NY	34.7	Ted Lyons	CHI	36.4
Normalized OPS			**Park Adjusted**			**Normalized ERA**			**Park Adjusted**		
Jimmie Foxx	PHI	181	Jimmie Foxx	PHI	186	Lefty Grove	BOS	165	Lefty Grove	BOS	181
Lou Gehrig	NY	170	Hank Greenberg	DET	178	Ted Lyons	CHI	148	Ted Lyons	CHI	157
Hank Greenberg	DET	165	Lou Gehrig	NY	172	Red Ruffing	NY	143	Ivy Andrews	STL	142
On Base Average			**Slugging Percentage**			**Percent of Team Wins**			**Wins Above Team**		
Lou Gehrig	NY	.466	Jimmie Foxx	PHI	.636	Wes Ferrell	BOS	.321	Johnny Marcum	PHI	7.1
Jimmie Foxx	PHI	.461	Hank Greenberg	DET	.628	Johnny Marcum	PHI	.293	Wes Ferrell	BOS	6.9
Mickey Cochrane	DET	.452	Lou Gehrig	NY	.583	Mel Harder	CLE	.268	Mel Harder	CLE	5.5
Isolated Power			**Players Overall**			**Pitchers Overall**			**Defensive Runs**		
Hank Greenberg	DET	.300	Jimmie Foxx	PHI	62.5	Wes Ferrell	BOS	71.9	Luke Appling	CHI	24.0
Jimmie Foxx	PHI	.290	Buddy Myer	WAS	62.0	Lefty Grove	BOS	61.5	Cecil Travis	WAS	18.9
Lou Gehrig	NY	.254	Hank Greenberg	DET	53.7	Red Ruffing	NY	37.6	Billy Werber	BOS	17.0

Club	W	L	R	OR	Avg	OBA	SLG	BPF	NOPS-A	BR	Adj	Wins	ERA	PPF	NERA-A	PR	Adj	Wins	Diff
DET	93	58	919	665	.290	.366	.435	93	115/124	96	153	14.3	3.82	88	117/102	96	14	1.3	1.8
NY	89	60	818	632	.280	.358	.416	99	108/109	42	51	4.8	3.60	96	124/118	126	97	9.1	.6
CLE	82	71	776	739	.284	.341	.421	97	105/108	3	29	2.7	4.15	96	107/103	47	17	1.5	1.2
BOS	78	75	718	732	.276	.353	.392	109	99/ 91	-9	-80	-7.6	4.05	110	110/121	62	129	12.1	-3.0
CHI	74	78	738	750	.275	.348	.382	105	95/ 90	-40	-82	-7.8	4.37	106	102/108	12	54	5.0	.7
WAS	67	86	823	903	.285	.357	.381	91	98/107	-17	51	4.8	5.25	92	85/ 78	-121	-174	-16.3	2.1
STL	65	87	718	930	.270	.344	.384	109	95/ 87	-49	-121	-11.5	5.26	113	85/ 96	-122	-31	-3.0	3.5
PHI	58	91	710	869	.279	.341	.406	97	100/103	-21	-0	-.1	5.12	100	87/ 87	-97	-98	-9.3	-7.1
			778		.280	.351	.402						4.45						

NATIONAL LEAGUE 1935

Batting Runs			Park Adjusted			Pitching Runs			Park Adjusted		
Arky Vaughan	PIT	71.9	Arky Vaughan	PIT	64.1	Cy Blanton	PIT	40.5	Cy Blanton	PIT	51.1
Mel Ott	NY	47.6	Mel Ott	NY	50.7	Hal Schumacher	NY	33.0	Bill Swift	PIT	38.7
Joe Medwick	STL	46.0	Joe Medwick	STL	43.7	Dizzy Dean	STL	32.7	Dizzy Dean	STL	31.9
Normalized OPS			**Park Adjusted**			**Normalized ERA**			**Park Adjusted**		
Arky Vaughan	PIT	193	Arky Vaughan	PIT	175	Cy Blanton	PIT	155	Cy Blanton	PIT	170
Mel Ott	NY	156	Mel Ott	NY	162	Bill Swift	PIT	149	Bill Swift	PIT	163
Joe Medwick	STL	155	Joe Medwick	STL	151	Hal Schumacher	NY	139	Syl Johnson	NY	135
On Base Average			**Slugging Percentage**			**Percent of Team Wins**			**Wins Above Team**		
Arky Vaughan	PIT	.491	Arky Vaughan	PIT	.607	Paul Derringer	CIN	.324	Paul Derringer	CIN	8.4
Mel Ott	NY	.407	Joe Medwick	STL	.576	Dizzy Dean	STL	.292	Fred Frankhouse	BOS	5.5
Stan Hack	CHI	.406	Mel Ott	NY	.555	Fred Frankhouse	BOS	.289	Van Mungo	BRO	4.9
Isolated Power			**Players Overall**			**Pitchers Overall**			**Defensive Runs**		
Wally Berger	BOS	.253	Billy Herman	CHI	65.0	Cy Blanton	PIT	46.8	Billy Jurges	CHI	27.7
Mel Ott	NY	.233	Arky Vaughan	PIT	59.4	Bill Swift	PIT	36.6	Billy Herman	CHI	22.1
Arky Vaughan	PIT	.222	Mel Ott	NY	49.0	Hal Schumacher	NY	35.3	Terry Moore	STL	13.1

Club	W	L	R	OR	Avg	OBA	SLG	BPF	NOPS-A	BR	Adj	Wins	ERA	PPF	NERA-A	PR	Adj	Wins	Diff
CHI	100	54	847	597	.288	.347	.414	98	114/116	81	97	9.4	3.26	93	123/115	118	76	7.4	6.2
STL	96	58	829	625	.284	.335	.405	103	108/105	33	13	1.2	3.54	99	114/113	74	71	6.9	10.9
NY	91	62	770	675	.286	.336	.416	96	111/116	54	82	8.0	3.78	94	106/100	38	2	.2	6.4
PIT	86	67	743	647	.285	.343	.402	110	109/ 99	49	-26	-2.6	3.42	109	117/129	91	148	14.4	-2.3
BRO	70	83	711	767	.277	.333	.376	95	99/103	-20	12	1.2	4.22	96	95/ 92	-29	-53	-5.3	-2.4
CIN	68	85	646	772	.265	.319	.378	91	96/105	-50	14	1.3	4.30	92	93/ 86	-42	-89	-8.7	-1.1
PHI	64	89	685	871	.269	.322	.378	115	96/ 84	-46	-154	-15.0	4.76	119	84/101	-112	5	.5	2.0
BOS	38	115	575	852	.263	.311	.362	94	89/ 94	-97	-55	-5.4	4.93	98	82/ 80	-133	-144	-14.0	-19.0
			726		.277	.331	.391						4.02						

AMERICAN LEAGUE   1936

Batting Runs			Park Adjusted			Pitching Runs			Park Adjusted		
Lou Gehrig	NY	82.2	Lou Gehrig	NY	87.8	Lefty Grove	BOS	62.6	Lefty Grove	BOS	62.5
Jimmie Foxx	BOS	56.8	Jimmie Foxx	BOS	56.5	Tommy Bridges	DET	47.1	Johnny Allen	CLE	43.2
Earl Averill	CLE	56.4	Earl Averill	CLE	55.5	Johnny Allen	CLE	43.0	Tommy Bridges	DET	39.2
Normalized OPS			**Park Adjusted**			**Normalized ERA**			**Park Adjusted**		
Lou Gehrig	NY	186	Lou Gehrig	NY	197	Lefty Grove	BOS	179	Lefty Grove	BOS	179
Jimmie Foxx	BOS	161	Bill Dickey	NY	164	Johnny Allen	CLE	146	Johnny Allen	CLE	146
Earl Averill	CLE	160	Jimmie Foxx	BOS	161	Pete Appleton	WAS	143	Pete Appleton	WAS	141
On Base Average			**Slugging Percentage**			**Percent of Team Wins**			**Wins Above Team**		
Lou Gehrig	NY	.478	Lou Gehrig	NY	.696	Harry Kelley	PHI	.283	Harry Kelley	PHI	6.9
Luke Appling	CHI	.474	Hal Trosky	CLE	.644	Tommy Bridges	DET	.277	Charlie Gehringer	CHI	6.1
Jimmie Foxx	BOS	.440	Jimmie Foxx	BOS	.631	Wes Ferrell	BOS	.270	Tommy Bridges	DET	6.0
Isolated Power			**Players Overall**			**Pitchers Overall**			**Defensive Runs**		
Lou Gehrig	NY	.342	Charlie Gehringer	DET	76.3	Lefty Grove	BOS	60.3	Jackie Hayes	CHI	18.3
Hal Trosky	CLE	.300	Lou Gehrig	NY	67.1	Johnny Allen	CLE	41.3	Charlie Gehringer	DET	18.1
Jimmie Foxx	BOS	.292	Luke Appling	CHI	65.5	Tommy Bridges	DET	40.3	Luke Appling	CHI	16.9

Club	W	L	R	OR	Avg	OBA	SLG	BPF	NOPS-A	BR	Adj	Wins	ERA	PPF	NERA-A	PR	Adj	Wins	Diff
NY	102	51	1065	731	.300	.381	.483	95	124/131	162	211	18.6	4.17	89	121/108	135	50	4.4	2.4
DET	83	71	921	871	.300	.377	.431	96	108/113	59	92	8.1	5.00	95	101/ 96	6	-30	-2.7	.6
CHI	81	70	920	873	.292	.374	.397	101	98/ 97	-8	-19	-1.7	5.06	101	99/100	-3	1	.1	7.2
WAS	82	71	889	799	.295	.365	.414	100	101/101	-6	-3	-.4	4.58	98	110/108	68	55	4.9	1.0
CLE	80	74	921	862	.304	.363	.461	101	113/112	70	62	5.4	4.83	100	104/104	32	33	2.9	-5.3
BOS	74	80	775	764	.276	.349	.400	100	93/ 93	-71	-74	-6.6	4.39	100	115/115	99	98	8.7	-5.1
STL	57	95	804	1064	.279	.355	.403	103	95/ 92	-50	-80	-7.1	6.24	108	81/ 87	-179	-122	-10.9	-1.0
PHI	53	100	714	1045	.269	.336	.376	101	83/ 82	-150	-160	-14.2	6.07	106	83/ 88	-154	-107	-9.5	.3
			876		.289	.363	.421						5.04						

NATIONAL LEAGUE   1936

Batting Runs			Park Adjusted			Pitching Runs			Park Adjusted		
Mel Ott	NY	61.9	Mel Ott	NY	63.4	Carl Hubbell	NY	57.7	Carl Hubbell	NY	51.8
Dolph Camilli	PHI	58.0	Paul Waner	PIT	52.1	Danny MacFayden	BOS	34.2	Danny MacFayden	BOS	31.7
Paul Waner	PIT	51.4	Dolph Camilli	PHI	50.9	Dizzy Dean	STL	29.6	Van Mungo	BRO	31.2
Normalized OPS			**Park Adjusted**			**Normalized ERA**			**Park Adjusted**		
Mel Ott	NY	177	Mel Ott	NY	180	Carl Hubbell	NY	174	Carl Hubbell	NY	166
Dolph Camilli	PHI	172	Johnny Mize	STL	165	Danny MacFayden	BOS	140	Danny MacFayden	BOS	137
Johnny Mize	STL	161	Paul Waner	PIT	161	Frank Gabler	NY	129	Claude Passeau	PHI	129
On Base Average			**Slugging Percentage**			**Percent of Team Wins**			**Wins Above Team**		
Arky Vaughan	PIT	.453	Mel Ott	NY	.588	Carl Hubbell	NY	.283	Carl Hubbell	NY	8.7
Mel Ott	NY	.448	Dolph Camilli	PHI	.577	Dizzy Dean	STL	.276	Red Lucas	PIT	5.3
Paul Waner	PIT	.446	Johnny Mize	STL	.577	Van Mungo	BRO	.269	Dizzy Dean	STL	4.1
Isolated Power			**Players Overall**			**Pitchers Overall**			**Defensive Runs**		
Dolph Camilli	PHI	.262	Dick Bartell	NY	58.2	Carl Hubbell	NY	52.7	Dick Bartell	NY	43.3
Mel Ott	NY	.260	Joe Medwick	STL	54.1	Danny MacFayden	BOS	29.6	Burgess Whitehead	NY	31.3
Johnny Mize	STL	.249	Billy Herman	CHI	53.0	Claude Passeau	PHI	29.4	Tony Cuccinello	BOS	22.0

Club	W	L	R	OR	Avg	OBA	SLG	BPF	NOPS-A	BR	Adj	Wins	ERA	PPF	NERA-A	PR	Adj	Wins	Diff
NY	92	62	742	621	.281	.337	.395	98	105/107	19	33	3.2	3.45	96	116/111	87	60	5.8	5.9
STL	87	67	795	794	.281	.336	.410	98	109/112	40	56	5.5	4.47	98	90/ 88	-70	-83	-8.2	12.7
CHI	87	67	755	603	.286	.349	.392	100	107/107	44	45	4.4	3.54	97	113/110	73	55	5.4	.2
PIT	84	70	804	718	.286	.349	.397	99	109/110	56	63	6.1	3.89	98	103/101	20	5	.5	.4
CIN	74	80	722	760	.274	.329	.388	94	101/107	-13	26	2.5	4.22	95	95/ 90	-30	-62	-6.2	.6
BOS	71	83	631	715	.265	.322	.356	97	89/ 92	-85	-61	-6.0	3.94	98	102/100	12	-1	-.2	.2
BRO	67	87	662	752	.272	.323	.353	104	89/ 85	-91	-119	-11.7	3.98	106	101/107	5	41	4.0	-2.3
PHI	54	100	726	874	.281	.339	.401	109	107/ 98	32	-32	-3.2	4.64	112	87/ 97	-94	-20	-2.0	-17.8
			730		.278	.335	.386						4.02						

AMERICAN LEAGUE 1937

Batting Runs			Park Adjusted			Pitching Runs			Park Adjusted		
Lou Gehrig	NY	73.3	Lou Gehrig	NY	76.3	Lefty Gomez	NY	70.7	Lefty Gomez	NY	58.7
Hank Greenberg	DET	67.2	Joe DiMaggio	NY	64.3	Lefty Grove	BOS	46.5	Lefty Grove	BOS	49.6
Joe DiMaggio	NY	61.3	Hank Greenberg	DET	58.5	Red Ruffing	NY	46.4	Monty Stratton	CHI	44.6
Normalized OPS			**Park Adjusted**			**Normalized ERA**			**Park Adjusted**		
Lou Gehrig	NY	178	Lou Gehrig	NY	184	Lefty Gomez	NY	198	Monty Stratton	CHI	201
Hank Greenberg	DET	174	Joe DiMaggio	NY	174	Monty Stratton	CHI	193	Lefty Gomez	NY	182
Joe DiMaggio	NY	169	Hank Greenberg	DET	159	Johnny Allen	CLE	181	Johnny Allen	CLE	165
On Base Average			**Slugging Percentage**			**Percent of Team Wins**			**Wins Above Team**		
Lou Gehrig	NY	.473	Joe DiMaggio	NY	.673	Harry Kelley	PHI	.241	Johnny Allen	CLE	7.1
Charlie Gehringer	DET	.458	Hank Greenberg	DET	.668	George Caster	PHI	.222	Lefty Grove	BOS	4.4
Hank Greenberg	DET	.436	Lou Gehrig	NY	.643	Lefty Grove	BOS	.213	Roxie Lawson	DET	4.2
Isolated Power			**Players Overall**			**Pitchers Overall**			**Defensive Runs**		
Hank Greenberg	DET	.332	Harlond Clift	STL	81.2	Lefty Gomez	NY	53.9	Harlond Clift	STL	41.9
Joe DiMaggio	NY	.327	Joe DiMaggio	NY	64.7	Lefty Grove	BOS	45.6	Jackie Hayes	CHI	24.2
Lou Gehrig	NY	.292	Bill Dickey	NY	56.1	Monty Stratton	CHI	44.4	Odell Hale	CLE	16.8

Club	W	L	R	OR	Avg	OBA	SLG	BPF	NOPS-A	BR	Adj	Wins	ERA	PPF	NERA-A	PR	Adj	Wins	Diff
NY	102	52	979	671	.283	.369	.456	97	118/121	113	139	12.8	3.65	92	127/116	150	90	8.3	3.9
DET	89	65	935	841	.292	.370	.452	109	116/107	106	28	2.6	4.88	109	95/103	-38	22	2.1	7.4
CHI	86	68	780	730	.280	.350	.400	105	97/ 92	-37	-76	-7.1	4.16	105	111/116	69	101	9.2	6.8
CLE	83	71	817	768	.280	.352	.423	93	104/112	4	61	5.6	4.38	91	105/ 96	36	-24	-2.3	2.7
BOS	80	72	821	775	.281	.357	.411	103	102/ 99	-0	-23	-2.2	4.48	102	103/106	21	37	3.4	2.8
WAS	73	80	757	841	.279	.351	.379	94	91/ 97	-73	-26	-2.5	4.57	95	101/ 96	7	-25	-1.4	1.4
PHI	54	97	699	854	.267	.341	.397	99	94/ 96	-62	-52	-4.8	4.85	101	95/ 96	-34	-27	-2.5	-14.1
STL	46	108	715	1023	.285	.348	.399	99	96/ 97	-46	-41	-3.9	6.01	104	77/ 80	-209	-178	-16.4	-10.7
				813	.281	.355	.415						4.62						

NATIONAL LEAGUE 1937

Batting Runs			Park Adjusted			Pitching Runs			Park Adjusted		
Joe Medwick	STL	68.8	Joe Medwick	STL	68.8	Jim Turner	BOS	43.7	Cliff Melton	NY	32.2
Johnny Mize	STL	57.4	Johnny Mize	STL	57.4	Cliff Melton	NY	35.8	Jim Turner	BOS	29.1
Dolph Camilli	PHI	54.9	Dolph Camilli	PHI	49.0	Lou Fette	BOS	29.6	Dizzy Dean	STL	25.9
Normalized OPS			**Park Adjusted**			**Normalized ERA**			**Park Adjusted**		
Joe Medwick	STL	183	Joe Medwick	STL	183	Jim Turner	BOS	164	Cliff Melton	NY	145
Dolph Camilli	PHI	179	Johnny Mize	STL	175	Cliff Melton	NY	150	Dizzy Dean	STL	144
Johnny Mize	STL	175	Dolph Camilli	PHI	164	Dizzy Dean	STL	145	Jim Turner	BOS	143
On Base Average			**Slugging Percentage**			**Percent of Team Wins**			**Wins Above Team**		
Dolph Camilli	PHI	.446	Joe Medwick	STL	.641	Jim Turner	BOS	.253	Lou Fette	BOS	5.5
Johnny Mize	STL	.427	Johnny Mize	STL	.595	Lou Fette	BOS	.253	Jim Turner	BOS	4.9
Joe Medwick	STL	.414	Dolph Camilli	PHI	.587	Wayne La Master	PHI	.246	Carl Hubbell	NY	4.0
Isolated Power			**Players Overall**			**Pitchers Overall**			**Defensive Runs**		
Joe Medwick	STL	.267	Dick Bartell	NY	61.2	Jim Turner	BOS	33.5	Dick Bartell	NY	31.9
Dolph Camilli	PHI	.248	Joe Medwick	STL	61.0	Cliff Melton	NY	30.4	Burgess Whitehead	NY	29.1
Johnny Mize	STL	.230	Dolph Camilli	PHI	46.3	Russ Bauers	PIT	25.4	Lew Riggs	CIN	17.8

Club	W	L	R	OR	Avg	OBA	SLG	BPF	NOPS-A	BR	Adj	Wins	ERA	PPF	NERA-A	PR	Adj	Wins	Diff
NY	95	57	732	602	.278	.334	.403	99	109/110	40	44	4.4	3.43	97	114/110	74	54	5.4	9.2
CHI	93	61	811	682	.287	.355	.416	104	118/114	117	88	8.8	3.98	102	98/100	-9	2	.2	7.1
PIT	86	68	704	646	.285	.343	.384	102	105/103	30	16	1.6	3.56	101	110/113	53	59	5.9	1.6
STL	81	73	789	733	.282	.331	.406	100	109/109	36	36	3.5	3.98	99	98/ 97	-9	-14	-1.5	2.0
BOS	79	73	579	556	.247	.314	.339	89	84/ 95	-108	-30	-3.1	3.21	87	122/106	106	28	2.8	3.3
BRO	62	91	616	772	.265	.327	.354	100	92/ 93	-56	-55	-5.5	4.13	103	95/ 97	-32	-16	-1.7	-7.3
PHI	61	92	724	869	.273	.334	.391	109	105/ 97	21	-39	-4.0	5.06	112	77/ 86	-175	-103	-10.3	-1.2
CIN	56	98	612	707	.254	.315	.360	95	91/ 96	-77	-41	-4.2	3.94	96	99/ 96	-3	-25	-2.5	-14.3
				696	.272	.332	.382						3.91						

AMERICAN LEAGUE   1938

Batting Runs			Park Adjusted			Pitching Runs			Park Adjusted		
Jimmie Foxx	BOS	78.2	Jimmie Foxx	BOS	79.2	Red Ruffing	NY	40.5	Thornton Lee	CHI	39.2
Hank Greenberg	DET	66.0	Hank Greenberg	DET	65.8	Lefty Gomez	NY	38.2	Red Ruffing	NY	37.3
Harlond Clift	STL	37.7	Harlond Clift	STL	36.0	Thornton Lee	CHI	35.4	Lefty Gomez	NY	35.2
Normalized OPS			**Park Adjusted**			**Normalized ERA**			**Park Adjusted**		
Jimmie Foxx	BOS	188	Jimmie Foxx	BOS	190	Lefty Grove	BOS	156	Lefty Grove	BOS	150
Hank Greenberg	DET	177	Hank Greenberg	DET	176	Red Ruffing	NY	145	Thornton Lee	CHI	141
Rudy York	DET	147	Rudy York	DET	146	Lefty Gomez	NY	143	Red Ruffing	NY	141
On Base Average			**Slugging Percentage**			**Percent of Team Wins**			**Wins Above Team**		
Jimmie Foxx	BOS	.462	Jimmie Foxx	BOS	.704	Buck Newsom	STL	.364	Buck Newsom	STL	9.1
Buddy Myer	WAS	.454	Hank Greenberg	DET	.683	George Caster	PHI	.302	Monty Stratton	CHI	5.3
Hank Greenberg	DET	.438	Jeff Heath	CLE	.602	Monty Stratton	CHI	.231	George Caster	PHI	4.5
Isolated Power			**Players Overall**			**Pitchers Overall**			**Defensive Runs**		
Hank Greenberg	DET	.369	Jimmie Foxx	BOS	63.4	Red Ruffing	NY	43.6	Frankie Crosetti	NY	19.2
Jimmie Foxx	BOS	.356	Harlond Clift	STL	53.1	Thornton Lee	CHI	43.6	Joe Gordon	NY	18.2
Rudy York	DET	.281	Joe Cronin	BOS	51.4	Lefty Gomez	NY	33.7	Harlond Clift	STL	12.3

Club	W	L	R	OR	Avg	OBA	SLG	BPF	NOPS-A	BR	Adj	Wins	ERA	PPF	NERA-A	PR	Adj	Wins	Diff
NY	99	53	966	710	.274	.366	.446	101	113/111	77	65	5.9	3.92	98	122/119	134	117	10.5	6.6
BOS	88	61	902	751	.299	.378	.434	99	113/114	87	96	8.6	4.45	96	108/104	50	24	2.2	2.7
CLE	86	66	847	782	.281	.350	.434	98	106/108	9	26	2.4	4.60	97	104/101	29	5	.5	7.2
DET	84	70	862	795	.272	.359	.411	100	102/101	0	-1	-.2	4.79	99	100/ 99	1	-4	-.5	7.6
WAS	75	76	814	873	.293	.362	.416	94	104/110	12	59	5.3	4.94	95	97/ 92	-22	-58	-5.3	-.5
CHI	65	83	709	752	.277	.344	.383	102	90/ 88	-89	-107	-9.7	4.36	103	110/113	63	83	7.5	-6.7
STL	55	97	755	962	.281	.355	.397	102	97/ 95	-37	-52	-4.8	5.80	105	83/ 87	-150	-113	-10.3	-6.0
PHI	53	99	726	956	.270	.348	.396	102	95/ 93	-56	-73	-6.7	5.49	106	87/ 93	-101	-59	-5.4	-10.9
			823		.281	.358	.415						4.79						

NATIONAL LEAGUE   1938

Batting Runs			Park Adjusted			Pitching Runs			Park Adjusted		
Mel Ott	NY	61.9	Mel Ott	NY	63.7	Bill Lee	CHI	36.4	Bill Lee	CHI	45.2
Johnny Mize	STL	59.2	Johnny Mize	STL	51.1	Paul Derringer	CIN	29.1	Clay Bryant	CHI	28.7
Arky Vaughan	PIT	36.4	Arky Vaughan	PIT	35.0	Clay Bryant	CHI	20.5	Paul Derringer	CIN	26.4
Normalized OPS			**Park Adjusted**			**Normalized ERA**			**Park Adjusted**		
Johnny Mize	STL	182	Mel Ott	NY	184	Bill Lee	CHI	142	Bill Lee	CHI	153
Mel Ott	NY	179	Johnny Mize	STL	163	Ira Hutchinson	BOS	138	Charlie Root	CHI	142
Ernie Lombardi	CIN	150	Ernie Lombardi	CIN	150	Charlie Root	CHI	133	Bill McGee	CHI	132
On Base Average			**Slugging Percentage**			**Percent of Team Wins**			**Wins Above Team**		
Mel Ott	NY	.442	Johnny Mize	STL	.614	Paul Derringer	CIN	.256	Bill Lee	CHI	4.8
Arky Vaughan	PIT	.433	Mel Ott	NY	.583	Bill Lee	CHI	.247	Vito Tamulis	BRO	4.2
Johnny Mize	STL	.422	Joe Medwick	STL	.536	Claude Passeau	PHI	.244	Bob Weiland	STL	4.0
Isolated Power			**Players Overall**			**Pitchers Overall**			**Defensive Runs**		
Johnny Mize	STL	.277	Arky Vaughan	PIT	61.7	Bill Lee	CHI	46.0	Pep Young	PIT	33.9
Mel Ott	NY	.271	Mel Ott	NY	59.7	Clay Bryant	CHI	29.0	Dick Bartell	NY	23.5
Ival Goodman	CIN	.241	Pep Young	PIT	38.7	Bill McGee	STL	24.1	Billy Herman	CHI	21.1

Club	W	L	R	OR	Avg	OBA	SLG	BPF	NOPS-A	BR	Adj	Wins	ERA	PPF	NERA-A	PR	Adj	Wins	Diff
CHI	89	63	713	598	.269	.338	.377	109	104/ 96	25	-35	-3.7	3.37	107	112/120	64	107	10.7	5.9
PIT	86	64	707	630	.279	.340	.388	102	108/106	48	34	3.5	3.47	101	109/110	49	52	5.2	2.3
NY	83	67	705	637	.271	.334	.396	98	109/112	45	61	6.2	3.62	99	104/100	24	2	.2	1.6
CIN	82	68	723	634	.277	.327	.406	100	110/111	43	46	4.6	3.62	98	104/102	24	12	1.2	1.2
BOS	77	75	561	618	.250	.309	.333	86	83/ 97	-118	-27	-2.8	3.40	86	111/ 96	58	-21	-2.3	6.1
STL	71	80	725	721	.279	.331	.407	112	112/100	56	-23	-2.4	3.84	112	98/110	-8	61	6.1	-8.2
BRO	69	80	704	710	.257	.338	.367	98	102/104	13	26	2.6	4.06	98	93/ 91	-41	-51	-5.2	-2.9
PHI	45	105	550	840	.254	.312	.333	99	84/ 85	-110	-102	-10.4	4.93	104	77/ 80	-168	-144	-14.6	-5.0
			674		.267	.329	.376						3.78						

AMERICAN LEAGUE 1939

Batting Runs			Park Adjusted			Pitching Runs			Park Adjusted		
Jimmie Foxx	BOS	66.1	Jimmie Foxx	BOS	58.5	Bob Feller	CLE	58.4	Bob Feller	CLE	55.4
Ted Williams	BOS	56.3	Joe DiMaggio	NY	57.8	Lefty Grove	BOS	44.0	Lefty Grove	BOS	52.8
Joe DiMaggio	NY	56.0	Bob Johnson	PHI	49.0	Red Ruffing	NY	43.6	Ted Lyons	CHI	34.8

Normalized OPS			Park Adjusted			Normalized ERA			Park Adjusted		
Jimmie Foxx	BOS	192	Joe DiMaggio	NY	187	Lefty Grove	BOS	182	Lefty Grove	BOS	198
Joe DiMaggio	NY	182	Jimmie Foxx	BOS	175	Ted Lyons	CHI	168	Ted Lyons	CHI	166
Ted Williams	BOS	164	Bob Johnson	PHI	158	Bob Feller	CLE	162	Bob Feller	CLE	159

On Base Average			Slugging Percentage			Percent of Team Wins			Wins Above Team		
Jimmie Foxx	BOS	.464	Jimmie Foxx	BOS	.694	Dutch Leonard	WAS	.308	Dutch Leonard	WAS	9.8
George Selkirk	NY	.452	Joe DiMaggio	NY	.671	Bob Feller	CLE	.276	Bob Feller	CLE	6.8
Joe DiMaggio	NY	.448	Hank Greenberg	DET	.622	Tommy Bridges	DET	.210	Tommy Bridges	DET	5.2

Isolated Power			Players Overall			Pitchers Overall			Defensive Runs		
Jimmie Foxx	BOS	.334	Joe DiMaggio	NY	58.0	Bob Feller	CLE	58.2	Bobby Doerr	BOS	30.2
Hank Greenberg	DET	.310	Bob Johnson	PHI	52.6	Lefty Grove	BOS	47.5	Mike Kreevich	CHI	12.0
Joe DiMaggio	NY	.290	Jimmie Foxx	BOS	49.9	Red Ruffing	NY	38.8	Ben Chapman	CLE	11.3

Club	W	L	R	OR	Avg	OBA	SLG	BPF	NOPS-A	BR	Adj	Wins	ERA	PPF	NERA-A	PR	Adj	Wins	Diff
NY	106	45	967	556	.287	.374	.451	98	120/123	134	154	14.2	3.31	91	139/127	196	132	12.2	4.1
BOS	89	62	890	795	.291	.363	.436	110	113/103	78	-1	-.1	4.56	109	101/110	9	71	6.6	7.1
CLE	87	67	797	700	.280	.350	.413	100	103/104	4	6	.5	4.08	98	113/111	82	68	6.3	3.2
CHI	85	69	755	737	.275	.349	.374	99	92/ 93	-61	-56	-5.3	4.30	99	107/106	48	40	3.7	9.5
DET	81	73	849	762	.279	.356	.426	111	109/ 98	43	-43	-4.0	4.29	110	108/118	50	120	11.0	-3.0
WAS	65	87	702	797	.278	.346	.379	86	93/108	-61	49	4.5	4.61	86	100/ 87	2	-92	-8.5	-7.0
PHI	55	97	711	1022	.271	.336	.400	96	96/100	-54	-25	-2.4	5.79	101	80/ 81	-173	-166	-15.4	-3.3
STL	43	111	733	1035	.268	.339	.381	101	92/ 91	-79	-86	-8.0	6.01	106	77/ 81	-211	-170	-15.8	-10.2
			801		.279	.352	.407						4.62						

NATIONAL LEAGUE 1939

Batting Runs			Park Adjusted			Pitching Runs			Park Adjusted		
Johnny Mize	STL	68.7	Johnny Mize	STL	66.2	Bucky Walters	CIN	57.8	Bucky Walters	CIN	61.1
Mel Ott	NY	45.9	Mel Ott	NY	47.2	Paul Derringer	CIN	33.0	Paul Derringer	CIN	36.1
Dolph Camilli	BRO	41.8	Dolph Camilli	BRO	36.6	Hugh Casey	BRO	24.8	Hugh Casey	BRO	30.7

Normalized OPS			Park Adjusted			Normalized ERA			Park Adjusted		
Johnny Mize	STL	185	Johnny Mize	STL	179	Bucky Walters	CIN	171	Bucky Walters	CIN	175
Mel Ott	NY	175	Mel Ott	NY	178	Junior Thompson	CIN	154	Junior Thompson	CIN	157
Dolph Camilli	BRO	149	Dolph Camilli	BRO	140	Bob Bowman	STL	150	Bob Bowman	STL	151

On Base Average			Slugging Percentage			Percent of Team Wins			Wins Above Team		
Mel Ott	NY	.449	Johnny Mize	STL	.626	Bucky Walters	CIN	.278	Paul Derringer	CIN	6.1
Johnny Mize	STL	.444	Mel Ott	NY	.581	Paul Derringer	CIN	.258	Bill Posedel	BOS	4.1
Dolph Camilli	BRO	.409	Dolph Camilli	BRO	.524	Curt Davis	STL	.239	Bucky Walters	CIN	4.1

Isolated Power			Players Overall			Pitchers Overall			Defensive Runs		
Johnny Mize	STL	.277	Johnny Mize	STL	43.7	Bucky Walters	CIN	76.8	Burgess Whitehead	NY	17.6
Mel Ott	NY	.273	Mel Ott	NY	39.9	Hugh Casey	BRO	32.9	Enos Slaughter	STL	15.4
Dolph Camilli	BRO	.234	Arky Vaughan	PIT	34.2	Paul Derringer	CIN	32.4	Hank Majeski	BOS	13.0

Club	W	L	R	OR	Avg	OBA	SLG	BPF	NOPS-A	BR	Adj	Wins	ERA	PPF	NERA-A	PR	Adj	Wins	Diff
CIN	97	57	767	595	.278	.343	.405	105	109·104	51	14	1.4	3.27	102	120·123	101	115	11.6	7.1
STL	92	61	779	633	.294	.354	.432	103	120·117	122	100	10.0	3.59	101	109·110	51	55	5.5	.0
BRO	84	69	708	645	.265	.338	.380	107	101· 95	0	-45	-4.6	3.64	106	107·114	42	79	7.9	4.2
CHI	84	70	724	678	.266	.336	.391	103	103·101	10	-8	-.9	3.80	102	103·105	17	30	3.0	4.9
NY	77	74	703	685	.272	.340	.396	98	106·108	28	42	4.2	4.06	97	96· 94	-21	-36	-3.7	.9
PIT	68	85	666	721	.276	.338	.384	103	102· 99	4	-15	-1.6	4.15	104	94· 98	-34	-10	-1.1	-5.8
BOS	63	88	572	659	.264	.314	.348	93	85· 91	-116	-69	-7.0	3.71	94	106· 99	31	-4	-.5	-5.0
PHI	45	106	553	856	.261	.318	.351	93	87 94	-97	-48	-4.9	5.17	98	76· 74	-184	-197	-19.8	-5.8
			684		.272	.335	.386						3.92						

Batting Runs				Park Adjusted				Pitching Runs				Park Adjusted		
Hank Greenberg	DET	68.8		Hank Greenberg	DET	60.3		Bob Feller	CLE	62.9		Buck Newsom	DET	55.8
Ted Williams	BOS	57.2		Ted Williams	BOS	56.2		Buck Newsom	DET	45.6		Johnny Rigney	CHI	55.2
Joe DiMaggio	NY	50.8		Joe DiMaggio	NY	55.1		Johnny Rigney	CHI	39.9		Bob Feller	CLE	46.2

Normalized OPS				Park Adjusted				Normalized ERA				Park Adjusted		
Hank Greenberg	DET	181		Joe DiMaggio	NY	179		Bob Feller	CLE	168		Buck Newsom	DET	167
Joe DiMaggio	NY	169		Hank Greenberg	DET	165		Buck Newsom	DET	155		Johnny Rigney	CHI	157
Ted Williams	BOS	166		Ted Williams	BOS	164		Johnny Rigney	CHI	141		Eddie Smith	CHI	152

On Base Average				Slugging Percentage				Percent of Team Wins				Wins Above Team		
Ted Williams	BOS	.442		Hank Greenberg	DET	.670		Bob Feller	CLE	.303		Buck Newsom	DET	7.0
Hank Greenberg	DET	.433		Joe DiMaggio	NY	.626		Sid Hudson	WAS	.266		Bob Feller	CLE	6.7
Charlie Gehringer	DET	.428		Ted Williams	BOS	.594		Johnny Babich	PHI	.259		Schoolboy Rowe	DET	5.6

Isolated Power				Players Overall				Pitchers Overall				Defensive Runs		
Hank Greenberg	DET	.330		Hank Greenberg	DET	49.5		Johnny Rigney	CHI	52.8		Bobby Doerr	BOS	18.2
Jimmie Foxx	BOS	.283		Lou Boudreau	CLE	47.0		Buck Newsom	DET	51.0		Joe Gordon	NY	15.7
Joe DiMaggio	NY	.274		Ted Williams	BOS	46.1		Bob Feller	CLE	42.6		Jeep Heffner	STL	15.2

Club	W	L	R	OR	Avg	OBA	SLG	BPF	NOPS-A	BR	Adj	Wins	ERA	PPF	NERA-A	PR	Adj	Wins	Diff
DET	90	64	888	717	.286	.366	.442	110	119/108	124	46	4.4	4.01	108	109/118	57	110	10.4	-1.7
CLE	89	65	710	637	.265	.332	.398	91	98/107	-42	22	2.1	3.63	89	121/108	115	43	4.1	5.8
NY	88	66	817	671	.259	.344	.418	94	107/113	25	70	6.6	3.89	91	113/102	75	15	1.4	3.0
CHI	82	72	735	672	.278	.340	.387	112	96/ 86	-40	-129	-12.3	3.74	111	117/130	100	175	16.6	.7
BOS	82	72	872	825	.286	.356	.449	101	119/117	109	100	9.5	4.89	101	90/ 90	-76	-73	-7.0	2.5
STL	67	87	757	882	.263	.333	.401	105	99/ 94	-34	-75	-7.2	5.12	108	86/ 92	-112	-59	-5.6	2.8
WAS	64	90	665	811	.271	.331	.374	94	91/ 96	-85	-41	-3.9	4.59	96	95/ 92	-30	-55	-5.3	-3.7
PHI	54	100	703	932	.262	.334	.387	95	95/100	-52	-18	-1.8	5.22	99	84/ 83	-124	-131	-12.4	-8.7
			768		.271	.342	.407						4.38						

Batting Runs				Park Adjusted				Pitching Runs				Park Adjusted		
Johnny Mize	STL	63.9		Johnny Mize	STL	60.9		Bucky Walters	CIN	46.5		Claude Passeau	CHI	40.7
Dolph Camilli	BRO	38.7		Elbie Fletcher	PIT	37.2		Claude Passeau	CHI	42.2		Bucky Walters	CIN	37.1
Elbie Fletcher	PIT	31.8		Arky Vaughan	PIT	34.6		Paul Derringer	CIN	26.0		Luke Hamlin	BRO	24.6

Normalized OPS				Park Adjusted				Normalized ERA				Park Adjusted		
Johnny Mize	STL	183		Johnny Mize	STL	176		Bucky Walters	CIN	155		Claude Passeau	CHI	152
Dolph Camilli	BRO	153		Elbie Fletcher	PIT	147		Claude Passeau	CHI	154		Bucky Walters	CIN	144
Bill Nicholson	CHI	145		Bill Nicholson	CHI	146		Rip Sewell	PIT	138		Luke Hamlin	BRO	140

On Base Average				Slugging Percentage				Percent of Team Wins				Wins Above Team		
Elbie Fletcher	PIT	.418		Johnny Mize	STL	.636		Kirby Higbe	PHI	.280		Freddie Fitzsimmons	BRO	6.4
Mel Ott	NY	.407		Bill Nicholson	CHI	.534		Claude Passeau	CHI	.267		Rip Sewell	PIT	6.2
Johnny Mize	STL	.404		Dolph Camilli	BRO	.529		Hugh Mulcahy	PHI	.260		Claude Passeau	CHI	5.0

Isolated Power				Players Overall				Pitchers Overall				Defensive Runs		
Johnny Mize	STL	.321		Arky Vaughan	PIT	64.0		Claude Passeau	CHI	47.0		Lonny Frey	CIN	25.8
Dolph Camilli	BRO	.242		Stan Hack	CHI	46.8		Bucky Walters	CIN	39.1		Billy Herman	CHI	24.9
Bill Nicholson	CHI	.236		Lonny Frey	CIN	43.3		Lon Warneke	STL	25.0		Stan Hack	CHI	21.1

Club	W	L	R	OR	Avg	OBA	SLG	BPF	NOPS-A	BR	Adj	Wins	ERA	PPF	NERA-A	PR	Adj	Wins	Diff
CIN	100	53	707	528	.266	.327	.379	97	103/107	6	29	2.9	3.05	93	126/117	126	82	8.3	12.3
BRO	88	65	697	621	.260	.327	.383	112	104/ 93	15	-70	-7.1	3.50	111	110/122	55	125	12.6	6.0
STL	84	69	747	699	.275	.336	.411	104	115/111	83	56	5.6	3.83	103	101/104	3	23	2.4	-.5
PIT	78	76	809	783	.276	.346	.394	93	113/122	84	135	13.7	4.36	92	88/ 81	-78	-128	-13.1	.4
CHI	75	79	681	636	.267	.331	.384	100	106/106	25	27	2.7	3.54	99	109/107	47	40	4.1	-8.8
NY	72	80	663	659	.267	.329	.374	102	102/100	3	-10	-1.2	3.79	102	101/104	8	22	2.2	-5.0
BOS	65	87	623	745	.256	.311	.349	94	90/ 95	-82	-43	-4.5	4.36	96	88/ 85	-77	-99	-10.1	3.6
PHI	50	103	494	750	.238	.300	.331	101	81/ 80	-132	-141	-14.4	4.40	106	87/ 93	-82	-45	-4.7	-7.4
				678	.264	.326	.376						3.85						

AMERICAN LEAGUE   1941

Batting Runs			Park Adjusted			Pitching Runs			Park Adjusted		
Ted Williams	BOS	102.0	Ted Williams	BOS	101.4	Thornton Lee	CHI	59.3	Thornton Lee	CHI	53.8
Joe DiMaggio	NY	64.3	Joe DiMaggio	NY	64.8	Bob Feller	CLE	38.1	Bob Feller	CLE	38.9
Charlie Keller	NY	45.7	Charlie Keller	NY	46.2	Eddie Smith	CHI	28.2	Al Benton	DET	24.2
Normalized OPS			**Park Adjusted**			**Normalized ERA**			**Park Adjusted**		
Ted Williams	BOS	237	Ted Williams	BOS	235	Thornton Lee	CHI	175	Thornton Lee	CHI	168
Joe DiMaggio	NY	183	Joe DiMaggio	NY	185	Al Benton	DET	140	Al Benton	DET	147
Charlie Keller	NY	161	Charlie Keller	NY	162	Charlie Wagner	BOS	135	Charlie Wagner	BOS	133
On Base Average			**Slugging Percentage**			**Percent of Team Wins**			**Wins Above Team**		
Ted Williams	BOS	.551	Ted Williams	BOS	.735	Bob Feller	CLE	.333	Bob Feller	CLE	8.6
Roy Cullenbine	STL	.452	Joe DiMaggio	NY	.643	Thornton Lee	CHI	.286	Thornton Lee	CHI	7.0
Joe DiMaggio	NY	.440	Jeff Heath	CLE	.586	Dutch Leonard	WAS	.257	Al Benton	DET	5.5
Isolated Power			**Players Overall**			**Pitchers Overall**			**Defensive Runs**		
Ted Williams	BOS	.329	Ted Williams	BOS	86.8	Thornton Lee	CHI	57.8	Ken Keltner	CLE	21.9
Joe DiMaggio	NY	.287	Joe DiMaggio	NY	61.6	Bob Feller	CLE	37.0	Jimmy Bloodworth	WAS	21.8
Charlie Keller	NY	.282	Charlie Keller	NY	40.1	Eddie Smith	CHI	27.4	Phil Rizzuto	NY	17.2

Club	W	L	R	OR	Avg	OBA	SLG	BPF	NOPS-A	BR	Adj	Wins	ERA	PPF	NERA-A	PR	Adj	Wins	Diff
NY	101	53	830	631	.269	.346	.419	99	112/113	64	70	6.7	3.52	96	118/113	97	70	6.8	10.5
BOS	84	70	865	750	.283	.366	.430	101	120/119	136	129	12.5	4.19	99	99/ 98	-5	-11	-1.2	-4.3
CHI	77	77	638	649	.255	.322	.343	96	83/ 87	-127	-99	-9.7	3.53	96	118/113	98	72	7.0	2.7
DET	75	79	686	743	.263	.340	.375	103	97/ 94	-25	-50	-5.0	4.17	105	99/104	-3	25	2.5	.5
CLE	75	79	677	668	.256	.323	.393	101	98/ 98	-39	-44	-4.4	3.90	101	107/107	39	42	4.1	-1.7
WAS	70	84	728	798	.272	.331	.376	98	95/ 97	-52	-37	-3.6	4.35	99	95/ 94	-31	-37	-3.7	.3
STL	70	84	765	823	.266	.359	.390	99	107/108	55	66	6.4	4.72	100	88/ 88	-87	-90	-8.8	-4.6
PHI	64	90	713	840	.268	.340	.387	102	101/ 99	-6	-19	-1.9	4.84	104	86/ 89	-103	-77	-7.6	-3.5
				738	.266	.341	.389						4.15						

NATIONAL LEAGUE   1941

Batting Runs			Park Adjusted			Pitching Runs			Park Adjusted		
Dolph Camilli	BRO	50.2	Dolph Camilli	BRO	49.6	Whit Wyatt	BRO	41.3	Whit Wyatt	BRO	37.3
Pete Reiser	BRO	48.0	Pete Reiser	BRO	47.4	Elmer Riddle	CIN	33.6	Ernie White	STL	36.1
Johnny Mize	STL	40.2	Stan Hack	CHI	38.1	Ernie White	STL	28.8	Elmer Riddle	CIN	32.0
Normalized OPS			**Park Adjusted**			**Normalized ERA**			**Park Adjusted**		
Pete Reiser	BRO	170	Pete Reiser	BRO	168	Elmer Riddle	CIN	162	Ernie White	STL	164
Dolph Camilli	BRO	169	Dolph Camilli	BRO	168	Whit Wyatt	BRO	155	Elmer Riddle	CIN	159
Johnny Mize	STL	164	Mel Ott	NY	149	Ernie White	STL	151	Whit Wyatt	BRO	150
On Base Average			**Slugging Percentage**			**Percent of Team Wins**			**Wins Above Team**		
Elbie Fletcher	PIT	.421	Pete Reiser	BRO	.558	Whit Wyatt	BRO	.220	Elmer Riddle	CIN	6.9
Stan Hack	CHI	.417	Dolph Camilli	BRO	.556	Kirby Higbe	BRO	.220	Howie Krist	STL	3.9
Dolph Camilli	BRO	.407	Johnny Mize	STL	.535	Bucky Walters	CIN	.216	Johnny Podgajny	PHI	3.6
						Elmer Riddle	CIN	.216			
Isolated Power			**Players Overall**			**Pitchers Overall**			**Defensive Runs**		
Dolph Camilli	BRO	.270	Pete Reiser	BRO	48.9	Whit Wyatt	BRO	42.2	Pinky May	PHI	21.5
Johnny Mize	STL	.218	Dixie Walker	BRO	30.3	Elmer Riddle	CIN	34.3	Lou Stringer	CHI	17.0
Pete Reiser	BRO	.215	Dolph Camilli	BRO	28.6	Ernie White	STL	32.3	Eddie Miller	BOS	11.6

Club	W	L	R	OR	Avg	OBA	SLG	BPF	NOPS-A	BR	Adj	Wins	ERA	PPF	NERA-A	PR	Adj	Wins	Diff
BRO	100	54	800	581	.272	.347	.405	101	120/119	128	123	12.6	3.14	97	116/112	78	58	6.0	4.4
STL	97	56	734	589	.272	.340	.377	111	110/ 99	62	-9	-1.1	3.19	109	114/124	70	119	12.2	9.4
CIN	88	66	616	564	.247	.312	.337	99	90/ 90	-72	-67	-7.0	3.17	98	115/113	72	61	6.3	11.7
PIT	81	73	690	643	.268	.338	.368	106	106/100	38	1	.1	3.49	105	104/110	22	51	5.3	-1.4
NY	74	79	667	706	.260	.326	.371	102	104/102	15	3	.3	3.95	103	92/ 95	-47	-32	-3.3	.6
CHI	70	84	666	670	.253	.327	.365	95	102/108	9	44	4.5	3.72	94	98/ 92	-12	-43	-4.5	-6.9
BOS	62	92	592	720	.251	.334	.334	94	89/ 95	-81	-39	-4.1	3.95	96	92/ 88	-47	-71	-7.4	-3.5
PHI	43	111	501	793	.244	.307	.331	97	86/ 89	-96	-77	-8.0	4.50	103	81/ 83	-131	-117	-12.1	-13.9
				658	.258	.326	.361						3.63						

AMERICAN LEAGUE 1942

Batting Runs			Park Adjusted			Pitching Runs			Park Adjusted		
Ted Williams	BOS	92.6	Ted Williams	BOS	91.5	Tiny Bonham	NY	34.8	Tex Hughson	BOS	31.1
Charlie Keller	NY	46.9	Charlie Keller	NY	50.0	Tex Hughson	BOS	33.2	Ted Lyons	CHI	30.5
Joe Gordon	NY	38.6	Joe Gordon	NY	41.5	Ted Lyons	CHI	31.1	Hal Newhouser	DET	29.3
Normalized OPS			**Park Adjusted**			**Normalized ERA**			**Park Adjusted**		
Ted Williams	BOS	221	Ted Williams	BOS	218	Ted Lyons	CHI	174	Ted Lyons	CHI	173
Charlie Keller	NY	161	Charlie Keller	NY	168	Tiny Bonham	NY	161	Hal Newhouser	DET	159
Wally Judnich	STL	156	Joe Gordon	NY	159	Spud Chandler	NY	154	Johnny Niggeling	STL	147
On Base Average			**Slugging Percentage**			**Percent of Team Wins**			**Wins Above Team**		
Ted Williams	BOS	.499	Ted Williams	BOS	.648	Phil Marchildon	PHI	.309	Phil Marchildon	PHI	7.4
Charlie Keller	NY	.417	Charlie Keller	NY	.513	Tex Hughson	BOS	.237	Tex Hughson	BOS	6.0
Wally Judnich	STL	.413	Wally Judnich	STL	.499	Jim Bagby	CLE	.227	Ted Lyons	CHI	5.9
Isolated Power			**Players Overall**			**Pitchers Overall**			**Defensive Runs**		
Ted Williams	BOS	.291	Ted Williams	BOS	84.3	Ted Lyons	CHI	35.0	Phil Rizzuto	NY	28.6
Chet Laabs	STL	.223	Joe Gordon	NY	54.6	Tex Hughson	BOS	33.2	Johnny Pesky	BOS	18.5
Charlie Keller	NY	.221	Johnny Pesky	BOS	42.9	Hal Newhouser	DET	30.7	Ken Keltner	CLE	16.0

Club	W	L	R	OR	Avg	OBA	SLG	BPF	NOPS-A	BR	Adj	Wins	ERA	PPF	NERA-A	PR	Adj	Wins	Diff
NY	103	51	801	507	.269	.346	.394	96	117/122	104	132	13.5	2.90	89	126/113	115	56	5.8	6.7
BOS	93	59	761	594	.276	.352	.403	101	121/119	132	123	12.6	3.45	98	106/104	31	21	2.2	2.2
STL	82	69	730	637	.259	.338	.385	108	112/103	68	14	1.4	3.59	107	102/109	9	48	4.9	.2
CLE	75	79	590	659	.253	.320	.345	94	94/101	-42	-1	-.2	3.59	95	102/ 96	10	-20	-2.2	.4
DET	73	81	589	587	.246	.314	.344	106	93/ 88	-57	-95	-9.8	3.13	106	117/124	82	117	12.0	-6.2
CHI	66	82	538	609	.246	.316	.318	98	85/ 87	-86	-74	-7.6	3.58	99	102/101	11	6	.6	-1.0
WAS	62	89	653	817	.258	.333	.341	95	97/102	-14	20	2.1	4.58	98	80/ 78	-138	-150	-15.4	-.1
PHI	55	99	549	801	.249	.309	.325	102	86/ 84	-100	-111	-11.5	4.44	107	82/ 88	-119	-81	-8.4	-2.1
			651		.257	.329	.357						3.66						

NATIONAL LEAGUE 1942

Batting Runs			Park Adjusted			Pitching Runs			Park Adjusted		
Enos Slaughter	STL	49.8	Mel Ott	NY	48.4	Mort Cooper	STL	47.7	Mort Cooper	STL	45.2
Mel Ott	NY	49.3	Enos Slaughter	STL	47.6	Johnny Beazley	STL	28.2	Johnny Vander Meer	CIN	29.7
Johnny Mize	NY	40.8	Johnny Mize	NY	39.9	Larry French	BRO	24.5	Ray Starr	CIN	26.6
Normalized OPS			**Park Adjusted**			**Normalized ERA**			**Park Adjusted**		
Mel Ott	NY	166	Mel Ott	NY	164	Mort Cooper	STL	187	Mort Cooper	STL	182
Enos Slaughter	STL	164	Johnny Mize	NY	160	Johnny Beazley	STL	155	Johnny Beazley	STL	151
Johnny Mize	NY	162	Enos Slaughter	STL	159	Curt Davis	BRO	140	Johnny Vander Meer	CIN	145
On Base Average			**Slugging Percentage**			**Percent of Team Wins**			**Wins Above Team**		
Elbie Fletcher	PIT	.417	Johnny Mize	NY	.521	Tommy Hughes	PHI	.286	Claude Passeau	CHI	5.6
Mel Ott	NY	.415	Mel Ott	NY	.497	Claude Passeau	CHI	.279	Tommy Hughes	PHI	4.6
Enos Slaughter	STL	.412	Enos Slaughter	STL	.494	Rip Sewell	PIT	.258	Johnny Vander Meer	CIN	3.7
Isolated Power			**Players Overall**			**Pitchers Overall**			**Defensive Runs**		
Dolph Camilli	BRO	.219	Enos Slaughter	STL	46.1	Mort Cooper	STL	40.9	Peewee Reese	BRO	18.0
Johnny Mize	NY	.216	Bill Nicholson	CHI	44.7	Bucky Walters	CIN	32.6	Vince DiMaggio	PIT	14.5
Mel Ott	NY	.202	Mel Ott	NY	40.6	Johnny Vander Meer	CIN	28.0	Al Glossop	PHI	14.0

Club	W	L	R	OR	Avg	OBA	SLG	BPF	NOPS-A	BR	Adj	Wins	ERA	PPF	NERA-A	PR	Adj	Wins	Diff
STL	106	48	755	482	.268	.338	.379	103	118/114	109	89	9.6	2.55	98	130/127	120	107	11.5	7.9
BRO	104	50	742	510	.264	.338	.362	103	112/109	79	59	6.3	2.84	99	117/115	73	66	7.1	13.6
NY	85	67	675	600	.254	.330	.361	101	110/108	59	51	5.4	3.31	100	100/100	0	0	.0	3.5
CIN	76	76	527	545	.231	.299	.321	106	88/ 83	-80	-115	-12.4	2.82	106	118/125	78	112	12.0	.4
PIT	66	81	585	631	.245	.320	.330	95	97/101	-16	11	1.1	3.59	96	92/ 89	-40	-59	-6.4	-2.2
CHI	68	86	591	665	.254	.321	.353	98	105/107	23	35	3.8	3.60	100	92/ 92	-44	-46	-5.0	-7.7
BOS	59	89	515	645	.240	.307	.329	95	93/ 97	-47	-21	-2.4	3.76	98	88/ 86	-64	-75	-8.2	-4.4
PHI	42	109	394	706	.232	.289	.306	99	81/ 81	-124	-120	-13.0	4.12	106	80/ 85	-120	-90	-9.7	-10.8
			598		.249	.318	.343						3.31						

AMERICAN LEAGUE 1943

Batting Runs			Park Adjusted			Pitching Runs			Park Adjusted		
Charlie Keller	NY	45.8	Charlie Keller	NY	49.3	Spud Chandler	NY	46.7	Spud Chandler	NY	38.5
Rudy York	DET	41.0	Rudy York	DET	39.6	Tiny Bonham	NY	25.8	Tex Hughson	BOS	23.7
Luke Appling	CHI	35.0	Roy Cullenbine	CLE	33.6	Dizzy Trout	DET	22.5	Dizzy Trout	DET	23.2
Normalized OPS			**Park Adjusted**			**Normalized ERA**			**Park Adjusted**		
Charlie Keller	NY	169	Charlie Keller	NY	179	Spud Chandler	NY	201	Spud Chandler	NY	184
Rudy York	DET	160	Jeff Heath	CLE	165	Tiny Bonham	NY	145	Mickey Haefner	WAS	148
Jeff Heath	CLE	148	Rudy York	DET	157	Mickey Haefner	WAS	144	Tommy Bridges	DET	139
On Base Average			**Slugging Percentage**			**Percent of Team Wins**			**Wins Above Team**		
Luke Appling	CHI	.419	Rudy York	DET	.527	Dizzy Trout	DET	.256	Spud Chandler	NY	5.6
Roy Cullenbine	CLE	.407	Charlie Keller	NY	.525	Jesse Flores	PHI	.245	Al Smith	CLE	4.9
Charlie Keller	NY	.396	Vern Stephens	STL	.482	Early Wynn	WAS	.214	Dizzy Trout	DET	4.8
Isolated Power			**Players Overall**			**Pitchers Overall**			**Defensive Runs**		
Rudy York	DET	.256	Lou Boudreau	CLE	57.0	Spud Chandler	NY	47.9	Lou Boudreau	CLE	27.6
Charlie Keller	NY	.254	Joe Gordon	NY	55.9	Dizzy Trout	DET	29.9	Joe Gordon	NY	24.7
Jeff Heath	CLE	.208	Rudy York	DET	53.0	Orval Grove	CHI	21.0	Rudy York	DET	17.8

Club	W	L	R	OR	Avg	OBA	SLG	BPF	NOPS-A	BR	Adj	Wins	ERA	PPF	NERA-A	PR	Adj	Wins	Diff
NY	98	56	669	542	.256	.337	.376	94	116/123	97	131	14.1	2.93	91	113/103	58	13	1.4	5.5
WAS	84	69	666	595	.254	.336	.347	104	106/102	45	21	2.2	3.18	103	104/106	17	32	3.4	1.8
CLE	82	71	600	577	.255	.329	.350	90	105/117	32	93	10.1	3.15	89	105/ 93	23	-35	-3.9	-.6
CHI	82	72	573	594	.247	.322	.320	107	94/ 88	-32	-74	-8.1	3.20	108	103/111	15	55	5.9	7.1
DET	78	76	632	560	.261	.324	.359	102	107/104	32	19	2.1	3.00	101	110/111	46	50	5.4	-6.5
STL	72	80	596	604	.245	.322	.349	96	103/107	14	35	3.8	3.41	96	97/ 93	-17	-36	-4.0	-3.8
BOS	68	84	563	607	.244	.308	.332	103	94/ 91	-47	-67	-7.4	3.45	104	95/100	-24	-1	-.2	-.4
PHI	49	105	497	717	.232	.294	.297	103	79/ 76	-136	-153	-16.7	4.05	108	81/ 88	-115	-76	-8.3	-3.0
			600		.249	.322	.341						3.30						

NATIONAL LEAGUE 1943

Batting Runs			Park Adjusted			Pitching Runs			Park Adjusted		
Stan Musial	STL	65.5	Stan Musial	STL	62.1	Max Lanier	STL	34.8	Nate Andrews	BOS	36.5
Bill Nicholson	CHI	47.7	Bill Nicholson	CHI	45.1	Mort Cooper	STL	32.6	Max Lanier	STL	35.4
Augie Galan	BRO	28.5	Augie Galan	BRO	31.4	Nate Andrews	BOS	25.4	Mort Cooper	STL	33.5
Normalized OPS			**Park Adjusted**			**Normalized ERA**			**Park Adjusted**		
Stan Musial	STL	184	Stan Musial	STL	176	Max Lanier	STL	177	Max Lanier	STL	179
Bill Nicholson	CHI	164	Bill Nicholson	CHI	158	Mort Cooper	STL	147	Mort Cooper	STL	148
Augie Galan	BRO	137	Mel Ott	NY	144	Whit Wyatt	BRO	136	Nate Andrews	BOS	145
On Base Average			**Slugging Percentage**			**Percent of Team Wins**			**Wins Above Team**		
Stan Musial	STL	.425	Stan Musial	STL	.562	Rip Sewell	PIT	.263	Rip Sewell	PIT	6.7
Augie Galan	BRO	.412	Bill Nicholson	CHI	.531	Al Javery	BOS	.250	Schoolboy Rowe	PHI	5.7
Billy Herman	BRO	.398	Walker Cooper	STL	.463	Hi Bithorn	CHI	.243	Ace Adams	NY	5.1
Isolated Power			**Players Overall**			**Pitchers Overall**			**Defensive Runs**		
Bill Nicholson	CHI	.222	Stan Musial	STL	52.5	Jim Tobin	BOS	39.1	Eddie Miller	CIN	24.9
Stan Musial	STL	.206	Bill Nicholson	CHI	43.0	Nate Andrews	BOS	37.5	Whitey Wietelmann	BOS	19.2
Mel Ott	NY	.184	Augie Galan	BRO	31.1	Max Lanier	STL	32.6	Eddie Stanky	CHI	18.0

Club	W	L	R	OR	Avg	OBA	SLG	BPF	NOPS-A	BR	Adj	Wins	ERA	PPF	NERA-A	PR	Adj	Wins	Diff
STL	105	49	679	475	.279	.334	.392	105	117/112	93	64	6.8	2.57	101	131/132	127	131	14.0	7.1
CIN	87	67	608	543	.256	.315	.340	105	96/ 91	-34	-67	-7.3	3.14	104	107/112	37	59	6.3	11.0
BRO	81	72	716	674	.272	.346	.357	95	110/115	71	100	10.6	3.88	94	87/ 82	-76	-106	-11.5	5.3
PIT	80	74	669	605	.262	.335	.357	100	107/106	44	42	4.5	3.06	99	110/109	48	42	4.5	-6.0
CHI	74	79	632	600	.261	.336	.351	104	105/101	38	15	1.6	3.24	103	104/107	20	37	4.0	-8.1
BOS	68	85	465	612	.233	.298	.309	107	81/ 76	-122	-163	-17.5	3.24	110	104/115	20	75	8.0	1.0
PHI	64	90	571	676	.249	.316	.335	95	94/100	-38	-6	-.7	3.78	97	89/ 86	-63	-81	-8.7	-3.6
NY	55	98	558	713	.247	.313	.335	93	93/101	-47	-3	-.4	4.09	95	82/ 79	-110	-133	-14.3	-6.7
			612		.258	.324	.347						3.37						

AMERICAN LEAGUE — 1944

Batting Runs			Park Adjusted			Pitching Runs			Park Adjusted		
Bob Johnson	BOS	52.9	Bob Johnson	BOS	54.9	Dizzy Trout	DET	51.2	Dizzy Trout	DET	71.6
Stan Spence	WAS	38.1	Stan Spence	WAS	46.2	Hal Newhouser	DET	41.9	Hal Newhouser	DET	60.0
Bobby Doerr	BOS	38.0	Bobby Doerr	BOS	39.7	Jack Kramer	STL	27.0	Hank Borowy	NY	25.5
Normalized OPS			**Park Adjusted**			**Normalized ERA**			**Park Adjusted**		
Bob Johnson	BOS	173	Bob Johnson	BOS	178	Dizzy Trout	DET	162	Dizzy Trout	DET	186
Bobby Doerr	BOS	163	Stan Spence	WAS	169	Hal Newhouser	DET	154	Hal Newhouser	DET	178
Stan Spence	WAS	150	Bobby Doerr	BOS	169	Tex Hughson	BOS	152	Tex Hughson	BOS	145
On Base Average			**Slugging Percentage**			**Percent of Team Wins**			**Wins Above Team**		
Bob Johnson	BOS	.431	Bobby Doerr	BOS	.528	Hal Newhouser	DET	.330	Hal Newhouser	DET	9.7
Lou Boudreau	CLE	.406	Bob Johnson	BOS	.528	Dizzy Trout	DET	.307	Tex Hughson	BOS	7.6
Bobby Doerr	BOS	.399	Johnny Lindell	NY	.500	Tex Hughson	BOS	.234	Dizzy Trout	DET	4.9
Isolated Power			**Players Overall**			**Pitchers Overall**			**Defensive Runs**		
Bob Johnson	BOS	.204	Lou Boudreau	CLE	64.7	Dizzy Trout	DET	88.3	Eddie Mayo	DET	24.3
Bobby Doerr	BOS	.203	Snuffy Stirnweiss	NY	56.8	Hal Newhouser	DET	62.3	Lou Boudreau	CLE	23.4
Johnny Lindell	NY	.200	Stan Spence	WAS	56.6	Jack Kramer	STL	29.6	Snuffy Stirnweiss	NY	18.9

Club	W	L	R	OR	Avg	OBA	SLG	BPF	NOPS-A	BR	Adj	Wins	ERA	PPF	NERA-A	PR	Adj	Wins	Diff
STL	89	65	684	587	.252	.323	.352	100	100/100	-3	-2	-.3	3.17	98	108/106	40	29	3.0	9.3
DET	88	66	658	581	.263	.332	.354	116	103/ 89	21	-79	-8.4	3.09	115	111/128	53	134	14.1	5.3
NY	83	71	674	617	.264	.333	.387	104	113/109	75	51	5.3	3.38	103	102/105	8	25	2.6	-1.9
BOS	77	77	739	676	.270	.336	.380	97	112/116	73	93	9.8	3.82	95	90/ 86	-60	-84	-8.9	-.9
PHI	72	82	525	594	.257	.314	.327	100	89/ 89	-72	-74	-7.9	3.26	102	105/107	26	34	3.6	-.7
CLE	72	82	643	677	.266	.331	.372	101	108/107	46	41	4.3	3.65	102	94/ 95	-34	-25	-2.7	-6.6
CHI	71	83	543	662	.247	.307	.320	98	86/ 88	-96	-81	-8.6	3.58	100	96/ 96	-22	-24	-2.6	5.2
WAS	64	90	592	664	.261	.324	.330	89	93/105	-40	30	3.1	3.49	89	98/ 87	-8	-66	-7.1	-9.1
				632	.260	.325	.353						3.43						

NATIONAL LEAGUE — 1944

Batting Runs			Park Adjusted			Pitching Runs			Park Adjusted		
Stan Musial	STL	60.9	Stan Musial	STL	59.7	Bucky Walters	CIN	38.4	Bucky Walters	CIN	35.6
Dixie Walker	BRO	51.2	Dixie Walker	BRO	54.0	Mort Cooper	STL	32.1	Red Munger	STL	28.8
Bill Nicholson	CHI	47.4	Augie Galan	BRO	49.5	Red Munger	STL	30.5	Mort Cooper	STL	28.4
Normalized OPS			**Park Adjusted**			**Normalized ERA**			**Park Adjusted**		
Stan Musial	STL	177	Dixie Walker	BRO	177	Ed Heusser	CIN	152	Ed Heusser	CIN	148
Mel Ott	NY	171	Stan Musial	STL	174	Bucky Walters	CIN	150	Bucky Walters	CIN	147
Dixie Walker	BRO	170	Mel Ott	NY	172	Mort Cooper	STL	147	Mort Cooper	STL	141
On Base Average			**Slugging Percentage**			**Percent of Team Wins**			**Wins Above Team**		
Stan Musial	STL	.440	Stan Musial	STL	.549	Bill Voiselle	NY	.313	Bill Voiselle	NY	6.5
Dixie Walker	BRO	.434	Bill Nicholson	CHI	.545	Jim Tobin	BOS	.277	Bucky Walters	CIN	6.4
Augie Galan	BRO	.426	Mel Ott	NY	.544	Bucky Walters	CIN	.258	Claude Passeau	CHI	3.9
Isolated Power			**Players Overall**			**Pitchers Overall**			**Defensive Runs**		
Bill Nicholson	CHI	.258	Stan Musial	STL	59.6	Bucky Walters	CIN	43.5	Buddy Kerr	NY	15.6
Mel Ott	NY	.256	Dixie Walker	BRO	36.6	Red Munger	STL	28.7	Woody Williams	CIN	14.3
Ron Northey	PHI	.209	Augie Galan	BRO	36.2	Jim Tobin	BOS	26.5	Stan Musial	STL	13.1

Club	W	L	R	OR	Avg	OBA	SLG	BPF	NOPS-A	BR	Adj	Wins	ERA	PPF	NERA-A	PR	Adj	Wins	Diff
STL	105	49	772	490	.275	.344	.402	102	118/116	110	99	10.1	2.68	96	135/130	148	127	13.0	4.9
PIT	90	63	744	662	.266	.338	.379	106	109/103	58	20	2.0	3.44	105	105/110	27	53	5.4	6.1
CIN	89	65	573	537	.254	.313	.338	98	90/ 91	-73	-63	-6.6	2.97	98	122/119	100	86	8.9	9.7
CHI	75	79	702	669	.261	.328	.360	103	101/ 98	-1	-22	-2.3	3.59	103	101/103	. 3	18	1.9	-1.6
NY	67	87	682	773	.263	.330	.370	99	104/105	22	28	2.8	4.29	101	84/ 85	-101	-96	-10.0	-2.8
BOS	65	89	593	674	.246	.308	.353	92	93/101	-60	-11	-1.2	3.67	93	98/ 92	-8	-45	-4.8	-6.0
BRO	63	91	690	832	.269	.331	.366	96	103/108	15	43	4.4	4.68	99	77/ 76	-161	-169	-17.4	-1.0
PHI	61	92	539	658	.251	.316	.336	103	90/ 87	-68	-87	-9.0	3.64	105	99/104	-4	25	2.6	-9.0
				662	.261	.326	.363						3.61						

Batting Runs			Park Adjusted			Pitching Runs			Park Adjusted		
Snuffy Stirnweiss	NY	39.1	Snuffy Stirnweiss	NY	37.1	Hal Newhouser	DET	54.1	Hal Newhouser	DET	63.4
Nick Etten	NY	29.9	Eddie Lake	BOS	31.9	Roger Wolff	WAS	34.6	Nels Potter	STL	38.0
Jeff Heath	CLE	29.4	Bobby Estalella	PHI	31.4	Dutch Leonard	WAS	29.8	Al Benton	DET	34.5
Normalized OPS			**Park Adjusted**			**Normalized ERA**			**Park Adjusted**		
Snuffy Stirnweiss	NY	149	Bobby Estalella	PHI	157	Hal Newhouser	DET	186	Hal Newhouser	DET	201
Bobby Estalella	PHI	142	Eddie Lake	BOS	147	Al Benton	DET	167	Al Benton	DET	180
Nick Etten	NY	139	Snuffy Stirnweiss	NY	145	Roger Wolff	WAS	159	Roger Wolff	WAS	155
On Base Average			**Slugging Percentage**			**Percent of Team Wins**			**Wins Above Team**		
Eddie Lake	BOS	.412	Snuffy Stirnweiss	NY	.476	Dave Ferriss	BOS	.296	Dave Ferriss	BOS	8.4
Bobby Estalella	PHI	.399	Vern Stephens	STL	.473	Hal Newhouser	DET	.284	Hal Newhouser	DET	7.0
Oscar Grimes	NY	.395	Nick Etten	NY	.437	Steve Gromek	CLE	.260	Steve Gromek	CLE	6.1
Isolated Power			**Players Overall**			**Pitchers Overall**			**Defensive Runs**		
Vern Stephens	STL	.184	Snuffy Stirnweiss	NY	64.9	Hal Newhouser	DET	69.0	Eddie Lake	BOS	26.7
Snuffy Stirnweiss	NY	.168	Eddie Lake	BOS	64.3	Nels Potter	STL	40.2	Snuffy Stirnweiss	NY	25.5
Pat Seerey	CLE	.164	Eddie Mayo	DET	27.0	Al Benton	DET	29.7	Irv Hall	PHI	24.7

Club	W	L	R	OR	Avg	OBA	SLG	BPF	NOPS-A	BR	Adj	Wins	ERA	PPF	NERA-A	PR	Adj	Wins	Diff
DET	88	65	633	565	.256	.324	.361	109	105/96	20	-30	-3.3	3.00	108	112/121	58	99	10.6	4.3
WAS	87	67	622	562	.258	.330	.334	99	98/99	-6	-0	-.1	2.93	98	115/112	69	56	6.0	4.1
STL	81	70	597	548	.249	.316	.341	113	96/85	-28	-106	-11.5	3.15	113	107/121	34	102	10.9	6.1
NY	81	71	676	606	.259	.343	.373	103	114/111	90	73	7.9	3.46	102	97/99	-12	-4	-.6	-2.3
CLE	73	72	557	548	.255	.326	.359	102	105/103	23	14	1.5	3.31	101	102/103	8	15	1.6	-2.6
CHI	71	78	596	633	.262	.326	.337	94	98/104	-12	19	2.1	3.70	95	91/86	-48	-74	-8.0	2.5
BOS	71	83	599	674	.260	.330	.346	95	102/108	12	45	4.8	3.80	96	89/85	-65	-87	-9.5	-1.3
PHI	52	98	494	638	.245	.306	.316	90	86/95	-95	-37	-4.1	3.63	92	93/86	-39	-78	-8.5	-10.4
				597	.255	.325	.346						3.37						

NATIONAL LEAGUE 1945

Batting Runs			Park Adjusted			Pitching Runs			Park Adjusted		
Tommy Holmes	BOS	62.9	Tommy Holmes	BOS	54.8	Hank Wyse	CHI	34.4	Hank Wyse	CHI	33.3
Phil Cavaretta	CHI	46.4	Phil Cavaretta	CHI	44.5	Claude Passeau	CHI	33.8	Claude Passeau	CHI	33.0
Augie Galan	BRO	36.8	Augie Galan	BRO	39.6	Ray Prim	CHI	25.7	Preacher Roe	PIT	25.1
Normalized OPS			**Park Adjusted**			**Normalized ERA**			**Park Adjusted**		
Tommy Holmes	BOS	175	Mel Ott	NY	162	Ray Prim	CHI	158	Ray Prim	CHI	157
Phil Cavaretta	CHI	164	Tommy Holmes	BOS	160	Claude Passeau	CHI	155	Claude Passeau	CHI	153
Mel Ott	NY	152	Phil Cavaretta	CHI	159	Harry Brecheen	STL	151	Harry Brecheen	STL	149
On Base Average			**Slugging Percentage**			**Percent of Team Wins**			**Wins Above Team**		
Phil Cavaretta	CHI	.449	Tommy Holmes	BOS	.577	Hank Wyse	CHI	.224	Harry Brecheen	STL	3.7
Augie Galan	BRO	.423	Whitey Kurowski	STL	.511	Hal Gregg	BRO	.207	Van Mungo	NY	3.7
Stan Hack	CHI	.420	Phil Cavaretta	CHI	.500	Nick Strincevich	PIT	.195	Andy Karl	PHI	3.6
Isolated Power			**Players Overall**			**Pitchers Overall**			**Defensive Runs**		
Tommy Holmes	BOS	.225	Tommy Holmes	BOS	47.2	Claude Passeau	CHI	36.3	Buddy Kerr	NY	28.3
Vince DiMaggio	PHI	.195	Eddie Stanky	BRO	43.5	Hank Wyse	CHI	32.9	Carden Gillenwater	BOS	20.4
Mel Ott	NY	.191	Stan Hack	CHI	39.8	Ray Prim	CHI	26.1	Stan Hack	CHI	15.7

Club	W	L	R	OR	Avg	OBA	SLG	BPF	NOPS-A	BR	Adj	Wins	ERA	PPF	NERA-A	PR	Adj	Wins	Diff
CHI	98	56	735	532	.277	.349	.372	103	108/105	54	34	3.4	2.98	99	127/126	124	119	11.9	5.7
STL	95	59	756	583	.273	.338	.371	102	104/102	25	12	1.2	3.24	99	117/116	87	80	8.0	8.8
BRO	87	67	795	724	.271	.350	.376	96	109/113	67	92	9.2	3.70	95	103/98	15	-13	-1.4	2.2
PIT	82	72	753	686	.267	.341	.377	102	107/105	45	31	3.1	3.76	101	101/102	6	11	1.1	.8
NY	78	74	668	700	.269	.336	.379	94	106/113	33	74	7.4	4.06	94	94/88	-39	-74	-7.5	2.0
BOS	67	85	721	728	.267	.334	.374	110	104/95	21	-46	-4.7	4.04	111	94/104	-36	25	2.6	-6.8
CIN	61	93	536	694	.249	.304	.333	92	84/91	-121	-67	-6.8	4.00	94	95/89	-29	-64	-6.5	-2.7
PHI	46	108	548	865	.246	.307	.326	101	83/82	-121	-126	-12.8	4.64	107	82/88	-125	-86	-8.7	-9.5
				689	.265	.333	.364						3.80						

AMERICAN LEAGUE   1946

Batting Runs		Park Adjusted		Pitching Runs		Park Adjusted	
Ted Williams	BOS 94.2	Ted Williams	BOS 88.5	Bob Feller	CLE 54.4	Hal Newhouser	DET 56.4
Hank Greenberg	DET 46.1	Charlie Keller	NY 45.8	Hal Newhouser	DET 51.1	Bob Feller	CLE 41.1
Charlie Keller	NY 45.9	Hank Greenberg	DET 41.6	Spud Chandler	NY 40.0	Dizzy Trout	DET 40.5

Normalized OPS		Park Adjusted		Normalized ERA		Park Adjusted	
Ted Williams	BOS 223	Ted Williams	BOS 207	Hal Newhouser	DET 181	Hal Newhouser	DET 190
Hank Greenberg	DET 170	Charlie Keller	NY 160	Spud Chandler	NY 167	Spud Chandler	NY 162
Charlie Keller	NY 160	Hank Greenberg	DET 159	Bob Feller	CLE 160	Jesse Flores	PHI 159

On Base Average		Slugging Percentage		Percent of Team Wins		Wins Above Team	
Ted Williams	BOS .497	Ted Williams	BOS .667	Bob Feller	CLE .382	Bob Feller	CLE 10.8
Charlie Keller	NY .405	Hank Greenberg	DET .604	Hal Newhouser	DET .283	Hal Newhouser	DET 6.6
Mickey Vernon	WAS .403	Charlie Keller	NY .533	Phil Marchildon	PHI .265	Earl Caldwell	CHI 5.4

Isolated Power		Base Stealing Runs		Relievers - Runs		Park Adjusted	
Hank Greenberg	DET .327	Bob Dillinger	STL 1.8	Earl Caldwell	CHI 14.4	Earl Caldwell	CHI 14.2
Ted Williams	BOS .325	Snuffy Stirnweiss	NY 1.8	Bob Lemon	CLE 10.6	Gordon Maltzberger	CHI 8.5
Charlie Keller	NY .258	George Case	CLE 1.8	Gordon Maltzberger	CHI 8.6	Bob Klinger	BOS 8.2

Defensive Runs		Players Overall		Pitchers Overall		Relief Points	
Bobby Doerr	BOS 28.4	Ted Williams	BOS 80.2	Hal Newhouser	DET 56.4	Earl Caldwell	CHI 38
Lou Boudreau	CLE 14.7	Bobby Doerr	BOS 47.4	Dizzy Trout	DET 47.4	Bob Klinger	BOS 22
Joe Gordon	NY 13.9	Hank Greenberg	DET 36.7	Spud Chandler	NY 40.9	Johnny Murphy	NY 20

Club	W	L	R	OR	Avg	OBA	SLG	BPF	NOPS-A	BR	Adj	Wins	ERA	PPF	NERA-A	PR	Adj	Wins	Diff
BOS	104	50	792	594	.271	.356	.402	108	120/112	136	85	8.9	3.38	105	104/109	19	45	4.7	13.4
DET	92	62	704	567	.258	.338	.374	107	107/100	43	-0	-.1	3.22	105	109/114	45	70	7.4	7.8
NY	87	67	684	547	.248	.334	.387	100	110/110	54	54	5.6	3.14	97	112/109	55	41	4.3	.1
WAS	76	78	608	706	.260	.328	.366	99	102/102	1	6	.6	3.74	101	94/ 95	-35	-28	-3.1	1.4
CHI	74	80	562	595	.257	.323	.333	90	90/ 91	-64	-57	-6.1	3.10	99	113/112	63	60	6.2	-3.2
CLE	68	86	537	638	.245	.313	.356	90	95/106	-48	15	1.6	3.62	91	97/ 88	-17	-66	-7.1	-3.5
STL	66	88	621	710	.251	.313	.356	98	95/ 96	-52	-40	-4.3	3.95	100	89/ 89	-68	-68	-7.3	.5
PHI	49	105	529	680	.253	.318	.338	102	90/ 89	-67	-79	-8.4	3.90	105	90/ 95	-59	-31	-3.3	-16.3
				630	.256	.328	.364						3.50						

NATIONAL LEAGUE   1946

Batting Runs		Park Adjusted		Pitching Runs		Park Adjusted	
Stan Musial	STL 70.9	Stan Musial	STL 63.8	Howie Pollet	STL 39.0	Howie Pollet	STL 45.9
Johnny Mize	NY 43.6	Johnny Mize	NY 44.4	Johnny Sain	BOS 35.6	Harry Brecheen	STL 29.7
Whitey Kurowski	STL 29.1	Phil Cavaretta	CHI 30.6	Harry Brecheen	STL 23.7	Johnny Sain	BOS 29.5

Normalized OPS		Park Adjusted		Normalized ERA		Park Adjusted	
Stan Musial	STL 188	Stan Musial	STL 172	Howie Pollet	STL 163	Howie Pollet	STL 174
Whitey Kurowski	STL 142	Del Ennis	PHI 147	Johnny Sain	BOS 155	Fireman Beggs	CIN 158
Del Ennis	PHI 140	Phil Cavaretta	CHI 144	Fireman Beggs	CIN 147	Ewell Blackwell	CIN 149

On Base Average		Slugging Percentage		Percent of Team Wins		Wins Above Team	
Eddie Stanky	BRO .436	Stan Musial	STL .587	Johnny Sain	BOS .247	Schoolboy Rowe	PHI 4.7
Stan Musial	STL .434	Del Ennis	PHI .485	Dave Koslo	NY .230	Fritz Ostermueller	PIT 4.2
Phil Cavaretta	CHI .401	Enos Slaughter	STL .465	Howie Pollet	STL .214	Emil Kush	CHI 3.3

Isolated Power		Base Stealing Runs		Relievers - Runs		Park Adjusted	
Stan Musial	STL .221	Caught Stealing		Hugh Casey	BRO 16.0	Junior Thompson	NY 14.9
Ron Northey	PHI .192	Not Available		Junior Thompson	NY 14.9	Hugh Casey	BRO 14.8
Ralph Kiner	PIT .183			Robert Malloy	CIN 5.3	Robert Malloy	CIN 7.3

Defensive Runs		Players Overall		Pitchers Overall		Relief Points	
Marty Marion	STL 21.9	Stan Musial	STL 48.2	Howie Pollet	STL 45.9	Hugh Casey	BRO 28
Bobby Adams	CIN 18.9	Del Ennis	PHI 36.4	Johnny Sain	BOS 38.4	Emil Kush	CHI 20
Lonny Frey	CIN 16.3	Johnny Mize	NY 35.3	Fireman Beggs	CIN 30.7	Hank Behrman	BRO 19

Club	W	L	R	OR	Avg	OBA	SLG	BPF	NOPS-A	BR	Adj	Wins	ERA	PPF	NERA-A	PR	Adj	Wins	Diff
STL	98	58	712	545	.265	.334	.381	110	110/100	54	-5	-.7	3.01	107	114/121	63	100	10.6	10.1
BRO	96	60	701	570	.260	.348	.361	100	108/108	61	64	6.8	3.04	97	112/109	59	42	4.4	6.8
CHI	82	71	626	581	.254	.331	.346	95	98/103	-8	19	2.1	3.25	94	105/ 99	26	-4	-.5	4.0
BOS	81	72	630	592	.264	.337	.353	95	102/107	16	46	4.9	3.38	94	101/ 95	6	-25	-2.7	2.3
PHI	69	85	560	705	.258	.315	.359	95	98/103	-28	-0	-.1	4.00	98	86/ 84	-87	-97	-10.4	2.6
CIN	67	87	523	570	.239	.307	.327	106	86/ 81	-100	-136	-14.5	3.08	107	111/119	53	91	9.7	-5.1
PIT	63	91	552	668	.250	.328	.344	102	97/ 95	-17	-31	-3.4	3.72	105	92/ 96	-44	-19	-2.1	-8.5
NY	61	93	612	685	.255	.328	.374	98	106/108	26	37	4.0	3.91	100	87/ 87	-73	-75	-8.1	-11.9
				615	.256	.329	.355						3.42						

Batting Runs			Park Adjusted			Pitching Runs			Park Adjusted		
Ted Williams	BOS	91.1	Ted Williams	BOS	90.0	Bob Feller	CLE	34.1	Hal Newhouser	DET	31.2
Joe DiMaggio	NY	36.5	Joe DiMaggio	NY	42.7	Hal Newhouser	DET	26.3	Bob Feller	CLE	28.7
Tommy Henrich	NY	26.4	Tommy Henrich	NY	32.9	Joe Haynes	CHI	25.9	Joe Haynes	CHI	27.5
Normalized OPS			**Park Adjusted**			**Normalized ERA**			**Park Adjusted**		
Ted Williams	BOS	214	Ted Williams	BOS	211	Joe Haynes	CHI	153	Joe Haynes	CHI	156
Joe DiMaggio	NY	153	Joe DiMaggio	NY	168	Bob Feller	CLE	138	Dick Fowler	PHI	138
Tommy Henrich	NY	137	Tommy Henrich	NY	151	Eddie Lopat	CHI	132	Eddie Lopat	CHI	135
On Base Average			**Slugging Percentage**			**Percent of Team Wins**			**Wins Above Team**		
Ted Williams	BOS	.499	Ted Williams	BOS	.634	Early Wynn	WAS	.266	Phil Marchildon	PHI	5.9
Ferris Fain	PHI	.414	Joe DiMaggio	NY	.522	Bob Feller	CLE	.250	Joe Haynes	CHI	5.6
Roy Cullenbine	DET	.401	Joe Gordon	CLE	.496	Phil Marchildon	PHI	.244	Bob Feller	CLE	4.9
Isolated Power			**Base Stealing Runs**			**Relievers - Runs**			**Park Adjusted**		
Ted Williams	BOS	.292	Bob Dillinger	STL	2.4	Joe Page	NY	19.0	Joe Page	NY	10.3
Jeff Heath	STL	.234	Elmer Valo	PHI	1.5	Russ Christopher	PHI	7.3	Russ Christopher	PHI	9.0
Joe Gordon	CLE	.224	Bingo Binks	PHI	1.2	Ed Klieman	CLE	6.9	Johnny Murphy	BOS	5.8
Defensive Runs			**Players Overall**			**Pitchers Overall**			**Relief Points**		
Bobby Doerr	BOS	21.5	Ted Williams	BOS	84.6	Hal Newhouser	DET	35.9	Joe Page	NY	55
George Kell	DET	16.1	Joe DiMaggio	NY	37.9	Bob Feller	CLE	32.2	Ed Klieman	CLE	40
Lou Boudreau	CLE	15.0	Lou Boudreau	CLE	36.2	Fred Hutchinson	DET	32.2	Russ Christopher	PHI	37

Club	W	L	R	OR	Avg	OBA	SLG	BPF	NOPS-A	BR	Adj	Wins	ERA	PPF	NERA-A	PR	Adj	Wins	Diff
NY	97	57	794	568	.271	.349	.407	91	118/130	110	171	17.7	3.39	85	109/ 93	48	-36	-3.8	6.1
DET	85	69	714	642	.258	.353	.377	105	110/105	77	42	4.4	3.57	104	104/108	21	45	4.7	-1.1
BOS	83	71	720	669	.265	.349	.382	101	111/109	71	62	6.4	3.81	101	97/ 98	-15	-12	-1.3	.9
CLE	80	74	687	588	.259	.324	.385	98	104/107	8	23	2.4	3.44	96	108/103	41	16	1.6	-1.0
PHI	78	76	633	614	.252	.333	.349	105	96/ 91	-23	-56	-5.9	3.51	105	105/111	30	59	6.1	.8
CHI	70	84	553	661	.256	.321	.342	100	91/ 91	-69	-69	-7.2	3.63	102	102/104	11	23	2.4	-2.2
WAS	64	90	496	675	.241	.313	.321	96	82/ 86	-116	-91	-9.5	3.97	99	93/ 93	-39	-43	-4.6	1.1
STL	59	95	564	744	.241	.320	.350	104	93/ 89	-54	-78	-8.2	4.33	108	86/ 92	-94	-51	-5.4	-4.5
				645		.256	.333	.364					3.70						

NATIONAL LEAGUE 1947

Batting Runs			Park Adjusted			Pitching Runs			Park Adjusted		
Ralph Kiner	PIT	61.4	Ralph Kiner	PIT	59.0	Warren Spahn	BOS	56.1	Warren Spahn	BOS	49.3
Johnny Mize	NY	48.3	Johnny Mize	NY	47.8	Ewell Blackwell	CIN	48.4	Ralph Branca	BRO	45.7
Whitey Kurowski	STL	41.8	Bob Elliott	BOS	39.3	Ralph Branca	BRO	43.6	Ewell Blackwell	CIN	36.2
Normalized OPS			**Park Adjusted**			**Normalized ERA**			**Park Adjusted**		
Ralph Kiner	PIT	177	Ralph Kiner	PIT	171	Warren Spahn	BOS	175	Warren Spahn	BOS	166
Johnny Mize	NY	161	Johnny Mize	NY	160	Ewell Blackwell	CIN	165	Ralph Branca	BRO	155
Whitey Kurowski	STL	154	Bob Elliott	BOS	150	Ralph Branca	BRO	152	Al Brazle	STL	151
On Base Average			**Slugging Percentage**			**Percent of Team Wins**			**Wins Above Team**		
Augie Galan	CIN	.449	Ralph Kiner	PIT	.639	Ewell Blackwell	CIN	.301	Ewell Blackwell	CIN	9.7
Whitey Kurowski	STL	.420	Johnny Mize	NY	.614	Dutch Leonard	PHI	.274	Larry Jansen	NY	8.8
Ralph Kiner	PIT	.417	Walker Cooper	NY	.586	Larry Jansen	NY	.259	Dutch Leonard	PHI	6.6
Isolated Power			**Base Stealing Runs**			**Relievers - Runs**			**Park Adjusted**		
Ralph Kiner	PIT	.326	Caught Stealing			Walter Lanfranconi	BOS	7.9	Emil Kush	CHI	10.2
Johnny Mize	NY	.312	Not Available			Emil Kush	CHI	7.1	Walter Lanfranconi	BOS	6.4
Walker Cooper	NY	.282				Gerry Staley	STL	4.1	Russ Meyer	CHI	4.9
Defensive Runs			**Players Overall**			**Pitchers Overall**			**Relief Points**		
Marty Marion	STL	16.7	Ralph Kiner	PIT	61.0	Warren Spahn	BOS	47.7	Hugh Casey	BRO	52
Emil Verban	PHI	14.8	Johnny Mize	NY	45.1	Ralph Branca	BRO	38.8	Ken Trinkle	NY	32
Frankie Gustine	PIT	14.6	Peewee Reese	BRO	33.4	Ewell Blackwell	CIN	37.1	Harry Gumbert	CIN	30

Club	W	L	R	OR	Avg	OBA	SLG	BPF	NOPS-A	BR	Adj	Wins	ERA	PPF	NERA-A	PR	Adj	Wins	Diff
BRO	94	60	774	668	.272	.364	.384	103	107/104	62	39	3.8	3.81	102	107/109	39	50	4.9	8.3
STL	89	65	780	634	.270	.347	.401	107	107/100	41	-11	-1.2	3.53	105	115/121	83	115	11.3	1.9
BOS	86	68	701	622	.275	.346	.390	97	104/108	19	42	4.1	3.61	95	113/107	69	37	3.6	1.2
NY	81	73	830	761	.271	.335	.454	101	120/119	95	91	9.0	4.44	100	92/ 91	-55	-59	-5.9	.9
CIN	73	81	681	755	.259	.330	.375	90	95/106	-47	26	2.5	4.41	90	92/ 93	-51	-113	-11.2	4.6
CHI	69	85	567	722	.259	.321	.361	105	89/ 85	-92	-124	-12.3	4.10	108	99/107	-5	41	4.0	.3
PIT	62	92	744	817	.261	.339	.406	103	107/104	31	9	.9	4.68	105	87/ 91	-92	-64	-6.4	-9.5
PHI	62	92	589	667	.258	.321	.352	96	86/ 89	-107	-82	-8.2	3.96	98	103/100	17	2	.2	-7.1
				708		.265	.338	.390					4.07						

AMERICAN LEAGUE 1948

Batting Runs			Park Adjusted			Pitching Runs			Park Adjusted		
Ted Williams	BOS	75.7	Ted Williams	BOS	80.3	Bob Lemon	CLE	47.9	Gene Bearden	CLE	41.4
Lou Boudreau	CLE	52.7	Lou Boudreau	CLE	53.3	Gene Bearden	CLE	47.4	Bob Lemon	CLE	40.2
Joe DiMaggio	NY	48.6	Joe DiMaggio	NY	47.4	Hal Newhouser	DET	38.4	Hal Newhouser	DET	36.8

Normalized OPS			Park Adjusted			Normalized ERA			Park Adjusted		
Ted Williams	BOS	195	Ted Williams	BOS	207	Gene Bearden	CLE	177	Gene Bearden	CLE	167
Joe DiMaggio	NY	162	Lou Boudreau	CLE	163	Bob Lemon	CLE	152	Ray Scarborough	WAS	161
Lou Boudreau	CLE	162	Joe DiMaggio	NY	160	Ray Scarborough	WAS	152	Bob Lemon	CLE	144

On Base Average			Slugging Percentage			Percent of Team Wins			Wins Above Team		
Ted Williams	BOS	.497	Ted Williams	BOS	.615	Hal Newhouser	DET	.269	Ray Scarborough	WAS	7.7
Lou Boudreau	CLE	.453	Joe DiMaggio	NY	.598	Ray Scarborough	WAS	.268	Hal Newhouser	DET	5.5
Elmer Valo	PHI	.432	Tommy Henrich	NY	.554	Bob Lemon	CLE	.206	Jack Kramer	BOS	4.4
						Gene Bearden	CLE	.206			

Isolated Power			Base Stealing Runs			Relievers - Runs			Park Adjusted		
Joe DiMaggio	NY	.278	Sherry Robertson	WAS	2.4	Ed Klieman	CLE	15.1	Ed Klieman	CLE	13.0
Tommy Henrich	NY	.247	Joe E. Tucker	CLE	2.1	Russ Christopher	CLE	9.1	Forrest Thompson	WAS	10.0
Ted Williams	BOS	.246	Dom DiMaggio	BOS	1.8	Forrest Thompson	WAS	6.3	Russ Christopher	CLE	7.5

Defensive Runs			Players Overall			Pitchers Overall			Relief Points		
Jerry Priddy	STL	22.8	Ted Williams	BOS	74.4	Bob Lemon	CLE	61.0	Joe Page	NY	38
Luke Appling	CHI	13.3	Lou Boudreau	CLE	64.2	Gene Bearden	CLE	49.2	Russ Christopher	CLE	38
Don Kolloway	CHI	13.0	Bobby Doerr	BOS	42.7	Hal Newhouser	DET	39.2	Earl Johnson	BOS	26

Club	W	L	R	OR	Avg	OBA	SLG	BPF	NOPS-A	BR	Adj	Wins	ERA	PPF	NERA-A	PR	Adj	Wins	Diff
CLE	97	58	840	568	.282	.360	.431	99	118/119	112	117	11.4	3.22	95	133/126	166	130	12.6	-4.4
BOS	96	59	907	720	.274	.374	.409	94	115/122	115	160	15.5	4.21	90	102/ 92	11	-51	-5.1	8.0
NY	94	60	857	633	.278	.356	.432	102	117/115	99	88	8.5	3.75	98	114/112	81	67	6.4	2.0
PHI	84	70	729	735	.260	.353	.362	107	96/ 89	-20	-73	-7.2	4.42	108	97/104	-21	29	2.8	11.4
DET	78	76	700	726	.267	.353	.375	98	100/102	0	11	1.1	4.15	99	103/102	20	11	1.1	-1.2
STL	59	94	671	849	.271	.345	.378	104	98/ 95	-20	-49	-4.9	5.00	108	86/ 92	-109	-59	-5.9	-6.8
WAS	56	97	578	796	.244	.322	.331	102	79/ 78	-150	-163	-15.9	4.66	106	92/ 97	-55	-18	-1.8	-2.8
CHI	51	101	559	814	.251	.329	.331	93	81/ 87	-133	-86	-8.4	4.89	97	88/ 85	-90	-107	-10.4	-6.1
			730		.266	.349	.382						4.28						

NATIONAL LEAGUE 1948

Batting Runs			Park Adjusted			Pitching Runs			Park Adjusted		
Stan Musial	STL	90.2	Stan Musial	STL	89.8	Johnny Sain	BOS	47.4	Johnny Sain	BOS	47.7
Johnny Mize	NY	44.6	Johnny Mize	NY	42.0	Harry Brecheen	STL	44.4	Harry Brecheen	STL	43.2
Ralph Kiner	PIT	38.5	Ralph Kiner	PIT	38.1	Dutch Leonard	PHI	36.3	Rex Barney	BRO	36.0

Normalized OPS			Park Adjusted			Normalized ERA			Park Adjusted		
Stan Musial	STL	208	Stan Musial	STL	207	Harry Brecheen	STL	177	Harry Brecheen	STL	174
Johnny Mize	NY	157	Andy Pafko	CHI	154	Dutch Leonard	PHI	158	Preacher Roe	BRO	168
Sid Gordon	NY	149	Johnny Mize	NY	152	Johnny Sain	BOS	152	Johnny Sain	BOS	152

On Base Average			Slugging Percentage			Percent of Team Wins			Wins Above Team		
Stan Musial	STL	.450	Stan Musial	STL	.702	Johnny Schmitz	CHI	.281	Johnny Schmitz	CHI	6.4
Bob Elliott	BOS	.423	Johnny Mize	NY	.564	Johnny Vander Meer	CIN	.266	Harry Brecheen	STL	6.2
Richie Ashburn	PHI	.410	Sid Gordon	NY	.537	Johnny Sain	BOS	.264	Johnny Vander Meer	CIN	5.1

Isolated Power			Base Stealing Runs			Relievers - Runs			Park Adjusted		
Stan Musial	STL	.326	Caught Stealing			Ted Wilks	STL	19.6	Ted Wilks	STL	18.9
Johnny Mize	NY	.275	Not Available			Andy Hansen	NY	10.9	Lefty Minner	BRO	13.9
Ralph Kiner	PIT	.268				Lefty Minner	BRO	10.7	Andy Hansen	NY	11.9

Defensive Runs			Players Overall			Pitchers Overall			Relief Points		
Richie Ashburn	PHI	12.3	Stan Musial	STL	81.9	Johnny Sain	BOS	47.9	Harry Gumbert	CIN	46
Frankie Gustine	PIT	11.9	Johnny Mize	NY	43.7	Harry Brecheen	STL	43.1	Ted Wilks	STL	31
Roy Smalley	CHI	11.7	Andy Pafko	CHI	39.3	Dutch Leonard	PHI	31.9	Kirby Higbe	PIT	28

Club	W	L	R	OR	Avg	OBA	SLG	BPF	NOPS-A	BR	Adj	Wins	ERA	PPF	NERA-A	PR	Adj	Wins	Diff
BOS	91	62	739	584	.275	.358	.399	103	113-110	94	73	7.3	3.37	100	117 117	90	91	9.1	-1.9
STL	85	69	742	646	.263	.340	.389	101	106-105	29	26	2.6	3.91	99	101 100	7	-0	-.1	5.5
BRO	84	70	744	667	.261	.339	.381	113	103- 91	13	-74	-7.5	3.75	112	105-117	31	102	10.1	4.4
PIT	83	71	706	699	.263	.338	.380	101	103-102	10	6	.6	4.15	110	95- 96	-29	-26	-2.7	8.0
NY	78	76	780	704	.256	.334	.408	103	110-106	45	22	2.2	3.93	102	101 103	4	18	1.8	-3.0
PHI	66	88	591	729	.259	.318	.368	92	94-102	-63	-9	-1.0	4.07	94	97 91	-17	-53	-5.4	-4.6
CIN	64	89	588	752	.247	.313	.365	101	92- 91	-75	-80	-8.1	4.47	104	88- 92	-76	-52	-5.3	1.0
CHI	64	90	597	706	.262	.322	.369	90	95-105	-51	14	1.4	4.01	92	99- 91	-7	-56	-5.7	-8.7
			686		.261	.333	.383						3.95						

AMERICAN LEAGUE 1949

Batting Runs			Park Adjusted			Pitching Runs			Park Adjusted		
Ted Williams	BOS	88.9	Ted Williams	BOS	89.4	Mel Parnell	BOS	46.6	Virgil Trucks	DET	55.3
Vern Stephens	BOS	36.9	Vern Stephens	BOS	37.5	Virgil Trucks	DET	42.2	Hal Newhouser	DET	41.1
Eddie Joost	PHI	31.9	Eddie Joost	PHI	35.7	Bob Lemon	CLE	37.6	Mel Parnell	BOS	40.0
Normalized OPS			**Park Adjusted**			**Normalized ERA**			**Park Adjusted**		
Ted Williams	BOS	201	Ted Williams	BOS	202	Mike Garcia	CLE	178	Mike Garcia	CLE	176
Tommy Henrich	NY	148	Tommy Henrich	NY	146	Mel Parnell	BOS	151	Virgil Trucks	DET	164
Vern Stephens	BOS	145	Vern Stephens	BOS	146	Virgil Trucks	DET	149	Fred Hutchinson	DET	157
On Base Average			**Slugging Percentage**			**Percent of Team Wins**			**Wins Above Team**		
Ted Williams	BOS	.490	Ted Williams	BOS	.650	Mel Parnell	BOS	.260	Mel Parnell	BOS	6.4
Luke Appling	CHI	.439	Vern Stephens	BOS	.539	Ray Scarborough	WAS	.260	Ray Scarborough	WAS	6.2
Eddie Joost	PHI	.429	Tommy Henrich	NY	.526	Bob Lemon	CLE	.247	Ellis Kinder	BOS	6.1
Isolated Power			**Base Stealing Runs**			**Relievers - Runs**			**Park Adjusted**		
Ted Williams	BOS	.307	Cliff Mapes	NY	1.8	Joe Page	NY	24.0	Joe Page	NY	22.9
Vern Stephens	BOS	.249	Birdie Tebbetts	BOS	1.8	Satchel Paige	CLE	10.7	Satchel Paige	CLE	10.1
Tommy Henrich	NY	.238	Ferris Fain	PHI	1.8	Frank Papish	CLE	6.9	Frank Papish	CLE	6.5
Defensive Runs			**Players Overall**			**Pitchers Overall**			**Relief Points**		
Bobby Doerr	BOS	22.4	Ted Williams	BOS	85.4	Bob Lemon	CLE	56.2	Joe Page	NY	72
Johnny Pesky	BOS	19.8	Bobby Doerr	BOS	49.7	Virgil Trucks	DET	44.6	Al Benton	CLE	28
Mickey Vernon	CLE	18.4	Eddie Joost	PHI	44.0	Hal Newhouser	DET	44.3	Tom Ferrick	STL	20

Club	W	L	R	OR	Avg	OBA	SLG	BPF	NOPS-A	BR	Adj	Wins	ERA	PPF	NERA-A	PR	Adj	Wins	Diff
NY	97	57	829	637	.269	.362	.400	102	109/107	58	46	4.5	3.70	98	114/112	76	65	6.4	9.1
BOS	96	58	896	667	.282	.381	.420	99	120/120	147	152	14.7	3.97	95	106/101	35	4	.4	3.9
CLE	89	65	675	574	.260	.339	.384	100	98/ 98	-29	-32	-3.2	3.35	98	125/123	130	120	11.7	3.5
DET	87	67	751	655	.267	.361	.378	111	102/ 92	20	-61	-6.0	3.77	110	111/123	66	132	12.9	3.1
PHI	81	73	726	725	.260	.361	.369	95	100/105	6	40	3.9	4.23	99	99/ 94	-4	-38	-3.8	3.9
CHI	63	91	648	737	.257	.347	.347	100	90/ 90	-65	-65	-6.4	4.30	102	98/ 99	-14	-4	-.5	-7.1
STL	53	101	667	913	.254	.339	.377	94	96/102	-40	-0	-.1	5.21	98	80/ 79	-151	-161	-15.8	-8.1
WAS	50	104	584	868	.254	.333	.356	99	89/ 89	-92	-86	-8.5	5.10	104	82/ 86	-134	-107	-10.6	-7.9
			722		.263	.353	.379						4.20						

NATIONAL LEAGUE 1949

Batting Runs			Park Adjusted			Pitching Runs			Park Adjusted		
Stan Musial	STL	72.4	Ralph Kiner	PIT	66.2	Dave Koslo	NY	36.2	Howie Pollet	STL	39.7
Ralph Kiner	PIT	70.0	Stan Musial	STL	64.8	Howie Pollet	STL	32.7	Dave Koslo	NY	32.5
Jackie Robinson	BRO	49.9	Jackie Robinson	BRO	44.9	Warren Spahn	BOS	32.6	Preacher Roe	BRO	31.9
Normalized OPS			**Park Adjusted**			**Normalized ERA**			**Park Adjusted**		
Ralph Kiner	PIT	188	Ralph Kiner	PIT	180	Dave Koslo	NY	161	Gerry Staley	STL	158
Stan Musial	STL	182	Stan Musial	STL	167	Gerry Staley	STL	148	Howie Pollet	STL	156
Jackie Robinson	BRO	156	Jackie Robinson	BRO	147	Howie Pollet	STL	146	Dave Koslo	NY	155
On Base Average			**Slugging Percentage**			**Percent of Team Wins**			**Wins Above Team**		
Stan Musial	STL	.438	Ralph Kiner	PIT	.658	Ken Raffensberger	CIN	.290	Warren Spahn	BOS	5.1
Jackie Robinson	BRO	.432	Stan Musial	STL	.624	Warren Spahn	BOS	.280	Ken Raffensberger	CIN	5.1
Ralph Kiner	PIT	.432	Jackie Robinson	BRO	.528	Vern Bickford	BOS	.213	Russ Meyer	PHI	4.5
Isolated Power			**Base Stealing Runs**			**Relievers - Runs**			**Park Adjusted**		
Ralph Kiner	PIT	.348	Caught Stealing			Eddie Erautt	CIN	8.7	Jim Konstanty	PHI	9.1
Stan Musial	STL	.286	Not Available			Jim Konstanty	PHI	8.6	Ted Wilks	STL	7.5
Del Ennis	PHI	.223				Bobby Hogue	BOS	7.3	Eddie Erautt	CIN	6.2
Defensive Runs			**Players Overall**			**Pitchers Overall**			**Relief Points**		
Red Schoendienst	STL	23.2	Stan Musial	STL	52.6	Howie Pollet	STL	38.6	Ted Wilks	STL	35
Richie Ashburn	PHI	18.3	Ralph Kiner	PIT	52.1	Dave Koslo	NY	33.5	Jim Konstanty	PHI	27
Roy Smalley	CHI	15.7	Jackie Robinson	BRO	49.9	Gerry Staley	STL	32.3	Gerry Staley	STL	18

Club	W	L	R	OR	Avg	OBA	SLG	BPF	NOPS-A	BR	Adj	Wins	ERA	PPF	NERA-A	PR	Adj	Wins	Diff
BRO	97	57	879	651	.274	.354	.419	106	116/110	103	59	5.9	3.80	102	106/109	38	53	5.2	8.9
STL	96	58	766	616	.277	.348	.404	109	110/101	61	-2	-.3	3.45	107	117/125	93	135	13.4	5.9
PHI	81	73	662	668	.254	.325	.388	101	99/ 98	-26	-33	-3.4	3.90	101	104/105	22	30	2.9	4.4
BOS	75	79	706	719	.258	.345	.374	96	101/104	6	33	3.2	3.99	96	101/ 97	8	-15	-1.5	-3.7
NY	73	81	736	693	.261	.340	.401	97	107/111	36	57	5.6	3.83	96	106/102	33	9	.9	-10.5
PIT	71	83	681	760	.259	.332	.384	105	100/ 96	-13	-45	-4.6	4.57	100	89/ 94	-78	-38	-3.9	2.5
CIN	62	92	627	770	.260	.316	.368	93	92/ 98	-83	-33	-3.4	4.34	95	93/ 89	-45	-76	-7.7	-4.0
CHI	61	93	593	773	.256	.312	.373	96	92/ 96	-82	-54	-5.4	4.50	99	90/ 89	-68	-75	-7.6	-3.0
			706		.262	.334	.389						4.04						

AMERICAN LEAGUE 1950

Batting Runs
Larry Doby	CLE	41.7
Ted Williams	BOS	40.9
Joe DiMaggio	NY	35.4

Park Adjusted
Larry Doby	CLE	43.3
Ted Williams	BOS	37.4
Al Rosen	CLE	36.1

Pitching Runs
Ned Garver	STL	34.2
Early Wynn	CLE	32.8
Art Houtteman	DET	31.8

Park Adjusted
Ned Garver	STL	49.9
Mel Parnell	BOS	31.3
Early Wynn	CLE	27.6

Normalized OPS
Larry Doby	CLE	151
Joe DiMaggio	NY	148
Hoot Evers	DET	144

Park Adjusted
Larry Doby	CLE	154
Joe DiMaggio	NY	146
Hoot Evers	DET	145

Normalized ERA
Early Wynn	CLE	143
Ned Garver	STL	135
Bob Feller	CLE	134

Park Adjusted
Ned Garver	STL	151
Early Wynn	CLE	136
Mel Parnell	BOS	131

On Base Average
Larry Doby	CLE	.442
Eddie Yost	WAS	.440
Johnny Pesky	BOS	.437

Slugging Percentage
Joe DiMaggio	NY	.585
Walt Dropo	BOS	.583
Hoot Evers	DET	.551

Percent of Team Wins
Bob Hooper	PHI	.288
Bob Lemon	CLE	.250
Ned Garver	STL	.224

Wins Above Team
Bob Hooper	PHI	7.8
Witto Aloma	CHI	3.7
Bob Lemon	CLE	3.5

Isolated Power
Joe DiMaggio	NY	.284
Walt Dropo	BOS	.261
Al Rosen	CLE	.256

Base Stealing Runs
Dom DiMaggio	BOS	2.1
Joe Collins	NY	1.5
Bobby Avila	CLE	1.5

Relievers - Runs
Howie Judson	CHI	7.9
Witto Aloma	CHI	7.7
Al Benton	CLE	7.0

Park Adjusted
Howie Judson	CHI	6.9
Howie Judson	CHI	6.9
Al Benton	CLE	5.5

Defensive Runs
Jerry Priddy	DET	32.5
Johnny Pesky	BOS	14.0
Irv Noren	WAS	13.3

Players Overall
Jerry Priddy	DET	42.2
Hoot Evers	DET	37.0
Phil Rizzuto	NY	35.4

Pitchers Overall
Ned Garver	STL	57.5
Bob Lemon	CLE	36.5
Mel Parnell	BOS	34.5

Relief Points
Mickey Harris	WAS	31
Joe Page	NY	25
Mickey McDermott	BOS	21

Club	W	L	R	OR	Avg	OBA	SLG	BPF	NOPS-A	BR	Adj	Wins	ERA	PPF	NERA-A	PR	Adj	Wins	Diff
NY	98	56	914	691	.282	.367	.441	102	115/113	92	80	7.5	4.15	98	110/108	65	50	4.7	8.8
DET	95	59	837	713	.282	.369	.417	99	109/110	60	69	6.5	4.12	97	111/107	71	48	4.5	7.0
BOS	94	60	1027	804	.302	.385	.464	106	126/118	185	132	12.4	4.88	104	94/ 97	-46	-20	-2.0	6.6
CLE	92	62	806	654	.269	.358	.422	98	108/110	39	55	5.1	3.74	95	122/116	128	94	8.8	1.1
WAS	67	87	690	813	.260	.347	.360	101	87/ 86	-92	-99	-9.4	4.65	103	98/101	-11	9	.9	-1.5
CHI	60	94	625	749	.260	.333	.364	96	85/ 88	-125	-98	-9.3	4.41	98	104/102	25	12	1.1	-8.9
STL	58	96	684	916	.246	.337	.370	108	88/ 81	-101	-158	-14.9	5.19	112	88/ 99	-93	-10	-1.1	-3.0
PHI	52	102	670	913	.261	.349	.378	90	93/103	-55	19	1.8	5.49	94	83/ 78	-135	-179	-16.9	-9.9
			782		.271	.356	.402						4.58						

NATIONAL LEAGUE 1950

Batting Runs
Stan Musial	STL	57.3
Ralph Kiner	PIT	48.7
Andy Pafko	CHI	41.5

Park Adjusted
Stan Musial	STL	60.1
Ralph Kiner	PIT	45.9
Sid Gordon	BOS	44.8

Pitching Runs
Robin Roberts	PHI	37.9
Larry Jansen	NY	34.6
Ewell Blackwell	CIN	34.1

Park Adjusted
Ewell Blackwell	CIN	41.2
Robin Roberts	PHI	31.8
Preacher Roe	BRO	31.4

Normalized OPS
Stan Musial	STL	169
Ralph Kiner	PIT	159
Andy Pafko	CHI	156

Park Adjusted
Stan Musial	STL	175
Sid Gordon	BOS	174
Bob Elliott	BOS	155

Normalized ERA
Sal Maglie	NY	153
Ewell Blackwell	CIN	140
Larry Jansen	NY	138

Park Adjusted
Ewell Blackwell	CIN	148
Sal Maglie	NY	147
Preacher Roe	BRO	134

On Base Average
Eddie Stanky	NY	.460
Stan Musial	STL	.437
Jackie Robinson	BRO	.423

Slugging Percentage
Stan Musial	STL	.596
Andy Pafko	CHI	.591
Ralph Kiner	PIT	.590

Percent of Team Wins
Ewell Blackwell	CIN	.258
Warren Spahn	BOS	.253
Johnny Sain	BOS	.241

Wins Above Team
Sal Maglie	NY	6.7
Dutch Hiller	CHI	5.5
Ewell Blackwell	CIN	4.0

Isolated Power
Ralph Kiner	PIT	.318
Andy Pafko	CHI	.288
Roy Campanella	BRO	.270

Base Stealing Runs
Caught Stealing		
Not Available		

Relievers - Runs
Jim Konstanty	PHI	25.0
Jack Kramer	NY	6.0
Milo Candini	PHI	4.8

Park Adjusted
Jim Konstanty	PHI	21.9
Johnny Vander Meer	CHI	6.2
Dutch Leonard	CHI	6.2

Defensive Runs
Roy Smalley	CHI	21.3
Billy Cox	BRO	15.5
Jackie Robinson	BRO	12.4

Players Overall
Stan Musial	STL	51.7
Eddie Stanky	NY	47.4
Sid Gordon	BOS	46.3

Pitchers Overall
Ewell Blackwell	CIN	40.2
Larry Jansen	NY	30.8
Sal Maglie	NY	30.7

Relief Points
Jim Konstanty	PHI	69
Dutch Leonard	CHI	19
Al Brazle	STL	18

Club	W	L	R	OR	Avg	OBA	SLG	BPF	NOPS-A	BR	Adj	Wins	ERA	PPF	NERA-A	PR	Adj	Wins	Diff
PHI	91	63	722	624	.265	.334	.396	98	101/104	-13	4	.4	3.50	96	119/113	101	73	7.1	6.5
BRO	89	65	847	724	.272	.349	.444	108	119/110	107	46	4.5	4.28	107	97/103	-21	22	2.1	5.3
NY	86	68	735	643	.258	.342	.392	98	102/104	4	19	1.9	3.71	96	112/107	66	40	3.9	3.2
BOS	83	71	785	736	.263	.342	.405	86	106/123	24	125	12.2	4.13	84	100/ 84	1	-98	-9.7	3.5
STL	78	75	693	670	.259	.339	.386	96	100/103	-14	11	1.1	3.97	95	104/ 99	26	-2	-.3	.7
CIN	66	87	654	734	.260	.327	.376	104	93/ 90	-64	-93	-9.2	4.32	106	96/102	-25	10	1.0	-2.3
CHI	64	89	643	772	.248	.315	.401	106	98/ 92	-52	-97	-9.6	4.28	109	97/106	-20	37	3.6	-6.5
PIT	57	96	681	857	.264	.338	.406	103	105/102	13	-11	-1.2	4.97	107	83/ 89	-124	-81	-8.0	-10.3
			720		.261	.336	.401						4.14						

Batting Runs			Park Adjusted			Pitching Runs			Park Adjusted		
Ted Williams	BOS	62.8	Ted Williams	BOS	63.3	Early Wynn	CLE	33.5	Billy Pierce	CHI	26.7
Larry Doby	CLE	37.0	Larry Doby	CLE	42.2	Eddie Lopat	NY	31.6	Ned Garver	STL	24.3
Ferris Fain	PHI	33.8	Eddie Yost	WAS	33.0	Billy Pierce	CHI	28.9	Early Wynn	CLE	20.3
Normalized OPS			Park Adjusted			Normalized ERA			Park Adjusted		
Ted Williams	BOS	175	Ted Williams	BOS	176	Eddie Lopat	NY	142	Billy Pierce	CHI	133
Larry Doby	CLE	153	Larry Doby	CLE	166	Early Wynn	CLE	136	Mel Parnell	BOS	124
Ferris Fain	PHI	149	Gil McDougald	NY	154	Billy Pierce	CHI	136	Ned Garver	STL	124
On Base Average			Slugging Percentage			Percent of Team Wins			Wins Above Team		
Ted Williams	BOS	.464	Ted Williams	BOS	.556	Ned Garver	STL	.385	Ned Garver	STL	11.6
Ferris Fain	PHI	.451	Larry Doby	CLE	.512	Bobby Shantz	PHI	.257	Bobby Shantz	PHI	6.4
Larry Doby	CLE	.428	Vic Wertz	DET	.511	Bob Feller	CLE	.237	Bob Feller	CLE	4.8
Isolated Power			Base Stealing Runs			Relievers - Runs			Park Adjusted		
Ted Williams	BOS	.237	Phil Rizzuto	NY	3.6	Ellis Kinder	BOS	22.2	Ellis Kinder	BOS	20.9
Vic Wertz	DET	.226	Chico Carrasquel	CHI	1.8	Witto Aloma	CHI	17.6	Witto Aloma	CHI	17.0
Larry Doby	CLE	.217	Jim Busby	CHI	1.2	Joe Ostrowski	NY	6.5	Ray Herbert	DET	4.5
Defensive Runs			Players Overall			Pitchers Overall			Relief Points		
Gil Coan	WAS	23.8	Ted Williams	BOS	63.9	Ned Garver	STL	32.9	Ellis Kinder	BOS	47
Vern Stephens	BOS	19.9	Eddie Joost	PHI	39.0	Billy Pierce	CHI	25.9	Carl Scheib	PHI	19
Chico Carrasquel	CHI	16.2	Yogi Berra	NY	38.0	Mel Parnell	BOS	24.3	Joe Ostrowski	NY	18
									Morrie Martin	PHI	18
									Witto Aloma	CHI	18

Club	W	L	R	OR	Avg	OBA	SLG	BPF	NOPS-A	BR	Adj	Wins	ERA	PPF	NERA-A	PR	Adj	Wins	Diff
NY	98	56	798	621	.269	.349	.408	89	111/124	61	136	13.4	3.56	85	116/ 98	85	-7	-.8	8.4
CLE	93	61	696	594	.256	.336	.389	92	102/111	0	55	5.4	3.39	89	122/109	114	47	4.6	6.0
BOS	87	67	804	725	.266	.358	.392	99	109/109	64	68	6.4	4.14	98	100/ 97	-2	-16	-1.6	4.9
CHI	81	73	714	644	.270	.348	.385	99	104/105	23	28	2.7	3.50	98	118/115	98	85	8.4	-7.1
DET	73	81	685	741	.265	.337	.380	108	100/ 92	-14	-72	-7.2	4.29	110	96/105	-25	35	3.5	-.3
PHI	70	84	736	745	.262	.349	.386	106	105/ 98	29	-15	-1.6	4.47	107	92/ 99	-51	-8	-.9	-4.5
WAS	62	92	672	764	.263	.336	.355	97	92/ 95	-59	-35	-3.5	4.48	98	92/ 90	-54	-66	-6.6	-4.9
STL	52	102	611	882	.247	.317	.357	107	88/ 82	-101	-148	-14.6	5.18	112	80/ 89	-160	-84	-8.3	-2.1
				715	.262	.342	.381						4.12						

Batting Runs			Park Adjusted			Pitching Runs			Park Adjusted		
Ralph Kiner	PIT	70.8	Stan Musial	STL	71.9	Sal Maglie	NY	34.0	Sal Maglie	NY	31.8
Stan Musial	STL	70.1	Ralph Kiner	PIT	65.2	Warren Spahn	BOS	33.7	Robin Roberts	PHI	30.1
Jackie Robinson	BRO	45.8	Jackie Robinson	BRO	46.5	Robin Roberts	PHI	32.5	Warren Spahn	BOS	27.2
Normalized OPS			Park Adjusted			Normalized ERA			Park Adjusted		
Ralph Kiner	PIT	187	Stan Musial	STL	187	Chet Nichols	BOS	137	Sal Maglie	NY	133
Stan Musial	STL	183	Ralph Kiner	PIT	175	Sal Maglie	NY	135	Chet Nichols	BOS	131
Roy Campanella	BRO	161	Roy Campanella	BRO	162	Warren Spahn	BOS	133	Robin Roberts	PHI	128
On Base Average			Slugging Percentage			Percent of Team Wins			Wins Above Team		
Ralph Kiner	PIT	.452	Ralph Kiner	PIT	.627	Murry Dickson	PIT	.313	Preacher Roe	BRO	7.8
Stan Musial	STL	.449	Stan Musial	STL	.614	Warren Spahn	BOS	.289	Murry Dickson	PIT	6.6
Jackie Robinson	BRO	.429	Roy Campanella	BRO	.590	Robin Roberts	PHI	.288	Sal Maglie	NY	6.0
Isolated Power			Base Stealing Runs			Relievers - Runs			Park Adjusted		
Ralph Kiner	PIT	.318	Sam Jethroe	BOS	7.5	Al Brazle	STL	14.7	Harry Perkowski	CIN	13.6
Bobby Thomson	NY	.268	Richie Ashburn	PHI	5.1	Monte Kennedy	NY	12.9	Al Brazle	STL	12.9
Roy Campanella	BRO	.265	Jackie Robinson	BRO	2.7	Harry Perkowski	CIN	12.8	Dutch Leonard	CHI	12.4
Defensive Runs			Players Overall			Pitchers Overall			Relief Points		
Richie Ashburn	PHI	25.8	Jackie Robinson	BRO	76.1	Robin Roberts	PHI	31.7	Clyde King	BRO	32
Jackie Robinson	BRO	21.2	Stan Musial	STL	73.0	Sal Maglie	NY	30.3	Frank Smith	CIN	27
Carl Furillo	BRO	13.6	Ralph Kiner	PIT	57.7	Warren Spahn	BOS	29.0	George Spencer	NY	26

Club	W	L	R	OR	Avg	OBA	SLG	BPF	NOPS-A	BR	Adj	Wins	ERA	PPF	NERA-A	PR	Adj	Wins	Diff
NY	98	59	781	641	.260	.347	.418	101	115/114	91	84	8.5	3.48	98	114/112	75	64	6.4	4.6
BRO	97	60	855	672	.275	.351	.434	99	121/123	130	137	13.7	3.87	96	102/ 98	13	-13	-1.4	6.2
STL	81	73	683	671	.264	.339	.382	98	102/105	7	23	2.3	3.95	97	100/ 97	1	-15	-1.6	3.3
BOS	76	78	723	662	.262	.336	.394	97	105/109	20	43	4.3	3.75	95	106/101	33	3	.3	-5.7
PHI	73	81	648	644	.260	.326	.375	98	97/ 98	-37	-27	-2.8	3.81	98	104/102	23	12	1.2	-2.4
CIN	68	86	559	667	.248	.304	.351	100	84/ 84	-133	-132	-13.3	3.70	102	107/109	40	50	5.1	-.7
PIT	64	90	689	845	.258	.331	.397	107	105/ 98	11	-37	-3.8	4.77	111	83/ 92	-124	-60	-6.1	-3.1
CHI	62	92	614	750	.250	.315	.364	99	90/ 92	-85	-76	-7.8	4.34	101	91/ 92	-58	-51	-5.2	-2.0
				694	.260	.331	.390						3.96						

AMERICAN LEAGUE   1952

Batting Runs			Park Adjusted			Pitching Runs			Park Adjusted		
Mickey Mantle	NY	40.3	Al Rosen	CLE	43.4	Allie Reynolds	NY	43.6	Bobby Shantz	PHI	52.3
Al Rosen	CLE	38.6	Larry Doby	CLE	42.4	Mike Garcia	CLE	42.2	Allie Reynolds	NY	41.0
Larry Doby	CLE	37.9	Mickey Mantle	NY	39.7	Bob Lemon	CLE	40.6	Billy Pierce	CHI	31.6

Normalized OPS			Park Adjusted			Normalized ERA			Park Adjusted		
Mickey Mantle	NY	156	Larry Doby	CLE	166	Allie Reynolds	NY	178	Allie Reynolds	NY	173
Larry Doby	CLE	155	Al Rosen	CLE	163	Mike Garcia	CLE	155	Bobby Shantz	PHI	168
Al Rosen	CLE	152	Mickey Mantle	NY	154	Bobby Shantz	PHI	148	Joe Dobson	CHI	147

On Base Average			Slugging Percentage			Percent of Team Wins			Wins Above Team		
Ferris Fain	PHI	.438	Larry Doby	CLE	.541	Bobby Shantz	PHI	.304	Bobby Shantz	PHI	10.1
Elmer Valo	PHI	.432	Mickey Mantle	NY	.530	Early Wynn	CLE	.247	Hal Newhouser	DET	3.6
Gene Woodling	NY	.397	Al Rosen	CLE	.524	Ted Gray	DET	.240	Allie Reynolds	NY	3.3
									Satchel Paige	STL	3.3
									Bob Cain	STL	3.3

Isolated Power			Base Stealing Runs			Relievers - Runs			Park Adjusted		
Larry Doby	CLE	.266	Phil Rizzuto	NY	1.5	Fritz Dorish	CHI	12.1	Fritz Dorish	CHI	12.3
Luke Easter	CLE	.249	Jay Porter	STL	1.2	Lefty Kennedy	CHI	7.0	Lefty Kennedy	CHI	7.1
Al Rosen	CLE	.222	Billy Goodman	BOS	1.2	Al Benton	BOS	5.5	Sandy Consuegra	WAS	5.7

Defensive Runs			Players Overall			Pitchers Overall			Relief Points		
Phil Rizzuto	NY	19.8	Larry Doby	CLE	42.7	Bobby Shantz	PHI	55.6	Fritz Dorish	CHI	32
Billy Goodman	BOS	19.2	Billy Goodman	BOS	34.7	Bob Lemon	CLE	39.9	Satchel Paige	STL	28
Gil McDougald	NY	18.1	Ferris Fain	PHI	32.1	Allie Reynolds	NY	38.7	Bob Hooper	PHI	20
									Sandy Consuegra	WAS	20

Club	W	L	R	OR	Avg	OBA	SLG	BPF	NOPS-A	BR	Adj	Wins	ERA	PPF	NERA-A	PR	Adj	Wins	Diff
NY	95	59	727	557	.267	.341	.403	101	115/115	88	83	8.6	3.14	97	117/114	81	67	6.9	2.4
CLE	93	61	763	606	.262	.342	.404	93	116/125	95	139	14.5	3.32	90	111/ 99	56	-3	-.4	1.9
CHI	81	73	610	568	.252	.327	.346	101	95/ 94	-34	-43	-4.6	3.25	101	113/114	66	69	7.2	1.4
PHI	79	75	664	723	.253	.343	.359	111	103/ 93	26	-44	-4.6	4.16	113	88/100	-73	0	-.0	6.7
WAS	78	76	598	608	.239	.317	.326	101	86/ 84	-97	-105	-11.0	3.37	102	109/111	48	57	5.9	6.2
BOS	76	78	668	658	.255	.328	.377	97	104/108	14	35	3.7	3.80	96	97/ 93	-18	-39	-4.2	-.5
STL	64	90	604	733	.250	.322	.356	95	96/101	-38	-5	-.6	4.12	97	89/ 87	-68	-84	-8.8	-3.6
DET	50	104	557	738	.243	.318	.352	101	94/ 93	-51	-54	-5.7	4.25	104	86/ 90	-88	-65	-6.8	-14.5
			649		.253	.330	.365						3.67						

NATIONAL LEAGUE   1952

Batting Runs			Park Adjusted			Pitching Runs			Park Adjusted		
Stan Musial	STL	55.7	Stan Musial	STL	59.8	Robin Roberts	PHI	41.8	Robin Roberts	PHI	40.9
Jackie Robinson	BRO	41.9	Jackie Robinson	BRO	38.0	Bob Rush	CHI	28.7	Bob Rush	CHI	30.0
Hank Sauer	CHI	32.8	Ralph Kiner	PIT	32.2	Karl Drews	PHI	26.0	Carl Erskine	BRO	26.0

Normalized OPS			Park Adjusted			Normalized ERA			Park Adjusted		
Stan Musial	STL	167	Stan Musial	STL	177	Warren Hacker	CHI	145	Warren Hacker	CHI	147
Jackie Robinson	BRO	151	Ted Kluszewski	CIN	150	Robin Roberts	PHI	144	Robin Roberts	PHI	143
Ted Kluszewski	CIN	145	Jackie Robinson	BRO	143	Billy Loes	BRO	138	Billy Loes	BRO	142

On Base Average			Slugging Percentage			Percent of Team Wins			Wins Above Team		
Jackie Robinson	BRO	.440	Stan Musial	STL	.538	Murry Dickson	PIT	.333	Robin Roberts	PHI	10.6
Stan Musial	STL	.432	Hank Sauer	CHI	.531	Robin Roberts	PHI	.322	Murry Dickson	PIT	5.8
Solly Hemus	STL	.392	Ted Kluszewski	CIN	.509	Ken Raffensberger	CIN	.246	Hoyt Wilhelm	NY	4.8

Isolated Power			Base Stealing Runs			Relievers - Runs			Park Adjusted		
Hank Sauer	CHI	.261	Peewee Reese	BRO	6.0	Joe Black	BRO	24.9	Joe Black	BRO	26.4
Ralph Kiner	PIT	.256	Jackie Robinson	BRO	3.0	Hoyt Wilhelm	NY	22.9	Hoyt Wilhelm	NY	23.5
Gil Hodges	BRO	.246	Sam Jethroe	BOS	3.0	Al Brazle	STL	12.2	Dutch Leonard	CHI	12.1

Defensive Runs			Players Overall			Pitchers Overall			Relief Points		
Red Schoendienst	STL	27.5	Jackie Robinson	BRO	57.7	Bob Rush	CHI	39.6	Joe Black	BRO	55
Hal Jeffcoat	CHI	14.7	Red Schoendienst	STL	47.6	Robin Roberts	PHI	39.2	Hoyt Wilhelm	NY	49
Richie Ashburn	PHI	14.1	Stan Musial	STL	45.2	Carl Erskine	BRO	27.6	Al Brazle	STL	45

Club	W	L	R	OR	Avg	OBA	SLG	BPF	NOPS-A	BR	Adj	Wins	ERA	PPF	NERA-A	PR	Adj	Wins	Diff
BRO	96	57	775	603	.262	.348	.399	106	117/110	107	70	7.3	3.53	102	106/108	31	46	4.7	7.5
NY	92	62	722	639	.256	.329	.399	102	112/109	56	41	4.3	3.59	101	104/105	21	26	2.7	8.0
STL	88	66	677	630	.267	.340	.391	95	112/118	68	104	10.7	3.66	93	102/ 95	11	-27	-2.9	3.2
PHI	87	67	657	552	.260	.332	.376	101	105/104	25	17	1.7	3.07	99	122/121	102	98	10.1	-1.9
CHI	77	77	628	631	.264	.321	.383	101	104/103	5	-2	-.3	3.58	101	104/106	23	31	3.2	-2.9
CIN	69	85	615	659	.249	.314	.366	97	97/100	-35	-15	-1.7	4.02	98	93/ 91	-42	-56	-5.9	-.4
BOS	64	89	569	651	.233	.301	.343	98	87/ 89	-101	-87	-9.1	3.78	99	99/ 98	-6	-10	-1.1	-2.3
PIT	42	112	515	793	.231	.300	.331	101	83/ 82	-124	-128	-13.3	4.65	106	80/ 85	-137	-102	-10.6	-11.0
			645		.253	.323	.374						3.73						

AMERICAN LEAGUE 1953

Batting Runs
- Al Rosen — CLE — 63.3
- Mickey Vernon — WAS — 39.7
- Minnie Minoso — CHI — 30.6

Park Adjusted
- Al Rosen — CLE — 72.8
- Mickey Vernon — WAS — 44.6
- Larry Doby — CLE — 34.5

Pitching Runs
- Billy Pierce — CHI — 38.3
- Eddie Lopat — NY — 31.0
- Mel Parnell — BOS — 24.9

Park Adjusted
- Billy Pierce — CHI — 47.0
- Mel Parnell — BOS — 35.4
- Mickey McDermott — BOS — 31.4

Normalized OPS
- Al Rosen — CLE — 176
- Mickey Vernon — WAS — 147
- Gus Zernial — PHI — 144

Park Adjusted
- Al Rosen — CLE — 200
- Mickey Vernon — WAS — 157
- Larry Doby — CLE — 152

Normalized ERA
- Eddie Lopat — NY — 165
- Billy Pierce — CHI — 147
- Johnny Sain — NY — 133
- Whitey Ford — NY — 133

Park Adjusted
- Billy Pierce — CHI — 157
- Eddie Lopat — NY — 145
- Mickey McDermott — BOS — 145

On Base Average
- Gene Woodling — NY — .429
- Al Rosen — CLE — .422
- Minnie Minoso — CHI — .410

Slugging Percentage
- Al Rosen — CLE — .613
- Gus Zernial — PHI — .559
- Yogi Berra — NY — .523

Percent of Team Wins
- Bob Porterfield — WAS — .289
- Mel Parnell — BOS — .250
- Bob Lemon — CLE — .228

Wins Above Team
- Bob Porterfield — WAS — 7.6
- Mel Parnell — BOS — 6.3
- Marlin Stuart — STL — 4.8

Isolated Power
- Al Rosen — CLE — .277
- Gus Zernial — PHI — .275
- Yogi Berra — NY — .227

Base Stealing Runs
- Cass Michaels — PHI — 2.1
- Gil Coan — WAS — 2.1
- Gus Zernial — PHI — 1.2

Relievers - Runs
- Ellis Kinder — BOS — 25.5
- Fritz Dorish — CHI — 9.8
- Bob Kuzava — NY — 6.8

Park Adjusted
- Ellis Kinder — BOS — 30.1
- Fritz Dorish — CHI — 14.5
- Satchel Paige — STL — 12.7

Defensive Runs
- Billy Hunter — STL — 17.2
- Jim Piersall — BOS — 16.3
- Bobby Avila — CLE — 15.3

Players Overall
- Al Rosen — CLE — 63.0
- Mickey Vernon — WAS — 35.1
- Yogi Berra — NY — 32.2

Pitchers Overall
- Billy Pierce — CHI — 39.7
- Mickey McDermott — BOS — 38.8
- Mel Parnell — BOS — 34.0

Relief Points
- Ellis Kinder — BOS — 68
- Fritz Dorish — CHI — 46
- Allie Reynolds — NY — 38

Club	W	L	R	OR	Avg	OBA	SLG	BPF	NOPS-A	BR	Adj	Wins	ERA	PPF	NERA-A	PR	Adj	Wins	Diff
NY	99	52	801	547	.273	.358	.417	94	118/126	113	157	15.7	3.20	88	125/110	120	50	5.0	2.8
CLE	92	62	770	627	.270	.349	.410	88	113/128	76	158	15.8	3.64	84	110/ 93	54	-40	-4.1	3.3
CHI	89	65	716	592	.258	.341	.364	109	97/ 89	-17	-80	-8.1	3.41	107	117/126	91	136	13.7	6.4
BOS	84	69	656	632	.264	.332	.384	110	101/ 92	-10	-76	-7.7	3.59	110	111/122	62	122	12.2	3.0
WAS	76	76	687	614	.263	.343	.368	94	99/106	-5	36	3.6	3.66	92	109/100	50	1	.1	-3.7
DET	60	94	695	923	.266	.331	.387	92	101/110	-9	45	4.5	5.25	96	76/ 73	-196	-220	-22.2	.7
PHI	59	95	632	799	.256	.321	.372	106	94/ 89	-60	-101	-10.2	4.66	109	86/ 94	-104	-46	-4.7	-3.0
STL	54	100	555	778	.249	.317	.363	109	90/ 83	-83	-140	-14.1	4.48	113	89/101	-74	5	.5	-9.4
			689		.262	.336	.383	0.10					3.99						

NATIONAL LEAGUE 1953

Batting Runs
- Stan Musial — STL — 62.7
- Duke Snider — BRO — 59.3
- Eddie Mathews — MIL — 54.9

Park Adjusted
- Stan Musial — STL — 65.5
- Duke Snider — BRO — 58.2
- Eddie Mathews — MIL — 57.1

Pitching Runs
- Warren Spahn — MIL — 64.6
- Robin Roberts — PHI — 59.2
- Harvey Haddix — STL — 34.4

Park Adjusted
- Warren Spahn — MIL — 57.8
- Robin Roberts — PHI — 54.1
- Harvey Haddix — STL — 28.9

Normalized OPS
- Stan Musial — STL — 169
- Duke Snider — BRO — 168
- Eddie Mathews — MIL — 164

Park Adjusted
- Stan Musial — STL — 175
- Eddie Mathews — MIL — 169
- Duke Snider — BRO — 166

Normalized ERA
- Warren Spahn — MIL — 204
- Robin Roberts — PHI — 156
- Bob Buhl — MIL — 144

Park Adjusted
- Warren Spahn — MIL — 193
- Robin Roberts — PHI — 151
- Bob Buhl — MIL — 136

On Base Average
- Stan Musial — STL — .437
- Jackie Robinson — BRO — .425
- Duke Snider — BRO — .419

Slugging Percentage
- Duke Snider — BRO — .627
- Eddie Mathews — MIL — .627
- Roy Campanella — BRO — .611

Percent of Team Wins
- Robin Roberts — PHI — .277
- Warren Spahn — MIL — .250
- Harvey Haddix — STL — .241

Wins Above Team
- Warren Spahn — MIL — 6.3
- Robin Roberts — STL — 5.4
- Frank Smith — CIN — 4.3

Isolated Power
- Eddie Mathews — MIL — .325
- Roy Campanella — BRO — .299
- Duke Snider — BRO — .292

Base Stealing Runs
- Peewee Reese — BRO — 3.0
- Jackie Robinson — BRO — 2.7
- Eddie Miksis — CHI — 1.5
- Hal Jeffcoat — CHI — 1.5

Relievers - Runs
- Hoyt Wilhelm — NY — 20.0
- Clem Labine — BRO — 18.4
- Ernie Johnson — MIL — 14.6

Park Adjusted
- Hoyt Wilhelm — NY — 17.9
- Clem Labine — BRO — 16.7
- Ernie Johnson — MIL — 12.5

Defensive Runs
- Johnny Logan — MIL — 24.2
- Red Schoendienst — STL — 23.6
- Richie Ashburn — PHI — 21.0

Players Overall
- Red Schoendienst — STL — 62.1
- Stan Musial — STL — 54.3
- Eddie Mathews — MIL — 53.5

Pitchers Overall
- Warren Spahn — MIL — 63.1
- Robin Roberts — PHI — 55.2
- Harvey Haddix — STL — 39.7

Relief Points
- Al Brazle — STL — 41
- Hoyt Wilhelm — NY — 36
- Lew Burdette — MIL — 32

Club	W	L	R	OR	Avg	OBA	SLG	BPF	NOPS-A	BR	Adj	Wins	ERA	PPF	NERA-A	PR	Adj	Wins	Diff
BRO	105	49	955	689	.285	.366	.474	101	130/129	188	178	17.1	4.10	97	105/101	28	7	.7	10.2
MIL	92	62	738	589	.266	.325	.415	97	103/106	-21	-2	-.3	3.30	95	130/123	152	117	11.2	4.1
STL	83	71	768	713	.273	.347	.424	97	111/114	53	77	7.5	4.22	95	102/ 97	10	-19	-1.9	.5
PHI	83	71	716	666	.265	.335	.396	98	100/102	-24	-10	-1.1	3.80	97	113/109	74	54	5.2	1.9
NY	70	84	768	747	.271	.336	.422	97	108/111	20	39	3.8	4.25	97	101/ 98	5	-14	-1.4	-9.3
CIN	68	86	714	788	.261	.325	.400	101	99/ 98	-41	-50	-4.9	4.63	103	93/ 95	-52	-35	-3.4	-.6
CHI	65	89	633	835	.260	.328	.399	102	99/ 97	-40	-52	-5.1	4.79	105	89/ 94	-75	-40	-4.0	-2.9
PIT	50	104	622	887	.247	.319	.356	104	84/ 81	-129	-155	-15.0	5.22	109	82/ 89	-140	-85	-8.3	-3.7
			739		.266	.335	.411						4.28						

Batting Runs

Ted Williams	BOS	70.9
Minnie Minoso	CHI	47.3
Mickey Mantle	NY	42.8

Park Adjusted

Ted Williams	BOS	71.1
Mickey Mantle	NY	44.2
Minnie Minoso	CHI	38.0

Pitching Runs

Mike Garcia	CLE	31.1
Early Wynn	CLE	30.1
Bob Lemon	CLE	28.7

Park Adjusted

Virgil Trucks	CHI	37.8
Mike Garcia	CLE	29.9
Early Wynn	CLE	28.9

Normalized OPS

Ted Williams	BOS	214
Minnie Minoso	CHI	159
Mickey Mantle	NY	155

Park Adjusted

Ted Williams	BOS	214
Mickey Mantle	NY	158
Al Rosen	CLE	145

Normalized ERA

Mike Garcia	CLE	141
Sandy Consuegra	CHI	138
Bob Lemon	CLE	137

Park Adjusted

Sandy Consuegra	CHI	151
Virgil Trucks	CHI	146
Mike Garcia	CLE	139

On Base Average

Ted Williams	BOS	.516
Minnie Minoso	CHI	.416
Al Rosen	CLE	.412

Slugging Percentage

Ted Williams	BOS	.635
Minnie Minoso	CHI	.535
Mickey Mantle	NY	.525

Percent of Team Wins

Steve Gromek	DET	.265
Bob Turley	BAL	.259
Joe Coleman	BAL	.241

Wins Above Team

Sandy Consuegra	CHI	5.0
Bob Turley	BAL	4.7
Steve Gromek	DET	3.8

Isolated Power

Ted Williams	BOS	.290
Mickey Mantle	NY	.225
Minnie Minoso	CHI	.215

Base Stealing Runs

Jim Busby	WAS	3.9
Spook Jacobs	PHI	3.3
Jackie Jensen	BOS	2.4

Relievers - Runs

Don Mossi	CLE	18.5
Ray Narleski	CLE	14.8
Fritz Dorish	CHI	12.1

Park Adjusted

Don Mossi	CLE	18.0
Fritz Dorish	CHI	16.3
Ray Narleski	CLE	14.4

Defensive Runs

Vic Power	PHI	14.6
Andy Carey	NY	14.1
Bobby Avila	CLE	13.8

Players Overall

Ted Williams	BOS	61.2
Bobby Avila	CLE	47.9
Minnie Minoso	CHI	44.7

Pitchers Overall

Bob Lemon	CLE	36.2
Virgil Trucks	CHI	36.1
Mike Garcia	CLE	27.6

Relief Points

Johnny Sain	NY	50
Ellis Kinder	BOS	36
Ray Narleski	CLE	30

Club	W	L	R	OR	Avg	OBA	SLG	BPF	NOPS-A	BR	Adj	Wins	ERA	PPF	NERA-A	PR	Adj	Wins	Diff
CLE	111	43	746	504	.262	.345	.403	103	113/109	76	53	5.5	2.78	99	134/132	149	142	14.7	13.8
NY	103	51	805	563	.268	.351	.408	98	116/119	101	114	11.8	3.26	93	114/106	70	30	3.1	11.0
CHI	94	60	711	521	.267	.350	.379	113	107/ 95	51	-30	-3.2	3.05	109	122/133	103	157	16.2	4.0
BOS	69	85	700	728	.266	.348	.395	100	111/112	73	75	7.8	4.01	100	93/ 93	-45	-43	-4.5	-11.3
DET	68	86	584	664	.258	.324	.367	100	97/ 97	-36	-33	-3.5	3.81	101	98/ 99	-12	-6	-.7	-4.8
WAS	66	88	632	680	.246	.328	.355	96	94/ 98	-43	-14	-1.6	3.84	96	97/ 93	-17	-39	-4.1	-5.3
BAL	54	100	483	668	.251	.316	.338	93	86/ 92	-103	-57	-6.0	3.88	96	96/ 92	-24	-48	-5.1	-12.0
PHI	51	103	542	875	.236	.307	.342	100	85/ 85	-116	-114	-11.9	5.18	106	72/ 77	-221	-184	-19.1	5.1
			650		.257	.334	.373						3.72						

Batting Runs

Duke Snider	BRO	64.6
Willie Mays	NY	61.8
Stan Musial	STL	60.7

Park Adjusted

Willie Mays	NY	63.6
Duke Snider	BRO	61.4
Stan Musial	STL	56.5

Pitching Runs

Johnny Antonelli	NY	51.2
Robin Roberts	PHI	41.5
Curt Simmons	PHI	35.5

Park Adjusted

Johnny Antonelli	NY	43.9
Robin Roberts	PHI	38.9
Curt Simmons	PHI	33.5

Normalized OPS

Willie Mays	NY	177
Duke Snider	BRO	175
Ted Kluszewski	CIN	169

Park Adjusted

Willie Mays	NY	181
Eddie Mathews	MIL	173
Duke Snider	BRO	169

Normalized ERA

Johnny Antonelli	NY	178
Lew Burdette	MIL	148
Curt Simmons	PHI	145

Park Adjusted

Johnny Antonelli	NY	167
Curt Simmons	PHI	142
Lew Burdette	MIL	137

On Base Average

Richie Ashburn	PHI	.442
Stan Musial	STL	.433
Eddie Mathews	MIL	.428

Slugging Percentage

Willie Mays	NY	.667
Duke Snider	BRO	.647
Ted Kluszewski	CIN	.642

Percent of Team Wins

Robin Roberts	PHI	.311
Harvey Haddix	STL	.250
Warren Spahn	MIL	.236

Wins Above Team

Robin Roberts	PHI	6.1
Brooks Lawrence	STL	5.9
Joe Nuxhall	CIN	4.3

Isolated Power

Willie Mays	NY	.322
Ted Kluszewski	CIN	.316
Eddie Mathews	MIL	.313

Base Stealing Runs

Dee Fondy	CHI	3.0
Bill Bruton	MIL	2.4
Johnny Temple	CIN	2.1

Relievers - Runs

Hoyt Wilhelm	NY	24.2
Marv Grissom	NY	23.2
Dave Jolly	MIL	20.2

Park Adjusted

Hoyt Wilhelm	NY	21.1
Marv Grissom	NY	19.8
Dave Jolly	MIL	16.6

Defensive Runs

Red Schoendienst	STL	30.0
Alex Grammas	STL	21.1
Johnny Logan	MIL	15.2

Players Overall

Willie Mays	NY	69.9
Eddie Mathews	MIL	55.2
Duke Snider	BRO	45.6

Pitchers Overall

Johnny Antonelli	NY	46.0
Robin Roberts	PHI	32.3
Curt Simmons	PHI	29.1

Relief Points

Jim Hughes	BRO	60
Marv Grissom	NY	49
Frank Smith	CIN	42

Club	W	L	R	OR	Avg	OBA	SLG	BPF	NOPS-A	BR	Adj	Wins	ERA	PPF	NERA-A	PR	Adj	Wins	Diff
NY	97	57	732	550	.264	.335	.424	98	107/110	17	34	3.4	3.09	94	132/123	151	112	11.1	5.5
BRO	92	62	778	740	.270	.353	.444	104	118/114	100	72	7.1	4.31	103	95/ 98	-35	-13	-1.4	9.3
MIL	89	65	670	670	.265	.330	.401	95	100/105	-32	1	.1	3.19	93	128/118	136	91	9.0	2.9
PHI	75	79	659	614	.267	.345	.395	99	102/102	-1	3	.3	3.59	98	113/111	73	62	6.2	-8.5
CIN	74	80	729	763	.262	.336	.406	103	102/ 99	-7	-29	-2.9	4.50	104	90/ 94	-64	-39	-4.0	3.9
STL	72	82	799	790	.281	.354	.421	105	111/106	65	29	2.9	4.49	105	91/ 95	-64	-32	-3.3	-4.6
CHI	64	90	700	766	.263	.327	.412	97	102/105	-20	-1	-.2	4.51	98	90/ 89	-65	-75	-7.6	-5.3
PIT	53	101	557	845	.248	.326	.350	96	84/ 88	-118	-88	-8.8	4.92	101	83/ 83	-126	-123	-12.3	-2.9
			703		.265	.338	.407						4.07						

AMERICAN LEAGUE 1955

Batting Runs		
Ted Williams	BOS	59.3
Mickey Mantle	NY	59.0
Al Kaline	DET	50.2

Park Adjusted		
Mickey Mantle	NY	63.2
Al Kaline	DET	54.2
Ted Williams	BOS	48.8

Pitching Runs		
Billy Pierce	CHI	45.6
Whitey Ford	NY	37.8
Frank Sullivan	BOS	30.4

Park Adjusted		
Frank Sullivan	BOS	55.0
Billy Pierce	CHI	43.1
Early Wynn	CLE	36.9

Normalized OPS		
Mickey Mantle	NY	178
Al Kaline	DET	160
Al Smith	CLE	137

Park Adjusted		
Mickey Mantle	NY	189
Al Kaline	DET	168
Roy Sievers	WAS	141

Normalized ERA		
Billy Pierce	CHI	201
Whitey Ford	NY	151
Early Wynn	CLE	141

Park Adjusted		
Billy Pierce	CHI	196
Frank Sullivan	BOS	165
Early Wynn	CLE	151

On Base Average		
Mickey Mantle	NY	.433
Al Kaline	DET	.425
Al Smith	CLE	.411

Slugging Percentage		
Mickey Mantle	NY	.611
Al Kaline	DET	.546
Larry Doby	CLE	.505

Percent of Team Wins		
Frank Sullivan	BOS	.214
Jim Wilson	BAL	.211
Billy Hoeft	DET	.203

Wins Above Team		
Billy Hoeft	DET	4.9
Billy Pierce	CHI	3.7
Mickey McDermott	WAS	3.6

Isolated Power		
Mickey Mantle	NY	.306
Roy Sievers	WAS	.218
Larry Doby	CLE	.214

Base Stealing Runs		
Earl Torgeson	DET	2.7
Billy Klaus	BOS	1.8
Mickey Mantle	NY	1.8

Relievers - Runs		
Sandy Consuegra	CHI	18.4
Don Mossi	CLE	14.1
Jim Konstanty	NY	13.6

Park Adjusted		
Leo Kiely	BOS	20.1
Sandy Consuegra	CHI	16.9
Don Mossi	CLE	16.8

Defensive Runs		
Nellie Fox	CHI	28.8
Willie Miranda	BAL	18.9
Gil McDougald	NY	16.8

Players Overall		
Mickey Mantle	NY	59.6
Al Kaline	DET	46.3
Nellie Fox	CHI	41.7

Pitchers Overall		
Frank Sullivan	BOS	49.7
Billy Pierce	CHI	41.3
Early Wynn	CLE	34.7

Relief Points		
Ray Narleski	CLE	53
Tom Gorman	KC	44
Ellis Kinder	BOS	41

Club	W	L	R	OR	Avg	OBA	SLG	BPF	NOPS-A	BR	Adj	Wins	ERA	PPF	NERA-A	PR	Adj	Wins	Diff
NY	96	58	762	569	.260	.343	.418	94	113/120	71	109	10.9	3.23	90	123/110	111	51	5.1	2.9
CLE	93	61	698	601	.257	.353	.394	109	109/100	60	-1	-.2	3.39	108	117/126	88	135	13.5	2.7
CHI	91	63	725	557	.268	.347	.388	100	106/105	31	28	2.9	3.37	97	118/114	90	73	7.4	3.8
BOS	84	70	755	652	.264	.354	.402	122	112/ 92	77	-74	-7.5	3.71	122	107/130	38	169	16.9	-2.4
DET	79	75	775	658	.266	.348	.394	95	108/114	46	81	8.1	3.80	92	104/ 96	25	-21	-2.2	-4.0
KC	63	91	638	911	.261	.323	.382	102	98/ 95	-39	-56	-5.8	5.35	108	74/ 80	-212	-163	-16.5	8.2
BAL	57	97	540	754	.240	.316	.320	90	77/ 85	-157	-92	-9.3	4.21	94	94/ 88	-38	-77	-7.8	-2.9
WAS	53	101	598	789	.248	.324	.351	91	89/ 97	-85	-28	-2.9	4.63	94	86/ 81	-100	-133	-13.4	-7.7
				686	.258	.339	.381						3.96						

NATIONAL LEAGUE 1955

Batting Runs		
Willie Mays	NY	62.3
Duke Snider	BRO	58.9
Eddie Mathews	MIL	50.0

Park Adjusted		
Willie Mays	NY	61.1
Duke Snider	BRO	55.3
Eddie Mathews	MIL	52.1

Pitching Runs		
Bob Friend	PIT	26.7
Robin Roberts	PHI	25.8
Don Newcombe	BRO	21.9

Park Adjusted		
Bob Friend	PIT	26.0
Robin Roberts	PHI	25.3
Don Newcombe	BRO	23.4

Normalized OPS		
Willie Mays	NY	176
Duke Snider	BRO	174
Eddie Mathews	MIL	166

Park Adjusted		
Willie Mays	NY	173
Eddie Mathews	MIL	171
Duke Snider	BRO	166

Normalized ERA		
Bob Friend	PIT	142
Don Newcombe	BRO	126
Bob Buhl	MIL	126

Park Adjusted		
Bob Friend	PIT	141
Don Newcombe	BRO	128
Robin Roberts	PHI	123

On Base Average		
Richie Ashburn	PHI	.449
Duke Snider	BRO	.421
Eddie Mathews	MIL	.417

Slugging Percentage		
Willie Mays	NY	.659
Duke Snider	BRO	.628
Eddie Mathews	MIL	.601

Percent of Team Wins		
Robin Roberts	PHI	.299
Bob Friend	PIT	.233
Joe Nuxhall	CIN	.227

Wins Above Team		
Bob Friend	PIT	5.9
Robin Roberts	PHI	5.9
Don Newcombe	BRO	4.8

Isolated Power		
Willie Mays	NY	.340
Duke Snider	BRO	.320
Eddie Mathews	MIL	.313

Base Stealing Runs		
Willie Mays	NY	4.8
Johnny Temple	CIN	3.3
Jackie Robinson	BRO	1.8

Relievers - Runs		
Hersh Freeman	CIN	19.2
Bob Miller	PHI	16.4
Paul La Palme	STL	13.2

Park Adjusted		
Hersh Freeman	CIN	20.9
Bob Miller	PHI	16.2
Paul La Palme	STL	14.2

Defensive Runs		
Roy McMillan	CIN	16.1
Willie Mays	NY	16.1
Dick Groat	PIT	15.5

Players Overall		
Willie Mays	NY	78.8
Ernie Banks	CHI	47.6
Duke Snider	BRO	46.1

Pitchers Overall		
Don Newcombe	BRO	41.0
Robin Roberts	PHI	34.0
Bob Friend	PIT	27.9

Relief Points		
Clem Labine	BRO	40
Jack Meyer	PHI	35
Hersh Freeman	CIN	32

Club	W	L	R	OR	Avg	OBA	SLG	BPF	NOPS-A	BR	Adj	Wins	ERA	PPF	NERA-A	PR	Adj	Wins	Diff
BRO	98	55	857	650	.271	.359	.448	105	124/118	142	109	10.8	3.68	101	110/111	54	63	6.2	4.4
MIL	85	69	743	668	.261	.329	.427	97	110/113	27	47	4.7	3.85	95	105/100	29	1	.1	3.2
NY	80	74	702	673	.260	.328	.402	102	102/101	-14	-24	-2.5	3.77	101	107/108	41	44	4.7	.8
PHI	77	77	675	666	.255	.343	.395	100	104/104	14	14	1.4	3.93	100	103/102	16	15	1.5	-2.9
CIN	75	79	761	684	.270	.344	.425	105	113/108	63	28	2.7	3.95	104	102/106	14	39	3.8	-8.6
CHI	72	81	626	713	.247	.307	.398	97	96/ 99	-74	-53	-5.4	4.17	98	97/ 95	-19	-30	-3.1	4.0
STL	68	86	654	757	.261	.324	.400	100	100/100	-29	-30	-3.1	4.56	102	89/ 90	-79	-66	-6.6	.7
PIT	60	94	560	767	.244	.310	.361	96	85/ 89	-126	-98	-9.9	4.39	99	92/ 91	-53	-57	-5.8	-1.4
				697	.259	.330	.407						4.04						

AMERICAN LEAGUE 1956

Batting Runs
Mickey Mantle NY 83.3
Ted Williams BOS 54.1
Minnie Minoso CHI 43.2

Park Adjusted
Mickey Mantle NY 84.7
Ted Williams BOS 53.4
Minnie Minoso CHI 41.8

Pitching Runs
Herb Score CLE 45.1
Early Wynn CLE 44.5
Whitey Ford NY 42.4

Park Adjusted
Herb Score CLE 44.8
Early Wynn CLE 44.2
Whitey Ford NY 36.3

Normalized OPS
Mickey Mantle NY 206
Ted Williams BOS 184
Minnie Minoso CHI 151

Park Adjusted
Mickey Mantle NY 209
Ted Williams BOS 182
Charlie Maxwell DET 148

Normalized ERA
Whitey Ford NY 168
Herb Score CLE 164
Early Wynn CLE 153

Park Adjusted
Herb Score CLE 164
Whitey Ford NY 159
Early Wynn CLE 153

On Base Average
Ted Williams BOS .479
Mickey Mantle NY .467
Minnie Minoso CHI .430

Slugging Percentage
Mickey Mantle NY .705
Ted Williams BOS .605
Charlie Maxwell DET .534

Percent of Team Wins
Frank Lary DET .256
Chuck Stobbs WAS .254
Billy Hoeft DET .244

Wins Above Team
Billy Pierce CHI 4.9
Tom Brewer BOS 4.6
Chuck Stobbs WAS 4.4

Isolated Power
Mickey Mantle NY .353
Ted Williams BOS .260
Vic Wertz CLE .245

Base Stealing Runs
Luis Aparicio CHI 3.9
Bobby Avila CLE 2.7
Mickey Mantle NY 2.4

Relievers - Runs
Ray Narleski CLE 17.3
Bob Grim NY 11.7
Tommy Byrne NY 9.8

Park Adjusted
Ray Narleski CLE 17.2
Bob Grim NY 9.6
Paul La Palme CHI 9.1

Defensive Runs
Al Kaline DET 17.2
Gil McDougald NY 11.2
Eddie Yost WAS 10.5

Players Overall
Mickey Mantle NY 80.2
Al Kaline DET 41.7
Ted Williams BOS 38.8

Pitchers Overall
Early Wynn CLE 48.2
Whitey Ford NY 43.1
Herb Score CLE 42.2

Relief Points
George Zuverink BAL 40
Ike Delock BOS 38
Don Mossi CLE 31

Club	W	L	R	OR	Avg	OBA	SLG	BPF	NOPS-A	BR	Adj	Wins	ERA	PPF	NERA-A	PR	Adj	Wins	Diff
NY	97	57	857	631	.270	.349	.434	98	115/117	80	93	9.1	3.63	94	115/108	81	44	4.3	6.6
CLE	88	66	712	581	.244	.337	.381	102	96/ 94	-34	-49	-4.9	3.32	100	125/125	129	127	12.5	3.5
CHI	85	69	776	634	.267	.352	.397	102	105/103	30	16	1.6	3.73	99	112/111	67	62	6.1	.3
BOS	84	70	780	751	.275	.365	.419	101	115/113	100	92	9.0	4.17	101	100/100	-0	3	.3	-2.3
DET	82	72	789	699	.279	.359	.420	101	113/112	84	74	7.2	4.06	100	102/102	15	14	1.4	-3.6
BAL	69	85	571	705	.244	.322	.350	91	84/ 91	-123	-63	-6.3	4.20	93	99/ 92	-5	-49	-4.9	3.2
WAS	59	95	652	924	.250	.343	.377	99	97/ 98	-26	-18	-1.9	5.33	104	78/ 81	-177	-152	-15.0	-1.1
KC	52	102	619	831	.252	.317	.370	104	88/ 85	-107	-135	-13.3	4.86	108	85/ 92	-106	-56	-5.6	-6.1
				720	.260	.343	.394						4.16						

NATIONAL LEAGUE 1956

Batting Runs
Duke Snider BRO 50.5
Frank Robinson CIN 39.1
Hank Aaron MIL 36.3

Park Adjusted
Duke Snider BRO 47.1
Stan Musial STL 38.3
Willie Mays NY 37.0

Pitching Runs
Warren Spahn MIL 30.8
Lew Burdette MIL 30.3
Johnny Antonelli NY 26.1

Park Adjusted
Warren Spahn MIL 28.2
Lew Burdette MIL 27.9
Johnny Antonelli NY 26.7

Normalized OPS
Duke Snider BRO 164
Frank Robinson CIN 148
Willie Mays NY 145

Park Adjusted
Duke Snider BRO 157
Willie Mays NY 147
Stan Musial STL 146

Normalized ERA
Lew Burdette MIL 139
Warren Spahn MIL 135
Johnny Antonelli NY 132

Park Adjusted
Lew Burdette MIL 136
Sal Maglie BRO 134
Johnny Antonelli NY 133

On Base Average
Duke Snider BRO .402
Jim Gilliam BRO .400
Stan Musial STL .390

Slugging Percentage
Duke Snider BRO .598
Joe Adcock MIL .597
Hank Aaron MIL .558

Percent of Team Wins
Johnny Antonelli NY .299
Don Newcombe BRO .290
Robin Roberts PHI .268

Wins Above Team
Don Newcombe BRO 8.3
Johnny Antonelli NY 7.2
Bob Rush CHI 4.7

Isolated Power
Duke Snider BRO .306
Joe Adcock MIL .306
Frank Robinson CIN .267

Base Stealing Runs
Willie Mays NY 6.0
Richie Ashburn PHI 2.4
Wally Post CIN 1.8
Eddie Mathews MIL 1.8

Relievers - Runs
Marv Grissom NY 19.9
Tom Acker CIN 13.2
Don Bessent BRO 11.1

Park Adjusted
Marv Grissom NY 20.1
Tom Acker CIN 15.6
Don Bessent BRO 11.9

Defensive Runs
Roy McMillan CIN 29.6
Richie Ashburn PHI 21.3
Randy Jackson BRO 19.8

Players Overall
Willie Mays NY 45.2
Hank Aaron MIL 36.6
Stan Musial STL 35.6

Pitchers Overall
Don Newcombe BRO 32.7
Warren Spahn MIL 32.0
Johnny Antonelli NY 29.5

Relief Points
Hersh Freeman CIN 59
Clem Labine BRO 50
Turk Lown CHI 36

Club	W	L	R	OR	Avg	OBA	SLG	BPF	NOPS-A	BR	Adj	Wins	ERA	PPF	NERA-A	PR	Adj	Wins	Diff
BRO	93	61	720	601	.258	.344	.419	105	115/110	81	52	5.3	3.57	102	105/108	30	44	4.5	6.2
MIL	92	62	709	569	.259	.325	.423	101	112/111	38	35	3.5	3.11	98	121/119	102	89	9.1	2.3
CIN	91	63	775	658	.266	.338	.441	108	121/111	103	47	4.8	3.84	107	98/105	-10	29	3.0	6.2
STL	76	78	678	698	.268	.335	.399	97	107/110	25	42	4.3	3.96	98	95/ 93	-29	-42	-4.5	-.9
PHI	71	83	668	738	.252	.331	.381	94	100/107	-11	30	3.1	4.20	95	90/ 85	-65	-97	-10.0	.9
NY	67	87	540	650	.244	.301	.382	99	93/ 95	-84	-75	-7.8	3.78	101	100/100	-0	2	.2	-2.4
PIT	66	88	588	653	.257	.310	.380	105	95/ 90	-69	-99	-10.2	3.74	106	101/107	5	40	4.1	-4.9
CHI	60	94	597	708	.244	.304	.382	94	94/100	-78	-40	-4.2	3.96	96	95/ 91	-28	-53	-5.5	-7.3
				659	.256	.324	.401						3.77						

AMERICAN LEAGUE 1957

Batting Runs
Ted Williams BOS 89.9
Mickey Mantle NY 88.9
Roy Sievers WAS 47.3

Park Adjusted
Mickey Mantle NY 90.9
Ted Williams BOS 85.1
Roy Sievers WAS 47.3

Pitching Runs
Jim Bunning DET 32.4
Frank Sullivan BOS 28.4
Tom Sturdivant NY 28.0

Park Adjusted
Frank Sullivan BOS 36.2
Jim Bunning DET 34.1
Tom Sturdivant NY 22.2

Normalized OPS
Ted Williams BOS 240
Mickey Mantle NY 220
Roy Sievers WAS 161

Park Adjusted
Mickey Mantle NY 227
Ted Williams BOS 223
Roy Sievers WAS 161

Normalized ERA
Bobby Shantz NY 155
Tom Sturdivant NY 149
Jim Bunning DET 140

Park Adjusted
Frank Sullivan BOS 150
Bobby Shantz NY 144
Jim Bunning DET 143

On Base Average
Ted Williams BOS .528
Mickey Mantle NY .515
Minnie Minoso CHI .413

Slugging Percentage
Ted Williams BOS .731
Mickey Mantle NY .665
Roy Sievers WAS .579

Percent of Team Wins
Jim Bunning DET .256
Billy Pierce CHI .222
Pedro Ramos WAS .218

Wins Above Team
Jim Bunning DET 7.1
Dick Donovan CHI 3.7
Ray Narleski CLE 3.4

Isolated Power
Ted Williams BOS .343
Mickey Mantle NY .300
Roy Sievers WAS .278

Base Stealing Runs
Jim Rivera CHI 4.2
Luis Aparicio CHI 3.6
Mickey Mantle NY 3.0

Relievers - Runs
Gerry Staley CHI 20.2
George Zuverink BAL 16.6
Virgil Trucks KC 9.8

Park Adjusted
Gerry Staley CHI 18.3
George Zuverink BAL 11.8
Virgil Trucks KC 9.7

Defensive Runs
Nellie Fox CHI 20.7
Gil McDougald NY 18.7
Billy Klaus BOS 16.9

Players Overall
Mickey Mantle NY 78.2
Ted Williams BOS 66.7
Nellie Fox CHI 53.5

Pitchers Overall
Frank Sullivan BOS 36.1
Jim Bunning DET 32.3
Bobby Shantz NY 29.3

Relief Points
Bob Grim NY 54
Ray Narleski CLE 42
Ike Delock BOS 34

Club	W	L	R	OR	Avg	OBA	SLG	BPF	NOPS-A	BR	Adj	Wins	ERA	PPF	NERA-A	PR	Adj	Wins	Diff
NY	98	56	723	534	.268	.341	.409	97	114/117	75	95	9.8	3.00	93	126/118	122	82	8.4	2.7
CHI	90	64	707	566	.260	.347	.375	99	105/107	39	48	5.0	3.35	96	113/108	69	44	4.5	3.5
BOS	82	72	721	668	.262	.343	.405	108	113/105	76	23	2.4	3.88	108	98/105	-13	30	3.1	-.5
DET	78	76	614	614	.257	.324	.378	101	100/ 99	-17	-26	-2.8	3.56	102	107/108	36	45	4.7	-.9
BAL	76	76	597	588	.252	.321	.353	91	91/101	-65	-7	-.8	3.46	90	109/ 98	51	-8	-.9	1.8
CLE	76	77	682	722	.252	.332	.382	105	103/ 98	9	-21	-2.2	4.06	106	93/ 99	-40	-5	-.6	2.4
KC	59	94	563	710	.244	.297	.394	97	98/100	-59	-41	-4.3	4.18	100	91/ 90	-59	-60	-6.3	-6.9
WAS	55	99	603	808	.244	.318	.363	100	94/ 94	-55	-55	-5.7	4.85	104	78/ 81	-162	-138	-14.4	-1.9
			651		.255	.328	.382						3.79						

NATIONAL LEAGUE 1957

Batting Runs
Willie Mays NY 60.5
Stan Musial STL 54.1
Hank Aaron MIL 48.3

Park Adjusted
Willie Mays NY 63.3
Stan Musial STL 51.7
Hank Aaron MIL 51.1

Pitching Runs
Warren Spahn MIL 35.8
Don Drysdale BRO 29.2
Bob Buhl MIL 27.5

Park Adjusted
Don Drysdale BRO 45.3
Johnny Podres BRO 40.7
Warren Spahn MIL 27.8

Normalized OPS
Stan Musial STL 175
Willie Mays NY 173
Hank Aaron MIL 158

Park Adjusted
Willie Mays NY 180
Stan Musial STL 169
Hank Aaron MIL 164

Normalized ERA
Johnny Podres BRO 146
Don Drysdale BRO 144
Warren Spahn MIL 144

Park Adjusted
Johnny Podres BRO 170
Don Drysdale BRO 169
Warren Spahn MIL 134

On Base Average
Stan Musial STL .428
Willie Mays NY .411
Ed Bouchee PHI .396

Slugging Percentage
Willie Mays NY .626
Stan Musial STL .612
Hank Aaron MIL .600

Percent of Team Wins
Jack Sanford PHI .247
Dick Drott CHI .242
Bob Friend PIT .226

Wins Above Team
Jack Sanford PHI 6.7
Dick Drott CHI 5.5
Turk Farrell PHI 4.3

Isolated Power
Duke Snider BRO .313
Ernie Banks CHI .295
Willie Mays NY .292

Base Stealing Runs
Johnny Temple CIN 2.7
Chico Fernandez PHI 2.4
Jim Gilliam BRO 1.8

Relievers - Runs
Turk Farrell PHI 13.8
Ed Roebuck BRO 12.4
Don McMahon MIL 12.2

Park Adjusted
Ed Roebuck BRO 19.4
Turk Farrell PHI 13.5
Clem Labine BRO 12.9

Defensive Runs
Richie Ashburn PHI 26.5
Don Blasingame STL 25.4
Johnny Logan MIL 22.1

Players Overall
Willie Mays NY 63.6
Hank Aaron MIL 54.3
Frank Robinson CIN 45.0

Pitchers Overall
Don Drysdale BRO 49.3
Johnny Podres BRO 40.8
Warren Spahn MIL 28.8

Relief Points
Turk Farrell PHI 38
Clem Labine BRO 37
Marv Grissom NY 32

Club	W	L	R	OR	Avg	OBA	SLG	BPF	NOPS-A	BR	Adj	Wins	ERA	PPF	NERA-A	PR	Adj	Wins	Diff
MIL	95	59	772	613	.269	.329	.442	96	118/123	82	108	11.0	3.47	93	112/104	64	23	2.3	4.7
STL	87	67	737	666	.274	.336	.405	104	109/105	36	10	1.0	3.77	103	103/105	16	32	3.3	5.7
BRO	84	70	690	591	.253	.328	.387	118	101/ 86	-10	-130	-13.3	3.35	117	116/135	81	183	18.6	1.7
CIN	80	74	747	781	.269	.341	.432	101	118/117	97	90	9.1	4.63	102	84/ 85	-115	-104	-10.6	4.5
PHI	77	77	623	656	.250	.325	.375	99	97/ 98	-38	-30	-3.1	3.79	99	102/101	13	9	.9	2.2
NY	69	85	643	701	.252	.313	.393	96	99/103	-41	-16	-1.8	4.01	97	97/ 94	-19	-36	-3.7	-2.5
PIT	62	92	586	696	.268	.318	.384	98	98/105	-45	0	-.0	3.87	95	100/ 95	0	-30	-3.2	-11.8
CHI	62	92	628	722	.244	.307	.380	97	94/ 97	-77	-54	-5.5	4.13	98	94/ 92	-39	-49	-5.1	-4.3
			678		.260	.325	.400						3.88						

AMERICAN LEAGUE   1958

Batting Runs			Park Adjusted			Pitching Runs			Park Adjusted		
Mickey Mantle	NY	64.2	Rocky Colavito	CLE	55.7	Whitey Ford	NY	42.7	Whitey Ford	NY	51.0
Ted Williams	BOS	53.6	Mickey Mantle	NY	55.4	Billy Pierce	CHI	29.6	Bob Turley	NY	30.8
Rocky Colavito	CLE	52.8	Ted Williams	BOS	52.7	Frank Lary	DET	24.9	Frank Lary	DET	27.0

Normalized OPS			Park Adjusted			Normalized ERA			Park Adjusted		
Ted Williams	BOS	186	Rocky Colavito	CLE	187	Whitey Ford	NY	187	Whitey Ford	NY	204
Mickey Mantle	NY	182	Ted Williams	BOS	183	Billy Pierce	CHI	141	Bob Turley	NY	138
Rocky Colavito	CLE	178	Mickey Mantle	NY	163	Jack Harshman	BAL	130	Billy Pierce	CHI	134

On Base Average			Slugging Percentage			Percent of Team Wins			Wins Above Team		
Ted Williams	BOS	.462	Rocky Colavito	CLE	.620	Pedro Ramos	WAS	.230	Dick Hyde	WAS	5.3
Mickey Mantle	NY	.445	Bob Cerv	KC	.592	Bob Turley	NY	.228	Bob Turley	NY	5.2
Pete Runnels	BOS	.418	Mickey Mantle	NY	.592	Cal McLish	CLE	.208	Cal McLish	CLE	4.7
						Frank Lary	DET	.208			

Isolated Power			Base Stealing Runs			Relievers - Runs			Park Adjusted		
Rocky Colavito	CLE	.317	Luis Aparicio	CHI	5.1	Dick Hyde	WAS	23.1	Dick Hyde	WAS	24.3
Bob Cerv	KC	.287	Jim Rivera	CHI	4.5	Ryne Duren	NY	14.8	Ryne Duren	NY	17.7
Mickey Mantle	NY	.287	Mickey Mantle	NY	3.6	Leo Kiely	BOS	6.9	Leo Kiely	BOS	7.5

Defensive Runs			Players Overall			Pitchers Overall			Relief Points		
Al Kaline	DET	22.2	Rocky Colavito	CLE	50.7	Whitey Ford	NY	52.3	Dick Hyde	WAS	53
Tony Kubek	NY	20.5	Mickey Mantle	NY	42.0	Jack Harshman	BAL	27.4	Ryne Duren	NY	48
Frank Malzone	BOS	16.3	Pete Runnels	BOS	40.8	Bob Turley	NY	26.6	Leo Kiely	BOS	32

Club	W	L	R	OR	Avg	OBA	SLG	BPF	NOPS-A	BR	Adj	Wins	ERA	PPF	NERA-A	PR	Adj	Wins	Diff
NY	92	62	759	577	.268	.338	.416	112	116/104	88	9	.9	3.22	109	117/128	84	136	14.1	.1
CHI	82	72	634	615	.257	.329	.367	96	99/103	-12	13	1.3	3.60	95	105/100	25	-1	-.3	3.9
BOS	79	75	697	691	.256	.340	.400	102	112/111	70	60	6.2	3.93	102	96/ 98	-23	-14	-1.5	-2.7
CLE	77	76	694	635	.258	.327	.403	95	110/115	38	67	6.9	3.72	94	101/ 95	7	-26	-2.8	-3.6
DET	77	77	659	606	.266	.329	.389	103	106/103	20	2	.3	3.59	102	105/107	27	39	4.0	-4.3
BAL	74	79	521	575	.241	.310	.350	93	89/ 96	-84	-41	-4.4	3.40	93	111/104	56	19	2.0	-.1
KC	73	81	642	713	.247	.309	.381	101	98/ 97	-39	-47	-5.0	4.15	103	91/ 93	-58	-42	-4.4	5.4
WAS	61	93	553	747	.240	.309	.357	99	91/ 92	-77	-71	-7.4	4.53	103	83/ 85	-116	-100	-10.4	1.8
				645	.254	.324	.383						3.77						

NATIONAL LEAGUE   1958

Batting Runs			Park Adjusted			Pitching Runs			Park Adjusted		
Willie Mays	SF	55.3	Willie Mays	SF	55.2	Lew Burdette	MIL	31.8	Sam Jones	STL	38.8
Ernie Banks	CHI	45.7	Hank Aaron	MIL	47.5	Stu Miller	SF	29.9	Stu Miller	SF	29.7
Hank Aaron	MIL	37.0	Ernie Banks	CHI	43.4	Sam Jones	STL	29.8	Robin Roberts	PHI	25.5

Normalized OPS			Park Adjusted			Normalized ERA			Park Adjusted		
Willie Mays	SF	163	Hank Aaron	MIL	166	Stu Miller	SF	160	Stu Miller	SF	159
Ernie Banks	CHI	155	Willie Mays	SF	163	Sam Jones	STL	137	Sam Jones	STL	148
Stan Musial	STL	150	Ernie Banks	CHI	150	Lew Burdette	MIL	136	Robin Roberts	PHI	126

On Base Average			Slugging Percentage			Percent of Team Wins			Wins Above Team		
Richie Ashburn	PHI	.441	Ernie Banks	CHI	.614	Bob Friend	PIT	.262	Robin Roberts	PHI	3.9
Stan Musial	STL	.426	Willie Mays	SF	.583	Robin Roberts	PHI	.246	Bob Purkey	CIN	3.9
Willie Mays	SF	.423	Hank Aaron	MIL	.546	Warren Spahn	MIL	.239	Red Witt	PIT	3.2

Isolated Power			Base Stealing Runs			Relievers - Runs			Park Adjusted		
Ernie Banks	CHI	.301	Willie Mays	SF	5.7	Don Elston	CHI	11.6	Don Elston	CHI	13.0
Frank Thomas	PIT	.247	Don Blasingame	STL	3.0	Roy Face	PIT	9.9	Bill Henry	CHI	10.8
Willie Mays	SF	.237	Don Zimmer	LA	3.0	Bill Henry	CHI	9.6	Willard Schmidt	CIN	10.5

Defensive Runs			Players Overall			Pitchers Overall			Relief Points		
Ken Boyer	STL	22.2	Willie Mays	SF	65.6	Sam Jones	STL	32.3	Roy Face	PIT	48
Roberto Clemente	PIT	21.5	Ernie Banks	CHI	52.4	Stu Miller	SF	30.6	Clem Labine	LA	33
Don Zimmer	LA	20.6	Hank Aaron	MIL	47.9	Robin Roberts	PHI	26.9	Don Elston	CHI	30

Club	W	L	R	OR	Avg	OBA	SLG	BPF	NOPS-A	BR	Adj	Wins	ERA	PPF	NERA-A	PR	Adj	Wins	Diff
MIL	92	62	675	541	.266	.331	.412	86	107/123	14	105	10.5	3.21	82	123/102	113	8	.8	3.7
PIT	84	70	662	607	.264	.319	.410	90	103/114	-20	43	4.3	3.56	88	111/ 98	59	-10	-1.1	3.8
SF	80	74	727	698	.263	.334	.422	100	110/110	37	36	3.6	3.98	100	99/ 99	-3	-5	-.6	-.0
CIN	76	78	695	621	.258	.333	.389	108	100/ 92	-18	-74	-7.6	3.72	107	106/114	35	80	8.0	-1.4
STL	72	82	619	704	.261	.331	.380	106	97/ 91	-37	-79	-8.0	4.11	108	96/104	-24	25	2.5	.5
CHI	72	82	709	725	.265	.332	.426	103	111/108	39	19	2.0	4.22	103	94/ 97	-39	-19	-2.0	-4.9
LA	71	83	668	761	.251	.319	.402	103	101/ 97	-31	-53	-5.4	4.48	105	88/ 93	-79	-48	-4.9	4.3
PHI	69	85	664	762	.266	.341	.400	101	105/104	20	11	1.1	4.33	103	91/ 94	-57	-37	-3.9	-5.2
				677	.262	.330	.405						3.95						

Batting Runs			Park Adjusted			Pitching Runs			Park Adjusted		
Al Kaline	DET	41.6	Mickey Mantle	NY	38.6	Hoyt Wilhelm	BAL	42.0	Hoyt Wilhelm	BAL	40.7
Eddie Yost	DET	37.9	Tito Francona	CLE	37.2	Camilo Pascual	WAS	32.5	Camilo Pascual	WAS	35.3
Tito Francona	CLE	36.6	Al Kaline	DET	35.4	Bob Shaw	CHI	30.1	Bob Shaw	CHI	24.7
Normalized OPS			**Park Adjusted**			**Normalized ERA**			**Park Adjusted**		
Al Kaline	DET	157	Mickey Mantle	NY	152	Hoyt Wilhelm	BAL	176	Hoyt Wilhelm	BAL	174
Harvey Kuenn	DET	147	Al Kaline	DET	144	Camilo Pascual	WAS	146	Camilo Pascual	WAS	150
Mickey Mantle	NY	146	Gene Woodling	BAL	139	Bob Shaw	CHI	144	Bob Shaw	CHI	136
On Base Average			**Slugging Percentage**			**Percent of Team Wins**			**Wins Above Team**		
Eddie Yost	DET	.437	Al Kaline	DET	.530	Camilo Pascual	WAS	.270	Camilo Pascual	WAS	7.2
Pete Runnels	BOS	.415	Harmon Killebrew	WAS	.516	Bud Daley	KC	.242	Don Mossi	DET	5.0
Al Kaline	DET	.414	Mickey Mantle	NY	.514	Early Wynn	CHI	.234	Frank Lary	DET	4.5
Isolated Power			**Base Stealing Runs**			**Relievers - Runs**			**Park Adjusted**		
Harmon Killebrew	WAS	.275	Luis Aparicio	CHI	9.0	Gerry Staley	CHI	20.8	Gerry Staley	CHI	18.1
Rocky Colavito	CLE	.255	Mickey Mantle	NY	4.5	Ryne Duren	NY	17.0	Ryne Duren	NY	15.5
Jim Lemon	WAS	.232	Jackie Jensen	BOS	3.0	Bobby Shantz	NY	15.8	Bobby Shantz	NY	13.9
Defensive Runs			**Players Overall**			**Pitchers Overall**			**Relief Points**		
Billy Gardner	BAL	15.3	Pete Runnels	BOS	39.1	Camilo Pascual	WAS	46.1	Turk Lown	CHI	46
Minnie Minoso	CLE	12.0	Tito Francona	CLE	38.4	Hoyt Wilhelm	BAL	33.8	Gerry Staley	CHI	39
Jim Landis	CHI	10.6	Mickey Mantle	NY	38.0	Whitey Ford	NY	24.7	Billy Loes	BAL	29
									Mike Fornieles	BOS	29

Club	W	L	R	OR	Avg	OBA	SLG	BPF	NOPS-A	BR	Adj	Wins	ERA	PPF	NERA-A	PR	Adj	Wins	Diff
CHI	94	60	669	588	.250	.330	.364	96	98/101	-19	4	.4	3.29	95	117/111	91	58	5.8	10.7
CLE	89	65	745	646	.263	.323	.408	99	110/111	32	40	4.1	3.75	97	103/100	18	-0	-.1	8.0
NY	79	75	687	647	.260	.321	.402	96	107/111	17	41	4.2	3.61	95	107/102	40	13	1.3	-3.4
DET	76	78	713	732	.258	.338	.400	109	111/101	57	-4	-.5	4.20	110	92/101	-50	9	.9	-1.4
BOS	75	79	726	696	.256	.338	.385	104	107/106	36	32	3.2	4.16	100	93/ 93	-45	-44	-4.5	-.7
BAL	74	80	551	621	.238	.312	.345	98	87/ 89	-95	-80	-8.2	3.56	99	108/107	46	38	3.9	1.3
KC	66	88	681	760	.263	.328	.390	99	105/107	16	25	2.5	4.35	100	89/ 89	-74	-73	-7.5	-6.0
WAS	63	91	619	701	.237	.310	.379	101	97/ 96	-42	-48	-5.0	4.01	103	96/ 99	-22	-6	-.7	-8.3
			674		.253	.325	.384						3.86						

Batting Runs			Park Adjusted			Pitching Runs			Park Adjusted		
Hank Aaron	MIL	62.8	Hank Aaron	MIL	67.1	Sam Jones	SF	33.8	Roger Craig	LA	34.3
Eddie Mathews	MIL	48.4	Eddie Mathews	MIL	52.6	Roger Craig	LA	32.1	Vern Law	PIT	32.2
Ernie Banks	CHI	44.4	Willie Mays	SF	45.7	Warren Spahn	MIL	32.0	Sam Jones	SF	26.2
Normalized OPS			**Park Adjusted**			**Normalized ERA**			**Park Adjusted**		
Hank Aaron	MIL	172	Hank Aaron	MIL	182	Sam Jones	SF	140	Vern Law	PIT	137
Eddie Mathews	MIL	157	Eddie Mathews	MIL	166	Bob Buhl	MIL	138	Gene Conley	PHI	134
Frank Robinson	CIN	157	Willie Mays	SF	160	Warren Spahn	MIL	133	Sam Jones	SF	131
On Base Average			**Slugging Percentage**			**Percent of Team Wins**			**Wins Above Team**		
Joe Cunningham	STL	.456	Hank Aaron	MIL	.636	Sam Jones	SF	.253	Roy Face	PIT	9.6
Hank Aaron	MIL	.406	Ernie Banks	CHI	.596	Warren Spahn	MIL	.244	Vern Law	PIT	5.2
Frank Robinson	CIN	.397	Eddie Mathews	MIL	.593	Lew Burdette	MIL	.244	Gene Conley	PHI	4.7
Isolated Power			**Base Stealing Runs**			**Relievers - Runs**			**Park Adjusted**		
Ernie Banks	CHI	.292	Willie Mays	SF	5.7	Stu Miller	SF	20.7	Bill Henry	CHI	19.1
Eddie Mathews	MIL	.286	Vada Pinson	CIN	2.7	Bill Henry	CHI	18.8	Stu Miller	SF	15.9
Hank Aaron	MIL	.281	Hank Aaron	MIL	2.4	Roy Face	PIT	12.8	Roy Face	PIT	14.0
Defensive Runs			**Players Overall**			**Pitchers Overall**			**Relief Points**		
Charlie Neal	LA	18.4	Hank Aaron	MIL	64.7	Don Newcombe	CIN	34.5	Roy Face	PIT	55
Don Blasingame	STL	18.2	Ernie Banks	CHI	53.6	Vern Law	PIT	33.8	Lindy McDaniel	STL	48
Bill Virdon	PIT	17.6	Willie Mays	SF	49.8	Roger Craig	LA	29.9	Don Elston	CHI	38

Club	W	L	R	OR	Avg	OBA	SLG	BPF	NOPS-A	BR	Adj	Wins	ERA	PPF	NERA-A	PR	Adj	Wins	Diff
LA	88	68	705	670	.257	.335	.396	104	105/101	17	-7	-.8	3.79	103	104/108	25	45	4.5	6.3
MIL	86	70	724	623	.265	.329	.417	95	110/116	32	69	6.9	3.51	92	113/104	68	21	2.1	-1.0
SF	83	71	705	613	.261	.324	.414	96	108/112	16	45	4.5	3.46	94	114/107	74	35	3.5	-2.0
PIT	78	76	651	680	.263	.322	.384	102	98/ 96	-38	-55	-5.6	3.90	103	101/104	7	26	2.6	4.0
CIN	74	80	764	738	.274	.340	.427	101	116/114	76	68	6.9	4.31	101	92/ 92	-54	-48	-4.9	-4.9
CHI	74	80	673	688	.249	.319	.398	100	101/101	-22	-24	-2.5	4.01	101	98/ 99	-9	-5	-.6	.2
STL	71	83	641	725	.269	.333	.400	101	106/104	14	4	.4	4.34	103	91/ 94	-58	-39	-4.1	-2.3
PHI	64	90	599	725	.242	.314	.362	100	89/ 89	-91	-90	-9.1	4.28	102	92/ 94	-49	-36	-3.8	-.1
			683		.260	.327	.400						3.95						

AMERICAN LEAGUE 1960

Batting Runs			Park Adjusted			Pitching Runs			Park Adjusted		
Mickey Mantle	NY	43.9	Mickey Mantle	NY	49.7	Jim Bunning	DET	30.4	Jim Bunning	DET	31.5
Ted Williams	BOS	43.5	Ted Williams	BOS	42.0	Frank Baumann	CHI	24.6	Frank Baumann	CHI	21.3
Roger Maris	NY	36.5	Roger Maris	NY	41.7	Art Ditmar	NY	18.0	Pedro Ramos	WAS	18.9

Normalized OPS			Park Adjusted			Normalized ERA			Park Adjusted		
Mickey Mantle	NY	157	Mickey Mantle	NY	170	Frank Baumann	CHI	145	Jim Bunning	DET	140
Roger Maris	NY	154	Roger Maris	NY	168	Jim Bunning	DET	139	Frank Baumann	CHI	139
Roy Sievers	CHI	150	Roy Sievers	CHI	152	Hal Brown	BAL	127	Hal Brown	BAL	131

On Base Average			Slugging Percentage			Percent of Team Wins			Wins Above Team		
Eddie Yost	DET	.416	Roger Maris	NY	.581	Bud Daley	KC	.276	Jim Perry	CLE	5.1
Gene Woodling	BAL	.403	Mickey Mantle	NY	.558	Ray Herbert	KC	.241	Bud Daley	KC	5.0
Pete Runnels	BOS	.403	Harmon Killebrew	WAS	.534	Jim Perry	CLE	.237	Bill Monbouquette	BOS	4.1

Isolated Power			Base Stealing Runs			Relievers - Runs			Park Adjusted		
Roger Maris	NY	.299	Luis Aparicio	CHI	10.5	Gerry Staley	CHI	18.5	Mike Fornieles	BOS	17.6
Mickey Mantle	NY	.283	Jim Landis	CHI	3.3	Mike Fornieles	BOS	14.9	Gerry Staley	CHI	16.4
Harmon Killebrew	WAS	.258	Al Kaline	DET	3.3	Dave Sisler	DET	12.4	Dave Sisler	DET	12.8

Defensive Runs			Players Overall			Pitchers Overall			Relief Points		
Luis Aparicio	CHI	31.2	Luis Aparicio	CHI	40.1	Jim Bunning	DET	28.7	Mike Fornieles	BOS	43
Vic Power	CLE	20.6	Mickey Mantle	NY	37.6	Frank Baumann	CHI	20.5	Gerry Staley	CHI	38
Clete Boyer	NY	14.8	Roger Maris	NY	37.1	Camilo Pascual	WAS	20.3	Johnny Klippstein	CLE	33

Club	W	L	R	OR	Avg	OBA	SLG	BPF	NOPS-A	BR	Adj	Wins	ERA	PPF	NERA-A	PR	Adj	Wins	Diff
NY	97	57	746	627	.260	.332	.426	92	114/124	63	117	11.9	3.52	89	110/ 98	55	-9	-1.0	9.2
BAL	89	65	682	606	.253	.334	.377	105	100/ 95	-7	-40	-4.1	3.51	104	110/114	55	77	7.8	8.4
CHI	87	67	741	617	.270	.348	.396	98	110/113	57	68	6.9	3.60	96	107/103	41	17	1.7	1.5
CLE	76	78	667	693	.267	.328	.388	101	102/101	-10	-16	-1.7	3.95	101	98/ 99	-12	-3	-.4	1.1
WAS	73	81	672	696	.244	.326	.384	104	100/ 96	-17	-47	-4.8	3.77	105	103/108	16	47	4.7	-3.9
DET	71	83	633	644	.239	.326	.375	101	97/ 96	-31	-36	-3.7	3.64	101	106/108	37	43	4.3	-6.6
BOS	65	89	658	775	.261	.336	.389	103	104/101	16	-5	-.6	4.62	106	84/ 89	-113	-79	-8.1	-3.4
KC	58	96	615	756	.249	.318	.366	95	93/ 97	-67	-37	-3.8	4.38	98	88/ 87	-76	-89	-9.1	-6.1
			677		.255	.331	.388						3.87						

NATIONAL LEAGUE 1960

Batting Runs			Park Adjusted			Pitching Runs			Park Adjusted		
Frank Robinson	CIN	47.8	Eddie Mathews	MIL	55.0	Mike McCormick	SF	29.8	Don Drysdale	LA	44.2
Eddie Mathews	MIL	46.0	Willie Mays	SF	51.4	Don Drysdale	LA	27.4	Johnny Podres	LA	31.5
Willie Mays	SF	43.5	Frank Robinson	CIN	47.1	Ernie Broglio	STL	25.5	Stan Williams	LA	30.4

Normalized OPS			Park Adjusted			Normalized ERA			Park Adjusted		
Frank Robinson	CIN	172	Eddie Mathews	MIL	179	Mike McCormick	SF	139	Don Drysdale	LA	152
Eddie Mathews	MIL	157	Willie Mays	SF	172	Ernie Broglio	STL	137	Stan Williams	LA	144
Willie Mays	SF	154	Frank Robinson	CIN	170	Don Drysdale	LA	132	Ernie Broglio	STL	141

On Base Average			Slugging Percentage			Percent of Team Wins			Wins Above Team		
Richie Ashburn	CHI	.416	Frank Robinson	CIN	.595	Glen Hobbie	CHI	.267	Bob Purkey	CIN	5.9
Frank Robinson	CIN	.413	Hank Aaron	MIL	.566	Bob Purkey	CIN	.254	Ernie Broglio	STL	5.3
Eddie Mathews	MIL	.401	Ken Boyer	STL	.562	Ernie Broglio	STL	.244	Turk Farrell	PHI	4.3

Isolated Power			Base Stealing Runs			Relievers - Runs			Park Adjusted		
Frank Robinson	CIN	.297	Maury Wills	LA	7.8	Lindy McDaniel	STL	21.5	Lindy McDaniel	STL	23.1
Ernie Banks	CHI	.283	Julian Javier	STL	3.3	Jim Brosnan	CIN	15.4	Ed Roebuck	LA	20.2
Hank Aaron	MIL	.275	Don Blasingame	SF	3.0	Ed Roebuck	LA	12.9	Turk Farrell	PHI	17.0

Defensive Runs			Players Overall			Pitchers Overall			Relief Points		
Bill Mazeroski	PIT	25.6	Willie Mays	SF	54.4	Don Drysdale	LA	49.6	Lindy McDaniel	STL	74
Maury Wills	LA	23.9	Ernie Banks	CHI	53.0	Ernie Broglio	STL	32.5	Roy Face	PIT	60
Ken Boyer	STL	16.2	Hank Aaron	MIL	50.7	Stan Williams	LA	30.6	Turk Farrell	PHI	36
										Jim Brosnan	CIN	36

Club	W	L	R	OR	Avg	OBA	SLG	BPF	NOPS-A	BR	Adj	Wins	ERA	PPF	NERA-A	PR	Adj	Wins	Diff
PIT	95	59	734	593	.276	.338	.407	100	114/113	72	69	7.2	3.49	98	108/105	42	29	3.0	7.9
MIL	88	66	724	658	.265	.327	.417	88	114/129	60	137	14.1	3.76	86	100/ 86	0	-82	-8.5	5.4
STL	86	68	639	616	.254	.323	.393	103	106/102	13	-8	-1.0	3.64	103	103/107	18	37	3.9	6.1
LA	82	72	662	593	.255	.327	.383	115	103/ 90	6	-93	-9.9	3.40	115	111/127	56	143	14.8	-.1
SF	79	75	671	631	.255	.319	.393	89	104/117	1	71	7.3	3.44	88	109/ 96	50	-21	-2.3	-2.9
CIN	67	87	640	692	.250	.320	.388	101	103/102	-0	-8	-.9	4.00	102	94/ 96	-36	-23	-2.5	-6.6
CHI	60	94	634	776	.243	.314	.369	98	96/ 97	-47	-37	-3.9	4.35	101	86/ 88	-91	-84	-8.7	-4.4
PHI	59	95	546	691	.239	.304	.351	108	88/ 81	-101	-153	-15.9	4.02	112	94/105	-38	28	2.9	-5.0
			656		.255	.322	.388						3.76						

AMERICAN LEAGUE 1961

Batting Runs / Park Adjusted / Pitching Runs / Park Adjusted

Batting Runs			Park Adjusted			Pitching Runs			Park Adjusted		
Norm Cash	DET	86.1	Norm Cash	DET	85.8	Billy Hoeft	BAL	30.7	Billy Hoeft	BAL	30.6
Mickey Mantle	NY	76.3	Mickey Mantle	NY	83.4	Dick Donovan	WAS	30.6	Dick Donovan	WAS	30.4
Jim Gentile	BAL	59.1	Jim Gentile	BAL	58.2	Bill Stafford	NY	29.2	Camilo Pascual	MIN	27.5
Normalized OPS			**Park Adjusted**			**Normalized ERA**			**Park Adjusted**		
Norm Cash	DET	204	Mickey Mantle	NY	220	Dick Donovan	WAS	168	Dick Donovan	WAS	168
Mickey Mantle	NY	199	Norm Cash	DET	203	Bill Stafford	NY	150	Don Mossi	DET	133
Jim Gentile	BAL	183	Jim Gentile	BAL	181	Don Mossi	DET	136	Milt Pappas	BAL	132
On Base Average			**Slugging Percentage**			**Percent of Team Wins**			**Wins Above Team**		
Norm Cash	DET	.488	Mickey Mantle	NY	.687	Whitey Ford	NY	.229	Whitey Ford	NY	6.7
Mickey Mantle	NY	.452	Norm Cash	DET	.662	Frank Lary	DET	.228	Don Schwall	BOS	5.4
Jim Gentile	BAL	.428	Jim Gentile	BAL	.646	Camilo Pascual	MIN	.214	Barry Latman	CLE	4.8
Isolated Power			**Base Stealing Runs**			**Relievers - Runs**			**Park Adjusted**		
Mickey Mantle	NY	.370	Luis Aparicio	CHI	8.1	Luis Arroyo	NY	24.2	Hoyt Wilhelm	BAL	21.1
Roger Maris	NY	.351	Dick Howser	KC	5.7	Hoyt Wilhelm	BAL	21.2	Tom Morgan	LA	20.3
Jim Gentile	BAL	.344	Chuck Hinton	WAS	3.6	Tom Morgan	LA	17.1	Luis Arroyo	NY	17.1
Defensive Runs			**Players Overall**			**Pitchers Overall**			**Relief Points**		
Clete Boyer	NY	31.8	Mickey Mantle	NY	75.9	Billy Hoeft	BAL	32.0	Luis Arroyo	NY	83
Al Kaline	DET	18.9	Norm Cash	DET	71.3	Dick Donovan	WAS	31.9	Hoyt Wilhelm	BAL	47
Tony Kubek	NY	17.4	Rocky Colavito	DET	51.2	Frank Lary	DET	29.1	Mike Fornieles	BOS	39

Club	W	L	R	OR	Avg	OBA	SLG	BPF	NOPS-A	BR	Adj	Wins	ERA	PPF	NERA-A	PR	Adj	Wins	Diff
NY	109	53	827	612	.263	.332	.442	91	117/129	82	149	14.8	3.46	87	116/101	90	4	.4	12.8
DET	101	61	841	671	.266	.349	.421	100	116/115	96	94	9.3	3.55	98	113/111	77	63	6.2	4.5
BAL	95	67	691	588	.254	.328	.390	101	101/100	-15	-24	-2.5	3.22	100	125/125	132	131	13.0	3.5
CHI	86	76	765	726	.265	.338	.395	93	105/113	17	70	6.9	4.06	92	99/91	-4	-59	-5.9	4.0
CLE	78	83	737	752	.266	.328	.406	102	106/103	9	-6	-.7	4.15	102	97/99	-19	-3	-.4	-1.4
BOS	76	86	729	792	.254	.336	.374	98	98/100	-21	-5	-.6	4.29	99	94/93	-41	-50	-5.1	.7
MIN	70	90	707	778	.250	.328	.397	109	103/94	-4	-69	-6.9	4.28	110	94/104	-40	26	2.6	-5.7
LA	70	91	744	784	.245	.333	.398	106	104/98	12	-33	-3.4	4.31	108	93/100	-44	3	.3	-7.4
WAS	61	100	618	776	.244	.317	.367	98	91/93	-82	-65	-6.6	4.23	100	95/95	-32	-33	-3.4	-9.5
KC	61	100	683	863	.247	.323	.354	101	89/88	-89	-98	-9.8	4.74	104	85/88	-112	-87	-8.7	-.9
			734		.256	.331	.395						4.02						

NATIONAL LEAGUE 1961

Batting Runs			Park Adjusted			Pitching Runs			Park Adjusted		
Frank Robinson	CIN	52.4	Hank Aaron	MIL	53.7	Warren Spahn	MIL	29.8	Jim O'Toole	CIN	35.7
Hank Aaron	MIL	46.0	Eddie Mathews	MIL	48.1	Jim O'Toole	CIN	26.3	Curt Simmons	STL	25.9
Willie Mays	SF	45.8	Willie Mays	SF	46.9	Mike McCormick	SF	23.0	Bob Gibson	STL	25.0
Normalized OPS			**Park Adjusted**			**Normalized ERA**			**Park Adjusted**		
Frank Robinson	CIN	166	Hank Aaron	MIL	171	Warren Spahn	MIL	134	Jim O'Toole	CIN	141
Willie Mays	SF	156	Eddie Mathews	MIL	162	Jim O'Toole	CIN	130	Curt Simmons	STL	138
Hank Aaron	MIL	155	Willie Mays	SF	158	Curt Simmons	STL	129	Bob Gibson	STL	133
On Base Average			**Slugging Percentage**			**Percent of Team Wins**			**Wins Above Team**		
Wally Moon	LA	.438	Frank Robinson	CIN	.611	Warren Spahn	MIL	.253	Johnny Podres	LA	5.5
Frank Robinson	CIN	.411	Orlando Cepeda	SF	.609	Don Cardwell	CHI	.234	Stu Miller	SF	4.0
Eddie Mathews	MIL	.405	Hank Aaron	MIL	.594	Art Mahaffey	PHI	.234	Don Cardwell	CHI	3.6
Isolated Power			**Base Stealing Runs**			**Relievers - Runs**			**Park Adjusted**		
Orlando Cepeda	SF	.297	Frank Robinson	CIN	4.8	Stu Miller	SF	18.6	Stu Miller	SF	16.6
Frank Robinson	CIN	.288	Lee Maye	MIL	2.4	Ron Perranoski	LA	14.2	Ron Perranoski	LA	15.0
Dick Stuart	PIT	.280	Billy Williams	CHI	1.8	Don McMahon	MIL	12.2	Bill Henry	CIN	12.7
Defensive Runs			**Players Overall**			**Pitchers Overall**			**Relief Points**		
Bill Mazeroski	PIT	31.4	Frank Robinson	CIN	48.8	Jim O'Toole	CIN	33.4	Stu Miller	SF	57
Roberto Clemente	PIT	13.4	Hank Aaron	MIL	42.6	Curt Simmons	STL	31.5	Jim Brosnan	CIN	48
Dick Groat	PIT	13.1	Roberto Clemente	PIT	41.5	Warren Spahn	MIL	28.9	Bill Henry	CIN	35

Club	W	L	R	OR	Avg	OBA	SLG	BPF	NOPS-A	BR	Adj	Wins	ERA	PPF	NERA-A	PR	Adj	Wins	Diff
CIN	93	61	710	653	.270	.328	.421	109	109/100	19	-41	-4.2	3.78	108	107/115	38	89	8.8	11.4
LA	89	65	735	697	.262	.340	.405	102	107/104	29	13	1.3	4.04	102	100/102	-0	11	1.1	9.7
SF	85	69	773	655	.264	.332	.423	98	110/112	35	45	4.5	3.77	96	107/103	40	16	1.6	1.9
MIL	83	71	712	656	.258	.330	.415	90	108/119	19	87	8.6	3.90	89	103/92	21	-50	-5.1	2.4
STL	80	74	703	668	.271	.336	.393	107	102/96	-2	-51	-5.1	3.74	107	108/115	44	86	8.5	-.4
PIT	75	79	694	675	.273	.330	.410	100	106/106	8	5	.5	3.92	100	103/103	17	18	1.8	-4.3
CHI	64	90	689	800	.255	.327	.418	95	108/113	16	49	4.8	4.48	97	90/87	-69	-87	-8.7	-9.1
PHI	47	107	584	796	.243	.311	.357	99	85/86	-122	-115	-11.6	4.61	103	88/90	-87	-70	-7.0	-11.5
			700		.262	.329	.405						4.03						

AMERICAN LEAGUE 1962

Batting Runs			Park Adjusted			Pitching Runs			Park Adjusted		
Mickey Mantle	NY	57.3	Mickey Mantle	NY	60.4	Hank Aguirre	DET	42.3	Hank Aguirre	DET	46.9
Norm Siebern	KC	41.1	Norm Siebern	KC	41.5	Whitey Ford	NY	30.8	Jim Kaat	MIN	31.3
Harmon Killebrew	MIN	33.1	Pete Runnels	BOS	28.3	Ralph Terry	NY	25.9	Dean Chance	LA	28.4
Normalized OPS			**Park Adjusted**			**Normalized ERA**			**Park Adjusted**		
Mickey Mantle	NY	192	Mickey Mantle	NY	203	Hank Aguirre	DET	180	Hank Aguirre	DET	188
Norm Siebern	KC	144	Norm Siebern	KC	145	Robin Roberts	BAL	143	Dean Chance	LA	142
Harmon Killebrew	MIN	142	Brooks Robinson	BAL	134	Whitey Ford	NY	137	Jim Kaat	MIN	133
On Base Average			**Slugging Percentage**			**Percent of Team Wins**			**Wins Above Team**		
Mickey Mantle	NY	.488	Mickey Mantle	NY	.605	Dick Donovan	CLE	.250	Dick Donovan	CLE	6.4
Norm Siebern	KC	.416	Harmon Killebrew	MIN	.545	Ralph Terry	NY	.240	Ray Herbert	CHI	5.8
Joe Cunningham	CHI	.415	Rocky Colavito	DET	.514	Ray Herbert	CHI	.235	Dave Wickersham	KC	4.8
Isolated Power			**Base Stealing Runs**			**Relievers - Runs**			**Park Adjusted**		
Harmon Killebrew	MIN	.303	Jake Wood	DET	5.4	Dick Radatz	BOS	24.2	Dick Radatz	BOS	25.1
Mickey Mantle	NY	.284	Dick Howser	KC	4.5	Dick Hall	BAL	22.1	Hoyt Wilhelm	BAL	17.0
Norm Cash	DET	.270	Ed Charles	KC	3.6	Hoyt Wilhelm	BAL	21.0	Dick Hall	BAL	17.0
Defensive Runs			**Players Overall**			**Pitchers Overall**			**Relief Points**		
Clete Boyer	NY	35.5	Mickey Mantle	NY	55.1	Jim Kaat	MIN	39.9	Dick Radatz	BOS	60
Zoilo Versalles	MIN	34.1	Clete Boyer	NY	41.1	Hank Aguirre	DET	35.6	Marshall Bridges	NY	48
Billy Moran	LA	22.2	Al Kaline	DET	34.6	Camilo Pascual	MIN	33.2	Terry Fox	DET	37

Club	W	L	R	OR	Avg	OBA	SLG	BPF	NOPS-A	BR	Adj	Wins	ERA	PPF	NERA-A	PR	Adj	Wins	Diff
NY	96	66	817	680	.267	.339	.426	95	116/123	87	126	12.7	3.70	92	107/ 99	44	-4	-.5	2.8
MIN	91	71	798	713	.260	.340	.412	106	112/105	67	20	2.0	3.89	106	102/108	13	49	4.9	3.0
LA	86	76	718	706	.250	.328	.380	105	99/ 94	-22	-61	-6.3	3.70	106	107/113	44	80	8.0	3.2
DET	85	76	758	692	.248	.332	.411	105	110/104	41	3	.3	3.81	105	104/109	26	57	5.7	-1.5
CHI	85	77	707	658	.257	.336	.372	100	99/ 99	-12	-9	-1.0	3.73	99	106/105	38	30	3.0	2.0
CLE	80	82	682	745	.245	.314	.388	95	98/103	-45	-13	-1.4	4.14	96	96/ 92	-26	-51	-5.2	5.6
BAL	77	85	652	680	.248	.316	.387	91	98/108	-41	23	2.4	3.70	90	107/ 97	45	-18	-1.9	-4.5
BOS	76	84	707	756	.258	.326	.403	101	106/105	9	2	.2	4.22	102	94/ 96	-39	-28	-2.9	-1.4
KC	72	90	745	837	.263	.334	.386	100	103/103	3	6	.6	4.79	101	83/ 84	-130	-124	-12.5	2.9
WAS	60	101	599	716	.250	.310	.373	103	93/ 90	-81	-105	-10.6	4.05	105	98/103	-11	21	2.1	-12.0
			718		.255	.328	.394						3.97						

NATIONAL LEAGUE 1962

Batting Runs			Park Adjusted			Pitching Runs			Park Adjusted		
Frank Robinson	CIN	66.6	Frank Robinson	CIN	73.2	Don Drysdale	LA	38.5	Bob Gibson	STL	41.5
Willie Mays	SF	54.1	Willie Mays	SF	59.6	Bob Purkey	CIN	36.1	Ernie Broglio	STL	35.5
Hank Aaron	MIL	54.0	Hank Aaron	MIL	56.1	Sandy Koufax	LA	28.6	Bob Friend	PIT	30.3
Normalized OPS			**Park Adjusted**			**Normalized ERA**			**Park Adjusted**		
Frank Robinson	CIN	177	Frank Robinson	CIN	192	Sandy Koufax	LA	155	Bob Gibson	STL	156
Hank Aaron	MIL	167	Willie Mays	SF	175	Bob Shaw	MIL	141	Ernie Broglio	STL	148
Willie Mays	SF	164	Hank Aaron	MIL	171	Bob Purkey	CIN	140	Bob Shaw	MIL	135
On Base Average			**Slugging Percentage**			**Percent of Team Wins**			**Wins Above Team**		
Frank Robinson	CIN	.424	Frank Robinson	CIN	.624	Roger Craig	NY	.250	Bob Purkey	CIN	7.3
Stan Musial	STL	.420	Hank Aaron	MIL	.618	Don Drysdale	LA	.245	Jack Sanford	SF	5.7
Bob Skinner	PIT	.397	Willie Mays	SF	.615	Bob Purkey	CIN	.235	Don Drysdale	LA	5.0
Isolated Power			**Base Stealing Runs**			**Relievers - Runs**			**Park Adjusted**		
Willie Mays	SF	.311	Maury Wills	LA	23.4	Roy Face	PIT	20.8	Roy Face	PIT	22.4
Hank Aaron	MIL	.296	Willie Davis	LA	5.4	Jim Umbricht	HOU	14.3	Don Elston	CHI	15.1
Frank Robinson	CIN	.282	Willie Mays	SF	4.2	Ron Perranoski	LA	12.8	Jim Umbricht	HOU	14.1
Defensive Runs			**Players Overall**			**Pitchers Overall**			**Relief Points**		
Bill Mazeroski	PIT	40.7	Frank Robinson	CIN	71.7	Bob Gibson	STL	47.9	Roy Face	PIT	65
Johnny Callison	PHI	25.3	Hank Aaron	MIL	60.3	Ernie Broglio	STL	33.7	Ron Perranoski	LA	46
Dick Groat	PIT	13.7	Willie Mays	SF	59.9	Warren Spahn	MIL	27.3	Jack Baldschun	PHI	43

Club	W	L	R	OR	Avg	OBA	SLG	BPF	NOPS-A	BR	Adj	Wins	ERA	PPF	NERA-A	PR	Adj	Wins	Diff
SF	103	62	878	690	.278	.344	.441	93	122/130	123	171	17.1	3.79	90	104/ 93	24	-39	-4.0	7.5
LA	102	63	842	697	.268	.339	.400	90	108/120	41	116	11.6	3.62	87	109/ 95	53	-30	-3.1	11.1
CIN	98	64	802	685	.270	.333	.417	92	112/121	54	112	11.2	3.75	90	105/ 94	31	-35	-3.6	9.4
PIT	93	68	706	626	.268	.323	.394	105	102/ 94	-14	-48	-4.8	3.38	104	117/121	90	115	11.5	5.9
MIL	86	76	730	665	.252	.328	.403	97	106/109	16	35	3.5	3.68	96	107/103	42	18	1.8	-.3
STL	84	78	774	664	.271	.337	.394	113	105/ 93	22	-76	-7.7	3.55	113	111/125	64	145	14.5	-3.7
PHI	81	80	705	759	.260	.332	.390	95	103/108	4	37	3.7	4.27	96	92/ 88	-52	-78	-7.9	4.7
HOU	64	96	592	717	.246	.312	.351	98	86/ 88	-116	-97	-9.8	3.82	99	103/102	19	13	1.3	-7.5
CHI	59	103	632	827	.253	.319	.377	111	96/ 86	-52	-131	-13.2	4.53	114	87/ 99	-94	-3	-.4	-8.4
NY	40	120	617	948	.240	.320	.361	106	91/ 86	-74	-121	-12.2	5.04	112	78/ 87	-174	-102	-10.3	-17.6
			728		.261	.329	.393						3.94						

AMERICAN LEAGUE 1963

Batting Runs			Park Adjusted			Pitching Runs			Park Adjusted		
Carl Yastrzemski	BOS	42.0	Carl Yastrzemski	BOS	42.0	Gary Peters	CHI	35.0	Gary Peters	CHI	36.6
Bob Allison	MIN	38.6	Bob Allison	MIN	35.7	Camilo Pascual	MIN	32.1	Camilo Pascual	MIN	34.0
Al Kaline	DET	34.3	Al Kaline	DET	35.6	Jim Bouton	NY	30.5	Juan Pizarro	CHI	31.1

Normalized OPS			Park Adjusted			Normalized ERA			Park Adjusted		
Bob Allison	MIN	153	Carl Yastrzemski	BOS	150	Gary Peters	CHI	156	Gary Peters	CHI	158
Carl Yastrzemski	BOS	150	Al Kaline	DET	150	Juan Pizarro	CHI	152	Juan Pizarro	CHI	155
Harmon Killebrew	MIN	150	Bob Allison	MIN	147	Camilo Pascual	MIN	147	Camilo Pascual	MIN	150

On Base Average			Slugging Percentage			Percent of Team Wins			Wins Above Team		
Carl Yastrzemski	BOS	.419	Harmon Killebrew	MIN	.555	Bill Monbouquette	BOS	.263	Bill Monbouquette	BOS	7.2
Albie Pearson	LA	.403	Bob Allison	MIN	.533	Steve Barber	BAL	.233	Dick Radatz	BOS	5.9
Norm Cash	DET	.388	Elston Howard	NY	.528	Camilo Pascual	MIN	.231	Camilo Pascual	MIN	5.0
						Whitey Ford	NY	.231			

Isolated Power			Base Stealing Runs			Relievers - Runs			Park Adjusted		
Harmon Killebrew	MIN	.297	Luis Aparicio	BAL	8.4	Dick Radatz	BOS	24.3	Dick Radatz	BOS	24.6
Bob Allison	MIN	.262	Jose Tartabull	KC	4.2	Bill Dailey	MIN	20.0	Bill Dailey	MIN	20.8
Jimmie Hall	MIN	.262	Al Weis	CHI	3.9	Stu Miller	BAL	17.2	Hoyt Wilhelm	CHI	15.7
			Bobby Richardson	NY	3.9						

Defensive Runs			Players Overall			Pitchers Overall			Relief Points		
Ron Hansen	CHI	27.1	Carl Yastrzemski	BOS	43.8	Gary Peters	CHI	44.1	Dick Radatz	BOS	74
Clete Boyer	NY	18.4	Bob Allison	MIN	30.1	Camilo Pascual	MIN	38.6	Stu Miller	BAL	56
Billy Moran	LA	17.4	Ron Hansen	CHI	28.8	Juan Pizarro	CHI	30.6	Bill Dailey	MIN	51

Club	W	L	R	OR	Avg	OBA	SLG	BPF	NOPS-A	BR	Adj	Wins	ERA	PPF	NERA-A	PR	Adj	Wins	Diff
NY	104	57	714	547	.252	.310	.403	97	110/114	27	48	5.1	3.08	94	118/111	89	52	5.5	12.9
CHI	94	68	683	544	.250	.325	.365	103	102/ 98	4	-19	-2.1	2.97	102	122/124	108	117	12.3	2.8
MIN	91	70	767	602	.255	.326	.430	104	123/118	116	88	9.3	3.28	102	111/113	57	68	7.1	-5.9
BAL	86	76	644	621	.249	.312	.380	92	103/112	-6	46	4.8	3.45	97	105/ 95	28	-24	-2.7	2.8
DET	79	83	700	703	.252	.329	.382	98	108/110	43	56	5.9	3.90	98	93/ 91	-43	-55	-5.9	-2.1
CLE	79	83	635	702	.239	.304	.381	99	101/102	-25	-19	-2.1	3.80	100	96/ 96	-26	-26	-2.8	2.9
BOS	76	85	666	704	.252	.313	.400	100	110/110	31	31	3.3	3.97	101	91/ 92	-54	-49	-5.3	-2.5
KC	73	89	615	704	.247	.316	.353	112	96/ 85	-40	-119	-12.6	3.92	114	93/106	-46	36	3.7	.8
LA	70	91	597	660	.250	.312	.354	95	95/ 99	-52	-20	-2.2	3.53	96	103/ 99	17	-7	-.8	-7.4
WAS	56	106	578	812	.227	.295	.351	100	89/ 89	-94	-95	-10.1	4.42	104	82/ 85	-126	-103	-11.0	-3.9
				660	.247	.314	.380						3.63						

NATIONAL LEAGUE 1963

Batting Runs			Park Adjusted			Pitching Runs			Park Adjusted		
Hank Aaron	MIL	63.0	Willie Mays	SF	59.7	Sandy Koufax	LA	48.6	Dick Ellsworth	CHI	47.8
Willie Mays	SF	55.8	Hank Aaron	MIL	55.8	Dick Ellsworth	CHI	38.3	Sandy Koufax	LA	43.5
Orlando Cepeda	SF	45.4	Orlando Cepeda	SF	49.1	Juan Marichal	SF	31.3	Bob Friend	PIT	31.4

Normalized OPS			Park Adjusted			Normalized ERA			Park Adjusted		
Hank Aaron	MIL	178	Willie Mays	SF	185	Sandy Koufax	LA	175	Dick Ellsworth	CHI	170
Willie Mays	SF	174	Orlando Cepeda	SF	174	Dick Ellsworth	CHI	156	Sandy Koufax	LA	167
Orlando Cepeda	SF	164	Hank Aaron	MIL	170	Bob Friend	PIT	140	Bob Friend	PIT	145

On Base Average			Slugging Percentage			Percent of Team Wins			Wins Above Team		
Eddie Mathews	MIL	.400	Hank Aaron	MIL	.586	Juan Marichal	SF	.284	Warren Spahn	MIL	9.1
Hank Aaron	MIL	.394	Willie Mays	SF	.582	Warren Spahn	MIL	.274	Juan Marichal	SF	8.9
Willie Mays	SF	.384	Willie McCovey	SF	.566	Dick Ellsworth	CHI	.268	Jim Maloney	CIN	8.7

Isolated Power			Base Stealing Runs			Relievers - Runs			Park Adjusted		
Willie McCovey	SF	.285	Hank Aaron	MIL	6.3	Ron Perranoski	LA	23.1	Ron Perranoski	LA	21.0
Willie Mays	SF	.268	Vada Pinson	CIN	3.3	Bob Veale	PIT	19.5	Bob Veale	PIT	20.4
Hank Aaron	MIL	.268	Tommy Harper	CIN	3.0	Johnny Klippstein	PHI	16.9	Johnny Klippstein	PHI	17.3

Defensive Runs			Players Overall			Pitchers Overall			Relief Points		
Bill Mazeroski	PIT	46.5	Willie Mays	SF	55.9	Dick Ellsworth	CHI	46.8	Ron Perranoski	LA	71
Ken Hubbs	CHI	23.4	Hank Aaron	MIL	55.7	Larry Jackson	CHI	36.0	Lindy McDaniel	CHI	63
Dick Schofield	PIT	17.7	Eddie Mathews	MIL	44.5	Sandy Koufax	LA	35.9	Jack Baldschun	PHI	47

Club	W	L	R	OR	Avg	OBA	SLG	BPF	NOPS-A	BR	Adj	Wins	ERA	PPF	NERA-A	PR	Adj	Wins	Diff
LA	99	63	640	550	.251	.311	.357	97	102/105	-2	14	1.6	2.86	95	115/110	71	47	5.1	11.4
STL	93	69	747	628	.271	.328	.403	102	121/119	119	106	11.5	3.32	100	99/ 99	-4	-3	-.4	.9
SF	88	74	725	641	.258	.318	.414	94	122/129	111	147	15.9	3.35	93	98/ 91	-9	-49	-5.5	-3.5
PHI	87	75	642	578	.252	.308	.381	102	108/106	28	16	1.7	3.09	101	106/108	32	38	4.1	.2
CIN	86	76	648	594	.246	.312	.371	105	107/102	25	-2	-.3	3.30	104	100/104	-1	20	2.2	3.1
MIL	84	78	677	603	.244	.314	.370	105	107/102	30	1	.1	3.26	104	101/104	4	24	2.6	.3
CHI	82	80	570	578	.238	.300	.363	108	100/ 93	-20	-70	-7.7	3.07	109	107/117	35	82	8.9	-.2
PIT	74	88	567	595	.250	.310	.359	103	102/100	-1	-17	-1.9	3.10	103	106/110	31	48	5.2	-10.3
HOU	66	96	464	640	.220	.284	.301	89	76/ 85	-158	-93	-10.2	3.44	91	96/ 87	-23	-71	-7.8	3.0
NY	51	111	501	774	.219	.286	.315	98	81/ 83	-128	-116	-12.7	4.12	102	80/ 82	-131	-118	-12.9	-4.4
				618	.245	.307	.364						3.29						

AMERICAN LEAGUE 1964

Batting Runs			Park Adjusted			Pitching Runs			Park Adjusted		
Mickey Mantle	NY	53.0	Mickey Mantle	NY	49.9	Dean Chance	LA	61.0	Dean Chance	LA	48.4
Bob Allison	MIN	45.0	Bob Allison	MIN	45.1	Joe Horlen	CHI	41.0	Whitey Ford	NY	43.6
Boog Powell	BAL	44.1	Harmon Killebrew	MIN	43.4	Whitey Ford	NY	40.7	Joe Horlen	CHI	32.3

Normalized OPS			Park Adjusted			Normalized ERA			Park Adjusted		
Mickey Mantle	NY	179	Mickey Mantle	NY	171	Dean Chance	LA	220	Dean Chance	LA	195
Boog Powell	BAL	175	Bob Allison	MIN	164	Joe Horlen	CHI	193	Whitey Ford	NY	175
Bob Allison	MIN	164	Boog Powell	BAL	164	Whitey Ford	NY	170	Joe Horlen	CHI	174

On Base Average			Slugging Percentage			Percent of Team Wins			Wins Above Team		
Mickey Mantle	NY	.426	Boog Powell	BAL	.606	Dean Chance	LA	.244	Dean Chance	LA	6.5
Bob Allison	MIN	.406	Mickey Mantle	NY	.591	Claude Osteen	WAS	.242	Dick Radatz	BOS	5.8
Boog Powell	BAL	.400	Tony Oliva	MIN	.557	Dave Wickersham	DET	.224	Wally Bunker	BAL	5.4

Isolated Power			Base Stealing Runs			Relievers - Runs			Park Adjusted		
Boog Powell	BAL	.316	Luis Aparicio	BAL	6.9	Bob Lee	LA	32.2	Bob Lee	LA	26.0
Mickey Mantle	NY	.288	Tom Tresh	NY	3.9	Hoyt Wilhelm	CHI	23.8	Dick Radatz	BOS	21.1
Harmon Killebrew	MIN	.277	Leon Wagner	CLE	3.0	Dick Radatz	BOS	23.3	Dick Hall	BAL	19.5

Defensive Runs			Players Overall			Pitchers Overall			Relief Points		
Bobby Knoop	LA	36.5	Tony Oliva	MIN	42.9	Whitey Ford	NY	46.0	Dick Radatz	BOS	81
Clete Boyer	NY	17.8	Jim Fregosi	LA	41.1	Dean Chance	LA	41.7	Hoyt Wilhelm	CHI	69
Carl Yastrzemski	BOS	15.4	Bob Allison	MIN	40.6	Joe Horlen	CHI	34.9	Stu Miller	BAL	53

Club	W	L	R	OR	Avg	OBA	SLG	BPF	NOPS-A	BR	Adj	Wins	ERA	PPF	NERA-A	PR	Adj	Wins	Diff
NY	99	63	730	577	.253	.319	.387	105	106/101	13	-20	-2.2	3.16	103	115/118	79	96	10.1	10.0
CHI	98	64	642	501	.247	.323	.353	93	96/104	-31	17	1.8	2.72	90	133/120	147	87	9.1	6.1
BAL	97	65	679	567	.248	.319	.387	107	106/ 99	14	-31	-3.4	3.16	106	115/121	76	109	11.5	7.9
DET	85	77	699	678	.253	.321	.395	96	109/114	35	63	6.6	3.84	95	94/ 90	-34	-61	-6.6	4.0
LA	82	80	544	551	.242	.306	.344	89	89/ 99	-88	-21	-2.3	2.91	89	124/110	115	49	5.1	-1.8
MIN	79	83	737	678	.252	.324	.427	100	120/120	98	100	10.5	3.58	99	101/100	8	1	.1	-12.6
CLE	79	83	689	693	.247	.315	.380	102	103/100	-7	-21	-2.3	3.75	102	97/ 99	-19	-6	-.7	1.0
BOS	72	90	688	793	.258	.324	.416	95	116/122	76	108	11.3	4.50	97	81/ 78	-137	-156	-16.6	-3.8
WAS	62	100	578	733	.231	.301	.348	103	89/ 86	-94	-114	-12.1	3.98	106	91/ 96	-56	-22	-2.4	-4.4
KC	57	105	621	836	.239	.313	.379	107	102/ 95	-13	-60	-6.4	4.70	111	77/ 85	-173	-109	-11.6	-6.0
			661		.247	.316	.382						3.63						

NATIONAL LEAGUE 1964

Batting Runs			Park Adjusted			Pitching Runs			Park Adjusted		
Willie Mays	SF	56.5	Willie Mays	SF	54.1	Don Drysdale	LA	48.1	Don Drysdale	LA	41.0
Ron Santo	CHI	55.1	Dick Allen	PHI	53.4	Sandy Koufax	LA	44.6	Sandy Koufax	LA	39.6
Dick Allen	PHI	50.5	Ron Santo	CHI	49.3	Chris Short	PHI	32.8	Bob Gibson	STL	34.9

Normalized OPS			Park Adjusted			Normalized ERA			Park Adjusted		
Willie Mays	SF	174	Rico Carty	MIL	171	Sandy Koufax	LA	204	Sandy Koufax	LA	192
Ron Santo	CHI	169	Willie Mays	SF	169	Don Drysdale	LA	162	Chris Short	PHI	153
Frank Robinson	CIN	164	Dick Allen	PHI	167	Chris Short	PHI	161	Don Drysdale	LA	153

On Base Average			Slugging Percentage			Percent of Team Wins			Wins Above Team		
Ron Santo	CHI	.401	Willie Mays	SF	.607	Larry Jackson	CHI	.316	Larry Jackson	CHI	9.7
Frank Robinson	CIN	.399	Ron Santo	CHI	.564	Sandy Koufax	LA	.238	Sandy Koufax	LA	8.4
Hank Aaron	MIL	.394	Dick Allen	PHI	.557	Juan Marichal	SF	.233	Bob Bruce	HOU	6.1

Isolated Power			Base Stealing Runs			Relievers - Runs			Park Adjusted		
Willie Mays	SF	.311	Maury Wills	LA	5.7	Al McBean	PIT	16.4	Bill Henry	CIN	16.2
Ron Santo	CHI	.252	Tommy Harper	CIN	5.4	Bill Henry	CIN	15.4	Al McBean	PIT	16.1
Frank Robinson	CIN	.241	Willie Davis	LA	4.8	Bob Miller	LA	14.2	Sammy Ellis	CIN	14.8

Defensive Runs			Players Overall			Pitchers Overall			Relief Points		
Bill Mazeroski	PIT	33.4	Willie Mays	SF	61.0	Don Drysdale	LA	46.5	Al McBean	PIT	57
Willie Davis	LA	19.4	Hank Aaron	MIL	56.2	Sandy Koufax	LA	34.5	Jack Baldschun	PHI	45
Johnny Callison	PHI	18.8	Ron Santo	CHI	54.0	Juan Marichal	SF	34.1	Hal Woodeshick	HOU	41

Club	W	L	R	OR	Avg	OBA	SLG	BPF	NOPS-A	BR	Adj	Wins	ERA	PPF	NERA-A	PR	Adj	Wins	Diff
STL	93	69	715	652	.272	.326	.392	116	112/ 97	62	-43	-4.7	3.43	116	103/120	16	108	11.4	5.3
PHI	92	70	693	632	.258	.317	.391	96	110/114	41	65	6.9	3.37	95	105/100	27	-0	-.1	4.2
CIN	92	70	660	566	.249	.310	.372	105	102/ 97	-9	-43	-4.6	3.07	104	115/120	75	98	10.3	5.3
SF	90	72	656	587	.246	.313	.382	103	106/102	16	-5	-.7	3.19	102	111/114	57	71	7.5	2.2
MIL	88	74	803	744	.272	.335	.418	96	123/130	132	166	17.5	4.12	94	86/ 81	-91	-126	-13.4	2.9
PIT	80	82	663	636	.264	.317	.389	100	109/110	35	38	4.0	3.52	99	100/100	2	-1	-.3	-4.7
LA	80	82	614	572	.250	.308	.340	95	91/ 96	-69	-39	-4.2	2.95	94	120/113	96	63	6.6	-3.4
CHI	76	86	649	724	.251	.316	.390	108	109/101	37	-13	-1.5	4.08	109	87/ 95	-86	-32	-3.5	.0
HOU	66	96	495	628	.229	.287	.315	94	77/ 82	-157	-120	-12.8	3.41	95	104/ 99	20	-5	-.6	-1.8
NY	53	109	569	776	.246	.297	.348	91	91/ 99	-83	-27	-3.0	4.25	94	83/ 78	-113	-147	-15.6	-9.4
			652		.254	.313	.374						3.54						

Batting Runs
Carl Yastrzemski	BOS	41.6
Tony Oliva	MIN	35.3
Rocky Colavito	CLE	34.6

Park Adjusted
Rocky Colavito	CLE	38.1
Carl Yastrzemski	BOS	37.5
Leon Wagner	CLE	32.6

Pitching Runs
Sam McDowell	CLE	38.9
Mel Stottlemyre	NY	26.9
Sonny Siebert	CLE	21.7

Park Adjusted
Sam McDowell	CLE	32.4
Mel Stottlemyre	NY	27.6
Denny McLain	DET	23.7

Normalized OPS
Carl Yastrzemski	BOS	163
Norm Cash	DET	149
Tony Oliva	MIN	147

Park Adjusted
Carl Yastrzemski	BOS	153
Leon Wagner	CLE	151
Rocky Colavito	CLE	149

Normalized ERA
Sam McDowell	CLE	159
Sonny Siebert	CLE	142
George Brunet	CAL	135

Park Adjusted
Sam McDowell	CLE	149
Pete Richert	WAS	137
Denny McLain	DET	137

On Base Average
Carl Yastrzemski	BOS	.398
Rocky Colavito	CLE	.387
Tony Oliva	MIN	.384

Slugging Percentage
Carl Yastrzemski	BOS	.536
Fred Whitfield	CLE	.513
Tony Conigliaro	BOS	.512

Percent of Team Wins
Mel Stottlemyre	NY	.260
Pete Richert	WAS	.214
Earl Wilson	BOS	.210

Wins Above Team
Mel Stottlemyre	NY	7.6
Denny McLain	DET	4.5
Jim Grant	MIN	4.1

Isolated Power
Norm Cash	DET	.246
Tony Conigliaro	BOS	.244
Carl Yastrzemski	BOS	.225

Base Stealing Runs
Zoilo Versalles	MIN	5.1
Bert Campaneris	KC	3.9
Vic Davalillo	CLE	3.6
Luis Aparicio	BAL	3.6

Relievers - Runs
Hoyt Wilhelm	CHI	26.4
Bob Lee	CAL	22.4
Stu Miller	BAL	20.7

Park Adjusted
Stu Miller	BAL	21.6
Hoyt Wilhelm	CHI	21.2
Bob Lee	CAL	19.6

Defensive Runs
Clete Boyer	NY	27.4
Bobby Knoop	CAL	18.6
Ron Hansen	CHI	18.3

Players Overall
Don Buford	CHI	42.8
Carl Yastrzemski	BOS	34.2
Tony Oliva	MIN	32.9

Pitchers Overall
Mel Stottlemyre	NY	33.8
Sam McDowell	CLE	31.4
Jim Kaat	MIN	30.1

Relief Points
Eddie Fisher	CHI	71
Stu Miller	BAL	69
Ron Kline	WAS	66

Club	W	L	R	OR	Avg	OBA	SLG	BPF	NOPS-A	BR	Adj	Wins	ERA	PPF	NERA-A	PR	Adj	Wins	Diff
MIN	102	60	774	600	.254	.327	.399	105	116/110	84	53	5.6	3.14	102	110/113	52	65	6.9	8.5
CHI	95	67	647	555	.246	.317	.364	93	102/110	1	49	5.2	2.99	91	116/105	77	23	2.5	6.3
BAL	94	68	641	578	.238	.309	.363	103	100/ 97	-20	-38	-4.2	2.98	102	116/118	78	89	9.5	7.7
DET	89	73	680	602	.238	.314	.374	105	104/100	10	-18	-2.0	3.35	104	103/107	17	38	4.0	6.0
CLE	87	75	663	613	.250	.317	.379	95	107/113	26	58	6.2	3.30	94	105/ 98	26	-7	-.9	.7
NY	77	85	611	604	.235	.300	.364	101	98/ 97	-41	-46	-5.0	3.29	101	105/106	28	32	3.4	-2.3
CAL	75	87	527	569	.239	.300	.341	95	90/ 95	-80	-46	-5.0	3.17	95	109/103	46	16	1.7	-2.7
WAS	70	92	591	721	.228	.306	.350	101	95/ 94	-46	-52	-5.7	3.93	103	88/ 91	-74	-57	-6.2	.9
BOS	62	100	669	791	.251	.329	.400	107	117/110	95	51	5.4	4.24	109	82/ 89	-124	-72	-7.8	-16.6
KC	59	103	585	755	.240	.311	.358	96	98/103	-24	2	.3	4.24	98	82/ 80	-123	-133	-14.3	-8.0
			639		.242	.313	.369						3.46						

Batting Runs
Willie Mays	SF	64.7
Billy Williams	CHI	49.3
Hank Aaron	MIL	45.9

Park Adjusted
Willie Mays	SF	56.1
Billy Williams	CHI	47.0
Frank Robinson	CIN	46.0

Pitching Runs
Sandy Koufax	LA	56.1
Juan Marichal	SF	46.0
Vern Law	PIT	33.3

Park Adjusted
Juan Marichal	SF	59.9
Sandy Koufax	LA	41.5
Bob Shaw	SF	34.5

Normalized OPS
Willie Mays	SF	189
Hank Aaron	MIL	162
Billy Williams	CHI	159

Park Adjusted
Willie Mays	SF	168
Frank Robinson	CIN	158
Billy Williams	CHI	154

Normalized ERA
Sandy Koufax	LA	174
Juan Marichal	SF	166
Vern Law	PIT	164

Park Adjusted
Juan Marichal	SF	186
Vern Law	PIT	162
Sandy Koufax	LA	155

On Base Average
Willie Mays	SF	.399
Frank Robinson	CIN	.388
Hank Aaron	MIL	.384

Slugging Percentage
Willie Mays	SF	.645
Hank Aaron	MIL	.560
Billy Williams	CHI	.552

Percent of Team Wins
Tony Cloninger	MIL	.279
Sandy Koufax	LA	.268
Bob Gibson	STL	.250

Wins Above Team
Sandy Koufax	LA	7.1
Tony Cloninger	MIL	6.9
Sammy Ellis	CIN	5.5

Isolated Power
Willie Mays	SF	.328
Willie McCovey	SF	.263
Mack Jones	MIL	.248

Base Stealing Runs
Jim Wynn	HOU	10.5
Maury Wills	LA	9.6
Tommy Harper	CIN	6.9

Relievers - Runs
Frank Linzy	SF	19.2
Billy O'Dell	MIL	16.7
Al McBean	PIT	15.8

Park Adjusted
Frank Linzy	SF	23.1
Billy O'Dell	MIL	18.7
Ted Abernathy	CHI	17.0

Defensive Runs
Bill Mazeroski	PIT	30.5
Maury Wills	LA	21.3
Gene Alley	PIT	20.1

Players Overall
Willie Mays	SF	58.7
Jim Wynn	HOU	50.3
Ron Santo	CHI	48.0

Pitchers Overall
Juan Marichal	SF	59.5
Sandy Koufax	LA	44.0
Vern Law	PIT	36.8

Relief Points
Ted Abernathy	CHI	64
Frank Linzy	SF	57
Billy McCool	CIN	52

Club	W	L	R	OR	Avg	OBA	SLG	BPF	NOPS-A	BR	Adj	Wins	ERA	PPF	NERA-A	PR	Adj	Wins	Diff
LA	97	65	608	521	.245	.314	.335	91	91/100	-60	-3	-.4	2.81	89	126/112	119	55	5.8	10.6
SF	95	67	682	593	.252	.315	.385	113	107/ 95	24	-56	-6.0	3.20	112	111/124	55	124	13.1	6.9
PIT	90	72	675	580	.265	.319	.382	100	107/107	27	26	2.8	3.01	98	117/116	86	78	8.2	-2.0
CIN	89	73	825	704	.273	.341	.439	100	132/132	188	189	20.0	3.89	98	91/ 89	-55	-66	-7.1	-4.9
MIL	86	76	708	633	.256	.311	.416	105	116/110	67	31	3.3	3.52	105	101/105	4	31	3.2	-1.5
PHI	85	76	654	667	.250	.315	.384	94	107/114	24	65	6.9	3.53	93	100/ 94	2	-35	-3.8	1.5
STL	80	81	707	674	.254	.316	.371	107	103/ 96	3	-43	-4.7	3.77	107	94/100	-37	3	.3	3.8
CHI	72	90	635	723	.238	.309	.358	103	97/ 94	-33	-53	-5.7	3.78	105	94/ 98	-38	-11	-1.3	-2.0
HOU	65	97	569	711	.237	.306	.340	90	90/100	-72	-7	-.8	3.84	91	92/ 84	-48	-97	-10.4	-4.8
NY	50	112	495	752	.221	.278	.327	100	79/ 79	-163	-161	-17.1	4.06	104	87/ 90	-83	-63	-6.7	-7.1
			656		.249	.313	.374						3.54						

AMERICAN LEAGUE 1966

Batting Runs			Park Adjusted			Pitching Runs			Park Adjusted		
Frank Robinson	BAL	73.6	Frank Robinson	BAL	72.0	Gary Peters	CHI	33.3	Jim Kaat	MIN	33.4
Harmon Killebrew	MIN	49.9	Harmon Killebrew	MIN	43.2	Joe Horlen	CHI	23.6	Gary Peters	CHI	25.5
Al Kaline	DET	42.1	Al Kaline	DET	37.6	Jim Kaat	MIN	23.4	Jim Perry	MIN	24.3
Normalized OPS			**Park Adjusted**			**Normalized ERA**			**Park Adjusted**		
Frank Robinson	BAL	197	Frank Robinson	BAL	193	Gary Peters	CHI	174	Gary Peters	CHI	157
Al Kaline	DET	165	Boog Powell	BAL	153	Joe Horlen	CHI	141	Jim Perry	MIN	147
Harmon Killebrew	MIN	165	Al Kaline	DET	153	Steve Hargan	CLE	138	Steve Hargan	CLE	146
On Base Average			**Slugging Percentage**			**Percent of Team Wins**			**Wins Above Team**		
Frank Robinson	BAL	.415	Frank Robinson	BAL	.637	Jim Kaat	MIN	.281	Jim Nash	KC	6.5
Al Kaline	DET	.396	Harmon Killebrew	MIN	.538	Denny McLain	DET	.227	Jim Kaat	MIN	5.4
Harmon Killebrew	MIN	.393	Al Kaline	DET	.534	Sonny Siebert	CLE	.198	Sonny Siebert	CLE	4.7
Isolated Power			**Base Stealing Runs**			**Relievers - Runs**			**Park Adjusted**		
Frank Robinson	BAL	.321	Bert Campaneris	KC	9.6	Jack Aker	KC	18.1	Jack Aker	KC	15.0
Harmon Killebrew	MIN	.257	Chico Salmon	CLE	2.4	Hoyt Wilhelm	CHI	15.9	Hoyt Wilhelm	CHI	12.9
Al Kaline	DET	.246	Tommie Agee	CHI	2.4	Stu Miller	BAL	12.1	Al Worthington	MIN	12.7
Defensive Runs			**Players Overall**			**Pitchers Overall**			**Relief Points**		
Bobby Knoop	CAL	23.3	Frank Robinson	BAL	61.3	Jim Kaat	MIN	37.1	Jack Aker	KC	76
Jim Fregosi	CAL	19.0	Al Kaline	DET	39.9	Gary Peters	CHI	32.7	Ron Kline	WAS	54
Clete Boyer	NY	18.6	Jim Fregosi	CAL	33.6	Jim Perry	MIN	27.0	Larry Sherry	DET	51

Club	W	L	R	OR	Avg	OBA	SLG	BPF	NOPS-A	BR	Adj	Wins	ERA	PPF	NERA-A	PR	Adj	Wins	Diff
BAL	97	63	755	601	.258	.325	.409	102	121/119	114	99	10.6	3.32	100	104/104	19	19	2.0	4.3
MIN	89	73	663	581	.249	.319	.382	109	111/101	49	-8	-1.0	3.14	109	110/119	-48	95	10.2	-1.2
DET	88	74	719	698	.251	.323	.406	108	119/111	102	53	5.7	3.84	108	89/ 96	-65	-22	-2.5	3.8
CHI	83	79	574	517	.231	.299	.331	92	89/ 97	-82	-31	-3.4	2.68	90	128/115	123	67	7.2	-1.8
CLE	81	81	574	586	.237	.299	.360	105	98/ 93	-37	-68	-7.4	3.23	105	106/112	34	64	6.9	.5
CAL	80	82	604	643	.232	.305	.354	100	98/ 98	-28	-27	-3.0	3.56	101	96/ 97	-19	-17	-1.9	3.9
KC	74	86	564	648	.236	.295	.337	92	90/ 97	-83	-33	-3.7	3.55	93	97/ 90	-18	-58	-6.3	4.0
WAS	71	88	557	659	.234	.296	.355	93	96/103	-49	-6	-.8	3.71	94	93/ 87	-42	-74	-8.1	.3
BOS	72	90	655	731	.240	.312	.376	108	107/ 99	24	-24	-2.7	3.92	110	88/ 96	-77	-24	-2.7	-3.6
NY	70	89	611	612	.235	.302	.374	93	103/112	-4	40	4.3	3.42	92	101/ 93	3	-38	-4.2	-9.6
				628	.240	.308	.369						3.44						

NATIONAL LEAGUE 1966

Batting Runs			Park Adjusted			Pitching Runs			Park Adjusted		
Dick Allen	PHI	56.8	Dick Allen	PHI	55.8	Sandy Koufax	LA	67.4	Sandy Koufax	LA	59.7
Ron Santo	CHI	51.2	Ron Santo	CHI	51.6	Juan Marichal	SF	47.0	Juan Marichal	SF	48.1
Willie McCovey	SF	46.9	Willie McCovey	SF	45.9	Jim Bunning	PHI	41.8	Jim Bunning	PHI	42.7
Normalized OPS			**Park Adjusted**			**Normalized ERA**			**Park Adjusted**		
Dick Allen	PHI	181	Dick Allen	PHI	178	Sandy Koufax	LA	209	Sandy Koufax	LA	196
Willie McCovey	SF	168	Willie McCovey	SF	165	Mike Cuellar	HOU	162	Juan Marichal	SF	163
Willie Stargell	PIT	164	Ron Santo	CHI	164	Juan Marichal	SF	162	Mike Cuellar	HOU	153
On Base Average			**Slugging Percentage**			**Percent of Team Wins**			**Wins Above Team**		
Ron Santo	CHI	.417	Dick Allen	PHI	.632	Sandy Koufax	LA	.284	Juan Marichal	SF	8.8
Joe Morgan	HOU	.412	Willie McCovey	SF	.586	Juan Marichal	SF	.269	Sandy Koufax	LA	7.6
Dick Allen	PHI	.398	Willie Stargell	PIT	.581	Bob Gibson	STL	.253	Phil Regan	LA	5.7
Isolated Power			**Base Stealing Runs**			**Relievers - Runs**			**Park Adjusted**		
Dick Allen	PHI	.315	Lou Brock	STL	11.4	Phil Regan	LA	25.9	Phil Regan	LA	23.1
Willie McCovey	SF	.291	Sonny Jackson	HOU	6.3	Clay Carroll	ATL	19.7	Joe Hoerner	STL	18.4
Willie Mays	SF	.268	Hank Aaron	ATL	4.5	Joe Hoerner	STL	17.5	Clay Carroll	ATL	18.2
Defensive Runs			**Players Overall**			**Pitchers Overall**			**Relief Points**		
Bill Mazeroski	PIT	40.8	Ron Santo	CHI	74.8	Juan Marichal	SF	54.8	Phil Regan	LA	69
Ron Santo	CHI	27.0	Dick Allen	PHI	51.3	Sandy Koufax	LA	50.6	Billy McCool	CIN	44
Dick Groat	PHI	18.2	Hank Aaron	ATL	49.7	Bob Gibson	STL	41.2	Roy Face	PIT	42

Club	W	L	R	OR	Avg	OBA	SLG	BPF	NOPS-A	BR	Adj	Wins	ERA	PPF	NERA-A	PR	Adj	Wins	Diff
LA	95	67	606	490	.256	.316	.362	96	97/101	-36	-12	-1.3	2.62	94	138/129	160	125	13.1	2.3
SF	93	68	675	626	.248	.304	.392	102	104/102	-13	-24	-2.6	3.24	101	111/112	61	66	6.9	8.2
PIT	92	70	759	641	.279	.331	.428	101	122/121	117	109	11.4	3.52	99	102/102	14	11	1.1	-1.5
PHI	87	75	696	640	.258	.323	.378	102	104/103	12	2	.2	3.57	101	101/102	5	9	1.0	4.8
ATL	85	77	782	683	.263	.329	.424	99	120/121	106	113	11.9	3.68	97	98/ 95	-12	-27	-2.9	-4.9
STL	83	79	571	577	.251	.300	.368	103	95/ 92	-65	-84	-8.9	3.12	103	116/119	80	98	10.2	.6
CIN	76	84	692	702	.260	.311	.395	106	106/100	7	-34	-3.7	4.08	107	88/ 95	-75	-33	-3.6	3.3
HOU	72	90	612	695	.255	.320	.365	94	99/106	-18	23	2.4	3.76	94	96/ 91	-23	-55	-5.9	-5.5
NY	66	95	587	761	.239	.303	.342	97	87/ 90	-98	-79	-8.4	4.17	100	87/ 86	-88	-90	-9.5	3.5
CHI	59	103	644	809	.254	.315	.380	100	103/103	-5	-1	-.2	4.34	102	83/ 85	-117	-105	-11.1	-10.7
				662	.256	.315	.384						3.61						

AMERICAN LEAGUE 1967

Batting Runs
Carl Yastrzemski	BOS	76.4
Harmon Killebrew	MIN	62.3
Frank Robinson	BAL	53.3

Park Adjusted
Carl Yastrzemski	BOS	64.7
Harmon Killebrew	MIN	60.7
Frank Robinson	BAL	53.5

Pitching Runs
Joe Horlen	CHI	33.6
Gary Peters	CHI	27.3
Jim Merritt	MIN	17.8

Park Adjusted
Joe Horlen	CHI	28.2
Gary Peters	CHI	21.9
Sonny Siebert	CLE	21.3

Normalized OPS
Carl Yastrzemski	BOS	205
Frank Robinson	BAL	188
Harmon Killebrew	MIN	184

Park Adjusted
Al Kaline	DET	189
Frank Robinson	BAL	188
Harmon Killebrew	MIN	180

Normalized ERA
Joe Horlen	CHI	157
Gary Peters	CHI	141
Sonny Siebert	CLE	136

Park Adjusted
Joe Horlen	CHI	148
Sonny Siebert	CLE	144
Lee Stange	BOS	136

On Base Average
Carl Yastrzemski	BOS	.421
Al Kaline	DET	.415
Harmon Killebrew	MIN	.413

Slugging Percentage
Carl Yastrzemski	BOS	.622
Frank Robinson	BAL	.576
Harmon Killebrew	MIN	.558

Percent of Team Wins
Earl Wilson	DET	.242
Jim Lonborg	BOS	.239
Dean Chance	MIN	.220

Wins Above Team
Joe Horlen	CHI	5.6
Jim Lonborg	BOS	5.4
Earl Wilson	DET	4.3

Isolated Power
Carl Yastrzemski	BOS	.295
Harmon Killebrew	MIN	.289
Frank Robinson	BAL	.265

Base Stealing Runs
Bert Campaneris	KC	6.9
Horace Clarke	NY	3.9
Fred Valentine	WAS	3.3

Relievers - Runs
Hoyt Wilhelm	CHI	18.9
Moe Drabowsky	BAL	17.1
Bob Locker	CHI	15.9

Park Adjusted
Hoyt Wilhelm	CHI	17.1
Moe Drabowsky	BAL	16.6
Dave Baldwin	WAS	13.6

Defensive Runs
Brooks Robinson	BAL	29.9
Paul Blair	BAL	16.1
Ken McMullen	WAS	12.0

Players Overall
Carl Yastrzemski	BOS	64.4
Al Kaline	DET	52.9
Brooks Robinson	BAL	49.9

Pitchers Overall
Joe Horlen	CHI	31.9
Gary Peters	CHI	31.0
Steve Hargan	CLE	22.1

Relief Points
Minnie Rojas	CAL	69
John Wyatt	BOS	53
Bob Locker	CHI	49

Club	W	L	R	OR	Avg	OBA	SLG	BPF	NOPS-A	BR	Adj	Wins	ERA	PPF	NERA-A	PR	Adj	Wins	Diff
BOS	92	70	722	614	.255	.323	.395	117	122/104	118	13	1.4	3.36	117	96/112	-20	67	7.4	2.2
MIN	91	71	671	590	.240	.310	.369	102	109/107	44	29	3.2	3.13	101	103/104	16	23	2.5	4.3
DET	91	71	683	587	.243	.327	.376	95	117/122	100	128	14.1	3.32	93	97/91	-13	-47	-5.3	1.1
CHI	89	73	531	491	.225	.293	.320	95	88/93	-79	-51	-5.8	2.45	94	132/124	129	98	10.8	3.0
CAL	84	77	567	587	.238	.302	.349	94	100/107	-11	22	2.4	3.20	94	101/95	5	-25	-2.8	3.9
WAS	76	85	550	637	.223	.289	.326	106	89/85	-80	-113	-12.6	3.38	108	96/103	-24	15	1.7	6.4
BAL	76	85	654	592	.240	.313	.372	100	111/111	54	56	6.2	3.32	99	97/96	-14	-21	-2.4	-8.2
CLE	75	87	559	613	.235	.295	.359	105	101/97	-14	-42	-4.7	3.25	106	99/105	-2	28	3.1	-4.4
NY	72	90	522	621	.225	.298	.317	91	89/98	-73	-16	-1.9	3.24	91	100/91	-1	-47	-5.3	-1.8
KC	62	99	533	660	.233	.297	.330	97	93/96	-53	-36	-4.1	3.68	99	88/87	-70	-75	-8.4	-6.0
				599	.236	.305	.351						3.23						

NATIONAL LEAGUE 1967

Batting Runs
Roberto Clemente	PIT	52.1
Hank Aaron	ATL	50.1
Ron Santo	CHI	47.9

Park Adjusted
Roberto Clemente	PIT	54.4
Hank Aaron	ATL	48.6
Rusty Staub	HOU	43.4

Pitching Runs
Jim Bunning	PHI	36.3
Phil Niekro	ATL	34.7
Gaylord Perry	SF	24.9

Park Adjusted
Jim Bunning	PHI	43.3
Phil Niekro	ATL	36.5
Dick Hughes	STL	27.0

Normalized OPS
Dick Allen	PHI	175
Roberto Clemente	PIT	171
Hank Aaron	ATL	167

Park Adjusted
Roberto Clemente	PIT	177
Rusty Staub	HOU	169
Dick Allen	PHI	164

Normalized ERA
Phil Niekro	ATL	181
Jim Bunning	PHI	147
Chris Short	PHI	141

Park Adjusted
Phil Niekro	ATL	185
Jim Bunning	PHI	156
Chris Short	PHI	150

On Base Average
Dick Allen	PHI	.404
Orlando Cepeda	STL	.403
Rusty Staub	HOU	.402

Slugging Percentage
Hank Aaron	ATL	.573
Dick Allen	PHI	.566
Roberto Clemente	PIT	.554

Percent of Team Wins
Tom Seaver	NY	.262
Mike McCormick	SF	.242
Claude Osteen	LA	.233

Wins Above Team
Tom Seaver	NY	6.2
Mike Cuellar	HOU	5.4
Mike McCormick	SF	5.0

Isolated Power
Hank Aaron	ATL	.267
Dick Allen	PHI	.259
Willie McCovey	SF	.259

Base Stealing Runs
Joe Morgan	HOU	5.7
Lou Brock	STL	4.8
Vada Pinson	CIN	3.0
Dick Allen	PHI	3.0

Relievers - Runs
Ted Abernathy	CIN	24.8
Frank Linzy	SF	20.0
Don Nottebart	CIN	12.6

Park Adjusted
Ted Abernathy	CIN	27.5
Frank Linzy	SF	20.5
Don Nottebart	CIN	14.7

Defensive Runs
Ron Santo	CHI	31.0
Bill Mazeroski	PIT	17.3
Hal Lanier	SF	16.1

Players Overall
Ron Santo	CHI	61.5
Roberto Clemente	PIT	46.4
Hank Aaron	ATL	42.4

Pitchers Overall
Jim Bunning	PHI	42.3
Phil Niekro	ATL	37.0
Gaylord Perry	SF	28.6

Relief Points
Ted Abernathy	CIN	65
Roy Face	PIT	43
Frank Linzy	SF	41

Club	W	L	R	OR	Avg	OBA	SLG	BPF	NOPS-A	BR	Adj	Wins	ERA	PPF	NERA-A	PR	Adj	Wins	Diff
STL	101	60	695	557	.263	.322	.379	113	110/97	52	-30	-3.4	3.05	112	111/124	53	117	12.7	11.2
SF	91	71	652	551	.245	.315	.372	103	106/103	24	5	.6	2.92	101	116/117	75	83	9.0	.4
CHI	87	74	702	624	.251	.319	.378	107	109/102	44	2	.2	3.48	106	97/103	-16	16	1.7	4.6
CIN	87	75	604	563	.239	.299	.372	107	101/95	-19	-62	-6.8	3.05	107	111/118	54	91	9.9	2.9
PHI	82	80	612	581	.242	.314	.357	106	101/95	-0	-39	-4.3	3.11	106	109/115	44	78	8.4	-3.1
PIT	81	81	679	693	.277	.327	.380	97	112/116	63	84	9.1	3.74	97	90/87	-58	-77	-8.4	-.7
ATL	77	85	631	640	.240	.309	.372	102	104/102	10	-2	-.3	3.48	102	97/99	-15	-2	-.3	-3.4
LA	73	89	519	595	.236	.303	.332	87	90/104	-73	9	1.0	3.21	87	105/91	27	-46	-5.1	-3.9
HOU	69	93	626	742	.249	.319	.364	89	104/118	20	91	9.8	4.03	89	84/75	-103	-160	-17.5	-4.3
NY	61	101	498	672	.238	.290	.325	98	84/86	-117	-107	-11.7	3.73	101	90/91	-56	-50	-5.6	-2.8
				622	.249	.312	.363						3.38						

AMERICAN LEAGUE   1968

Batting Runs			Park Adjusted			Pitching Runs			Park Adjusted		
Carl Yastrzemski	BOS	57.5	Carl Yastrzemski	BOS	59.2	Luis Tiant	CLE	39.4	Luis Tiant	CLE	43.9
Frank Howard	WAS	45.1	Frank Howard	WAS	51.7	Denny McLain	DET	38.2	Denny McLain	DET	41.8
Willie Horton	DET	41.1	Ken Harrelson	BOS	41.7	Sam McDowell	CLE	35.1	Sam McDowell	CLE	39.8
Normalized OPS			**Park Adjusted**			**Normalized ERA**			**Park Adjusted**		
Carl Yastrzemski	BOS	179	Frank Howard	WAS	187	Luis Tiant	CLE	186	Luis Tiant	CLE	196
Willie Horton	DET	170	Carl Yastrzemski	BOS	184	Sam McDowell	CLE	165	Sam McDowell	CLE	174
Frank Howard	WAS	167	Ken Harrelson	BOS	168	Dave McNally	BAL	153	Tommy John	CHI	158
On Base Average			**Slugging Percentage**			**Percent of Team Wins**			**Wins Above Team**		
Carl Yastrzemski	BOS	.429	Frank Howard	WAS	.552	Denny McLain	DET	.301	Denny McLain	DET	9.7
Frank Robinson	BAL	.391	Willie Horton	DET	.543	Mel Stottlemyre	NY	.253	Luis Tiant	CLE	6.1
Mickey Mantle	NY	.387	Ken Harrelson	BOS	.518	Luis Tiant	CLE	.244	Mel Stottlemyre	NY	5.1
Isolated Power			**Base Stealing Runs**			**Relievers - Runs**			**Park Adjusted**		
Frank Howard	WAS	.278	Bert Campaneris	OAK	5.4	Wilbur Wood	CHI	19.6	Wilbur Wood	CHI	22.4
Willie Horton	DET	.258	Tommy McCraw	CHI	3.0	Hoyt Wilhelm	CHI	13.1	Hoyt Wilhelm	CHI	14.7
Ken Harrelson	BOS	.243	Joe Foy	BOS	3.0	Vicente Romo	CLE	12.5	Vicente Romo	CLE	13.9
Defensive Runs			**Players Overall**			**Pitchers Overall**			**Relief Points**		
Horace Clarke	NY	30.3	Carl Yastrzemski	BOS	63.5	Denny McLain	DET	42.0	Wilbur Wood	CHI	45
Luis Aparicio	CHI	28.1	Frank Howard	WAS	42.2	Sam McDowell	CLE	38.8	Al Worthington	MIN	39
Brooks Robinson	BAL	16.1	Bill Freehan	DET	37.3	Luis Tiant	CLE	36.7	Sparky Lyle	BOS	33

Club	W	L	R	OR	Avg	OBA	SLG	BPF	NOPS-A	BR	Adj	Wins	ERA	PPF	NERA-A	PR	Adj	Wins	Diff
DET	103	59	671	492	.235	.309	.385	106	120/113	102	67	7.7	2.71	103	110/113	44	60	6.9	7.4
BAL	91	71	579	497	.225	.306	.352	103	108/105	41	26	2.9	2.66	101	112/113	51	58	6.6	.4
CLE	86	75	516	504	.234	.294	.327	105	96/ 91	-33	-62	-7.2	2.66	105	112/118	52	78	9.0	3.8
BOS	86	76	614	611	.236	.316	.352	97	111/114	62	77	8.9	3.33	97	89/ 87	-56	-69	-8.1	4.2
NY	83	79	536	531	.214	.293	.318	102	93/ 91	-45	-57	-6.7	2.79	102	107/109	30	41	4.7	4.0
OAK	82	80	569	544	.240	.306	.343	95	105/111	19	49	5.6	2.94	94	101/ 95	6	-22	-2.7	-1.9
MIN	79	83	562	546	.237	.301	.350	106	106/100	20	-12	-1.5	2.90	106	103/109	13	43	5.0	-5.4
CAL	67	95	498	615	.227	.293	.318	97	93/ 96	-50	-32	-3.8	3.43	99	87/ 86	-72	-78	-9.0	-1.2
CHI	67	95	463	527	.228	.286	.311	104	88/ 85	-81	-102	-11.8	2.75	105	108/114	38	63	7.2	-9.4
WAS	65	96	524	665	.224	.289	.336	89	97/109	-30	28	3.2	3.64	91	82/ 75	-105	-147	-17.0	-1.8
			553		.230	.299	.339						2.98						

NATIONAL LEAGUE   1968

Batting Runs			Park Adjusted			Pitching Runs			Park Adjusted		
Willie McCovey	SF	48.8	Willie McCovey	SF	51.4	Bob Gibson	STL	63.1	Bob Gibson	STL	57.9
Pete Rose	CIN	44.5	Hank Aaron	ATL	43.9	Jerry Koosman	NY	26.5	Jerry Koosman	NY	35.7
Hank Aaron	ATL	39.0	Jim Wynn	HOU	42.6	Bob Veale	PIT	25.2	Tom Seaver	NY	33.8
Normalized OPS			**Park Adjusted**			**Normalized ERA**			**Park Adjusted**		
Willie McCovey	SF	176	Willie McCovey	SF	184	Bob Gibson	STL	266	Bob Gibson	STL	252
Dick Allen	PHI	160	Hank Aaron	ATL	167	Bobby Bolin	SF	150	Jerry Koosman	NY	159
Pete Rose	CIN	158	Jim Wynn	HOU	166	Bob Veale	PIT	145	Tom Seaver	NY	150
On Base Average			**Slugging Percentage**			**Percent of Team Wins**			**Wins Above Team**		
Pete Rose	CIN	.394	Willie McCovey	SF	.545	Juan Marichal	SF	.295	Juan Marichal	SF	8.9
Willie McCovey	SF	.383	Dick Allen	PHI	.520	Jerry Koosman	NY	.260	Steve Blass	PIT	7.2
Jim Wynn	HOU	.378	Billy Williams	CHI	.500	Chris Short	PHI	.250	Jerry Koosman	NY	6.2
Isolated Power			**Base Stealing Runs**			**Relievers - Runs**			**Park Adjusted**		
Dick Allen	PHI	.257	Lou Brock	STL	11.4	Ron Kline	PIT	16.5	Ron Kline	PIT	17.2
Willie McCovey	SF	.252	Hank Aaron	ATL	5.4	Hank Aguirre	LA	9.9	Ted Abernathy	CIN	15.9
Ernie Banks	CHI	.223	Willie Davis	LA	4.8	Frank Linzy	SF	9.5	Cal Koonce	NY	9.5
						Jim Grant	LA	9.5			
Defensive Runs			**Players Overall**			**Pitchers Overall**			**Relief Points**		
Bill Mazeroski	PIT	23.5	Hank Aaron	ATL	54.1	Bob Gibson	STL	59.2	Joe Hoerner	STL	48
Don Kessinger	CHI	19.8	Willie McCovey	SF	45.9	Tom Seaver	NY	35.2	Jim Brewer	LA	41
Jim Wynn	HOU	17.9	Jim Wynn	HOU	44.3	Jerry Koosman	NY	31.0	Ted Abernathy	CIN	39

Club	W	L	R	OR	Avg	OBA	SLG	BPF	NOPS-A	BR	Adj	Wins	ERA	PPF	NERA-A	PR	Adj	Wins	Diff
STL	97	65	583	472	.249	.300	.346	97	103/106	2	19	2.2	2.49	95	120/114	82	56	6.5	7.3
SF	88	74	599	529	.239	.310	.341	95	104/108	20	46	5.2	2.71	94	110/103	45	15	1.8	.0
CHI	84	78	612	611	.242	.300	.366	107	109/102	36	-3	-.5	3.41	108	87/ 94	-68	-29	-3.5	7.0
CIN	83	79	690	673	.273	.322	.389	117	123/105	130	29	3.3	3.56	118	84/ 99	-95	-5	-.7	-.6
ATL	81	81	514	549	.252	.308	.339	92	102/111	9	53	6.1	2.92	92	102/ 94	11	-26	-3.2	-2.9
PIT	80	82	583	532	.252	.309	.343	103	104/101	19	3	.3	2.74	102	109/111	40	50	5.8	-7.1
PHI	76	86	543	615	.233	.297	.333	97	97/100	-23	-6	-.8	3.36	98	89/ 87	-59	-68	-8.0	3.7
LA	76	86	470	509	.230	.291	.319	90	91/101	-60	-14	-.6	2.69	90	111/100	48	-1	-.2	-4.2
NY	73	89	473	499	.228	.283	.315	110	87/ 79	-90	-144	-16.6	2.72	110	110/121	43	95	10.9	-2.3
HOU	72	90	510	588	.231	.300	.317	93	93/100	-40	-1	-.3	3.25	94	92/ 86	-43	-72	-8.4	-.3
			558		.243	.302	.341						2.98						

Batting Runs
			Park Adjusted				Pitching Runs				Park Adjusted		
Harmon Killebrew	MIN	65.6	Harmon Killebrew	MIN	65.9	Mike Cuellar	BAL	40.2	Mike Cuellar	BAL	38.9
Reggie Jackson	OAK	62.0	Reggie Jackson	OAK	65.1	Fritz Peterson	NY	32.6	Denny McLain	DET	34.8
Frank Howard	WAS	56.8	Frank Howard	WAS	61.4	Dick Bosman	WAS	30.8	Andy Messersmith	CAL	28.9

Normalized OPS / Park Adjusted / **Normalized ERA** / Park Adjusted
Reggie Jackson	OAK	182	Reggie Jackson	OAK	190	Dick Bosman	WAS	165	Dick Bosman	WAS	154
Harmon Killebrew	MIN	181	Harmon Killebrew	MIN	182	Jim Palmer	BAL	155	Jim Palmer	BAL	153
Rico Petrocelli	BOS	176	Frank Howard	WAS	182	Mike Cuellar	BAL	152	Mike Cuellar	BAL	151

On Base Average / Slugging Percentage / **Percent of Team Wins** / Wins Above Team
Harmon Killebrew	MIN	.430	Reggie Jackson	OAK	.608	Sam McDowell	CLE	.290	Denny McLain	DET	7.1
Frank Robinson	BAL	.417	Rico Petrocelli	BOS	.589	Denny McLain	DET	.267	Sam McDowell	CLE	7.1
Mike Epstein	WAS	.416	Harmon Killebrew	MIN	.584	Mel Stottlemyre	NY	.250	Diego Segui	SEA	5.5

Isolated Power / Base Stealing Runs / **Relievers - Runs** / Park Adjusted
Reggie Jackson	OAK	.333	Bert Campaneris	OAK	13.8	Ken Tatum	CAL	21.6	Ken Tatum	CAL	21.0
Harmon Killebrew	MIN	.308	Tommy Harper	SEA	11.1	Ron Perranoski	MIN	20.3	Ron Perranoski	MIN	19.1
Rico Petrocelli	BOS	.292	Jose Cardenal	CLE	7.2	Eddie Watt	BAL	15.6	Sparky Lyle	BOS	15.7

Defensive Runs / Players Overall / **Pitchers Overall** / Relief Points
Leo Cardenas	MIN	31.1	Reggie Jackson	OAK	64.8	Mike Cuellar	BAL	35.5	Ron Perranoski	MIN	70
Luis Aparicio	CHI	30.3	Rico Petrocelli	BOS	57.5	Andy Messersmith	CAL	29.7	Ken Tatum	CAL	56
Brooks Robinson	BAL	18.9	Harmon Killebrew	MIN	51.8	Denny McLain	DET	28.7	Sparky Lyle	BOS	47

East	W	L	R	OR	Avg	OBA	SLG	BPF	NOPS-A	BR	Adj	Wins	ERA	PPF	NERA-A	PR	Adj	Wins	Diff
BAL	109	53	779	517	.265	.346	.414	102	122/119	135	120	12.6	2.83	99	128/127	130	124	13.0	2.5
DET	90	72	701	601	.242	.318	.387	105	106/101	17	-13	-1.4	3.32	104	109/113	50	71	7.5	2.9
BOS	87	75	743	736	.251	.335	.415	107	119/111	110	61	6.4	3.94	108	92/ 99	-49	-4	-.5	.1
WAS	86	76	694	644	.251	.332	.378	94	107/114	38	78	8.2	3.49	93	104/ 96	21	-19	-2.1	-1.1
NY	80	81	562	587	.235	.310	.344	91	90/100	-73	-14	-1.5	3.23	90	112/101	64	6	.7	.4
CLE	62	99	573	717	.237	.309	.345	96	90/ 94	-76	-51	-5.4	3.94	98	92/ 90	-49	-62	-6.6	-6.4

West	W	L	R	OR	Avg	OBA	SLG	BPF	NOPS-A	BR	Adj	Wins	ERA	PPF	NERA-A	PR	Adj	Wins	Diff
MIN	97	65	790	618	.268	.342	.408	100	119/120	118	121	12.7	3.25	97	111/109	62	46	4.9	-1.5
OAK	88	74	740	678	.249	.330	.376	96	106/110	32	60	6.3	3.72	95	97/ 92	-15	-46	-4.9	5.6
CAL	71	91	528	652	.230	.302	.319	97	80/ 83	-136	-116	-12.3	3.55	98	102/100	13	2	.2	2.1
KC	69	93	586	688	.240	.311	.338	106	89/ 84	-83	-123	-13.0	3.72	108	98/105	-14	31	3.3	-2.3
CHI	68	94	625	723	.247	.322	.357	107	98/ 91	-24	-73	-7.7	4.20	109	86/ 94	-91	-37	-4.0	-1.3
SEA	64	98	639	799	.234	.317	.346	99	93/ 94	-52	-47	-5.0	4.34	101	83/ 85	-116	-108	-11.4	-.6
				663	.246	.323	.369						3.63						

NATIONAL LEAGUE 1969

Batting Runs / Park Adjusted / **Pitching Runs** / Park Adjusted
Willie McCovey	SF	76.1	Willie McCovey	SF	75.8	Juan Marichal	SF	50.0	Bob Gibson	STL	54.7
Hank Aaron	ATL	56.3	Pete Rose	CIN	58.4	Bob Gibson	STL	49.6	Juan Marichal	SF	49.4
Pete Rose	CIN	56.0	Jim Wynn	HOU	54.8	Bill Singer	LA	44.4	Tom Seaver	NY	42.6

Normalized OPS / Park Adjusted / **Normalized ERA** / Park Adjusted
Willie McCovey	SF	209	Willie McCovey	SF	208	Juan Marichal	SF	171	Steve Carlton	STL	172
Hank Aaron	ATL	178	Jim Wynn	HOU	174	Steve Carlton	STL	166	Bob Gibson	STL	172
Roberto Clemente	PIT	166	Hank Aaron	ATL	173	Bob Gibson	STL	165	Juan Marichal	SF	171

On Base Average / Slugging Percentage / **Percent of Team Wins** / Wins Above Team
Willie McCovey	SF	.458	Willie McCovey	SF	.656	Tom Seaver	NY	.250	Tom Seaver	NY	6.5
Jim Wynn	HOU	.440	Hank Aaron	ATL	.607	Phil Niekro	ATL	.247	Bob Moose	PIT	5.3
Pete Rose	CIN	.432	Dick Allen	PHI	.573	Larry Dierker	HOU	.247	Rick Wise	PHI	5.0

Isolated Power / Base Stealing Runs / **Relievers - Runs** / Park Adjusted
Willie McCovey	SF	.336	Bobby Bonds	SF	11.1	Tug McGraw	NY	15.0	Tug McGraw	NY	15.1
Hank Aaron	ATL	.307	Lou Brock	STL	7.5	Wayne Granger	CIN	13.0	Wayne Granger	CIN	11.0
Dick Allen	PHI	.285	Joe Morgan	HOU	6.3	Jim Brewer	LA	10.2	Cecil Upshaw	ATL	8.9

Defensive Runs / Players Overall / **Pitchers Overall** / Relief Points
Don Kessinger	CHI	25.2	Willie McCovey	SF	57.7	Bob Gibson	STL	59.6	Wayne Granger	CIN	66
Hal Lanier	SF	19.8	Pete Rose	CIN	53.1	Juan Marichal	SF	52.0	Cecil Upshaw	ATL	62
Don Money	PHI	17.0	Jim Wynn	HOU	52.4	Tom Seaver	NY	44.6	Fred Gladding	HOU	58

East	W	L	R	OR	Avg	OBA	SLG	BPF	NOPS-A	BR	Adj	Wins	ERA	PPF	NERA-A	PR	Adj	Wins	Diff
NY	100	62	632	541	.242	.313	.351	102	94/ 93	-51	-60	-6.4	2.99	100	121/121	100	102	10.7	14.7
CHI	92	70	720	611	.253	.326	.384	106	108/102	37	0	-.0	3.34	105	108/113	41	68	7.1	3.9
PIT	88	74	725	652	.277	.336	.398	100	115/115	84	87	9.1	3.61	99	100/ 98	-0	-8	-1.0	-1.1
STL	87	75	595	540	.253	.318	.359	105	98/ 94	-24	-55	-5.9	2.94	104	122/127	106	130	13.7	-1.8
PHI	63	99	645	745	.241	.314	.372	100	101/101	-11	-13	-1.4	4.17	102	86/ 88	-90	-80	-8.5	-8.1
MON	52	110	582	791	.240	.312	.359	104	96/ 92	-39	-66	-7.1	4.33	107	83/ 89	-115	-74	-7.9	-14.0

West	W	L	R	OR	Avg	OBA	SLG	BPF	NOPS-A	BR	Adj	Wins	ERA	PPF	NERA-A	PR	Adj	Wins	Diff
ATL	93	69	691	631	.258	.323	.380	103	106/103	20	3	.3	3.54	102	102/104	10	22	2.3	9.4
SF	90	72	713	636	.242	.336	.361	100	104/103	29	29	3.1	3.26	100	110/110	55	52	5.5	.5
CIN	89	73	798	768	.277	.338	.422	97	123/127	133	153	16.1	4.13	96	87/ 84	-86	-107	-11.3	3.2
LA	85	77	645	561	.254	.316	.359	95	98/103	-30	2	.2	3.09	93	116/109	82	44	4.6	-.8
HOU	81	81	676	668	.240	.332	.352	95	100/106	7	43	4.5	3.60	94	100/ 94	0	-34	-3.7	-.9
SD	52	110	468	746	.225	.286	.329	95	80/ 84	-152	-122	-12.9	4.23	98	85/ 84	-99	-108	-11.4	-4.7
				658	.250	.321	.369						3.60						

AMERICAN LEAGUE 1970

Batting Runs			Park Adjusted			Pitching Runs			Park Adjusted		
Carl Yastrzemski	BOS	71.7	Carl Yastrzemski	BOS	66.6	Jim Palmer	BAL	34.0	Sam McDowell	CLE	42.9
Frank Howard	WAS	54.1	Frank Howard	WAS	55.7	Sam McDowell	CLE	27.0	Tommy John	CHI	28.1
Harmon Killebrew	MIN	49.5	Boog Powell	BAL	52.7	Clyde Wright	CAL	25.8	Ray Culp	BOS	25.5
Normalized OPS			**Park Adjusted**			**Normalized ERA**			**Park Adjusted**		
Carl Yastrzemski	BOS	186	Carl Yastrzemski	BOS	174	Diego Segui	OAK	145	Sam McDowell	CLE	143
Boog Powell	BAL	164	Boog Powell	BAL	173	Jim Palmer	BAL	137	Diego Segui	OAK	135
Frank Howard	WAS	164	Frank Howard	WAS	168	Clyde Wright	CAL	131	Ray Culp	BOS	130
On Base Average			**Slugging Percentage**			**Percent of Team Wins**			**Wins Above Team**		
Carl Yastrzemski	BOS	.453	Carl Yastrzemski	BOS	.592	Sam McDowell	CLE	.263	Sam McDowell	CLE	6.2
Frank Howard	WAS	.420	Boog Powell	BAL	.549	Clyde Wright	CAL	.256	Clyde Wright	CAL	5.0
Boog Powell	BAL	.417	Harmon Killebrew	MIN	.546	Jim Perry	MIN	.245	Steve Hargan	CLE	4.9
Isolated Power			**Base Stealing Runs**			**Relievers - Runs**			**Park Adjusted**		
Harmon Killebrew	MIN	.275	Amos Otis	KC	8.7	Jim Grant	OAK	25.8	Jim Grant	OAK	22.2
Carl Yastrzemski	BOS	.263	Bert Campaneris	OAK	6.6	Darold Knowles	WAS	22.2	Darold Knowles	WAS	21.4
Frank Howard	WAS	.263	Ed Stroud	WAS	3.9	Stan Williams	MIN	21.7	Stan Williams	MIN	20.1
			Alex Johnson	CAL	3.9						
Defensive Runs			**Players Overall**			**Pitchers Overall**			**Relief Points**		
Ed Brinkman	WAS	31.5	Carl Yastrzemski	BOS	58.1	Sam McDowell	CLE	36.6	Ron Perranoski	MIN	74
Graig Nettles	CLE	25.9	Tommy Harper	MIL	49.4	Tommy John	CHI	32.8	Lindy McDaniel	NY	71
Bobby Knoop	CHI	22.7	Tony Oliva	MIN	43.0	Jim Grant	OAK	24.6	Tom Timmermann	DET	59

East	W	L	R	OR	Avg	OBA	SLG	BPF	NOPS-A	BR	Adj	Wins	ERA	PPF	NERA-A	PR	Adj	Wins	Diff
BAL	108	54	792	574	.257	.346	.401	95	115/121	101	136	14.1	3.15	92	118/108	94	44	4.5	8.4
NY	93	69	680	612	.251	.327	.365	91	99/109	-17	45	4.7	3.25	89	115/102	77	11	1.2	6.2
BOS	87	75	786	722	.262	.338	.428	107	121/113	119	73	7.6	3.90	107	95/102	-28	10	1.1	-2.7
DET	79	83	666	731	.238	.325	.374	105	101/96	-3	-35	-3.8	4.09	106	91/96	-59	-23	-2.5	4.3
CLE	76	86	649	675	.249	.316	.394	111	105/94	3	-72	-7.6	3.90	113	95/107	-29	45	4.7	-2.1
WAS	70	92	626	689	.238	.323	.358	98	96/98	-35	-21	-2.2	3.80	98	98/96	-12	-21	-2.2	-6.5

West	W	L	R	OR	Avg	OBA	SLG	BPF	NOPS-A	BR	Adj	Wins	ERA	PPF	NERA-A	PR	Adj	Wins	Diff
MIN	98	64	744	605	.262	.329	.403	99	111/113	50	60	6.2	3.23	97	115/111	78	58	6.1	4.7
OAK	89	73	678	593	.249	.327	.392	95	107/113	28	65	6.3	3.30	93	113/105	66	24	2.5	-1.2
CAL	86	76	631	630	.251	.311	.363	92	94/102	-63	-12	-1.4	3.48	92	107/98	39	-10	-1.1	7.5
MIL	65	97	613	751	.242	.321	.358	98	95/97	-41	-27	-2.9	4.21	100	88/88	-78	-80	-8.4	-4.7
KC	65	97	611	705	.244	.311	.348	98	89/91	-87	-75	-7.9	3.78	99	98/98	-9	-14	-1.5	-6.6
CHI	56	106	633	822	.253	.317	.362	110	95/86	-49	-118	-12.4	4.56	113	82/93	-133	-53	-5.6	-7.0
				676	.250	.324	.379						3.72						

NATIONAL LEAGUE 1970

Batting Runs			Park Adjusted			Pitching Runs			Park Adjusted		
Willie McCovey	SF	62.0	Willie McCovey	SF	66.5	Tom Seaver	NY	40.1	Ferguson Jenkins	CHI	49.5
Rico Carty	ATL	54.4	Rico Carty	ATL	52.4	Gaylord Perry	SF	31.2	Bob Gibson	STL	48.1
Tony Perez	CIN	52.1	Tony Perez	CIN	51.9	Bob Gibson	STL	30.4	Ken Holtzman	CHI	46.1
Normalized OPS			**Park Adjusted**			**Normalized ERA**			**Park Adjusted**		
Willie McCovey	SF	180	Willie McCovey	SF	192	Tom Seaver	NY	144	Bob Gibson	STL	147
Rico Carty	ATL	176	Rico Carty	ATL	171	Wayne Simpson	CIN	134	Tom Seaver	NY	144
Jim Hickman	CHI	166	Tony Perez	CIN	163	Luke Walker	PIT	133	Ken Holtzman	CHI	143
On Base Average			**Slugging Percentage**			**Percent of Team Wins**			**Wins Above Team**		
Rico Carty	ATL	.456	Willie McCovey	SF	.612	Bob Gibson	STL	.303	Bob Gibson	STL	11.0
Willie McCovey	SF	.446	Tony Perez	CIN	.589	Gaylord Perry	SF	.267	Carl Morton	MON	6.0
Dick Dietz	SF	.430	Johnny Bench	CIN	.587	Ferguson Jenkins	CHI	.262	Gaylord Perry	SF	5.0
Isolated Power			**Base Stealing Runs**			**Relievers - Runs**			**Park Adjusted**		
Willie McCovey	SF	.323	Bobby Bonds	SF	8.4	Dick Selma	PHI	19.3	Chuck Taylor	STL	20.3
Johnny Bench	CIN	.294	Lou Brock	STL	6.3	Clay Carroll	CIN	16.8	Dick Selma	PHI	16.9
Dick Allen	STL	.281	Bobby Tolan	CIN	5.1	Don Gullett	CIN	14.1	Clay Carroll	CIN	16.4
Defensive Runs			**Players Overall**			**Pitchers Overall**			**Relief Points**		
Dal Maxvill	STL	29.3	Willie McCovey	SF	62.3	Bob Gibson	STL	58.3	Wayne Granger	CIN	77
Gene Alley	PIT	27.2	Joe Morgan	HOU	53.3	Ken Holtzman	CHI	46.7	Dave Giusti	PIT	67
Doug Rader	HOU	22.2	Bobby Bonds	SF	48.4	Tom Seaver	NY	46.3	Jim Brewer	LA	56

East	W	L	R	OR	Avg	OBA	SLG	BPF	NOPS-A	BR	Adj	Wins	ERA	PPF	NERA-A	PR	Adj	Wins	Diff
PIT	89	73	729	664	.270	.328	.406	96	106/110	12	38	3.8	3.71	95	109/104	56	25	2.5	1.7
CHI	84	78	806	679	.259	.335	.415	119	111/93	49	-88	-8.8	3.76	119	108/128	47	169	16.8	-5.0
NY	83	79	695	630	.249	.336	.370	101	97/96	-22	-28	-2.9	3.46	100	117/117	97	98	9.7	-4.8
STL	76	86	744	747	.263	.333	.379	113	99/88	-19	-115	-11.5	4.06	113	100/113	-0	88	8.7	-2.2
PHI	73	88	594	730	.238	.307	.356	95	86/90	-124	-87	-8.7	4.17	96	97/93	-18	-45	-4.6	5.8
MON	73	89	687	807	.237	.324	.365	97	93/96	-61	-38	-3.9	4.51	98	90/88	-71	-84	-8.5	4.4

West	W	L	R	OR	Avg	OBA	SLG	BPF	NOPS-A	BR	Adj	Wins	ERA	PPF	NERA-A	PR	Adj	Wins	Diff
CIN	102	60	775	681	.270	.339	.436	100	118/118	94	92	9.2	3.71	99	109/108	56	50	5.0	6.8
LA	87	74	749	684	.270	.337	.382	95	101/107	-3	32	3.2	3.82	94	106/100	37	-1	-.2	3.5
SF	86	76	831	826	.262	.353	.409	94	113/121	92	137	13.6	4.50	94	90/84	-71	-113	-11.4	2.7
HOU	79	83	744	763	.259	.334	.391	94	103/110	5	52	5.1	4.23	93	96/89	-28	-72	-7.2	.1
ATL	76	86	736	772	.270	.337	.404	103	108/104	32	10	1.0	4.35	104	93/97	-46	-23	-2.4	-3.6
SD	63	99	681	788	.246	.314	.391	100	98/98	-48	-46	-4.7	4.38	101	93/93	-50	-45	-4.5	-8.8
				731	.258	.332	.392						4.05						

Batting Runs			Park Adjusted			Pitching Runs			Park Adjusted		
Bobby Murcer	NY	53.7	Bobby Murcer	NY	57.8	Wilbur Wood	CHI	57.7	Vida Blue	OAK	64.3
Merv Rettenmund	BAL	34.8	Roy White	NY	37.8	Vida Blue	OAK	57.2	Wilbur Wood	CHI	62.8
Don Buford	BAL	34.5	Norm Cash	DET	36.4	Jim Palmer	BAL	24.7	Bert Blyleven	MIN	27.3
Normalized OPS			**Park Adjusted**			**Normalized ERA**			**Park Adjusted**		
Bobby Murcer	NY	174	Bobby Murcer	NY	185	Vida Blue	OAK	191	Vida Blue	OAK	202
Tony Oliva	MIN	157	Norm Cash	DET	169	Wilbur Wood	CHI	181	Wilbur Wood	CHI	189
Norm Cash	DET	154	Al Kaline	DET	164	Jim Palmer	BAL	129	Bert Blyleven	MIN	131
On Base Average			**Slugging Percentage**			**Percent of Team Wins**			**Wins Above Team**		
Bobby Murcer	NY	.429	Tony Oliva	MIN	.546	Wilbur Wood	CHI	.278	Wilbur Wood	CHI	6.3
Merv Rettenmund	BAL	.424	Bobby Murcer	NY	.543	Mickey Lolich	DET	.275	Andy Messersmith	CAL	5.7
Al Kaline	DET	.421	Norm Cash	DET	.531	Andy Messersmith	CAL	.263	Dave McNally	BAL	5.2
Isolated Power			**Base Stealing Runs**			**Relievers - Runs**			**Park Adjusted**		
Norm Cash	DET	.248	Amos Otis	KC	10.8	Ken Sanders	MIL	23.4	Ken Sanders	MIL	25.5
Reggie Jackson	OAK	.231	Freddie Patek	KC	6.3	Tom Burgmeier	KC	16.9	Tom Burgmeier	KC	16.2
Frank Robinson	BAL	.229	Bert Campaneris	OAK	6.0	Steve Mingori	CLE	13.0	Steve Mingori	CLE	14.7
Defensive Runs			**Players Overall**			**Pitchers Overall**			**Relief Points**		
Graig Nettles	CLE	39.5	Graig Nettles	CLE	47.0	Wilbur Wood	CHI	59.7	Ken Sanders	MIL	64
Bill Melton	CHI	24.2	Bobby Murcer	NY	44.3	Vida Blue	OAK	57.5	Fred Scherman	DET	54
Gene Michael	NY	21.5	Bill Melton	CHI	40.8	Jim Hunter	OAK	29.6	Ted Abernathy	KC	48

East	W	L	R	OR	Avg	OBA	SLG	BPF	NOPS-A	BR	Adj	Wins	ERA	PPF	NERA-A	PR	Adj	Wins	Diff
BAL	101	57	742	530	.261	.349	.398	100	120/120	133	131	14.1	3.00	97	116/112	73	58	6.3	1.6
DET	91	71	701	645	.254	.327	.405	91	116/128	88	144	15.5	3.64	90	95/ 86	-26	-85	-9.2	3.7
BOS	85	77	691	667	.252	.325	.397	102	113/112	68	58	6.3	3.83	101	91/ 92	-57	-49	-5.4	3.1
NY	82	80	648	641	.254	.331	.360	94	104/111	22	62	6.7	3.45	93	101/ 94	3	-34	-3.8	-1.9
WAS	63	96	537	660	.230	.309	.326	89	87/ 98	-85	-16	-1.9	3.70	90	94/ 84	-35	-91	-9.9	-4.8
CLE	60	102	543	747	.238	.302	.342	105	90/ 86	-82	-111	-12.0	4.27	108	81/ 87	-128	-85	-9.3	.3

West	W	L	R	OR	Avg	OBA	SLG	BPF	NOPS-A	BR	Adj	Wins	ERA	PPF	NERA-A	PR	Adj	Wins	Diff
OAK	101	60	691	564	.252	.323	.384	107	109/101	42	-3	-.5	3.06	106	113/120	66	99	10.7	10.3
KC	85	76	603	566	.250	.316	.353	99	97/ 99	-27	-19	-2.1	3.26	98	107/104	33	22	2.4	4.2
CHI	79	83	617	597	.250	.327	.373	104	106/102	33	7	.8	3.13	104	111/115	55	77	8.3	-11.1
CAL	76	86	511	576	.231	.292	.329	104	83/ 80	-130	-154	-16.7	3.10	105	112/117	61	89	9.5	2.1
MIN	74	86	654	670	.260	.326	.372	106	106/100	28	-9	-1.0	3.82	107	91/ 97	-54	-18	-2.0	-2.9
MIL	69	92	534	609	.229	.306	.329	103	87/ 85	-87	-104	-11.3	3.38	104	103/107	14	36	3.9	-4.0
				623	.247	.320	.364						3.47						

Batting Runs			Park Adjusted			Pitching Runs			Park Adjusted		
Hank Aaron	ATL	65.2	Joe Torre	STL	60.5	Tom Seaver	NY	54.3	Tom Seaver	NY	50.0
Joe Torre	STL	62.5	Willie Stargell	PIT	60.3	Dave Roberts	SD	41.1	Dave Roberts	SD	40.9
Willie Stargell	PIT	58.7	Hank Aaron	ATL	59.8	Don Wilson	HOU	30.3	Ferguson Jenkins	CHI	34.5
Normalized OPS			**Park Adjusted**			**Normalized ERA**			**Park Adjusted**		
Hank Aaron	ATL	202	Willie Stargell	PIT	192	Tom Seaver	NY	197	Tom Seaver	NY	189
Willie Stargell	PIT	187	Hank Aaron	ATL	185	Dave Roberts	SD	165	Dave Roberts	SD	165
Joe Torre	STL	175	Joe Torre	STL	171	Don Wilson	HOU	142	Don Wilson	HOU	136
On Base Average			**Slugging Percentage**			**Percent of Team Wins**			**Wins Above Team**		
Willie Mays	SF	.429	Hank Aaron	ATL	.669	Ferguson Jenkins	CHI	.289	Ferguson Jenkins	CHI	6.5
Joe Torre	STL	.424	Willie Stargell	PIT	.628	Rick Wise	PHI	.254	Don Gullett	CIN	6.1
Hank Aaron	ATL	.414	Joe Torre	STL	.555	Clay Kirby	SD	.246	Tom Seaver	NY	5.7
Isolated Power			**Base Stealing Runs**			**Relievers - Runs**			**Park Adjusted**		
Hank Aaron	ATL	.341	Lou Brock	STL	7.8	Tug McGraw	NY	21.8	Tug McGraw	NY	20.1
Willie Stargell	PIT	.333	Joe Morgan	HOU	7.2	Danny Frisella	NY	15.1	Jim Brewer	LA	14.0
Lee May	CIN	.253	Willie Mays	SF	5.1	Jim Brewer	HOU	14.8	Danny Frisella	NY	13.7
Defensive Runs			**Players Overall**			**Pitchers Overall**			**Relief Points**		
Dal Maxvill	STL	18.4	Willie Stargell	PIT	49.6	Tom Seaver	NY	54.8	Dave Giusti	PIT	64
Tommy Helms	CIN	14.0	Hank Aaron	ATL	49.2	Ferguson Jenkins	CHI	48.6	Jerry Johnson	SF	51
Tony Perez	CIN	11.4	Joe Torre	STL	44.3	Dave Roberts	SD	43.2	Jim Brewer	LA	51

East	W	L	R	OR	Avg	OBA	SLG	BPF	NOPS-A	BR	Adj	Wins	ERA	PPF	NERA-A	PR	Adj	Wins	Diff
PIT	97	65	788	599	.274	.333	.416	97	121/125	121	138	14.8	3.31	95	105/ 99	26	-2	-.4	1.6
STL	90	72	739	699	.275	.342	.385	103	114/111	91	74	7.9	3.86	102	90/ 92	-63	-50	-5.5	6.5
NY	83	79	588	550	.249	.321	.351	97	98/101	-19	-0	-.1	3.00	96	116/111	77	55	5.9	-3.8
CHI	83	79	637	648	.258	.327	.378	107	108/101	41	0	-.0	3.61	99	96/103	-22	18	1.9	.1
MON	71	90	622	729	.246	.325	.343	102	96/ 95	-19	-29	-3.2	4.12	103	84/ 87	-102	-85	-9.2	2.9
PHI	67	95	558	688	.233	.300	.350	103	92/ 89	-73	-92	-10.0	3.72	105	93/ 98	-39	-11	-1.3	-2.7

West	W	L	R	OR	Avg	OBA	SLG	BPF	NOPS-A	BR	Adj	Wins	ERA	PPF	NERA-A	PR	Adj	Wins	Diff
SF	90	72	706	644	.247	.331	.378	95	109/114	57	86	9.2	3.33	94	104/ 98	22	-9	-1.1	.8
LA	89	73	663	587	.266	.328	.370	100	106/105	28	26	2.8	3.24	99	107/107	37	34	3.6	1.6
ATL	82	80	643	699	.257	.314	.385	109	107/ 98	18	-39	-4.3	3.75	111	93/102	-44	15	1.6	3.7
HOU	79	83	585	567	.240	.308	.340	96	90/ 93	-80	-58	-6.3	3.13	96	111/106	55	33	3.5	.8
CIN	79	83	586	581	.241	.301	.366	91	97/107	-44	11	1.2	3.35	90	103/ 93	19	-35	-3.9	.7
SD	61	100	486	610	.233	.294	.332	98	85/ 86	-116	-105	-11.4	3.23	100	108/107	39	38	4.0	-12.1
				633	.252	.318	.366						3.47						

AMERICAN LEAGUE 1972

Batting Runs			Park Adjusted			Pitching Runs			Park Adjusted		
Dick Allen	CHI	66.1	Dick Allen	CHI	67.3	Gaylord Perry	CLE	44.1	Gaylord Perry	CLE	53.3
Bobby Murcer	NY	44.9	Bobby Murcer	NY	50.3	Jim Hunter	OAK	33.7	Mickey Lolich	DET	41.9
John Mayberry	KC	43.0	John Mayberry	KC	43.4	Jim Palmer	BAL	30.5	Jim Palmer	BAL	34.5
Normalized OPS			**Park Adjusted**			**Normalized ERA**			**Park Adjusted**		
Dick Allen	CHI	203	Dick Allen	CHI	207	Luis Tiant	BOS	161	Gaylord Perry	CLE	173
Pudge Fisk	BOS	168	Bobby Murcer	NY	181	Gaylord Perry	CLE	160	Luis Tiant	BOS	162
John Mayberry	KC	167	John Mayberry	KC	168	Jim Hunter	OAK	150	Jim Palmer	BAL	155
On Base Average			**Slugging Percentage**			**Percent of Team Wins**			**Wins Above Team**		
Dick Allen	CHI	.422	Dick Allen	CHI	.603	Gaylord Perry	CLE	.333	Gaylord Perry	CLE	7.4
Carlos May	CHI	.408	Pudge Fisk	BOS	.538	Wilbur Wood	CHI	.276	Jim Palmer	BAL	6.1
John Mayberry	KC	.396	Bobby Murcer	NY	.537	Jim Palmer	BAL	.263	Jim Hunter	OAK	5.1
Isolated Power			**Base Stealing Runs**			**Relievers - Runs**			**Park Adjusted**		
Dick Allen	CHI	.294	Bert Campaneris	OAK	7.2	Sparky Lyle	NY	13.9	Darold Knowles	OAK	12.0
Pudge Fisk	BOS	.245	Don Baylor	BAL	6.0	Darold Knowles	OAK	12.5	Jerry Bell	MIL	11.4
Bobby Murcer	NY	.244	Freddie Patek	KC	5.7	Jerry Bell	MIL	11.2	Sparky Lyle	NY	10.2
Defensive Runs			**Players Overall**			**Pitchers Overall**			**Relief Points**		
Freddie Patek	KC	27.5	Dick Allen	CHI	47.7	Gaylord Perry	CLE	56.2	Sparky Lyle	NY	83
Aurelio Rodriguez	DET	24.0	Bobby Murcer	NY	45.2	Jim Palmer	BAL	37.4	Terry Forster	CHI	65
Gene Michael	NY	22.7	Pudge Fisk	BOS	40.6	Mickey Lolich	DET	35.9	Rollie Fingers	OAK	55

East	W	L	R	OR	Avg	OBA	SLG	BPF	NOPS-A	BR	Adj	Wins	ERA	PPF	NERA-A	PR	Adj	Wins	Diff
DET	86	70	558	514	.237	.306	.356	119	105/ 89	18	-80	-9.3	2.96	119	104/124	17	108	12.2	5.0
BOS	85	70	640	620	.248	.320	.376	101	116/115	83	80	9.1	3.51	101	88/ 88	-66	-63	-7.3	5.7
BAL	80	74	519	430	.229	.304	.339	105	99/ 94	-11	-40	-4.7	2.54	104	121/126	81	101	11.5	-3.8
NY	79	76	557	527	.249	.318	.357	91	109/119	45	92	10.5	3.05	90	101/ 91	4	-42	-4.9	-4.1
CLE	72	84	472	519	.234	.295	.330	107	93/ 87	-53	-89	-10.2	2.97	108	103/112	16	54	6.2	-2.0
MIL	65	91	493	595	.235	.303	.328	99	95/ 96	-34	-29	-3.4	3.45	101	89/ 89	-58	-56	-6.4	-3.2

West	W	L	R	OR	Avg	OBA	SLG	BPF	NOPS-A	BR	Adj	Wins	ERA	PPF	NERA-A	PR	Adj	Wins	Diff
OAK	93	62	604	457	.240	.308	.366	100	109/108	37	36	4.1	2.58	98	119/116	77	67	7.6	3.9
CHI	87	67	566	538	.238	.311	.346	98	103/105	14	26	2.9	3.12	97	98/ 96	-6	-20	-2.3	9.4
MIN	77	77	537	539	.244	.311	.344	104	103/ 99	9	-12	-1.4	2.86	104	108/112	33	54	6.1	-4.7
KC	76	78	580	545	.255	.329	.353	99	110/111	65	68	7.8	3.24	99	95/ 94	-25	-31	-3.6	-5.1
CAL	75	80	454	533	.242	.294	.330	87	93/107	-54	13	1.5	3.06	87	100/ 88	1	-57	-6.6	2.7
TEX	54	100	461	628	.217	.292	.290	95	79/ 83	-114	-89	-10.2	3.53	98	87/ 85	-70	-81	-9.3	-3.5
				537	.239	.308	.343						3.07						

NATIONAL LEAGUE 1972

Batting Runs			Park Adjusted			Pitching Runs			Park Adjusted		
Billy Williams	CHI	61.1	Billy Williams	CHI	55.5	Steve Carlton	PHI	56.9	Steve Carlton	PHI	55.5
Johnny Bench	CIN	44.0	Johnny Bench	CIN	45.1	Don Sutton	LA	41.8	Bob Gibson	STL	34.7
Cesar Cedeno	HOU	42.9	Willie Stargell	PIT	38.4	Bob Gibson	STL	30.8	Don Sutton	LA	31.0
Normalized OPS			**Park Adjusted**			**Normalized ERA**			**Park Adjusted**		
Billy Williams	CHI	183	Billy Williams	CHI	169	Steve Carlton	PHI	175	Steve Carlton	PHI	173
Willie Stargell	PIT	162	Johnny Bench	CIN	163	Gary Nolan	CIN	173	Gary Nolan	CIN	167
Johnny Bench	CIN	161	Willie Stargell	PIT	159	Don Sutton	LA	166	Don Sutton	LA	149
On Base Average			**Slugging Percentage**			**Percent of Team Wins**			**Wins Above Team**		
Joe Morgan	CIN	.419	Billy Williams	CHI	.606	Steve Carlton	PHI	.458	Steve Carlton	PHI	17.1
Billy Williams	CHI	.403	Willie Stargell	PIT	.558	Bob Gibson	STL	.253	Bob Gibson	STL	5.7
Ron Santo	CHI	.397	Johnny Bench	CIN	.541	Tom Seaver	NY	.253	Dave Roberts	SD	5.5
Isolated Power			**Base Stealing Runs**			**Relievers - Runs**			**Park Adjusted**		
Billy Williams	CHI	.274	Bobby Bonds	SF	9.6	Mike Marshall	MON	21.5	Mike Marshall	MON	24.7
Johnny Bench	CIN	.271	Lou Brock	STL	8.1	Tug McGraw	NY	20.7	Tug McGraw	NY	18.1
Willie Stargell	PIT	.265	Joe Morgan	CIN	7.2	Jim Brewer	LA	19.0	Jim Brewer	LA	15.9
Defensive Runs			**Players Overall**			**Pitchers Overall**			**Relief Points**		
Dave Cash	PIT	26.8	Joe Morgan	CIN	52.1	Steve Carlton	PHI	58.5	Clay Carroll	CIN	82
Tommy Helms	HOU	25.1	Johnny Bench	CIN	49.6	Bob Gibson	STL	42.9	Tug McGraw	NY	64
Don Money	PHI	21.0	Billy Williams	CHI	40.0	Don Sutton	LA	30.2	Mike Marshall	MON	56

East	W	L	R	OR	Avg	OBA	SLG	BPF	NOPS-A	BR	Adj	Wins	ERA	PPF	NERA-A	PR	Adj	Wins	Diff
PIT	96	59	691	512	.274	.327	.397	102	115/113	74	62	6.7	2.81	100	123/122	101	99	10.7	1.2
CHI	85	70	685	567	.257	.332	.387	108	113/104	74	23	2.4	3.22	107	107/115	37	77	8.2	-3.2
NY	83	73	528	538	.225	.309	.332	93	89/ 96	-67	-26	-2.9	3.27	93	106/ 99	29	-5	-.7	8.6
STL	75	81	568	600	.260	.319	.355	103	99/ 96	-13	-32	-3.5	3.42	104	101/105	5	25	2.7	-2.2
MON	70	86	513	609	.234	.304	.325	105	86/ 82	-91	-123	-13.3	3.59	107	96/103	-20	16	1.8	3.6
PHI	59	97	503	635	.236	.304	.344	97	92/ 94	-64	-48	-5.3	3.67	99	94/ 93	-32	-37	-4.1	-9.6

West	W	L	R	OR	Avg	OBA	SLG	BPF	NOPS-A	BR	Adj	Wins	ERA	PPF	NERA-A	PR	Adj	Wins	Diff
CIN	95	59	707	557	.251	.333	.380	98	111/113	67	77	8.2	3.21	96	108/104	39	18	1.9	7.8
HOU	84	69	708	636	.258	.329	.393	110	114/107	75	15	1.6	3.77	109	92/100	-48	1	.2	-5.8
LA	85	70	584	527	.256	.321	.360	91	102/111	1	54	5.8	2.78	90	124/111	105	50	5.3	-3.7
ATL	70	84	628	730	.258	.330	.382	104	111/107	60	38	4.1	4.28	105	81/ 85	-125	-96	-10.4	-.7
SF	69	86	662	649	.244	.311	.384	98	106/108	17	27	2.9	3.70	98	93/ 92	-37	-47	-5.2	-6.2
SD	58	95	488	665	.227	.284	.332	90	83/ 92	-129	-71	-7.7	3.78	92	92/ 84	-49	-92	-10.0	-.8
				605	.248	.317	.365						3.46						

Batting Runs
John Mayberry	KC	41.0
Reggie Jackson	OAK	40.5
Rod Carew	MIN	38.9

Park Adjusted
Reggie Jackson	OAK	48.6
Sal Bando	OAK	44.3
John Mayberry	KC	35.6

Pitching Runs
Bert Blyleven	MIN	47.1
Jim Palmer	BAL	46.7
Nolan Ryan	CAL	34.5

Park Adjusted
Bert Blyleven	MIN	65.2
Jim Palmer	BAL	49.9
Bill Lee	BOS	41.0

Normalized OPS
Reggie Jackson	OAK	156
Reggie Smith	BOS	156
John Mayberry	KC	152

Park Adjusted
Reggie Jackson	OAK	176
Sal Bando	OAK	163
Gene Tenace	OAK	152

Normalized ERA
Jim Palmer	BAL	159
Bert Blyleven	MIN	152
Bill Lee	BOS	139

Park Adjusted
Bert Blyleven	MIN	172
Jim Palmer	BAL	163
Bill Lee	BOS	147

On Base Average
John Mayberry	KC	.420
Rod Carew	MIN	.415
Carl Yastrzemski	BOS	.411

Slugging Percentage
Reggie Jackson	OAK	.531
Reggie Smith	BOS	.515
Sal Bando	OAK	.498

Percent of Team Wins
Wilbur Wood	CHI	.312
Joe Coleman	DET	.271
Jim Colborn	MIL	.270

Wins Above Team
Jim Hunter	OAK	7.0
Jim Colborn	MIL	6.7
Rogelio Moret	BOS	5.2

Isolated Power
Reggie Jackson	OAK	.237
Frank Robinson	CAL	.223
Reggie Smith	BOS	.213

Base Stealing Runs
Tommy Harper	BOS	7.8
Bert Campaneris	OAK	4.2
Don Baylor	BAL	4.2

Relievers - Runs
John Hiller	DET	33.1
Rollie Fingers	OAK	27.0
Bob Reynolds	BAL	23.2

Park Adjusted
John Hiller	DET	33.4
Bob Reynolds	BAL	24.3
Ray Corbin	MIN	21.1

Defensive Runs
Freddie Patek	KC	34.4
Graig Nettles	NY	25.9
Buddy Bell	CLE	24.4

Players Overall
Rod Carew	MIN	52.4
Reggie Jackson	OAK	45.9
Bobby Grich	BAL	40.5

Pitchers Overall
Bert Blyleven	MIN	63.6
Jim Palmer	BAL	48.8
Bill Lee	BOS	42.2

Relief Points
John Hiller	DET	91
Sparky Lyle	NY	55
Rollie Fingers	OAK	52

East	W	L	R	OR	Avg	OBA	SLG	BPF	NOPS-A	BR	Adj	Wins	ERA	PPF	NERA-A	PR	Adj	Wins	Diff
BAL	97	65	754	561	.266	.348	.389	105	109/104	61	29	2.9	3.08	102	124/127	120	136	13.9	-.8
BOS	89	73	738	647	.267	.340	.401	106	110/104	58	14	1.4	3.66	106	104/110	26	61	6.3	.3
DET	85	77	642	674	.254	.322	.390	100	102/102	-8	-10	-1.1	3.90	101	98/ 99	-11	-7	-.8	5.9
NY	80	82	641	610	.261	.324	.378	91	99/109	-24	37	3.8	3.34	89	114/102	76	12	1.2	-6.0
MIL	74	88	708	731	.253	.327	.388	95	103/108	2	35	3.6	3.98	95	96/ 91	-25	-54	-5.7	-4.9
CLE	71	91	680	826	.256	.317	.387	100	100/ 98	-26	-44	-4.6	4.58	105	83/ 87	-123	-94	-9.7	4.3

West	W	L	R	OR	Avg	OBA	SLG	BPF	NOPS-A	BR	Adj	Wins	ERA	PPF	NERA-A	PR	Adj	Wins	Diff
OAK	94	68	758	615	.260	.336	.389	89	106/119	28	108	11.0	3.29	86	116/100	87	0	.0	2.0
KC	88	74	755	752	.261	.342	.381	108	105/ 98	31	-21	-2.3	4.21	108	91/ 98	-61	-11	-1.2	10.5
MIN	81	81	738	692	.270	.344	.393	113	109/ 96	56	-36	-3.8	3.77	113	101/115	8	89	9.1	-5.3
CAL	79	83	629	657	.253	.320	.348	93	89/ 96	-83	-35	-3.7	3.56	93	107/ 99	42	-2	-.3	2.0
CHI	77	85	652	705	.256	.326	.372	105	98/ 93	-27	-62	-6.5	3.86	106	99/105	-5	31	3.2	-.7
TEX	57	105	619	844	.255	.320	.361	95	93/ 97	-62	-31	-3.3	4.64	98	82/ 81	-128	-141	-14.5	-6.2
				693	.259	.331	.381						3.82						

Batting Runs
Willie Stargell	PIT	57.8
Darrell Evans	ATL	54.8
Hank Aaron	ATL	45.5

Park Adjusted
Willie Stargell	PIT	64.2
Darrell Evans	ATL	44.7
Joe Morgan	CIN	44.4

Pitching Runs
Tom Seaver	NY	51.3
Don Sutton	LA	35.4
Steve Rogers	MON	31.7

Park Adjusted
Tom Seaver	NY	49.5
Wayne Twitchell	PHI	36.1
Don Sutton	LA	33.7

Normalized OPS
Willie Stargell	PIT	184
Willie McCovey	SF	167
Darrell Evans	ATL	164

Park Adjusted
Willie Stargell	PIT	203
Willie McCovey	SF	160
Tony Perez	CIN	159

Normalized ERA
Tom Seaver	NY	177
Don Sutton	LA	151
Wayne Twitchell	PHI	147

Park Adjusted
Tom Seaver	NY	174
Wayne Twitchell	PHI	158
Don Sutton	LA	149

On Base Average
Ken Singleton	MON	.429
Willie McCovey	SF	.425
Ron Fairly	MON	.422

Slugging Percentage
Willie Stargell	PIT	.646
Darrell Evans	ATL	.556
Willie McCovey	SF	.546

Percent of Team Wins
Dave Roberts	SD	.283
Ron Bryant	SF	.273
Tom Seaver	NY	.232

Wins Above Team
Dave Roberts	SD	8.0
Ron Bryant	SF	5.7
Tom Seaver	NY	5.2

Isolated Power
Willie Stargell	PIT	.347
Willie McCovey	SF	.279
Darrell Evans	ATL	.276

Base Stealing Runs
Joe Morgan	CIN	11.1
Lou Brock	STL	9.0
Cesar Cedeno	HOU	7.8

Relievers - Runs
Pedro Borbon	CIN	20.3
Mike Marshall	MON	20.0
Dave Giusti	PIT	14.4

Park Adjusted
Mike Marshall	MON	21.5
Bob Locker	CHI	18.3
Pedro Borbon	CIN	17.6

Defensive Runs
Ron Cey	LA	21.5
Don Kessinger	CHI	17.8
Tim Foli	MON	15.3

Players Overall
Joe Morgan	CIN	58.6
Willie Stargell	PIT	56.6
Cesar Cedeno	HOU	46.4

Pitchers Overall
Tom Seaver	NY	52.0
Steve Rogers	MON	33.0
Steve Renko	MON	32.3

Relief Points
Mike Marshall	MON	79
Dave Giusti	PIT	56
Tug McGraw	NY	54

East	W	L	R	OR	Avg	OBA	SLG	BPF	NOPS-A	BR	Adj	Wins	ERA	PPF	NERA-A	PR	Adj	Wins	Diff
NY	82	79	608	588	.246	.317	.338	99	89/ 90	-80	-73	-7.7	3.27	98	112/110	64	56	5.8	3.4
STL	81	81	643	603	.259	.328	.357	90	97/108	-23	42	4.4	3.25	89	113/100	68	1	.1	-4.5
PIT	80	82	704	693	.261	.317	.405	90	109/121	30	94	9.8	3.74	90	98/ 88	-10	-71	-7.5	-3.3
MON	79	83	668	702	.251	.341	.364	101	103/102	30	20	2.1	3.73	102	99/101	-8	3	.3	-4.4
CHI	77	84	614	655	.247	.322	.357	110	96/ 87	-34	-104	-10.9	3.66	112	100/112	2	71	7.4	-.0
PHI	71	91	642	717	.249	.312	.371	106	98/ 92	-39	-82	-8.7	4.00	108	92/ 99	-52	-6	-.7	-.6

West	W	L	R	OR	Avg	OBA	SLG	BPF	NOPS-A	BR	Adj	Wins	ERA	PPF	NERA-A	PR	Adj	Wins	Diff
CIN	99	63	741	621	.254	.335	.383	96	107/111	43	68	7.1	3.43	95	107/101	39	6	.6	10.3
LA	95	66	675	565	.263	.326	.371	100	101/101	-4	-3	-.4	3.00	98	122/120	111	102	10.6	4.3
SF	88	74	739	702	.262	.337	.407	104	115/110	87	58	6.0	3.80	104	97/101	-20	4	.4	.6
HOU	82	80	681	672	.251	.314	.376	98	99/101	-27	-14	-1.5	3.79	98	97/ 95	-18	-32	-3.4	5.9
ATL	76	85	799	774	.266	.341	.427	113	122/109	134	46	4.8	4.25	113	86/ 98	-93	-13	-1.5	-7.8
SD	60	102	548	770	.244	.298	.351	93	88/ 94	-109	-64	-6.7	4.16	95	88/ 84	-77	-103	-10.8	-3.4
				672	.254	.324	.376						3.67						

AMERICAN LEAGUE 1974

Batting Runs			Park Adjusted			Pitching Runs			Park Adjusted		
Jeff Burroughs	TEX	45.3	Rod Carew	MIN	48.0	Jim Hunter	OAK	40.1	Gaylord Perry	CLE	41.4
Rod Carew	MIN	44.7	Jeff Burroughs	TEX	46.1	Gaylord Perry	CLE	39.7	Jim Hunter	OAK	36.6
Reggie Jackson	OAK	41.0	Reggie Jackson	OAK	41.5	Bert Blyleven	MIN	30.2	Luis Tiant	BOS	31.4

Normalized OPS			Park Adjusted			Normalized ERA			Park Adjusted		
Dick Allen	CHI	168	Dick Allen	CHI	167	Jim Hunter	OAK	146	Gaylord Perry	CLE	146
Reggie Jackson	OAK	160	Jeff Burroughs	TEX	162	Gaylord Perry	CLE	144	Jim Hunter	OAK	142
Jeff Burroughs	TEX	160	Reggie Jackson	OAK	161	Andy Hassler	CAL	139	Al Fitzmorris	KC	139

On Base Average			Slugging Percentage			Percent of Team Wins			Wins Above Team		
Rod Carew	MIN	.435	Dick Allen	CHI	.563	Nolan Ryan	CAL	.324	Nolan Ryan	CAL	7.9
Carl Yastrzemski	BOS	.421	Reggie Jackson	OAK	.514	Ferguson Jenkins	TEX	.301	Ferguson Jenkins	TEX	7.4
Jeff Burroughs	TEX	.405	Jeff Burroughs	TEX	.504	Steve Busby	KC	.286	Steve Busby	KC	6.3

Isolated Power			Base Stealing Runs			Relievers - Runs			Park Adjusted		
Dick Allen	CHI	.262	Reggie Jackson	OAK	4.5	Sparky Lyle	NY	24.9	Tom Murphy	MIL	23.8
Reggie Jackson	OAK	.225	Rich Coggins	BAL	4.2	Tom Murphy	MIL	23.5	Sparky Lyle	NY	23.2
Frank Robinson	CLE	.211	Vada Pinson	KC	3.3	Steve Foucault	TEX	22.0	John Hiller	DET	21.6

Defensive Runs			Players Overall			Pitchers Overall			Relief Points		
Brooks Robinson	BAL	22.7	Rod Carew	MIN	73.8	Gaylord Perry	CLE	42.5	Terry Forster	CHI	55
Aurelio Rodriguez	DET	21.9	Reggie Jackson	OAK	49.0	Jim Hunter	OAK	33.0	Tom Murphy	MIL	50
Rod Carew	MIN	17.0	Bobby Grich	BAL	45.0	Luis Tiant	BOS	28.9	Rollie Fingers	OAK	49

East	W	L	R	OR	Avg	OBA	SLG	BPF	NOPS-A	BR	Adj	Wins	ERA	PPF	NERA-A	PR	Adj	Wins	Diff
BAL	91	71	659	612	.256	.325	.370	98	101/103	-1	9	.9	3.29	98	110/108	55	41	4.2	4.8
NY	89	73	671	623	.263	.328	.368	97	101/105	1	21	2.2	3.32	96	109/105	50	28	2.9	3.0
BOS	84	78	696	661	.264	.336	.377	106	106/101	39	1	.1	3.72	106	97/103	-14	18	1.9	1.0
CLE	77	85	662	694	.255	.312	.370	101	98/ 97	-35	-40	-4.3	3.82	101	95/ 96	-31	-23	-2.5	2.9
MIL	76	86	647	660	.244	.310	.369	101	97/ 97	-40	-43	-4.6	3.77	101	96/ 97	-22	-19	-2.0	1.7
DET	72	90	620	768	.247	.304	.366	106	94/ 89	-63	-104	-11.0	4.17	109	87/ 94	-87	-37	-4.0	6.0

West	W	L	R	OR	Avg	OBA	SLG	BPF	NOPS-A	BR	Adj	Wins	ERA	PPF	NERA-A	PR	Adj	Wins	Diff
OAK	90	72	689	551	.247	.324	.373	99	102/103	1	7	.7	2.96	97	122/119	106	91	9.5	-1.2
TEX	84	76	690	698	.272	.338	.377	99	107/108	42	50	5.2	3.84	99	94/ 93	-33	-40	-4.3	3.0
MIN	82	80	673	669	.272	.338	.378	95	107/112	44	75	7.8	3.64	95	100/ 95	-2	-31	-3.3	-3.5
CHI	80	80	684	721	.268	.333	.389	100	109/109	49	46	4.8	3.94	101	92/ 93	-50	-44	-4.7	-.1
KC	77	85	667	662	.259	.329	.364	107	101/ 94	-1	-48	-5.2	3.51	107	103/111	18	62	6.5	-5.3
CAL	68	94	618	657	.254	.323	.356	91	96/106	-29	30	3.1	3.52	91	103/ 93	16	-36	-3.9	-12.2
				665		.258	.325	.371					3.63						

NATIONAL LEAGUE 1974

Batting Runs			Park Adjusted			Pitching Runs			Park Adjusted		
Mike Schmidt	PHI	48.3	Mike Schmidt	PHI	49.9	Phil Niekro	ATL	41.8	Phil Niekro	ATL	49.2
Joe Morgan	CIN	45.2	Willie Stargell	PIT	47.6	Jon Matlack	NY	35.9	Jon Matlack	NY	39.8
Willie Stargell	PIT	44.4	Joe Morgan	CIN	47.4	Andy Messersmith	LA	33.8	Buzz Capra	ATL	37.8

Normalized OPS			Park Adjusted			Normalized ERA			Park Adjusted		
Willie Stargell	PIT	163	Willie Stargell	PIT	171	Buzz Capra	ATL	159	Buzz Capra	ATL	169
Mike Schmidt	PHI	162	Mike Schmidt	PHI	166	Phil Niekro	ATL	152	Phil Niekro	ATL	162
Joe Morgan	CIN	158	Joe Morgan	CIN	163	Jon Matlack	NY	151	Jon Matlack	NY	156

On Base Average			Slugging Percentage			Percent of Team Wins			Wins Above Team		
Joe Morgan	CIN	.430	Mike Schmidt	PHI	.546	Phil Niekro	ATL	.227	Mike Caldwell	SF	6.3
Willie Stargell	PIT	.409	Willie Stargell	PIT	.537	Jim Lonborg	PHI	.213	Larry Hardy	SD	4.6
Bob Bailey	MON	.400	Reggie Smith	STL	.528	Jerry Koosman	NY	.211	Mike Torrez	MON	4.3

Isolated Power			Base Stealing Runs			Relievers - Runs			Park Adjusted		
Mike Schmidt	PHI	.264	Lou Brock	STL	15.6	Mike Marshall	LA	27.9	Tom House	ATL	22.1
Willie Stargell	PIT	.236	Larry Lintz	MON	10.8	Dale Murray	MON	20.2	Dale Murray	MON	21.0
Johnny Bench	CIN	.227	Joe Morgan	CIN	10.2	Tom House	ATL	19.5	Mike Marshall	LA	20.7

Defensive Runs			Players Overall			Pitchers Overall			Relief Points		
Dave Cash	PHI	29.9	Mike Schmidt	PHI	71.9	Phil Niekro	ATL	49.4	Mike Marshall	LA	60
Mike Schmidt	PHI	26.0	Joe Morgan	CIN	59.9	Jim Barr	SF	40.9	Pedro Borbon	CIN	41
Bud Harrelson	NY	17.5	Dave Concepcion	CIN	47.0	Jon Matlack	NY	35.7	Randy Moffitt	SF	33
Tim Foli	MON	17.5							Al Hrabosky	STL	33
									Dave Giusti	PIT	33

East	W	L	R	OR	Avg	OBA	SLG	BPF	NOPS-A	BR	Adj	Wins	ERA	PPF	NERA-A	PR	Adj	Wins	Diff
PIT	88	74	751	657	.274	.338	.391	95	111/111	66	99	10.3	3.51	94	103/ 97	20	-17	-1.9	-1.4
STL	86	75	677	643	.265	.334	.365	104	102/ 98	9	-17	-1.9	3.49	104	104/108	23	46	4.8	2.7
PHI	80	82	676	701	.261	.322	.373	98	102/104	-6	7	.8	3.92	98	93/ 91	-46	-58	-6.2	4.4
MON	79	82	662	657	.254	.338	.350	104	99/ 96	2	-15	-1.6	3.59	103	101/104	6	22	2.3	-2.2
NY	71	91	572	646	.235	.314	.329	103	86/ 84	-98	-116	-12.2	3.42	104	106/110	34	56	5.9	-3.7
CHI	66	96	669	826	.251	.329	.365	102	101/ 99	-0	-13	-1.5	4.29	104	85/ 88	-107	-83	-8.7	-4.8

West	W	L	R	OR	Avg	OBA	SLG	BPF	NOPS-A	BR	Adj	Wins	ERA	PPF	NERA-A	PR	Adj	Wins	Diff
LA	102	60	798	561	.272	.346	.401	95	116/123	103	139	14.4	2.98	91	122/111	105	55	5.7	.8
CIN	98	64	776	631	.260	.345	.394	97	114/118	95	117	12.2	3.42	95	106/101	34	3	.3	4.5
ATL	88	74	661	563	.249	.321	.363	107	98/ 92	-22	-69	-7.3	3.05	106	119/126	94	131	13.6	.7
HOU	81	81	653	632	.263	.324	.378	96	104/108	6	32	3.3	3.48	95	104/ 99	24	-2	-.3	-3.0
SF	72	90	634	723	.252	.323	.358	111	97/ 88	-27	-100	-10.5	3.80	113	96/108	-26	47	4.9	-3.4
SD	60	102	541	830	.229	.304	.330	92	83/ 90	-121	-69	-7.3	4.61	95	79/ 75	-157	-185	-19.4	5.7
				673		.255	.328	.367					3.63						

AMERICAN LEAGUE 1975

Batting Runs			Park Adjusted			Pitching Runs			Park Adjusted		
John Mayberry	KC	56.0	John Mayberry	KC	58.5	Jim Palmer	BAL	61.0	Jim Palmer	BAL	47.1
Fred Lynn	BOS	49.4	Ken Singleton	BAL	47.7	Jim Hunter	NY	44.1	Jim Hunter	NY	40.2
Rod Carew	MIN	44.4	Fred Lynn	BOS	41.4	Frank Tanana	CAL	33.2	Bert Blyleven	MIN	33.8
Normalized OPS			**Park Adjusted**			**Normalized ERA**			**Park Adjusted**		
Fred Lynn	BOS	172	John Mayberry	KC	177	Jim Palmer	BAL	181	Jim Palmer	BAL	163
John Mayberry	KC	171	Gene Tenace	OAK	163	Jim Hunter	NY	147	Dennis Eckersley	CLE	151
Rod Carew	MIN	161	Ken Singleton	BAL	159	Dennis Eckersley	CLE	146	Jim Hunter	NY	143
On Base Average			**Slugging Percentage**			**Percent of Team Wins**			**Wins Above Team**		
Rod Carew	MIN	.428	Fred Lynn	BOS	.566	Jim Hunter	NY	.277	Frank Tanana	CAL	5.7
John Mayberry	KC	.419	John Mayberry	KC	.547	Jim Kaat	CHI	.267	Jim Kaat	CHI	5.3
Ken Singleton	BAL	.418	Boog Powell	CLE	.524	Jim Palmer	BAL	.256	Jim Palmer	BAL	5.0
Isolated Power			**Base Stealing Runs**			**Relievers - Runs**			**Park Adjusted**		
Reggie Jackson	OAK	.258	Mickey Rivers	CAL	12.6	Rich Gossage	CHI	30.8	Rich Gossage	CHI	31.9
John Mayberry	KC	.256	Freddie Patek	KC	5.4	Jim Todd	OAK	20.4	Dave La Roche	CLE	15.8
Bobby Bonds	NY	.242	Amos Otis	KC	5.1	Dave La Roche	CLE	14.5	John Hiller	DET	15.1
			Rod Carew	MIN	5.1						
Defensive Runs			**Players Overall**			**Pitchers Overall**			**Relief Points**		
Mark Belanger	BAL	29.4	Toby Harrah	TEX	62.0	Jim Palmer	BAL	49.4	Rich Gossage	CHI	62
Bucky Dent	CHI	29.2	Rod Carew	MIN	57.7	Bert Blyleven	MIN	36.3	Rollie Fingers	OAK	62
Bobby Grich	BAL	25.4	Bobby Grich	BAL	56.3	Jim Hunter	NY	35.8	Dave La Roche	CLE	41

East	W	L	R	OR	Avg	OBA	SLG	BPF	NOPS-A	BR	Adj	Wins	ERA	PPF	NERA-A	PR	Adj	Wins	Diff
BOS	95	65	796	709	.275	.347	.417	112	118/106	108	27	2.7	3.99	111	95/106	-31	36	3.7	8.5
BAL	90	69	682	553	.252	.328	.373	92	100/108	-13	42	4.3	3.17	90	120/107	100	38	3.8	2.4
NY	83	77	681	588	.264	.328	.382	98	102/104	-2	8	.8	3.29	97	115/112	79	62	6.4	-4.1
CLE	79	80	688	703	.261	.329	.392	103	105/102	17	-5	-.6	3.84	104	99/102	-7	15	1.5	-1.4
MIL	68	94	675	792	.250	.323	.389	99	103/104	-1	5	.5	4.34	100	87/88	-84	-84	-8.7	-4.8
DET	57	102	570	786	.249	.303	.366	104	91/87	-93	-121	-12.5	4.29	107	88/95	-77	-34	-3.6	-6.4

West	W	L	R	OR	Avg	OBA	SLG	BPF	NOPS-A	BR	Adj	Wins	ERA	PPF	NERA-A	PR	Adj	Wins	Diff
OAK	98	64	758	606	.254	.335	.391	87	107/122	34	121	12.4	3.29	84	115/97	80	-14	-1.5	6.1
KC	91	71	710	649	.261	.336	.394	97	108/111	40	62	6.4	3.49	96	108/104	48	22	2.2	1.4
TEX	79	83	714	733	.256	.332	.371	101	100/99	-9	-14	-1.5	3.90	101	97/98	-11	-11	-1.2	.8
MIN	76	83	724	736	.271	.343	.386	108	108/100	46	-6	-.7	4.06	108	93/101	-42	7	.7	-3.5
CHI	75	86	655	703	.255	.334	.358	101	97/96	-23	-31	-3.3	3.93	102	96/98	-22	-11	-1.2	-1.0
CAL	72	89	628	723	.246	.324	.328	96	85/88	-98	-72	-7.4	3.89	97	97/95	-15	-32	-3.4	2.3
				690	.258	.330	.379						3.79						

NATIONAL LEAGUE 1975

Batting Runs			Park Adjusted			Pitching Runs			Park Adjusted		
Joe Morgan	CIN	57.1	Joe Morgan	CIN	58.3	Andy Messersmith	LA	47.9	Randy Jones	SD	47.2
Greg Luzinski	PHI	47.5	Greg Luzinski	PHI	50.1	Randy Jones	SD	44.0	Tom Seaver	NY	38.4
Ted Simmons	STL	35.7	Mike Schmidt	PHI	36.2	Tom Seaver	NY	39.0	Andy Messersmith	LA	32.3
Normalized OPS			**Park Adjusted**			**Normalized ERA**			**Park Adjusted**		
Joe Morgan	CIN	172	Joe Morgan	CIN	174	Randy Jones	SD	162	Randy Jones	SD	167
Greg Luzinski	PHI	159	Greg Luzinski	PHI	164	Andy Messersmith	LA	158	Tom Seaver	NY	152
Dave Parker	PIT	147	Mike Schmidt	PHI	150	Tom Seaver	NY	153	Jerry Reuss	PIT	141
On Base Average			**Slugging Percentage**			**Percent of Team Wins**			**Wins Above Team**		
Joe Morgan	CIN	.471	Dave Parker	PIT	.541	Randy Jones	SD	.282	Tom Seaver	NY	7.8
Jim Wynn	LA	.407	Greg Luzinski	PHI	.540	Tom Seaver	NY	.268	Randy Jones	SD	7.4
Pete Rose	CIN	.407	Mike Schmidt	PHI	.523	Carl Morton	ATL	.254	Al Hrabosky	STL	5.4
Isolated Power			**Base Stealing Runs**			**Relievers - Runs**			**Park Adjusted**		
Mike Schmidt	PHI	.274	Davey Lopes	LA	15.9	Al Hrabosky	STL	21.1	Al Hrabosky	STL	22.8
Dave Kingman	NY	.263	Joe Morgan	CIN	14.1	Bob Apodaca	NY	20.3	Bob Apodaca	NY	20.1
Greg Luzinski	PHI	.240	Lou Brock	STL	7.2	Tom Hilgendorf	PHI	16.1	Tom Hilgendorf	PHI	14.5
Defensive Runs			**Players Overall**			**Pitchers Overall**			**Relief Points**		
Mike Schmidt	PHI	26.1	Joe Morgan	CIN	83.4	Randy Jones	SD	50.7	Al Hrabosky	STL	67
Darrell Evans	ATL	23.4	Mike Schmidt	PHI	63.9	Tom Seaver	NY	42.8	Rawly Eastwick	CIN	51
Manny Trillo	CHI	19.0	Darrell Evans	ATL	38.3	Bob Forsch	STL	34.8	Dale Murray	MON	40
									Tug McGraw	PHI	40
									Dave Giusti	PIT	40

East	W	L	R	OR	Avg	OBA	SLG	BPF	NOPS-A	BR	Adj	Wins	ERA	PPF	NERA-A	PR	Adj	Wins	Diff
PIT	92	69	712	565	.263	.325	.402	100	110/110	44	41	4.3	3.01	98	120/119	99	90	9.4	-2.2
PHI	86	76	735	694	.269	.344	.402	96	115/120	98	122	12.7	3.82	96	95/91	-29	-55	-5.8	-1.9
NY	82	80	646	625	.256	.321	.361	100	97/97	-35	-33	-3.6	3.39	100	107/107	40	37	3.9	.7
STL	82	80	662	689	.273	.329	.375	103	103/100	7	-15	-1.6	3.57	104	102/106	10	34	3.6	-.9
MON	75	87	601	690	.244	.319	.348	111	92/83	-62	-133	-14.0	3.73	112	97/109	16	57	6.0	2.0
CHI	75	87	712	827	.259	.341	.368	103	104/101	33	13	1.4	4.56	105	80/83	-148	-121	-12.7	5.4

West	W	L	R	OR	Avg	OBA	SLG	BPF	NOPS-A	BR	Adj	Wins	ERA	PPF	NERA-A	PR	Adj	Wins	Diff
CIN	108	54	840	586	.271	.355	.401	98	118/120	128	140	14.6	3.37	95	108/102	43	13	1.4	11.1
LA	88	74	648	534	.248	.328	.365	91	100/110	-9	54	5.6	2.92	88	124/109	116	45	4.7	-3.3
SF	80	81	659	671	.259	.336	.365	101	102/100	11	1	.1	3.74	102	97/99	-16	-7	-.8	.2
SD	71	91	552	683	.244	.313	.335	101	86/85	-99	-106	-11.2	3.50	103	104/107	21	38	3.9	-2.7
ATL	67	94	583	739	.244	.315	.346	97	91/93	-74	-53	-5.7	3.92	99	93/91	-46	-53	-5.6	-2.2
HOU	64	97	664	711	.254	.322	.359	97	96/99	-35	-14	-1.6	4.05	97	90/87	-67	-83	-8.8	-6.2
				668	.257	.329	.369						3.63						

AMERICAN LEAGUE 1976

Batting Runs			Park Adjusted			Pitching Runs			Park Adjusted		
Rod Carew	MIN	40.0	Hal McRae	KC	41.3	Vida Blue	OAK	38.6	Mark Fidrych	DET	41.8
Hal McRae	KC	38.8	Rod Carew	MIN	40.8	Jim Palmer	BAL	35.3	Jim Palmer	BAL	40.0
George Brett	KC	36.2	George Brett	KC	39.1	Frank Tanana	CAL	34.7	Vida Blue	OAK	32.6

Normalized OPS			Park Adjusted			Normalized ERA			Park Adjusted		
Hal McRae	KC	155	Hal McRae	KC	162	Mark Fidrych	DET	151	Mark Fidrych	DET	164
Rod Carew	MIN	152	Rod Carew	MIN	153	Vida Blue	OAK	150	Jim Palmer	BAL	145
Reggie Jackson	BAL	148	George Brett	KC	152	Frank Tanana	CAL	145	Vida Blue	OAK	142

On Base Average			Slugging Percentage			Percent of Team Wins			Wins Above Team		
Hal McRae	KC	.412	Reggie Jackson	BAL	.502	Mark Fidrych	DET	.257	Mark Fidrych	DET	7.4
Mike Hargrove	TEX	.401	Jim Rice	BOS	.482	Luis Tiant	BOS	.253	Frank Tanana	CAL	6.6
Rod Carew	MIN	.398	Graig Nettles	NY	.475	Frank Tanana	CAL	.250	Wayne Garland	BAL	6.4
						Jim Palmer	BAL	.250			

Isolated Power			Base Stealing Runs			Relievers - Runs			Park Adjusted		
Reggie Jackson	BAL	.225	Bert Campaneris	OAK	9.0	Mark Littell	KC	16.7	John Hiller	DET	19.7
Graig Nettles	NY	.221	Mickey Rivers	NY	8.7	Rollie Fingers	OAK	15.8	Sparky Lyle	NY	15.5
Gene Tenace	OAK	.209	Don Baylor	OAK	8.4	John Hiller	DET	15.4	Mark Littell	KC	14.4

Defensive Runs			Players Overall			Pitchers Overall			Relief Points		
Mark Belanger	BAL	16.8	George Brett	KC	41.0	Mark Fidrych	DET	46.5	Bill Campbell	MIN	69
Graig Nettles	NY	16.5	Graig Nettles	NY	36.5	Jim Palmer	BAL	41.3	Rollie Fingers	OAK	55
Aurelio Rodriguez	DET	15.9	Rod Carew	MIN	36.0	Vida Blue	OAK	29.5	Sparky Lyle	NY	52

East	W	L	R	OR	Avg	OBA	SLG	BPF	NOPS-A	BR	Adj	Wins	ERA	PPF	NERA-A	PR	Adj	Wins	Diff
NY	97	62	730	575	.269	.330	.389	104	112/108	64	40	4.2	3.19	102	111/113	54	65	6.9	6.4
BAL	88	74	619	598	.243	.311	.358	104	97/ 93	-32	-58	-6.2	3.31	104	106/110	34	56	6.0	7.2
BOS	83	79	716	660	.263	.327	.402	106	115/108	80	38	4.0	3.52	106	100/106	0	36	3.8	-5.8
CLE	81	78	615	615	.263	.324	.359	98	101/103	-1	12	1.2	3.48	98	101/ 99	7	-5	-.7	1.0
DET	74	87	609	709	.257	.318	.365	107	101/ 94	-6	-52	-5.7	3.87	109	91/ 99	-54	-3	-.5	-.4
MIL	66	95	570	655	.246	.314	.340	98	92/ 94	-56	-42	-4.5	3.65	99	97/ 95	-19	-26	-2.9	-7.1

West	W	L	R	OR	Avg	OBA	SLG	BPF	NOPS-A	BR	Adj	Wins	ERA	PPF	NERA-A	PR	Adj	Wins	Diff
KC	90	72	713	611	.269	.331	.371	96	107/111	37	63	6.7	3.21	94	110/104	52	19	2.0	.3
OAK	87	74	686	598	.246	.327	.361	96	102/106	13	36	3.9	3.26	95	108/103	43	13	1.4	1.2
MIN	85	77	743	704	.274	.343	.375	99	111/113	78	86	9.1	3.72	98	95/ 93	-32	-41	-4.5	-.6
TEX	76	86	616	652	.250	.323	.341	102	95/ 94	-31	-42	-4.5	3.47	102	102/104	9	22	2.3	-2.8
CAL	76	86	550	631	.235	.309	.318	93	84/ 91	-104	-57	-6.2	3.36	93	105/ 98	27	-12	-1.3	2.5
CHI	64	97	586	745	.255	.317	.349	98	96/ 98	-35	-25	-2.7	4.25	100	83/ 83	-117	-114	-12.2	-1.5
				646	.256	.323	.361						3.52						

NATIONAL LEAGUE 1976

Batting Runs			Park Adjusted			Pitching Runs			Park Adjusted		
Joe Morgan	CIN	61.4	Joe Morgan	CIN	56.2	Tom Seaver	NY	27.6	John Denny	STL	26.1
Mike Schmidt	PHI	42.7	Mike Schmidt	PHI	40.9	Randy Jones	SD	26.7	Phil Niekro	ATL	24.5
Bill Madlock	CHI	40.6	Bob Watson	HOU	39.9	J.R. Richard	HOU	24.3	Andy Messersmith	ATL	24.3

Normalized OPS			Park Adjusted			Normalized ERA			Park Adjusted		
Joe Morgan	CIN	191	Joe Morgan	CIN	176	John Denny	STL	139	John Denny	STL	145
Bill Madlock	CHI	159	Bob Watson	HOU	163	Doug Rau	LA	136	Doug Rau	LA	136
Mike Schmidt	PHI	155	Cesar Cedeno	HOU	153	Tom Seaver	NY	135	Pat Zachry	CIN	135

On Base Average			Slugging Percentage			Percent of Team Wins			Wins Above Team		
Joe Morgan	CIN	.453	Joe Morgan	CIN	.576	Randy Jones	SD	.301	Randy Jones	SD	7.4
Bill Madlock	CHI	.415	George Foster	CIN	.530	J.R. Richard	HOU	.250	Phil Niekro	ATL	5.9
Pete Rose	CIN	.406	Mike Schmidt	PHI	.524	Jerry Koosman	NY	.244	Jerry Koosman	NY	5.6

Isolated Power			Base Stealing Runs			Relievers - Runs			Park Adjusted		
Dave Kingman	NY	.268	Davey Lopes	LA	12.9	Charlie Hough	LA	20.7	Charlie Hough	LA	20.6
Mike Schmidt	PHI	.262	Joe Morgan	CIN	12.6	Rawly Eastwick	CIN	17.1	Rawly Eastwick	CIN	19.5
Joe Morgan	CIN	.256	Frank Taveras	PIT	10.8	Ron Reed	PHI	14.9	Randy Moffitt	SF	15.8

Defensive Runs			Players Overall			Pitchers Overall			Relief Points		
Mike Schmidt	PHI	24.6	Mike Schmidt	PHI	48.0	John Denny	STL	26.1	Rawly Eastwick	CIN	69
Rennie Stennett	PIT	18.5	Joe Morgan	CIN	52.7	Jim Barr	SF	26.2	Charlie Hough	LA	52
Garry Maddox	PHI	17.6	George Foster	CIN	38.0	Phil Niekro	ATL	25.8	Skip Lockwood	NY	51

East	W	L	R	OR	Avg	OBA	SLG	BPF	NOPS-A	BR	Adj	Wins	ERA	PPF	NERA-A	PR	Adj	Wins	Diff
PHI	101	61	770	557	.272	.342	.395	102	117/114	106	91	9.6	3.09	100	113/113	67	65	6.9	3.4
PIT	92	70	708	630	.267	.323	.391	98	111/113	50	62	6.6	3.37	97	104/101	22	4	.5	3.9
NY	86	76	615	538	.246	.320	.352	94	98/104	-18	17	1.8	2.94	93	119/111	91	51	5.4	-2.1
CHI	75	87	611	728	.251	.316	.356	105	98/ 93	-26	-59	-6.4	3.94	107	89/ 95	-70	-29	-3.2	3.6
STL	72	90	629	671	.260	.325	.359	103	101/ 98	1	-21	-2.3	3.61	104	97/101	-16	7	.8	-7.5
MON	55	107	531	734	.235	.293	.340	102	87/ 85	-108	-122	-13.0	4.00	105	88/ 92	-79	-50	-5.4	-7.6

West	W	L	R	OR	Avg	OBA	SLG	BPF	NOPS-A	BR	Adj	Wins	ERA	PPF	NERA-A	PR	Adj	Wins	Diff
CIN	102	60	857	633	.280	.360	.424	108	131/121	214	158	16.8	3.51	106	100/106	0	33	3.5	.8
LA	92	70	608	543	.251	.315	.349	101	96/ 95	-38	-43	-4.7	3.01	100	116/116	81	80	8.5	7.2
HOU	80	82	625	657	.256	.325	.347	85	97/115	-19	79	8.4	3.55	84	99/ 83		-98	-10.5	1.1
SF	74	88	595	686	.246	.314	.345	103	94/ 91	-47	-65	-7.0	3.53	104	99/103	-3	20	2.1	-2.1
SD	73	89	570	662	.247	.313	.337	87	91/104	-64	14	1.5	3.65	88	96/ 84	-22	-91	-9.8	.3
ATL	70	92	620	700	.245	.322	.334	115	93/ 81	-43	-139	-14.9	3.87	117	91/106	-57	37	4.0	-.1
				645	.255	.323	.361						3.51						

AMERICAN LEAGUE 1977

Batting Runs			Park Adjusted			Pitching Runs			Park Adjusted		
Rod Carew	MIN	67.0	Rod Carew	MIN	68.7	Nolan Ryan	CAL	43.4	Bert Blyleven	TEX	50.6
Jim Rice	BOS	51.4	Ken Singleton	BAL	58.2	Jim Palmer	BAL	41.4	Nolan Ryan	CAL	35.9
Ken Singleton	BAL	48.5	Mitchell Page	OAK	42.0	Frank Tanana	CAL	41.1	Frank Tanana	CAL	35.1

Normalized OPS			Park Adjusted			Normalized ERA			Park Adjusted		
Rod Carew	MIN	176	Ken Singleton	BAL	180	Frank Tanana	CAL	160	Bert Blyleven	TEX	171
Jim Rice	BOS	160	Rod Carew	MIN	180	Bert Blyleven	TEX	150	Frank Tanana	CAL	152
Ken Singleton	BAL	158	Mitchell Page	OAK	161	Nolan Ryan	CAL	147	Dave Rozema	DET	142

On Base Average			Slugging Percentage			Percent of Team Wins			Wins Above Team		
Rod Carew	MIN	.452	Jim Rice	BOS	.593	Nolan Ryan	CAL	.257	Dave Rozema	DET	5.7
Ken Singleton	BAL	.442	Rod Carew	MIN	.570	Dave Lemanczyk	TOR	.241	Frank Tanana	CAL	4.7
Mike Hargrove	TEX	.424	Reggie Jackson	NY	.550	Dave Goltz	MIN	.238	Dave Goltz	MIN	4.7

Isolated Power			Base Stealing Runs			Relievers - Runs			Park Adjusted		
Jim Rice	BOS	.273	Mitchell Page	OAK	9.6	Sparky Lyle	NY	29.0	Bill Campbell	BOS	27.9
Reggie Jackson	NY	.265	Freddie Patek	KC	8.1	Pablo Torrealba	OAK	19.0	Sparky Lyle	NY	25.7
Andy Thornton	CLE	.263	Toby Harrah	TEX	5.1	Lerrin La Grow	CHI	17.8	Pete Vuckovich	TOR	17.0

Defensive Runs			Players Overall			Pitchers Overall			Relief Points		
Mark Belanger	BAL	30.7	Rod Carew	MIN	63.3	Bert Blyleven	TEX	51.0	Bill Campbell	BOS	79
Chet Lemon	CHI	26.3	Mitchell Page	OAK	49.5	Frank Tanana	CAL	36.3	Sparky Lyle	NY	73
Bert Campaneris	TEX	24.1	Ken Singleton	BAL	46.8	Nolan Ryan	CAL	35.9	Lerrin La Grow	CHI	61

East	W	L	R	OR	Avg	OBA	SLG	BPF	NOPS-A	BR	Adj	Wins	ERA	PPF	NERA-A	PR	Adj	Wins	Diff
NY	100	62	831	651	.281	.347	.444	97	119/123	104	128	12.8	3.63	95	112/106	72	37	3.7	2.6
BAL	97	64	719	653	.261	.332	.393	88	100/114	-22	68	6.7	3.75	86	109/ 93	53	-39	-3.9	13.7
BOS	97	64	859	712	.281	.349	.465	117	126/107	144	20	2.0	4.16	117	98/114	-12	94	9.4	5.2
DET	74	88	714	751	.264	.321	.410	107	102/ 96	-24	-76	-7.6	4.14	108	98/106	-9	42	4.1	-3.5
CLE	71	90	676	739	.269	.337	.380	97	97/100	-30	-9	-1.0	4.10	98	99/ 97	-3	-19	-2.0	-6.5
MIL	67	95	639	765	.258	.316	.389	99	95/ 99	-72	-45	-4.5	4.33	97	94/ 91	-40	-58	-5.8	-3.6
TOR	54	107	605	822	.252	.318	.365	108	88/ 82	-105	-161	-16.1	4.58	110	89/ 98	-79	-11	-1.2	-9.2

West	W	L	R	OR	Avg	OBA	SLG	BPF	NOPS-A	BR	Adj	Wins	ERA	PPF	NERA-A	PR	Adj	Wins	Diff
KC	102	60	822	651	.277	.343	.436	101	116/115	82	76	7.5	3.54	99	115/114	86	80	7.9	5.5
TEX	94	68	767	657	.270	.345	.405	115	107/ 93	34	-74	-7.4	3.55	114	115/131	85	180	17.9	2.6
CHI	90	72	844	771	.278	.347	.444	97	119/123	107	128	12.7	4.25	96	96/ 92	-28	-52	-5.3	1.6
MIN	84	77	867	776	.282	.351	.417	98	112/114	71	86	8.5	4.38	97	93/ 90	-48	-68	-6.8	1.8
CAL	74	88	675	695	.255	.327	.386	95	97/102	-46	-8	-.9	3.76	95	108/102	50	14	1.4	-7.5
SEA	64	98	624	855	.256	.314	.381	98	92/ 94	-90	-74	-7.5	4.85	100	84/ 84	-122	-122	-12.2	2.7
OAK	63	98	605	749	.240	.311	.352	93	82/ 88	-144	-97	-9.7	4.05	94	101/ 95	4	-32	-3.3	-4.5
				732	.266	.333	.405						4.07						

NATIONAL LEAGUE 1977

Batting Runs			Park Adjusted			Pitching Runs			Park Adjusted		
George Foster	CIN	56.0	George Foster	CIN	55.1	John Candelaria	PIT	40.5	Rick Reuschel	CHI	45.7
Reggie Smith	LA	50.3	Greg Luzinski	PHI	52.4	Steve Carlton	PHI	40.1	John Candelaria	PIT	38.7
Greg Luzinski	PHI	48.6	Reggie Smith	LA	50.4	Burt Hooton	LA	32.0	Steve Carlton	PHI	30.5

Normalized OPS			Park Adjusted			Normalized ERA			Park Adjusted		
Reggie Smith	LA	168	Greg Luzinski	PHI	171	John Candelaria	PIT	167	Rick Reuschel	CHI	164
George Foster	CIN	168	Reggie Smith	LA	168	Burt Hooton	LA	149	Rick Reuschel	CHI	159
Greg Luzinski	PHI	162	Mike Schmidt	PHI	166	Steve Carlton	PHI	148	Burt Hooton	LA	146

On Base Average			Slugging Percentage			Percent of Team Wins			Wins Above Team		
Reggie Smith	LA	.432	George Foster	CIN	.631	Phil Niekro	ATL	.262	Bob Forsch	STL	7.4
Joe Morgan	CIN	.420	Greg Luzinski	PHI	.594	Rick Reuschel	CHI	.247	Rick Reuschel	CHI	6.1
Gene Tenace	SD	.417	Reggie Smith	LA	.576	Bob Forsch	STL	.241	John Candelaria	PIT	6.1

Isolated Power			Base Stealing Runs			Relievers - Runs			Park Adjusted		
George Foster	CIN	.311	Frank Taveras	PIT	10.2	Rich Gossage	PIT	33.8	Bruce Sutter	CHI	36.5
Mike Schmidt	PHI	.300	Cesar Cedeno	HOU	9.9	Bruce Sutter	CHI	30.5	Rich Gossage	PIT	32.8
Greg Luzinski	PHI	.285	Gene Richards	SD	9.6	Gary Lavelle	SF	24.3	Gary Lavelle	SF	28.6

Defensive Runs			Players Overall			Pitchers Overall			Relief Points		
Ivan DeJesus	CHI	36.1	Mike Schmidt	PHI	76.8	Rick Reuschel	CHI	50.9	Rollie Fingers	SD	77
Manny Trillo	CHI	35.9	Dave Parker	PIT	61.1	John Candelaria	PIT	41.5	Bruce Sutter	CHI	73
Dave Parker	PIT	31.1	George Foster	CIN	59.5	Steve Carlton	PHI	40.8	Rich Gossage	PIT	65

East	W	L	R	OR	Avg	OBA	SLG	BPF	NOPS-A	BR	Adj	Wins	ERA	PPF	NERA-A	PR	Adj	Wins	Diff
PHI	101	61	847	668	.279	.351	.448	95	124/131	143	180	18.2	3.71	92	105/ 97	33	-16	-1.7	3.5
PIT	96	66	734	665	.274	.334	.413	99	109/110	37	43	4.4	3.61	98	108/106	50	39	3.9	6.7
STL	83	79	737	688	.270	.332	.388	93	101/108	-12	34	3.4	3.81	92	103/ 95	16	-31	-3.2	1.8
CHI	81	81	692	739	.266	.333	.387	111	101/ 91	-10	-92	-9.4	4.02	113	97/110	-16	66	6.7	2.8
MON	75	87	665	736	.260	.320	.402	101	102/101	-19	-29	-3.1	4.01	102	98/100	-15	-0	-.1	-2.8
NY	64	98	587	663	.244	.315	.346	97	84/ 87	-124	-100	-10.2	3.78	97	104/101	22	4	.4	-7.2

West	W	L	R	OR	Avg	OBA	SLG	BPF	NOPS-A	BR	Adj	Wins	ERA	PPF	NERA-A	PR	Adj	Wins	Diff
LA	98	64	769	582	.266	.338	.418	100	112/112	58	59	6.0	3.22	98	121/118	113	98	9.9	1.1
CIN	88	74	802	725	.274	.348	.436	101	120/118	115	106	10.8	4.22	100	93/ 93	-47	-45	-4.7	.9
HOU	81	81	680	650	.254	.322	.385	95	98/103	-40	-4	-.5	3.54	94	111/104	61	23	2.4	-1.9
SF	75	87	673	711	.263	.326	.383	107	98/ 91	-34	-87	-8.9	3.76	108	104/113	25	78	7.9	-5.0
SD	69	93	692	834	.249	.325	.375	91	96/105	-48	13	1.3	4.45	93	88/ 81	-86	-133	-13.6	.2
ATL	61	101	678	895	.254	.322	.376	109	95/ 87	-57	-119	-12.1	4.86	112	80/ 90	-152	-74	-7.6	-.2
				713	.262	.330	.396						3.91						

AMERICAN LEAGUE 1978

Batting Runs
Jim Rice	BOS	58.7
Larry Hisle	MIL	36.3
Andy Thornton	CLE	35.6

Park Adjusted
Jim Rice	BOS	52.4
Andy Thornton	CLE	41.1
Ken Singleton	BAL	40.5

Pitching Runs
Ron Guidry	NY	62.0
Mike Caldwell	MIL	46.0
Jon Matlack	TEX	45.4

Park Adjusted
Ron Guidry	NY	63.9
Mike Caldwell	MIL	49.8
Jon Matlack	TEX	41.6

Normalized OPS
Jim Rice	BOS	169
Amos Otis	KC	154
Larry Hisle	MIL	153

Park Adjusted
Amos Otis	KC	167
Andy Thornton	CLE	163
Ken Singleton	BAL	160

Normalized ERA
Ron Guidry	NY	217
Jon Matlack	TEX	167
Mike Caldwell	MIL	160

Park Adjusted
Ron Guidry	NY	221
Mike Caldwell	MIL	165
Jon Matlack	TEX	161

On Base Average
Rod Carew	MIN	.415
Ken Singleton	BAL	.410
Mike Hargrove	TEX	.391

Slugging Percentage
Jim Rice	BOS	.600
Larry Hisle	MIL	.533
Doug De Cinces	BAL	.526

Percent of Team Wins
Ron Guidry	NY	.250
Mike Caldwell	MIL	.237
Jim Palmer	BAL	.233

Wins Above Team
Ron Guidry	NY	9.4
Enrique Romo	SEA	5.3
Larry Gura	KC	5.3

Isolated Power
Jim Rice	BOS	.285
Gorman Thomas	MIL	.270
Andy Thornton	CLE	.254

Base Stealing Runs
Julio Cruz	SEA	11.7
Ron LeFlore	DET	10.8
Bump Wills	TEX	7.2

Relievers - Runs
Rich Gossage	NY	26.3
Bob Stanley	BOS	18.6
John Hiller	DET	14.6

Park Adjusted
Rich Gossage	NY	27.2
Bob Stanley	BOS	22.6
John Hiller	DET	18.6

Defensive Runs
Mark Belanger	BAL	26.0
Buddy Bell	CLE	24.6
Bump Wills	TEX	24.1

Players Overall
Jim Rice	BOS	53.4
Roy Smalley	MIN	50.9
Amos Otis	KC	50.0

Pitchers Overall
Ron Guidry	NY	65.7
Mike Caldwell	MIL	50.6
Jon Matlack	TEX	42.0

Relief Points
Rich Gossage	NY	63
Dave La Roche	CAL	61
Don Stanhouse	BAL	51

East	W	L	R	OR	Avg	OBA	SLG	BPF	NOPS-A	BR	Adj	Wins	ERA	PPF	NERA-A	PR	Adj	Wins	Diff
NY	100	63	735	582	.267	.332	.388	103	104/101	14	-6	-.7	3.18	102	119/121	97	107	11.0	8.2
BOS	99	64	796	657	.267	.339	.424	108	117/109	96	43	4.5	3.54	107	107/114	39	80	8.3	4.8
MIL	93	69	804	650	.276	.342	.432	104	120/115	114	84	8.7	3.65	103	103/107	20	39	4.0	-.7
BAL	90	71	659	633	.258	.329	.396	90	106/117	18	83	8.6	3.57	89	106/ 94	33	-30	-3.2	4.1
DET	86	76	714	653	.271	.341	.392	110	108/ 98	45	-25	-2.7	3.67	110	103/113	17	79	8.1	-.5
CLE	69	90	639	694	.261	.326	.379	92	100/108	-18	34	3.5	3.99	92	95/ 87	-32	-81	-8.5	-5.6
TOR	59	102	590	775	.250	.310	.359	98	90/ 91	-92	-81	-8.4	4.53	100	83/ 83	-119	-118	-12.3	-.8

West	W	L	R	OR	Avg	OBA	SLG	BPF	NOPS-A	BR	Adj	Wins	ERA	PPF	NERA-A	PR	Adj	Wins	Diff
KC	92	70	743	634	.268	.333	.399	92	108/117	34	88	9.0	3.44	90	110/ 99	54	-5	-.6	2.6
TEX	87	75	692	632	.253	.335	.381	98	103/105	11	28	2.9	3.42	97	110/107	58	37	3.9	-.7
CAL	87	75	691	666	.259	.333	.370	101	99/ 98	-13	-23	-2.4	3.66	101	103/104	19	26	2.7	5.8
MIN	73	89	666	678	.267	.342	.375	98	103/106	21	37	3.8	3.71	98	102/ 99	12	-3	-.4	-11.4
CHI	71	90	634	731	.264	.320	.379	102	98/ 97	-34	-44	-4.7	4.23	103	89/ 92	-70	-54	-5.7	.8
OAK	69	93	532	690	.245	.305	.351	103	86/ 84	-116	-133	-13.8	3.62	105	104/109	25	52	5.4	-3.5
SEA	56	104	614	834	.248	.317	.359	104	91/ 88	-72	-97	-10.1	4.72	106	80/ 85	-148	-109	-11.3	-2.5
				679	.261	.329	.385						3.78						

NATIONAL LEAGUE 1978

Batting Runs
Dave Parker	PIT	52.9
Jeff Burroughs	ATL	50.1
Greg Luzinski	PHI	40.9

Park Adjusted
Dave Parker	PIT	51.6
Jeff Burroughs	ATL	43.5
Jack Clark	SF	42.0

Pitching Runs
Bob Knepper	SF	27.4
Steve Rogers	MON	27.1
Craig Swan	NY	26.4

Park Adjusted
Phil Niekro	ATL	43.0
Craig Swan	NY	30.8
Steve Rogers	MON	26.8

Normalized OPS
Dave Parker	PIT	171
Jeff Burroughs	ATL	168
Reggie Smith	LA	163

Park Adjusted
Dave Parker	PIT	168
Jack Clark	SF	163
Reggie Smith	LA	160

Normalized ERA
Craig Swan	NY	147
Steve Rogers	MON	145
Pete Vuckovich	STL	141

Park Adjusted
Craig Swan	NY	155
Steve Rogers	MON	145
Phil Niekro	ATL	140

On Base Average
Jeff Burroughs	ATL	.436
Dave Parker	PIT	.395
Gene Tenace	SD	.394

Slugging Percentage
Dave Parker	PIT	.585
Reggie Smith	LA	.559
George Foster	CIN	.546

Percent of Team Wins
Phil Niekro	ATL	.275
Ross Grimsley	MON	.263
Gaylord Perry	SD	.250

Wins Above Team
Gaylord Perry	SD	8.4
Ross Grimsley	MON	6.7
J.R. Richard	HOU	5.8

Isolated Power
George Foster	CIN	.265
Reggie Smith	LA	.264
Greg Luzinski	PHI	.261

Base Stealing Runs
Davey Lopes	LA	11.1
Omar Moreno	PIT	8.1
Bake McBride	PHI	6.6

Relievers - Runs
Kent Tekulve	PIT	18.7
Doug Bair	CIN	17.8
Ron Reed	PHI	16.4

Park Adjusted
Kent Tekulve	PIT	19.5
Doug Bair	CIN	19.3
Ron Reed	PHI	15.8

Defensive Runs
Manny Trillo	CHI	33.8
Ozzie Smith	SD	32.7
Garry Templeton	STL	27.1

Players Overall
Dave Parker	PIT	52.1
Jack Clark	SF	44.3
Ted Simmons	STL	38.5

Pitchers Overall
Phil Niekro	ATL	50.0
Craig Swan	NY	30.5
Steve Carlton	PHI	27.2

Relief Points
Rollie Fingers	SD	73
Kent Tekulve	PIT	71
Doug Bair	CIN	64

East	W	L	R	OR	Avg	OBA	SLG	BPF	NOPS-A	BR	Adj	Wins	ERA	PPF	NERA-A	PR	Adj	Wins	Diff
PHI	90	72	708	586	.258	.331	.388	100	109/109	49	46	4.9	3.33	99	107/106	40	33	3.5	.6
PIT	88	73	684	637	.257	.323	.385	102	106/104	21	9	1.0	3.42	102	105/106	26	34	3.6	2.9
CHI	79	83	664	724	.264	.334	.361	110	102/ 93	11	-52	-5.6	4.05	111	88/ 98	-76	-10	-1.1	4.8
MON	76	86	633	611	.254	.308	.379	100	100/100	-26	-26	-2.8	3.42	100	105/104	26	23	2.3	-4.7
STL	69	93	600	657	.249	.306	.358	99	93/ 94	-65	-58	-6.2	3.59	100	100/ 99	-1	-4	-.5	-5.3
NY	66	96	607	690	.245	.317	.352	104	94/ 91	-46	-72	-7.7	3.87	105	93/ 98	-46	-14	-1.6	-5.6

West	W	L	R	OR	Avg	OBA	SLG	BPF	NOPS-A	BR	Adj	Wins	ERA	PPF	NERA-A	PR	Adj	Wins	Diff
LA	95	67	727	573	.264	.340	.402	102	116/114	98	87	9.2	3.13	100	115/114	73	71	7.6	-2.7
CIN	92	69	710	688	.256	.337	.393	104	112/108	75	51	5.4	3.83	104	94/ 97	-38	-17	-1.9	8.0
SF	89	73	613	594	.248	.320	.374	91	102/112	-0	56	5.9	3.30	90	109/ 98	46	-11	-1.2	3.4
SD	84	78	591	598	.252	.323	.348	91	95/105	-36	23	2.4	3.28	90	109/ 98	48	-10	-1.1	1.7
HOU	74	88	605	634	.258	.315	.355	88	95/107	-48	25	2.7	3.63	87	99/ 86	-7	-79	-8.5	-1.2
ATL	69	93	600	750	.244	.317	.363	110	98/ 89	-27	-91	-9.7	4.10	113	87/ 99	-82	-9	-1.0	-1.2
				645	.254	.323	.372						3.58						

AMERICAN LEAGUE 1979

Batting Runs
Fred Lynn	BOS	62.1
Jim Rice	BOS	50.2
Sixto Lezcano	MIL	44.6

Park Adjusted
Fred Lynn	BOS	59.4
Ken Singleton	BAL	47.7
Jim Rice	BOS	47.2

Pitching Runs
Tommy John	NY	38.7
Ron Guidry	NY	37.9
Dennis Eckersley	BOS	34.1

Park Adjusted
Jerry Koosman	MIN	39.1
Dennis Eckersley	BOS	37.1
Ron Guidry	NY	31.5

Normalized OPS
Fred Lynn	BOS	182
Sixto Lezcano	MIL	165
Jim Rice	BOS	160

Park Adjusted
Fred Lynn	BOS	176
Steve Kemp	DET	164
Sixto Lezcano	MIL	160

Normalized ERA
Ron Guidry	NY	152
Tommy John	NY	143
Dennis Eckersley	BOS	142

Park Adjusted
Dennis Eckersley	BOS	145
Ron Guidry	NY	143
Jerry Koosman	MIN	139

On Base Average
Darrell Porter	KC	.429
Fred Lynn	BOS	.426
Brian Downing	CAL	.420

Slugging Percentage
Fred Lynn	BOS	.637
Jim Rice	BOS	.596
Sixto Lezcano	MIL	.573

Percent of Team Wins
Jerry Koosman	MIN	.244
Tommy John	NY	.236
Mike Flanagan	BAL	.225

Wins Above Team
Ron Davis	NY	5.7
Tommy John	NY	5.3
Jack Morris	DET	5.2

Isolated Power
Fred Lynn	BOS	.303
Gorman Thomas	MIL	.294
Jim Rice	BOS	.271

Base Stealing Runs
Willie Wilson	KC	17.7
Ron LeFlore	DET	15.0
Julio Cruz	SEA	9.3

Relievers - Runs
Jim Kern	TEX	42.2
Sid Monge	CLE	26.6
Aurelio Lopez	DET	25.7

Park Adjusted
Jim Kern	TEX	41.7
Mike Marshall	MIN	32.8
Sid Monge	CLE	28.8

Defensive Runs
Roy Smalley	MIN	32.7
Bucky Dent	NY	30.9
Rick Burleson	BOS	26.3

Players Overall
Roy Smalley	MIN	57.0
George Brett	KC	55.6
Fred Lynn	BOS	54.1

Pitchers Overall
Jerry Koosman	MIN	41.0
Jim Kern	TEX	41.0
Dennis Eckersley	BOS	37.8

Relief Points
Jim Kern	TEX	79
Mike Marshall	MIN	70
Aurelio Lopez	DET	57

East	W	L	R	OR	Avg	OBA	SLG	BPF	NOPS-A	BR	Adj	Wins	ERA	PPF	NERA-A	PR	Adj	Wins	Diff
BAL	102	57	757	582	.261	.339	.419	96	107/112	25	54	5.3	3.27	94	129/121	153	112	10.9	6.3
MIL	95	66	807	722	.280	.347	.448	103	118/114	95	72	7.0	4.04	102	105/107	30	46	4.5	2.9
BOS	91	69	841	711	.283	.347	.456	103	120/116	108	81	8.0	4.03	103	105/108	32	49	4.8	-1.8
NY	89	71	734	672	.266	.331	.406	95	101/106	-20	15	1.4	3.86	94	110/103	59	20	2.0	5.6
DET	85	76	770	738	.269	.342	.415	93	107/115	24	77	7.6	4.28	92	99/ 91	-6	-61	-6.0	2.9
CLE	81	80	760	805	.258	.344	.384	103	98/ 95	-20	-41	-4.1	4.58	104	92/ 96	-54	-30	-3.0	7.7
TOR	53	109	613	862	.251	.313	.363	101	84/ 83	-141	-147	-14.5	4.85	103	87/ 90	-96	-73	-7.3	-6.2

West	W	L	R	OR	Avg	OBA	SLG	BPF	NOPS-A	BR	Adj	Wins	ERA	PPF	NERA-A	PR	Adj	Wins	Diff
CAL	88	74	866	768	.282	.354	.429	93	114/123	83	140	13.7	4.34	91	98/ 89	-16	-76	-7.5	.8
KC	85	77	851	816	.282	.347	.422	105	110/105	51	12	1.2	4.45	105	95/100	-35	0	-.0	2.9
TEX	83	79	750	698	.278	.337	.409	100	104/104	-0	1	.1	3.87	99	109/108	58	52	5.1	-3.2
MIN	82	80	764	725	.278	.344	.402	111	103/ 93	7	-75	-7.4	4.15	111	102/113	12	89	8.7	-.2
CHI	73	87	730	748	.275	.335	.410	111	104/ 93	-3	-87	-8.5	4.13	112	102/115	16	95	9.3	-7.7
SEA	67	95	711	820	.269	.334	.404	100	102/101	-16	-18	-1.9	4.61	101	92/ 93	-60	-50	-5.0	-7.1
OAK	54	108	573	860	.239	.304	.346	89	77/ 86	-188	-111	-10.9	4.77	91	89/ 81	-86	-143	-14.1	-2.0
				752	.270	.337	.408						4.23						

NATIONAL LEAGUE 1979

Batting Runs
Dave Winfield	SD	47.4
Keith Hernandez	STL	47.1
Mike Schmidt	PHI	44.9

Park Adjusted
Dave Winfield	SD	48.5
Mike Schmidt	PHI	46.9
Keith Hernandez	STL	41.0

Pitching Runs
J.R. Richard	HOU	33.4
Joe Niekro	HOU	21.7
Steve Rogers	MON	20.5

Park Adjusted
Phil Niekro	ATL	32.8
John Fulgham	STL	24.4
Jim Bibby	PIT	18.3

Normalized OPS
Mike Schmidt	PHI	157
Dave Winfield	SD	157
Dave Kingman	CHI	157

Park Adjusted
George Foster	CIN	163
Mike Schmidt	PHI	162
Dave Winfield	SD	160

Normalized ERA
J.R. Richard	HOU	138
Dan Schatzeder	MON	132
Burt Hooton	LA	126

Park Adjusted
Phil Niekro	ATL	125
Dan Schatzeder	MON	125
Bruce Kison	PIT	125

On Base Average
Keith Hernandez	STL	.421
Pete Rose	PHI	.421
Gene Tenace	SD	.407

Slugging Percentage
Dave Kingman	CHI	.613
Mike Schmidt	PHI	.564
George Foster	CIN	.561

Percent of Team Wins
Phil Niekro	ATL	.318
Joe Niekro	HOU	.236
Rick Reuschel	CHI	.225

Wins Above Team
Phil Niekro	ATL	5.5
Rick Sutcliffe	LA	4.6
Tom Seaver	CIN	4.3

Isolated Power
Dave Kingman	CHI	.325
Mike Schmidt	PHI	.311
George Foster	CIN	.259

Base Stealing Runs
Davey Lopes	LA	10.8
Omar Moreno	PIT	10.5
Jerry Royster	ATL	5.7

Relievers - Runs
Joe Sambito	HOU	19.8
Elias Sosa	MON	19.8
Tom Hume	CIN	17.8

Park Adjusted
Bruce Sutter	CHI	19.2
Kent Tekulve	PIT	18.5
Elias Sosa	MON	17.3

Defensive Runs
Garry Templeton	STL	26.2
Ozzie Smith	SD	23.8
Ted Sizemore	CHI	23.4

Players Overall
Mike Schmidt	PHI	60.3
Dave Winfield	SD	53.8
Keith Hernandez	STL	49.1

Pitchers Overall
Phil Niekro	ATL	36.9
John Fulgham	STL	23.2
Bruce Sutter	CHI	20.7

Relief Points
Bruce Sutter	CHI	80
Kent Tekulve	PIT	74
Joe Sambito	HOU	53

East	W	L	R	OR	Avg	OBA	SLG	BPF	NOPS-A	BR	Adj	Wins	ERA	PPF	NERA-A	PR	Adj	Wins	Diff
PIT	98	64	775	643	.272	.333	.416	108	114/105	68	12	1.3	3.42	107	109/117	53	95	9.8	5.9
MON	95	65	701	581	.264	.321	.408	97	108/112	23	45	4.6	3.14	95	119/113	97	66	6.8	3.6
STL	86	76	731	693	.278	.335	.401	108	110/102	46	-8	-.9	3.74	108	100/108	1	49	5.0	.9
PHI	84	78	683	718	.266	.343	.396	97	110/113	63	81	8.4	4.16	98	90/ 88	-66	-81	-8.5	3.1
CHI	80	82	706	707	.269	.331	.403	105	109/104	40	7	.7	3.90	105	96/101	-24	6	.7	-2.4
NY	63	99	593	706	.250	.315	.350	100	89/ 89	-89	-92	-9.6	3.84	102	97/ 99	-15	-4	-.6	-7.8

West	W	L	R	OR	Avg	OBA	SLG	BPF	NOPS-A	BR	Adj	Wins	ERA	PPF	NERA-A	PR	Adj	Wins	Diff
CIN	90	71	731	644	.264	.340	.396	96	109/114	55	83	8.6	3.60	95	104/ 98	22	-10	-1.1	2.0
HOU	89	73	583	582	.256	.317	.344	87	88/100	-92	-9	-1.0	3.20	86	117/101	87	4	8.6	8.6
LA	79	83	739	717	.263	.333	.412	96	112/117	62	90	9.2	3.87	95	97/ 92	-20	-47	-4.9	-6.3
SF	71	91	672	751	.246	.322	.365	96	95/ 98	-45	-21	-2.3	4.16	97	90/ 87	-67	-83	-8.6	.9
SD	68	93	603	681	.242	.313	.348	99	88/ 89	-96	-87	-9.0	3.70	100	101/101	6	3	.4	-3.8
ATL	66	94	669	763	.256	.320	.377	112	98/ 88	-30	-109	-11.3	4.18	114	89/102	-68	12	1.2	-3.9
				682	.261	.327	.385						3.74						

AMERICAN LEAGUE 1980

Batting Runs			Park Adjusted			Pitching Runs			Park Adjusted		
George Brett	KC	64.8	George Brett	KC	65.6	Mike Norris	OAK	47.6	Mike Norris	OAK	34.0
Reggie Jackson	NY	49.1	Reggie Jackson	NY	50.5	Larry Gura	KC	34.1	Britt Burns	CHI	31.1
Cecil Cooper	MIL	42.9	Cecil Cooper	MIL	43.4	Britt Burns	CHI	31.9	Larry Gura	KC	30.9
Normalized OPS			**Park Adjusted**			**Normalized ERA**			**Park Adjusted**		
George Brett	KC	204	George Brett	KC	207	Rudy May	NY	164	Rudy May	NY	158
Reggie Jackson	NY	169	Reggie Jackson	NY	172	Mike Norris	OAK	159	Mike Norris	OAK	143
Cecil Cooper	MIL	152	Cecil Cooper	MIL	153	Britt Burns	CHI	143	Britt Burns	CHI	141
On Base Average			**Slugging Percentage**			**Percent of Team Wins**			**Wins Above Team**		
George Brett	KC	.461	George Brett	KC	.664	Mike Norris	OAK	.265	Mike Norris	OAK	7.6
Willie Randolph	NY	.429	Reggie Jackson	NY	.597	Steve Stone	BAL	.250	Steve Stone	BAL	6.5
Rickey Henderson	OAK	.422	Ben Oglivie	MIL	.563	Len Barker	CLE	.241	Danny Darwin	TEX	5.6
Isolated Power			**Base Stealing Runs**			**Relievers - Runs**			**Park Adjusted**		
Reggie Jackson	NY	.298	Willie Wilson	KC	17.7	Doug Corbett	MIN	31.1	Doug Corbett	MIN	32.0
George Brett	KC	.274	Rickey Henderson	OAK	14.4	Tom Burgmeier	BOS	22.5	Tom Burgmeier	BOS	26.8
Ben Oglivie·	MIL	.258	Julio Cruz	SEA	9.3	Rich Gossage	NY	19.5	Rich Gossage	NY	17.8
Defensive Runs			**Players Overall**			**Pitchers Overall**			**Relief Points**		
Rick Burleson	BOS	31.0	George Brett	KC	68.9	Mike Norris	OAK	37.3	Dan Quisenberry	KC	83
Roy Smalley	MIN	25.7	Rickey Henderson	OAK	65.8	Doug Corbett	MIN	35.0	Rich Gossage	NY	76
Doug DeCinces	BAL	23.4	Ben Oglivie	MIL	49.5	Larry Gura	KC	31.6	Ed Farmer	CHI	65

East	W	L	R	OR	Avg	OBA	SLG	BPF	NOPS-A	BR	Adj	Wins	ERA	PPF	NERA-A	PR	Adj	Wins	Diff
NY	103	59	820	662	.267	.346	.425	98	114/116	79	94	9.4	3.59	96	113/108	74	49	4.9	7.7
BAL	100	62	805	640	.273	.344	.413	102	110/107	52	36	3.6	3.64	101	111/112	65	70	7.0	8.5
MIL	86	76	811	682	.275	.332	.448	99	117/118	78	82	8.2	3.73	98	108/106	50	38	3.7	-7.0
BOS	83	77	757	767	.283	.343	.436	109	116/106	86	18	1.8	4.39	110	92/101	-55	7	.7	.5
DET	84	78	830	757	.273	.351	.409	107	110/104	66	16	1.6	4.25	106	95/101	-33	7	.7	.6
CLE	79	81	738	807	.277	.355	.381	98	103/105	26	43	4.3	4.69	98	86/ 85	-102	-114	-11.4	6.1
TOR	67	95	624	762	.251	.310	.383	101	92/ 91	-91	-101	-10.2	4.19	103	96/ 99	-24	-4	-.5	-3.3

West	W	L	R	OR	Avg	OBA	SLG	BPF	NOPS-A	BR	Adj	Wins	ERA	PPF	NERA-A	PR	Adj	Wins	Diff
KC	97	65	809	694	.286	.348	.413	99	111/112	62	72	7.2	3.85	98	105/102	31	15	1.5	7.3
OAK	83	79	686	642	.259	.324	.385	91	96/106	-50	17	1.7	3.49	89	116/104	90	20	2.0	-1.7
MIN	77	84	670	724	.265	.322	.381	101	95/ 94	-64	-70	-7.1	3.93	102	103/104	18	28	2.8	.9
TEX	76	85	756	752	.284	.342	.405	100	107/107	30	33	3.3	4.04	100	100/100	0	-1	-.2	-7.5
CHI	70	90	587	722	.259	.314	.370	98	89/ 91	-103	-89	-8.9	3.91	99	103/102	20	15	1.5	-2.6
CAL	65	95	698	797	.265	.335	.378	100	97/ 97	-32	-30	-3.1	4.52	101	89/ 90	-75	-70	-7.1	-4.8
SEA	59	103	610	793	.248	.311	.356	98	84/ 86	-135	-122	-12.3	4.38	100	92/ 92	-54	-54	-5.5	-4.2
				729	.269	.334	.399						4.04						

NATIONAL LEAGUE 1980

Batting Runs			Park Adjusted			Pitching Runs			Park Adjusted		
Mike Schmidt	PHI	56.5	Mike Schmidt	PHI	50.4	Steve Carlton	PHI	42.9	Steve Carlton	PHI	53.0
Keith Hernandez	STL	43.3	Keith Hernandez	STL	40.8	Don Sutton	LA	33.0	Don Sutton	LA	29.8
Mike Easler	PIT	37.2	Mike Easler	PIT	35.5	Jerry Reuss	LA	27.8	Jerry Reuss	LA	24.3
Normalized OPS			**Park Adjusted**			**Normalized ERA**			**Park Adjusted**		
Mike Schmidt	PHI	178	Mike Schmidt	PHI	164	Don Sutton	LA	163	Steve Carlton	PHI	167
Keith Hernandez	STL	151	Jack Clark	SF	156	Steve Carlton	PHI	154	Don Sutton	LA	157
Jack Clark	SF	151	Keith Hernandez	STL	146	Jerry Reuss	LA	143	Jerry Reuss	LA	138
On Base Average			**Slugging Percentage**			**Percent of Team Wins**			**Wins Above Team**		
Keith Hernandez	STL	.410	Mike Schmidt	PHI	.624	Steve Carlton	PHI	.264	Jim Bibby	PIT	7.3
Cesar Cedeno	HOU	.390	Bob Horner	ATL	.529	Jim Bibby	PIT	.229	Steve Carlton	PHI	6.9
Jack Clark	SF	.390	Jack Clark	SF	.517	Joe Niekro	HOU	.215	Jerry Reuss	LA	5.2
Isolated Power			**Base Stealing Runs**			**Relievers - Runs**			**Park-Adjusted**		
Mike Schmidt	PHI	.338	Ron LeFlore	MON	17.7	Tug McGraw	PHI	21.9	Bill Caudill	CHI	25.5
Bob Horner	ATL	.261	Jerry Mumphrey	SD	12.6	Bill Caudill	CHI	20.3	Tug McGraw	PHI	25.0
Jack Clark	SF	.233	Dave Collins	CIN	11.1	Rick Camp	ATL	20.3	Tom Hume	CIN	20.2
Defensive Runs			**Players Overall**			**Pitchers Overall**			**Relief Points**		
Ozzie Smith	SD	42.8	Mike Schmidt	PHI	71.4	Steve Carlton	PHI	51.3	Rollie Fingers	SD	59
Garry Templeton	STL	31.2	Ozzie Smith	SD	52.3	Don Sutton	LA	25.6	Bruce Sutter	CHI	58
Mike Schmidt	PHI	20.9	Garry Templeton	STL	49.0	Tug McGraw	PHI	25.5	Tom Hume	CIN	58

East	W	L	R	OR	Avg	OBA	SLG	BPF	NOPS-A	BR	Adj	Wins	ERA	PPF	NERA-A	PR	Adj	Wins	Diff
PHI	91	71	728	639	.270	.330	.400	109	112/103	61	2	.2	3.46	108	104/113	24	74	7.8	2.0
MON	90	72	694	629	.257	.327	.388	98	108/110	36	52	5.5	3.48	96	104/100	21	0	-.0	3.5
PIT	83	79	666	646	.266	.325	.388	104	107/103	27	3	.3	3.58	104	101/105	5	27	2.8	-1.1
STL	74	88	738	710	.275	.331	.400	103	112/109	83	40	4.3	3.93	108	92/ 95	-50	-31	-3.4	-7.9
NY	67	95	611	702	.257	.322	.345	98	93/ 95	-51	-37	-4.1	3.89	99	93/ 92	-44	-50	-5.4	-4.6
CHI	64	98	614	728	.251	.311	.365	108	96/ 89	-46	-99	-10.5	3.89	110	93/102	-45	14	1.5	-7.9

West	W	L	R	OR	Avg	OBA	SLG	BPF	NOPS-A	BR	Adj	Wins	ERA	PPF	NERA-A	PR	Adj	Wins	Diff
HOU	93	70	637	589	.261	.328	.367	94	101/108	2	40	4.3	3.11	93	116/108	82	40	4.2	3.0
LA	92	71	663	591	.263	.325	.388	97	107/110	29	47	4.9	3.26	96	111/106	57	34	3.6	2.0
CIN	89	73	707	670	.262	.330	.386	108	108/100	39	-11	-1.3	3.85	108	94/101	-38	7	.7	8.6
ATL	81	80	630	660	.250	.308	.380	95	100/105	-26	2	.2	3.77	95	96/ 91	-24	-51	-5.5	5.8
SF	75	86	573	634	.244	.311	.342	97	89/ 92	-81	-60	-6.4	3.45	97	104/102	25	9	.9	-.0
SD	73	89	591	654	.255	.326	.342	91	93/102	-47	14	1.5	3.66	91	99/ 89	-7	-62	-6.7	-2.8
				654	.259	.323	.374						3.61						

AMERICAN LEAGUE   1981

Batting Runs			Park Adjusted			Pitching Runs			Park Adjusted		
Dwight Evans	BOS	39.8	Dwight Evans	BOS	37.8	Steve McCatty	OAK	27.7	Dave Stieb	TOR	24.7
Tom Paciorek	SEA	28.4	Rickey Henderson	OAK	31.6	Dave Righetti	NY	18.7	Larry Gura	KC	22.6
Bobby Grich	CAL	28.3	Tom Paciorek	SEA	28.7	Larry Gura	KC	18.0	Dennis Leonard	KC	20.6
Normalized OPS			**Park Adjusted**			**Normalized ERA**			**Park Adjusted**		
Dwight Evans	BOS	169	Eddie Murray	BAL	163	Steve McCatty	OAK	158	Dennis Lamp	CHI	153
Bobby Grich	CAL	164	Dwight Evans	BOS	163	Sammy Stewart	BAL	157	Sammy Stewart	BAL	149
Tom Paciorek	SEA	156	Rickey Henderson	OAK	158	Dennis Lamp	CHI	152	Larry Gura	KC	143
On Base Average			**Slugging Percentage**			**Percent of Team Wins**			**Wins Above Team**		
Mike Hargrove	CLE	.432	Bobby Grich	CAL	.543	Dave Stieb	TOR	.297	Dave Stieb	TOR	4.6
Dwight Evans	BOS	.418	Eddie Murray	BAL	.534	Dennis Leonard	KC	.260	Pete Vuckovich	MIL	4.5
Rickey Henderson	OAK	.411	Dwight Evans	BOS	.522	Dennis Martinez	BAL	.237	Dennis Martinez	BAL	4.1
Isolated Power			**Base Stealing Runs**			**Relievers - Runs**			**Park Adjusted**		
Eddie Murray	BAL	.241	Julio Cruz	SEA	8.1	Rollie Fingers	MIL	22.7	Rollie Fingers	MIL	22.3
Bobby Grich	CAL	.239	Rick Manning	CLE	5.7	Rich Gossage	NY	15.1	Doug Corbett	MIN	14.9
Gorman Thomas	MIL	.234	Willie Wilson	KC	5.4	Dan Quisenberry	KC	13.2	Dan Quisenberry	KC	14.9
Defensive Runs			**Players Overall**			**Pitchers Overall**			**Relief Points**		
Buddy Bell	TEX	27.0	Rickey Henderson	OAK	51.1	Dave Stieb	TOR	26.5	Rollie Fingers	MIL	65
Robin Yount	MIL	25.2	Buddy Bell	TEX	48.0	Larry Gura	KC	22.9	Rich Gossage	NY	44
Bump Wills	TEX	23.9	Bobby Grich	CAL	37.7	Rollie Fingers	MIL	22.5	Lamarr Hoyt	CHI	35

East	W	L	R	OR	Avg	OBA	SLG	BPF	NOPS-A	BR	Adj	Wins	ERA	PPF	NERA-A	PR	Adj	Wins	Diff
MIL	62	47	493	459	.257	.317	.391	99	106/107	9	13	1.3	3.91	98	94/ 92	-26	-32	-3.5	9.6
BAL	59	46	429	437	.251	.331	.379	95	106/111	22	42	4.4	3.70	95	99/ 94	-2	-21	-2.3	4.4
NY	59	48	421	343	.252	.328	.391	97	109/112	29	40	4.2	2.90	96	126/121	81	64	6.7	-5.5
DET	60	49	427	404	.256	.334	.368	98	103/106	14	25	2.6	3.54	97	104/101	14	2	.2	2.7
BOS	59	49	519	481	.275	.343	.399	104	115/111	65	48	5.1	3.83	103	96/ 99	-17	-3	-.4	.4
CLE	52	51	431	442	.263	.331	.351	94	97/104	-10	15	1.6	3.88	94	94/ 88	-22	-46	-4.9	3.8
TOR	37	69	329	466	.226	.288	.330	117	79/ 68	-102	-172	-18.2	3.81	120	96/115	-15	61	6.4	-4.2

West	W	L	R	OR	Avg	OBA	SLG	BPF	NOPS-A	BR	Adj	Wins	ERA	PPF	NERA-A	PR	Adj	Wins	Diff
OAK	64	45	458	403	.247	.314	.379	91	102/111	-7	31	3.2	3.30	90	111/ 99	40	-1	-.2	6.5
TEX	57	48	452	389	.270	.329	.369	84	103/122	5	72	7.6	3.40	82	108/ 89	28	-40	-4.3	1.2
CHI	54	52	476	423	.272	.338	.387	101	111/109	41	35	3.7	3.50	101	105/105	17	20	2.1	-4.8
KC	50	53	397	405	.267	.327	.383	106	106/100	17	-8	-.9	3.56	106	103/110	11	35	3.7	-4.3
CAL	51	59	476	453	.256	.332	.380	104	107/102	26	7	.7	3.71	104	99/103	-4	11	1.2	-5.8
SEA	44	65	426	521	.251	.316	.368	99	99/ 99	-18	-15	-1.6	4.24	101	86/ 87	-63	-60	-6.4	-2.5
MIN	41	68	378	486	.240	.295	.338	109	84/ 77	-86	-126	-13.4	3.98	112	92/103	-33	12	1.2	-1.4
				437															
					.256	.323	.373						3.66						

NATIONAL LEAGUE   1981

Batting Runs			Park Adjusted			Pitching Runs			Park Adjusted		
Mike Schmidt	PHI	50.1	Mike Schmidt	PHI	46.0	Nolan Ryan	HOU	29.8	Steve Carlton	PHI	29.7
Andre Dawson	MON	28.8	Andre Dawson	MON	34.0	Bob Knepper	HOU	22.9	Tom Seaver	CIN	22.0
George Foster	CIN	27.7	George Foster	CIN	24.1	Steve Carlton	PHI	22.7	F. Valenzuela	LA	19.6
Normalized OPS			**Park Adjusted**			**Normalized ERA**			**Park Adjusted**		
Mike Schmidt	PHI	204	Mike Schmidt	PHI	187	Nolan Ryan	HOU	206	Nolan Ryan	HOU	169
Andre Dawson	MON	159	Andre Dawson	MON	180	Bob Knepper	HOU	160	Steve Carlton	PHI	158
George Foster	CIN	152	Tim Raines	MON	153	Burt Hooton	LA	153	Vida Blue	SF	151
On Base Average			**Slugging Percentage**			**Percent of Team Wins**			**Wins Above Team**		
Mike Schmidt	PHI	.439	Mike Schmidt	PHI	.644	Mike Krukow	CHI	.237	Tom Seaver	CIN	5.0
Keith Hernandez	STL	.405	Andre Dawson	MON	.553	Steve Carlton	PHI	.220	Steve Carlton	PHI	4.3
Gary Matthews	PHI	.404	George Foster	CIN	.519	Tom Seaver	CIN	.212	Rick Camp	ATL	3.8
Isolated Power			**Base Stealing Runs**			**Relievers - Runs**			**Park Adjusted**		
Mike Schmidt	PHI	.328	Tim Raines	MON	14.7	Gary Lucas	SD	14.9	Rick Camp	ATL	15.5
Andre Dawson	MON	.251	Lon Lacy	PIT	5.4	Rick Camp	ATL	14.5	Gary Lucas	SD	14.4
Dave Kingman	NY	.235	Andre Dawson	MON	5.4	Al Holland	SF	12.2	Al Holland	SF	14.4
Defensive Runs			**Players Overall**			**Pitchers Overall**			**Relief Points**		
Ozzie Smith	SD	27.8	Mike Schmidt	PHI	64.7	Steve Carlton	PHI	27.9	Bruce Sutter	STL	51
Mike Schmidt	PHI	22.4	Andre Dawson	MON	48.3	Tom Seaver	CIN	24.4	Rick Camp	ATL	49
Garry Templeton	STL	13.7	Tim Raines	MON	31.3	F. Valenzuela	LA	24.2	Greg Minton	SF	45

East	W	L	R	OR	Avg	OBA	SLG	BPF	NOPS-A	BR	Adj	Wins	ERA	PPF	NERA-A	PR	Adj	Wins	Diff
STL	59	43	464	417	.265	.339	.377	104	110/106	43	25	2.7	3.63	104	96/100	-14	-0	-.2	5.5
MON	60	48	443	394	.246	.319	.370	89	102/116	3	51	5.4	3.30	86	106/ 92	21	-29	-3.2	3.8
PHI	59	48	491	472	.273	.344	.389	109	115/106	66	27	3.0	4.05	109	86/ 94	-58	-23	-2.6	5.1
PIT	46	56	407	425	.257	.314	.369	101	101/100	-8	-11	-1.2	3.56	101	98/ 99	-7	-3	-.4	-3.3
NY	41	62	348	432	.248	.311	.356	99	96/ 97	-25	-20	-2.2	3.55	100	98/ 99	-5	-4	-.5	-7.8
CHI	38	65	370	483	.236	.306	.340	96	90/ 93	-51	-35	-3.9	4.02	98	87/ 86	-55	-61	-6.6	-3.0

West	W	L	R	OR	Avg	OBA	SLG	BPF	NOPS-A	BR	Adj	Wins	ERA	PPF	NERA-A	PR	Adj	Wins	Diff
CIN	66	42	464	440	.267	.339	.385	107	113/105	53	22	2.4	3.72	107	94/101	-24	2	.2	9.4
LA	63	47	450	356	.262	.325	.374	100	105/106	16	18	2.0	3.01	98	116/113	53	44	4.7	1.3
HOU	61	49	394	331	.257	.321	.356	85	99/116	-11	52	5.6	2.66	82	131/107	92	21	2.3	-1.9
SF	56	55	427	414	.250	.322	.357	106	99/ 94	-7	-31	-3.4	3.28	106	106/113	24	46	5.0	-1.9
ATL	50	56	395	416	.243	.308	.349	103	93/ 90	-40	-52	-5.7	3.45	103	101/105	4	18	1.9	.8
SD	41	69	382	455	.256	.316	.346	100	94/ 94	-32	-32	-3.5	3.73	101	94/ 95	-25	-20	-2.2	-8.2
				420															
					.255	.322	.364						3.49						

AMERICAN LEAGUE 1982

Batting Runs
Robin Yount MIL 50.3
Dwight Evans BOS 48.5
Eddie Murray BAL 42.5

Park Adjusted
Robin Yount MIL 55.7
Hal McRae KC 43.5
Eddie Murray BAL 43.0

Pitching Runs
Rick Sutcliffe CLE 26.9
Dave Stieb TOR 26.6
Jim Palmer BAL 23.9

Park Adjusted
Dave Stieb TOR 42.5
Floyd Bannister SEA 32.8
Rick Sutcliffe CLE 31.0

Normalized OPS
Robin Yount MIL 160
Eddie Murray BAL 156
Dwight Evans BOS 155

Park Adjusted
Robin Yount MIL 171
Hal McRae KC 158
Eddie Murray BAL 157

Normalized ERA
Rick Sutcliffe CLE 138
Bob Stanley BOS 131
Jim Palmer BAL 130

Park Adjusted
Rick Sutcliffe CLE 144
Dave Stieb TOR 141
Bob Stanley BOS 140

On Base Average
Dwight Evans BOS .403
Toby Harrah CLE .400
Rickey Henderson OAK .399

Slugging Percentage
Robin Yount MIL .578
Dave Winfield NY .560
Eddie Murray BAL .549

Percent of Team Wins
Charlie Hough TEX .250
Lamarr Hoyt CHI .218
Dave Stieb TOR .218

Wins Above Team
Charlie Hough TEX 5.5
MIN 4.8
Pete Vuckovich MIL 4.6

Isolated Power
Dave Winfield NY .280
Gorman Thomas MIL .261
Reggie Jackson CAL .257

Base Stealing Runs
Rickey Henderson OAK 13.8
Paul Molitor MIL 6.9
Miguel Dilone CLE 6.9

Relievers - Runs
Dan Spillner CLE 23.7
Dan Quisenberry KC 23.1
Tom Burgmeier BOS 20.2

Park Adjusted
Dan Spillner CLE 26.3
Bill Caudill SEA 24.3
Tom Burgmeier BOS 23.4

Defensive Runs
Buddy Bell TEX 36.9
Tony Bernazard CHI 27.5
Doug De Cinces CAL 23.8

Players Overall
Robin Yount MIL 69.4
Doug De Cinces CAL 57.8
Buddy Bell TEX 55.8

Pitchers Overall
Dave Stieb TOR 46.2
Floyd Bannister SEA 32.2
Rick Sutcliffe CLE. 31.8

Relief Points
Dan Quisenberry KC 81
Bill Caudill SEA 67
Rich Gossage NY 63

East	W	L	R	OR	Avg	OBA	SLG	BPF	NOPS-A	BR	Adj	Wins	ERA	PPF	NERA-A	PR	Adj	Wins	Diff
MIL	95	67	891	717	.279	.337	.455	93	121/130	109	158	15.8	3.98	91	103/ 94	16	-41	-4.2	2.4
BAL	94	68	774	687	.266	.344	.419	99	112/113	67	72	7.2	3.99	98	102/101	15	5	.5	5.3
BOS	89	73	753	713	.274	.342	.407	107	108/101	39	-9	-1.0	4.04	107	101/108	6	51	5.1	3.9
DET	83	79	729	685	.266	.326	.418	99	108/108	15	20	2.0	3.81	99	107/106	43	35	3.5	-3.5
NY	79	83	709	716	.256	.331	.398	98	103/105	-4	10	1.0	3.99	98	102/100	14	0	-.0	-3.0
TOR	78	84	651	701	.262	.317	.383	110	95/ 85	-69	-147	-14.8	3.95	112	103/116	22	102	10.2	1.7
CLE	78	84	683	748	.262	.343	.373	103	98/ 95	-12	-36	-3.7	4.11	104	99/103	-5	22	2.2	-1.5

West	W	L	R	OR	Avg	OBA	SLG	BPF	NOPS-A	BR	Adj	Wins	ERA	PPF	NERA-A	PR	Adj	Wins	Diff
CAL	93	69	814	670	.274	.350	.433	100	118/118	108	108	10.8	3.82	98	107/105	43	33	3.3	-2.1
KC	90	72	784	717	.285	.340	.428	93	114/123	67	118	11.8	4.08	92	100/ 92	1	-53	-5.4	2.6
CHI	87	75	786	710	.273	.340	.413	98	109/111	43	56	5.6	3.87	97	106/103	34	17	1.7	-1.3
SEA	76	86	651	712	.254	.313	.381	112	93/ 83	-83	-169	-17.0	3.90	113	105/118	30	118	11.8	.2
OAK	68	94	691	819	.236	.312	.367	99	89/ 90	-105	-97	-9.8	4.55	100	90/ 90	-74	-73	-7.4	4.2
TEX	64	98	590	749	.249	.309	.359	93	86/ 93	-127	-74	-7.5	4.30	93	95/ 89	-34	-76	-7.7	-1.8
MIN	60	102	657	819	.257	.319	.396	99	99/100	-42	-35	-3.6	4.76	101	86/ 86	-108	-104	-10.5	-6.9
			726		.264	.330	.402						4.08						

NATIONAL LEAGUE 1982

Batting Runs
Mike Schmidt PHI 47.2
Al Oliver MON 42.7
Jason Thompson PIT 40.7

Park Adjusted
Mike Schmidt PHI 51.0
Pedro Guerrero LA 42.7
Leon Durham CHI 35.8

Pitching Runs
Steve Rogers MON 37.0
Joe Niekro HOU 34.2
Joaquin Andujar STL 33.6

Park Adjusted
Steve Rogers MON 52.9
Joaquin Andujar STL 40.3
Joe Niekro HOU 29.8

Normalized OPS
Mike Schmidt PHI 165
Pedro Guerrero LA 153
Leon Durham CHI 152

Park Adjusted
Mike Schmidt PHI 174
Pedro Guerrero LA 163
Sixto Lezcano SD 158

Normalized ERA
Steve Rogers MON 150
Joe Niekro HOU 146
Joaquin Andujar STL 146

Park Adjusted
Steve Rogers MON 172
Joaquin Andujar STL 155
Joe Niekro HOU 140

On Base Average
Mike Schmidt PHI .407
Keith Hernandez STL .404
Joe Morgan SF .402

Slugging Percentage
Mike Schmidt PHI .547
Pedro Guerrero LA .536
Leon Durham CHI .521

Percent of Team Wins
Steve Carlton PHI .258
Mario Soto CIN .230
Steve Rogers MON .221

Wins Above Team
Phil Niekro ATL 6.3
Steve Rogers MON 5.6
Steve Carlton PHI 5.5

Isolated Power
Mike Schmidt PHI .267
Bob Horner ATL .240
Pedro Guerrero LA .231

Base Stealing Runs
Tim Raines MON 13.8
Mookie Wilson NY 7.8
Dickie Thon HOU 6.3

Relievers - Runs
Greg Minton SF 24.3
Rod Scurry PIT 21.7
Jeff Reardon MON 18.7

Park Adjusted
Rod Scurry PIT 26.9
Jeff Reardon MON 24.9
Steve Bedrosian ATL 23.5

Defensive Runs
Ozzie Smith STL 33.9
Glenn Hubbard ATL 22.9
Rafael Ramirez ATL 18.4

Players Overall
Mike Schmidt PHI 65.8
Pedro Guerrero LA 53.5
Sixto Lezcano SD 44.3

Pitchers Overall
Steve Rogers MON 52.1
Joaquin Andujar STL 40.2
Rod Scurry PIT 27.5

Relief Points
Bruce Sutter STL 82
Greg Minton SF 76
Gene Garber ATL 66

East	W	L	R	OR	Avg	OBA	SLG	BPF	NOPS-A	BR	Adj	Wins	ERA	PPF	NERA-A	PR	Adj	Wins	Diff
STL	92	70	685	609	.264	.337	.364	107	103/ 97	24	-21	-2.3	3.37	106	107/114	39	76	8.0	5.4
PHI	89	73	664	654	.260	.325	.376	95	104/110	14	50	5.2	3.61	94	100/ 94	0	-34	-3.7	6.5
MON	86	76	697	616	.262	.327	.396	115	111/ 96	52	-45	-4.3	3.82	114	109/124	46	130	13.6	-3.8
PIT	84	78	724	696	.273	.330	.408	112	115/103	80	-0	-.1	3.82	113	94/106	-34	39	4.1	-1.0
CHI	73	89	676	709	.260	.319	.375	103	102/ 99	-3	-22	-2.4	3.93	103	92/ 95	-51	-30	-3.3	-2.3
NY	65	97	609	723	.247	.307	.350	97	91/ 94	-74	-56	-6.0	3.88	99	93/ 92	-43	-51	-5.4	-4.6

West	W	L	R	OR	Avg	OBA	SLG	BPF	NOPS-A	BR	Adj	Wins	ERA	PPF	NERA-A	PR	Adj	Wins	Diff
ATL	89	73	739	702	.256	.327	.383	109	107/ 98	33	-28	-3.0	3.83	109	94/103	-34	19	2.0	9.0
LA	88	74	691	612	.264	.330	.388	94	109/115	47	86	9.0	3.27	93	110/103	56	14	1.5	-3.5
SF	87	75	673	687	.253	.329	.376	92	105/114	27	81	8.5	3.67	91	98/ 90	-9	-59	-6.3	3.7
SD	81	81	675	658	.257	.313	.359	89	95/107	-48	22	2.3	3.53	88	102/ 90	13	-55	-5.9	3.6
HOU	77	85	569	620	.247	.305	.349	96	90/ 94	-83	-55	-5.9	3.42	96	106/101	31	7	.7	1.2
CIN	61	101	545	661	.251	.313	.350	98	93/ 95	-62	-46	-4.9	3.67	99	98/ 97	-9	-15	-1.7	-13.4
				662	.258	.322	.373						3.61						

AMERICAN LEAGUE 1983

Batting Runs			Park Adjusted			Pitching Runs			Park Adjusted		
Wade Boggs	BOS	50.6	Wade Boggs	BOS	48.4	Rick Honeycutt	TEX	32.3	Dave Stieb	TOR	42.5
Eddie Murray	BAL	44.9	Robin Yount	MIL	41.9	Dave Stieb	TOR	31.9	Rick Honeycutt	TEX	32.7
Jim Rice	BOS	38.7	Eddie Murray	BAL	41.4	Mike Boddicker	BAL	26.1	Scott McGregor	BAL	29.5

Normalized OPS			Park Adjusted			Normalized ERA			Park Adjusted		
George Brett	KC	158	George Brett	KC	161	Rick Honeycutt	TEX	169	Rick Honeycutt	TEX	170
Wade Boggs	BOS	157	Robin Yount	MIL	157	Mike Boddicker	BAL	147	Mike Boddicker	BAL	152
Eddie Murray	BAL	155	Wade Boggs	BOS	153	Dave Stieb	TOR	134	Dave Stieb	TOR	145

On Base Average			Slugging Percentage			Percent of Team Wins			Wins Above Team		
Wade Boggs	BOS	.449	George Brett	KC	.563	Rick Sutcliffe	CLE	.243	Rick Sutcliffe	CLE	5.9
Rickey Henderson	OAK	.415	Jim Rice	BOS	.550	Lamarr Hoyt	CHI	.242	Ken Schrom	MIN	5.9
Rod Carew	CAL	.411	Eddie Murray	BAL	.538	Ron Guidry	NY	.231	Richard Dotson	CHI	5.2

Isolated Power			Base Stealing Runs			Relievers - Runs			Park Adjusted		
George Brett	KC	.252	Rickey Henderson	OAK	21.0	Dan Quisenberry	KC	33.0	Dan Quisenberry	KC	32.0
Ron Kittle	CHI	.250	Rudy Law	CHI	15.9	Bob Stanley	BOS	19.7	Bob Stanley	BOS	21.9
Greg Luzinski	CHI	.247	Willie Wilson	KC	12.9	Tippy Martinez	BAL	19.7	Tippy Martinez	BAL	21.1

Defensive Runs			Players Overall			Pitchers Overall			Relief Points		
Gary Ward	MIN	22.2	Wade Boggs	BOS	58.0	Dave Stieb	TOR	42.9	Dan Quisenberry	KC	97
Bobby Grich	CAL	20.8	Rickey Henderson	OAK	57.4	Rick Honeycutt	TEX	35.6	Bob Stanley	BOS	72
Cal Ripken	BAL	15.9	Cal Ripken	BAL	57.2	Dan Quisenberry	KC	33.5	Rich Gossage	NY	65

East	W	L	R	OR	Avg	OBA	SLG	BPF	NOPS-A	BR	Adj	Wins	ERA	PPF	NERA-A	PR	Adj	Wins	Diff
BAL	98	64	799	652	.269	.343	.421	104	113/108	70	38	3.8	3.64	103	112/116	70	91	9.1	4.1
DET	92	70	789	679	.274	.338	.427	97	114/118	67	91	9.1	3.82	95	107/102	41	10	1.0	.9
NY	91	71	770	703	.273	.339	.416	96	111/116	52	84	8.4	3.90	95	105/ 99	29	-5	-.6	2.2
TOR	89	73	795	726	.277	.341	.436	109	117/108	89	27	2.7	4.12	108	99/107	-6	48	4.7	.6
MIL	87	75	764	708	.277	.336	.418	91	110/121	44	111	11.0	4.02	90	101/ 91	9	-59	-6.0	.9
BOS	78	84	724	775	.270	.337	.409	103	108/105	35	15	1.5	4.34	103	94/ 97	-42	-20	-2.1	-2.4
CLE	70	92	704	785	.265	.341	.369	110	97/ 88	-22	-92	-9.2	4.43	111	92/102	-56	15	1.4	-3.2

West	W	L	R	OR	Avg	OBA	SLG	BPF	NOPS-A	BR	Adj	Wins	ERA	PPF	NERA-A	PR	Adj	Wins	Diff
CHI	99	63	800	650	.262	.332	.413	98	108/110	26	37	3.7	3.68	97	111/107	63	41	4.1	10.1
KC	79	83	696	767	.271	.322	.397	98	101/103	-30	-15	-1.6	4.26	98	96/ 94	-29	-39	-4.0	3.6
TEX	77	85	639	609	.255	.312	.366	101	89/ 88	-107	-113	-11.4	3.32	101	123/124	124	128	12.8	-5.4
OAK	74	88	708	782	.262	.330	.381	96	98/102	-33	-6	-.6	4.35	96	94/ 90	-43	-66	-6.7	.2
CAL	70	92	722	779	.260	.325	.393	94	100/107	-28	17	1.7	4.32	94	94/ 89	-39	-80	-8.0	-4.6
MIN	70	92	709	822	.261	.321	.401	105	102/ 97	-22	-58	-5.9	4.68	106	87/ 93	-95	-53	-5.4	.2
SEA	60	102	558	740	.240	.303	.360	102	85/ 83	-134	-146	-14.7	4.20	104	97/101	-18	4	.4	-6.7
			727		.266	.330	.401						4.08						

NATIONAL LEAGUE 1983

Batting Runs			Park Adjusted			Pitching Runs			Park Adjusted		
Dale Murphy	ATL	46.2	Mike Schmidt	PHI	43.2	John Denny	PHI	34.2	John Denny	PHI	33.4
Mike Schmidt	PHI	43.2	Dale Murphy	ATL	41.9	Mario Soto	CIN	28.7	Mario Soto	CIN	33.3
Pedro Guerrero	LA	37.8	Pedro Guerrero	LA	36.1	Atlee Hammaker	SF	26.5	Atlee Hammaker	SF	29.2

Normalized OPS			Park Adjusted			Normalized ERA			Park Adjusted		
Dale Murphy	ATL	157	Mike Schmidt	PHI	155	Atlee Hammaker	SF	162	Atlee Hammaker	SF	168
Mike Schmidt	PHI	155	Jose Cruz	HOU	149	John Denny	PHI	153	John Denny	PHI	152
Pedro Guerrero	LA	149	Dale Murphy	ATL	149	Bob Welch	LA	137	Mario Soto	CIN	141

On Base Average			Slugging Percentage			Percent of Team Wins			Wins Above Team		
Mike Schmidt	PHI	.402	Dale Murphy	ATL	.540	Mario Soto	CIN	.230	John Denny	PHI	6.0
Dale Murphy	ATL	.396	Andre Dawson	MON	.539	John Denny	PHI	.211	Jesse Orosco	NY	5.3
Tim Raines	MON	.395	Pedro Guerrero	LA	.531	Steve Rogers	MON	.207	Mario Soto	CIN	4.0
						Bill Gullickson	MON	.207			

Isolated Power			Base Stealing Runs			Relievers - Runs			Park Adjusted		
Mike Schmidt	PHI	.270	Tim Raines	MON	18.6	Jesse Orosco	NY	26.5	Jesse Orosco	NY	25.8
Andre Dawson	MON	.240	Al Wiggins	SD	12.0	Lee Smith	CHI	22.6	Kent Tekulve	PIT	24.5
Darrell Evans	SF	.239	Willie McGee	STL	6.9	Kent Tekulve	PIT	22.0	Lee Smith	CHI	21.0

Defensive Runs			Players Overall			Pitchers Overall			Relief Points		
Ryne Sandberg	CHI	40.4	Dickie Thon	HOU	57.1	John Denny	PHI	34.3	Al Holland	PHI	62
Mike Schmidt	PHI	21.5	Mike Schmidt	PHI	55.9	Mario Soto	CIN	29.8	Lee Smith	CHI	56
Larry Bowa	CHI	21.3	Tim Raines	MON	53.4	Atlee Hammaker	SF	28.1	Jesse Orosco	NY	53

East	W	L	R	OR	Avg	OBA	SLG	BPF	NOPS-A	BR	Adj	Wins	ERA	PPF	NERA-A	PR	Adj	Wins	Diff
PHI	90	72	696	635	.249	.331	.373	100	103/103	17	17	1.7	3.35	99	109/108	47	43	4.5	2.8
PIT	84	78	659	648	.264	.327	.383	106	105/ 99	17	-23	-2.5	3.55	106	102/109	14	51	5.3	.1
MON	82	80	677	646	.264	.329	.386	103	107/103	30	7	.8	3.59	103	101/105	9	28	2.9	-2.6
STL	79	83	679	710	.270	.337	.384	100	108/108	47	47	4.9	3.84	100	95/ 95	-32	-30	-3.2	-3.7
CHI	71	91	701	719	.261	.322	.401	96	109/114	35	61	6.3	4.08	96	89/ 86	-69	-92	-9.7	-6.6
NY	68	94	575	680	.241	.301	.344	97	86/ 89	-111	-95	-10.0	3.68	98	99/ 97	-6	-15	-1.6	-1.4

West	W	L	R	OR	Avg	OBA	SLG	BPF	NOPS-A	BR	Adj	Wins	ERA	PPF	NERA-A	PR	Adj	Wins	Diff
LA	91	71	654	609	.250	.320	.370	102	102/100	-5	-20	-2.2	3.11	102	117/119	86	97	10.1	2.0
ATL	88	74	746	640	.272	.344	.400	106	115/109	91	53	5.5	3.67	105	99/104	-5	22	2.3	-.8
HOU	85	77	643	646	.257	.323	.375	90	101/113	-6	60	6.2	3.45	89	106/ 94	31	-32	-3.4	1.2
SD	81	81	653	653	.250	.313	.351	94	91/ 97	-73	-36	-3.8	3.61	94	101/ 94	4	-31	-3.4	7.2
SF	79	83	687	697	.247	.328	.375	103	103/ 99	11	-12	-1.3	3.73	104	98/101	-14	8	.8	-1.5
CIN	74	88	623	710	.239	.317	.356	103	94/ 92	-48	-66	-7.0	4.01	104	91/ 94	-59	-35	-3.7	3.8
			666		.255	.324	.376						3.64						

AMERICAN LEAGUE 1984

Batting Runs			Park Adjusted			Pitching Runs			Park Adjusted		
Eddie Murray	BAL	46.9	Eddie Murray	BAL	50.4	Mike Boddicker	BAL	35.0	Bert Blyleven	CLE	37.8
Dwight Evans	BOS	46.7	Don Mattingly	NY	48.4	Dave Stieb	TOR	34.6	Dave Stieb	TOR	34.9
Don Mattingly	NY	41.4	Dave Winfield	NY	45.2	Bert Blyleven	CLE	30.8	Mike Boddicker	BAL	29.4

Normalized OPS			Park Adjusted			Normalized ERA			Park Adjusted		
Eddie Murray	BAL	154	Don Mattingly	NY	168	Mike Boddicker	BAL	143	Bert Blyleven	CLE	148
Dwight Evans	BOS	153	Dave Winfield	NY	165	Dave Stieb	TOR	141	Dave Stieb	TOR	141
Don Mattingly	NY	153	Eddie Murray	BAL	161	Bert Blyleven	CLE	140	Geoff Zahn	CAL	138

On Base Average			Slugging Percentage			Percent Of Team Wins			Wins Above Team		
Eddie Murray	BAL	.415	Harold Baines	CHI	.541	Bert Blyleven	CLE	.253	Bert Blyleven	CLE	8.3
Wade Boggs	BOS	.409	Don Mattingly	NY	.537	Mike Boddicker	BAL	.235	Mark Langston	SEA	5.6
Rickey Henderson	OAK	.401	Dwight Evans	BOS	.532	Charlie Hough	TEX	.232	Doyle Alexander	TOR	5.1

Isolated Power			Base Stealing Runs			Relievers - Runs			Park Adjusted		
Tony Armas	BOS	.263	Willie Wilson	KC	11.1	Willie Hernandez	DET	32.2	Willie Hernandez	DET	28.0
Steve Balboni	KC	.253	Dave Collins	TOR	9.6	Dan Quisenberry	KC	19.3	Doug Corbett	CAL	20.6
Ron Kittle	CHI	.238	Rickey Henderson	OAK	9.0	Doug Corbett	CAL	17.8	Ernie Camacho	CLE	20.3

Defensive Runs			Players Overall			Pitchers Overall			Relief Points		
Cal Ripken	BAL	38.7	Cal Ripken	BAL	93.7	Bert Blyleven	CLE	38.2	Dan Quisenberry	KC	97
Wade Boggs	BOS	25.9	Buddy Bell	TEX	50.8	Dave Stieb	TOR	35.9	Bill Caudill	OAK	83
Buddy Bell	TEX	24.0	Wade Boggs	BOS	48.8	Mike Boddicker	BAL	35.0	Willie Hernandez	DET	79

East	W	L	R	OR	Avg	OBA	SLG	BPF	NOPS-A	BR	Adj	Wins	ERA	PPF	NERA-A	PR	Adj	Wins	Diff
DET	104	58	829	643	.271	.345	.432	95	118/124	104	138	13.9	3.49	93	115/107	82	38	3.8	5.3
TOR	89	73	750	696	.273	.333	.421	101	112/111	53	47	4.8	3.87	100	103/104	21	22	2.2	1.0
NY	87	75	758	679	.276	.342	.404	91	109/120	47	115	11.6	3.80	89	105/94	33	-36	-3.8	-1.8
BOS	86	76	810	764	.283	.343	.441	110	120/109	113	40	4.0	4.19	110	95/105	-30	35	3.5	-2.5
BAL	85	77	681	667	.254	.329	.388	96	101/105	-12	16	1.6	3.71	95	108/103	46	15	1.6	.8
CLE	75	87	761	766	.265	.339	.384	106	102/96	5	-38	-3.9	4.27	106	94/100	-44	-2	-.3	-1.8
MIL	67	94	641	734	.262	.319	.370	93	93/100	-74	-21	-2.2	4.06	96	98/94	-9	-35	-3.6	-7.7

West	W	L	R	OR	Avg	OBA	SLG	BPF	NOPS-A	BR	Adj	Wins	ERA	PPF	NERA-A	PR	Adj	Wins	Diff
KC	84	78	673	673	.268	.320	.399	98	102/104	-21	-9	-1.0	3.92	98	102/100	13	1	.1	3.9
CAL	81	81	696	697	.249	.322	.381	107	97/90	-44	-94	-9.6	3.96	107	101/108	6	54	5.4	4.2
MIN	81	81	673	675	.265	.321	.385	102	98/96	-43	-59	-6.0	3.86	102	104/106	23	38	3.8	2.2
OAK	77	85	738	796	.259	.332	.404	90	106/119	19	93	9.4	4.47	89	89/80	-74	-141	-14.3	.9
SEA	74	88	682	774	.258	.326	.384	103	99/96	-30	-50	-5.2	4.33	104	92/96	-52	-26	-2.7	.9
CHI	74	88	679	736	.247	.316	.395	107	100/93	-35	-85	-8.7	4.14	108	97/104	-22	29	2.9	-1.3
TEX	69	92	656	714	.261	.315	.377	102	94/91	-74	-91	-9.3	3.91	103	102/105	13	34	3.4	-5.6
			716		.264	.329	.398						4.00						

NATIONAL LEAGUE 1984

Batting Runs			Park Adjusted			Pitching Runs			Park Adjusted		
Dale Murphy	ATL	43.8	Keith Hernandez	NY	37.5	Alejandro Pena	LA	24.5	Alejandro Pena	LA	30.7
Mike Schmidt	PHI	41.2	Tim Raines	MON	37.1	Dwight Gooden	NY	24.0	Orel Hershiser	LA	25.9
Ryne Sandberg	CHI	37.9	Mike Schmidt	PHI	37.1	Rick Rhoden	PIT	23.0	Ricky Mahler	ATL	25.8

Normalized OPS			Park Adjusted			Normalized ERA			Park Adjusted		
Mike Schmidt	PHI	157	Chili Davis	SF	151	Alejandro Pena	LA	144	Alejandro Pena	LA	156
Dale Murphy	ATL	156	Gary Carter	MON	151	Dwight Gooden	NY	138	Orel Hershiser	LA	146
Ryne Sandberg	CHI	147	Mike Schmidt	PHI	148	Orel Hershiser	LA	135	Rick Honeycutt	LA	137

On Base Average			Slugging Percentage			Percent Of Team Wins			Wins Above Team		
Gary Matthews	CHI	.417	Dale Murphy	ATL	.547	Mario Soto	CIN	.257	Mario Soto	CIN	8.5
Keith Hernandez	NY	.415	Mike Schmidt	PHI	.536	Joaquin Andujar	STL	.238	Rick Sutcliffe	CHI	6.6
Tony Gwynn	SD	.411	Ryne Sandberg	CHI	.520	Joe Niekro	HOU	.200	Frank Williams	SF	4.0

Isolated Power			Base Stealing Runs			Relievers - Runs			Park Adjusted		
Mike Schmidt	PHI	.259	Tim Raines	MON	16.5	Bruce Sutter	STL	28.1	Bruce Sutter	STL	26.4
Dale Murphy	ATL	.257	Juan Samuel	PHI	12.6	Bill Dawley	HOU	18.1	Steve Bedrosian	ATL	16.9
Leon Durham	CHI	.226	Mookie Wilson	NY	8.4	Craig Lefferts	SD	17.3	Bill Dawley	HOU	15.3
			Al Wiggins	SD	8.4						

Defensive Runs			Players Overall			Pitchers Overall			Relief Points		
Ozzie Smith	STL	30.8	Ryne Sandberg	CHI	64.0	Ricky Mahler	ATL	32.8	Bruce Sutter	STL	93
Ryne Sandberg	CHI	23.0	Ozzie Smith	STL	51.6	F. Valenzuela	LA	31.5	Lee Smith	CHI	77
Glenn Hubbard	ATL	22.1	Tim Raines	MON	48.1	Rick Rhoden	PIT	28.7	Jesse Orosco	NY	76

East	W	L	R	OR	Avg	OBA	SLG	BPF	NOPS-A	BR	Adj	Wins	ERA	PPF	NERA-A	PR	Adj	Wins	Diff
CHI	96	65	762	658	.260	.333	.397	110	114/104	78	15	1.5	3.76	109	96/104	-25	25	2.6	11.3
NY	90	72	652	676	.257	.322	.369	97	102/105	1	20	2.1	3.63	97	99/96	-4	-21	-2.3	9.2
STL	84	78	652	645	.252	.319	.351	97	96/99	-34	-13	-1.5	3.59	96	100/97	1	-18	-2.0	6.5
PHI	81	81	720	690	.266	.335	.407	106	117/111	102	61	6.4	3.62	106	99/105	-3	31	3.3	-9.6
MON	78	83	593	585	.251	.314	.362	91	98/107	-31	23	2.4	3.31	91	109/99	45	-6	-.8	-4.2
PIT	75	87	615	567	.255	.312	.363	95	97/103	-36	-3	-.4	3.12	94	115/108	78	42	4.4	-10.0

West	W	L	R	OR	Avg	OBA	SLG	BPF	NOPS-A	BR	Adj	Wins	ERA	PPF	NERA-A	PR	Adj	Wins	Diff
SD	92	70	686	634	.259	.320	.371	96	102/106	-0	25	2.6	3.48	95	103/98	19	-9	-1.1	9.5
HOU	80	82	693	630	.264	.326	.371	94	104/110	14	53	5.5	3.33	93	108/100	43	1	.2	-6.7
ATL	80	82	632	655	.247	.319	.361	115	99/86	-17	-115	-12.2	3.57	116	101/117	4	97	10.2	1.0
LA	79	83	580	600	.244	.308	.348	107	92/86	-67	-114	-12.1	3.17	108	113/122	68	114	12.0	-2.0
CIN	70	92	627	747	.244	.316	.356	100	96/96	-33	-33	-3.6	4.16	102	86/88	-92	-82	-8.8	1.4
SF	66	96	682	807	.265	.330	.375	95	106/111	31	64	6.7	4.40	96	82/79	-129	-149	-15.8	-5.9
			658		.255	.321	.369						3.59						

Detroit 104- 58 .642(1)

Player	Pos	G	HR	RBI	Avg	OBA	SLG	NOPS-A		Runs-A		DEF	BSR	Wins
Lance Parrish	*CD	147	33	98	.237	.290	.443	100	104	-2	1	2	-1.2	.8
Dave Bergman	*1/O	120	7	44	.273	.358	.417	114	119	6	7	8	-1.5	.7
Lou Whitaker	*2	143	13	56	.289	.360	.407	112	117	10	13	-0	-1.2	1.8
Alan Trammell	*SD	139	14	69	.314	.383	.468	134	141	27	30	-1	-2.1	3.4
Howard Johnson	*3/S1O	116	12	50	.248	.326	.394	98	103	-0	1	-8	-.6	-.7
Kirk Gibson	*O	149	27	91	.282	.367	.516	141	148	30	33	-11	3.3	2.0
Chet Lemon	*O	141	20	76	.287	.360	.495	134	140	23	26	7	-1.5	2.8
Larry Herndon	*O	125	7	43	.280	.336	.400	103	108	1	4	-7	.6	-.7
Darrell Evans	D13	131	16	63	.232	.356	.384	105	110	5	7	3	-.6	.5
Tom Brookens	3S2	113	5	26	.246	.307	.397	93	98	-1	-0	-5	-1.8	-.4
Barbaro Garbey	13DO/2	110	5	52	.287	.327	.391	98	102	-0	1	3	-2.4	-.2
John Grubb	OD	86	8	17	.267	.397	.432	129	136	9	10	0	.3	.7
Rusty Kuntz	OD	84	2	22	.286	.398	.414	125	131	6	7	-0	-.6	.3
Ruppert Jones	O	79	12	37	.284	.347	.516	135	142	10	11	-0	-1.8	.6
Marty Castillo	C3	70	4	17	.234	.285	.383	83	87	-3	-2	-1	.3	-.1
Others		122	2	27	.242	.289	.316	68	71	-10	-9	0	1.8	-.4

Pitcher	P	G	IP	W	L	ERA	NERA-A		Runs-A		Wins
Jack Morris	S	35	240	19	11	3.60	111	104	11	3	.5
Dan Petry	S	35	233	18	8	3.24	123	115	20	12	1.6
Milt Wilcox	S	33	194	17	8	3.99	100	93	0	-5	-.4
Juan Berenguer	S	31	168	11	10	3.48	115	107	10	5	.3
Willie Hernandez	R	80	140	9	3	1.93	207	193	32	28	2.7
Aurelio Lopez	R	71	138	10	1	2.93	136	127	16	12	1.1
Doug Bair	R	47	94	5	3	3.73	107	100	3	0	.0
Dave Rozema	M	29	101	7	6	3.74	107	100	3	0	.0
Others		69	156	8	8	4.67	86	80	-11	-15	-1.8

Toronto 89- 73 .549(2)

Player	Pos	G	HR	RBI	Avg	OBA	SLG	NOPS-A		Runs-A		DEF	BSR	Wins
Ernie Whitt	*C	124	15	46	.238	.331	.425	108	107	3	3	2	-1.8	1.0
Willie Upshaw	*1	152	19	84	.278	.347	.464	122	121	17	16	-0	.6	.5
Damaso Garcia	*2	152	5	46	.284	.312	.374	89	88	-9	-10	-6	6.6	-.4
Alfredo Griffin	*S2	140	4	30	.241	.250	.298	51	51	-27	-27	-5	1.5	-2.3
Rance Mulliniks	*3/S2	125	3	42	.324	.385	.440	128	127	13	13	-2	-1.2	1.1
George Bell	*O/3	159	26	87	.292	.328	.498	125	124	18	17	-1	2.1	1.2
Lloyd Moseby	*O	158	18	92	.280	.372	.470	131	130	26	26	8	6.3	3.4
Dave Collins	*O/1	128	2	44	.308	.369	.444	124	123	15	14	-0	9.6	1.8
Cliff Johnson	*D/1	127	16	61	.304	.393	.507	147	146	24	24	0	-.6	2.0
Garth Iorg	*3/2S	121	1	25	.227	.245	.304	51	51	-16	-16	-2	-1.5	-1.9
Jesse Barfield	O	110	14	49	.284	.359	.466	126	125	11	11	1	1.2	1.0
Buck Martinez	C	102	5	37	.220	.312	.349	83	82	-4	-4	1	-1.8	-.1
Willie Aikens	D/1	93	11	26	.205	.298	.376	85	85	-4	-4	0	0.0	-.7
Tony Fernandez	S3	88	3	19	.270	.320	.356	87	86	-3	-3	-2	-2.7	-.4
Rick Leach	O1/P	65	0	7	.261	.323	.375	92	92	-0	-0	0	0.0	-.3
Others		69	1	7	.179	.193	.304	35	35	-4	-4	0	-.6	-.6

Pitcher	P	G	IP	W	L	ERA	NERA-A		Runs-A		Wins
Dave Stieb	S	35	267	16	8	2.83	141	141	35	35	3.6
Doyle Alexander	S	36	262	17	6	3.13	128	128	25	26	2.6
Luis Leal	S	35	222	13	8	3.89	103	103	3	3	.3
Jim Clancy	S	36	220	13	15	5.11	78	78	-26	-26	-2.7
Roy Lee Jackson	R	54	86	7	8	3.56	112	113	4	4	.4
Dennis Lamp	R	56	85	8	8	4.55	88	88	-4	-4	-.4
Jim Acker	R	32	72	3	5	4.38	91	92	-2	-2	-.3
Jimmy Key	R	63	62	4	5	4.65	86	86	-3	-3	-.3
Jim Gott	M	35	110	7	6	4.01	100	100	0	0	-.1
Others		37	78	1	4	4.38	91	91	-2	-2	-.2

New York 87- 75 .537(3)

Player	Pos	G	HR	RBI	Avg	OBA	SLG	NOPS-A		Runs-A		DEF	BSR	Wins
Butch Wynegar	*C	129	6	45	.267	.361	.342	96	105	0	5	0	-2.1	1.1
Don Mattingly	*1O	153	23	110	.343	.386	.537	153	168	41	48	6	-.3	4.4
Willie Randolph	*2	142	2	31	.287	.382	.348	103	114	7	14	13	-.6	3.4
Bob Meacham	S/2	99	2	25	.253	.319	.328	79	87	-8	-4	1	-.3	.3
Toby Harrah	3/2O	88	1	26	.217	.333	.296	76	84	-6	-3	7	.9	.5
Ken Griffey	O1	120	7	56	.273	.324	.381	94	104	-2	1	5	-.6	.1
Omar Moreno	*O	117	4	38	.259	.297	.361	81	89	-9	-5	-2	-.6	-1.3
Dave Winfield	*O	141	19	100	.340	.397	.515	150	165	38	45	2	-.6	4.2
Don Baylor	*D/O	134	27	89	.262	.343	.489	127	140	18	24	0	-.3	1.9
Steve Kemp	OD	94	7	41	.291	.373	.403	115	126	7	11	-0	.6	.8
Roy Smalley	3S/1	67	7	26	.239	.290	.388	86	94	-3	-1	2	0.0	.1
Mike Pagliarulo	3	67	7	34	.239	.292	.448	101	112	0	2	6	0.0	.9
Brian Dayett	O	64	4	23	.244	.299	.402	92	101	-1	0	0	0.0	-.3
Tim Foli	S23/1	61	0	16	.252	.265	.319	61	67	-8	-6	3	0.0	-.1
Oscar Gamble	DO	54	10	27	.184	.320	.440	108	119	1	3	0	.3	.2
Andre Robertson	S/2	52	0	6	.214	.236	.264	38	42	-11	-9	1	0.0	-.6
Others		137	4	32	.259	.307	.346	80	89	-8	-4	0	0.0	-.4

Pitcher	P	G	IP	W	L	ERA	NERA-A		Runs-A		Wins
Phil Niekro	S	32	216	16	8	3.08	130	116	22	12	1.3
Ron Guidry	S	29	196	10	11	4.50	89	79	-10	-19	-2.1
Dennis Rasmussen	S	24	148	9	6	4.56	88	78	-8	-15	-1.8
Joe Cowley	S	16	83	9	2	3.58	112	100	4	0	.0
John Montefusco	S	11	55	5	3	3.60	111	99	2	0	-.1
Bob Shirley	R	41	114	3	3	3.39	118	105	8	2	.2
Jay Howell	R	61	104	9	4	2.68	149	133	15	10	1.2
Dave Righetti	R	64	96	5	6	2.34	171	152	18	13	1.3
Jose Rijo	R	24	62	2	8	4.79	83	74	-4	-7	-.8
Mike Armstrong	R	36	54	3	2	3.50	114	102	3	0	.0
Ray Fontenot	M	35	169	8	9	3.62	110	98	7	-0	-.1
Others		76	168	8	13	5.09	79	70	-19	-27	-2.9

Boston 86- 76 .531(4)

Player	Pos	G	HR	RBI	Avg	OBA	SLG	NOPS-A		Runs-A		DEF	BSR	Wins
Rich Gedman	*C	133	24	72	.269	.315	.506	123	112	12	6	-2	0.0	1.1
Bill Buckner	*1	114	11	67	.278	.323	.410	101	92	0	-4	7	-.6	-.7
Marty Barrett	*2	139	3	45	.303	.361	.383	106	96	5	-0	-1	-.3	.3
Jackie Gutierrez	*S	151	2	29	.263	.287	.316	67	61	-19	-25	-24	.6	-4.0
Wade Boggs	*3	158	6	55	.325	.409	.416	129	117	29	20	26	-.3	4.9
Jim Rice	*O	159	28	122	.280	.326	.467	117	106	13	5	7	1.2	.7
Tony Armas	*OD	157	43	123	.268	.304	.531	126	114	18	10	-5	-1.5	-.3
Dwight Evans	*O	162	32	104	.295	.392	.532	153	139	47	38	-3	.3	2.9
Mike Easler	*D1	156	27	91	.313	.377	.516	144	131	36	28	2	-.3	2.2
Rick Miller	O/1	95	0	12	.260	.350	.317	86	78	-1	-2	0	-.3	-.6
Reid Nichols	O	74	1	14	.226	.309	.306	71	65	-4	-5	-0	0.0	-.9
Glenn Hoffman	S/32	64	0	4	.189	.241	.243	34	31	-6	-7	-3	-.6	-.8
Others		154	4	29	.240	.310	.311	73	66	-12	-17	1	-1.8	-1.5

Pitcher	P	G	IP	W	L	ERA	NERA-A		Runs-A		Wins
Bruce Hurst	S	33	218	12	12	3.92	102	112	2	12	1.2
Bobby Ojeda	S	33	217	12	12	3.98	100	111	0	10	1.1
Oil Can Boyd	S	29	198	12	12	4.36	92	101	-7	1	.3
Al Nipper	S	29	183	11	6	3.89	103	114	2	11	1.3
Roger Clemens	S	21	133	9	4	4.33	92	102	-4	1	.1
Dennis Eckersley	S	9	65	4	4	4.98	80	88	-6	-3	-.3
Bob Stanley	R	57	107	9	10	3.53	113	125	6	10	1.3
Mark Clear	R	47	67	8	3	4.03	99	109	0	3	.3
John Henry Johnson	R	30	64	1	2	3.52	114	125	3	6	.6
Steve Crawford	R	35	62	5	0	3.34	120	132	5	7	.7
Mike Brown	M	15	67	1	8	6.85	58	64	-20	-17	-1.8
Others		25	62	2	3	5.08	79	87	-6	-4	-.5

Baltimore 85- 77 .525(5)

Player	Pos	G	HR	RBI	Avg	OBA	SLG	NOPS-A		Runs-A		DEF	BSR	Wins
Rick Dempsey	*C	109	11	34	.230	.315	.364	87	91	-4	-3	1	-.9	.3
Eddie Murray	*1	162	29	110	.306	.415	.509	154	161	47	50	7	1.8	4.7
Rich Dauer	*2/3	127	2	24	.254	.297	.335	75	78	-13	-11	-8	-1.5	-1.6
Cal Ripken	*S	162	27	86	.304	.375	.510	142	149	37	40	39	0.0	9.4
Ron Jackson	3	12	0	2	.286	.286	.357	77	80	-0	-0	-3	-1.2	-.6
John Shelby	*O	128	6	30	.209	.248	.313	54	57	-23	-21	1	1.2	-2.4
Al Bumbry	*O	119	3	24	.270	.320	.337	82	86	-7	-5	-9	-.3	-2.0
Mike Young	*O	123	17	52	.252	.356	.431	117	122	10	13	-8	.6	.1
Ken Singleton	*D	111	6	36	.215	.288	.289	60	63	-18	-16	0	0.0	-2.1
Gary Roenicke	*O	121	10	44	.224	.348	.380	101	106	2	4	-1	-.9	-.3
John Lowenstein	OD/1	105	8	28	.237	.322	.374	92	96	-2	-0	-0	.3	-.5
Todd Cruz	3/P	96	3	9	.218	.265	.310	58	61	-7	-6	-6	-2.1	-1.5
Floyd Rayford	C3/1	86	4	27	.256	.298	.360	81	85	-6	-4	-4	-1.8	-.7
Lenn Sakata	2/O	81	3	11	.191	.221	.255	31	33	-14	-13	-2	.6	-1.3
Jim Dwyer	O	76	2	21	.255	.348	.360	96	101	0	1	-0	-1.2	-.3
Benny Ayala	DO	60	4	24	.212	.262	.364	71	74	-4	-3	0	-.3	-.6
Others		94	3	21	.279	.346	.372	99	103	0	1	-0	.3	-.0

Pitcher	P	G	IP	W	L	ERA	NERA-A		Runs-A		Wins
Mike Boddicker	S	34	261	20	11	2.79	143	136	35	29	3.5
Mike Flanagan	S	34	227	13	13	3.53	113	108	12	7	.6
Storm Davis	S	35	225	14	9	3.12	128	122	22	17	1.5
Scott McGregor	S	30	196	15	12	3.95	101	96	1	-2	-.2
Sammy Stewart	R	60	93	7	4	3.29	122	116	7	5	.5
Tippy Martinez	R	55	90	4	9	3.90	103	98	1	-0	-.1
Tom Underwood	R	37	72	1	0	3.50	114	109	4	2	.3
Bill Swaggerty	R	23	57	3	2	5.21	77	73	-7	-8	-1.0
Dennis Martinez	M	34	142	6	9	5.01	80	76	-15	-18	-1.8
Others		27	77	2	8	5.49	73	69	-12	-13	-1.5

Cleveland 75- 87 .463(6)

Player	Pos	G	HR	RBI	Avg	OBA	SLG	NOPS-A		Runs-A		DEF	BSR	Wins
Jerry Willard	C	87	10	37	.224	.298	.386	88	83	-3	-5	-0	.3	-.2
Mike Hargrove	*1	133	2	44	.267	.363	.335	95	89	0	-2	2	-1.2	-1.0
Tony Bernazard	*2	140	2	38	.221	.293	.287	61	58	-21	-25	5	-1.8	-1.6
Julio Franco	*S	160	3	79	.286	.335	.348	89	84	-8	-12	3	-.3	.2
Brook Jacoby	*3/S	126	7	40	.264	.319	.369	90	85	-5	-8	-16	-.3	-2.5
George Vukovich	*O	134	9	60	.304	.356	.439	119	112	11	8	15	-2.1	1.6
Brett Butler	*O	159	3	49	.269	.364	.355	100	95	3	-1	5	2.4	.1
Mel Hall	O	83	7	30	.257	.350	.397	106	100	3	1	-0	-.3	-.3
Andy Thornton	*D1	155	33	99	.271	.371	.484	134	127	29	24	0	-1.2	1.7
Pat Tabler	1O3/2	144	10	68	.290	.358	.410	112	106	8	5	-2	.3	-.2
Carmelo Castillo	O	87	10	36	.261	.333	.464	118	112	5	3	3	-1.5	.2
Mike Fischlin	23S	85	1	14	.226	.290	.308	66	62	-5	-6	1	-.6	-.3
Chris Bando	C/13	75	12	41	.291	.383	.505	143	135	13	12	-0	-.9	1.4
Joe Carter	O/1	66	13	41	.275	.309	.467	111	105	3	1	3	-1.8	-.0
Broderick Perkins	D/1	58	0	4	.197	.284	.212	40	37	-4	-5	0	0.0	-.6
Otis Nixon	O	49	0	1	.154	.222	.154	6	6	-10	-11	2	0.0	-1.2
Others		92	1	23	.255	.320	.314	76	72	-5	-6	-2	.6	-.6

Pitcher	P	G	IP	W	L	ERA	NERA-A		Runs-A		Wins
Bert Blyleven	S	33	245	19	7	2.87	140	148	31	38	3.8
Neal Heaton	S	38	199	12	15	5.20	77	82	-26	-20	-2.4
Steve Comer	S	22	117	4	8	5.69	70	75	-21	-18	-1.8
Rick Sutcliffe	S	15	94	4	5	5.17	77	82	-11	-9	-1.0
Ernie Camacho	R	69	100	5	9	2.43	165	175	17	20	2.0
Tom Waddell	R	58	97	7	4	3.06	131	139	10	13	1.3
Mike Jeffcoat	R	63	75	5	2	3.00	133	142	8	10	1.1
Jamie Easterly	R	26	69	3	1	3.39	118	125	5	7	.7
Luis Aponte	R	25	50	1	0	4.14	97	103	-0	1	.0
Steve Farr	M	31	116	3	11	4.58	87	93	-6	-3	-.4
Roy Smith	M	22	86	5	5	4.60	87	92	-5	-2	-.5
Don Schulze	M	19	86	3	6	4.81	83	88	-7	-4	-.6
Dan Spillner	M	14	51	0	5	5.65	71	75	-8	-7	-.7
Others		36	81	4	9	6.67	60	64	-23	-21	-2.3

Milwaukee   67- 94   .416(7)

Player	Pos	G	HR	RBI	Avg	OBA	SLG	NOPS-A		Runs-A		DEF	BSR	Wins
Jim Sundberg	*C	110	7	43	.261	.334	.399	102	110	1	5	3	-.3	1.5
Cecil Cooper	*1D	148	11	67	.275	.309	.386	91	98	-7	-2	3	1.2	-.9
Jim Gantner	*2	153	3	56	.282	.319	.344	84	90	-13	-7	12	-1.2	1.1
Robin Yount	*SD	160	16	80	.298	.367	.441	122	132	20	26	5	1.8	4.2
Ed Romero	3S2/1O	116	1	31	.252	.310	.294	68	74	-14	-10	7	-.9	-.1
Ben Oglivie	*O	131	12	60	.262	.328	.384	96	104	-1	2	-2	-3.6	-.9
Dion James	*O	128	1	30	.295	.353	.377	102	111	2	6	-15	-3.0	-1.8
Rick Manning	*O	119	7	31	.249	.319	.370	90	97	-3	-0	-4	-2.7	-1.3
Ted Simmons	D13	132	4	52	.221	.270	.300	57	62	-28	-23	2	.9	-2.7
Charlie Moore	O/C	70	2	17	.234	.276	.314	63	68	-9	-7	-1	-2.4	-1.3
Roy Howell	3/1	68	4	17	.232	.284	.348	74	80	-5	-3	3	-.6	-.2
Mark Brouhard	O	66	6	22	.239	.302	.365	84	91	-3	-2	-1	-1.8	-.8
Bill Schroeder	C/1	61	14	25	.257	.291	.486	111	120	2	4	1	-.6	.8
Bob Clark	O	58	2	16	.260	.328	.361	90	98	-1	0	-1	-2.7	-.7
Others		116	6	51	.236	.321	.358	88	95	-4	-1	6	-2.7	.2

Pitcher	P	G	IP	W	L	ERA	NERA-A		Runs-A		Wins
Don Sutton	S	33	213	14	12	3.76	106	102	6	2	.0
Moose Haas	S	31	189	9	11	4.00	100	96	0	-2	-.1
Jaime Cocanower	S	33	175	8	16	4.01	100	96	0	-2	-.2
Peter Ladd	R	54	91	4	9	5.24	76	73	-12	-13	-1.6
Tom Tellmann	R	50	81	6	3	2.78	144	138	11	10	1.0
Rick Waits	R	47	73	2	4	3.58	112	107	3	2	.2
Bob McClure	M	39	140	4	8	4.37	91	88	-5	-7	-.8
Mike Caldwell	M	26	126	6	13	4.64	86	83	-8	-10	-1.1
Chuck Porter	M	17	81	6	4	3.89	103	99	1	0	-.0
Bob Gibson	M	18	69	2	5	4.96	81	77	-6	-8	-.8
Others		96	195	6	9	3.83	104	100	4	0	-.0

AMERICAN LEAGUE   1984   WEST DIVISION

Kansas City   84- 78   .519(1)

Player	Pos	G	HR	RBI	Avg	OBA	SLG	NOPS-A		Runs-A		DEF	BSR	Wins
Don Slaught	*C	124	4	42	.264	.302	.379	87	89	-7	-6	-5	0.0	-.5
Steve Balboni	*1	126	28	77	.244	.320	.498	123	125	12	13	-3	0.0	.1
Frank White	*2	129	17	56	.271	.313	.445	107	109	3	4	16	-1.5	2.5
Onix Concepcion	S/23	90	1	23	.282	.322	.338	83	85	-5	-5	3	-.9	.3
Greg Pryor	*32/S1	123	4	25	.263	.302	.356	81	83	-6	-5	7	-1.8	.2
Darryl Motley	*O	146	15	70	.284	.321	.441	108	110	5	6	0	-3.6	-.2
Willie Wilson	*O	128	2	44	.301	.352	.390	105	107	4	5	2	11.1	1.4
Pat Sheridan	*O	138	8	53	.283	.340	.399	104	106	3	4	-2	2.1	-.2
Hal McRae	D	106	3	42	.303	.372	.397	113	115	7	7	0	-1.8	.2
Jorge Orta	DO/2	122	9	50	.298	.346	.457	120	122	10	11	0	-.6	.7
George Brett	*3	104	13	69	.284	.349	.459	122	124	11	12	6	-.6	1.8
John Wathan	C1/O	97	2	10	.181	.271	.269	50	51	-10	-10	-2	-1.8	-1.4
Dane Iorg	1O/3	78	5	30	.255	.294	.404	91	93	-2	-2	-1	-.6	-.9
Buddy Biancalana	S2	66	2	9	.194	.229	.299	45	45	-9	-9	2	-.9	-.5
U L Washington	S	63	1	10	.224	.283	.276	55	56	-9	-9	2	-2.4	-.6
Others		159	3	29	.210	.263	.285	52	52	-19	-19	-0	-2.1	-2.5

Pitcher	P	G	IP	W	L	ERA	NERA-A		Runs-A		Wins
Buddy Black	S	35	257	17	12	3.12	128	126	25	23	2.6
Mark Gubicza	S	29	189	10	14	4.05	99	97	-0	-2	-.1
Larry Gura	S	31	169	12	9	5.17	77	76	-21	-22	-2.3
Charlie Leibrandt	S	23	144	11	7	3.63	110	108	6	5	.3
Danny Jackson	S	15	76	2	6	4.26	94	92	-1	-2	-.3
Dan Quisenberry	R	72	129	6	3	2.65	151	148	19	18	2.1
Joe Beckwith	R	49	101	8	4	3.39	118	116	7	6	.6
Mark Huismann	R	38	75	3	3	4.20	95	93	-1	-1	-.2
Bret Saberhagen	M	38	158	10	11	3.47	115	113	9	8	.8
Mike Jones	M	23	81	2	3	4.89	82	80	-7	-8	-1.0
Others		22	65	3	6	6.23	64	63	-15	-16	-1.6

California 81- 81   .500(2)

Player	Pos	G	HR	RBI	Avg	OBA	SLG	NOPS-A		Runs-A		DEF	BSR	Wins
Bob Boone	*C	139	3	32	.202	.244	.262	40	38	-35	-39	0	-.9	-3.3
Rod Carew	1	93	3	31	.295	.371	.353	102	95	2	-0	-2	-.6	-1.0
Bobby Grich	213	116	18	58	.256	.360	.452	123	115	12	9	5	-2.4	1.5
Dick Schofield	*S	140	4	21	.193	.264	.263	46	43	-28	-31	7	.3	-1.6
Doug DeCinces	*3	146	20	82	.269	.333	.431	110	103	7	2	-1	.6	.2
Fred Lynn	*O	142	23	79	.271	.367	.474	131	122	23	18	9	-.6	2.1
Gary Pettis	*O	140	2	29	.227	.333	.300	77	72	-10	-14	-5	4.8	-2.1
Brian Downing	*OD	156	23	91	.275	.365	.462	127	119	21	16	-2	-2.4	.5
Reggie Jackson	*D/O	143	25	81	.223	.300	.406	93	87	-5	-9	0	0.0	-1.6
Juan Beniquez	O	110	8	39	.336	.373	.452	127	119	13	10	0	-1.8	.5
Rob Wilfong	2/S	108	6	33	.248	.298	.362	81	76	-7	-10	5	-.3	-.1
Rob Picciolo	S3/2O	87	1	9	.202	.202	.277	31	29	-10	-11	2	-.6	-.6
Jerry Narron	C/1	69	3	17	.247	.289	.340	73	69	-4	-6	0	0.0	-.4
Mike Brown	O	62	7	22	.284	.342	.520	135	126	6	5	0	-1.8	.2
Daryl Sconiers	1	57	4	17	.244	.301	.344	78	73	-4	-5	-0	-.9	-1.1
Others		71	0	8	.158	.223	.194	17	16	-18	-19	-0	0.0	-2.4

Pitcher	P	G	IP	W	L	ERA	NERA-A		Runs-A		Wins
Mike Witt	S	34	247	15	11	3.46	116	124	15	23	2.2
Ron Romanick	S	33	230	12	12	3.76	106	114	6	14	1.2
Geoff Zahn	S	28	199	13	10	3.12	128	138	19	26	2.7
Tommy John	S	32	181	7	13	4.52	88	95	-10	-4	-.4
Jim Slaton	S	32	163	7	10	4.97	80	86	-17	-11	-1.2
Doug Corbett	R	45	85	5	1	2.12	189	203	18	21	2.1
Luis Sanchez	R	49	84	9	7	3.32	120	129	6	9	.9
Curt Kaufman	R	29	69	2	3	4.57	88	94	-3	-1	-.2
Bruce Kison	M	20	65	4	5	5.40	74	80	-9	-7	-.9
Others		63	135	7	9	5.07	79	85	-15	-11	-1.1

Minnesota 81- 81   .500(2)

Player	Pos	G	HR	RBI	Avg	OBA	SLG	NOPS-A		Runs-A		DEF	BSR	Wins
Dave Engle	CD	109	4	38	.266	.312	.353	84	82	-8	-9	-0	-.6	-.7
Kent Hrbek	*1	149	27	107	.311	.387	.522	149	146	37	36	-6	-.3	1.8
Tim Teufel	*2	157	14	61	.262	.351	.400	107	105	7	5	-10	-1.5	.2
Houston Jimenez	*S	108	0	19	.201	.240	.245	34	34	-25	-26	-17	-.6	-3.8
Gary Gaetti	*3/OS	162	5	65	.262	.318	.350	85	83	-11	-13	17	.3	.6
Mickey Hatcher	*OD1/3	152	5	69	.302	.346	.406	107	105	6	4	10	-.6	.7
Kirby Puckett	*O	128	0	31	.296	.321	.336	82	80	-13	-14	22	0.0	.2
Tom Brunansky	*O	155	32	85	.254	.322	.460	114	111	9	8	0	-1.8	.0
Bob Bush	D/1	113	11	43	.222	.301	.389	89	87	-4	-5	0	-.9	-1.0
Darrell Brown	OD	95	1	19	.273	.310	.342	80	79	-6	-7	3	.6	-.6
Ron Washington	S/23	88	3	23	.294	.312	.447	107	105	1	1	-7	-.3	-.2
Tim Laudner	C	87	10	35	.206	.260	.389	77	75	-8	-9	-0	0.0	-.6
Dave Meier	O/3	59	0	13	.238	.273	.306	60	59	-7	-7	2	-.6	-.9
Others		128	2	28	.221	.294	.274	58	57	-14	-15	-2	0.0	-1.8

Pitcher	P	G	IP	W	L	ERA	NERA-A		Runs-A		Wins
Frank Viola	S	35	258	18	12	3.21	125	127	23	25	2.3
Mike Smithson	S	36	252	15	13	3.68	109	111	9	12	1.1
John Butcher	S	34	225	13	11	3.44	116	119	14	16	1.5
Ken Schrom	S	25	137	5	11	4.47	90	92	-6	-5	-.8
Pete Filson	R	55	119	6	5	4.08	98	100	-0	0	-.1
Ron Davis	R	64	83	7	11	4.55	88	90	-4	-3	-.5
Rick Lysander	R	36	57	4	3	3.47	115	118	3	4	.4
Ed Hodge	M	25	100	4	3	4.77	84	86	-8	-7	-.9
Albert Williams	M	17	69	3	5	5.74	70	71	-12	-12	-1.3
Others		84	138	6	7	3.39	118	121	9	11	1.1

Oakland 77- 85 .475(4)

Player	Pos	G	HR	RBI	Avg	OBA	SLG	NOPS-A		Runs-A		DEF	BSR	Wins
Mike Heath	*CO/3S	140	13	64	.248	.289	.396	87	97	-8	-2	6	-.3	.8
Bruce Bochte	*1	148	5	52	.264	.338	.345	90	100	-4	1	-12	-2.4	-2.4
Joe Morgan	*2	116	6	43	.244	.361	.351	98	109	1	7	-19	.6	-.9
Tony Phillips	S2/O	154	4	37	.266	.329	.359	90	101	-4	1	-20	-.6	-1.1
Carney Lansford	*3	151	14	74	.300	.347	.439	116	129	12	20	-0	.9	2.3
Rickey Henderson	*O	142	16	58	.293	.401	.458	137	153	29	36	5	9.0	4.6
Dwayne Murphy	*O	153	33	88	.256	.346	.472	124	138	18	26	10	-1.8	2.9
Mike Davis	*O	134	9	46	.230	.290	.364	80	89	-10	-5	1	-1.2	-1.1
Dave Kingman	*D/1	147	35	118	.268	.329	.505	127	142	18	25	-0	0.0	2.0
Bill Almon	O1D/3CS	105	7	16	.223	.258	.374	73	81	-7	-5	-0	-2.7	-1.4
Mark Wagner	S3/2P	83	0	12	.230	.287	.310	65	73	-3	-2	-3	.6	-.2
Don Hill	S/23	73	2	16	.230	.251	.299	52	58	-11	-8	-6	-.3	-1.2
Davey Lopes	O2/3	72	9	36	.257	.347	.430	114	127	5	8	-0	3.6	1.0
Jim Essian	C/3	63	2	10	.235	.350	.346	93	104	0	1	2	-.3	.7
Jeff Burroughs	D/O	58	2	8	.211	.371	.310	91	101	0	1	0	0.0	.0
Garry Hancock	O/1P	51	0	8	.217	.217	.250	29	32	-5	-4	0	0.0	-.6
Others		76	1	11	.246	.311	.370	88	98	-1	0	0	0.0	.2

Pitcher	P	G	IP	W	L	ERA	NERA-A		Runs-A		Wins
Ray Burris	S	34	212	13	10	3.14	127	114	20	10	.7
Steve McCatty	S	33	180	8	14	4.75	84	75	-14	-22	-2.5
Bill Krueger	S	26	142	10	10	4.75	84	75	-11	-18	-2.0
Curt Young	S	20	109	9	4	4.05	99	88	-0	-5	-.6
Keith Atherton	R	57	104	7	6	4.33	92	83	-3	-8	-1.1
Bill Caudill	R	68	96	9	7	2.72	147	132	14	9	.8
Tim Conroy	R	38	93	1	6	5.23	77	68	-12	-16	-1.8
Lary Sorensen	M	46	183	6	13	4.92	81	73	-18	-26	-2.9
Mike Warren	M	24	90	3	6	4.90	82	73	-8	-12	-1.5
Chris Codiroli	M	28	89	6	4	5.87	68	61	-17	-22	-2.3
Others		68	130	5	5	5.40	74	66	-19	-25	-2.6

Seattle 74- 88 .457(5)

Player	Pos	G	HR	RBI	Avg	OBA	SLG	NOPS-A		Runs-A		DEF	BSR	Wins
Bob Kearney	*C	133	7	43	.225	.259	.334	63	61	-21	-23	3	-.9	-1.3
Alvin Davis	*1	152	27	116	.284	.395	.497	145	141	38	35	-2	-.9	2.0
Jack Perconte	*2	155	0	31	.294	.359	.346	96	94	-0	-2	5	5.1	1.5
Spike Owen	*S	152	3	43	.245	.309	.326	76	74	-16	-18	17	0.0	1.1
Jim Presley	3	70	10	36	.227	.248	.402	77	75	-8	-9	-5	-.3	-1.5
Phil Bradley	*O	124	0	24	.301	.373	.363	105	102	3	2	-1	1.5	-.2
Dave Henderson	OD	112	14	43	.280	.321	.466	115	112	6	5	2	-1.5	.2
Al Cowens	*O	139	15	78	.277	.315	.435	105	102	2	0	-5	-.3	-1.0
Ken Phelps	D/1	101	24	51	.241	.382	.521	147	143	20	19	0	-.9	1.5
Barry Bonnell	O3/1	110	8	48	.264	.315	.394	95	92	-2	-3	-0	.3	-.9
Steve Henderson	OD	109	10	35	.262	.341	.409	106	104	3	2	0	-1.8	-.4
Larry Milbourne	32/S	79	1	22	.265	.305	.313	71	69	-7	-8	-1	-1.2	-1.0
Pat Putnam	DO/1	64	2	16	.200	.257	.277	48	47	-10	-10	0	.9	-1.2
Domingo Ramos	3S/12	59	0	2	.185	.233	.210	23	23	-7	-8	-0	-.6	-.9
Others		218	8	47	.208	.301	.290	64	63	-24	-26	-2	-1.8	-2.9

Pitcher	P	G	IP	W	L	ERA	NERA-A		Runs-A		Wins
Mark Langston	S	35	225	17	10	3.40	118	122	15	19	2.0
Mike Moore	S	34	212	7	17	4.97	80	84	-22	-18	-1.6
Jim Beattie	S	32	211	12	16	3.41	117	122	14	17	1.9
Matt Young	S	22	113	6	8	5.73	70	73	-21	-19	-1.9
Salome Barojas	S	19	95	6	5	3.98	100	105	0	2	.2
Ed Vande Berg	R	50	130	8	12	4.78	84	87	-10	-8	-.9
Bob Stoddard	R	27	79	2	3	5.13	78	81	-9	-7	-.8
Dave Beard	R	43	76	3	2	5.80	69	72	-14	-13	-1.4
Ed Nunez	R	37	68	2	2	3.18	126	131	6	7	.7
Paul Mirabella	R	52	68	2	5	4.37	92	95	-2	-1	-.1
Mike Stanton	R	54	61	4	4	3.54	113	117	3	4	.4
Roy Thomas	R	21	50	3	2	5.22	77	80	-6	-5	-.6
Others		28	53	2	2	4.08	98	102	0	0	-.0

Chicago 74- 88 .457(5)

Player	Pos	G	HR	RBI	Avg	OBA	SLG	NOPS-A		Runs-A		DEF	BSR	Wins
Pudge Fisk	C	102	21	43	.231	.292	.468	107	100	2	-1	-1	1.8	.4
Greg Walker	*1D	136	24	75	.294	.349	.532	140	131	22	18	-5	-.6	.6
Julio Cruz	*2	143	5	43	.222	.298	.311	69	64	-16	-20	20	.6	.6
Scott Fletcher	*S2/3	149	3	35	.250	.329	.311	78	73	-11	-15	16	.6	1.1
Vance Law	*32/OS	151	17	59	.252	.312	.403	96	90	-2	-6	-9	.6	-1.4
Harold Baines	*O	147	29	94	.304	.364	.541	147	137	35	30	0	-.9	2.3
Rudy Law	*O	136	6	37	.251	.310	.345	81	76	-11	-16	-9	-1.5	-3.3
Ron Kittle	*O	139	32	74	.215	.298	.453	104	98	1	-2	-0	-2.7	-1.1
Greg Luzinski	*D	125	13	58	.238	.333	.364	93	87	-2	-6	0	.9	-1.0
Jerry Hairston	OD	115	5	19	.260	.375	.401	115	107	6	4	-0	-.6	.0
Tom Paciorek	1O	111	4	29	.256	.311	.358	85	79	-7	-10	-4	1.8	-1.9
Mike Squires	13/OP	104	0	6	.183	.239	.195	22	20	-8	-8	-1	-.6	-1.6
Jerry Dybzinski	S3/2	94	1	10	.235	.313	.311	73	68	-4	-5	5	1.5	.6
Marc Hill	C/1	77	5	20	.233	.275	.373	77	72	-5	-7	-0	-.6	-.6
Dave Stegman	O	55	2	11	.261	.306	.380	89	83	-1	-1	-0	.9	-.4
Others		161	5	27	.179	.255	.260	43	40	-25	-28	-3	2.1	-2.9

Pitcher	P	G	IP	W	L	ERA	NERA-A		Runs-A		Wins
Richard Dotson	S	32	246	14	15	3.59	112	121	11	20	2.0
Tom Seaver	S	34	237	11	11	3.95	101	109	1	10	1.1
Lamarr Hoyt	S	34	236	13	18	4.46	90	97	-11	-3	-.3
Floyd Bannister	S	34	218	14	11	4.83	83	89	-19	-11	-1.4
Ron Reed	R	51	73	0	6	3.08	130	140	7	10	1.1
Juan Agosto	R	49	55	2	1	3.11	129	139	5	7	1.0
Britt Burns	M	34	117	4	12	5.00	80	86	-12	-8	-.9
Gene Nelson	M	20	75	3	5	4.44	90	97	-3	-0	-.1
Others		111	197	9	9	3.97	101	109	1	8	1.0

Texas 69- 92 .429(7)

Player	Pos	G	HR	RBI	Avg	OBA	SLG	NOPS-A		Runs-A		DEF	BSR	Wins
Donnie Scott	C	81	3	20	.221	.282	.298	61	59	-11	-12	-0	-.6	-1.0
Pete O'Brien	*1/O	142	18	80	.287	.353	.448	120	117	14	12	0	-2.1	-.0
Wayne Tolleson	*2/S3O	118	0	9	.213	.277	.251	48	46	-22	-24	-15	4.2	-3.1
Curt Wilkerson	*S2	153	1	26	.248	.283	.279	56	55	-27	-29	-13	-2.4	-3.5
Buddy Bell	*3	148	11	83	.315	.388	.458	133	130	26	24	24	0.0	5.1
Bill Sample	*O	130	5	33	.247	.290	.327	70	69	-19	-20	4	1.8	-2.0
Gary Ward	*O	155	21	79	.284	.344	.447	117	114	13	11	-4	-.9	-.0
George Wright	OD	101	9	48	.243	.275	.384	80	78	-10	-11	1	-1.2	-1.7
Larry Parrish	OD3	156	22	101	.285	.337	.465	120	117	15	13	-0	-1.8	.5
Mickey Rivers	DO	102	4	33	.300	.320	.387	94	92	-2	-3	0	-1.5	-.9
Ed Yost	C	80	6	25	.182	.202	.273	30	29	-22	-23	-0	-.9	-2.1
Bob Jones	O1	64	4	22	.259	.312	.371	88	86	-1	-2	0	-.3	-.6
Marv Foley	C/13	63	6	19	.217	.313	.391	94	91	-0	-0	0	0.0	-.0
Jeff Kunkel	S	50	3	7	.204	.221	.324	49	47	-9	-10	-3	-.6	-1.2
Others		196	7	33	.242	.331	.325	82	80	-7	-8	-6	.6	-1.5

Pitcher	P	G	IP	W	L	ERA	NERA-A		Runs-A		Wins
Charlie Hough	S	36	266	16	14	3.76	106	110	7	11	1.4
Frank Tanana	S	35	246	15	15	3.26	123	127	20	24	2.5
Danny Darwin	S	35	224	8	12	3.94	102	105	2	5	.3
Dave Stewart	S	32	192	7	14	4.73	84	87	-15	-12	-1.4
Mike Mason	S	36	184	9	13	3.62	110	114	8	10	.9
Dave Schmidt	R	43	70	6	6	2.57	155	160	11	12	1.3
Odell Jones	R	33	59	2	4	3.66	109	113	2	3	.4
Dickie Noles	M	18	58	2	3	5.12	78	81	-6	-5	-.8
Others		83	138	4	11	4.89	82	84	-13	-11	-1.1

Chicago   96- 65   .596(1)

Player	Pos	G	HR	RBI	Avg	OBA	SLG	NOPS-A		Runs-A		DEF	BSR	Wins
Jody Davis	*C	150	19	94	.256	.319	.421	106	97	3	-2	1	-2.1	.1
Leon Durham	*1	137	23	96	.279	.372	.505	144	132	28	22	4	0.0	2.2
Ryne Sandberg	*2	156	19	84	.314	.369	.520	147	135	38	31	23	5.4	6.7
Larry Bowa	*S	133	0	17	.223	.274	.269	53	48	-23	-28	5	.6	-1.0
Ron Cey	*3	146	25	97	.240	.329	.442	115	105	9	3	-20	-.3	-2.0
Gary Matthews	*O	147	14	82	.291	.417	.428	138	126	30	24	-12	.3	.6
Bob Dernier	*O	143	3	32	.278	.356	.362	102	93	3	-2	-2	3.3	-.8
Keith Moreland	*O1/3C	140	16	80	.279	.329	.422	110	100	5	0	-8	-2.1	-1.7
Henry Cotto	O	105	0	8	.274	.325	.308	79	72	-3	-5	-0	.9	-.9
Gary Woods	O/2	87	3	10	.235	.336	.388	103	94	1	-0	0	0.0	-.3
Thad Bosley	O	55	2	14	.296	.378	.418	124	113	3	2	0	.9	.2
Jay Johnstone	O	52	0	3	.288	.350	.370	102	93	0	-0	0	0.0	-.1
Others		297	12	61	.243	.307	.359	86	79	-10	-17	1	.3	-1.1
Pitchers	P	435	0	25	.145	.187	.168	96	87	-0	-4	4	-.6	-1.0

Pitcher	P	G	IP	W	L	ERA	NERA-A		Runs-A		Wins
Steve Trout	S	32	190	13	7	3.41	105	115	4	11	1.3
Dennis Eckersley	S	24	160	10	8	3.04	118	129	10	16	1.4
Rick Sutcliffe	S	20	150	16	1	2.70	133	145	15	20	2.6
Scott Sanderson	S	24	141	8	5	3.13	115	125	7	12	1.2
Dick Ruthven	S	23	127	6	10	5.03	71	78	-19	-15	-1.7
Chuck Rainey	S	17	88	5	7	4.30	84	91	-6	-3	-.5
Lee Smith	R	69	101	9	7	3.65	98	107	-0	3	.2
Tim Stoddard	R	58	92	10	6	3.82	94	103	-1	1	-.1
Rich Bordi	R	31	83	5	2	3.47	104	113	1	4	.2
George Frazier	R	37	64	6	3	4.08	88	96	-2	-0	-.2
Warren Brusstar	R	41	64	1	1	3.09	116	126	4	6	.7
Dickie Noles	R	21	51	2	2	5.12	70	76	-8	-6	-.9
Rick Reuschel	M	19	92	5	5	5.18	69	75	-15	-12	-1.0
Others		16	30	0	1	6.00	60	65	-7	-6	-.7

New York   90- 72   .556(2)

Player	Pos	G	HR	RBI	Avg	OBA	SLG	NOPS-A		Runs-A		DEF	BSR	Wins
Mike Fitzgerald	*C	112	2	33	.242	.291	.306	68	70	-14	-13	2	.3	-.9
Keith Hernandez	*1	154	15	94	.311	.415	.449	143	147	35	38	18	-1.2	4.9
Wally Backman	*2/S	128	1	26	.280	.362	.339	98	101	1	2	-2	4.2	.8
Jose Oquendo	S	81	0	10	.222	.286	.249	51	53	-11	-10	-2	2.4	-.5
Hubie Brooks	*3S	153	16	73	.283	.342	.417	112	116	9	11	-19	-1.2	-.9
George Foster	*O	146	24	86	.269	.314	.443	111	114	6	7	3	-.6	.3
Mookie Wilson	*O	154	10	54	.276	.309	.409	100	103	-1	0	10	8.4	1.2
Darryl Strawberry	*O	147	26	97	.251	.345	.467	126	130	18	20	5	3.3	2.3
Danny Heep	O1	99	1	12	.231	.326	.312	80	82	-4	-3	2	.3	-.5
Rusty Staub	/1	78	1	18	.264	.303	.361	86	88	-0	-0	0	0.0	-.1
Kelvin Chapman	2/3	75	3	23	.289	.358	.401	113	116	4	4	-1	-1.8	.2
Ron Gardenhire	S2/3	74	1	10	.246	.278	.304	63	65	-9	-9	-2	1.2	-.6
Ron Hodges	C	64	1	11	.208	.354	.264	76	78	-1	-1	0	-.3	-.1
Rafael Santana	S	51	1	12	.276	.317	.382	95	98	-0	-0	-2	-1.8	.0
Jerry Martin	O/1	51	3	5	.154	.206	.264	31	32	-8	-7	1	0.0	-1.0
Others		133	1	29	.224	.285	.280	59	61	-15	-14	-2	-1.2	-2.0
Pitchers	P	446	1	14	.165	.183	.188	104	108	0	1	-2	.3	.3

Pitcher	P	G	IP	W	L	ERA	NERA-A		Runs-A		Wins
Dwight Gooden	S	31	218	17	9	2.60	138	134	24	22	2.4
Walt Terrell	S	33	215	11	12	3.52	102	99	2	-0	-.4
Ron Darling	S	33	206	12	9	3.80	95	92	-4	-6	-.6
Bruce Berenyi	S	19	115	9	6	3.76	96	93	-1	-2	-.3
Sid Fernandez	S	15	90	6	6	3.50	103	100	1	0	-.2
Jesse Orosco	R	60	87	10	6	2.59	139	135	10	9	1.0
Brent Gaff	R	47	84	3	2	3.64	99	96	0	-0	-.2
Doug Sisk	R	50	78	1	3	2.08	173	168	13	12	1.3
Tom Gorman	R	36	58	6	0	2.95	122	118	4	4	.3
Tim Leary	R	20	54	3	3	4.00	90	87	-1	-2	-.3
Ed Lynch	M	40	124	9	8	4.50	80	78	-11	-13	-1.5
Others		56	115	3	8	6.50	55	54	-36	-37	-4.0

St Louis   84- 78   .519(3)

Player	Pos	G	HR	RBI	Avg	OBA	SLG	NOPS-A		Runs-A		DEF	BSR	Wins
Darrell Porter	*C	127	11	68	.232	.335	.363	96	99	-0	0	-1	-.3	.3
David Green	*1O	126	15	65	.268	.300	.416	99	103	-1	-0	-3	-.3	-1.0
Tom Herr	*2	145	4	49	.276	.337	.346	92	95	-4	-2	11	-.3	1.2
Ozzie Smith	*S	124	1	44	.257	.349	.337	94	97	-1	0	31	6.3	5.4
Terry Pendleton	3	67	1	33	.324	.363	.420	119	123	6	7	11	3.0	2.2
George Hendrick	*O/1	120	9	69	.277	.327	.406	105	108	2	4	-4	-1.2	-.7
Willie McGee	*O	145	6	50	.291	.326	.394	101	105	0	2	4	6.9	.7
Lonnie Smith	*O	145	6	49	.250	.352	.341	95	99	-0	1	-10	7.2	-.8
Andy Van Slyke	O31	137	7	50	.244	.356	.368	104	107	4	5	2	5.4	.8
Tito Landrum	O	105	3	26	.272	.311	.387	95	98	-0	-0	-1	-1.5	-.8
Art Howe	31/2S	89	2	12	.216	.306	.295	69	72	-4	-4	4	-1.2	-.2
Steve Braun	O/3	86	0	16	.276	.383	.327	101	104	1	1	0	0.0	.0
Mike Jorgensen	1	59	1	12	.245	.315	.357	88	91	-1	-0	-0	0.0	-.3
Ken Oberkfell	3/2S	50	0	11	.309	.379	.395	118	121	4	5	5	-.9	.9
Others		223	7	35	.199	.253	.292	53	55	-27	-25	10	.6	-.8
Pitchers	P	480	2	21	.116	.162	.148	71	73	-4	-3	5	-.3	-.5

Pitcher	P	G	IP	W	L	ERA	NERA-A		Runs-A		Wins
Joaquin Andujar	S	36	261	20	14	3.34	107	104	7	4	.9
Dave LaPoint	S	33	193	12	10	3.96	91	87	-7	-10	-1.6
Danny Cox	S	29	156	9	11	4.04	89	86	-7	-9	-1.0
Kurt Kepshire	S	17	109	6	5	3.30	109	105	4	2	-.2
Bruce Sutter	R	71	123	5	7	1.54	234	226	28	26	2.8
Neil Allen	R	57	119	9	6	3.55	101	98	1	-0	.1
Jeff Lahti	R	63	85	4	2	3.71	97	94	-0	-1	-.2
Dave Rucker	R	50	73	2	3	2.10	171	165	12	11	1.1
Ricky Horton	M	37	126	9	4	3.43	105	101	2	1	.2
John Stuper	M	15	61	3	5	5.31	68	65	-11	-11	-1.4
Bob Forsch	M	16	52	2	5	6.06	59	57	-13	-14	-1.4
Others		51	90	3	6	4.60	78	75	-9	-10	-1.1

Philadelphia   81- 81   .500(4)

Player	Pos	G	HR	RBI	Avg	OBA	SLG	NOPS-A		Runs-A		DEF	BSR	Wins
Ozzie Virgil	*C	141	18	68	.261	.334	.434	114	108	8	5	-5	-.3	.3
Len Matuszek	1/O	101	12	43	.248	.354	.458	127	119	10	8	0	-.6	.4
Juan Samuel	*2	160	15	69	.272	.307	.442	108	102	4	0	-15	12.6	.1
Ivan DeJesus	*S	144	0	35	.257	.327	.306	79	74	-10	-13	-6	.6	-.6
Mike Schmidt	*3/1S	151	36	106	.277	.388	.536	157	148	41	37	13	-2.7	4.9
Von Hayes	*O	152	16	67	.292	.360	.447	126	118	19	15	8	6.6	2.4
Garry Maddox	O	77	5	19	.282	.319	.390	98	93	-0	-2	-2	-.3	-.9
Sixto Lezcano	O	109	14	40	.277	.371	.480	138	130	13	11	-3	-.6	.3
Glenn Wilson	*O/3	132	6	31	.240	.279	.372	81	77	-8	-11	0	1.5	-1.6
Greg Gross	O1	112	0	16	.322	.396	.376	118	111	6	4	0	.3	.1
Tim Corcoran	1O	102	5	36	.341	.443	.486	161	151	19	17	0	-.6	1.5
John Wockenfuss	1C/3	86	6	24	.289	.390	.417	127	120	7	6	-0	.3	.5
Luis Aguayo	32S	58	3	11	.278	.350	.458	125	118	2	2	-0	0.0	.3
Kiko Garcia	S3/2	57	0	5	.233	.281	.267	55	51	-3	-3	0	0.0	-.1
Joe Lefebvre	O/3	52	3	18	.250	.351	.363	101	95	1	0	0	-1.2	-.4
Jeff Stone	O	51	1	15	.362	.398	.465	142	134	10	9	0	5.1	1.2
Others		166	6	48	.249	.318	.363	91	86	-4	-7	-1	-1.5	-1.0
Pitchers	P	451	1	22	.132	.146	.156	66	62	-6	-9	-4	.6	-1.2

Pitcher	P	G	IP	W	L	ERA	NERA-A		Runs-A		Wins
Steve Carlton	S	33	229	13	7	3.58	100	107	0	6	.5
Jerry Koosman	S	36	224	14	15	3.25	110	117	8	14	1.0
Charles Hudson	S	30	174	9	11	4.03	89	94	-8	-3	-1.0
John Denny	S	22	154	7	7	2.45	146	155	19	23	2.8
Shane Rawley	S	18	120	10	6	3.83	94	100	-2	0	-.3
Marty Bystrom	S	11	57	4	4	5.05	71	75	-8	-7	-.9
Kevin Gross	R	44	129	8	5	4.12	87	93	-6	-3	-.6
Al Holland	R	68	98	5	10	3.40	106	112	2	4	.2
Larry Andersen	R	64	91	3	7	2.37	151	160	12	15	1.5
Bill Campbell	R	57	81	6	5	3.44	104	111	1	3	.2
Others		66	101	2	4	5.35	67	71	-19	-16	-1.7

Montreal 78- 83 .484(5)

Player	Pos	G	HR	RBI	Avg	OBA	SLG	NOPS-A		Runs-A		DEF	BSR	Wins
Gary Carter	*C1	159	27	106	.294	.368	.487	138	151	30	36	2	-.6	4.3
Terry Francona	1/O	58	1	18	.346	.364	.467	132	144	8	10	3	0.0	1.1
Doug Flynn	2S	124	0	17	.243	.267	.281	54	59	-21	-18	-16	0.0	-3.2
Angel Salazar	S	80	0	12	.155	.179	.201	7	7	-21	-19	-11	-.3	-2.7
Tim Wallach	*3/S	160	18	72	.246	.313	.395	98	107	-2	3	17	-3.3	1.7
Andre Dawson	*O	138	17	86	.248	.304	.409	99	108	-2	3	11	.9	.9
Tim Raines	*O/2	160	8	60	.309	.395	.437	134	146	30	37	1	16.5	5.1
Jim Wohlford	O/3	95	5	29	.300	.344	.451	122	133	5	8	-4	.9	.1
Derrell Thomas	SO2/31	108	0	20	.255	.312	.321	78	85	-6	-3	-10	-2.4	-1.5
Pete Rose	1O	95	0	23	.259	.335	.295	78	86	-6	-3	2	-.3	-.5
Miguel Dilone	O	88	1	10	.278	.348	.367	101	110	0	2	0	6.9	.8
Bryan Little	2/S	85	0	9	.244	.332	.293	77	84	-6	-3	-10	-1.2	-1.5
Mike Stenhouse	O1	80	4	16	.183	.292	.297	66	72	-7	-5	1	0.0	-.8
Dan Driessen	1	51	9	32	.254	.323	.479	123	134	4	6	2	-.6	.6
Others		272	5	32	.212	.267	.272	52	56	-29	-25	-7	.6	-3.0
Pitchers	P	434	1	11	.136	.170	.174	89	98	-1	1	-5	-.6	-.0

Pitcher	P	G	IP	W	L	ERA	NERA-A		Runs-A		Wins
Bill Gullickson	S	32	227	12	9	3.61	100	90	0	-8	-1.5
Charlie Lea	S	30	224	15	10	2.89	124	113	17	9	.8
Bryn Smith	S	28	179	12	13	3.32	108	98	5	-0	.2
Steve Rogers	S	31	169	6	15	4.31	83	76	-13	-19	-2.0
David Palmer	S	20	105	7	3	3.86	93	85	-2	-6	-.6
Bob James	R	62	96	6	6	3.66	98	89	-0	-3	-.6
Jeff Reardon	R	68	87	7	7	2.90	124	113	7	4	.1
Gary Lucas	R	55	53	0	3	2.72	132	120	5	3	.4
Dan Schatzeder	M	36	136	7	7	2.71	132	120	13	8	1.1
Others		71	155	6	10	2.73	132	120	15	9	.8

Pittsburgh 75- 87 .463(6)

Player	Pos	G	HR	RBI	Avg	OBA	SLG	NOPS-A		Runs-A		DEF	BSR	Wins
Tony Pena	*C	147	15	78	.286	.334	.425	112	118	8	11	7	-1.2	2.3
Jason Thompson	*1	154	17	74	.254	.359	.389	110	116	9	13	-8	0.0	-.3
Johnny Ray	*2	155	6	67	.312	.358	.434	122	128	15	18	-6	-.3	1.6
Dale Berra	*S/3	136	9	52	.222	.278	.318	67	71	-19	-16	5	-1.5	.3
Bill Madlock	3/1	103	4	44	.253	.300	.323	75	79	-13	-10	-2	.3	-1.5
Leon Lacy	*O/2	138	12	70	.321	.364	.464	131	138	18	21	15	-.3	3.2
Marvell Wynne	*O	154	0	39	.266	.311	.337	82	86	-15	-11	-1	-4.2	-2.6
Doug Frobel	*O	126	12	28	.203	.272	.388	83	88	-6	-5	-0	-.9	-1.3
Lee Mazzilli	O/1	111	4	21	.237	.339	.331	89	94	-2	-0	2	1.8	-.1
Jim Morrison	32/S1	100	11	45	.286	.332	.454	119	125	7	8	-2	-.6	.6
Milt May	C	50	1	8	.177	.255	.240	40	42	-7	-6	1	-.6	-.6
Others		270	5	41	.212	.266	.297	58	61	-31	-27	3	0.0	-2.8
Pitchers	P	418	2	19	.183	.208	.219	135	143	8	10	0	-.9	.4

Pitcher	P	G	IP	W	L	ERA	NERA-A		Runs-A		Wins
Rick Rhoden	S	33	238	14	9	2.72	132	124	23	17	3.0
Larry McWilliams	S	34	227	12	11	2.93	122	115	17	11	.9
John Tudor	S	32	212	12	11	3.27	110	103	8	2	.5
Jose DeLeon	S	30	192	7	13	3.75	96	90	-2	-7	-1.5
John Candelaria	S	33	185	12	11	2.72	132	124	18	13	1.3
Don Robinson	R	51	122	6	8	3.02	119	112	8	5	1.0
Kent Tekulve	R	72	88	3	9	2.66	135	127	9	7	.9
Lee Tunnell	R	26	68	1	7	5.29	68	64	-12	-14	-1.5
Others		97	135	9	10	2.80	128	121	12	9	.7

San Diego　92- 70　.568(1)

Player	Pos	G	HR	RBI	Avg	OBA	SLG	NOPS-A		Runs-A		DEF	BSR	Wins
Terry Kennedy	*C	148	14	57	.240	.287	.353	79	82	-15	-13	-6	-.9	-1.8
Steve Garvey	*1	161	8	86	.284	.312	.373	91	95	-7	-4	-4	-.9	-1.9
Al Wiggins	*2	158	3	34	.258	.344	.329	90	94	-5	-2	-28	8.4	-1.9
Garry Templeton	*S	148	2	35	.258	.313	.320	78	82	-13	-10	-22	.6	-1.8
Graig Nettles	*3	124	20	65	.228	.334	.413	109	113	5	7	-2	0.0	.3
Tony Gwynn	*O	158	5	71	.351	.411	.444	140	146	34	37	12	-.9	4.2
Kevin McReynolds	*O	147	20	75	.278	.322	.465	119	124	11	13	11	-2.7	1.6
Carmelo Martinez	*O/1	149	13	66	.250	.346	.395	108	112	6	9	14	-1.5	1.5
Luis Salazar	3O/S	93	3	17	.241	.261	.329	65	67	-10	-9	-0	-.9	-1.3
Tim Flannery	23S	86	2	10	.273	.350	.391	108	112	1	2	-2	.6	.2
Bobby Brown	O	85	3	29	.251	.297	.368	86	89	-3	-2	3	2.4	-.0
Kurt Bevacqua	13/O	59	1	9	.200	.326	.275	70	73	-2	-1	0	0.0	-.4
Others		164	9	41	.202	.276	.368	80	83	-6	-5	-4	.6	-.7
Pitchers	P	452	6	34	.180	.212	.255	157	164	12	14	-6	0.0	1.3

Pitcher	P	G	IP	W	L	ERA	NERA-A		Runs-A		Wins
Eric Show	S	32	207	15	9	3.39	106	101	5	1	.7
Tim Lollar	S	31	196	11	13	3.90	92	87	-6	-10	-.6
Eddie Whitson	S	31	189	14	8	3.24	111	105	7	4	.1
Mark Thurmond	S	32	179	14	8	2.97	121	115	12	9	1.3
Craig Lefferts	R	62	106	3	4	2.12	169	161	17	15	1.6
Rich Gossage	R	62	102	10	6	2.91	123	117	8	6	.5
Greg Booker	R	32	57	1	1	3.32	108	103	2	1	.2
Dave Dravecky	M	50	157	9	8	2.92	123	117	12	9	.7
Andy Hawkins	M	36	146	8	9	4.68	77	73	-17	-20	-2.2
Others		79	123	7	4	5.05	71	68	-19	-21	-2.4

Houston　80- 82　.494(2)

Player	Pos	G	HR	RBI	Avg	OBA	SLG	NOPS-A		Runs-A		DEF	BSR	Wins
Mark Bailey	*C	108	9	34	.212	.321	.343	87	92	-4	-2	0	-.6	-.0
Enos Cabell	*1	127	8	44	.310	.343	.417	113	120	6	9	-2	-4.2	-.2
Bill Doran	*2S	147	4	41	.261	.343	.356	96	102	-0	3	9	-.9	1.7
Craig Reynolds	*S/3	146	6	60	.260	.290	.364	83	88	-12	-9	16	1.5	2.6
Phil Garner	32	128	4	45	.278	.359	.388	110	117	6	8	11	-.3	2.1
Jose Cruz	*O	160	12	95	.312	.386	.462	137	146	31	35	4	1.8	3.5
Jerry Mumphrey	*O	151	9	83	.290	.359	.391	111	117	8	12	-6	.3	-.1
Terry Puhl	*O	132	9	55	.301	.383	.434	129	137	19	22	-2	-.9	1.2
Kevin Bass	O	121	2	29	.260	.279	.360	78	83	-9	-7	-1	-1.5	-1.6
Ray Knight	31	88	2	29	.223	.263	.281	53	56	-17	-15	4	-1.8	-1.5
Dennis Walling	31/O	87	3	31	.281	.327	.402	104	110	1	2	4	1.5	.7
Alan Ashby	C	66	4	27	.262	.335	.361	96	102	-0	0	0	0.0	.2
Jim Pankovitz	2/SO	53	1	14	.284	.301	.407	97	104	-0	0	1	0.0	.1
Others		149	5	31	.224	.294	.335	76	81	-8	-6	2	-.6	-.3
Pitchers	P	474	1	22	.137	.177	.172	92	98	-0	1	-6	.6	.4

Pitcher	P	G	IP	W	L	ERA	NERA-A		Runs-A		Wins
Joe Niekro	S	38	248	16	12	3.05	118	110	15	8	.7
Bob Knepper	S	35	234	15	10	3.19	113	105	10	4	.6
Nolan Ryan	S	30	184	12	11	3.03	118	110	11	6	.2
Bill Dawley	R	60	98	11	4	1.93	186	173	18	15	1.7
Vern Ruhle	R	40	90	1	9	4.60	78	73	-9	-12	-1.3
Dave Smith	R	53	77	5	4	2.22	162	150	12	10	.9
Frank DiPino	R	57	75	4	9	3.36	107	99	2	0	-.1
Julio Solano	R	31	51	1	3	1.94	185	172	9	8	.8
Mike Scott	M	31	154	5	11	4.68	77	71	-18	-22	-2.4
Mike LaCoss	M	39	132	7	5	4.02	89	83	-5	-9	-1.1
Others		60	106	3	4	3.65	98	91	-0	-3	-.6

Atlanta 80- 82 .494(2)

Player	Pos	G	HR	RBI	Avg	OBA	SLG	NOPS-A		Runs-A		DEF	BSR	Wins
Bruce Benedict	C	95	4	25	.223	.304	.297	69	60	-11	-16	1	-.9	-1.5
Chris Chambliss	*1	135	9	44	.257	.355	.362	102	89	3	-4	-0	-.9	-1.2
Glenn Hubbard	*2	120	9	43	.234	.333	.380	100	87	1	-6	22	.6	2.1
Rafael Ramirez	*S	145	2	48	.266	.298	.327	75	65	-19	-29	1	-6.0	-1.7
Randy Johnson	3	91	5	30	.279	.329	.374	97	84	-0	-5	-0	-3.0	-1.2
Brad Komminsk	O	90	8	36	.203	.277	.316	66	58	-13	-18	-4	.6	-2.9
Dale Murphy	*O	162	36	100	.290	.374	.547	156	135	44	33	-7	1.5	2.0
Claudell Washington	*O	120	17	61	.286	.376	.469	136	118	21	13	-8	.9	-.0
Gerald Perry	1O	122	7	47	.265	.378	.372	111	97	8	1	-2	-2.7	-1.0
Albert Hall	O	87	1	9	.261	.309	.338	82	71	-2	-5	-1	-.6	-1.2
Jerry Royster	23SO	81	1	21	.207	.259	.295	55	48	-13	-17	4	-.6	-1.3
Alex Trevino	C	79	3	28	.244	.290	.338	76	66	-8	-12	1	.3	-1.0
Ken Oberkfell	3/2	50	1	10	.233	.294	.308	69	60	-6	-9	-0	-1.5	-1.3
Bob Watson	1	49	2	12	.212	.287	.329	73	63	-2	-4	0	0.0	-.6
Others		207	6	50	.241	.311	.322	78	68	-14	-23	1	2.7	-2.3
Pitchers	P	446	0	14	.160	.199	.192	116	101	3	-2	11	.6	-.2

Pitcher	P	G	IP	W	L	ERA	NERA-A		Runs-A		Wins
Ricky Mahler	S	38	222	13	10	3.12	115	134	12	26	3.4
Pascual Perez	S	30	212	14	8	3.74	96	112	-2	10	1.1
Len Barker	S	21	126	7	8	3.86	93	108	-3	4	.6
Gene Garber	R	62	106	3	6	3.06	118	136	6	13	1.4
Steve Bedrosian	R	40	84	9	6	2.36	152	177	12	17	1.6
Jeff Dedmon	R	54	81	4	3	3.78	95	110	-1	4	.4
Donnie Moore	R	47	64	4	5	2.95	122	141	5	9	.9
Craig Mc Murtry	M	37	183	9	17	4.33	83	96	-14	-2	-.1
Rick Camp	M	31	149	8	6	3.26	110	128	5	15	1.4
Pete Falcone	M	35	120	5	7	4.13	87	101	-6	1	-.0
Others		45	101	4	6	4.01	90	104	-4	2	.4

Los Angeles 79- 83 .488(4)

Player	Pos	G	HR	RBI	Avg	OBA	SLG	NOPS-A		Runs-A		DEF	BSR	Wins
Mike Scioscia	*C	114	5	38	.273	.371	.370	108	101	6	3	1	0.0	.8
Greg Brock	1	88	14	34	.225	.323	.402	103	96	1	-1	4	2.4	.2
Steve Sax	*2	145	1	35	.243	.301	.304	71	66	-21	-26	13	-1.2	-1.2
Dave Anderson	*S3	121	3	34	.251	.335	.329	87	81	-4	-8	15	1.5	2.1
German Rivera	3	94	2	17	.260	.325	.357	91	85	-1	-3	7	.3	.2
Ken Landreaux	*O	134	11	47	.251	.299	.374	88	82	-7	-11	-6	-2.4	-2.7
Mike Marshall	*O1	134	21	65	.257	.316	.438	110	102	5	1	-14	-.6	-2.1
Candy Maldonado	*O/3	116	5	28	.268	.321	.382	97	90	-0	-3	-13	-1.8	-2.5
Pedro Guerrero	3O1	144	16	72	.303	.362	.462	130	121	21	16	2	-2.1	1.3
Bill Russell	SO/2	89	0	19	.267	.331	.321	84	78	-4	-6	6	-1.2	.4
Terry Whitfield	O	87	4	18	.244	.313	.356	87	82	-2	-4	-1	-2.1	-1.2
Franklin Stubbs	1O	87	8	17	.194	.274	.341	72	67	-8	-9	2	-.6	-1.3
Steve Yeager	C	74	4	29	.228	.300	.310	71	67	-6	-8	1	-.9	-.8
R.J. Reynolds	O	73	2	24	.258	.302	.350	83	77	-5	-7	-2	-.9	-1.5
Bob Bailor	23S	65	0	8	.275	.317	.305	75	70	-3	-4	3	.3	.1
Rafael Landestoy	23/O	53	1	2	.185	.200	.241	23	22	-5	-5	1	0.0	-.5
Others		178	2	26	.167	.238	.218	29	27	-23	-25	0	.3	-3.0
Pitchers	P	423	3	17	.146	.180	.190	104	97	1	-1	5	.3	-.0

Pitcher	P	G	IP	W	L	ERA	NERA-A		Runs-A		Wins
Fernando Valenzuela	S	34	261	12	17	3.03	118	128	16	24	3.3
Alejandro Pena	S	28	199	12	6	2.49	144	156	24	31	2.9
Rick Honeycutt	S	29	184	10	9	2.84	127	137	15	21	2.3
Bob Welch	S	31	179	13	13	3.77	95	103	-3	2	.1
Burt Hooton	R	54	110	3	6	3.44	105	113	2	5	.4
Pat Zachry	R	58	83	5	6	3.80	95	102	-1	1	.2
Ken Howell	R	32	51	5	5	3.35	107	116	1	3	.3
Orel Hershiser	M	45	190	11	8	2.65	135	146	20	26	2.9
Jerry Reuss	M	30	99	5	7	3.82	94	102	-1	1	.1
Others		80	105	3	6	3.86	93	101	-2	0	-.2

Cincinnati 70-92 .432(5)

Player	Pos	G	HR	RBI	Avg	OBA	SLG	NOPS-A		Runs-A		DEF	BSR	Wins
Brad Gulden	*C	107	4	33	.226	.309	.308	74	74	-9	-9	2	-.6	-.5
Dan Driessen	1	81	7	28	.280	.384	.436	130	130	10	10	-4	0.0	.2
Ron Oester	*2/S	150	3	38	.242	.296	.316	72	72	-20	-20	-12	.9	-2.9
Dave Concepcion	*S3/1	154	4	58	.245	.312	.320	78	78	-14	-14	-20	3.0	-2.2
Nick Esasky	31	113	10	45	.193	.305	.348	83	83	-6	-6	-4	-.9	-1.5
Dave Parker	*O	156	16	94	.285	.331	.410	107	107	5	5	1	-2.7	-.4
Eddie Milner	*O	117	7	29	.232	.337	.342	91	91	-2	-2	5	-1.5	-.4
Gary Redus	*O	123	7	22	.254	.342	.376	102	102	2	2	-1	7.8	.2
Cesar Cedeno	O1	110	10	47	.276	.323	.429	110	110	4	4	0	3.9	.3
Tom Foley	S2/3	106	5	27	.253	.312	.357	88	88	-4	-4	-12	-.3	-.8
Wayne Krenchicki	3/12	97	6	22	.298	.365	.470	133	133	8	8	-1	-.6	.5
Duane Walker	O	83	10	28	.292	.395	.528	157	157	15	15	2	.3	1.5
Tony Perez	1	71	2	15	.241	.297	.343	79	79	-3	-3	-1	0.0	-.7
Dann Bilardello	C	68	2	10	.209	.287	.280	60	60	-8	-8	2	-.6	-.7
Eric Davis	O	57	10	30	.224	.322	.466	119	119	4	4	2	1.8	.6
Others		173	2	27	.244	.319	.298	74	74	-11	-11	-3	-.3	-1.7
Pitchers	P	490	1	25	.147	.178	.185	100	100	0	0	-5	0.0	.2

Pitcher	P	G	IP	W	L	ERA	NERA-A		Runs-A		Wins
Mario Soto	S	33	237	18	7	3.53	102	103	2	3	.4
Jeff Russell	S	33	182	6	18	4.25	84	86	-12	-11	-1.2
Joe Price	S	30	172	7	13	4.19	86	87	-10	-9	-1.4
Jay Tibbs	S	14	101	6	2	2.85	126	128	8	9	.7
Tom Hume	R	54	113	4	13	5.65	64	65	-25	-24	-2.6
Ted Power	R	78	109	9	7	2.81	128	130	10	10	1.1
Bob Owchinko	R	49	94	3	5	4.12	87	89	-4	-4	-.5
Johnny Franco	R	54	79	6	2	2.62	137	139	9	9	1.0
Bill Scherrer	R	36	52	1	1	5.02	72	73	-7	-7	-.8
Frank Pastore	M	24	98	3	8	6.52	55	56	-31	-30	-3.4
Bruce Berenyi	M	13	51	3	7	6.00	60	61	-13	-12	-1.5
Others		71	173	4	9	4.16	86	88	-10	-9	-1.2

San Francisco 66-96 .407(6)

Player	Pos	G	HR	RBI	Avg	OBA	SLG	NOPS-A		Runs-A		DEF	BSR	Wins
Bob Brenly	*C1/O	145	20	80	.291	.355	.464	129	135	19	22	-1	-3.6	2.0
Al Oliver	1	91	0	34	.298	.339	.366	98	103	-0	1	-0	-.6	-.4
Manny Trillo	2/3	98	4	36	.254	.303	.342	81	85	-10	-7	3	0.0	-.3
Johnnie LeMaster	*S	132	4	32	.217	.268	.282	54	57	-26	-24	13	2.1	.5
Joel Youngblood	*3O/2	134	10	51	.254	.328	.358	93	98	-3	-0	-15	-2.1	-2.1
Chili Davis	*O	137	21	81	.315	.369	.507	144	151	28	31	13	-1.2	3.8
Dan Gladden	O	86	4	31	.351	.411	.447	141	148	20	22	5	-.3	2.4
Jeff Leonard	*O	136	21	86	.302	.360	.484	135	142	23	26	5	.9	2.7
Scot Thompson	1/O	120	1	31	.306	.382	.355	108	114	4	5	-0	-.3	.1
Dusty Baker	O	100	3	32	.292	.392	.374	116	122	7	9	0	.6	.7
Brad Wellman	2S3	93	2	25	.226	.278	.291	60	63	-13	-12	3	0.0	-.5
Gene Richards	O	87	0	4	.252	.340	.281	76	80	-3	-2	0	-.3	-.4
Duane Kuiper	2/1	83	0	11	.200	.276	.209	38	40	-8	-7	1	-.6	-.8
Fran Mullins	S3/2	57	2	10	.218	.277	.345	74	78	-3	-2	1	.3	.1
Jack Clark	O/1	57	11	44	.320	.439	.537	173	182	22	23	0	-.3	2.1
John Rabb	1/OC	54	3	9	.195	.283	.317	68	72	-2	-2	0	-.3	-.4
Others		141	6	37	.249	.317	.369	92	97	-3	-1	-5	-2.1	-.9
Pitchers	P	522	0	12	.115	.160	.129	59	62	-7	-5	-0	0.0	-.8

Pitcher	P	G	IP	W	L	ERA	NERA-A		Runs-A		Wins
Bill Laskey	S	35	208	9	14	4.33	83	80	-16	-19	-2.7
Mike Krukow	S	35	199	11	12	4.57	79	76	-21	-23	-2.8
Jeff Robinson	S	34	172	7	15	4.55	79	76	-17	-20	-2.3
Greg Minton	R	74	124	4	9	3.77	95	92	-1	-3	-.4
Frank Williams	R	61	106	9	4	3.57	101	97	0	-0	.4
Gary Lavelle	R	77	101	5	4	2.76	130	126	9	8	.7
Randy Lerch	R	37	72	5	3	4.25	85	82	-4	-5	-.6
Bob Lacey	R	34	51	1	3	3.88	93	89	-1	-1	-.2
Mark Davis	M	46	175	5	17	5.35	67	65	-33	-36	-4.0
Mark Grant	M	11	54	1	4	6.33	57	55	-15	-16	-2.0
Others		77	198	9	11	4.64	78	75	-22	-25	-2.8

APPENDIX TO THE 2015 EDITION

Top 500 Players of All Time (through 2014)

#	PLAYER	OVERALL WINS
1	Barry Bonds	130.1
2	Babe Ruth	129.0
3	Nap Lajoie	95.3
4	Walter Johnson	89.9
5	Ted Williams	86.4
6	Rogers Hornsby	86.2
7	Ty Cobb	86.1
8	Willie Mays	84.3
9	Henry Aaron	83.7
10	Tris Speaker	82.9
11	Honus Wagner	82.2
12	Mike Schmidt	77.3
13	Cy Young	77.0
14	Stan Musial	75.9
15	Roger Clemens	73.6
16	Eddie Collins	73.1
17	Mickey Mantle	71.6
18	Rickey Henderson	71.2
19	Lou Gehrig	71.0
20	Albert Pujols*	69.6
21	Joe Morgan	68.5
22	Alex Rodriguez*	67.3
23	Frank Robinson	65.4
24	Greg Maddux	64.2

#	PLAYER	OVERALL WINS
25	Grover Alexander	62.9
26	Mariano Rivera	61.9
27	Mel Ott	60.7
28	Lefty Grove	59.1
29	Jimmie Foxx	58.7
30	Christy Mathewson	56.3
31	Kid Nichols	56.2
32	Jeff Bagwell	52.9
33	Eddie Mathews	52.5
34	Warren Spahn	51.4
35	Bobby Grich	50.6
36	Wade Boggs	50.5
37	George Davis	50.0
38	Frank Thomas	49.8
39	Pedro Martinez	49.5
	Tom Seaver	49.5
41	Manny Ramirez	49.4
42	Randy Johnson	49.1
43	Bill Dahlen	48.5
44	Joe DiMaggio	45.9
45	Ken Griffey	45.7
46	Gabby Hartnett	45.4
47	Charlie Gehringer	45.1
48	Bob Gibson	45.0
49	Ozzie Smith	44.8
50	Ron Santo	44.7
51	Gary Sheffield	44.3
52	Edgar Martinez	44.0
53	Al Kaline	43.7
54	Roger Connor	43.3
	Jim Thome	43.3
	Arky Vaughan	43.3
57	Lou Boudreau	43.2
58	Ed Delahanty	42.8
	Cal Ripken	42.8
60	George Brett	42.7
	Chipper Jones	42.7
	Carl Yastrzemski	42.7
63	Luke Appling	42.6
64	John Clarkson	42.5
	Johnny Bench	42.5
	Dan Brouthers	42.5
67	Yogi Berra	42.0
	Scott Rolen	42.0
69	Mike Piazza	41.8
70	Miguel Cabrera*	41.7
71	Barry Larkin	41.5
	Ivan Rodriguez	41.5

#	PLAYER	OVERALL WINS
73	Darrell Evans	41.4
74	Tom Glavine	40.8
75	Bobby Doerr	40.4
76	Carl Hubbell	40.2
77	Robin Yount	40.0
78	Joe Cronin	39.6
79	Vladimir Guerrero	39.5
80	Rod Carew	39.4
81	Bill Dickey	39.3
	Willie McCovey	39.3
83	Tony Gwynn	39.0
84	Dick Allen	38.9
85	Gary Carter	38.8
	Carlton Fisk	38.8
87	Reggie Jackson	38.7
88	Tim Raines	38.5
89	Joe Jackson	38.4
90	Bid McPhee	38.3
91	John Smoltz	38.0
92	Bobby Wallace	37.9
93	Johnny Mize	37.8
94	Jeff Kent	37.7
95	Ed Walsh	37.6
96	Frankie Frisch	37.3
97	Whitey Ford	37.2
	Hal Newhouser	37.2
99	Hoyt Wilhelm	37.1
100	Jim Edmonds	36.8
101	Amos Rusie	36.7
	Ryne Sandberg	36.7
103	Mike Mussina	36.5
	Jack Glasscock	36.5
105	Dave Bancroft	36.2
106	Roberto Clemente	36.1
	Paul Molitor	36.1
108	Willie Randolph	36.0
109	Rafael Palmeiro	35.8
110	Bob Johnson	35.7
	Troy Tulowitzki*	35.7
112	Tim Keefe	35.6
113	Jackie Robinson	35.4
114	Cap Anson	35.3
	Chase Utley*	35.3
116	Roberto Alomar	35.2
	Lou Whitaker	35.2
118	Mark McGwire	35.1
	Joe Sewell	35.1
120	Jim Palmer	35.0

#	PLAYER	OVERALL WINS
	Eddie Murray	35.0
122	Lance Berkman	34.9
123	Mickey Cochrane	34.8
	Bill Mazeroski	34.8
125	Robinson Cano*	34.4
126	Steve Carlton	34.0
	Bob Lemon	34.0
128	Bobby Abreu	33.9
	Todd Helton	33.9
130	Larry Walker	33.8
	Paul Waner	33.8
132	Kevin Brown	33.7
	Frank Baker	33.7
134	Bob Caruthers	33.6
135	Ted Lyons	33.5
	Keith Hernandez	33.5
137	Harry Heilmann	33.1
138	Gaylord Perry	32.9
139	Mordecai Brown	32.7
140	Hank Greenberg	32.6
141	Don Drysdale	32.4
142	Curt Schilling	32.1
143	Bert Blyleven	31.8
144	Carl Mays	31.7
145	Bob Feller	31.6
	Billy Herman	31.6
	Harmon Killebrew	31.6
	Reggie Smith	31.6
149	Charley Radbourn	31.5
150	Red Ruffing	31.4
151	Billy Wagner	31.3
	Roy Halladay	31.3
153	Wes Ferrell	31.2
154	Alan Trammell	31.1
155	Joe Mauer*	30.9
156	Brian Giles	30.8
157	Carlos Beltran*	30.7
158	Dennis Eckersley	30.5
	Cupid Childs	30.5
160	Fergie Jenkins	30.4
	Norm Cash	30.4
	Heinie Groh	30.4
163	Eddie Plank	30.3
	Robin Roberts	30.3
	Albert Belle	30.3
166	David Wright*	30.2
167	David Ortiz*	30.1
168	Tony Mullane	30.0

#	PLAYER	OVERALL WINS
169	Dave Winfield	29.9
170	Sam Crawford	29.8
	Miguel Tejada	29.8
172	Rich Gossage	29.7
	Dizzy Trout	29.7
	Elmer Flick	29.7
175	Jack Clark	29.6
	Tim Hudson*	29.6
	Willie Stargell	29.6
	Jimmy Wynn	29.6
179	Sam Thompson	29.5
180	Stan Hack	29.4
181	Buck Ewing	29.3
182	Trevor Hoffman	29.2
	Dazzy Vance	29.2
184	Phil Niekro	29.1
185	Joey Votto*	29.0
186	Joe Gordon	28.8
187	King Kelly	28.4
188	Clark Griffith	28.3
189	Dick Bartell	28.2
190	Craig Biggio	28.1
191	Buddy Bell	28.0
	Fred Dunlap	28.0
193	Bret Saberhagen	27.9
	Johan Santana*	27.9
	Fred Clarke	27.9
196	Ernie Banks	27.8
197	Juan Marichal	27.7
	Pete Browning	27.7
199	Tommy Bridges	27.3
	Jorge Posada	27.3
201	Bobby Bonds	27.2
	Art Fletcher	27.2
	Billy Hamilton	27.2
204	Urban Shocker	27.0
205	Joe Nathan*	26.9
206	Hughie Jennings	26.7
	Bill Terry	26.7
208	Roy Campanella	26.6
	Hardy Richardson	26.6
210	Harlond Clift	26.3
	George Sisler	26.3
212	John Franco	26.2
	Dolf Luque	26.2
	Sherry Magee	26.2
215	Sammy Sosa	26.1
	Gene Tenace	26.1

#	PLAYER	OVERALL WINS
217	Stan Coveleski	26.0
218	Jesse Burkett	25.9
219	Billy Pierce	25.6
	Dave Stieb	25.6
	Joe Medwick	25.6
222	Jim Fregosi	25.5
223	Gil McDougald	25.3
224	Roy Oswalt	25.1
225	Thurman Munson	25.0
226	Harry Brecheen	24.9
227	Bucky Walters	24.8
228	Dwight Evans	24.7
	Rafael Furcal*	24.7
	Andy Pettitte	24.7
231	Billy Williams	24.6
232	Jack Fournier	24.5
	Robin Ventura	24.5
	Bernie Williams	24.5
235	John Olerud	24.4
	Pete Rose	24.4
237	Eddie Cicotte	24.3
238	Eppa Rixey	24.2
	Will Clark	24.2
240	Zack Wheat	24.1
241	Charlie Bennett	23.9
	Clayton Kershaw*	23.9
	C. C. Sabathia*	23.9
244	Adrian Gonzalez*	23.8
245	Jack Stivetts	23.7
246	Matt Holliday*	23.6
	Dan Quisenberry	23.6
248	Kirby Puckett	23.5
249	Red Faber	23.4
250	Jimmy Key	23.3
	Goose Goslin	23.3
252	David Cone	23.2
	Rick Reuschel	23.2
254	Lenny Dykstra	23.1
	Ralph Kiner	23.1
	Rusty Staub	23.1
257	Addie Joss	23.0
	Silver King	23.0
	Eddie Rommel	23.0
	Jimmy Collins	23.0
	Ian Kinsler*	23.0
	Al Simmons	23.0
263	Lee Smith	22.9
	Jose Canseco	22.9
	Fred Tenney	22.9

#	PLAYER	OVERALL WINS
	Joe Torre	22.9
267	Richie Ashburn	22.8
	Roger Bresnahan	22.8
269	Rollie Fingers	22.7
	Rocky Colavito	22.7
	Roy Thomas	22.7
272	Lon Warneke	22.6
	Jake Beckley	22.6
	Frank Chance	22.6
275	Darryl Strawberry	22.5
276	Ryan Braun*	22.4
	Miller Huggins	22.4
	Charlie Keller	22.4
	Minnie Minoso	22.4
	Tim Salmon	22.4
281	Dizzy Dean	22.3
	Sandy Koufax	22.3
	Cesar Cedeno	22.3
	Lance Parrish	22.3
285	Mark Buehrle*	22.2
	Roberto Hernandez	22.2
	Dutch Leonard	22.2
	Nolan Ryan	22.2
	Rube Waddell	22.2
	Fred McGriff	22.2
291	Ernie Lombardi	22.1
292	Joe Wood	22.0
	Travis Jackson	22.0
	Duke Snider	22.0
	Adam Wainwright*	22.0
296	Adrian Beltre*	21.9
	Guy Hecker	21.9
	Nomar Garciaparra	21.9
	Frank Howard	21.9
	Del Pratt	21.9
301	Tom Henke	21.7
	Evan Longoria*	21.7
	Francisco Rodriguez*	21.7
	Ted Simmons	21.7
305	Luis Tiant	21.6
	Bob Elliott	21.6
307	Jonathan Papelbon*	21.5
	Kent Tekulve	21.5
	Harry Stovey	21.5
310	Ron Cey	21.2
	Ed Konetchy	21.2
	Graig Nettles	21.2
313	Jim McCormick	21.1
	Carlos Delgado	21.1

#	PLAYER	OVERALL WINS
	Chuck Klein	21.1
316	John Hiller	21.0
	John Wetteland	21.0
	Wally Berger	21.0
	Andrew McCutchen*	21.0
320	Javy Lopez	20.9
321	Red Lucas	20.8
	Fred Pfeffer	20.8
	Hanley Ramirez*	20.8
324	Eric Davis	20.7
	Felix Hernandez*	20.7
	Joe Kelley	20.7
	Jimmy Sheckard	20.7
	Vern Stephens	20.7
	John Ward	20.7
	Chief Zimmer	20.7
331	Tony Fernandez	20.6
332	Ron Guidry	20.5
	Jose Cruz	20.5
	Glenn Hubbard	20.5
	Orlando Hudson	20.5
	Kevin Mitchell	20.5
	Tony Oliva	20.5
	Darrell Porter	20.5
	Eddie Stanky	20.5
340	Nig Cuppy	20.4
	Jason Giambi*	20.4
	Tommy John	20.4
343	Kevin Appier	20.3
344	Chuck Finley	20.2
	Ken Caminiti	20.2
346	Lefty Gomez	20.1
	Ken Boyer	20.1
	Phil Rizzuto	20.1
349	Yadier Molina	20.0
	Jesse Tannehill	20.0
	Vic Willis	20.0
	Art Devlin	20.0
	Brian Downing	20.0
	Wally Schang	20.0
355	Ken Singleton	19.8
	Mike Trout*	19.8
357	Wilbur Cooper	19.7
358	Doug Jones	19.6
	Denny Lyons	19.6
360	Robb Nen	19.5
	Early Wynn	19.5
	John McGraw	19.5
	Joe Tinker	19.5

#	PLAYER	OVERALL WINS
364	Orel Hershiser	19.4
	Sparky Lyle	19.4
	Mike Marshall	19.4
	Jack Quinn	19.4
	Luis Gonzalez	19.4
	Charley Jones	19.4
370	Doc White	19.3
	Carlos Zambrano	19.3
	Bill Joyce	19.3
373	Ferris Fain	19.2
	Toby Harrah	19.2
	Fred Lynn	19.2
376	Spud Chandler	19.1
	Gene Alley	19.1
378	Freddie Fitzsimmons	19.0
379	Joe McGinnity	18.9
	Jesse Orosco	18.9
	Mickey Welch	18.9
	Tony Phillips	18.9
383	Schoolboy Rowe	18.8
	Roy Cullenbine	18.8
	Bill Freehan	18.8
386	Bobby Shantz	18.7
	Juan Gonzalez	18.7
	Pee Wee Reese	18.7
	Matt Williams	18.7
390	Ed Lopat	18.6
	Don Newcombe	18.6
	Moises Alou	18.6
	George Foster	18.6
	Andruw Jones	18.6
395	Bruce Sutter	18.5
	Orlando Cepeda	18.5
	Chet Lemon	18.5
398	Doc Gooden	18.4
	Gavvy Cravath	18.4
	Jim Rice	18.4
	Pie Traynor	18.4
402	Burleigh Grimes	18.3
	Mike Griffin	18.3
404	Curt Davis	18.2
	Larry Doby	18.2
	Mike Scioscia	18.2
	Hack Wilson	18.2
408	Rico Carty	18.1
409	Max Carey	18.0
	Dave Concepcion	18.0
411	Ray Lankford	17.9
	Red Schoendienst	17.9

#	PLAYER	OVERALL WINS
413	Tony Lazzeri	17.8
	Bobby Veach	17.8
	Ryan Zimmerman*	17.8
416	Babe Adams	17.7
417	Eric Chavez	17.6
	Andy Messersmith	17.6
	Brandon Webb	17.6
	Ken Williams	17.6
421	Victor Martinez*	17.5
	Don Sutton	17.5
	Babe Herman	17.5
424	Tom Gordon	17.4
	Jeff Montgomery	17.4
	Pedro Guerrero	17.4
427	Rick Aguilera	17.3
	Ed Reulbach	17.3
	Earl Averill	17.3
	Andre Dawson	17.3
	Bobby Knoop	17.3
	Boog Powell	17.3
	Al Rosen	17.3
	Justin Verlander*	17.3
435	Francisco Cordero	17.2
	Jeff Heath	17.2
	Chief Meyers	17.2
	Bill Nicholson	17.2
439	Jose Bautista*	17.1
	Cole Hamels*	17.1
	Paul O'Neill	17.1
	Jayson Werth*	17.1
443	John Candelaria	17.0
	Noodles Hahn	17.0
	Mel Harder	17.0
	Kenny Lofton	17.0
	Jake Peavy*	17.0
	Johnny Pesky	17.0
	John Valentin	17.0
	Jered Weaver*	17.0
451	John Tudor	16.9
	Roy Smalley	16.9
453	Larry French	16.8
	Virgil Trucks	16.8
	Ellis Burks	16.8
	Rick Burleson	16.8
	Johnny Logan	16.8
	Andy Van Slyke	16.8
459	Russell Martin*	16.7

#	PLAYER	OVERALL WINS
	Bob Shawkey	16.7
	Mel Stottlemyre	16.7
462	Ray Chapman	16.6
463	Zach Greinke*	16.5
	Brian McCann*	16.5
465	Jim Bunning	16.4
	Jim Kaat	16.4
	Bobby Bonilla	16.4
	Lave Cross	16.4
469	Hippo Vaughn	16.3
	David Justice	16.3
471	Charlie Buffinton	16.2
	Murry Dickson	16.2
473	Keith Foulke	16.1
	Sam Leever	16.1
	Jim O'Rourke	16.1
	Ray Schalk	16.1
477	Larry Jackson	16.0
478	George Uhle	15.9
	Johnny Evers	15.9
	Dustin Pedroia*	15.9
481	Thornton Lee	15.8
	Brett Butler	15.8
	Mike Hargrove	15.8
484	Mike Garcia	15.7
	Waite Hoyt	15.7
	Max Lanier	15.7
	Troy Percival	15.7
	Jack Taylor	15.7
	Frank Viola	15.7
	Earl Battey	15.7
	Ichiro Suzuki*	15.7
492	Jerry Koosman	15.6
	Jon Lester	15.6
	Deacon Phillippe	15.6
	Mike Timlin	15.6
	Luis Aparicio	15.6
	Willie Kamm	15.6
	Rico Petrocelli	15.6
499	Mel Parnell	15.5
	Steve Rogers	15.5
	Nap Rucker	15.5
	Bob Wickman	15.5
	J. D. Drew	15.5
	Ron Hansen	15.5

* active player

BIBLIOGRAPHY

Adams, Dallas. "The Probability of the League Leader Batting .400." *Baseball Research Journal*, 1981.

———. "Average Pitching and Fielding Skills." *Baseball Research Journal*, 1982.

Adelman, Melvin L. "The First Baseball Game, the First Newspaper References to Baseball, and The New York Club: A Note on the Early History of Baseball." *Journal of Sport History*, 7: 3, Winter 1980.

Allen, Lee. *100 Years of Baseball*. New York: Bartholomew House, 1950.

———. *The Hot Stove League*. New York: A.S. Barnes, 1955.

The Baseball Encyclopedia. New York: Macmillan and Information Concepts, Inc., 1969.

Baseball Guide. Various years and publishers (Beadle, DeWitt, Spalding, Reach, The Sporting News).

Boswell, Thomas. "Baseball's Best Stat," *Inside Sports*, January 31, 1981.

———. "Dwight Evans Should Be MVP." *Inside Sports*, February 1982.

Briggs, W.G., et al. "Notes on Digital Computer Simulation." Including "Baseball-O-Mation" simulation study prepared for presentation before the Operations Research Society of America in 1960. Harbridge House, 1964, 1965.

Brown, Melvin. "Optimal Batting Order in Baseball: An Application of a Performance Evaluation Methodology Based on a Systems Approach." Unpublished paper prepared for the MITRE Corporation, Westgate Research Park, McLean, Va., November 10, 1972.

          . "Performance Evaluation: A Systems Approach, with Particular Attention to Baseball." Publication data as above.

Chadwick, Henry. *The Game of Baseball: How to Learn It, How to Play It, and How to Teach It; with Sketches of Noted Players*. New York: George Munro, 1868.

Clifton, Merritt. *Relative Baseball*. Brigham, Quebec: Samisdat, 1979.

Codell, Barry F. "The Base-Out Percentage: Baseball's Newest Yardstick." *Baseball Research Journal*, 1979.

Cook, Earnshaw. *Percentage Baseball*. Baltimore: Waverly Press, 1964; Cambridge, Mass.: MIT Press, 1966.

Cover, Thomas M. and Keilers, Carroll W. "An Offensive Earned Run Average for Baseball." *Operations Research*, 25: 5, September-October 1977.

Cramer, Richard D. "Do Clutch Hitters Exist?" *Baseball Research Journal*, 1977.

          . "Average Batting Skill Through Major League History." *Baseball Research Journal*, 1980.

          . "Response to Commentary." *Baseball Research Journal*, 1981. (See entry for William D. Rubinstein.)

Cramer, Richard D. and Palmer, Pete. "The Batter's Run Average." *Baseball Research Journal*, 1974.

Davids, L. Robert. "Modern Base-Stealing Proficiency." *Baseball Research Journal*, 1981.

Deane, Bill. "The Best Fielders of the Century." *The National Pastime*, 1983.

D'Esopo, Donato A. and Lefkowitz, Benjamin. "The Distribution of Runs in the Game of Baseball." Unpublished paper prepared for the Stanford Research Institute and delivered at the annual meeting of the American Statistical Association, Stanford University, August 24, 1960.

Durso, Joe. "Major Changes in the Road to the Big Leagues." New York *Times*, June 6, 1983.

          . "The Players' Choice: Dawson." New York *Times*, July 4, 1983.

Efron, Bradley and Morris Carl . "Stein's Paradox in Statistics." *Scientific American*, May 1977.

Freeze, R. Allan. "An Analysis of Baseball Batting Order by Monte Carlo Simulation." *Operations Research*, 22: 4, July-August 1974.

Gammons, Peter. "The Search for That Ultimate Batting Statistic." Boston *Globe*, May 12, 1978.

Haas, Alex J. "Hit by Pitch." *Baseball Research Journal*, 1972.

Hecht, Henry. "A Box Full of Goodies," *Sports Illustrated*, April 4, 1983.

James, Bill. "The Relief Pitcher's ERA Advantage." *Baseball Research Journal*, 1977.

          . *The Baseball Abstract*. Lawrence, Kansas: 1980, 1981; New York: Ballantine Books, 1982, 1983.

Kingsley, Robert H. "Where They Are Doing the Swinging." *Sports Illustrated*, July 16, 1973.

          . "Lots of Home Runs in Atlanta." *Baseball Research Journal*, 1980.

Klein, Joe. "Computerball Is Here." *Sport*, April 1983.

Koppett, Leonard. *A Thinking Man's Guide to Baseball*. New York: E. P. Dutton, 1967.

Lanigan, Ernest J. *Baseball Cyclopedia*. New York: Baseball Magazine Company, 1922.

Lindsey, George R. "Statistical Data Useful for the Operation of a Baseball Team." *Operations Research*, 7: 3, May-June 1959.

———. "The Progress of a Score During a Baseball Game." *American Statistical Association Journal*, September 1961.

———. "An Investigation of Strategies in Baseball." *Operations Research*, 11: 4, July-August 1963.

McKean, Kevin. "Turning Baseball into a Science," *Discovery*, June 1982.

Mann, Stephen L. "The Run Productivity Average: The Central Formula for a Science of Baseball." Unpublished paper, 1977.

Maywar, James P. "Who Are the Most Impressive Strikeout Pitchers?" *Baseball Research Journal*, 1981.

Mills, Eldon G. and Harlan D. *Player Win Averages*. South Brunswick, N.J.: A. S. Barnes, 1970.

Moreland, George. *Balldom*. New York: Balldom Publishing, 1914.

The Official Encyclopedia of Baseball. Edited by Hy Turkin and S.C. Thompson. New York: A. S. Barnes, 1951.

Okrent, Dan. "He Does It by the Numbers." *Sports Illustrated*, May 25, 1981.

Oliver, Ted. *Kings of the Mound: A Pitchers' Rating Manual*. Los Angeles: Self-published, 1944.

Orem, Preston D. *Baseball (1845–1881) from the Newspaper Accounts*. Altadena, Calif.: Self-published, 1961.

———. *Baseball from the Newspaper Accounts, 1882–1891*. Altadena, Calif.: Self-published, 1966, 1967. Ten xerographed pamphlets, one for each year.

Pankin, Mark. "Evaluating Offensive Performance in Baseball." *Operations Research*, 26: 4, July-August 1978.

Peterson, A. V., Jr. "Comparing the Run-Scoring Abilities of Two Different Batting Orders." In *Optimal Strategies in Sports*, ed. S. P. Ladany and R. E. Machol. New York: North-Holland Publishing, 1977.

Rickey, Branch. "Goodby to Some Old Baseball Ideas." *Life*, August 2, 1954.

Rubinstein, William D. "Average Batting Skill Through Major League History: A Commentary." *Baseball Research Journal*, 1981.

Schwarzenbart, Paul. "Ballpark Effects on Fielding Statistics—American League Parks." *Baseball Analyst*, April 1983.

Shoebotham, David. "Relative Batting Averages." *Baseball Research Journal*, 1976.

Skipper, James K., Jr. "Is Pitching 75% of Baseball?" *Baseball Research Journal*, 1980.

Smith, David, "Maury Wills and the Value of a Stolen Base." *Baseball Research Journal*, 1980.

Soolman, Arnold. Untitled paper on runs and wins. Unpublished, 1971.

Sudyk, Bob. "Computer Picks Top Clutch-Hitters." *The Sporting News,* April 16, 1966.

Tattersall, John. "Hitting Homers at Home and on the Road." *Baseball Research Journal,* 1976.

Trueman, Richard E. "Hitting Performance Under Pressure: Are Clutch Hitters for Real?" Presented at the Joint National ORSA/TIMS Meeting, Atlanta, November 8, 1977.

Waggoner, Glen. "The Best Player in Baseball." *Sport,* June 1983.

Wiley, George T. "Computers in Baseball Research." *Baseball Research Journal,* 1976.